# Non-Antibiotic Therapies for Helicobacter pylori

Jie Gao • Yiqi Du
Editors

# Non-Antibiotic Therapies for Helicobacter pylori

## From Mechanisms to Clinical Translation

Springer

*Editors*
Jie Gao
Shanghai Changhai Hospital
Naval Medical University
Shanghai, China

Yiqi Du
Department of Gastroenterology
Shanghai Changhai Hospital
Naval Medical University
Shanghai, China

ISBN 978-981-95-9597-6 ISBN 978-981-95-9598-3 (eBook)
https://doi.org/10.1007/978-981-95-9598-3



This Springer imprint is published by the registered company Springer Nature Singapore Pte Ltd.
The registered company address is: 152 Beach Road, #21-01/04 Gateway East, Singapore 189721, Singapore

# Preface

*Helicobacter pylori* (*H. pylori*) infection continues to pose a substantial global public health challenge, being intricately associated with chronic gastritis, peptic ulcers, and gastric cancer. For many years, antibiotic-based eradication regimens have served as the cornerstone of clinical management. However, the growing crisis of antibiotic resistance, increasing rates of treatment failure, and the adverse effects of therapy-induced dysbiosis of the gut microbiota have necessitated the urgent exploration of effective and safe non-antibiotic alternative or adjunctive strategies. This pursuit has emerged as a leading and dynamic area of research within the field of gastroenterology.

The book *Non-Antibiotic Therapies for Helicobacter pylori* arises from a pivotal moment in medical research. Its objective is not to replace traditional treatments but to furnish researchers, clinicians, and pharmaceutical scientists with a thorough, systematic, and detailed examination of the non-antibiotic therapeutic domain. Our aim is to amalgamate recent advancements from fundamental science to clinical application, incorporating multidisciplinary viewpoints to highlight innovative investigations across a range of advanced methodologies, including nanotechnology, immunomodulation, probiotics, natural products, and gene editing.

The book is organized into seven thematic sections to provide a logical progression for the reader. Chapters 1–6 revisit the fundamental science of *H. pylori* infection, establishing the essential biological and clinical foundation. Chapters 7–22 explore specialized areas in depth: the targeted delivery and antimicrobial strategies of nanomaterials; immunotherapy based on host-pathogen interactions; microbiome-modulating probiotic therapies; and the knowledge-rich domain of Traditional Chinese Medicine and natural products. Chapters 23–26 concentrate on highly promising emerging fields, such as phage therapy and gene editing technology. Finally, Chapters 27–29 ground the discussion in clinical application, case analyses, and future perspectives, aiming to bridge the gap between laboratory research and clinical practice.

This publication represents the culmination of a highly collaborative endeavor, uniting eminent experts from renowned medical schools and hospitals throughout China, alongside international contributors from esteemed institutions such as the Technical University of Munich. Each chapter is authored by a distinguished scholar in their respective field, whose dedicated contributions ensure the authority and foresight of the content presented herein. On behalf of the editorial team, I express our profound gratitude to all contributors, colleagues involved in the supporting research, and everyone who has facilitated the publication of this book. We earnestly hope that this book will inspire new ideas, promote broader international collaboration, and contribute significantly to the ongoing global efforts against *H. pylori* infection.

Shanghai, China
Dec 27, 2025

Jie Gao

# Contents

**Part I Fundamentals and Pathogenesis of *Helicobacter pylori***

1 **History and Definition of *H. pylori*** .... 3
Yiqi Du, Xianzhu Zhou, Shihan Xu, and Jiayue Wang

2 **Biological Characteristics of *H. pylori*** .... 15
Chuan Yin, Jing Wang, and Te Shi

3 **Epidemiology of *H. pylori* Infection** .... 23
Yiqi Du, Xianzhu Zhou, Yifan Qiu, and Jiayue Wang

4 **Mechanisms of *H. pylori* Colonization and Pathogenesis** .... 41
Wei Hu and Jieyi Ye

5 **The Role of *H. pylori* in Gastric Carcinogenesis** .... 61
Wei Hu and Zhouyingyi Liu

6 **Managing *H. pylori* in the Era of Antibiotic Resistance** .... 85
Chuan Yin, Te Shi, and Jing Wang

**Part II Nanomaterials for Non-antibiotic Treatment**

7 **Nonresponsive Nanomaterials** .... 95
Min Lu, Huizhen Fan, Chenxi Wang, and Qiang Yu

8 **Externally Responsive Nanomaterials** .... 117
Min Lu, Huizhen Fan, Chenxi Wang, and Qiang Yu

9 **Internally Environment-Responsive Nanomaterials** .... 127
Min Lu, Chenxi Wang, Huizhen Fan, and Qiang Yu

10 **Multi-stimuli-Responsive Nanomaterials** .... 151
Xin Ding, Xiangyu Luo, Li Yang, and Xinyi Zhong

11 **Antimicrobial Mechanisms of Nanomaterials** .... 173
Xin Ding, Xiangyu Luo, Li Yang, and Xinyi Zhong

**Part III Immune-Based Strategies Against *H. pylori***

**12 Host Immune Response to *H. pylori* Infection** . . . . . 191
Yuan Zhuang

**13 Immune Evasion Mechanisms of *H. pylori*** . . . . . 209
Yuan Zhuang

**14 Immunotherapeutic Strategies Against *H. pylori*** . . . . . 229
Xiaobo He

**15 Clinical Research and Challenges of Immunotherapy** . . . . . 249
Verena Friedrich, Dietmar Pfeiffer, Paolo Palma, Pierre van Damme, and Markus Gerhard

**16 The Future of Immunotherapy** . . . . . 281
Xiaobo He

**Part IV Probiotic Therapies and Microbial Interventions**

**17 Probiotic Mechanisms in *H. pylori* Infection** . . . . . 305
Weiliang Hou and Xueting Li

**18 Probiotics in *H. pylori* Adjunctive Therapy** . . . . . 319
Jiulong Zhao and Yunyi Yang

**19 The Future of Probiotic Therapy** . . . . . 331
Weiliang Hou and Xueting Li

**Part V Traditional Medicine and Natural Product Based Therapies**

**20 The Theoretical Basis of Traditional Chinese Medicine for Treating *Helicobacter pylori*** . . . . . 351
Ziyou Zhong and Meicun Yao

**21 Traditional Chinese Medicine Pharmacology and Clinical Practice in *Helicobacter pylori*** . . . . . 365
Chuqiu Zhang and Meicun Yao

**22 Natural Product Therapy Against *H. pylori*** . . . . . 387
Yanyan Shi and Meicun Yao

**Part VI Innovative Non-antibiotic Therapies and Clinical Translation**

**23 Phage Therapy for *Helicobacter pylori*** . . . . . 409
Jinlong Ru, Jinling Xue, Raquel Mejías Luque, Markus Gerhard, and Li Deng

**24 Antibacterial Peptides Against *Helicobacter pylori*** . . . . . 429
Zelin Zheng, Yejiao Shi, and Honggang Hu

**25 Gene Editing Technology Against *Helicobacter pylori*** . . . . . . . . . . . . . . 447
Mengfan Li, Jie Gao, and Tinglin Zhang

**26 Potassium-Competitive Acid Blockers Against *Helicobacter pylori*** . . . 457
Zhiqiang Song

**27 Nonantibiotic Clinical Trials for *Helicobacter pylori*** . . . . . . . . . . . . . . . 483
Yingli Gong, Jie Gao, and Tinglin Zhang

**28 Policy and Market Prospects for Nonantibiotic Therapy** . . . . . . . . . . . 493
Haowei Wu, Jie Gao, and Tinglin Zhang

**29 The Future of Nonantibiotic Therapy** . . . . . . . . . . . . . . . . . . . . . . . . . . 503
Jie Gao, Jing Wang, and Chuan Yin

# Editors and Contributors

## About the Editor-in-Chief

**Jie Gao** He is a distinguished scientist, Principal Investigator, and doctoral supervisor at Shanghai Changhai Hospital (China), where he also serves as Deputy Director of the Shanghai Key Laboratory of Nautical Medicine and Translation of Drugs and Medical Devices. He has made pioneering contributions to the field of *H. pylori* treatment. By successfully applying his original "**Clearance-Remodeling-Repair (3R)**" **integrated strategy**, he has helped shift the treatment paradigm from simple antibacterial therapy toward a holistic model that combines pathogen eradication, microbiome preservation, and mucosal repair. His team is dedicated to developing precise, antibiotic-free therapeutic strategies that avoid disrupting the gut flora. A landmark study published as a cover article in *Acta Pharmaceutica Sinica B*—which was simultaneously recognized as both an ESI Hot Paper and an ESI Highly Cited Paper—demonstrates his influential work in antibiotic-free *H. pylori* clearance. Further high-impact studies have appeared in journals such as *Journal of Nanobiotechnology* (also an ESI Hot Paper), *Gut Microbes*, *Advanced Science*, *Journal of Controlled Release*, collectively forming an innovative research system in *H. pylori* treatment.

His international academic standing is reinforced by his tenure as a Senior Visiting Scholar at the University of Michigan, Ann Arbor, and his election as a **Fellow of the Royal Society of Biology (FRSB)**. He has authored or co-authored 125 high-impact SCI papers in leading journals, with a cumulative Impact Factor exceeding 1130 and over 5,100 citations (H-index 42, i10-index

72), including 2 ESI Hot Papers, 8 ESI Highly Cited Papers, and 9 cover features. He has secured 13 competitive national and provincial grants and holds 12 patents.

He holds editorial roles for 21 prestigious journals including *Nano-Micro Letters* and *iMeta*, and occupies leadership positions in several academic societies. His contributions have been honored with awards such as the Shanghai Natural Science Award and the Shanghai "Youth Science and Technology Rising Star" Award, solidifying his reputation as a pioneer in integrated gastrointestinal therapeutics.

**Yiqi Du** Deputy Director of the Department of Gastroenterology at Changhai Hospital, Naval Medical University, Chief Physician & Professor. He concurrently serves as a member of the Chinese Gastroenterology Society, Member of the *H. pylori* Group, Deputy Director of the Small Bowel Endoscopy Group of the Chinese Digestive Endoscopy Society, Council Member of the Chinese Medical Doctor Association, and Deputy Director of the Shanghai Gastroenterology Society. His research focuses on early cancer screening of the digestive tract, *H. pylori* infection, treatment of severe acute pancreatitis, and endoscopic diagnosis and treatment of small intestinal diseases. The research area covers the screening of *H. pylori* and familial infection, novel treatment on *H. pylori* as non-antibiotic regimens. He has led two National Key R&D Programs and seven National Natural Science Foundation projects, with 100 SCI-indexed papers published as first or corresponding author, including one with a peak impact factor of 25.8 (Gut). As the third contributor, he has won two National Science and Technology Progress Awards (2007, 2013), and as the second contributor, two Shanghai Science and Technology Progress Awards. He is recognized as a Special Government Allowance Expert by the State Council, a Shanghai Leading Talent, and a Military Discipline Top Talent.

## Deputy Editors

**Markus Gerhard** TUM School of Medicine and Health, Department Preclinical Medicine, Institute of Medical Microbiology, Immunology and Hygiene, Technical University of Munich, Munich, Germany

German Center for Infection Research (DZIF), Partner Site Munich, Munich, Germany

**Honggang Hu** Institute of Translational Medicine, Shanghai University, Shanghai, China

School of Medicine, Shanghai University, Shanghai, China

**Weiliang Hou** Department of Gastroenterology, Shanghai Institute of Pancreatic Diseases, National Key Laboratory of Immunity and Inflammation, Changhai Clinical Research Unit, Changhai Hospital, Naval Medical University, Shanghai, China

**Yongkang Lai** Department of Gastroenterology, Jiangxi Provincial Key Laboratory of Digestive Diseases, Jiangxi Clinical Research Center for Gastroenterology, Digestive Disease Hospital, The First Affiliated Hospital, Jiangxi Medical College, Nanchang University, Nanchang, Jiangxi, China

**Min Lu** Department of Orthopaedics, Shanghai Key Laboratory for Prevention and Treatment of Bone and Joint Diseases, Shanghai Institute of Traumatology and Orthopaedics, Ruijin Hospital, Shanghai Jiao Tong University School of Medicine, Shanghai, China

**Meicun Yao** School of Pharmaceutical Sciences (Shenzhen), Sun Yat-sen University, Shenzhen, China

**Chuan Yin** Department of Gastroenterology, Shanghai Changzheng Hospital, Naval Medical University, Shanghai, China

**Jiulong Zhao** Department of Gastroenterology, Shanghai Institute of Pancreatic Diseases, National Key Laboratory of Immunity and Inflammation, Changhai Clinical Research Unit, Changhai Hospital, Naval Medical University, Shanghai, China

**Yuan Zhuang** National Engineering Research Center of Immunological Products, Department of Microbiology and Biochemical Pharmacy, College of Pharmacy and Laboratory Medicine, Third Military Medical University, Chongqing, China

## Editorial Board

**Li Deng** Technical University of Munich, School of Life Sciences, Freising, Germany

TUM Central Institute of Infection Prevention (ZIP), Freising, Germany

Institute of Virology, Helmholtz Centre Munich—German Research Centre for Environmental Health, Neuherberg, Germany

German Center for Infection Research (DZIF), Munich, Germany

**Xin Ding** School of Medicine, Shenzhen Campus of Sun Yat-sen University, Shenzhen, China

State Key Laboratory of Anti-Infective Drug Discovery and Development, School of Pharmaceutical Sciences, Sun Yat-sen University, Guangzhou, China

**Yiqi Du** Department of Gastroenterology, Shanghai Changhai Hospital, Naval Medical University, Shanghai, China

**Jie Gao** Changhai Clinical Research Unit, Shanghai Changhai Hospital, Naval Medical University, Shanghai, China

Shanghai Key Laboratory of Nautical Medicine and Translation of Drugs and Medical Devices, Shanghai, China

College of Life Science, Mudanjiang Medical University, Mudanjiang, China

**Markus Gerhard** TUM School of Medicine and Health, Department Preclinical Medicine, Institute of Medical Microbiology, Immunology and Hygiene, Technical University of Munich, Munich, Germany

German Center for Infection Research (DZIF), Partner Site Munich, Munich, Germany

**Xiaobo He** National Key Laboratory of Immunity and Inflammation & Institute of Immunology, College of Basic Medical Sciences, Naval Medical University, Shanghai, China

**Wei Hu** Department of Gastroenterology, Shenzhen Hospital, Southern Medical University, Shenzhen, Guangdong, China

Shenzhen School of Clinical Medicine, Southern Medical University, Shenzhen, Guangdong, China

**Honggang Hu** Institute of Translational Medicine, Shanghai University, Shanghai, China

School of Medicine, Shanghai University, Shanghai, China

**Weiliang Hou** Department of Gastroenterology, Shanghai Institute of Pancreatic Diseases, National Key Laboratory of Immunity and Inflammation, Changhai Clinical Research Unit, Changhai Hospital, Naval Medical University, Shanghai, China

**Min Lu** Department of Orthopaedics, Shanghai Key Laboratory for Prevention and Treatment of Bone and Joint Diseases, Shanghai Institute of Traumatology and Orthopaedics, Ruijin Hospital, Shanghai Jiao Tong University School of Medicine, Shanghai, China

**Yanyan Shi** Peking University Third Hospital, Peking, China

**Yejiao Shi** Institute of Translational Medicine, Shanghai University, Shanghai, China

**Zhiqiang Song** Department of Gastroenterology, Peking University Third Hospital, Beijing, China

**Meicun Yao** School of Pharmaceutical Sciences (Shenzhen), Sun Yat-sen University, Shenzhen, China

**Chuan Yin** Department of Gastroenterology, Shanghai Changzheng Hospital, Naval Medical University, Shanghai, China

**Tinglin Zhang** Changhai Clinical Research Unit, Shanghai Changhai Hospital, Naval Medical University, Shanghai, China

Shanghai Key Laboratory of Nautical Medicine and Translation of Drugs and Medical Devices, Shanghai, China

**Jiulong Zhao** Department of Gastroenterology, Shanghai Institute of Pancreatic Diseases, National Key Laboratory of Immunity and Inflammation, Changhai Clinical Research Unit, Changhai Hospital, Naval Medical University, Shanghai, China

**Xianzhu Zhou** Department of Gastroenterology, Shanghai Changhai Hospital, Naval Medical University, Shanghai, China

**Yuan Zhuang** National Engineering Research Center of Immunological Products, Department of Microbiology and Biochemical Pharmacy, College of Pharmacy and Laboratory Medicine, Third Military Medical University, Chongqing, China

## Managing Editor

**Yingli Gong** Changhai Clinical Research Unit, Shanghai Changhai Hospital, Naval Medical University, Shanghai, China

College of Life Science, Mudanjiang Medical University, Mudanjiang, China

# Part I

# Fundamentals and Pathogenesis of *Helicobacter pylori*

# History and Definition of *H. pylori* 1

Yiqi Du, Xianzhu Zhou, Shihan Xu, and Jiayue Wang

*Helicobacter pylori (H. pylori)* is a microaerophilic, spiral-shaped Gram-negative bacterium first identified in the early 1980s. Since then, its role as a major etiological factor in gastric diseases has been widely recognized [1, 2]. Early studies demonstrated that the organism can persist in highly acidic environments and is closely associated with chronic gastritis and peptic ulcer disease. These observations overturned the prevailing concept that excessive gastric acid was the sole cause of ulcer formation [1, 2]. Marshall's self-infection experiment subsequently fulfilled Koch's postulates and provided direct evidence of *H. pylori* pathogenicity [3]. In 1994, the US National Institutes of Health convened a consensus conference that confirmed the etiological role of *H. pylori* in peptic ulcer disease, a milestone considered a turning point that brought *H. pylori* research into mainstream medicine [4]. In the same year, the International Agency for Research on Cancer classified *H. pylori* as a Group I carcinogen [5].

Over the past four decades, research on *H. pylori* has made leaps and bounds. From a microbiological perspective, the bacterium demonstrates remarkable environmental adaptability through the activity of urease, flagella, and key virulence factors such as CagA and VacA [6–9]. From an epidemiological perspective, prospective studies have established *H. pylori* infection as a major risk factor for both peptic ulcer disease and gastric cancer [10, 11]. Globally, an estimated 89% of non-cardia gastric cancers are attributable to *H. pylori* [12], and large-scale studies have confirmed that eradication reduces gastric cancer risk by 36–46% [13, 14]. Meanwhile, diagnostic approaches have been refined, with the urea breath test, stool antigen test, and endoscopic histological examination now widely used in clinical practice. Furthermore, international consensus reports, including Maastricht VI and

Y. Du (✉) · X. Zhou · S. Xu · J. Wang
Department of Gastroenterology, Shanghai Changhai Hospital, Naval Medical University, Shanghai, China



J. Gao, Y. Du (eds.), *Non-Antibiotic Therapies for Helicobacter pylori*,
https://doi.org/10.1007/978-981-95-9598-3_1

the American College of Gastroenterology (ACG) guidelines, have provided standardized definitions and management strategies for *H. pylori* infection [15, 16].

Research on *H. pylori* infection traces the path of a newly identified pathogen from discovery to broad acceptance and, at the same time, marks a paradigm shift in gastric cancer prevention and the management of gastrointestinal diseases. This chapter reviews the definition and historical development of *H. pylori* infection and provides the theoretical and historical foundation for later discussions on nonantibiotic treatment strategies.

## 1.1 Definition of *H. pylori* Infection

### 1.1.1 Microbiological Definition

*H. pylori* is a Gram-negative, microaerophilic, spiral-shaped bacterium with 4–6 flagella, conferring strong motility and the ability to penetrate gastric mucus [6, 17]. One of its defining biological features is the production of abundant urease, enabling survival in the acidic stomach by hydrolyzing urea to ammonia and carbon dioxide, thereby buffering the local microenvironment [18].

### 1.1.2 Conceptual and Clinical Definition

The term "infection" refers specifically to the persistent colonization of the gastric mucosa by *H. pylori*, regardless of symptoms or associated diseases [15, 19]. While most infected individuals remain asymptomatic, some progress to chronic gastritis, peptic ulcer disease, or gastric malignancies. This distinction highlights infection as a biological state, with clinical significance determined by subsequent pathological consequences.

Historically, *H. pylori* was first recognized as the etiological basis of peptic ulcer disease at the 1994 NIH Consensus Meeting [4]. More recently, the 2015 Kyoto Global Consensus expanded this view, explicitly defining *H. pylori* infection as an infectious disease in its own right, warranting both diagnosis and treatment, rather than merely being a risk factor for gastritis or ulceration. Current international guidelines consistently emphasize *H. pylori* infection as the principal target of gastric cancer prevention strategies [15, 16, 20, 53].

### 1.1.3 Diagnostic Approaches

Diagnosis has evolved from invasive endoscopic methods (histology, culture, rapid urease test) to noninvasive approaches. The urea breath test is now the preferred tool due to its high sensitivity and specificity [21]. Stool antigen testing is particularly suitable for children and patients unable to undergo endoscopy and can be integrated into colorectal cancer screening protocols [22]. Serology, though useful for epidemiological studies, cannot differentiate current from past infection and is no longer recommended for routine diagnosis [23].

## 1.2 Discovery and Early Research History

### 1.2.1 Initial Observations and Hypothesis

In the late 1970s, Australian pathologist Robin Warren repeatedly observed curved bacilli in gastric biopsy specimens from chronic gastritis patients, hypothesizing a microbial role in gastric inflammation [2]. At that time, however, the stomach was widely considered a sterile environment, and these observations initially received little attention. In 1983, Warren and clinician Barry Marshall reported their observations of these "unidentified curved bacilli" in patients with gastritis in a correspondence published in *The Lancet* [2]. In the subsequent year, they demonstrated a high prevalence of the organism in patients with peptic ulcer disease and were the first to propose that *H. pylori* could serve as the etiological agent of peptic ulceration [1]. This hypothesis fundamentally challenged the prevailing view that excessive gastric acid was the sole cause of peptic ulcer disease and provoked substantial debate within the medical community.

### 1.2.2 Experimental and Clinical Evidence

To prove causality, Marshall ingested a broth containing *H. pylori* in 1985, subsequently developing acute gastritis confirmed by endoscopy and histology [3]. Though ulcers did not develop, this experiment fulfilled Koch's postulates in principle. Clinical evidence soon followed: a pivotal 1990 Lancet study by Rauws et al. demonstrated that duodenal ulcers healed permanently only when *H. pylori* was eradicated, while persistence of infection led to recurrence [24].

### 1.2.3 Acceptance and Milestones

Accumulating evidence shifted the scientific community from skepticism to acceptance. The 1994 NIH Consensus formally recognized *H. pylori* infection as the primary cause of peptic ulcer disease [4]. Taxonomically, Goodwin et al. reclassified the organism from *Campylobacter pyloridis* to *H. pylori* in 1989 based on phylogenetic analysis [6].

Beyond ulcer disease, prospective cohort studies established *H. pylori* as a carcinogenic factor, linking infection to gastric cancer [10, 11]. In recognition of their groundbreaking work, Marshall and Warren received the 2005 Nobel Prize in Physiology or Medicine, underscoring the global impact of this discovery [25].

## 1.3 Structure and Molecular Features

*H. pylori* exhibits a range of distinctive morphological and molecular characteristics that allow it to persist in the highly acidic gastric environment and to maintain long-term interactions with its human host.

### 1.3.1 Genomic Characteristics

The complete *H. pylori* genome, published in Nature in 1997, revealed a size of 1.67 Mb encoding 1590 genes—remarkably small compared with other bacteria [7]. This compact yet diverse genome underlies its high adaptability, genetic variability, and propensity to develop antibiotic resistance.

### 1.3.2 Virulence Factors

The pathogenicity of *H. pylori* is mediated by multiple virulence determinants, among which the cag pathogenicity island (cagPAI) and the vacuolating cytotoxin VacA are the most extensively characterized and most closely linked to disease development. *Cag*PAI encodes a type IV secretion system that translocates CagA into host epithelial cells [8, 26]. Phosphorylated CagA activates *SHP*2/ERK signaling, causing cytoskeletal rearrangements, abnormal proliferation, and increased oncogenic potential [9]. The vacuolating cytotoxin VacA, with allelic polymorphisms in its signal (s) and mid (m) regions, induces apoptosis via mitochondrial pathways [27, 28]. Other outer membrane proteins (such as BabA, SabA) mediate adhesion, while enzymes such as γ-glutamyl transferase and lipopolysaccharide contribute to immune modulation [17].

Overall, strains harboring the cag pathogenicity island (cagPAI) elicit more pronounced gastric mucosal inflammation and are strongly associated with an increased risk of severe clinical outcomes, including gastric cancer, thereby representing the principal pathogenic determinant of *H. pylori*–related gastric carcinogenesis [9, 10]. In contrast, the distinct genotypes of *vacA*—particularly high-toxicity alleles—modulate disease phenotype and risk primarily by influencing the extent of epithelial injury and inflammatory responses [28, 29]. Moreover, the extensive genomic polymorphism of *H. pylori* results in marked geographic and host-specific variation in virulence profiles. Acting in concert with host genetic susceptibility and environmental exposures, this bacterial heterogeneity provides a biological explanation for the paradox that *H. pylori* infection is highly prevalent worldwide, whereas only a minority of infected individuals ultimately progress to peptic ulcer disease or gastric cancer [30, 31].

### 1.3.3 Survival and Colonization Mechanisms

Long-term persistence of *H. pylori* in the highly acidic gastric environment is supported by several specialized adaptive mechanisms.

#### 1.3.3.1 Urease Activity

*H. pylori* possesses a pH-dependent urease system that hydrolyzes urea under acidic conditions to generate ammonia, thereby neutralizing local acidity and creating a favorable microenvironment for survival; this mechanism constitutes the biological basis of the urea breath test and the rapid urease test [18, 32].

#### 1.3.3.2 Flagellar Motility

*H. pylori* is equipped with multiple flagella, and the resulting motility enables traversal of the gastric mucus layer and access to relatively low-acidity niches adjacent to epithelial cells, thereby facilitating initial colonization [6, 33].

#### 1.3.3.3 Adhesion

*H. pylori* achieves stable attachment by binding host adhesion molecules through outer membrane adhesins, such as BabA [34] and SabA [35], thereby limiting clearance by gastric contents and sustaining persistent infection.

## 1.4 *H. pylori*-Associated Diseases

### 1.4.1 Peptic Ulcer Disease

*H. pylori* is the leading cause of peptic ulcer disease. Since Warren and Marshall identified its role in the 1980s, extensive evidence has shown that eradication therapy dramatically reduces ulcer recurrence. In a landmark trial, duodenal ulcer patients who achieved eradication had a 12-month recurrence rate of 0%, compared with 89% in those who remained infected [24]. Current international guidelines [15, 16] consistently recommend routine testing for *H. pylori* infection and eradication therapy in patients with active peptic ulcer disease or a documented history of ulceration, with antimicrobial treatment now recognized as a cornerstone of peptic ulcer management.

### 1.4.2 Gastric Cancer

The correlation between *H. pylori* and gastric cancer(GC) was first demonstrated by Parsonnet et al. in 1991 [11]. In 1994, the International Agency for Research on Cancer classified *H. pylori* as a Group I carcinogen, a designation reaffirmed in 2022 [5]. The 2020 Taipei Consensus further estimated that nearly 89% of non-cardia gastric cancers worldwide are attributable to *H. pylori* [12].

Carcinogenesis follows the Correa cascade: chronic gastritis progresses to atrophy, intestinal metaplasia, dysplasia, and eventually carcinoma [36]. CagA, delivered by the cag PAI, activates multiple signaling pathways that drive abnormal proliferation and loss of polarity [8, 9]. VacA toxin promotes persistence by inducing apoptosis and suppressing host immunity [27, 28], while impaired DNA repair accelerates transformation. Interactions between bacterial virulence genotypes and host polymorphisms, such as IL-1B, further explain variation in cancer risk [9].

Clinical data provide compelling evidence that eradication lowers gastric cancer risk. In 2018, a randomized controlled trial conducted by Choi et al. in Korea demonstrated that *H. pylori* eradication reduced the risk of metachronous gastric cancer after endoscopic resection by 50% (hazard ratio, 0.50; 95% confidence interval, 0.26–0.94; $P = 0.03$) [37]. In addition, a large population-based randomized trial of 148,000 individuals reported about a one-third reduction in cancer risk after eradication, with the greatest benefit in younger individuals [38]. These findings underpin the "test-and-treat" strategy and have prompted high-incidence countries to include *H. pylori* screening and eradication in primary prevention. As the only GC risk factor amenable to reversal through anti-infective therapy, *H. pylori* has been designated by multiple international consensus statements—including the Maastricht [15], ACG [16], and Taipei [20] consensus reports—as a primary target for population-level prevention, and represents the most important preventable and controllable risk factor for gastric cancer.

### 1.4.3 Gastric Mucosa-Associated Lymphoid Tissue (MALT) Lymphoma

*H. pylori* infection is also causally linked to gastric MALT lymphoma. Chronic colonization drives persistent B-cell activation and clonal expansion. In 1993, Wotherspoon et al. reported complete regression of low-grade gastric MALT lymphoma after *H. pylori* eradication [39]. Bayerdörffer et al. later confirmed remission or stabilization in most patients [40]. These studies established eradication as both infection control and tumor therapy, making MALT lymphoma a paradigm of infection-related malignancy. Current guidelines, including Maastricht VI, recommend eradication as first-line therapy for early gastric MALT lymphoma [15].

### 1.4.4 Extra-Gastric Diseases

#### 1.4.4.1 Iron-Deficiency Anemia and Immune Thrombocytopenic Purpura

Beyond the stomach, *H. pylori* has been implicated in systemic conditions, most consistently iron-deficiency anemia (IDA) and immune thrombocytopenic purpura (ITP). In patients with unexplained IDA, eradication often improves hemoglobin levels, likely by restoring acid secretion and iron absorption [21]. In ITP, molecular mimicry and cross-reactive immune responses may contribute to platelet destruction, and several studies show platelet recovery after eradication [23]. Consequently, international guidelines recommend *H. pylori* testing and eradication in patients with unexplained IDA or ITP [15, 16].

#### 1.4.4.2 Colorectal Cancer (CRC)

The possible link between *H. pylori* and colorectal cancer remains debated. A 2024 community-based RCT in China (>180,000 participants) found no association between *H. pylori* infection and CRC incidence or mortality [38]. In contrast, a US veterans cohort study (>800,000 participants) reported increased CRC risk and mortality in *H. pylori*-positive individuals, with risk significantly reduced after eradication [41]. A meta-analysis of nearly 49 million participants further supported an association between *H. pylori* and CRC and adenoma, with eradication lowering the risk of incident CRC [42]. Mechanistic studies suggest *H. pylori* may promote CRC through immune modulation, STAT3 activation, and disruption of the mucosal barrier [43]. While these findings are suggestive, causality remains uncertain and requires confirmation in well-designed prospective studies.

## 1.5 Evolution of *H. pylori* Management Strategies

Management strategies for *H. pylori* have shifted profoundly from individual clinical treatment toward population-level prevention. In the early years following its discovery, the primary goal of *H. pylori* management was to control gastritis and prevent recurrence of peptic ulcers. Entering the twenty-first century, high-quality prospective studies demonstrated that eradication of *H. pylori* not only reduces ulcer relapse but also halts progression of precancerous lesions such as atrophy and intestinal metaplasia, thereby lowering both gastric cancer incidence and mortality [13, 37, 38]. Thus, *H. pylori* management has moved beyond clinical care and become a cornerstone of primary prevention for gastric cancer.

### 1.5.1 "Test-and-Treat" Strategy

As evidence of the causal role of *H. pylori* in gastric cancer accumulated, the Maastricht II Consensus (2002) proposed the "test-and-treat" strategy [44]. The guideline recommended *H. pylori* testing and eradication in dyspeptic patients under 45 years of age without alarm symptoms, particularly in high-prevalence regions. However, because this approach relied on healthcare-seeking behavior and was largely limited to hospital-based settings, it proved insufficient for reducing overall infection levels in high-burden populations.

### 1.5.2 "Screen-and-Treat" Strategy

In 2012, the Maastricht IV Consensus proposed the "screen-and-treat" strategy [45]. This approach expanded the target population from symptomatic patients to the general public in high-risk regions, marking a transition toward proactive, population-based prevention. Its effectiveness has been supported by large-scale studies across Europe and East Asia. For example, a nationwide cohort in Korea (2009–2021) involving 916,438 infected individuals demonstrated that eradication therapy reduced standardized gastric cancer incidence and mortality by 38% and 69%, respectively [46]. Furthermore, health economic analysis suggests that, the "screen-and-treat" strategy is the most cost-effective gastric cancer prevention measure compared to no screening or endoscopic screening [47]. The above evidence provides strong support for incorporating the "screen-and-treat" strategy into government public health management measures.

### 1.5.3 "Family-Based" Strategy

*H. pylori* infection is highly prevalent within families, with parents identified as the primary source of transmission to children, and significant cross-transmission observed between spouses and siblings [48]. Based on this, Chinese researchers proposed the "family-based management" strategy in 2022, recommending that household members of diagnosed patients be tested and treated simultaneously [49].

A large-scale nationwide cohort study in China reported a household-level infection rate of 71.2%, substantially exceeding the individual infection rate of 40.7%. In approximately 20% of households, all family members are infected [50]. These findings indicate that the "family-based" strategy may offer greater efficiency in identifying individuals with *H. pylori* infection [51]. In addition, meta-analyses showed that "family-based" eradication reduces the recurrence rate of *H. pylori* infection from 18.4% to 7.2% and is also associated with favorable cost-effectiveness [52]. The "family-based' strategy therefore demonstrates advantages in case detection, interruption of transmission, and economic efficiency, supporting its potential role as an important public health strategy in regions with a high prevalence of *H. pylori* infection.

## References

1. Marshall B, Warren JR. Unidentified curved bacilli in the stomach of patients with gastritis and peptic ulceration. Lancet. 1984;323(8390):1311–5. https://doi.org/10.1016/S0140-6736(84)91816-6.
2. Warren JR, Marshall B. Unidentified curved bacilli on gastric epithelium in active chronic gastritis. Lancet. 1983;1(8336):1273–5.
3. Marshall BJ, Armstrong JA, McGeochie DB, Glancy RJ. Attempt to fulfil Koch's postulates for pyloric campylobacter. Med J Aust. 1985;142:436–9.
4. NIH Consensus Development Panel on H. pylori in Peptic Ulcer Disease. H. pylori in peptic ulcer disease. NIH Consensus Dev Conf Statement 1994;12(1):1–23.

5. IARC. IARC Monographs on the Identification of Carcinogenic Hazards to Humans. https://monographs.iarc.who.int/list-of-classifications. Accessed 2025 Oct 8
6. Goodwin CS, Armstrong JA, Chilvers T, Peters M, Collins MD, Sly L, et al. Transfer of campylobacter pylori and campylobacter mustelae to helicobacter gen. Nov. as H. pylori comb. nov. and helicobacter mustelae comb. nov., respectively. Int J Syst Evol Microbiol. 1989;39(4):397–405. https://doi.org/10.1099/00207713-39-4-397.
7. Tomb JF, White O, Kerlavage AR, Clayton RA, Sutton GG, Fleischmann RD, et al. The complete genome sequence of the gastric pathogen H. pylori. Science. 1997;277(5323):1770–8. https://doi.org/10.1126/science.277.5323.1770.
8. Censini S, Lange C, Xiang Z, Crabtree JE, Ghiara P, Borodovsky M, et al. cag, a pathogenicity Island of H. pylori, encodes type I-specific and disease-associated virulence factors. Proc Natl Acad Sci USA. 1996;93(11):5788–92.
9. Hatakeyama M. H. pylori CagA and gastric cancer: a paradigm for hit-and-run carcinogenesis. Cell Host Microbe. 2014;15(3):306–16. https://doi.org/10.1016/j.chom.2014.02.008.
10. Uemura N, Okamoto S, Yamamoto S, Matsumura N, Yamaguchi S, Yamakido M, et al. H. pylori infection and the development of gastric cancer. N Engl J Med. 2001;345(11):784–9.
11. Parsonnet J, Friedman GD, Vandersteen DP, Chang Y, Vogelman JH, Orentreich N, et al. H. pylori infection and the risk of gastric carcinoma. N Engl J Med. 1991;325(16):1127–31.
12. Liou JM, Malfertheiner P, Lee YC, Sheu BS, Sugano K, Cheng HC, et al. Screening and eradication of H. pylori for gastric cancer prevention: the Taipei global consensus. Gut. 2020;69(12):2093–112. https://doi.org/10.1136/gutjnl-2020-322368.
13. Ford AC, Yuan Y, Park JY, Forman D, Moayyedi P. Eradication therapy to prevent gastric cancer in helicobacter pylori-positive individuals: systematic review and meta-analysis of randomized controlled trials and observational studies. Gastroenterology. 2025;169(2):261–76. https://doi.org/10.1053/j.gastro.2024.12.033.
14. Ford AC, Yuan Y, Moayyedi P. H. pylori eradication therapy to prevent gastric cancer: systematic review and meta-analysis. Gut. 2020;69(12):2113–21. https://doi.org/10.1136/gutjnl-2020-320839.
15. Malfertheiner P, Megraud F, Rokkas T, Gisbert JP, Liou JM, Schulz C, et al. Management of H. pylori infection: the Maastricht VI/Florence consensus report. Gut. 2022;71(9):1893–908. https://doi.org/10.1136/gutjnl-2022-327745.
16. Chey WD, Howden CW, Moss SF, Morgan DR, Greer KB, Grover S, et al. ACG clinical guideline: treatment of H. pylori infection. Am J Gastroenterol. 2024;119(9):1730–53. https://doi.org/10.14309/ajg.0000000000002968.
17. Windsor HM, O'Rourke J. Bacteriology and taxonomy of H. pylori. Gastroenterol Clin North Am. 2000;29(3):633–648. https://doi.org/10.1016/S0889-8553(05)70134-X.
18. Eaton KA, Brooks CL, Morgan DR, Krakowka S. Essential role of urease in pathogenesis of gastritis induced by H. pylori in gnotobiotic piglets. Infect Immun. 1991;59(10):3615–22.
19. Malfertheiner P, Camargo MC, El-Omar E, Liou JM, Peek R, Schulz C, et al. H. pylori infection. Nat Rev Dis Primers. 2023;9(1):19. https://doi.org/10.1038/s41572-023-00431-8.
20. Liou JM, Malfertheiner P, Hong TC, Cheng HC, Sugano K, Shah S, et al. Screening and eradication of H. pylori for gastric cancer prevention: Taipei global consensus II. Gut. 2025;74(11):1767–91. https://doi.org/10.1136/gutjnl-2025-336027.
21. Shatila M, Thomas AS. Current and future perspectives in the diagnosis and management of H. pylori infection. J Clin Med. 2022;11(17):5086. https://doi.org/10.3390/jcm11175086.
22. Lee YC, Chiang TH, Chiu HM, Su WW, Chou KC, Chen SL, et al. Screening for H. pylori to prevent gastric cancer: a pragmatic randomized clinical trial. JAMA. 2024;332(19):1642–51. https://doi.org/10.1001/jama.2024.14887.
23. Saleem N, Howden CW. Update on the management of H. pylori infection. Curr Treat Options Gastroenterol. 2020;18(3):476–87. https://doi.org/10.1007/s11938-020-00300-3.
24. Rauws EA, Tytgat GN. Cure of duodenal ulcer associated with eradication of H. pylori. Lancet. 1990;335(8690):1233–5.
25. Pincock S. Nobel prize winners Robin Warren and Barry Marshall. Lancet. 2005;366(9495):1429. https://doi.org/10.1016/S0140-6736(05)67587-3.

26. Odenbreit S, Püls J, Sedlmaier B, Gerland E, Fischer W, Haas R. Translocation of H. pylori CagA into gastric epithelial cells by type IV secretion. Science. 2000;287(5457):1497–500. https://doi.org/10.1126/science.287.5457.1497.
27. Atherton JC, Cao P, Peek RM Jr, Tummuru MK, Blaser MJ, Cover TL. Mosaicism in vacuolating cytotoxin alleles of H. pylori. Association of specific vacA types with cytotoxin production and peptic ulceration. J Biol Chem. 1995;270(30):17771–7. https://doi.org/10.1074/jbc.270.30.17771.
28. Foegeding NJ, Caston RR, McClain MS, Ohi MD, Cover TL. An overview of H. pylori VacA toxin biology. Toxins (Basel). 2016;8(6):173. https://doi.org/10.3390/toxins8060173.
29. Rhead JL, Letley DP, Mohammadi M, Hussein N, Mohagheghi MA, Eshagh Hosseini M, et al. A new H. pylori vacuolating cytotoxin determinant, the intermediate region, is associated with gastric cancer. Gastroenterology. 2007;133(3):926–36. https://doi.org/10.1053/j.gastro.2007.06.056.
30. Falush D, Wirth T, Linz B, Pritchard JK, Stephens M, Kidd M, et al. Traces of human migrations in H. pylori populations. Science. 2003;299(5612):1582–5. https://doi.org/10.1126/science.1080857.
31. Linz B, Balloux F, Moodley Y, Manica A, Liu H, Roumagnac P, et al. An African origin for the intimate association between humans and H. pylori. Nature. 2007;445(7130):915–918. https://doi.org/10.1038/nature05562.
32. Scott DR, Marcus EA, Weeks DL, Sachs G. Mechanisms of acid resistance due to the urease system of H. pylori. Gastroenterology. 2002;123(1):187–95. https://doi.org/10.1053/gast.2002.34218.
33. Ottemann KM, Lowenthal AC. H. pylori uses motility for initial colonization and to attain robust infection. Infect Immun. 2002;70(4):1984–90. https://doi.org/10.1128/iai.70.4.1984-1990.2002.
34. Ilver D, Arnqvist A, Ogren J, Frick IM, Kersulyte D, Incecik ET, et al. H. pylori adhesin binding fucosylated histo-blood group antigens revealed by retagging. Science. 1998;279(5349):373–7. https://doi.org/10.1126/science.279.5349.373.
35. Mahdavi J, Sondén B, Hurtig M, Olfat FO, Forsberg L, Roche N, et al. H. pylori SabA adhesin in persistent infection and chronic inflammation. Science. 2002;297(5581):573–8. https://doi.org/10.1126/science.1069076.
36. Correa P. Human gastric carcinogenesis: a multistep and multifactorial process – first American Cancer Society award lecture on cancer epidemiology and prevention. Cancer Res. 1992;52(24):6735–40.
37. Choi IJ, Kook MC, Kim YI, Cho SJ, Lee JY, Kim CG, et al. H. pylori therapy for the prevention of Metachronous gastric cancer. N Engl J Med. 2018;378(12):1085–95. https://doi.org/10.1056/NEJMoa1708423.
38. Pan KF, Li WQ, Zhang L, Liu WD, Ma JL, Zhang Y, et al. Gastric cancer prevention by community eradication of H. pylori: a cluster-randomized controlled trial. Nat Med. 2024;30(11):3250–60. https://doi.org/10.1038/s41591-024-03153-w.
39. Wotherspoon AC, Doglioni C, Diss TC, Pan L, Moschini A, de Boni M, et al. Regression of primary low-grade B-cell gastric lymphoma of mucosa-associated lymphoid tissue type after eradication of H. pylori. Lancet. 1993;342(8879):1008–12.
40. Bayerdörffer E, Neubauer A, Rudolph B, Thiede C, Lehn N, Eidt S, et al. Regression of primary gastric lymphoma of mucosa-associated lymphoid tissue type after cure of H. pylori infection. MALT lymphoma study group. Lancet. 1995;345(8959):1591–4.
41. Shah SC, Camargo MC, Lamm M, Bustamante R, Roumie CL, Wilson O, et al. Impact of H. pylori infection and treatment on colorectal cancer in a large, nationwide cohort. J Clin Oncol. 2024;42(16):1881–9. https://doi.org/10.1200/JCO.23.00703.
42. Li X, Tao HQ, Zhao JE, Zhu J, Du LB, Gerhard M, et al. H. pylori infection, anti-H. pylori treatment and risk of colorectal cancer and adenoma: an observational study and a meta-analysis. EClinicalMedicine. 2025;84:103299. https://doi.org/10.1016/j.eclinm.2025.103299.

43. Ralser A, Dietl A, Jarosch S, Engelsberger V, Wanisch A, Janssen KP, et al. H. pylori promotes colorectal carcinogenesis by deregulating intestinal immunity and inducing a mucus-degrading microbiota signature. Gut. 2023;72(7):1258–70. https://doi.org/10.1136/gutjnl-2022-328075.
44. Malfertheiner P, Mégraud F, O'Morain C, Hungin APS, Jones R, Axon A, et al. Current concepts in the management of H. pylori infection – the Maastricht 2-2000 consensus report. Aliment Pharmacol Ther. 2002;16(Suppl 1):S1–S16.
45. Malfertheiner P, Megraud F, O'Morain CA, Atherton J, Axon AT, Bazzoli F, et al. Management of H. pylori infection – the Maastricht IV/Florence consensus report. Gut. 2012;61(5):646–64. https://doi.org/10.1136/gutjnl-2012-302084.
46. Jung YS, Tran MTX, Park B, Moon CM. Preventive effect of H. pylori treatment on gastric cancer incidence and mortality: a Korean population study. Gastroenterology. 2025;170(1):284–93. https://doi.org/10.1053/j.gastro.2025.03.036.
47. Wang Z, Han W, Xue F, Zhao Y, Wu P, Chen Y, et al. Nationwide gastric cancer prevention in China, 2021-2035: a decision analysis on effect, affordability and cost-effectiveness optimisation. BMJ Open Gastroenterol. 2024;1(1):e000215. https://doi.org/10.1136/bmjgast-2024-000215.
48. Ding SZ. Global whole family based-H. pylori eradication strategy to prevent its related diseases and gastric cancer. World J Gastroenterol. 2020;26(10):995–1004. https://doi.org/10.3748/wjg.v26.i10.995.
49. Ding SZ, Du YQ, Lu H, Wang WH, Cheng H, Chen SY, et al. Chinese consensus report on family-based H. pylori infection control and management (2021 edition). Gut. 2022;71(2):238–53. https://doi.org/10.1136/gutjnl-2021-325630.
50. Zhou XZ, Lyu NH, Zhu HY, Cai QC, Kong XY, Xie P, et al. Large-scale, national, family-based epidemiological study on H. pylori infection in China: the time to change practice for related disease prevention. Gut. 2023;72(5):855–69. https://doi.org/10.1136/gutjnl-2022-328965.
51. Zhao J-B, Yuan L, Yu X-C, Shao Q-Q, Ma J, Yu M, et al. Whole family-based H. pylori eradication is a superior strategy to single-infected patient treatment approach: a systematic review and meta-analysis. Helicobacter. 2021;26(3):e12793. https://doi.org/10.1111/hel.12793.
52. Ma Y, Zhou X, Liu Y, Xu S, Ma A, Du Y, et al. An economic evaluation of family-based versus traditional H. pylori screen-and-treat strategy: based on real-world data and microsimulation model. Helicobacter. 2024;29(4):e13123. https://doi.org/10.1111/hel.13123.
53. Sugano K, Tack J, Kuipers EJ, Graham DY, El-Omar EM, Miura S, Haruma K, Asaka M, Uemura N, Malfertheiner P; faculty members of Kyoto Global Consensus Conference. Kyoto global consensus report on Helicobacter pylori gastritis. Gut. 2015 Sep;64(9):1353–1367. https://doi.org/10.1136/gutjnl-2015-309252. Epub 2015 Jul 17. PMID: 26187502;PMCID: PMC4552923./p.

# Biological Characteristics of *H. pylori* 2

Chuan Yin, Jing Wang, and Te Shi

*Helicobacter pylori* is a Gram-negative, microaerophilic, spiral-shaped bacterium. Since its discovery by Marshall and Warren in 1982, it has represented a milestone breakthrough in the research of gastrointestinal diseases. This bacterium can colonize the highly acidic environment of the human stomach and is closely associated with conditions such as chronic gastritis, peptic ulcers, and gastric cancer [1]. This chapter will systematically elaborate on its biological characteristics, including fundamental morphology, ultrastructure, physiological properties, and molecular biological mechanisms, to deepen the understanding of its pathogenic processes.

## 2.1 The Morphology of *H. pylori*

### 2.1.1 Microscopic Morphology of *H. pylori*

*H. pylori* appears as a typical spiral, S-shaped, or curved rod under the light microscope. In wet mounts prepared from fresh cultures, the bacteria exhibit rapid corkscrew or tumbling motility, facilitated by their flagella, which enables them to

C. Yin (✉)
Department of Gastroenterology, Shanghai Changzheng Hospital, Naval Medical University, Shanghai, China
e-mail: chuanyinmd@smmu.edu.cn

J. Wang
Changhai Clinical Research Unit, Shanghai Changhai Hospital, Naval Medical University, Shanghai, China

College of Life Science, Mudanjiang Medical University, Mudanjiang, China

Te Shi
Department of Gastroenterology, Chinese People's Liberation Army of Naval Medical Center, Shanghai, China


J. Gao, Y. Du (eds.), *Non-Antibiotic Therapies for Helicobacter pylori*,
https://doi.org/10.1007/978-981-95-9598-3_2

penetrate the mucus layer covering the gastric mucosa. They are Gram-negative, measuring approximately 0.5–1.0 μm in width and 2.0–4.0 μm in length, and lack both a capsule and the ability to form spores [2].

Under the electron microscope, the bacterial surface appears smooth with blunt, rounded ends and a spiral curvature. Most bacterial cells possess a tuft of 2–6 sheathed flagella at one end (although bipolar flagellation has also been observed). The flagella are approximately 2.5 μm in length and 30 nm in diameter [3]. Filamentous appendages resembling pili on the bacterial surface are considered crucial for mediating adhesion to gastric mucosal epithelial cells. The propulsion generated by the flagella allows *H. pylori* to traverse the mucus layer and ultimately colonize the surface of epithelial cells.

In aged cultures or under unfavorable conditions, *H. pylori* can transform into coccoid forms. When subjected to Gram staining in this state, the bacteria exhibit variation in size, uneven staining intensity, and poorly defined boundaries [4]. In direct smears from gastric biopsy specimens, *H. pylori* is often found clustered within lightly stained mucus bands, typically arranged in a "file-like" pattern. Bacterial colonies are usually dense with few contaminating microbes, and a microcapsule-like structure can sometimes be observed surrounding the bacterial cells. In histopathological sections, *H. pylori* is primarily located on the surface of gastric mucosal epithelial cells, particularly within the gastric pits. It is rarely found in areas of intestinal metaplasia within the stomach but can be present in areas of gastric metaplasia in the duodenum.

When cultured under microaerophilic conditions at 37 °C for approximately 72 h, *H. pylori* forms small, raised colonies with neat edges on solid media, each measuring less than 1 mm in diameter and appearing gray-white. Opening the culture plate reveals a distinctive odor [5]. Wet mounts prepared from fresh colonies allow observation of the characteristic spiral and tumbling motility under dark-field microscopy [6].

### 2.1.2 Ultrastructure of *H. pylori*

*H. pylori* exhibits highly specialized ultrastructural features under electron microscopy, particularly in its flagellar apparatus, which possesses a complex and precise structure closely related to its motility and colonization ability. Under transmission electron microscopy, a circular basal body is visible at the root of each flagellum, embedded within the inner bacterial cell wall. This region is often accompanied by a distinct electron-lucent zone, speculated to be involved in energy storage or conversion for flagellar movement. The flagellum has a diameter of approximately 30 nm, and its base connects to a circular structure about 50 nm in diameter on the cell membrane. Below the basal body lies an electron-transparent zone, 20–40 nm wide, which lacks a clear membrane boundary separating it from the cytoplasmic matrix [2]. The overall flagellar structure can be divided into three main parts: the basal body, the hook, and the filament, similar to most Gram-negative bacteria.

However, its unique feature is a sheath-like structure, the "flagellar sheath," enveloping the filament.

#### 2.1.2.1 Basal Body

Embedded within the cell wall, the basal body consists of a central rod and two pairs of concentric rings. The proximal pair, the M and S rings, are anchored to the cell membrane; the distal pair, the P and L rings, connect to the peptidoglycan layer and the lipopolysaccharide of the outer membrane, respectively. As *H. pylori* is Gram-negative, the presence of P and L rings is a key feature distinguishing it from Gram-positive bacteria, which possess only M and S rings.

#### 2.1.2.2 Hook

The hook is the curved section where the flagellum extends out of the bacterium. It is hook-shaped, approximately 17 nm in diameter and 45 nm long, acting as a flexible joint connecting the basal body to the filament and composed of various polypeptide polymers.

#### 2.1.2.3 Filament

The filament is the fibrous part extending beyond the bacterium. It is a hollow helical structure formed by the polymerization of flagellin subunits, with a diameter of about 13.5 nm and a length reaching up to 20 μm. Its chemical composition shares similarities with skeletal muscle actin, relating to its motor function.

#### 2.1.2.4 Flagellar Sheath

The flagellar sheath is a lipoprotein bilayer structure formed by the extension of the bacterial outer membrane. It originates from the basal body, enveloping the hook and part of the filament. This structure may protect the filament and help resist the gastric acidic environment. Furthermore, after tannic acid treatment, a layer of glycocalyx about 40 nm thick can be observed covering the bacterial surface. This glycocalyx appears filamentous or reticulated under EM, connects to gastric mucosal epithelial cells, and is a crucial material basis mediating bacterial adhesion. The colonization of *H. pylori* relies not only on flagella and the glycocalyx; electron microscopy also frequently reveals direct attachment between the bacterial cell wall and the host epithelial cell membran [3].

## 2.2 Physiological and in vitro Culture Characteristics of *H. pylori*

*H. pylori* is a microaerophilic, Gram-negative bacterium with unique physiological traits, enabling it to colonize the highly acidic human stomach. Its adaptation relies on specialized enzyme systems and metabolic pathways. Below, we systematically elaborate on its physiological characteristics from four aspects: gas requirements, culture medium properties, growth conditions, and energy metabolism.

### 2.2.1 Gas Requirements and Cultivation Environment

*H. pylori* is a microaerophile, requiring an oxygen concentration of 5–8%, along with 10% $CO_2$ and 85% $N_2$, to maintain a low-oxygen environment. High humidity (>95%) enhances its survival. Growth is inhibited under anaerobic or atmospheric oxygen conditions; thus, a triple-gas incubator or continuous agitation (for liquid culture) is necessary to preserve microaerophilic conditions [7].

### 2.2.2 Nutritional Needs and Culture Medium Properties

#### 2.2.2.1 Basic Medium Components

*H. pylori* depends on blood- or serum-supplemented media for growth factors. Common media include Columbia blood agar and Brucella agar, requiring additions of whole blood or fetal bovine serum [8]. Commercial media typically contain peptone, yeast extract, inorganic salts, and inhibitors (e.g., vancomycin, amphotericin B) to suppress contaminants.

#### 2.2.2.2 Additives and Optimization Strategies

Adding sodium pyruvate or ferrous sulfate can increase bacterial yield, while soluble starch (1%) or activated charcoal (0.2%) helps adsorb harmful oxygen radicals. To inhibit fungi and Gram-positive bacteria, antibiotics like vancomycin (6 mg/L) and amphotericin B (2 mg/L) are often incorporated [9].

#### 2.2.2.3 Colony Morphology and Growth Cycle

After 3–7 days of incubation at 37 °C, *H. pylori* forms pinprick-sized (diameter < 1 mm), grayish-white, translucent colonies with slight hemolysis on blood agar. Subculture cycles take 2–3 days, and aging cultures may transform into coccoid forms (metabolically dormant state).

### 2.2.3 Growth Conditions and Stability

#### 2.2.3.1 Temperature and pH Adaptability

The optimal growth temperature is 37 °C (range: 30–37 °C). No growth occurs at 25 °C, and few strains survive at 42 °C. The preferred pH range is 6.6–7.2. However, its urease hydrolyzes urea to produce ammonia, creating an alkaline microclimate that enables survival in stomach acid below pH 3.5.

#### 2.2.3.2 Preservation and Stability

Long-term storage requires ultra-low temperatures (−80 °C or liquid nitrogen). Under ambient conditions, survival is limited to 2–14 days (depending on $O_2$ and humidity). Liquid cultures need agitation to maintain oxygen distribution and remove metabolic waste.

### 2.2.4 Energy Metabolism and Carbon Source Utilization

*H. pylori* primarily utilizes organic acids (e.g., succinate) and amino acids as carbon sources. Essential amino acids include arginine, isoleucine, histidine, methionine, leucine, valine, and phenylalanine, with some strains requiring serine or alanin [10]. Although glucose is nonessential, it synergizes with alanine to significantly promote growth. Energy generation depends on a unique respiratory quinone system (distinct from Campylobacter's menaquinone-6), potentially linked to the pentose phosphate pathway and organic acid oxidation [11].

In conclusion, the physiological features of *H. pylori*—microaerophilic dependence, complex nutritional demands, and versatile energy metabolism—underscore its adaptability. Urease activity is central to gastric colonization. Successful clinical cultivation necessitates strict control of gas, humidity, and medium composition, leveraging inherent resistance (e.g., to vancomycin) for selective culture. Deeper insights into its metabolic pathways may inform novel targeted therapies.

## 2.3 Molecular Biological Characteristics of *H. pylori*

*H. pylori* exhibits distinct molecular biological features characterized by high genomic diversity and complex functions of virulence genes. These traits collectively determine the pathogenicity, adaptability, and host–bacterium interactions of the strain. The following sections systematically elaborate on its DNA composition, genetic diversity, and key virulence genes.

### 2.3.1 DNA Composition and Genetic Diversity

The genome of *H. pylori* is relatively small but demonstrates significant genetic polymorphism. For example, the standard strain 26,695 has a circular genome of 1.66 Mb with 1552 coding regions and an average GC content of 39% [12]. Genomic regions with heterogeneous GC content contain insertion sequences (e.g., IS605 and IS606), whose copy numbers and distribution vary markedly among strains. These sequences are often adjacent to 5S rRNA genes or repetitive sequences and are associated with variations in pathogenicity islands (e.g., cag-PAI). The length and sequence of cag-PAI are highly variable across strains, directly influencing bacterial virulence. Comparative genomics between strains 26,695 and J99 reveal substantial differences in genome length (26,695: 1.67 Mb; J99: 1.64 Mb) and insertion sequence content (e.g., strain 26,695 carries five full copies of IS605, while J99 lacks them). Such diversity arises through five primary mechanisms: point mutations, chimera recombination, gain or loss of non-conserved genes, genomic rearrangements, and horizontal gene transfer via mobile genetic elements (e.g., plasmids and transposons). These mechanisms contribute to *H. pylori* being recognized as one of the most genetically diverse microorganisms known [13].

### 2.3.2 Key Virulence Genes and Their Functions

The pathogenicity of *H. pylori* relies on multiple virulence genes, whose expression and regulation determine infection outcomes (e.g., gastritis, ulcers, or gastric cancer). Based on virulence gene profiles, strains are classified into three types: Type I (carrying both cag-PAI and vacA), Type II (lacking both), and intermediate type (carrying one of the two) [14, 15]. Other genes such as dupA, iceA, urease genes, and oipA also contribute to pathogenesis through diverse mechanisms.

#### 2.3.2.1 Cag Pathogenicity Island (Cag-PAI)

The cag-PAI is a ~ 40 kb gene cluster on the chromosome encoding a type IV secretion system (T4SS). This system injects the CagA protein into host cells via interactions between CagL and host integrin receptors. CagA, encoded by the cagAgene (120–145 kDa), is one of the most extensively studied toxins of *H. pylori*. Strains carrying cag-PAI exhibit stronger virulence than those without it. The C-terminal region of CagA contains repeated EPIYA motifs (Glu-Pro-Ile-Tyr-Ala), which can be categorized into A, B, C, and D types based on sequence variations. Combinations of these motifs generate diverse CagA proteins associated with different strain genotypes.

After injection into host cells via T4SS, CagA undergoes tyrosine phosphorylation, activating SHP-2 phosphatase and triggering the MAPK/ERK signaling pathway. This leads to "hummingbird phenotype" changes: loss of cell polarity, disruption of tight junctions, and inhibition of dendritic cell maturation and Th1 immune responses, facilitating immune evasion [9]. Strains with intact cag-PAI (e.g., Type I) are highly virulent and strongly associated with severe gastric diseases.

#### 2.3.2.2 Vacuolating Cytotoxin Gene (*VacA*)

The *vacA* gene encodes a vacuolating cytotoxin with four structural domains: signal (s1/s2), middle (m1/m2), deletion (d1/d2), and intermediate (i1/i2) regions [16]. The toxin forms anion channels in host cell membranes, inducing vacuolation and mitochondrial apoptosis. Its activity depends on allele combinations; for example, s1/m1 strains exhibit strong vacuolating activity, while s2/m2 strains lack this ability. *vacA* synergizes with cag-PAI to disrupt epithelial barriers and suppress T-cell function, exacerbating inflammation.

#### 2.3.2.3 Other Virulence Factors

##### Duodenal Ulcer-Promoting Gene A (DupA)

Located in the plasticity region of the chromosome, it is fused from jhp0917-jhp0918(VirB4 homologs) to form a 1.8 kb ORF. dupA1 is linked to duodenal ulcer risk and may form a T4SS-like structure that upregulates IL-8, amplifying inflammation [17].

#### Induced by Contact with Epithelium (IceA)

The *iceA1* allele is induced upon contact with gastric epithelium and activates the NF-κB/ERK pathway, enhancing IL-8/IL-6 secretion and neutrophil infiltration. IceA2 is associated with asymptomatic or mild gastritis, indicating a low-inflammatory phenotype [18].

#### Urease Gene

Urease hydrolyzes urea to ammonia, neutralizing gastric acid and enabling colonization [19]. However, excess ammonia damages epithelial cells, generates carcinogenic nitrosamines, and disrupts vesicular transport, promoting malignant transformation [20].

#### Outer Inflammatory Protein A (OipA)

Encoded by *hopH*, its expression is regulated by CT dinucleotide repeats in the 5′ signal region. The "ON" state of OipA is significantly correlated with peptic ulcers and gastric cancer risk, acting as a molecular switch for virulence [21].

The molecular characteristics of *H. pylori* are defined by high genomic diversity and virulence gene polymorphism. Key virulence factors such as cag-PAI and vacA promote disease progression by disrupting host signaling, inducing cell damage, and modulating immune responses. Additional genes like dupA, iceA, urease, and oipA further refine strain-specific pathogenicity. Understanding these mechanisms provides a foundation for targeted therapies and personalized management strategies.

## 2.4 Conclusion

The biological characteristics of *H. pylori* (*H. pylori*) are closely associated with its pathogenicity. Its spiral shape and flagella enable it to penetrate the gastric mucus layer, its urease activity protects it from gastric acid, and its adhesins mediate colonization, while the diversity of virulence genes (such as cagA and vacA) determines the disease outcome. These characteristics not only explain the persistent survival of *H. pylori* in the gastric mucosa but also provide insights for targeted therapeutic strategies (e.g., urease inhibitors, flagellar assembly disruptors). Future research needs to focus on the correlation between genotypes and clinical phenotypes to advance personalized prevention and treatment strategies.

## References


1. Uemura N, Okamoto S, Yamamoto S, Matsumura N, Yamaguchi S, Yamakido M, et al. *H. pylori* infection and the development of gastric cancer. N Engl J Med. 2001;345(11):784–9. https://doi.org/10.1056/NEJMoa001999.
2. Sato F. *H.* pylori in culture: an ultrastructural study. Hokkaido Igaky Zasshi. 2000;75(3):187–96.
3. Liu Y, Hidaka E, Kaneko Y, Akamatsu T, Ota H. Ultrastructure of *H. pylori* in human gastric mucosa and *H. pylori*-infected human gastric mucosa using transmission electron

microscopy and the high-pressure freezing–freeze substitution technique. J Gastroenterol. 2006;41(6):569–74. https://doi.org/10.1007/s00535-006-1813-2.

4. Saito N, Sato F, Kato M, Takeda H, Sugiyama T, et al. Detection of coccoid *H. pylori*: light microscopical immunogold silver enhancing stain. Helicobacter. 1998;3(3):170–3.
5. Bukholm G, Tannæs T, Nedenskov P, Esbensen Y, Grav HJ, et al. Colony variation of *H. pylori*: pathogenic potential is correlated to cell wall lipid composition. Scand J Gastroenterol. 1997;32(5):445–54. https://doi.org/10.3109/00365529709025079.
6. Ottemann KM, Lowenthal AC. *H. pylori* uses motility for initial colonization and to attain robust infection. Infect Immun. 2002;70(4):1984–90. https://doi.org/10.1128/IAI.70.4.1984-1990.2002.
7. Bury-Moné S, Kaakoush NO, Asencio C, Mégraud F, Thibonnier M, et al. Is *H. pylori* a true Microaerophile? Helicobacter. 2006;11(4):296–303. https://doi.org/10.1111/j.1523-5378.2006.00413.x.
8. Whitmire JM, Merrell DS. Successful culture techniques for Helicobacter species: general culture techniques for *H. pylori*. Methods Mol Biol. 2012;921:17–27. https://doi.org/10.1007/978-1-62703-005-2_4.
9. Jiang X, Doyle MP. Growth supplements for *H. pylori*. J Clin Microbiol. 2000;38(5):1984–7. https://doi.org/10.1128/JCM.38.5.1984-1987.2000.
10. Mendz GL, Hazell SL. Aminoacid utilization by *H. pylori*. Int J Biochem Cell Biol. 1995;27(10):1085–93. https://doi.org/10.1016/1357-2725(95)00069-2.
11. Quaglia NC, Dambrosio A. *H. pylori*: a foodborne pathogen? World J Gastroenterol. 2018;24(31):3472–87. https://doi.org/10.3748/wjg.v24.i31.3472.
12. Tomb JF, White O, Kerlavage AR, Clayton RA, Sutton GG, et al. The complete genome sequence of the gastric pathogen *H. pylori*. Nature. 1997;388(6642):539–47. https://doi.org/10.1038/41483.
13. Denic M, Touati E, De Reuse H. Review: pathogenesis of *H. pylori* infection. Helicobacter. 2020;25(Suppl 1):e12736. https://doi.org/10.1111/hel.12736.
14. Blaser MJ, Atherton JC. *H. pylori* persistence: biology and disease. J Clin Invest. 2004;113(3):321–33. https://doi.org/10.1172/JCI20925.
15. Montecucco C, Rappuoli R. Living dangerously: how *H. pylori* survives in the human stomach. Nat Rev Mol Cell Biol. 2001;2(6):457–66. https://doi.org/10.1038/35073084.
16. Rhead JL, Letley DP, Mohammadi M, Hussein N, Mohagheghi MA, et al. A new *H. pylori* vacuolating cytotoxin determinant, the intermediate region, is associated with gastric cancer. Gastroenterology. 2007;133(3):926–36. https://doi.org/10.1053/j.gastro.2007.06.056.
17. Fischer W, Windhager L, Rohrer S, Zeiller M, Karnholz A, Hoffmann R, et al. Strain-specific genes of *H. pylori*: genome evolution driven by a novel type IV secretion system and genomic Island transfer. Nucleic Acids Res. 2010;38(18):6089–101. https://doi.org/10.1093/nar/gkq378.
18. Peek RM Jr, Crabtree JE. Helicobacter infection and gastric neoplasia. J Pathol. 2006;208(2):233–48. https://doi.org/10.1002/path.1868.
19. Mobley HL, Island MD, Hausinger RP. Molecular biology of microbial ureases. Microbiol Rev. 1995;59(3):451–80. https://doi.org/10.1128/mr.59.3.451-480.1995.
20. Smoot DT, Mobley HL, Chippendale GR, Lewison JF, Resau JH. *H. pylori* urease activity is toxic to human gastric epithelial cells. Infect Immun. 1990;58(6):1992–4. https://doi.org/10.1128/iai.58.6.1992-1994.1990.
21. Yamaoka Y, Ojo O, Fujimoto S, Odenbreit S, Haas R, et al. *H. pylori* outer membrane proteins and gastroduodenal disease. Gut. 2006;55(6):775–81. https://doi.org/10.1136/gut.2005.083014.

# Epidemiology of *H. pylori* Infection 3

Yiqi Du, Xianzhu Zhou, Yifan Qiu, and Jiayue Wang

*Helicobacter pylori* infection is one of the most common chronic bacterial infections globally, affecting nearly half of the world's population. Although the global infection rate has shown an overall declining trend over the past few decades, its epidemiological characteristics still demonstrate significant heterogeneity across different regions and populations. A comprehensive understanding of the epidemiological status of *H. pylori* infection, including its geographical distribution, transmission patterns, strain characteristics, age distribution, antibiotic resistance, and associated risk factors, is of great significance for formulating effective prevention and control strategies. This chapter will systematically elaborate on the epidemiological characteristics of *H. pylori* infection worldwide, analyze key factors influencing infection distribution, and explore current prevention and control strategies and their effectiveness.

## 3.1 Global Infection Status

### 3.1.1 Global and Regional Infection Rate

Overall, among the global adult population, the *H. pylori* infection rate has decreased from 52.6% before 1990 to 43.9% during 2015–2022 [1]. *H. pylori* infection exhibits significant geographical distribution differences worldwide, reflecting enormous disparities in socioeconomic development levels, sanitary infrastructure, and healthcare accessibility across different regions. According to the latest global systematic epidemiological studies, this regional gradient distribution pattern is undergoing dynamic changes [2].

Y. Du (✉) · X. Zhou · Y. Qiu · J. Wang
Department of Gastroenterology, Shanghai Changhai Hospital, Naval Medical University, Shanghai, China



J. Gao, Y. Du (eds.), *Non-Antibiotic Therapies for Helicobacter pylori*,
https://doi.org/10.1007/978-981-95-9598-3_3

The infection rate in Asia remains the highest globally. In East Asia (China, Japan, South Korea), the adult population infection rate maintains a high level of 50–70%, showing a significant age cohort effect—the infection rate in older cohorts is notably higher than in younger cohorts. Meta-analysis of mainland China shows an overall infection rate of approximately 55–60%, with rural areas having significantly higher prevalence than urban areas, primarily for disparities in sanitary conditions and socioeconomic status. In South Asia (India, Pakistan, Bangladesh), prevalence typically exceeds 70%, closely related to the region's high population density, relatively inadequate sanitary facilities, and water quality.

Europe presents clear east–west and north–south gradient distribution characteristics. In developed Western European countries (Germany, France, United Kingdom, Netherlands), adult prevalence has decreased to 20–40% and continues to show a declining trend. In contrast, Eastern European countries maintain relatively high prevalence of 50–70%. Mediterranean coastal countries fall between these two extremes, with rates of approximately 35–55%. Nordic countries show the best performance, with adult population prevalence reduced to below 30%.

The Americas demonstrate significant north–south and racial disparities. In North America (United States, Canada), the overall infection rate is 25–50%, but significant differences exist among different racial groups: the infection rate among African Americans (50–60%) is markedly higher than among white populations (20–30%), while the infection rate among Latino immigrants can reach 60–80%. Latin American countries generally have higher prevalence, with adult prevalence in Mexico, Brazil, Peru, and other countries mostly ranging between 60% and 90%, showing a strong negative correlation with socioeconomic status.

The overall infection rate in Africa continues to remain high. In sub-Saharan Africa, prevalence typically exceeds 70%, with some areas reaching over 90%. North African countries have relatively lower prevalence but still maintain levels of 60–80%. Notably, despite extremely high *H. pylori* prevalence in Africa, gastric cancer incidence is relatively low—a phenomenon called the "African enigma" [3], which may be related to host genetic factors, environmental factors, and regional differences in strain virulence.

### 3.1.2 Temporal Trends in Prevalence

Recent large-scale meta-analysis studies show that over the past 40+ years, global *H. pylori* prevalence has shown an overall declining trend [4].

Developed countries demonstrate a sustained and significant declining trend. This decline is primarily attributed to the synergistic effects of multiple factors: significant improvement in sanitary conditions, particularly the establishment of safe drinking water supply systems and the popularization of sewage treatment technology; improvement in household living conditions, increased per capita living space, and reduced living density; nonspecific pathogen clearance effects due to widespread antibiotic use; and significant reduction in childhood prevalence. Birth cohort studies show that prevalence can reach 60–80% among those born before 1950, while rates have decreased to 10–20% among those born after 2000 [5, 6].

The temporal trends in developing countries are more complex and diverse. Some rapidly developing middle-income countries, particularly urban areas, have begun to show declining trends similar to developed countries. However, prevalence in rural and impoverished areas remains high. In some regions, prevalence has even shown an upward trend due to population concentration caused by rapid urbanization and lagging sanitary infrastructure development.

Particular attention should be paid to the fact that although prevalence in adult populations is declining, the decline in prevalence among children and adolescents is relatively modest [7], suggesting that we need to pay more attention to childhood infection prevention measures.

### 3.1.3 Impact of Socioeconomic Factors

There exists a strong and stable negative correlation between *H. pylori* infection and socioeconomic status, an association that has been consistently confirmed across all global regions. Childhood socioeconomic circumstances are considered the strongest predictor of adult *H. pylori* infection status, a finding that profoundly reflects the epidemiological characteristic that infection occurs mainly in early life. Key influencing factors include: household income level—children from low-income families have prevalence two to three times higher than those from high-income families; parental education level—particularly maternal education level shows a significant negative correlation with childhood prevalence; living conditions—residential crowding (more than two persons per room) increases infection risk two- to fourfold; sanitary facilities—lack of indoor running water and modern sanitation facilities increases infection risk three- to fivefold.

Urban–rural disparities are particularly prominent in developing countries, where rural prevalence is typically 10–20 percentage points higher than urban rates. The fundamental reason for this difference lies in relatively weak sanitary infrastructure in rural areas, incomplete safe drinking water supply systems, scarce medical resources, and lower levels of health education penetration.

## 3.2 *H. pylori* Transmission Routes

### 3.2.1 Molecular Epidemiological Evidence of Person-to-Person Transmission

Current molecular epidemiological studies strongly confirm that person-to-person direct transmission is the primary route of *H. pylori* infection, accounting for 60–80% of all infection transmission events. This conclusion is based on extensive DNA fingerprinting analysis and multilocus sequence typing (MLST) studies.

Oral–oral transmission is considered one of the most important transmission routes. Evidence supporting this view includes: saliva detection studies showing that *H. pylori* DNA can be detected in saliva of 10–60% of infected individuals, with approximately 30% yielding viable bacteria in culture; dental plaque studies finding that 20–80% of infected individuals have *H. pylori* in their dental plaque, and genotyping analysis shows these strains are highly consistent with the patients' gastric strains. However, there remains some controversy regarding the oral cavity's capacity as a long-term colonization site for *H. pylori*, with some scholars suggesting that the special nature of the oral environment may not favor long-term *H. pylori* survival.

Fecal–oral transmission may play a more important role in developing countries with poor sanitary conditions. Related evidence shows: PCR detection finds *H. pylori* DNA in feces of 60–90% of infected individuals, but viable bacteria isolation rates are relatively low, only 10–30%; epidemiological surveys show that *H. pylori* prevalence is strongly correlated with environmental sanitary conditions; in areas with improper sewage treatment, prevalence is significantly elevated. It is noteworthy that *H. pylori* in feces has limited survival capacity and typically needs to complete transmission in a short time to cause successful infection.

### 3.2.2 Molecular Mechanisms of Intrafamilial Transmission

Intrafamilial transmission is the most important transmission mode of *H. pylori* infection, a conclusion strongly supported by numerous high-quality molecular epidemiological studies [8]. Modern molecular biological techniques, such as DNA fingerprinting analysis, multilocus sequence typing, and whole-genome sequencing, have provided us with unprecedented precision tools to track strain transmission trajectories within families.

Mother-to-child transmission is widely recognized as the main source of childhood *H. pylori* infection. Rothenbacher et al.'s large-scale population study showed that children of infected mothers have an infection risk 7–16 times higher than children of uninfected mothers. A 5-year longitudinal follow-up study in Japan found through DNA fingerprinting analysis that 69% of infected children carried strains completely identical to their mothers' strains, while only 17% matched their fathers' strains [9]. A population study in Germany further confirmed that maternal infection status is the strongest predictor of childhood infection [10]. The main mechanisms of this mother-to-child transmission may include: pre-chewing food for feeding (relatively common in certain cultural contexts), sharing utensils, intimate physical contact such as kissing, and close contact during daily care. Infection predominantly occurs in the critical age range of 2–5 years, suggesting transmission is achieved mainly through close contact rather than perinatal vertical transmission.

Spousal transmission, though relatively uncommon, does exist and provides important clues for understanding adult *H. pylori* acquisition. Studies show spousal infection concordance rates of 20–40%, increasing with marriage duration. Molecular typing analysis shows that 40–70% of infected spouses carry identical or highly similar strains. This relatively low spousal transmission rate may reflect adults' relative resistance to *H. pylori* infection.

Sibling transmission is also a component of intrafamilial horizontal transmission. Studies find sibling infection concordance rates of 30–70%, with shared bedrooms significantly increasing transmission risk, and siblings closer in age showing higher infection concordance rates. Older children often act as infection sources for younger children, a phenomenon more pronounced in large families.

### 3.2.3 Limited Role of Environmental Transmission

Although environmental transmission plays a relatively limited role in *H. pylori* infection, under specific conditions, particularly in areas with poor sanitary conditions, it may play an important supplementary role.

Evidence for waterborne transmission is relatively weak but still worthy of attention. Studies show that using untreated well water is associated with *H. pylori* infection, with an odds ratio (OR) of 2–4. Although *H. pylori* DNA can be detected in water sources, viable bacteria isolation is quite difficult. *H. pylori* survival time in aquatic environments is typically less than 7 days, but may extend survival time by forming coccoid forms, although the infectivity of this morphology remains unclear. Municipal water supply systems pose extremely low transmission risk due to chlorination disinfection.

Evidence for foodborne transmission is even weaker. Although some studies show raw vegetables may slightly increase infection risk, evidence for animal products as transmission vectors is insufficient. Food processing and proper cooking can effectively eliminate potential contamination, while *H. pylori* survival capacity under refrigeration conditions is poor.

The role of animal reservoir hosts in human *H. pylori* infection is extremely limited. Although *H. pylori*-like bacteria can be detected in pets (such as cats and dogs), phylogenetic analysis shows these are mostly species-specific strains, with rare human–animal transmission. Occupational exposure studies (such as veterinarians, slaughterhouse workers) have also not shown significant increases in prevalence.

## 3.3 Strain Typing of Infection

### 3.3.1 Major Virulence Gene Typing Systems

The pathogenicity of *H. pylori* largely depends on the combination of virulence factors it carries. The polymorphism of these virulence genes not only causes significant differences in strain pathogenicity but also forms an important molecular basis for understanding geographical distribution differences in *H. pylori*-related diseases.

#### 3.3.1.1 cagA Gene and Its EPIYA Motif Polymorphism

The cagA gene-encoded cytotoxin-associated protein A (CagA) is recognized as the most important virulence factor of *H. pylori* and the first bacterial oncoprotein confirmed in human cancer [11]. The global cagA positivity rate ranges from 60% to 90%, but with significant geographical distribution differences: over 95% in East Asia, 60–80% in Western countries, and 40–80% in Africa.

The geographical specificity of EPIYA motifs constitutes the core of CagA typing. The types and numbers of these motifs directly determine the pathogenic intensity of the CagA protein. Western-type CagA contains EPIYA-A, B, and C motifs, with a typical pattern of ABC, and some strains contain multiple C motif repeats (ABCC, ABCCC). Molecular biological studies show that the number of C motif repeats is positively correlated with disease severity, with each additional C repeat increasing gastric cancer risk two- to threefold. East Asian-type CagA is characterized by containing EPIYA-A, B, and D motifs, with a typical pattern of ABD. A key finding is that a single D motif has pathogenic potential equivalent to 3–5 C motifs, a molecular mechanism that powerfully explains the fundamental reason for high gastric cancer incidence in East Asia [12]. Recent structural biology research reveals that the D motif of East Asian-type CagA, through stronger binding capacity to host protein SHP-2 and more sustained ERK signaling pathway activation, leads to more severe gastric mucosal atrophy and intestinal metaplasia processes.

#### 3.3.1.2 Mosaic Structure of vacA Gene

The vacA gene-encoded vacuolating toxin, although present in all *H. pylori* strains, exhibits enormous variation in activity. This variation stems from the unique mosaic structure of the vacA gene, initially discovered and described by Atherton et al. [13].

The vacA gene structure contains multiple variable regions: the signal region (s region) determines toxin secretion and activation; the middle region (m region) affects cell specificity and toxic intensity; and the later-discovered intermediate region (i region), among others. In s region typing, the s1 type encodes active toxin and can be further subdivided into s1a (common in East Asia, Northern Europe), s1b (common in Western Europe, Latin America), s1c (unique to East Asia), and other subtypes, while the s2 type encodes toxin with very low or no activity. In m region typing, the m1 type has broad toxic effects on various cell lines, while the m2 type has relatively limited toxic effects. The relationship between genotype combinations and disease risk has been widely confirmed: the s1m1 type combination has the strongest virulence, with a gastric cancer odds ratio of 5–8; s1m2 type represents medium virulence level, mainly increasing peptic ulcer risk; s2m2 type is a low-virulence combination, rarely causing severe disease and mostly seen in asymptomatic infections.

## 3.3.2 Geographically Specific Strain Distribution Patterns

### 3.3.2.1 High Virulence Characteristics of East Asian Strains

*H. pylori* strains in East Asia (China, Japan, South Korea) exhibit unique high virulence characteristic profiles, corresponding clearly with the region's globally highest gastric cancer incidence. Characteristics of strains in this region include: cagA gene positivity rate exceeding 95%, almost entirely East Asian-type CagA (EPIYA-ABD pattern); vacA gene predominantly high-virulence s1m1 type (accounting for 70–90%); i1 type proportion significantly higher than other regions. This high-virulence gene combination (cagA positive/vacA s1m1/babA2 positive triple-positive) detection rate exceeds 60% in East Asia, far higher than other regions worldwide. From clinical manifestations, East Asia not only has the highest incidence of gastric cancer globally but also relatively early age of onset, with intestinal-type gastric cancer as the main type.

### 3.3.2.2 Diversified Characteristics of Western Strains

Strain characteristics in Europe, North America, Australia, and other regions form sharp contrast with East Asia: cagA positivity rate at medium level (60–80%), but entirely Western-type CagA (EPIYA-ABC/CC/CCC pattern); vacA genotypes show diversified distribution, with s1m1 type (40–60%), s1m2 type (20–30%), and s2m2 type (20–30%) coexisting; adhesin gene babA2 positivity rate is actually higher than in Asia. From the disease spectrum perspective, duodenal ulcers are relatively more common in Western regions, with diffuse-type gastric cancer as the main type, possibly related to the relative proportions and mechanisms of action of different virulence genotypes.

### 3.3.2.3 Unique Diversity of African Strains

The African continent has the richest *H. pylori* genetic diversity, a phenomenon reflecting the long history of human–*H. pylori* coevolution. Unique features of strains in this region include: presence of rare EPIYA motif combinations (such as B-B, B-D hybrid types); cagA positivity rate varies greatly (40–80%), with some regions even dominated by cagA-negative strains; overall relatively moderate virulence genotype expression. These strain characteristics well explain the "African enigma" phenomenon—despite extremely high *H. pylori* prevalence in the region (often exceeding 80%), gastric cancer incidence is relatively low, suggesting the complexity of host-pathogen-environment interactions far exceeds our current understanding.

## 3.3.3 Clinical Application Value of Molecular Typing

### 3.3.3.1 Disease Risk Stratification Based on Virulence Genotypes

Modern precision medicine concepts have begun incorporating *H. pylori* virulence genotyping into disease risk assessment systems. High-risk genotype combinations (cagA positive/vacA s1m1) increase gastric cancer risk 6–12-fold, with East

Asian-type cagA having two to three times higher carcinogenic risk compared to Western-type cagA. When three high-virulence markers are simultaneously present (cagA positive/vacA s1m1/babA2 positive), gastric cancer risk can increase 8–15-fold.

Medium-risk genotypes (cagA positive/vacA s1m2) mainly increase gastric cancer risk two- to fourfold, or vacA s1m1 type alone increases risk three- to fivefold. Low-risk genotypes (cagA negative/vacA s2m2) are mainly asymptomatic infections, rarely leading to severe disease.

#### 3.3.3.2 Formulation of Individualized Treatment Strategies

Virulence genotyping is gradually guiding clinical treatment decision-making. For infected individuals carrying high-virulence strains, active eradication treatment is recommended even in asymptomatic states. When high-risk genotypes coexist with other risk factors (such as family history of gastric cancer), closer monitoring protocols are needed. Different regions also need to adjust treatment regimens based on local strain prevalence. The latest Maastricht VI/Florence consensus report emphasizes the importance of individualized treatment strategies against the background of rising antibiotic resistance [14].

#### 3.3.3.3 Optimization of Population Screening Strategies

Different regions should formulate differentiated screening strategies based on local strain prevalence. The East Asian model adopts universal screening and treatment strategies because high-virulence strains are prevalent in the region, with clear preventive benefits. The Western model is more suitable for risk stratification strategies, focusing on populations carrying high-risk genotypes. Developing countries, under limited medical resource conditions, may consider selective screening, prioritizing high-risk populations.

## 3.4 Age of Infection

*H. pylori* infection occurs mainly in early life, and infection during this period has a decisive impact on an individual's lifelong *H. pylori* infection status. According to systematic analysis of high-quality literature published from 2014 to 2024, a large meta-analysis covering 224 studies from 71 countries with 2,979,179 people showed that global *H. pylori* infection rate decreased from 58.2% in 1980–1990 to 43.1% in 2011–2022, with an average annual decline rate of approximately 0.66% [4]. This declining trend is particularly evident in adult populations. However, it is noteworthy that prevalence in children and adolescents remained at a relatively high level of 35.1% during 2015–2022, with no significant decline, a phenomenon further confirmed in a comprehensive analysis of 1,748 articles from 111 countries [1]. This difference in infection rate change trends between adults and children suggests that prevention strategies for childhood infection still need further strengthening.

Regarding the exact age of *H. pylori* acquisition, the classic Bogalusa Heart Study provided detailed data through 21 years of prospective follow-up of Louisiana children. The study clearly showed that 4–5 years is the peak period for *H. pylori* infection, with an annual incidence rate of 2.1%, followed by a stepwise decline: 1.5% at 7–9 years, and only 0.3% in early adulthood at 21–23 years [15]. The study found that the vast majority of new infections (approximately 90%) occur before age 10, with new infections after age 10 being relatively rare. The situation is more severe in developing countries; a birth cohort study in Bangladesh showed that in areas with poor sanitary conditions, 50% of children are already infected before age 2, and prevalence can reach over 80% by age 5 [16].

Birth cohort effect studies provide a unique perspective for understanding intergenerational changes in *H. pylori* infection. A multigenerational cohort study in Rotterdam, Netherlands, showed a trend of decreasing prevalence across generations: the infection rate for 6–8-year-old children born in 1978 was 19%, declining to 9% by 1993, but remaining at around 9% in 2005–2006, indicating that childhood infection rate decline in developed countries has entered a plateau phase [17]. More importantly, the proportion of CagA-positive strains dramatically decreased from 41% in adults in 1992 to 0.9% in children in 2005, and this rapid disappearance of high-virulence strains may have profound implications for future gastric cancer incidence. A multiethnic cohort study in the United States also confirmed that prevalence was as high as 60–80% among those born before 1950, while rates have decreased to 10–20% among those born after 2000.

## 3.5 Pediatric Prevalence

Childhood is a critical period for *H. pylori* infection. A global systematic review and meta-analysis published in 2022 included data from 152,650 children in 198 studies, revealing a clear age-dependent infection pattern: the overall infection rate for preschool children aged 0–6 years was 26.0% (95% CI: 21.4–31.0), increasing to 33.9% (95% CI: 28.6–39.7) for school-age children aged 7–12 years, and further rising to 41.6% (95% CI: 35.6–47.8) for adolescents aged 13–18 years, approaching adult levels [7]. This pattern of prevalence gradually increasing with age has universality globally, but the rate of increase and final plateau levels vary significantly across different regions and socioeconomic conditions.

Geographical distribution differences are an important feature of pediatric *H. pylori* infection epidemiology. According to World Bank income classifications, the overall infection rate for children in high-income countries was 21.7% (95% CI: 16.9–27.4), while in low- and middle-income countries the childhood infection rate reached 43.2% (95% CI: 36.5–50.2), fully reflecting the profound impact of socioeconomic development levels on *H. pylori* transmission [7]. More detailed regional analysis shows: African children have the highest infection rate globally, averaging approximately 70%, with some areas in Nigeria reaching 87.7%; Oceania has the lowest rate at only 24.4%; Asia shows significant internal heterogeneity, with Japanese childhood prevalence declining to 1.8%, while some areas in South Asia still exceed 60%. Europe also has an obvious gradient, with Dutch childhood prevalence of 1–9%, while Portuguese 13-year-old children still have prevalence as high as 66.2%.

Temporal trend analysis shows that although global adult *H. pylori* prevalence has continued to decline over the past few decades, the decline in childhood prevalence has been relatively limited. A global systematic review and meta-analysis published in 2024 covering 1980–2022 found that adult *H. pylori* prevalence significantly decreased from 52.6% (95% CI 49.6–55.6%) before 1990 to 43.9% (95% CI 42.3–45.5%) in 2015–2022, while childhood and adolescent prevalence during the same period remained high at 35.1% (95% CI 30.5–40.1%), showing the characteristic of relatively slow decline in childhood prevalence [1]. This difference may reflect that childhood is an active period for *H. pylori* transmission, and even under improved sanitary conditions, close contact among children and relatively poor personal hygiene habits still provide opportunities for *H. pylori* transmission. In some rapidly urbanizing developing countries, childhood prevalence has even rebounded due to increased residential density from population influx.

## 3.6 Adult Prevalence

The epidemiological characteristics of adult *H. pylori* infection show patterns distinctly different from children. According to a large meta-analysis published in 2024 covering 1,748 articles from 111 countries, global adult *H. pylori* prevalence decreased from 52.6% before 1990 to 43.9% in 2015–2022, a decline of 15.9% (95% CI: −20.5% to −11.3%) over 30 years. This declining trend shows obvious differences across regions: the Western Pacific region had the largest decline, from 65.3% to 42.1%; Southeast Asia was next, from 58.7% to 44.2%; while the African region declined only from 72.8% to 70.1%, almost maintaining high levels.

Age-stratified analysis reveals an "age cohort effect" where adult prevalence gradually increases with age. In developed countries, prevalence is 25–35% for young adults aged 18–30, 30–40% for ages 31–40, 35–45% for ages 41–50, 40–50% for ages 51–60, and can reach 45–55% for those over 60. This increase mainly reflects the different sanitary conditions experienced by different birth cohorts during childhood. In contrast, developing countries have significantly higher prevalence across all age groups, reaching 50–60% for ages 18–30 and 75–85% for those over 60, with some areas even exceeding 90%.

Racial and ethnic differences are another important characteristic of adult *H. pylori* infection. A study of 913,328 adults in the US Veterans Health Administration system showed an overall infection rate of 25.8%, but with significant differences among racial groups: African Americans 40.2%, Hispanics 36.7% (first-generation immigrants 52.3%, second generation and beyond 28.6%), Asian Americans 35.8%, and whites only 20.1% [1]. The impact of socioeconomic status is equally significant; a 2018 systematic review showed that education level is negatively correlated with prevalence: uneducated individuals 68.4%, primary school 55.2%, secondary school 42.7%, college and above 31.5% [18]. Occupational exposure is also an important factor, with gastroenterologists and other healthcare workers having infection risks 1.8–2.5 times higher than the general population.

## 3.7 Reprevalence

Reprevalence after *H. pylori* eradication varies significantly across different populations and regions. A 2017 systematic review including 132 studies with 53,934 patient-years of follow-up data found a global annual reinfection rate of 3.1% (95% CI: 2.0–5.0%) [19]. However, this average masks enormous regional differences: developed countries have annual reprevalence of only 1.0–2.0% (Japan 0.22%, United States 1.1%, Western Europe 1.5%), while developing countries are significantly higher (Latin America 7.9%, Africa 10.9%, South Asia 12.5%). This difference is negatively correlated with the Human Development Index (HDI): countries with very high HDI 3.1%, high HDI countries 6.2%, medium HDI countries 8.7%, low HDI countries 10.9%.

Age is the most important factor affecting reprevalence. Children under 5 years are a high-risk group for reinfection, with annual reprevalence of 5–15%, significantly higher than adults. An Irish prospective pediatric cohort study found that 5 years is a critical age threshold: children over 5 years have an annual reinfection rate of 2.0%, approaching adult levels; while children under 5 years, especially those aged 2–3 years, can have annual reprevalence of 8–12% [20].

Worth mentioning is that a Chinese multicenter prospective study including 5,193 successfully eradicated patients found an annual reinfection rate of 1.5% (95% CI: 1.2–1.8%), significantly lower than the global average [21]. Multivariate analysis showed main risk factors including: ethnic minorities (HR = 4.7), low education level (HR = 1.7), family history of gastric cancer (HR = 9.9), use of well or river water (HR = 2.3), living with infected individuals (HR = 3.1), and frequent dining out (HR = 1.8). Time distribution shows reinfection mainly occurs in the first year after eradication (accounting for 40–50%), 25–30% in the second year, then declining annually thereafter, suggesting the importance of early close follow-up.

## 3.8 Household Prevalence

Intrafamilial transmission is the most important transmission mode of *H. pylori* infection, with infection status among family members showing high clustering. A German study of 1,221 preschool children and their parents found that when mothers were infected, children's prevalence significantly increased, with an adjusted OR of 7.9 (95% CI: 4.0–15.7); paternal infection had a relatively smaller impact, with an OR of 3.8 (95% CI: 0.8–19.1) [22]. This maternal–infant transmission advantage may be related to mothers' primary role in daily child care, more frequent physical contact, and feeding practices in certain regions.

Molecular epidemiological studies provide more precise evidence. A Swedish MLST study found that 81% of siblings carried identical strains, higher than mother–child pairs (56%) and spousal pairs (22%) [23]. A study of Hispanic populations in the southwestern United States showed an OR of 3.2 (95% CI: 1.14–9.1) when having one infected sibling, rising to 9.4 (95% CI: 3.1–28.5) when having both infected siblings and other family members [24].

Family clustering studies further confirmed this phenomenon. An Italian survey of 3,289 residents found children's prevalence was 44% when both parents were infected, 30% when one parent was infected, and only 21% when neither parent was infected ($P < 0.001$) [25]. A prospective cohort study in the Texas-Mexico border region of the United States showed that persistent *H. pylori* infection in older siblings had a significant impact on younger siblings, with a hazard ratio (HR) of 7.6 (95% CI: 1.6–37), with even higher risk when siblings were ≤ 3 years apart in age, HR reaching 16 (95% CI: 2.5–112). Additionally, family socioeconomic factors also significantly affect prevalence: low-income families have markedly higher prevalence than high-income families; increased residential density is positively correlated with prevalence; and families lacking indoor running water facilities have higher prevalence than those with running water. Maternal education level is negatively correlated with family prevalence, with mothers having higher education showing significantly lower family member prevalence [26, 27].

## 3.9 Epidemiology of *H. pylori* Antibiotic Resistance

*H. pylori* antibiotic resistance has become a global public health challenge. According to a 2018 systematic review and meta-analysis covering WHO regions worldwide, global primary clarithromycin resistance rates range from 10% to 34%, with the Americas at 10% (95% CI: 4–16%), while the Eastern Mediterranean and Western Pacific regions are as high as 29% and 34%, respectively. Secondary resistance rates are generally higher than primary resistance rates, reaching 50–90%. The latest 2024 global study reported clarithromycin resistance rates reaching 33.3%, showing a temporal increasing trend, from 29.1% in 2015–2019 to 36.5% in 2020–2023 [28].

Metronidazole resistance rates are highest globally, with the Eastern Mediterranean region showing primary resistance rates of 56% and secondary resistance rates of 65% [29]. Levofloxacin resistance rates have significantly increased in the Western Pacific region, from 18% in 2006–2015 to 35% in 2016–2022. In contrast, amoxicillin and tetracycline resistance rates remain low, at only 1–6% in most regions globally. The latest meta-analysis from the Asia-Pacific region shows that during 2019–2022, clarithromycin resistance rates reached 30%, metronidazole resistance rates were as high as 61%, and levofloxacin resistance rates were 35%, reflecting the severe antibiotic resistance situation in the region [30]. National surveillance data from Japan reported primary clarithromycin resistance rates of 35.5% in 2018–2020, highlighting that even in developed countries, resistance problems are equally serious [31].

Over the past 10–15 years, antibiotic resistance of *H. pylori* has shown an obvious upward trend. Clarithromycin resistance rates increased from 24.28% in

2010–2017 to 32.14% in 2018–2021, and globally from 29.1% in 2015–2023 to 36.5%. The increase in levofloxacin resistance rates is particularly significant, rising in South Korea from 12% in 2003–2008 to 31% in 2012–2018. This trend is closely related to community antibiotic use patterns, particularly the overuse of macrolide antibiotics during the COVID-19 pandemic, which has accelerated the development of resistance [32].

Differences between primary and secondary resistance are obvious. Kocsmár et al.'s study clearly distinguished differences between primary and secondary resistance. Primary clarithromycin resistance rate was 5.5%, while secondary resistance rates after macrolide antibiotic treatment were significantly higher. Secondary resistance rates after eradication treatment failure are generally as high as 50–90%, 2–10 times higher than primary resistance rates [33]. This difference has important guiding significance for clinical treatment strategy selection, emphasizing the importance of first-line treatment success.

Resistance is closely related to treatment failure. Meta-analysis shows clarithromycin resistance has an odds ratio (OR) of 6.97 (95% CI: 5.23–9.28, $P < 0.001$), with treatment failure rates as high as 55%, while sensitive strains have failure rates below 10%. Metronidazole resistance increases treatment failure risk two- to threefold, with failure rates of 38%. In patients with dual resistance (clarithromycin + metronidazole), 77.6% are resistant to both antibiotics, with multidrug-resistant strains using standard therapy having success rates of only 30–50% [29].

## 3.10 Epidemiological Prevention and Control Strategies

The test-and-treat strategy has become an important approach for *H. pylori* infection management. The 2024 American College of Gastroenterology (ACG) guidelines recommend 14-day bismuth quadruple therapy as the first-line treatment regimen when antibiotic susceptibility is unknown [34]. Effectiveness data for this strategy shows that for functional dyspepsia patients, the number needed to treat (NNT) is 13 (95% CI: 8–20). Eradicating *H. pylori* can significantly reduce gastric cancer risk, with a reduction of 35% (RR 0.65, 95% CI 0.46–0.95), and in high-risk populations, the NNT to prevent one gastric cancer case is 72 (95% CI: 56–119).

From an economic perspective, this strategy also demonstrates good cost-effectiveness. Finnish research shows that compared to non-screening, the incremental cost per case for screening is only $26; a New Zealand modeling study shows the incremental cost-effectiveness ratio (ICER) per quality-adjusted life year (QALY) gained is $13,800 (95% UI: 7,700-26,300), far below the usual cost-effectiveness threshold. A Chinese modeling study further indicates that *H. pylori* screening every 3–5 years is most cost-effective, with ICERs ranging from $1,163-1,238/QALY.

In regions with high *H. pylori* prevalence, population screening-treatment strategies have shown significant success. Since 2013, Japan has included *H. pylori*-associated chronic gastritis in national health insurance, reducing gastric cancer mortality by 30% through endoscopic screening combined with eradication treatment. Since

1999, South Korea has implemented the National Cancer Screening Program (NCSP), providing biennial endoscopic examinations for adults over 40, reducing gastric cancer mortality by 47%. A trial conducted in Linqu County, China, from 2011 to 2013 covering over 180,000 people showed endoscopic screening reduced gastric cancer mortality by 57%, with an ICER of only 1,370 RMB per QALY. Modeling studies also show that implementing population screening in high-risk areas can significantly reduce gastric cancer incidence and mortality risks with significant cost-effectiveness.

Family-based infection control strategies are important pathways for interrupting *H. pylori* transmission. Intrafamilial transmission, especially mother-to-child transmission, is the main mode of infection [35]. The global annual reinfection rate is approximately 3.1%, with higher reinfection risk if family members are not treated. A Chinese multicenter prospective study including 5,193 successfully eradicated patients found an annual reinfection rate of 1.5% (95% CI: 1.2–1.8%), significantly lower than the global average. A 2023 large-scale family epidemiological survey in China covering 10,735 families with 31,098 people showed 9.34% of families had universal infection [8]. Family intervention strategies have multiple advantages: can improve treatment compliance, reduce reinfection risk, centrally manage high-risk populations, and have high cost-effectiveness. Studies show that patients treated on a family basis have a 9-month reinfection rate of 7.1%, while those with only individual treatment have reprevalence as high as 38.6% [35].

Globally, population-level prevention strategies are receiving increasing attention. The World Health Organization recommends countries consider conducting *H. pylori* screening to prevent gastric cancer, especially in high-prevalence areas and high-risk populations for gastric cancer. The 2022 Maastricht VI/Florence consensus officially classified *H. pylori* as an infectious disease (ICD-11) and recommended a "test and treat all infected individuals" strategy [14]. Asia-Pacific regional guidelines recommend family screening for relatives of gastric cancer patients and implementing eradication treatment before gastric mucosal atrophy occurs, with high-prevalence countries actively promoting population screening. Currently, *H. pylori* vaccines remain in laboratory or preclinical research stages, with clinical application expected to be difficult within the next 5–6 years. Although multiple candidate vaccines are under development, bacterial genetic diversity and immune escape mechanisms remain the main challenges currently faced.

## 3.11 Risk Factors for Infection

Socioeconomic factors have a significant impact on *H. pylori* prevalence. A global meta-analysis including 224 studies from 71 countries covering nearly 3 million people showed that global *H. pylori* prevalence has decreased from 58.2% in 1980–1990 to 43.1% in 2011–2022 [4]. High-income countries, younger populations, and countries with higher universal health coverage generally have lower prevalence. A Japanese study involving 3,423 subjects further revealed that socioeconomic status shows a significant linear negative correlation with infection risk:

medium socioeconomic status relative to low status has an OR of 0.67 (95% CI: 0.53–0.84), high relative to low has an OR of 0.43 (95% CI: 0.34–0.54), with trend P-value = $9.7 \times 10^{-17}$. Additionally, number of siblings is positively correlated with infection risk: having three to four siblings relative to ≤2 has an OR of 1.74 (95% CI: 1.47–2.06), ≥5 siblings have an OR of 2.54 (95% CI: 2.12–3.04) [7].

Occupational risk is also an important factor affecting prevalence. Systematic reviews indicate that healthcare workers (especially those in gastroenterology), agricultural workers, sewage treatment workers, and miners have significantly higher prevalence than the general population [36]. Such occupational exposures increase infection risk, suggesting the need for targeted occupational protection measures.

Sanitary and environmental conditions also play a key role in infection transmission. Analysis based on NHANES 1999–2000 data of 7,493 participants showed that well water use in children aged 3–19 had an adjusted OR of 1.7 (95% CI: 1.1–2.6); residential crowding in children had an aOR of 2.3 (95% CI: 1.5–3.7); adults engaged in soil-related occupations had an aOR of 1.9 (95% CI: 1.2–2.9); with combined well water use and soil occupational exposure, adults had an aOR of 2.7 (95% CI: 1.3–5.6) [37].

Diet and lifestyle also have potential impacts on infection risk. Systematic reviews indicate that foods with protective effects include broccoli, garlic, green tea, cranberries, and curcumin; risk factors include high-salt diets, processed foods, and contaminated water or food. Consumption of fresh fruits and vegetables has protective effects (OR 0.6–0.8), with vitamin C showing particularly significant protective effects. Smoking has an aOR of 1.8 (95% CI: 1.5–2.2), while evidence for alcohol effects is inconsistent, with high-quality studies still limited [38].

Host genetic factors also play important roles in infection susceptibility and disease progression. Studies show that IL-1β gene polymorphisms are associated with increased gastric cancer risk (OR 1.5–6.5); TNF-α variants can affect inflammatory response intensity; IL-10 polymorphisms participate in immune regulation; CYP2C19 polymorphisms affect drug metabolism and treatment efficacy. Additionally, blood type and Lewis antigens may be related to bacterial adhesion, but further verification is still needed. These genetic mechanism discoveries provide theoretical basis for advancing individualized prevention and treatment strategies.

## 3.12 Conclusion

Epidemiological research on *H. pylori* infection reveals the complexity and diversity of this global health problem. From geographical distribution, significant gaps still exist between developing and developed countries, and between urban and rural areas; from age distribution, childhood infection remains the main pattern, with childhood infection rate decline lagging significantly behind adults; from strain characteristics, differences in virulence gene distribution across regions are closely related to regional disease spectrum characteristics; from the resistance situation, continuously rising antibiotic resistance rates have become a global challenge.

Intrafamilial transmission as the most important transmission route and socioeconomic factors as the most important influencing factors—these epidemiological characteristics provide a scientific basis for formulating precise prevention and control strategies. With the promotion and implementation of test-and-treat strategies, population screening programs, and family-based intervention measures, *H. pylori* infection prevention and control is transitioning from individual treatment to population prevention, which will lay the foundation for ultimately achieving effective control of *H. pylori*-related diseases.

## References


1. Chen YC, Malfertheiner P, Yu HT, Kuo CL, Chang YY, Meng FT, et al. Global prevalence of *H. pylori* infection and incidence of gastric Cancer between 1980 and 2022. Gastroenterology. 2024;166(4):605–19. https://doi.org/10.1053/j.gastro.2023.12.022.
2. Hooi JKY, Lai WY, Ng WK, Suen MMY, Underwood FE, Tanyingoh D, et al. Global prevalence of *H. pylori* infection: systematic review and meta-analysis. Gastroenterology. 2017;153(2):420–9. https://doi.org/10.1053/j.gastro.2017.04.022.
3. Holcombe C. *H. pylori*: the African enigma. Gut. 1992;33(4):429–31. https://doi.org/10.1136/gut.33.4.429.
4. Li Y, Choi H, Leung K, Jiang F, Graham DY, Leung WK. Global prevalence of *H. pylori* infection between 1980 and 2022: a systematic review and meta-analysis. Lancet. Gastroenterol Hepatol. 2023;8(6):553–64. https://doi.org/10.1016/s2468-1253(23)00070-5.
5. Leontiadis GI, Moayyedi P, Ford AC. *H. pylori* infection. BMJ Clin Evid. 2009:0406
6. Wang C, Nishiyama T, Kikuchi S, Inoue M, Sawada N, Tsugane S, et al. Changing trends in the prevalence of *H. pylori* infection in Japan (1908-2003): a systematic review and meta-regression analysis of 170,752 individuals. Sci Rep. 2017;7(1):15491. https://doi.org/10.1038/s41598-017-15490-7.
7. Yuan C, Adeloye D, Luk TT, Huang L, He Y, Xu Y, et al. The global prevalence of and factors associated with *H. pylori* infection in children: a systematic review and meta-analysis. Lancet child Adolesc. Health. 2022;6(3):185–94. https://doi.org/10.1016/s2352-4642(21)00400-4.
8. Zhou XZ, Lyu NH, Zhu HY, Cai QC, Kong XY, Xie P, et al. Large-scale, national, family-based epidemiological study on *H. pylori* infection in China: the time to change practice for related disease prevention. Gut. 2023;72(5):855–69. https://doi.org/10.1136/gutjnl-2022-328965.
9. Konno M, Fujii N, Yokota S, Sato K, Takahashi M, Sato K, et al. Five-year follow-up study of mother-to-child transmission of *H. pylori* infection detected by a random amplified polymorphic DNA fingerprinting method. J Clin Microbiol. 2005;43(5):2246–50. https://doi.org/10.1128/jcm.43.5.2246-2250.2005.
10. Rothenbacher D, Winkler M, Gonser T, Adler G, Brenner H. Role of infected parents in transmission of *H. pylori* to their children. Pediatr Infect Dis J. 2002;21(7):674–9. https://doi.org/10.1097/00006454-200207000-00014.
11. Rappuoli R, Bagnoli F, Blaser MJ. The story behind the science: the discovery of *H. pylori* CagA, the first tumorigenic bacterial protein. MBio. 2025;16(9):e0364124. https://doi.org/10.1128/mbio.03641-24.
12. Hayashi T, Senda M, Suzuki N, Nishikawa H, Ben C, Tang C, Nagase L, Inoue K, Senda T, Hatakeyama M. Differential mechanisms for SHP2 binding and activation are exploited by geographically distinct Helicobacter pylori CagA oncoproteins. Cell Rep. 2017;20(12):2876–2890. https://doi.org/10.1016/j.celrep.2017.08.080.
13. Atherton JC, Cao P, Peek RM Jr, Tummuru MK, Blaser MJ, Cover TL. Mosaicism in vacuolating cytotoxin alleles of *H. pylori*. Association of specific vacA types with

cytotoxin production and peptic ulceration. J Biol Chem. 1995;270(30):17771–7. https://doi.org/10.1074/jbc.270.30.17771.
14. Malfertheiner P, Megraud F, Rokkas T, Gisbert JP, Liou JM, Schulz C, et al. Management of *H. pylori* infection: the Maastricht VI/Florence consensus report. Gut. 2022; https://doi.org/10.1136/gutjnl-2022-327745.
15. Malaty HM, El-Kasabany A, Graham DY, Miller CC, Reddy SG, Srinivasan SR, et al. Age at acquisition of *H. pylori* infection: a follow-up study from infancy to adulthood. Lancet. 2002;359(9310):931–5. https://doi.org/10.1016/s0140-6736(02)08025-x.
16. Kienesberger S, Perez-Perez GI, Olivares AZ, Bardhan P, Sarker SA, Hasan KZ, et al. When is *H. pylori* acquired in populations in developing countries? A birth-cohort study in Bangladeshi children. Gut Microbes. 2018;9(3):252–63. https://doi.org/10.1080/19490976.2017.1421887.
17. den Hoed CM, Vila AJ, Holster IL, Perez-Perez GI, Blaser MJ, de Jongste JC, et al. *H. pylori* and the birth cohort effect: evidence for stabilized colonization rates in childhood. Helicobacter. 2011;16(5):405–9. https://doi.org/10.1111/j.1523-5378.2011.00854.x.
18. Zamani M, Ebrahimtabar F, Zamani V, Miller WH, Alizadeh-Navaei R, Shokri-Shirvani J, et al. Systematic review with meta-analysis: the worldwide prevalence of *H. pylori* infection. Aliment Pharmacol Ther. 2018;47(7):868–76. https://doi.org/10.1111/apt.14561.
19. Hu Y, Wan JH, Li XY, Zhu Y, Graham DY, Lu NH. Systematic review with meta-analysis: the global recurrence rate of *H. pylori*. Aliment Pharmacol Ther. 2017;46(9):773–9. https://doi.org/10.1111/apt.14319.
20. Rowland M, Kumar D, Daly L, O'Connor P, Vaughan D, Drumm B. Low rates of *H. pylori* reinfection in children. Gastroenterology. 1999;117(2):336–41. https://doi.org/10.1053/gast.1999.0029900336.
21. Xie Y, Song C, Cheng H, Xu C, Zhang Z, Wang J, et al. Long-term follow-up of *H. pylori* reinfection and its risk factors after initial eradication: a large-scale multicentre, prospective open cohort, observational study. Emerg Microbes Infect. 2020;9(1):548–57. https://doi.org/10.1080/22221751.2020.1737579.
22. Rothenbacher D, Bode G, Berg G, Knayer U, Gonser T, Adler G, et al. *H. pylori* among preschool children and their parents: evidence of parent-child transmission. J Infect Dis. 1999;179(2):398–402. https://doi.org/10.1086/314595.
23. Kivi M, Tindberg Y, Sörberg M, Casswall TH, Befrits R, Hellström PM, et al. Concordance of *H. pylori* strains within families. J Clin Microbiol. 2003;41(12):5604–8. https://doi.org/10.1128/jcm.41.12.5604-5608.2003.
24. Garg PK, Perry S, Sanchez L, Parsonnet J. Concordance of *H. pylori* infection among children in extended-family homes. Epidemiol Infect. 2006;134(3):450–9. https://doi.org/10.1017/s0950268805005352.
25. Dominici P, Bellentani S, Di Biase AR, Saccoccio G, Le Rose A, Masutti F, et al. Familial clustering of *H. pylori* infection: population based study. BMJ. 1999;319(7209):537–40. https://doi.org/10.1136/bmj.319.7209.537.
26. Rothenbacher D, Bode G, Berg G, Gommel R, Gonser T, Adler G, et al. Prevalence and determinants of *H. pylori* infection in preschool children: a population-based study from Germany. Int J Epidemiol. 1998;27(1):135–41. https://doi.org/10.1093/ije/27.1.135.
27. Nouraie M, Latifi-Navid S, Rezvan H, Radmard AR, Maghsudlu M, Zaer-Rezaii H, et al. Childhood hygienic practice and family education status determine the prevalence of *H. pylori* infection in Iran. Helicobacter. 2009;14(1):40–6. https://doi.org/10.1111/j.1523-5378.2009.00657.x.
28. Yu Y, Xue J, Lin F, Liu D, Zhang W, Ru S, et al. Global primary antibiotic resistance rate of *H. pylori* in recent 10 years: a systematic review and meta-analysis. Helicobacter. 2024;29(3):e13103. https://doi.org/10.1111/hel.13103.
29. Savoldi A, Carrara E, Graham DY, Conti M, Tacconelli E. Prevalence of antibiotic resistance in *H. pylori*: a systematic review and meta-analysis in World Health Organization regions. Gastroenterology. 2018;155(5):1372–82.e17. https://doi.org/10.1053/j.gastro.2018.07.007.

30. Kuo YT, Liou JM, El-Omar EM, Wu JY, Leow AHR, Goh KL, et al. Primary antibiotic resistance in *H. pylori* in the Asia-Pacific region: a systematic review and meta-analysis. Lancet Gastroenterol Hepatol. 2017;2(10):707–15. https://doi.org/10.1016/s2468-1253(17)30219-4.
31. Okimoto T, Ando T, Sasaki M, Ono S, Kobayashi I, Shibayama K, et al. Antimicrobial-resistant *H. pylori* in Japan: report of nationwide surveillance for 2018-2020. Helicobacter. 2024;29(1):e13028. https://doi.org/10.1111/hel.13028.
32. Sholeh M, Khoshnood S, Azimi T, Mohamadi J, Kaviar VH, Hashemian M, et al. The prevalence of clarithromycin-resistant *H. pylori* isolates: a systematic review and meta-analysis. PeerJ. 2023;11:e15121. https://doi.org/10.7717/peerj.15121.
33. Kocsmár É, Buzás GM, Szirtes I, Kocsmár I, Kramer Z, Szijártó A, et al. Primary and secondary clarithromycin resistance in *H. pylori* and mathematical modeling of the role of macrolides. Nat Commun. 2021;12(1):2255. https://doi.org/10.1038/s41467-021-22557-7.
34. Chey WD, Howden CW, Moss SF, Morgan DR, Greer KB, Grover S, et al. ACG clinical guideline: treatment of *H. pylori* infection. Am J Gastroenterol. 2024;119(9):1730–53. https://doi.org/10.14309/ajg.0000000000002968.
35. Ding SZ. Global whole family based-*H. pylori* eradication strategy to prevent its related diseases and gastric cancer. World J Gastroenterol. 2020;26(10):995–1004. https://doi.org/10.3748/wjg.v26.i10.995.
36. Kheyre H, Morais S, Ferro A, Costa AR, Norton P, Lunet N, et al. The occupational risk of *H. pylori* infection: a systematic review. Int Arch Occup Environ Health. 2018;91(6):657–74. https://doi.org/10.1007/s00420-018-1315-6.
37. Krueger WS, Hilborn ED, Converse RR, Wade TJ. Environmental risk factors associated with *H. pylori* seroprevalence in the United States: a cross-sectional analysis of NHANES data. Epidemiol Infect. 2015;143(12):2520–31. https://doi.org/10.1017/s0950268814003938.
38. Mard SA, Khadem Haghighian H, Sebghatulahi V, Ahmadi B. Dietary factors in relation to *H. pylori* infection. Gastroenterol Res Pract. 2014;2014:826910. https://doi.org/10.1155/2014/826910.

# Mechanisms of *H. pylori* Colonization and Pathogenesis

4

Wei Hu and Jieyi Ye

## 4.1 Introduction

*H. pylori* demonstrates a remarkable capacity not merely to persist but to actively thrive within the hostile environment of the human stomach—characterized by extreme acidity, potent enzymatic activity, and continual immune challenge—highlighting its sophisticated adaptation and intricate virulence mechanisms. The infectious process, driven by the coordinated action of multiple factors, unfolds through two sequential stages: colonization and damage. Colonization is essential for establishing and maintaining a persistent infection, requiring the bacterium to first traverse the acidic gastric lumen and adhere firmly to the gastric epithelium. Upon successful colonization, *H. pylori* deploys a suite of virulence factors that interact with host cells, provoking direct or indirect cytopathic effects, triggering substantial inflammatory responses, and disrupting immune homeostasis. The progressive accumulation of cellular injury over time can evolve into a range of gastrointestinal pathologies. This chapter systematically examines the roles and mechanisms of *H. pylori* adhesion and virulence factors operative during these critical phases of infection.

W. Hu (✉) · J. Ye
Department of Gastroenterology, Shenzhen Hospital, Southern Medical University, Shenzhen, Guangdong, China

Shenzhen School of Clinical Medicine, Southern Medical University, Shenzhen, Guangdong, China
e-mail: huwei1683013@smu.edu.cn



J. Gao, Y. Du (eds.), *Non-Antibiotic Therapies for Helicobacter pylori*, https://doi.org/10.1007/978-981-95-9598-3_4

## 4.2 Colonization of the Gastric Niche by *H. pylori*

*H. pylori* primarily resides within the gastric mucus layer or adheres directly to epithelial cells, while a smaller subpopulation colonizes the gastric glands as microcolonies [ 1, 2]. Successful colonization constitutes the critical first step in establishing a chronic infection. This is a remarkable feat, given that the gastric lumen presents a profoundly hostile environment for most microbes, with bactericidal acidity levels typically ranging from pH 0.9 to 1.5. Nonetheless, *H. pylori* has evolved sophisticated adaptive mechanisms that permit not just survival, but effective colonization within this extreme milieu.

### 4.2.1 Helical Cell Shape and Flagella

Bacterial motility is a well-established virulence factor that underpins successful colonization. Motile *H. pylori* strains colonize the gastric mucosa more extensively and sustain colonization longer than nonmotile strains [3]. This capacity for movement is critically dependent on the bacterium's characteristic helical shape. Straight-rod mutants exhibit significantly impaired motility, displaying reduced speed and a lower proportion of motile cells compared to the wild type [4]. The spiral morphology is essential for penetrating the viscous gastric mucus layer, and its loss results in deficient colonization [5]. Consistent with these findings, murine infection models confirm that shape mutants display attenuated stomach colonization, underscoring the necessity of the helical structure for robust infection [6].

The motility apparatus of *H. pylori* is composed of 4–8 sheathed flagella situated at a single pole, each consisting of two flagellin subunits, FlaA and FlaB. These flagella are indispensable for bacterial propulsion and colonization. Notably, although mutants deficient in the motB gene retain structurally normal flagella, they are nonmotile and exhibit a marked reduction in the ability to colonize mouse stomachs, highlighting the functional necessity of flagellar rotation [7]. Both FlaA and FlaB are essential for motility; deletion of either flagellin compromises movement and attenuates colonization [8]. Moreover, motility efficiency is influenced by flagellar number: bacteria possessing four flagella demonstrate a 19% increase in speed over those with only three [ 4].

### 4.2.2 Chemotactic System

*H. pylori* employs a sophisticated chemotaxis system that enables it to detect and respond to environmental signals by modulating flagellar rotation. This capability allows the bacterium to escape the highly acidic gastric lumen and migrate toward favorable niches for colonization. Sensing is mediated by three transmembrane chemoreceptors—TlpA, TlpB, and TlpC—and a cytoplasmic receptor, TlpD. TlpA recognizes arginine and bicarbonate [9, 10]; TlpB is involved in pH- and urea-taxis and also perceives the quorum-sensing molecule autoinducer-2 (AI-2) [11–14]; TlpC responds to environmental lactate [15]; and TlpD mediates repellent chemotaxis from reactive oxygen species (ROS) and hypochlorous acid (HClO), while also monitoring cellular energy status [16–18].

Structurally, these chemoreceptors share a conserved modular architecture, consisting of a ligand-binding sensory domain and a signaling MA domain. The MA domain recruits adapter proteins CheW and CheV, which together complex with the histidine kinase CheA [19]. Upon activation, CheA autophosphorylates and transfers the phosphate group to the response regulator CheY. Phosphorylated CheY then binds to the flagellar switch protein FliM, inducing a change in swimming direction [20, 21]. Signal termination is primarily mediated by the phosphatase CheZ, with evidence suggesting that ChePep may also contribute to CheY dephosphorylation (Fig. 4.1) [22, 23].

The functional integrity of the chemotaxis system is essential for effective colonization. Competitive coinfection experiments in mice reveal that strains lacking TlpA or TlpC are significantly outcompeted by the wild type [24]. Similarly,

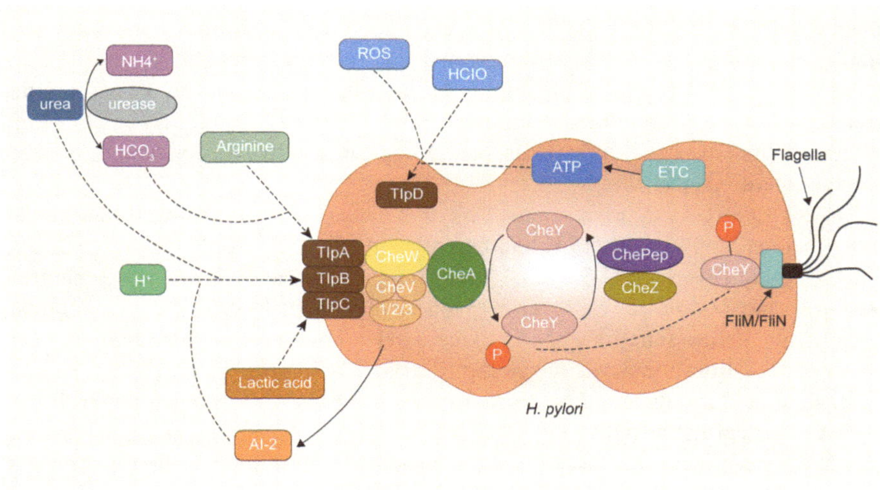


**Fig. 4.1** *Helicobacter pylori* chemotaxis system. Chemical and environmental sensing in *H. pylori* is facilitated by three transmembrane chemoreceptors (TlpA, TlpB, TlpC) and one cytoplasmic receptor (TlpD), each with specialized functions. TlpA serves as a receptor for arginine and bicarbonate; TlpB is implicated in pH- and urea-taxis and also perceives AI-2; TlpC is responsive to lactate; TlpD functions both in mediating repellent chemotaxis away from ROS and HClO, and in monitoring the internal energy status of the cell. The perception of signals through these TlpA-D receptors thus initiates the downstream chemotactic transduction pathway. Signal relay to CheA occurs via the coupling proteins CheW or CheV1/2/3, resulting in the inhibition of CheA autophosphorylation by attractants or its promotion by repellents. Following phosphotransfer to CheY, the phosphorylated response regulator then binds to the flagellar switch protein FliM and alters flagellar motor activity. The signal is terminated by the dephosphorylation of CheY-P, a reaction catalyzed by the CheZ phosphatase in conjunction with ChePep. AI-2, autoinducer-2; ROS, reactive oxygen species; HClO, hypochlorous acid. (Original illustration created by Wei HU. All rights reserved)

non-chemotactic mutants deficient in central signaling components (e.g., CheW, CheA, or CheY) exhibit reduced colonization in mono-infection and are severely disadvantaged in competitive settings [25]. These findings underscore the critical role of directed motility in *H. pylori* persistence within the gastric environment.

### 4.2.3 Urease

*H. pylori* survives the extreme acidity of the gastric lumen through constitutive production of urease. Although predominantly intracellular, a fraction of urease associates with the bacterial surface, largely through lytic release of cytoplasmic content [26, 27]. Urease hydrolyzes urea to yield ammonia and carbamate—the latter decomposing spontaneously into additional ammonia and carbonic acid. Protonation of ammonia generates a localized "ammonia cloud," neutralizing the immediate environment and creating a pH-neutral microniche conducive to bacterial survival.

Urease production is governed by a seven-gene cluster. The ureA and ureB genes encode the enzyme's structural subunits, while ureE, ureF, ureG, and ureH encode chaperones essential for nickel insertion and apoenzyme activation [28]. A key regulatory component is UreI, a proton-gated urea channel encoded by ureI. Under acidic conditions, UreI opens to permit urea entry into the cytoplasm, where internal urease hydrolyzes it to generate ammonia. This pH-dependent gating ensures urea uptake occurs only in acidic milieus, preventing futile substrate consumption under neutral conditions [28].

Urease activity is tightly regulated by ambient pH [27, 29]. Intracellular urease shows low activity at neutral pH but increases up to tenfold as pH drops to 5.5–6.5, remaining high even at pH 3.0 [27]. This acid-triggered activation enables neutralization of the periplasm, safeguarding intracellular metabolism [29]. In contrast, extracellular urease functions optimally near pH 7.5, with activity declining sharply below pH 5.0, questioning its physiological role in acid protection [27].

Beyond acid resistance, urease acts as a multifunctional virulence factor. It disrupts gastric epithelial tight junctions [30] and promotes angiogenesis, supporting gastric cancer progression [31]. Urease also activates TLR2 signaling via the PI3K pathway, stabilizing HIF-1$\alpha$ [32], and binds MHC-II molecules to induce apoptosis in gastric epithelial cells [33]. Additionally, it stimulates monocytes to release pro-inflammatory cytokines such as IL-1$\beta$, IL-6, IL-8, and TNF-$\alpha$, thereby amplifying local inflammation [34].

### 4.2.4 Outer Membrane Proteins

The outer membrane proteins (OMPs) of *H. pylori* play indispensable roles in host colonization and disease progression. Functioning primarily as adhesins, they mediate high-affinity binding to specific host receptors, thereby facilitating bacterial persistence. Beyond adhesion, numerous OMPs act as bona fide virulence factors by modulating host cell signaling, provoking inflammatory responses, and ultimately contributing to the transition from asymptomatic colonization to overt pathology.

AlpA and AlpB (also designated HopC and HopB) promote gastric colonization by binding laminin within the host extracellular matrix [35]. Gerbil infection models confirm that ΔalpA or ΔalpB mutants exhibit severe adherence defects and significantly reduced colonization, underscoring their critical role [35]. In addition to mediating attachment, AlpA/B induce the secretion of pro-inflammatory cytokines IL-8 and IL-6, and activate signaling pathways—including MAPKs, AP-1, and NF-κB—thereby exacerbating gastric inflammation and tissue damage [36]. AlpB has also been implicated in biofilm formation, potentially aiding long-term persistence [37].

BabA (HopS) binds host blood group antigens, enabling epithelial adherence. Based on binding specificity, strains are classified as "specialists" (binding Leb/H1 antigens of blood group O) or "generalists" (additionally recognizing A-Leb and B-Leb) [38, 39]. Structural studies reveal that antigen recognition involves cooperative interactions between the CL2 loop, DL1 region, and a conserved Asp-Ser-Ser triad in DL2, with binding stability maintained by a CL2 disulfide bond [39]. BabA–Leb engagement enhances T4SS activity, upregulates pro-inflammatory and metaplastic factors, and promotes carcinogenesis [40]. Notably, BabA-mediated adhesion is acid-sensitive, allowing reversible attachment and bacterial dispersal under fluctuating acidity [41].

SabA (HopP) recognizes sialylated Lewis X (sLex) antigens on host cells [42]. Chronic infection upregulates β3GnT5, increasing sLex expression and reinforcing SabA-dependent adhesion [43, 44]. Inflammation further amplifies sLex synthesis via TNF/NF-κB-driven transcription of B3GNT5, B3GALT5, and FUT3 [45]. SabA binding to neutrophils triggers reactive oxygen species (ROS) production, causing oxidative damage to gastric epithelium and worsening tissue injury [46].

HopQ binds CEACAM receptors (CEACAM1, 3, 5, 6) via their N-terminal IgV-like domains [47, 48]. This interaction is essential for T4SS-mediated delivery of CagA into host cells [48] and contributes to noncanonical NF-κB activation [ 49]. HopQ–CEACAM1 engagement also suppresses T cell and NK cell activity, facilitating immune evasion [50].

OipA (HopH) exists in functional (IL-8-inducing) and nonfunctional forms [51]. Functional OipA synergizes with CagA to augment virulence and is clinically associated with higher bacterial load, severe inflammation, and elevated IL-8 [51–53]. It also promotes apoptosis, indicated by an increased Bax/Bcl-2 ratio and caspase-3 cleavage [54].

HopZ facilitates epithelial adhesion, as HopZ mutants show significant binding defects [55]. Its expression undergoes phase variation via CT-dinucleotide repeat slippage in the signal peptide region, enabling ON/OFF switching. The "ON" state is selected during early infection to aid colonization, while ON/OFF status stabilizes during chronicity, suggesting stage-specific adaptive roles [56].

HomB contributes to epithelial adhesion and IL-8 induction; knockout mutants are severely impaired in both functions [57]. The presence of HomB is linked to an increased risk of gastric cancer, positioning it as a potential discriminatory marker between gastric cancer and duodenal ulcer [58].

### 4.2.5 LacdiNAc-Specific Adhesin

LacdiNAc-specific adhesin (LabA) functions as a key adhesin that enables *H. pylori* to colonize the gastric epithelium by specifically recognizing the LacdiNAc motif (GalNAcβ1-4GlcNAc, LDN) present on host MUC5AC mucins, thereby facilitating bacterial attachment [59].

### 4.2.6 Heat Shock Proteins

Heat shock proteins (HSPs) are a class of highly conserved molecular chaperones induced under stress conditions, with a primary role in maintaining proper protein folding, stability, and function. In *H. pylori*, major HSPs—including GroEL, GroES, DnaK, DnaJ, GrpE, and CbpA—are tightly regulated at the transcriptional level by the repressors HspR and HrcA [60–62]. These proteins are essential for bacterial survival and colonization, enabling *H. pylori* to withstand gastric challenges such as high acidity and oxidative stress.

Beyond facilitating bacterial stress adaptation, HSPs contribute to pathogenesis by mediating *H. pylori* adhesion to sulfated glycolipid complexes (SGCs), a process implicated in the establishment of initial infection [63]. Notably, the transcriptional repressors HrcA and HspR—key regulators of the HSP regulon—are upregulated during biofilm formation, correlating with enhanced bacterial antibiotic resistance [64]. HSPs also exacerbate host inflammatory responses: HSP60 (GroEL) binds Toll-like receptors (TLR-2/4), activating NF-κB, ERK, and p38 MAPK signaling pathways to stimulate TNF-α and IL-8 production [65]. Additionally, GroEL promotes IL-6 secretion from macrophages via NF-κB activation [ 66]. Furthermore, HSP60 enhances angiogenesis through the CXCR2/PLCβ2/$Ca2^{+}$ signaling axis in endothelial cells, thereby supporting tumor progression [67].

### 4.2.7 Gamma-Glutamyl Transpeptidase

Gamma-glutamyl transpeptidase (GGT) in *H. pylori* catalyzes the extracellular hydrolysis of glutamine to glutamate and ammonia, as well as the breakdown of glutathione to glutamate and cysteinylglycine (Cys-Gly). The resulting glutamate is imported into the bacterial cell via the GltS transporter, where it partly serves as a substrate for glutamine synthesis, thereby integrating into bacterial nitrogen metabolism [68, 69]. Furthermore, persistent GGT-mediated glutathione depletion in the host disrupts cellular redox homeostasis, increasing susceptibility to oxidative stress, DNA damage, cell death, and carcinogenesis [68]. GGT synthesis itself is an energy-demanding process that consumes ATP; its overproduction in host cells can lead to substantial energy depletion, exacerbating cellular dysfunction [68]. The ammonia generated from these reactions not only neutralizes gastric acid to protect the bacterium but also exerts cytotoxic effects that damage gastric epithelial cells [70].

GGT acts as a critical virulence factor that enhances *H. pylori* stomach colonization. This is evidenced in infection models where GGT-deficient mutants, despite establishing infection, achieve significantly lower bacterial loads than wild-type strains [71]. GGT further supports persistence by helping maintain urease activity, crucial for survival in gastric acid [70]. On the immune front, GGT-derived glutamate activates dendritic cell glutamate receptors, thereby inhibiting cAMP signaling and IL-6 secretion and promoting a tolerogenic DC phenotype. These DCs facilitate regulatory T cell (Treg) expansion, which suppresses effector T cells, thus creating an immune-tolerant niche that permits long-term bacterial persistence [72, 73].

In addition to its role in colonization, GGT contributes significantly to the pathogenesis of *H. pylori*. It stimulates the production of $H_2O_2$ in host cells, which in turn activates the NF-κB signaling pathway to promote the secretion of IL-8. The resultant oxidative stress can also cause DNA damage and induce apoptosis [74]. Furthermore, GGT triggers the mitochondrial apoptotic pathway, evidenced by the release of cytochrome c and subsequent activation of caspase-9 and caspase-3 in AGS cells, both of which are recognized as key executors in GGT-mediated apoptosis [75].

### 4.2.8 Neutrophil-Activating Protein

*H. pylori* neutrophil-activating protein (Hp-NAP) supports bacterial colonization by binding to sulfated oligosaccharides on salivary mucins, thereby mediating adhesion to host surfaces [76]. In environments with limited sialylation, it also enhances SabA-dependent attachment, further promoting bacterial persistence [77]. Beyond its role in adhesion, Hp-NAP triggers reactive oxygen species (ROS) production in host cells. Paradoxically, it also serves a protective function for the bacterium by sequestering free iron, thereby reducing iron-catalyzed oxidative DNA damage, and improving bacterial survival under oxidative stress [78].

The mechanism of Hp-NAP-induced ROS generation involves a multistep signaling cascade. Through the activation of pertussis toxin-sensitive G proteins, phosphatidylinositol 3-kinase (PI3-K), Src family tyrosine kinases, and elevated cytosolic calcium, Hp-NAP stimulates NADPH oxidase in neutrophils and monocytes, leading to robust ROS production [79]. Concurrently, Hp-NAP enhances neutrophil adhesion to the endothelium. After internalization and transcellular transport, it upregulates β2 integrin expression and primes its high-affinity conformation via the same signaling pathway, facilitating firm neutrophil adhesion [79, 80]. Although the resulting oxidative burst can damage both host and bacterial cells, Hp-NAP enables an adaptive response by promoting biofilm formation. In the absence of

Hp-NAP, bacteria fail to form protective biofilms under $H_2O_2$-induced stress, underscoring its critical role in oxidative stress adaptation [81].

In addition to orchestrating oxidative and adhesive events, Hp-NAP acts as a potent immunomodulator that drives a broad inflammatory response across multiple immune cell types. In polymorphonuclear cells (PMNs), it stimulates the release of IL-8 (via TLR2 and G protein signaling), MIP-1α (CCL3), and MIP-1β (CCL4) [80, 82]. In blood mononuclear cells (MNCs), it induces the production of TNF-α, tissue factor (TF), and plasminogen activator inhibitor-2 (PAI-2) [83]. Hp-NAP also activates mast cells to secrete histamine and IL-6 [84]. Furthermore, by prompting neutrophils and monocytes to express IL-12 and IL-23, it skews the immune microenvironment toward a T helper 1 (Th1)-polarized response, reinforcing chronic inflammatory pathology [85].

## 4.3 From Colonization to Damage: The Pathogenic Phase of *H. pylori*

Upon securing stable colonization, *H. pylori* transitions to a damaging lifestyle, orchestrating a direct assault on host cells through the coordinated release of virulence factors. These factors disrupt cellular structure and function, thereby driving the initiation and progression of *H. pylori*-related disease.

### 4.3.1 Cytotoxin-Associated Gene A

*H. pylori* strains are stratified into two major groups based on the presence or absence of the cytotoxin-associated gene A (CagA). CagA-positive strains demonstrate superior adaptability and enhanced pathogenic potential relative to CagA-negative isolates. This advantage is partly attributable to the ability of CagA to modulate the expression of flagellin genes flaA and flaB, thereby fine-tuning bacterial motility—a critical factor in successful colonization [86]. Such functional regulation underscores the increased virulence associated with CagA-positive *H. pylori*.

The molecular machinery responsible for CagA delivery is encoded within the approximately 40 kb cag pathogenicity island (CagPAI), which includes the type IV secretion system (T4SS) and the CagA effector itself. Upon bacterial contact with host gastric epithelial cells, the T4SS is activated, translocating CagA directly into the host cytoplasm. Inside the host cell, CagA is phosphorylated on specific tyrosine residues within its C-terminal EPIYA motifs by host kinases such as c-Src, Fyn, and Abl [80]. Both phosphorylated and non-phosphorylated forms of CagA interact with a range of host proteins, effectively hijacking normal cellular signaling [87]. In addition to CagA, the T4SS delivers nonprotein bacterial components—including LPS metabolites, peptidoglycan, and bacterial DNA—into host cells, where they further modulate inflammatory and immune signaling pathways [87].

Once delivered, CagA acts as a multifunctional virulence effector that disrupts host cell homeostasis through several distinct mechanisms. $CagA^+$ infection induces a loss of cellular polarity, disrupts intercellular junctions, promotes pseudopod extension, and facilitates basement membrane degradation, collectively enabling an invasive cellular phenotype [88]. By inhibiting the serine/threonine kinase Par1, CagA impairs DNA repair and contributes to double-strand break accumulation [89]. Furthermore, via its CRPIA motif, non-phosphorylated CagA binds the hepatocyte growth factor receptor Met, leading to sustained PI3K/Akt signaling and subsequent activation of β-catenin and NF-κB pathways, which drive aberrant epithelial proliferation and inflammation [90]. CagA also interferes with E-cadherin/β-catenin complex formation, resulting in β-catenin nuclear accumulation and promoting intestinal metaplasia [91]. Additionally, by disrupting the interaction between the tumor suppressor p53 and its activator ASPP2, CagA impairs apoptosis and facilitates malignant transformation (Fig. 4.2) [92].

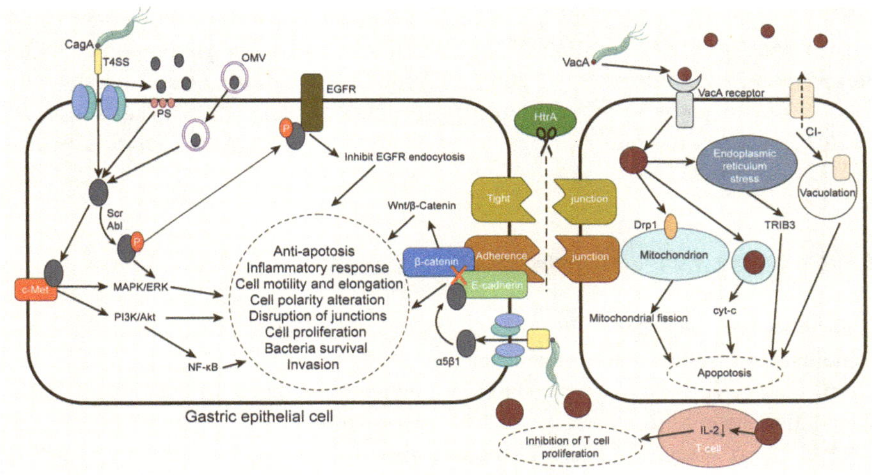


**Fig. 4.2** Multifaceted mechanisms of *Helicobacter pylori* virulence factors CagA and VacA in modulating host epithelial and immune responses. (Left panel) Upon attachment to the host cell, *H. pylori* delivers the CagA oncoprotein directly into the gastric epithelial cytoplasm via the T4SS. Following translocation, CagA is phosphorylated by Src/Abl family kinases and engages with host signaling molecules, notably c-Met. This interaction initiates the activation of multiple downstream pathways—including MAPK/ERK, PI3K/Akt, NF-κB, and Wnt/β-catenin—collectively promoting pro-inflammatory responses, enhancing cell survival, and disrupting epithelial integrity. In parallel, OMVs may serve as an auxiliary mechanism for CagA delivery. (Right panel) The VacA toxin binds to specific cell-surface receptors and is internalized. Inside the cell, VacA triggers endoplasmic reticulum stress, induces mitochondrial damage and fission, and stimulates cytoplasmic vacuolation, culminating in apoptosis. Beyond these direct cytotoxic effects, VacA also modulates immune responses by suppressing T-cell proliferation, a process potentially mediated through the IL-2/IL-21 signaling axis. T4SS, type IV secretion system; OMVs, outer membrane vesicles. (Original illustration created by Wei HU. All rights reserved)

### 4.3.2 Vacuolating Cytotoxin A

Vacuolating cytotoxin A (VacA) is a pore-forming toxin secreted by *H. pylori*, composed of two functional subunits: p33 and p55. The p33 subunit forms the inner transmembrane channel responsible for chloride ion transport, while the p55 subunit mediates binding to host cell receptors [93]. VacA interacts with several host transmembrane receptors, including receptor protein tyrosine phosphatases α and β (RPTPα/RPTPβ) and low-density lipoprotein receptor-related protein 1 (LRP1). These receptors facilitate toxin internalization and concurrently activate intracellular signaling pathways that promote cell death and contribute to gastric ulcer formation [94, 95].

VacA is secreted through the type V autotransporter system and enters host cells via endocytosis [93]. Upon internalization, it exerts multiple pathogenic effects. VacA induces cytoplasmic vacuolation through a process requiring rab7 GTPase and ATPase proton pump activity [96]. It also triggers mitochondrial fission via dynamin-related protein 1 (Drp1), leading to apoptosis of gastric epithelial cells [97]. Under endoplasmic reticulum (ER) stress, VacA upregulates pseudokinase TRIB3, promoting autophagic cell death [98]. By downregulating connexin 43 (Cx43), VacA causes its cytoplasmic accumulation and subsequent apoptosis [99]. Furthermore, VacA suppresses T cell proliferation by disrupting the T cell receptor/IL-2 signaling pathway in a calcineurin-dependent manner [100]. Additionally, VacA contributes to respiratory pathogenesis by stimulating the secretion of IL-8 and IL-6 (Fig. 4.2) [101].

### 4.3.3 Lipopolysaccharide

The lipopolysaccharide (LPS) of *H. pylori* comprises three distinct structural domains: lipid A, core oligosaccharide, and O antigen [102]. The lipid A moiety exhibits weak Toll-like receptor 4 (TLR4) activation capacity due to its low phosphorylation level and long fatty acid chains, distinguishing it from LPS of other Gram-negative bacteria [102]. The core oligosaccharide, divided into inner and outer regions, contributes to colonization through different mechanisms: the inner core binds laminin, whereas the outer core influences colonization via α-1,6-glucan [102, 103]. The O antigen facilitates immune evasion by molecular mimicry of host Lewis blood group antigens, thereby supporting persistent gastric colonization [102]. Additionally, LPS promotes bacterial survival in the stomach by interacting with gastric trefoil factor 1 (TFF1) in the mucus layer [104].

LPS acts as a multifunctional activator in *H. pylori*-triggered gastric pathogenesis. It upregulates TLR2 in AGS cells, thereby triggering STAT3 and ERK1/2 activation and enhancing claudin-4, −6, −7, and −9 expression in a time-dependent manner [105]. Through CD14 engagement [106], LPS induces monocytes to release IL-8, ENA-78, MCP-1, TNF-α, and regulatory cytokines such as IL-10, IL-18, and IL-12 [107, 108]. LPS further impairs mucosal integrity by suppressing mucin synthesis and promotes apoptosis via caspase-3 activation dependent on ERK/p38 signaling [109]. Moreover, it enhances epithelial proliferation and inflammation via the MEK1/2-ERK1/2 pathway, suggesting a role in promoting gastric tumorigenesis [110].

### 4.3.4 High Temperature Requirement A

High temperature requirement A (HtrA) is a serine protease produced by *H. pylori*. Deletion of HtrA renders the bacterium more susceptible to stressors such as heat, pH fluctuation, and osmotic shock, indicating its role in cellular protection [111]. As a key virulence factor, HtrA cleaves host tight and adherens junction proteins—including Occludin, Claudin-8, and E-cadherin—thereby disrupting the gastric epithelial barrier [112, 113]. This cleavage creates paracellular openings that allow *H. pylori* to invade host tissues, after which HtrA facilitates CagA translocation by interacting with α5β1 integrin receptors [114].

### 4.3.5 Outer Membrane Vesicle

Outer membrane vesicles (OMVs) released by *H. pylori* are spherical nanoparticles typically measuring 20–300 nm in diameter, with their size and abundance influenced by environmental conditions such as pH and temperature [115]. These vesicles encapsulate a diverse array of biomolecules, including lipids, lipopolysaccharides (LPS), proteins, peptidoglycans, and nucleic acids, which collectively contribute to their functional versatility [115].

Beyond their structural role, OMVs function as key vehicles for the delivery of bacterial virulence factors into host cells. This entry is facilitated through both clathrin-dependent and -independent endocytic pathways, enabling OMVs to exert significant pathogenic effects [116]. In addition to direct host cell modulation, OMVs play a critical role in biofilm formation and antibiotic resistance. For instance, OMVs are abundant within the biofilm matrix of the high-biofilm-forming strain TK1402, and exogenous addition of purified OMVs enhances biofilm development in vitro [117]. OMVs further stabilize biofilms by protecting extracellular DNA (eDNA)—a major structural component—from nuclease degradation. Surface-associated eDNA also promotes inter-vesicle aggregation, thereby reinforcing the biofilm architecture [118].

Within the gastric environment, OMVs emerge as critical mediators of *H. pylori*-driven disease progression. At elevated concentrations, they inhibit the growth of gastric epithelial cells, induce cytotoxicity, and stimulate the secretion of IL-8, amplifying local inflammatory responses [119]. Cell death is triggered through a non-mitochondrial apoptotic pathway, underscoring a distinct mechanism of OMV-induced damage [120]. Moreover, OMVs enhance the carcinogenic potential of *H. pylori*; exposure to OMVs from toxigenic strains leads to increased micronucleus formation in gastric epithelial cells, indicating genomic instability and supporting their role in gastric carcinogenesis [121].

Notably, the impact of OMVs extends beyond the gastric niche. Emerging evidence links these vesicles to extra-gastric pathologies. For example, hepatocyte-derived exosomes pretreated with *H. pylori* OMVs can activate hematopoietic stem cells and upregulate fibrosis-related markers (β-catenin, vimentin, α-SMA, TIMP-1), thereby accelerating liver fibrosis [122]. Furthermore, *H. pylori* OMVs have been shown to cross the blood–brain barrier, where they are internalized by astrocytes. This uptake activates glial cells via the C3-C3aR signaling pathway, promoting neuroinflammation and neuronal dysfunction, and ultimately contributing to the pathogenesis of Alzheimer's disease [123].

## 4.4 Conclusion

In summary, *H. pylori* employs a sophisticated pathogenic strategy reliant on the coordinated action of numerous virulence factors. Initial colonization ensures bacterial survival within the hostile gastric niche, while subsequent pathogenesis—marked by the induction of inflammatory and immune responses, disruption of key physiological functions, and promotion of cell death—collectively drives disease progression. A deep mechanistic understanding of these adhesins and virulence factors is therefore essential for developing advanced diagnostic, therapeutic, and prophylactic interventions.

## References

1. Hazell SL, Lee A, Brady L, Hennessy W. Campylobacter pyloridis and gastritis: association with intercellular spaces and adaptation to an environment of mucus as important factors in colonization of the gastric epithelium. J Infect Dis. 1986;153(4):658–63. https://doi.org/10.1093/infdis/153.4.658.
2. Sigal M, Rothenberg ME, Logan CY, Lee JY, Honaker RW, Cooper RL, et al. *H. pylori* activates and expands Lgr5(+) stem cells through direct colonization of the gastric glands. Gastroenterology. 2015;148(7):1392–404 e21. https://doi.org/10.1053/j.gastro.2015.02.049.
3. Eaton KA, Morgan DR, Krakowka S. Motility as a factor in the colonisation of gnotobiotic piglets by *H. pylori*. J Med Microbiol. 1992;37(2):123–7. https://doi.org/10.1099/00222615-37-2-123.
4. Martinez LE, Hardcastle JM, Wang J, Pincus Z, Tsang J, Hoover TR, et al. *H. pylori* strains vary cell shape and flagellum number to maintain robust motility in viscous environments. Mol Microbiol. 2016;99(1):88–110. https://doi.org/10.1111/mmi.13218.
5. Sycuro LK, Wyckoff TJ, Biboy J, Born P, Pincus Z, Vollmer W, et al. Multiple peptidoglycan modification networks modulate *H. pylori*'s cell shape, motility, and colonization potential. PLoS Pathog. 2012;8(3):e1002603. https://doi.org/10.1371/journal.ppat.1002603.
6. Sycuro LK, Pincus Z, Gutierrez KD, Biboy J, Stern CA, Vollmer W, et al. Peptidoglycan crosslinking relaxation promotes *H. pylori*'s helical shape and stomach colonization. Cell. 2010;141(5):822–33. https://doi.org/10.1016/j.cell.2010.03.046.
7. Ottemann KM, Lowenthal AC. *H. pylori* uses motility for initial colonization and to attain robust infection. Infect Immun. 2002;70(4):1984–90. https://doi.org/10.1128/IAI.70.4.1984-1990.2002.

8. Eaton KA, Suerbaum S, Josenhans C, Krakowka S. Colonization of gnotobiotic piglets by *H. pylori* deficient in two flagellin genes. Infect Immun. 1996;64(7):2445–8. https://doi.org/10.1128/iai.64.7.2445-2448.1996.
9. Cerda O, Rivas A, Toledo H. *H. pylori* strain ATCC700392 encodes a methyl-accepting chemotaxis receptor protein (MCP) for arginine and sodium bicarbonate. FEMS Microbiol Lett. 2003;224(2):175–81. https://doi.org/10.1016/S0378-1097(03)00423-3.
10. Cerda OA, Nunez-Villena F, Soto SE, Ugalde JM, Lopez-Solis R, Toledo H. tlpA gene expression is required for arginine and bicarbonate chemotaxis in *H. pylori*. Biol Res. 2011;44(3):277–82.
11. Croxen MA, Sisson G, Melano R, Hoffman PS. The *H. pylori* chemotaxis receptor TlpB (HP0103) is required for pH taxis and for colonization of the gastric mucosa. J Bacteriol. 2006;188(7):2656–65. https://doi.org/10.1128/JB.188.7.2656-2665.2006.
12. Huang JY, Sweeney EG, Sigal M, Zhang HC, Remington SJ, Cantrell MA, et al. Chemodetection and destruction of host urea allows *H. pylori* to locate the epithelium. Cell Host Microbe. 2015;18(2):147–56. https://doi.org/10.1016/j.chom.2015.07.002.
13. Rader BA, Wreden C, Hicks KG, Sweeney EG, Ottemann KM, Guillemin K. *H. pylori* perceives the quorum-sensing molecule AI-2 as a chemorepellent via the chemoreceptor TlpB. Microbiology (Reading). 2011;157(Pt 9):2445–55. https://doi.org/10.1099/mic.0.049353-0.
14. Hanyu H, Engevik KA, Matthis AL, Ottemann KM, Montrose MH, Aihara E. *H. pylori* uses the TlpB receptor to sense sites of gastric injury. Infect Immun. 2019;87(9) https://doi.org/10.1128/IAI.00202-19.
15. Machuca MA, Johnson KS, Liu YC, Steer DL, Ottemann KM, Roujeinikova A. *H. pylori* chemoreceptor TlpC mediates chemotaxis to lactate. Sci Rep. 2017;7(1):14089. https://doi.org/10.1038/s41598-017-14372-2.
16. Schweinitzer T, Mizote T, Ishikawa N, Dudnik A, Inatsu S, Schreiber S, et al. Functional characterization and mutagenesis of the proposed behavioral sensor TlpD of *H. pylori*. J Bacteriol. 2008;190(9):3244–55. https://doi.org/10.1128/JB.01940-07.
17. Perkins A, Tudorica DA, Amieva MR, Remington SJ, Guillemin K. *H. pylori* senses bleach (HOCl) as a chemoattractant using a cytosolic chemoreceptor. PLoS Biol. 2019;17(8):e3000395. https://doi.org/10.1371/journal.pbio.3000395.
18. Collins KD, Andermann TM, Draper J, Sanders L, Williams SM, Araghi C, et al. The *H. pylori* CZB cytoplasmic chemoreceptor TlpD forms an autonomous polar chemotaxis signaling complex that mediates a tactic response to oxidative stress. J Bacteriol. 2016;198(11):1563–75. https://doi.org/10.1128/JB.00071-16.
19. Alexander RP, Lowenthal AC, Harshey RM, Ottemann KM. CheV: CheW-like coupling proteins at the core of the chemotaxis signaling network. Trends Microbiol. 2010;18(11):494–503. https://doi.org/10.1016/j.tim.2010.07.004.
20. Lowenthal AC, Hill M, Sycuro LK, Mehmood K, Salama NR, Ottemann KM. Functional analysis of the *H. pylori* flagellar switch proteins. J Bacteriol. 2009;191(23):7147–56. https://doi.org/10.1128/JB.00749-09.
21. Lertsethtakarn P, Ottemann KM, Hendrixson DR. Motility and chemotaxis in campylobacter and helicobacter. Ann Rev Microbiol. 2011;65:389–410. https://doi.org/10.1146/annurev-micro-090110-102908.
22. Howitt MR, Lee JY, Lertsethtakarn P, Vogelmann R, Joubert LM, Ottemann KM, et al. ChePep controls *H. pylori* infection of the gastric glands and chemotaxis in the Epsilonproteobacteria. MBio. 2011;2(4) https://doi.org/10.1128/mBio.00098-11.
23. Lertsethtakarn P, Howitt MR, Castellon J, Amieva MR, Ottemann KM. *H. pylori* CheZ(HP) and ChePep form a novel chemotaxis-regulatory complex distinct from the core chemotaxis signaling proteins and the flagellar motor. Mol Microbiol. 2015;97(6):1063–78. https://doi.org/10.1111/mmi.13086.
24. Andermann TM, Chen YT, Ottemann KM. Two predicted chemoreceptors of *H. pylori* promote stomach infection. Infect Immun. 2002;70(10):5877–81. https://doi.org/10.1128/IAI.70.10.5877-5881.2002.

25. Terry K, Williams SM, Connolly L, Ottemann KM. Chemotaxis plays multiple roles during *H. pylori* animal infection. Infect Immun. 2005;73(2):803–11. https://doi.org/10.1128/IAI.73.2.803-811.2005.
26. Hawtin PR, Stacey AR, Newell DG. Investigation of the structure and localization of the urease of *H. pylori* using monoclonal antibodies. J Gen Microbiol. 1990;136(10):1995–2000. https://doi.org/10.1099/00221287-136-10-1995.
27. Scott DR, Weeks D, Hong C, Postius S, Melchers K, Sachs G. The role of internal urease in acid resistance of *H. pylori*. Gastroenterology. 1998;114(1):58–70. https://doi.org/10.1016/s0016-5085(98)70633-x.
28. Weeks DL, Eskandari S, Scott DR, Sachs G. A H+-gated urea channel: the link between *H. pylori* urease and gastric colonization. Science. 2000;287(5452):482–5. https://doi.org/10.1126/science.287.5452.482.
29. Rektorschek M, Weeks D, Sachs G, Melchers K. Influence of pH on metabolism and urease activity of *H. pylori*. Gastroenterology. 1998;115(3):628–41. https://doi.org/10.1016/s0016-5085(98)70142-8.
30. Wroblewski LE, Shen L, Ogden S, Romero-Gallo J, Lapierre LA, Israel DA, et al. *H. pylori* dysregulation of gastric epithelial tight junctions by urease-mediated myosin II activation. Gastroenterology. 2009;136(1):236–46. https://doi.org/10.1053/j.gastro.2008.10.011.
31. Olivera-Severo D, Uberti AF, Marques MS, Pinto MT, Gomez-Lazaro M, Figueiredo C, et al. A new role for *H. pylori* urease: contributions to angiogenesis. Front Microbiol. 2017;8:1883. https://doi.org/10.3389/fmicb.2017.01883.
32. Valenzuela-Valderrama M, Cerda-Opazo P, Backert S, Gonzalez MF, Carrasco-Veliz N, Jorquera-Cordero C, et al. The *H. pylori* urease virulence factor is required for the induction of hypoxia-induced factor-1alpha in gastric cells. Cancers (Basel). 2019;11(6) https://doi.org/10.3390/cancers11060799.
33. Fan X, Gunasena H, Cheng Z, Espejo R, Crowe SE, Ernst PB, et al. *H. pylori* urease binds to class II MHC on gastric epithelial cells and induces their apoptosis. J Immunol. 2000;165(4):1918–24. https://doi.org/10.4049/jimmunol.165.4.1918.
34. Harris PR, Mobley HL, Perez-Perez GI, Blaser MJ, Smith PD. *H. pylori* urease is a potent stimulus of mononuclear phagocyte activation and inflammatory cytokine production. Gastroenterology. 1996;111(2):419–25. https://doi.org/10.1053/gast.1996.v111.pm8690207.
35. Senkovich OA, Yin J, Ekshyyan V, Conant C, Traylor J, Adegboyega P, et al. *H. pylori* AlpA and AlpB bind host laminin and influence gastric inflammation in gerbils. Infect Immun. 2011;79(8):3106–16. https://doi.org/10.1128/IAI.01275-10.
36. Lu H, Wu JY, Beswick EJ, Ohno T, Odenbreit S, Haas R, et al. Functional and intracellular signaling differences associated with the *H. pylori* AlpAB adhesin from Western and east Asian strains. J Biol Chem. 2007;282(9):6242–54. https://doi.org/10.1074/jbc.M611178200.
37. Yonezawa H, Osaki T, Fukutomi T, Hanawa T, Kurata S, Zaman C, et al. Diversification of the AlpB outer membrane protein of *H. pylori* affects biofilm formation and cellular adhesion. J Bacteriol. 2017;199(6) https://doi.org/10.1128/JB.00729-16.
38. Aspholm-Hurtig M, Dailide G, Lahmann M, Kalia A, Ilver D, Roche N, et al. Functional adaptation of BabA, the *H. pylori* ABO blood group antigen binding adhesin. Science. 2004;305(5683):519–22. https://doi.org/10.1126/science.1098801.
39. Moonens K, Gideonsson P, Subedi S, Bugaytsova J, Romao E, Mendez M, et al. Structural insights into polymorphic ABO glycan binding by *H. pylori*. Cell Host Microbe. 2016;19(1):55–66. https://doi.org/10.1016/j.chom.2015.12.004.
40. Ishijima N, Suzuki M, Ashida H, Ichikawa Y, Kanegae Y, Saito I, et al. BabA-mediated adherence is a potentiator of the *H. pylori* type IV secretion system activity. J Biol Chem. 2011;286(28):25256–64. https://doi.org/10.1074/jbc.M111.233601.
41. Bugaytsova JA, Bjornham O, Chernov YA, Gideonsson P, Henriksson S, Mendez M, et al. *H. pylori* adapts to chronic infection and gastric disease via pH-responsive BabA-mediated adherence. Cell Host Microbe. 2017;21(3):376–89. https://doi.org/10.1016/j.chom.2017.02.013.

42. Pang SS, Nguyen STS, Perry AJ, Day CJ, Panjikar S, Tiralongo J, et al. The three-dimensional structure of the extracellular adhesion domain of the sialic acid-binding adhesin SabA from *H. pylori*. J Biol Chem. 2014;289(10):6332–40. https://doi.org/10.1074/jbc.M113.513135.
43. Marcos NT, Magalhaes A, Ferreira B, Oliveira MJ, Carvalho AS, Mendes N, et al. *H. pylori* induces beta3GnT5 in human gastric cell lines, modulating expression of the SabA ligand sialyl-Lewis x. J Clin Invest. 2008;118(6):2325–36. https://doi.org/10.1172/JCI34324.
44. Mahdavi J, Sonden B, Hurtig M, Olfat FO, Forsberg L, Roche N, et al. *H. pylori* SabA adhesin in persistent infection and chronic inflammation. Science. 2002;297(5581):573–8. https://doi.org/10.1126/science.1069076.
45. Magalhaes A, Marcos-Pinto R, Nairn AV, Dela Rosa M, Ferreira RM, Junqueira-Neto S, et al. *H. pylori* chronic infection and mucosal inflammation switches the human gastric glycosylation pathways. Biochim Biophys Acta. 2015;1852(9):1928–39. https://doi.org/10.1016/j.bbadis.2015.07.001.
46. Unemo M, Aspholm-Hurtig M, Ilver D, Bergstrom J, Boren T, Danielsson D, et al. The sialic acid binding SabA adhesin of *H. pylori* is essential for nonopsonic activation of human neutrophils. J Biol Chem. 2005;280(15):15390–7. https://doi.org/10.1074/jbc.M412725200.
47. Javaheri A, Kruse T, Moonens K, Mejias-Luque R, Debraekeleer A, Asche CI, et al. *H. pylori* adhesin HopQ engages in a virulence-enhancing interaction with human CEACAMs. Nat Microbiol. 2016;2:16189. https://doi.org/10.1038/nmicrobiol.2016.189.
48. Koniger V, Holsten L, Harrison U, Busch B, Loell E, Zhao Q, et al. *H. pylori* exploits human CEACAMs via HopQ for adherence and translocation of CagA. Nat Microbiol. 2016;2:16188. https://doi.org/10.1038/nmicrobiol.2016.188.
49. Maubach G, Sokolova O, Tager C, Naumann M. CEACAMs interaction with *H. pylori* HopQ supports the type 4 secretion system-dependent activation of non-canonical NF-kappaB. Int J Med Microbiol. 2020;310(6):151444. https://doi.org/10.1016/j.ijmm.2020.151444.
50. Gur C, Maalouf N, Gerhard M, Singer BB, Emgard J, Temper V, et al. The *H. pylori* HopQ outermembrane protein inhibits immune cell activities. Onco Targets Ther. 2019;8(4):e1553487. https://doi.org/10.1080/2162402X.2018.1553487.
51. Yamaoka Y, Kwon DH, Graham DY. A M(r) 34,000 proinflammatory outer membrane protein (oipA) of *H. pylori*. Proc Natl Acad Sci USA. 2000;97(13):7533–8. https://doi.org/10.1073/pnas.130079797.
52. Yamaoka Y, Kikuchi S, et al. Importance of *H. pylori* oipA in clinical presentation, gastric inflammation, and mucosal interleukin 8 production. Gastroenterology. 2002;123(2):414–24. https://doi.org/10.1053/gast.2002.34781.
53. Farzi N, Yadegar A, Aghdaei HA, Yamaoka Y, Zali MR. Genetic diversity and functional analysis of oipA gene in association with other virulence factors among *H. pylori* isolates from Iranian patients with different gastric diseases. Infect Genet Evol. 2018;60:26–34. https://doi.org/10.1016/j.meegid.2018.02.017.
54. Teymournejad O, Mobarez AM, et al. Binding of the *H. pylori* OipA causes apoptosis of host cells via modulation of Bax/Bcl-2 levels. Sci Rep. 2017;7(1):8036. https://doi.org/10.1038/s41598-017-08176-7.
55. Peck B, Ortkamp M, Diehl KD, Hundt E, Knapp B. Conservation, localization and expression of HopZ, a protein involved in adhesion of *H. pylori*. Nucleic Acids Res. 1999;27(16):3325–33. https://doi.org/10.1093/nar/27.16.3325.
56. Kennemann L, Brenneke B, Andres S, Engstrand L, Meyer TF, Aebischer T, et al. In vivo sequence variation in HopZ, a phase-variable outer membrane protein of *H. pylori*. Infect Immun. 2012;80(12):4364–73. https://doi.org/10.1128/IAI.00977-12.
57. Oleastro M, Cordeiro R, Ferrand J, Nunes B, Lehours P, Carvalho-Oliveira I, et al. Evaluation of the clinical significance of homB, a novel candidate marker of *H. pylori* strains associated with peptic ulcer disease. J Infect Dis. 2008;198(9):1379–87. https://doi.org/10.1086/592166.
58. Jung SW, Sugimoto M, Graham DY, Yamaoka Y. homB status of *H. pylori* as a novel marker to distinguish gastric cancer from duodenal ulcer. J Clin Microbiol. 2009;47(10):3241–5. https://doi.org/10.1128/JCM.00293-09.

59. Rossez Y, Gosset P, Boneca IG, Magalhaes A, Ecobichon C, Reis CA, et al. The lacdiNAc-specific adhesin LabA mediates adhesion of *H. pylori* to human gastric mucosa. J Infect Dis. 2014;210(8):1286–95. https://doi.org/10.1093/infdis/jiu239.
60. Manna OM, Caruso Bavisotto C, Gratie MI, Damiani P, Tomasello G, Cappello F. The role of *H. pylori* heat shock proteins in gastric diseases' pathogenesis. Int J Mol Sci. 2025;26(11) https://doi.org/10.3390/ijms26115065.
61. Roncarati D, Danielli A, Spohn G, Delany I, Scarlato V. Transcriptional regulation of stress response and motility functions in *H. pylori* is mediated by HspR and HrcA. J Bacteriol. 2007;189(20):7234–43. https://doi.org/10.1128/JB.00626-07.
62. Roncarati D, Scarlato V. The interplay between two transcriptional repressors and chaperones orchestrates *H. pylori* eat-shock response. Int J Mol Sci. 2018;19(6) https://doi.org/10.3390/ijms19061702.
63. Huesca M, Borgia S, Hoffman P, Lingwood CA. Acidic pH changes receptor binding specificity of *H. pylori*: a binary adhesion model in which surface heat shock (stress) proteins mediate sulfatide recognition in gastric colonization. Infect Immun. 1996;64(7):2643–8. https://doi.org/10.1128/iai.64.7.2643-2648.1996.
64. Hathroubi S, Zerebinski J, Ottemann KM. *H. pylori* biofilm involves a multigene stress-biased response, including a structural role for flagella. MBio. 2018;9(5) https://doi.org/10.1128/mBio.01973-18.
65. Lin CY, Huang YS, Li CH, Hsieh YT, Tsai NM, He PJ, et al. Characterizing the polymeric status of *H. pylori* heat shock protein 60. Biochem Biophys Res Commun. 2009;388(2):283–9. https://doi.org/10.1016/j.bbrc.2009.07.159.
66. Gobert AP, Bambou JC, Werts C, Balloy V, Chignard M, Moran AP, et al. *H. pylori* heat shock protein 60 mediates interleukin-6 production by macrophages via a toll-like receptor (TLR)-2-, TLR-4-, and myeloid differentiation factor 88-independent mechanism. J Biol Chem. 2004;279(1):245–50. https://doi.org/10.1074/jbc.M307858200.
67. Lin CS, He PJ, Hsu WT, Wu MS, Wu CJ, Shen HW, et al. *H. pylori*-derived heat shock protein 60 enhances angiogenesis via a CXCR2-mediated signaling pathway. Biochem Biophys Res Commun. 2010;397(2):283–9. https://doi.org/10.1016/j.bbrc.2010.05.101.
68. Shibayama K, Wachino J, Arakawa Y, Saidijam M, Rutherford NG, Henderson PJ. Metabolism of glutamine and glutathione via gamma-glutamyltranspeptidase and glutamate transport in *H. pylori*: possible significance in the pathophysiology of the organism. Mol Microbiol. 2007;64(2):396–406. https://doi.org/10.1111/j.1365-2958.2007.05661.x.
69. Leduc D, Gallaud J, Stingl K, de Reuse H. Coupled amino acid deamidase-transport systems essential for *H. pylori* colonization. Infect Immun. 2010;78(6):2782–92. https://doi.org/10.1128/IAI.00149-10.
70. Fuchs S, Fiedler MK, Heiduk N, Wanisch A, Mibus C, Singh D, et al. *H. pylori* gamma-glutamyltransferase is linked to proteomic adaptions important for colonization. Gut Microbes. 2025;17(1):2488048. https://doi.org/10.1080/19490976.2025.2488048.
71. McGovern KJ, Blanchard TG, Gutierrez JA, Czinn SJ, Krakowka S, Youngman P. Gamma-Glutamyltransferase is a *H. pylori* virulence factor but is not essential for colonization. Infect Immun. 2001;69(6):4168–73. https://doi.org/10.1128/IAI.69.6.4168-4173.2001.
72. Kabisch R, Semper RP, Wustner S, Gerhard M, Mejias-Luque R. *H. pylori* gamma-Glutamyltranspeptidase induces tolerogenic human dendritic cells by activation of glutamate receptors. J Immunol. 2016;196(10):4246–52. https://doi.org/10.4049/jimmunol.1501062.
73. Oertli M, Noben M, Engler DB, Semper RP, Reuter S, Maxeiner J, et al. *H. pylori* gamma-glutamyl transpeptidase and vacuolating cytotoxin promote gastric persistence and immune tolerance. Proc Natl Acad Sci USA. 2013;110(8):3047–52. https://doi.org/10.1073/pnas.1211248110.
74. Gong M, Ling SS, Lui SY, Yeoh KG, Ho B. *H. pylori* gamma-glutamyl transpeptidase is a pathogenic factor in the development of peptic ulcer disease. Gastroenterology. 2010;139(2):564–73. https://doi.org/10.1053/j.gastro.2010.03.050.

75. Kim KM, Lee SG, Park MG, Song JY, Kang HL, Lee WK, et al. Gamma-glutamyltranspeptidase of *H. pylori* induces mitochondria-mediated apoptosis in AGS cells. Biochem Biophys Res Commun. 2007;355(2):562–7. https://doi.org/10.1016/j.bbrc.2007.02.021.
76. Namavar F, Sparrius M, Veerman EC, Appelmelk BJ, Vandenbroucke-Grauls CM. Neutrophil-activating protein mediates adhesion of *H. pylori* to sulfated carbohydrates on high-molecular-weight salivary mucin. Infect Immun. 1998;66(2):444–7. https://doi.org/10.1128/IAI.66.2.444-447.1998.
77. Petersson C, Forsberg M, Aspholm M, Olfat FO, Forslund T, Boren T, et al. *H. pylori* SabA adhesin evokes a strong inflammatory response in human neutrophils which is down-regulated by the neutrophil-activating protein. Med Microbiol Immunol. 2006;195(4):195–206. https://doi.org/10.1007/s00430-006-0018-x.
78. Wang G, Hong Y, Olczak A, Maier SE, Maier RJ. Dual roles of *H. pylori* NapA in inducing and combating oxidative stress. Infect Immun. 2006;74(12):6839–46. https://doi.org/10.1128/IAI.00991-06.
79. Satin B, Del Giudice G, Della Bianca V, Dusi S, Laudanna C, Tonello F, et al. The neutrophil-activating protein (HP-NAP) of *H. pylori* is a protective antigen and a major virulence factor. J Exp Med. 2000;191(9):1467–76. https://doi.org/10.1084/jem.191.9.1467.
80. Polenghi A, Bossi F, Fischetti F, Durigutto P, Cabrelle A, Tamassia N, et al. The neutrophil-activating protein of *H. pylori* crosses endothelia to promote neutrophil adhesion in vivo. J Immunol. 2007;178(3):1312–20. https://doi.org/10.4049/jimmunol.178.3.1312.
81. Zhao Y, Cai Y, Chen Z, Li H, Xu Z, Li W, et al. SpoT-mediated NapA upregulation promotes oxidative stress-induced *H. pylori* biofilm formation and confers multidrug resistance. Antimicrob Agents Chemother. 2023;65(5) https://doi.org/10.1128/AAC.00152-21.
82. Wen SH, Hong ZW, Chen CC, Chang HW, Fu HW. *H. pylori* neutrophil-activating protein directly interacts with and activates toll-like receptor 2 to induce the secretion of Interleukin-8 from neutrophils and ATRA-induced differentiated HL-60 cells. Int J Mol Sci. 2021;22(21) https://doi.org/10.3390/ijms222111560.
83. Montemurro P, Barbuti G, Dundon WG, Del Giudice G, Rappuoli R, Colucci M, et al. *H. pylori* neutrophil-activating protein stimulates tissue factor and plasminogen activator inhibitor-2 production by human blood mononuclear cells. J Infect Dis. 2001;183(7):1055–62. https://doi.org/10.1086/319280.
84. Tsai CC, Kuo TY, Hong ZW, Yeh YC, Shih KS, Du SY, et al. *H. pylori* neutrophil-activating protein induces release of histamine and interleukin-6 through G protein-mediated MAPKs and PI3K/Akt pathways in HMC-1 cells. Virulence. 2015;6(8):755–65. https://doi.org/10.1080/21505594.2015.1043505.
85. Amedei A, Cappon A, Codolo G, Cabrelle A, Polenghi A, Benagiano M, et al. The neutrophil-activating protein of *H. pylori* promotes Th1 immune responses. J Clin Invest. 2006;116(4):1092–101. https://doi.org/10.1172/JCI27177.
86. Figura N, Trabalzini L, Mini R, Bernardini G, Scaloni A, Talamo F, et al. Inactivation of *H. pylori* cagA gene affects motility. Helicobacter. 2004;9(3):185–93. https://doi.org/10.1111/j.1083-4389.2004.00224.x.
87. Tran SC, Bryant KN, Cover TL. The *H. pylori* cag pathogenicity Island as a determinant of gastric cancer risk. Gut Microbes. 2024;16(1):2314201. https://doi.org/10.1080/19490976.2024.2314201.
88. Bagnoli F, Buti L, Tompkins L, Covacci A, Amieva MR. *H. pylori* CagA induces a transition from polarized to invasive phenotypes in MDCK cells. Proc Natl Acad Sci USA. 2005;102(45):16339–44. https://doi.org/10.1073/pnas.0502598102.
89. Zamperone A, Cohen D, Stein M, Viard C, Musch A. Inhibition of polarity-regulating kinase PAR1b contributes to *H. pylori* inflicted DNA double Strand breaks in gastric cells. Cell Cycle. 2019;18(3):299–311. https://doi.org/10.1080/15384101.2018.1560121.
90. Suzuki M, Mimuro H, Kiga K, Fukumatsu M, Ishijima N, Morikawa H, et al. *H. pylori* CagA phosphorylation-independent function in epithelial proliferation and inflammation. Cell Host Microbe. 2009;5(1):23–34. https://doi.org/10.1016/j.chom.2008.11.010.

91. Murata-Kamiya N, Kurashima Y, Teishikata Y, Yamahashi Y, Saito Y, Higashi H, et al. *H. pylori* CagA interacts with E-cadherin and deregulates the beta-catenin signal that promotes intestinal transdifferentiation in gastric epithelial cells. Oncogene. 2007;26(32):4617–26. https://doi.org/10.1038/sj.onc.1210251.
92. Buti L, Spooner E, Van der Veen AG, Rappuoli R, Covacci A, Ploegh HL. *H. pylori* cytotoxin-associated gene A (CagA) subverts the apoptosis-stimulating protein of p53 (ASPP2) tumor suppressor pathway of the host. Proc Natl Acad Sci USA. 2011;108(22):9238–43. https://doi.org/10.1073/pnas.1106200108.
93. Nejati S, Karkhah A, Darvish H, Validi M, Ebrahimpour S, Nouri HR. Influence of *H. pylori* virulence factors CagA and VacA on pathogenesis of gastrointestinal disorders. Microb Pathog. 2018;117:43–8. https://doi.org/10.1016/j.micpath.2018.02.016.
94. Yahiro K, Wada A, Nakayama M, Kimura T, Ogushi K, Niidome T, et al. Protein-tyrosine phosphatase alpha, RPTP alpha, is a *H. pylori* VacA receptor. J Biol Chem. 2003;278(21):19183–9. https://doi.org/10.1074/jbc.M300117200.
95. Chauhan N, Tay ACY, Marshall BJ, Jain U. *H. pylori* VacA, a distinct toxin exerts diverse functionalities in numerous cells: an overview. Helicobacter. 2019;24(1):e12544. https://doi.org/10.1111/hel.12544.
96. de Bernard M, Moschioni M, Habermann A, Griffiths G, Montecucco C. Cell vacuolization induced by *H. pylori* VacA cytotoxin does not depend on late endosomal SNAREs. Cell Microbiol. 2002;4(1):11–8. https://doi.org/10.1046/j.1462-5822.2002.00163.x.
97. Jain P, Luo ZQ, Blanke SR. *H. pylori* vacuolating cytotoxin A (VacA) engages the mitochondrial fission machinery to induce host cell death. Proc Natl Acad Sci USA. 2011;108(38):16032–7. https://doi.org/10.1073/pnas.1105175108.
98. Zhu P, Xue J, Zhang ZJ, Jia YP, Tong YN, Han D, et al. *H. pylori* VacA induces autophagic cell death in gastric epithelial cells via the endoplasmic reticulum stress pathway. Cell Death Dis. 2017;8(12):3207. https://doi.org/10.1038/s41419-017-0011-x.
99. Yahiro K, Akazawa Y, Nakano M, Suzuki H, Hisatune J, Isomoto H, et al. *H. pylori* VacA induces apoptosis by accumulation of connexin 43 in autophagic vesicles via a Rac1/ERK-dependent pathway. Cell Death Discov. 2015;1:15035. https://doi.org/10.1038/cddiscovery.2015.35.
100. Gebert B, Fischer W, Weiss E, Hoffmann R, Haas R. *H. pylori* vacuolating cytotoxin inhibits T lymphocyte activation. Science. 2003;301(5636):1099–102. https://doi.org/10.1126/science.1086871.
101. Nakashima S, Kakugawa T, Yura H, Tomonaga M, Harada T, Hara A, et al. Identification of *H. pylori* VacA in human lung and its effects on lung cells. Biochem Biophys Res Commun. 2015;460(3):721–6. https://doi.org/10.1016/j.bbrc.2015.03.096.
102. Li H, Liao T, Debowski AW, Tang H, Nilsson HO, Stubbs KA, et al. Lipopolysaccharide structure and biosynthesis in *H. pylori*. Helicobacter. 2016;21(6):445–61. https://doi.org/10.1111/hel.12301.
103. Altman E, Chandan V, Larocque S, Aubry A, Logan SM, Vinogradov E, et al. Effect of the HP0159 ORF mutation on the lipopolysaccharide structure and colonizing ability of *H. pylori*. FEMS Immunol Med Microbiol. 2008;53(2):204–13. https://doi.org/10.1111/j.1574-695X.2008.00416.x.
104. Reeves EP, Ali T, Leonard P, Hearty S, O'Kennedy R, May FE, et al. *H. pylori* lipopolysaccharide interacts with TFF1 in a pH-dependent manner. Gastroenterology. 2008;135(6):2043-54. https://doi.org/10.1053/j.gastro.2008.08.049.
105. Chavarria-Velazquez CO, Torres-Martinez AC, Montano LF, Rendon-Huerta EP. TLR2 activation induced by *H. pylori* LPS promotes the differential expression of claudin-4, −6, −7 and −9 via either STAT3 and ERK1/2 in AGS cells. Immunobiology. 2018;223(1):38–48. https://doi.org/10.1016/j.imbio.2017.10.016.
106. Bliss CM Jr, Golenbock DT, Keates S, Linevsky JK, Kelly CP. *H. pylori* lipopolysaccharide binds to CD14 and stimulates release of interleukin-8, epithelial neutrophil-activating peptide 78, and monocyte chemotactic protein 1 by human monocytes. Infect Immun. 1998;66(11):5357–63. https://doi.org/10.1128/IAI.66.11.5357-5363.1998.

107. Taylor JM, Ziman ME, Huff JL, Moroski NM, Vajdy M, Solnick JV. *H. pylori* lipopolysaccharide promotes a Th1 type immune response in immunized mice. Vaccine. 2006;24(23):4987–94. https://doi.org/10.1016/j.vaccine.2006.03.043.
108. Shimoyama A, Saeki A, Tanimura N, Tsutsui H, Miyake K, Suda Y, et al. Chemical synthesis of *H. pylori* lipopolysaccharide partial structures and their selective proinflammatory responses. Chemistry. 2011;17(51):14464–74. https://doi.org/10.1002/chem.201003581.
109. Slomiany BL, Slomiany A. Disruption in gastric mucin synthesis by *H. pylori* lipopolysaccharide involves ERK and p38 mitogen-activated protein kinase participation. Biochem Biophys Res Commun. 2002;294(2):220–4. https://doi.org/10.1016/S0006-291X(02)00463-1.
110. Yokota S, Okabayashi T, Rehli M, Fujii N, Amano K. *H. pylori* lipopolysaccharides upregulate toll-like receptor 4 expression and proliferation of gastric epithelial cells via the MEK1/2-ERK1/2 mitogen-activated protein kinase pathway. Infect Immun. 2010;78(1):468–76. https://doi.org/10.1128/IAI.00903-09.
111. Zarzecka U, Modrak-Wojcik A, Figaj D, Apanowicz M, Lesner A, Bzowska A, et al. Properties of the HtrA protease from bacterium *H. pylori* whose activity is indispensable for growth under stress conditions. Front Microbiol. 2019;10:961. https://doi.org/10.3389/fmicb.2019.00961.
112. Hoy B, Lower M, Weydig C, Carra G, Tegtmeyer N, Geppert T, et al. *H. pylori* HtrA is a new secreted virulence factor that cleaves E-cadherin to disrupt intercellular adhesion. EMBO Rep. 2010;11(10):798–804. https://doi.org/10.1038/embor.2010.114.
113. Zhang Z, Huang Q, Tao X, Song G, Zheng P, Li H, et al. The unique trimeric assembly of the virulence factor HtrA from *H. pylori* occurs via N-terminal domain swapping. J Biol Chem. 2019;294(20):7990–8000. https://doi.org/10.1074/jbc.RA119.007387.
114. Tegtmeyer N, Wessler S, Necchi V, Rohde M, Harrer A, Rau TT, et al. *H. pylori* employs a unique basolateral type IV secretion mechanism for CagA delivery. Cell Host Microbe. 2017;22(4):552–60. https://doi.org/10.1016/j.chom.2017.09.005.
115. Li J, Liao T, Chua EG, Zhang M, Shen Y, Song X, et al. *H. pylori* outer membrane vesicles: biogenesis, composition, and biological functions. Int J Biol Sci. 2024;20(10):4029–43. https://doi.org/10.7150/ijbs.94156.
116. Olofsson A, Nygard Skalman L, Obi I, Lundmark R, Arnqvist A. Uptake of *H. pylori* vesicles is facilitated by clathrin-dependent and clathrin-independent endocytic pathways. MBio. 2014;5(3):e00979–14. https://doi.org/10.1128/mBio.00979-14.
117. Yonezawa H, Osaki T, Kurata S, Fukuda M, Kawakami H, Ochiai K, et al. Outer membrane vesicles of *H. pylori* TK1402 are involved in biofilm formation. BMC Microbiol. 2009;9:197. https://doi.org/10.1186/1471-2180-9-197.
118. Grande R, Di Marcantonio MC, Robuffo I, Pompilio A, Celia C, Di Marzio L, et al. *H. pylori* ATCC 43629/NCTC 11639 outer membrane vesicles (OMVs) from biofilm and planktonic phase associated with extracellular DNA (eDNA). Front Microbiol. 2015;6:1369. https://doi.org/10.3389/fmicb.2015.01369.
119. Ismail S, Hampton MB, Keenan JI. *H. pylori* outer membrane vesicles modulate proliferation and interleukin-8 production by gastric epithelial cells. Infect Immun. 2003;71(10):5670–5. https://doi.org/10.1128/IAI.71.10.5670-5675.2003.
120. Ayala G, Torres L, Espinosa M, Fierros-Zarate G, Maldonado V, Melendez-Zajgla J. External membrane vesicles from *H. pylori* induce apoptosis in gastric epithelial cells. FEMS Microbiol Lett. 2006;260(2):178–85. https://doi.org/10.1111/j.1574-6968.2006.00305.x.
121. Chitcholtan K, Hampton MB, Keenan JI. Outer membrane vesicles enhance the carcinogenic potential of *H. pylori*. Carcinogenesis. 2008;29(12):2400–5. https://doi.org/10.1093/carcin/bgn218.
122. Bolori S, Shegefti S, Baghaei K, Yadegar A, Moon KM, Foster LJ, et al. The Effects of *H pylori*-Derived Outer Membrane Vesicles on Hepatic Stellate Cell Activation and Liver Fibrosis In Vitro. Biomed Res Int. 2023;2023:4848643. https://doi.org/10.1155/2023/4848643.
123. Xie J, Cools L, Van Imschoot G, Van Wonterghem E, Pauwels MJ, Vlaeminck I, et al. *H. pylori*-derived outer membrane vesicles contribute to Alzheimer's disease pathogenesis via C3-C3aR signalling. J Extracell Vesicles. 2023;12(2):e12306. https://doi.org/10.1002/jev2.12306.

# 5 The Role of *H. pylori* in Gastric Carcinogenesis

Wei Hu and Zhouyingyi Liu

## 5.1 Introduction

Gastric cancer represents a significant global health burden, characterized by high incidence and mortality rates worldwide [1]. Its pathogenesis involves a complex interplay of environmental, genetic, and microbial factors, among which *H. pylori* (*H. pylori*) stands out as a major risk factor [2]. As a predominant pathogen of the human stomach, *H. pylori* not only drives a range of gastric pathologies—such as chronic gastritis and peptic ulcers—but also significantly promotes gastric carcinogenesis. This bacterium is notably one of the most successful human pathogens. Typically acquired in childhood, it can persist for decades, often throughout the host's lifetime unless treated. The ultimate clinical outcome of *H. pylori* infection is determined by dynamic interactions between bacterial virulence determinants, host genetic susceptibility, and environmental factors within specific population settings [3].

Building on this foundation, the present review aims to integrate foundational research with clinical insights by systematically tracing the pathogenic cascade of *H. pylori*—from its genomic features to the molecular and cellular events leading to gastric cancer.

W. Hu (✉) · Z. Liu
Department of Gastroenterology, Shenzhen Hospital, Southern Medical University, Shenzhen, Guangdong, China

Shenzhen School of Clinical Medicine, Southern Medical University, Shenzhen, Guangdong, China
e-mail: huwei1683013@smu.edu.cn



J. Gao, Y. Du (eds.), *Non-Antibiotic Therapies for Helicobacter pylori*,
https://doi.org/10.1007/978-981-95-9598-3_5

## 5.2 Genomic Characteristics and Virulence Factors

The genome of *H. pylori* shows high genetic plasticity and significant geographic diversity, features that are closely linked to its successful colonization of the human stomach and its role in gastric disease pathogenesis [4]. This notable heterogeneity is reflected in a repertoire of virulence genes associated with gastric cancer development, notably including the *cag* pathogenicity island (cagPAI), vacuolating cytotoxin A (vacA), urease, flagella, and adhesin genes such as hopQ, sabA, oipA, and babA (Fig. 5.1) [5, 6].

### 5.2.1 CagPAI and Cytotoxin-Associated Gene A (CagA)

The *H. pylori* cagPAI, spanning approximately 40 kb, comprises around 28 genes that encode components of the type IV secretion system (T4SS) and associated effector proteins [7]. Strains carrying a functional cagPAI are strongly linked to severe clinical outcomes, including gastric atrophy and cancer [8, 9]. A key effector,

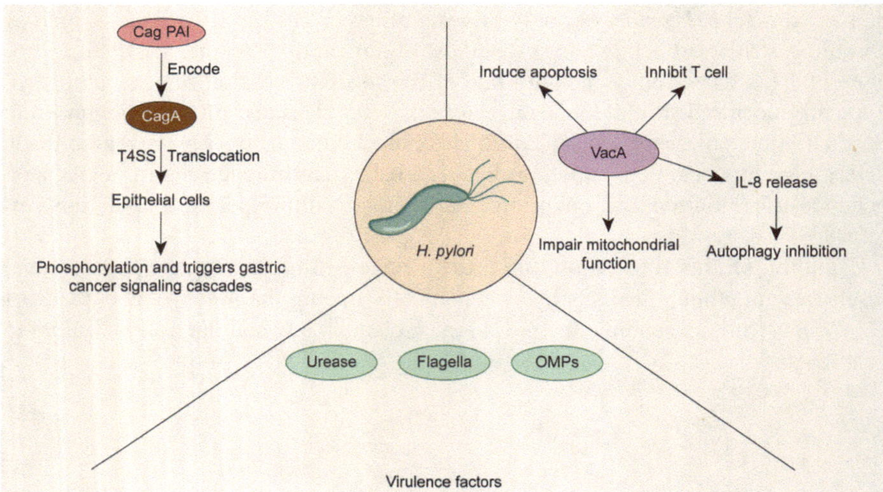


**Fig. 5.1** Key virulence factors of *Helicobacter pylori* and their functional roles in pathogenesis. CagPAI encodes the CagA oncoprotein, which is translocated into host epithelial cells via the T4SS, triggering signal transduction pathways that contribute to gastritis and carcinogenesis. VacA exerts pleiotropic effects, including the induction of apoptosis, inhibition of T-cell proliferation, disruption of mitochondrial function, and modulation of IL-8-mediated inflammation and autophagy. Additional critical factors for colonization and persistence—Urease, Flagella, and OMPs—are collectively highlighted for their roles in acid neutralization, motility, and immune evasion. cagPAI, the *cag* pathogenicity island; *CagA* cytotoxin-associated gene A, *T4SS* type IV secretion system, *VacA* vacuolating cytotoxin A, *OMPs* outer membrane proteins. (Original illustration created by Wei HU. All rights reserved)

CagA (120–140 kDa), possesses distinct amino-terminal domains and is delivered into gastric epithelial cells via the T4SS [10, 11].

Upon entry, CagA undergoes phosphorylation by host tyrosine kinases such as c-SRC and c-ABL at specific Glu-Pro-Ile-Tyr-Ala (EPIYA) motifs [12, 13], enabling interaction with SH2 domain-containing protein tyrosine phosphatase-2 (SHP2) and activation of downstream signaling pathways [14]. The virulence potential of CagA is influenced by the type and number of EPIYA repeats it contains. Four EPIYA types (A, B, C, D) have been identified, with EPIYA-A and EPIYA-B present in nearly all cag-positive strains [15]. Strains harboring the EPIYA-D motif exhibit a heightened risk of gastric cancer, attributed to their stronger binding affinity for SHP2 [16, 17]. Additionally, recent studies indicate that an increased number of EPIYA-C repeats may upregulate the transcription of host genes involved in oncogenesis—such as erbB2, HGF-R, FGFR4, and TGF-β [18]. Beyond altering epithelial cell behavior, CagA also activates inflammatory signaling pathways that recruit immune cells and reshape the tumor microenvironment [ 19]. A recent study has elucidated a specific mechanism by which CagA promotes pyroptosis in gastric cancer: activation of the miRNA-1290/NKD1/NLRP3 axis [20].

### 5.2.2 Vacuolating Cytotoxin A (VacA)

*H. pylori* secretes VacA, a pore-forming toxin that enables bacterial evasion of host defenses by targeting both epithelial and immune cells. Structurally, the N-terminal p33 subunit forms anion-selective channels, while the C-terminal p55 subunit mediates cellular binding and apoptosis induction [21, 22]. Functionally, VacA disrupts mitochondrial integrity, impairs autophagy, alters cell polarity, and suppresses T-cell activation and proliferation [13, 22]. It also promotes excessive interleukin-8 (IL-8) release and disrupts cellular amino acid homeostasis, further contributing to autophagic inhibition [23, 24].

The virulence of VacA is influenced by genetic polymorphisms within its signal (s), intermediate (i), middle (m), and deletion (d) regions, which define its activity and disease association [25, 26]. Specific vacA genotypes—particularly s1, i1, and m1 forms—are significantly associated with an elevated risk of gastric cancer and may correlate with distinct histological subtypes [27–29]. Moreover, VacA synergizes with other virulence factors such as CagA: it promotes CagA accumulation in host cells, and together they subvert host iron-uptake machinery to support bacterial replication [30, 31]. Recent evidence also suggests that VacA modulates the gastric microenvironment by influencing inflammatory signaling, potentially via STAT3 pathway suppression, and reduces STAT3, Bcl-2, and Bcl-XL expression in a dose-dependent manner [32].

### 5.2.3 Other Virulence Factors

Beyond CagA and VacA, *H. pylori* employs a diverse arsenal of molecular tools to sustain infection and enhance pathogenicity [33]. Urease, for example, hydrolyzes urea to neutralize gastric acid, thereby creating a localized niche with a survivable pH [34–36]. Flagella power bacterial motility and support biofilm formation, while also inducing IL-8 release to dysregulate host immune responses [35, 37, 38]. Outer membrane proteins (OMPs)—including BabA, SabA, and OipA—bind host cell receptors, facilitating long-term colonization, chronic inflammation, and elevated IL-8 secretion [4, 33, 38, 39]. Together, these mechanisms not only promote bacterial persistence but also collectively drive the pathogenesis of gastric cancer.

## 5.3 Molecular Mechanism of *H. pylori* in Gastric Cancer Carcinogenesis

The pathogenesis of gastric cancer involves complex biological processes and molecular mechanisms that remain incompletely understood. Dysregulation of multiple signaling pathways is known to critically contribute to gastric carcinogenesis. This chapter focuses on key molecular mechanisms through which *H. pylori* subverts host cellular processes to promote gastric cancer, with particular emphasis on the JAK-STAT, NF-κB, and Wnt/β-catenin signaling pathways (Fig. 5.2). These pathways are consistently disrupted during *H. pylori*-induced gastric tumor development.

### 5.3.1 JAK-STAT Signaling Pathway

STAT3 is a transcription factor that plays a pivotal role in *H. pylori*-induced gastric cancer progression. Upon cytokine binding—such as IL-6, IL-11, TNF, and IL-17—to their cognate receptors, receptor-associated tyrosine kinases including JAK1, JAK2, Src, and EGFR become activated and phosphorylated [40, 41]. This in turn triggers STAT3 phosphorylation (p-STAT3), facilitating its nuclear translocation and subsequent regulation of downstream genes involved in inflammation, proliferation, epithelial–mesenchymal transition (EMT), invasion, and metastasis [42, 43].

*H. pylori* activates the JAK-STAT pathway through multiple distinct mechanisms. The bacterium sustains STAT3 activation by upregulating interleukin-6 (IL-6) [44]. In addition, the virulence factor CagA, once delivered into host epithelial cells, promotes persistent STAT3 activation via the SHP-2 signaling cascade, thereby stimulating transcription of downstream target genes [45]. *H. pylori* also engages Toll-like receptor 2 (TLR2), amplifying inflammatory responses and further enhancing STAT3 activation [46, 47]. Recent studies have elucidated that STAT3 drives gastric carcinogenesis by inducing DAB2 overexpression through the SRC–YAP1 axis [48]. Moreover, *H. pylori* infection upregulates FGFR4 in a STAT3-dependent manner, contributing to key processes in gastric cancer progression [49]. In certain contexts, *H. pylori* has also been reported to downregulate other components within the JAK-STAT pathway [50, 51].

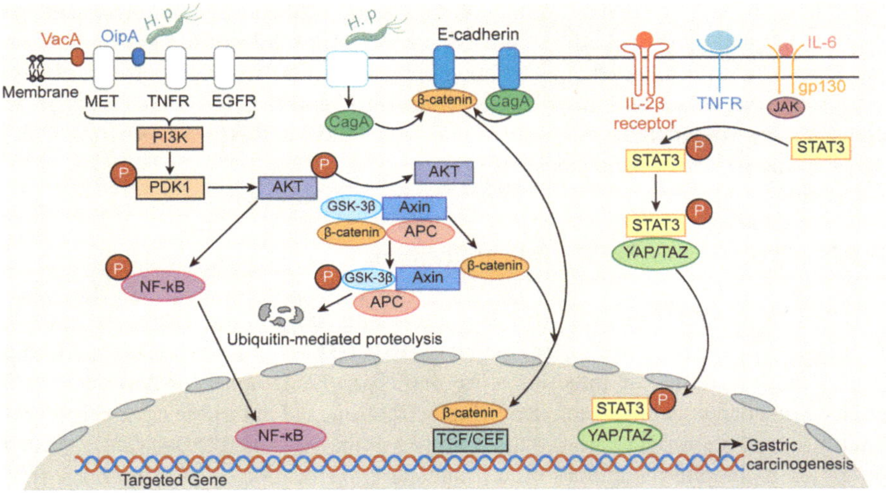


**Fig. 5.2** *Helicobacter pylori* virulence factors disrupt host signaling pathways to promote gastric carcinogenesis. The schematic diagram illustrates the key molecular mechanisms by which *H. pylori* virulence factors, including the secreted toxins CagA, VacA, and OipA, dysregulate critical host cell signaling pathways to drive oncogenic transformation. A central event is the disruption of the Wnt/β-catenin pathway by the injected CagA oncoprotein. CagA interacts with and stabilizes β-catenin by subverting its regulatory complex (containing Axin, GSK-3β, and APC), inhibiting ubiquitin-mediated proteolysis. This leads to β-catenin accumulation and nuclear translocation, where it binds the TCF/LEF transcription factor to activate pro-carcinogenic genes. Concurrently, other pathways are activated. The JAK-STAT signaling cascade is triggered, exemplified by IL-6/gp130-mediated JAK activation, resulting in STAT3 phosphorylation and nuclear translocation. The NF-κB pathway is activated via a PI3K-PDK1-AKT signaling axis, leading to NF-κB nuclear translocation. The involvement of the YAP/TAZ transcriptional co-activators is also indicated, suggesting a role for Hippo pathway dysregulation. The concerted dysregulation of these interconnected pathways—JAK-STAT, NF-κB, and Wnt/β-catenin—by *H. pylori* effectors converges to modulate the expression of target genes fundamental to gastric carcinogenesis. OipA, outer inflammatory protein A. (Original illustration created by Wei HU. All rights reserved)

### 5.3.2 NF-κB Signaling Pathway

NF-κB serves as a pivotal transcription factor in *H. pylori*-associated gastric carcinogenesis. The bacterium activates NF-κB through several distinct mechanisms: the CagA protein directly triggers NF-κB activation and nuclear translocation [52]; *H. pylori* also upregulates pro-inflammatory factors such as TNF-α and IL-1β, which stimulate the IKK complex to initiate NF-κB signaling [53, 54]. Furthermore, NF-κB induces expression of IL-32, creating a positive feedback loop that amplifies its own activation [55].

Notably, NF-κB regulates gastric cancer development by driving the overexpression of key factors including HNF4α, PRDX2, and CDX2 following *H. pylori* infection [56]. Recent studies reveal that NF-κB activation promotes tumor progression via the PIEZO1–YAP1–CTGF axis, which remodels the tumor microenvironment

[57], and through RASAL2 upregulation, which enhances carcinogenesis via β-catenin signaling [58]. Additionally, NF-κB activation transcribes genes that promote cell cycle progression, inhibit apoptosis, and stimulate angiogenesis, collectively fostering a tumor-supportive environment [59, 60]. NF-κB also engages in cross talk with other signaling pathways such as STAT3 and HIF-1α, synergistically accelerating gastric carcinogenesis [61].

### 5.3.3 Wnt/β-catenin Signaling Pathway

The Wnt/β-catenin signaling pathway represents a critical regulatory axis in *H. pylori*-associated gastric carcinogenesis [62, 63]. This pathway is implicated from the early stages of infection, where *H. pylori* disrupts Wnt-dependent cell polarity mechanisms linked to stem cell proliferation and enteroendocrine differentiation [64]. A major activator of the pathway is the virulence factor CagA, which promotes β-catenin accumulation and nuclear translocation. CagA interacts with LRP8 to form a CagA/LRP8/β-catenin complex that amplifies β-catenin nuclear translocation and drives oncogenesis [65].

In the nucleus, β-catenin functions as a transcriptional co-activator, upregulating genes that perturb cell-cycle control, inhibit apoptosis, and promote stem-cell renewal, collectively fostering a tumor-permissive milieu [64, 66]. *H. pylori* also activates Wnt/β-catenin signaling via ASCL1-mediated expression of aquaporin-5 (AQP5), contributing to inflammation and gastritis [67]. The pathway further intersects with epithelial–mesenchymal transition (EMT): nuclear β-catenin induces EMT-related transcription factors, conferring stem-like properties and enhancing tumor progression [68, 69]. Additionally, pathway activation stimulates tumor cell proliferation, migration, and invasion [70], and β-catenin–TCF/LEF complexes upregulate oncogenes such as MYC, CCND1, and LGR5to propel cell-cycle progression [71].

The Wnt/β-catenin cascade does not act in isolation; it engages in extensive cross talk with other signaling networks modulated by *H. pylori*. For instance, infection triggers malignant transformation by regulating Wnt/β-catenin to activate NF-κB-dependent inflammation and proliferation [72]. Moreover, *H. pylori* enhances the nuclear accumulation and transcriptional activity of both YAP and β-catenin through their physical interaction, synergistically promoting carcinogenesis [73].

### 5.3.4 Other Signaling Pathways

In addition to the major pathways discussed above, *H. pylori* activates several other signaling cascades—such as the MAPK (ERK, JNK, p38), PI3K/Akt, Hippo, HGF/Met, TGF-β, Hedgehog, and Notch pathways. These pathways collectively regulate essential cellular processes including proliferation, survival, migration, invasion, differentiation, apoptosis, stemness, and microRNA expression. Moreover, they engage in intricate cross-regulatory interactions, not only among themselves, but also with classical signaling axes, forming a complex network that underpins *H. pylori*-driven pathogenesis.

MAPK pathway: *H. pylori* targets the key cellular MAPK pathways (ERK, JNK, p38), each playing distinct roles in gastric cancer. The ERK pathway is sustainably activated to enhance proliferation and survival [74, 75]. Bacterial T4SS and CagA leverage the JNK pathway to modulate stress and inflammatory responses [75], while the MEK/ERK/SRF axis mediates CagA-induced ferroptosis susceptibility via ether lipid biosynthesis [76]. These pathways further integrate with broader signaling networks, as ERK activates NF-κB and JNK enhances STAT3 activity, illustrating a sophisticated cross-regulatory framework underlying pathogenesis [77].

Hippo pathway: The Hippo signaling pathway operates through a sequential phosphorylation cascade. Central to this pathway, mammalian sterile 20-like kinases 1 and 2 (MST1/2) bind to Salvador homolog 1 (SAV1), which then phosphorylates and activates large tumor suppressors 1 and 2 (LATS1/2). In conjunction with MOB kinase activator 1A and 1B (MOB1A/B), activated LATS1/2 phosphorylates Yes-associated protein and transcriptional co-activator with PDZ-binding motif (YAP/TAZ), thereby inhibiting their activity by restricting nuclear translocation [78]. Conversely, when YAP/TAZ are not phosphorylated, they translocate into the nucleus and interact with transcription factors of the TEAD family, driving the expression of downstream target genes [79]. YAP activation represents a key mechanism through which *H. pylori* links chronic gastritis to gastric carcinogenesis. YAP signaling induces inflammatory responses and recruits immune cells, leading to elevated production of IL-6 and IL-1, which in turn facilitate tumor development [80]. Furthermore, the virulence factor CagA disrupts tight junctions by dysregulating partitioning-defective 1 (PAR1)—a serine/threonine kinase critical for cell polarity—and subsequently promotes YAP/TAZ-mediated gene transcription [81]. Additionally, CagA binds to SHP2, triggering its phosphatase activity [ 82]. SHP2 is essential for RAS-ERK pathway activation, and YAP/TAZ can physically interact with SHP2 to promote its nuclear translocation [83]. Nuclear SHP2 further enhances β-catenin transcriptional activity by facilitating the parafibromin/β-catenin complex, leading to upregulation of Wnt target genes [84]. Thus, CagA-induced SHP2 activation serves as a critical node for cross-regulation between the Hippo and Wnt/β-catenin pathways.

PI3K/Akt pathway: *H. pylori* virulence factors activate the PI3K/Akt pathway, which plays a central role in promoting proliferation, survival, migration, and invasion in gastric cancer [77]. Pathway activation upregulates cell cycle regulators such as CCND1 and concurrently suppresses apoptotic processes [85]. Additionally, Akt facilitates epithelial–mesenchymal transition (EMT) by activating key transcription factors including Snail and Twist [86]. The virulence factor CagA further enhances Akt-driven EMT by promoting HDM2 phosphorylation, leading to CK2β degradation and subsequent Akt activation [87]. The PI3K/Akt axis engages in cross-regulation with upstream receptors (e.g., EGFR, HER2) and downstream effectors such as mTOR and NF-κB, forming an expanded signaling network [88]. Infection with *H. pylori* activates the PI3K/Akt/mTOR cascade, driving the malignant transformation of gastric epithelial cells [89]. Moreover, CagA confers cancer stem cell-like properties through activation of the PI3K/Akt/FOXO3a pathway, further underscoring its role in tumor initiation and progression [90].

Other pathways: *H. pylori* activates multiple signaling pathways—including HGF/Met, TGF-β, Hedgehog, and Notch—all of which are closely linked to tumorigenesis [91–94]. These pathways collectively regulate gastric epithelial cell behaviors such as proliferation, differentiation, apoptosis, and stemness. Importantly, a complex cross-regulatory network is sustained, involving interactions among pathways such as Wnt/β-catenin, Notch, and Hedgehog, which synergistically amplify carcinogenic processes [95].

## 5.4 Cellular Pathogenesis Induced by *H. pylori* in Gastric Cancer

Next, we focus on key cellular mechanisms triggered by *H. pylori*—such as oxidative stress, DNA damage, chronic inflammation, and reprogrammed cell death—which collectively contribute to the initiation and progression of gastric cancer (Fig. 5.3).

### 5.4.1 Oxidative Stress

*H. pylori* infection initiates the production of reactive oxygen and nitrogen species (ROS/RNS) in the gastric mucosa, resulting in persistent oxidative stress—a well-established driver of gastric carcinogenesis [96–98]. Beyond inducing diverse forms of DNA damage, this process disrupts antioxidant defenses and metabolic homeostasis in gastric epithelial cells, fostering a tumor-promoting microenvironment.

The generation of ROS/RNS by *H. pylori* occurs through both direct and indirect pathways. Direct mechanisms include bacterial stimulation of ROS accumulation in gastric epithelial cells [99, 100], with virulence factors such as VacA and CagA playing regulatory roles; VacA, for instance, influences autophagy in a manner associated with ROS production [101, 102]. Indirectly, *H. pylori* incites a sustained inflammatory response that recruits neutrophils and macrophages, which contribute to ROS/RNS generation [ 103, 104]. The bacterium also upregulates enzymes like NADPH oxidase (NOX) and inducible nitric oxide synthase (iNOS), amplifying oxidative stress [105]. Another mechanism involves spermine oxidase, which produces $H_2O_2$ via spermine oxidation, further promoting oxidative injury in gastric cells [106].

Elevated ROS/RNS levels lead to various forms of DNA damage, including double-strand breaks and base oxidation. Unrepaired lesions accumulate, promoting gene mutations and chromosomal rearrangements that accelerate genomic instability [107, 108]. This instability—partly attributable to *H. pylori*-induced DNA strand breaks—facilitates cellular transformation and gastric cancer progression [109, 110]. Moreover, *H. pylori* impairs DNA repair pathways, for example, by downregulating repair genes such as OGG1 [111]. Gastric epithelial cells

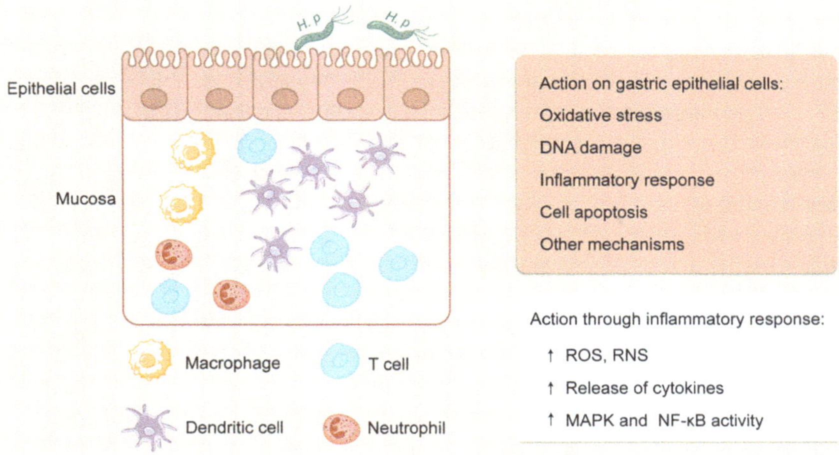


**Fig. 5.3** Cellular pathogenesis induced by *Helicobacter pylori* in gastric cancer. The schematic diagram provides an integrated overview of the pathogenic cascade triggered by *H. pylori* infection in the gastric mucosa. *H. pylori* attaches to the apical surface of gastric epithelial cells, where it directly induces cellular damage, initiating key oncogenic processes such as oxidative stress, DNA damage, inflammatory signaling, and dysregulated apoptosis. Infiltration of immune cells—including macrophages, dendritic cells, T cells, and neutrophils—into the lamina propria further amplifies the tissue response. These cells contribute to a pro-inflammatory microenvironment characterized by elevated ROS/RNS, enhanced cytokine release, and sustained activation of MAPK and NF-κB signaling pathways. Together, these direct and immune-mediated mechanisms drive the progressive damage that underlies gastric carcinogenesis. ROS/RNS, reactive oxygen and nitrogen species. (Original illustration created by Wei HU. )

with preexisting DNA repair deficiencies become especially susceptible to oxidative damage, significantly increasing the risk of malignant transformation [112, 113].

*H. pylori* also undermines host antioxidant defenses by altering the expression and activity of key molecules including superoxide dismutase (SOD), thioredoxin (Trx), AP endonuclease 1 (APE1), and glutathione (GSH) [103, 114]. Furthermore, it depletes antioxidants such as vitamins C and E from the gastric mucosa, thereby weakening the overall antioxidant capacity [115].

Oxidative stress additionally disrupts cellular metabolism, suppressing pyruvate utilization and the tricarboxylic acid (TCA) cycle while enhancing glycolysis—a metabolic shift that further raises intracellular ROS levels. Mitochondrial dysfunction exacerbates this cycle, creating a self-sustaining loop of oxidative damage [116]. Neutrophils contribute significantly to this process by upregulating NOX and iNOS, thereby enhancing ROS/RNS production and perpetuating the pathogenic cycle [117, 118].

### 5.4.2 DNA Damage

*H. pylori* infection promotes gastric carcinogenesis through the induction of diverse forms of DNA damage, including oxidative DNA lesions, double-strand breaks (DSBs), ribosome-related damage, and microsatellite instability (MSI) [119, 120]. Oxidative stress triggered by the bacterium leads to the formation of DNA adducts such as 8-hydroxy-2′-deoxyguanosine (8-OHdG), fostering mutagenesis and genomic instability [117]. Virulence factors secreted by *H. pylori* also induce DSBs, and faulty repair of these breaks can lead to chromosomal rearrangements [118]. Additionally, *H. pylori* activates ribosomal enzymes that cleave RNA, further compounding genomic instability [119, 120].

These forms of DNA damage contribute significantly to genomic instability—characterized by gene mutations and chromosomal aberrations—which serves as a key driver of gastric cancer progression [121]. *H. pylori* infection is associated with frequent genomic alterations in gastric tissues, including copy number variations, duplications, and deletions. These alterations disrupt the normal function of tumor suppressors and oncogenes, thereby promoting cellular transformation [121]. The severity of genomic instability varies among bacterial strains; notably, CagA-positive *H. pylori* induces more pronounced microsatellite instability and chromosomal deletions [122]. The bacterium also impairs the tumor suppressor p53, enhancing proliferation in CagA-translocated cells and further amplifying genomic instability [123].

Compounding these effects, *H. pylori* compromises cellular DNA repair capacity through multiple mechanisms. It downregulates key DNA damage response genes—including ATM, p53, and p21—impairing the detection and repair of DNA lesions [124]. The bacterium also disrupts specific repair pathways such as nucleotide excision repair (NER) and interferes with the repair of double-strand breaks by hampering both homologous recombination and nonhomologous end-joining activities [61]. Furthermore, the chronic inflammatory environment induced by *H. pylori*, marked by immune cell infiltration and sustained ROS/RNS release, perpetuates DNA damage and promotes continuous cell proliferation, establishing a self-reinforcing cycle of genomic instability [125].

### 5.4.3 Inflammatory Response

*H. pylori* infection initiates a persistent inflammatory response in the gastric mucosa through multiple molecular mechanisms. Following adhesion to gastric epithelial cells, the bacterium secretes CagA, which reprograms bile acid metabolism to promote inflammation and support carcinogenesis [126]. Additionally, *H. pylori* activates pattern recognition receptors (PRRs), including NOD1 and NOD2 [127], thereby stimulating downstream signaling cascades such as MAPK and NF-κB [117]. These pathways elevate the expression of inflammatory cytokines (e.g., IL-8, IL-6), chemokines, adhesion molecules, and enzymes like COX-2. The resulting inflammatory milieu facilitates the recruitment and activation of immune cells—particularly neutrophils and macrophages—establishing a self-sustaining cycle of chronic inflammation.

Infection persistently activates both innate and adaptive immune responses, leading to the accumulation of diverse immune cells within the gastric mucosa [128, 129]. These cells release cytokines, chemical mediators, and reactive oxygen species, collectively inflicting cumulative damage on gastric epithelial cells and fostering a tumor-promoting microenvironment.

The roles of specific immune cells in this process are distinct and multifaceted. Neutrophils, among the first responders, are recruited by chemokines and generate substantial reactive oxygen species, inducing oxidative stress and DNA damage in epithelial cells. *H. pylori* exploits neutrophil plasticity by promoting N1-like polarization, enhancing its virulence, and contributing to gastric cancer development [130]. Tumor-associated macrophages (TAMs) exhibit dual functions in inflammation and cancer. *H. pylori* upregulates indoleamine 2,3-dioxygenase (IDO) in macrophages, driving M2 polarization [131], while its phospholipase A modulates the p38 signaling pathway to manipulate macrophage autophagy, facilitating immune evasion and bacterial survival [132]. Dendritic cells experience impaired function upon infection, including compromised antigen presentation, leading to dysregulated adaptive immunity. This impairment supports immune evasion and sustains chronic inflammation, thereby promoting carcinogenesis [133]. T lymphocytes, including $CD4^+$ and $CD8^+$ subsets, are recruited in response to infection. Th1 and Th17 $CD4^+$ T cells produce cytokines, such as IFN-$\gamma$ and IL-17, amplifying macrophage and neutrophil activity. However, chronic T cell activation can result in tissue damage and tumor promotion [134]. A recent study further highlights that NLRP3 modulates the infiltration of $CD4^+$ T cells and M2 macrophages within the gastric cancer microenvironment [20].

Inflammatory mediators further contribute to carcinogenesis through various mechanisms. Cytokines such as IL-6 and TNF-$\alpha$ activate oncogenic pathways including STAT3 and NF-$\kappa$B, promoting the expression of tumor-related genes [135]. Immune cells recruited by chemokines generate substantial ROS and RNS, inducing DNA damage and genomic instability [136]. Matrix metalloproteinases (MMPs) secreted by TAMs enhance the invasive capacity of tumor cells [137].

The inflammatory response dynamically interacts with other carcinogenic pathways, forming a self-reinforcing network. Persistent immune cells continuously produce ROS/RNS, exacerbating oxidative stress and upregulating inflammatory factors such as IL-8 [138]. Chronic inflammation also promotes *H. pylori*-induced DNA damage and mutations, while DNA damage reciprocally amplifies inflammatory signaling [139]. Inflammatory pathways such as NF-$\kappa$B engage in cross-regulation with other oncogenic cascades—including STAT3 and Wnt/$\beta$-catenin—collectively accelerating gastric carcinogenesis.

### 5.4.4 Cell Apoptosis

Dysregulation of apoptosis represents a key event in *H. pylori*-induced gastric carcinogenesis, as the bacterium manipulates apoptotic processes in gastric epithelial cells to favor tumor survival and expansion. Although *H. pylori* has been shown to induce apoptosis in vitro [140, 141], its role in vivo is context-dependent and varies with the stage of infection.

The dual role of *H. pylori* in apoptosis regulation follows a temporal pattern: a proapoptotic response predominates during early infection, whereas antiapoptotic mechanisms prevail as malignancy develops. Initially, *H. pylori* promotes apoptosis as part of the host defense strategy. Virulence factors such as VacA activate the mitochondria-mediated intrinsic apoptotic pathway [142], while oxidative stress and DNA damage further contribute to programmed cell death, aiding the elimination of infected cells, and limiting bacterial colonization. As infection progresses, however, *H. pylori* shifts toward suppressing apoptosis—a phenomenon frequently observed in gastric cancer tissues and cell lines [69]. This inhibition supports tumor cell survival, proliferation, and dissemination [143].

*H. pylori* modulates apoptosis in gastric epithelial cells through several interconnected pathways: (1) PI3K/Akt signaling: Activation of this pathway leads to Akt-mediated phosphorylation and regulation of Bcl-2 family proteins [144]; (2) Bcl-2 family proteins: The bacterium upregulates antiapoptotic members (e.g., Bcl-2, Bcl-xL) and downregulates proapoptotic factors (e.g., Bax, Bad) [145]; (3) p53-dependent pathway: *H. pylori* suppresses p53 expression and activity, thereby obstructing p53-mediated apoptosis [146]; (4) NF-κB pathway: NF-κB activation upregulates antiapoptotic genes such as c-IAP2 and Survivin [147]; Notably, PPARγ activation can inhibit *H. pylori*-induced NF-κB signaling and attenuate gastric epithelial apoptosis [148]. Additionally, *H. pylori*-induced DNA damage and oxidative stress serve as upstream regulators that indirectly influence these apoptotic pathways [149, 150].

### 5.4.5 Other Mechanisms

Beyond its well-characterized roles in oxidative stress, DNA damage, inflammatory response, and apoptosis regulation, *H. pylori* employs additional strategies to promote gastric cancer development. These include interference with cell cycle progression and modulation of the autophagic process. Moreover, emerging research has begun to elucidate the significance of epigenetic alterations, intestinal microbiota dysbiosis, tumor microenvironment immune cells, extracellular vesicles, and noncoding RNAs in *H. pylori*-driven carcinogenesis.

*H. pylori* infection disrupts cell cycle checkpoints, leading to uncontrolled proliferation. Key mechanisms involve the upregulation of cyclins (e.g., cyclin D1, cyclin E) and cyclin-dependent kinases, coupled with downregulation of cell cycle inhibitors such as p21 and p27 [151, 152]. These effects are mediated primarily through oncogenic pathways including STAT3, Wnt/β-catenin, and NF-κB. Cell cycle dysregulation not only directly accelerates tumor growth but also contributes to genomic instability and apoptosis resistance. For example, compromised DNA checkpoint control impairs damage repair, leading to mutation accumulation [153, 154].

Autophagy, a cellular self-degradation process critical for maintaining homeostasis, plays a dual role in cancer [155]. *H. pylori* has evolved strategies to evade or exploit host autophagy: while it initially induces autophagy to mitigate cellular stress, persistent infection leads to sustained autophagic alterations that influence cell survival, growth, and immune evasion [40, 156, 157]. The bacterium promotes YAP-dependent autophagy by upregulating Beclin 1 and LC3B-II and enhancing autophagosome formation [158]. Conversely, CAPZA1 acts as a negative regulator by inhibiting autolysosome formation, thereby reducing CagA degradation and carcinogenic risk [159]. VacA targets lysosomal calcium channel TRPML1 to disrupt endolysosomal trafficking, creating survival niches for the bacterium [160]. Additionally, the virulence factor gamma-glutamyl transpeptidase (HpGGT) disrupts lysosomal membrane integrity, impairing autophagic flux [22, 161].

Cutting-edge research continues to uncover novel mechanisms in *H. pylori*-associated gastric cancer. These include epigenetic modifications such as RNA methylation and histone alterations [162, 163], dysbiosis of the gut microbiota that perturbs host metabolism and immunity [164, 165], and the roles of immune cells [166], extracellular vesicles [167], microRNAs [168], and long noncoding RNAs [169] within the tumor microenvironment. These emerging areas highlight the multifaceted nature of *H. pylori* carcinogenesis and offer promising directions for future investigation.

## 5.5 Conclusion

By examining the diverse mechanisms through which *H. pylori* induces gastric cancer, this chapter aims to offer deeper mechanistic insights to guide future research. Although approximately half of the global population is infected with *H. pylori*, only a small subset progresses to gastric malignancy. With advancing understanding of these mechanisms, identifying high-risk individuals has become a priority for developing targeted therapeutic and preventive strategies. Furthermore, as ongoing research continues to reveal the complexity of *H. pylori*–host interactions, these advances are expected to translate into more effective approaches for gastric cancer prevention and treatment, ultimately improving patient prognosis worldwide.

However, the mechanisms underlying *H. pylori*-induced gastric carcinogenesis remain incompletely elucidated, necessitating further investigation. Future studies should focus on: (1) Clarifying the precise mechanisms by which *H. pylori* virulence factors interact with host cells to initiate carcinogenesis; (2) Investigating the influence of bacterial genotype diversity on disease progression and treatment response; and (3) Developing novel therapeutic targets and immunotherapies based on molecular pathways dysregulated by *H. pylori*.

## References


1. Bray F, Laversanne M, Sung H, Ferlay J, Siegel RL, Soerjomataram I, et al. Global cancer statistics 2022: GLOBOCAN estimates of incidence and mortality worldwide for 36 cancers in 185 countries. CA Cancer J Clin. 2024;74(3):229–63. https://doi.org/10.3322/caac.21834.
2. Li Y, Choi H, Leung K, Jiang F, Graham DY, Leung WK. Global prevalence of *H. pylori* infection between 1980 and 2022: a systematic review and meta-analysis. Lancet Gastroenterol Hepatol. 2023;8(6):553–64. https://doi.org/10.1016/S2468-1253(23)00070-5.
3. Lee YC, Chen TH, Chiu HM, Shun CT, Chiang H, Liu TY, et al. The benefit of mass eradication of *H. pylori* infection: a community-based study of gastric cancer prevention. Gut. 2013;62(5):676–82. https://doi.org/10.1136/gutjnl-2012-302240.
4. Salama NR, Hartung ML, Muller A. Life in the human stomach: persistence strategies of the bacterial pathogen *H. pylori*. Nat Rev Microbiol. 2013;11(6):385–99. https://doi.org/10.1038/nrmicro3016.
5. Suerbaum S, Josenhans C. *H. pylori* evolution and phenotypic diversification in a changing host. Nat Rev Microbiol. 2007;5(6):441–52. https://doi.org/10.1038/nrmicro1658.
6. Thorell K, Munoz-Ramirez ZY, Wang D, Sandoval-Motta S, Boscolo Agostini R, Ghirotto S, et al. The *H. pylori* genome project: insights into *H. pylori* population structure from analysis of a worldwide collection of complete genomes. Nat Commun. 2023;14(1):8184. https://doi.org/10.1038/s41467-023-43562-y.
7. Tran SC, Bryant KN, Cover TL. The *H. pylori* cag pathogenicity Island as a determinant of gastric cancer risk. Gut Microbes. 2024;16(1):2314201. https://doi.org/10.1080/19490976.2024.2314201.
8. Amieva MR, El-Omar EM. Host-bacterial interactions in *H. pylori* infection. Gastroenterology. 2008;134(1):306–23. https://doi.org/10.1053/j.gastro.2007.11.009.
9. Hatakeyama M. SagA of CagA in *H. pylori* pathogenesis. Curr Opin Microbiol. 2008;11(1):30–7. https://doi.org/10.1016/j.mib.2007.12.003.
10. Ding SZ, Zheng PY. *H. pylori* infection induced gastric cancer; advance in gastric stem cell research and the remaining challenges. Gut Pathog. 2012;4(1):18. https://doi.org/10.1186/1757-4749-4-18.
11. Wu X, Zhao Y, Zhang H, Yang W, Yang J, Sun L, et al. Mechanism of regulation of the *H. pylori* Cagbeta ATPase by CagZ. Nat Commun. 2023;14(1):479. https://doi.org/10.1038/s41467-023-36218-4.
12. Poppe M, Feller SM, Romer G, Wessler S. Phosphorylation of *H. pylori* CagA by c-Abl leads to cell motility. Oncogene. 2007;26(24):3462–72. https://doi.org/10.1038/sj.onc.1210139.
13. Padda J, Khalid K, Cooper AC, Jean-Charles G. Association between *H. pylori* and gastric carcinoma. Cureus. 2021;13(5):e15165. https://doi.org/10.7759/cureus.15165.
14. Alipour M. Molecular mechanism of *H. pylori*-induced gastric cancer. J Gastrointest Cancer. 2021;52(1):23–30. https://doi.org/10.1007/s12029-020-00518-5.
15. Hatakeyama M. Oncogenic mechanisms of the *H. pylori* CagA protein. Nat Rev Cancer. 2004;4(9):688–94. https://doi.org/10.1038/nrc1433.
16. Shiota S, Suzuki R, Yamaoka Y. The significance of virulence factors in *H. pylori*. J Dig Dis. 2013;14(7):341–9. https://doi.org/10.1111/1751-2980.12054.
17. Li Q, Liu J, Gong Y, Yuan Y. Association of CagA EPIYA-D or EPIYA-C phosphorylation sites with peptic ulcer and gastric cancer risks: a meta-analysis. Medicine (Baltimore). 2017;96(17):e6620. https://doi.org/10.1097/MD.0000000000006620.
18. Soutto M, Zhang X, Bhat N, Chen Z, Zhu S, Maacha S, et al. Fibroblast growth factor receptor-4 mediates activation of nuclear factor erythroid 2-related Factor-2 in gastric tumorigenesis. Redox Biol. 2024;69:102998. https://doi.org/10.1016/j.redox.2023.102998.
19. Wang H, Zhao M, Shi F, Zheng S, Xiong L, Zheng L. A review of signal pathway induced by virulent protein CagA of *H. pylori*. Front Cell Infect Microbiol. 2023;13:1062803. https://doi.org/10.3389/fcimb.2023.1062803.

20. Wan C, Wang P, Xu Y, Zhu Y, Chen H, Cao X, et al. Mechanism and role of *H. pylori* CagA-induced NLRP3 inflammasome in gastric cancer immune cell infiltration. Sci Rep. 2025;15(1):14335. https://doi.org/10.1038/s41598-025-98301-8.
21. Chauhan N, Tay ACY, Marshall BJ, Jain U. *H. pylori* VacA, a distinct toxin exerts diverse functionalities in numerous cells: an overview. Helicobacter. 2019;24(1):e12544. https://doi.org/10.1111/hel.12544.
22. Amieva M, Peek RM Jr. Pathobiology of *H. pylori*-induced gastric cancer. Gastroenterology. 2016;150(1):64–78. https://doi.org/10.1053/j.gastro.2015.09.004.
23. Baj J, Forma A, Sitarz M, Portincasa P, Garruti G, Krasowska D, et al. *H. pylori* virulence factors-mechanisms of bacterial pathogenicity in the gastric microenvironment. Cells. 2020;10(1) https://doi.org/10.3390/cells10010027.
24. Kim IJ, Lee J, Oh SJ, Yoon MS, Jang SS, Holland RL, et al. *H. pylori* infection modulates host cell metabolism through VacA-dependent inhibition of mTORC1. Cell Host Microbe. 2018;23(5):583–93 e8. https://doi.org/10.1016/j.chom.2018.04.006.
25. Bridge DR, Merrell DS. Polymorphism in the *H. pylori* CagA and VacA toxins and disease. Gut Microbes. 2013;4(2):101–17. https://doi.org/10.4161/gmic.23797.
26. Atherton JC, Cao P, Peek RM Jr, Tummuru MK, Blaser MJ, Cover TL. Mosaicism in vacuolating cytotoxin alleles of *H. pylori*. Association of specific vacA types with cytotoxin production and peptic ulceration. J Biol Chem. 1995;270(30):17771–7. https://doi.org/10.1074/jbc.270.30.17771.
27. Ferreira RM, Machado JC, Letley D, Atherton JC, Pardo ML, Gonzalez CA, et al. A novel method for genotyping the *H. pylori* vacA intermediate region directly in gastric biopsy specimens. J Clin Microbiol. 2012;50(12):3983–9. https://doi.org/10.1128/JCM.02087-12.
28. Yamaoka Y, Kodama T, Gutierrez O, Kim JG, Kashima K, Graham DY. Relationship between *H. pylori* iceA, cagA, and vacA status and clinical outcome: studies in four different countries. J Clin Microbiol. 1999;37(7):2274–9. https://doi.org/10.1128/JCM.37.7.2274-2279.1999.
29. Sterbenc A, Jarc E, Poljak M, Homan M. *H. pylori* virulence genes. World J Gastroenterol. 2019;25(33):4870–84. https://doi.org/10.3748/wjg.v25.i33.4870.
30. Abdullah M, Greenfield LK, Bronte-Tinkew D, Capurro MI, Rizzuti D, Jones NL. VacA promotes CagA accumulation in gastric epithelial cells during *H. pylori* infection. Sci Rep. 2019;9(1):38. https://doi.org/10.1038/s41598-018-37095-4.
31. Tan S, et al. *H. pylori* perturbs iron trafficking in the epithelium to grow on the cell surface. PLoS Pathog. 2011;7(5):e1002050. https://doi.org/10.1371/journal.ppat.1002050.
32. Matsumoto A, Isomoto H, Nakayama M, Hisatsune J, Nishi Y, Nakashima Y, et al. *H. pylori* VacA reduces the cellular expression of STAT3 and pro-survival Bcl-2 family proteins, Bcl-2 and Bcl-XL, leading to apoptosis in gastric epithelial cells. Dig Dis Sci. 2011;56(4):999–1006. https://doi.org/10.1007/s10620-010-1420-1.
33. Sedarat Z, Taylor-Robinson AW. *H. pylori* outer membrane proteins and virulence factors: potential targets for novel therapies and vaccines. Pathogens. 2024;13(5) https://doi.org/10.3390/pathogens13050392.
34. Debowski AW, Walton SM, Chua EG, Tay AC, Liao T, Lamichhane B, et al. *H. pylori* gene silencing in vivo demonstrates urease is essential for chronic infection. PLoS Pathog. 2017;13(6):e1006464. https://doi.org/10.1371/journal.ppat.1006464.
35. Gu H. Role of flagella in the pathogenesis of *H. pylori*. Curr Microbiol. 2017;74(7):863–9. https://doi.org/10.1007/s00284-017-1256-4.
36. Celli JP, Turner BS, Afdhal NH, Keates S, Ghiran I, Kelly CP, et al. *H. pylori* moves through mucus by reducing mucin viscoelasticity. Proc Natl Acad Sci USA. 2009;106(34):14321–6. https://doi.org/10.1073/pnas.0903438106.
37. Kaplan M, Oikonomou CM, Wood CR, Chreifi G, Subramanian P, Ortega DR, et al. Novel transient cytoplasmic rings stabilize assembling bacterial flagellar motors. EMBO J. 2022;41(10):e109523. https://doi.org/10.15252/embj.2021109523.
38. Liu X, Tachiyama S, Zhou X, Mathias RA, Bonny SQ, Khan MF, et al. Bacterial flagella hijack type IV pili proteins to control motility. Proc Natl Acad Sci USA. 2024;121(4):e2317452121. https://doi.org/10.1073/pnas.2317452121.

39. Sallas ML, Dos Santos MP, Orcini WA, David EB, Peruquetti RL, Payao SLM, et al. Status (on/off) of oipA gene: their associations with gastritis and gastric cancer and geographic origins. Arch Microbiol. 2019;201(1):93–7. https://doi.org/10.1007/s00203-018-1580-5.
40. Hu W, Zhang L, Li MX, Shen J, Liu XD, Xiao ZG, et al. Vitamin D3 activates the autolysosomal degradation function against *H. pylori* through the PDIA3 receptor in gastric epithelial cells. Autophagy. 2019;15(4):707–25. https://doi.org/10.1080/15548627.2018.1557835.
41. Heinrich PC, Behrmann I, Haan S, Hermanns HM, Muller-Newen G, Schaper F. Principles of interleukin (IL)-6-type cytokine signalling and its regulation. Biochem J. 2003;374(Pt 1):1–20. https://doi.org/10.1042/BJ20030407.
42. Kitamura H, Ohno Y, Toyoshima Y, Ohtake J, Homma S, Kawamura H, et al. Interleukin-6/STAT3 signaling as a promising target to improve the efficacy of cancer immunotherapy. Cancer Sci. 2017;108(10):1947–52. https://doi.org/10.1111/cas.13332.
43. Jackson CB, Judd LM, Menheniott TR, Kronborg I, Dow C, Yeomans ND, et al. Augmented gp130-mediated cytokine signalling accompanies human gastric cancer progression. J Pathol. 2007;213(2):140–51. https://doi.org/10.1002/path.2218.
44. Yang J, Chatterjee-Kishore M, Staugaitis SM, Nguyen H, Schlessinger K, Levy DE, et al. Novel roles of unphosphorylated STAT3 in oncogenesis and transcriptional regulation. Cancer Res. 2005;65(3):939–47.
45. Bronte-Tinkew DM, Terebiznik M, Franco A, Ang M, Ahn D, Mimuro H, et al. *H. pylori* cytotoxin-associated gene a activates the signal transducer and activator of transcription 3 pathway in vitro and in vivo. Cancer Res. 2009;69(2):632–9. https://doi.org/10.1158/0008-5472.CAN-08-1191.
46. Tye H, Kennedy CL, Najdovska M, McLeod L, McCormack W, Hughes N, et al. STAT3-driven upregulation of TLR2 promotes gastric tumorigenesis independent of tumor inflammation. Cancer Cell. 2012;22(4):466–78. https://doi.org/10.1016/j.ccr.2012.08.010.
47. Tye H, Jenkins BJ. Tying the knot between cytokine and toll-like receptor signaling in gastrointestinal tract cancers. Cancer Sci. 2013;104(9):1139–45. https://doi.org/10.1111/cas.12205.
48. Duan Y, Kong P, Huang M, Yan Y, Dou Y, Huang B, et al. STAT3-mediated up-regulation of DAB2 via SRC-YAP1 signaling axis promotes *H. pylori*-driven gastric tumorigenesis. Biomark Res. 2024;12(1):33. https://doi.org/10.1186/s40364-024-00577-x.
49. Zhang X, Soutto M, Chen Z, Bhat N, Zhu S, Eissmann MF, et al. Induction of fibroblast growth factor receptor 4 by *H. pylori* via signal transducer and activator of transcription 3 with a feedforward activation loop involving SRC signaling in gastric cancer. Gastroenterology. 2022;163(3):620–36 e9. https://doi.org/10.1053/j.gastro.2022.05.016.
50. Mitchell DJ, Huynh HQ, Ceponis PJ, Jones NL, Sherman PM. *H. pylori* disrupts STAT1-mediated gamma interferon-induced signal transduction in epithelial cells. Infect Immun. 2004;72(1):537–45. https://doi.org/10.1128/IAI.72.1.537-545.2004.
51. Morey P, Pfannkuch L, Pang E, Boccellato F, Sigal M, Imai-Matsushima A, et al. *H. pylori* depletes cholesterol in gastric glands to prevent interferon gamma signaling and escape the inflammatory response. Gastroenterology. 2018;154(5):1391–404 e9. https://doi.org/10.1053/j.gastro.2017.12.008.
52. Lamb A, Chen LF. Role of the *H. pylori*-induced inflammatory response in the development of gastric cancer. J Cell Biochem. 2013;114(3):491–7. https://doi.org/10.1002/jcb.24389.
53. Brandt S, Kwok T, Hartig R, Konig W, Backert S. NF-kappaB activation and potentiation of proinflammatory responses by the *H. pylori* CagA protein. Proc Natl Acad Sci USA. 2005;102(26):9300–5. https://doi.org/10.1073/pnas.0409873102.
54. Yasumoto K, Okamoto S, Mukaida N, Murakami S, Mai M, Matsushima K. Tumor necrosis factor alpha and interferon gamma synergistically induce interleukin 8 production in a human gastric cancer cell line through acting concurrently on AP-1 and NF-kB-like binding sites of the interleukin 8 gene. J Biol Chem. 1992;267(31):22506–11.
55. Sakitani K, Hirata Y, Hayakawa Y, Serizawa T, Nakata W, Takahashi R, et al. Role of interleukin-32 in *H. pylori*-induced gastric inflammation. Infect Immun. 2012;80(11):3795–803. https://doi.org/10.1128/IAI.00637-12.

56. Wang S, Chen Z, Zhu S, Lu H, Peng D, Soutto M, et al. PRDX2 protects against oxidative stress induced by *H. pylori* and promotes resistance to cisplatin in gastric cancer. Redox Biol. 2020;28:101319. https://doi.org/10.1016/j.redox.2019.101319.
57. Chen B, Liu X, Yu P, Xie F, Kwan JSH, Chan WN, et al. *H. pylori*-induced NF-kappaB-PIEZO1-YAP1-CTGF axis drives gastric cancer progression and cancer-associated fibroblast-mediated tumour microenvironment remodelling. Clin Transl Med. 2023;13(11):e1481. https://doi.org/10.1002/ctm2.1481.
58. Cao L, Zhu S, Lu H, Soutto M, Bhat N, Chen Z, et al. *H. pylori*-induced RASAL2 hrough activation of nuclear factor-kappaB promotes gastric tumorigenesis via beta-catenin signaling Axis. Gastroenterology. 2022;162(6):1716–31 e17. https://doi.org/10.1053/j.gastro.2022.01.046.
59. Huang S, Bucana CD, Van Arsdall M, Fidler IJ. Stat1 negatively regulates angiogenesis, tumorigenicity and metastasis of tumor cells. Oncogene. 2002;21(16):2504–12. https://doi.org/10.1038/sj.onc.1205341.
60. Karin M. Nuclear factor-kappaB in cancer development and progression. Nature. 2006;441(7092):431–6. https://doi.org/10.1038/nature04870.
61. Zhong W, Wu K, Long Z, Zhou X, Zhong C, Wang S, et al. Gut dysbiosis promotes prostate cancer progression and docetaxel resistance via activating NF-kappaB-IL6-STAT3 axis. Microbiome. 2022;10(1):94. https://doi.org/10.1186/s40168-022-01289-w.
62. Koushyar S, Powell AG, Vincan E, Phesse TJ. Targeting Wnt signaling for the treatment of gastric cancer. Int J Mol Sci. 2020;21(11) https://doi.org/10.3390/ijms21113927.
63. Liu J, Xiao Q, Xiao J, Niu C, Li Y, Zhang X, et al. Wnt/beta-catenin signalling: function, biological mechanisms, and therapeutic opportunities. Signal Transduct Target Ther. 2022;7(1):3. https://doi.org/10.1038/s41392-021-00762-6.
64. Takahashi-Kanemitsu A, Lu M, Knight CT, Yamamoto T, Hayashi T, Mii Y, et al. The *H. pylori* CagA oncoprotein disrupts Wnt/PCP signaling and promotes hyperproliferation of pyloric gland base cells. Sci Signal. 2023;16(794):eabp9020. https://doi.org/10.1126/scisignal.abp9020.
65. Liu B, Bukhari I, Li F, Ren F, Xia X, Hu B, et al. Enhanced LRP8 expression induced by *H. pylori* drives gastric cancer progression by facilitating beta-catenin nuclear translocation. J Adv Res. 2025;69:299–312. https://doi.org/10.1016/j.jare.2024.04.002.
66. Song P, Gao Z, Bao Y, Chen L, Huang Y, Liu Y, et al. Wnt/beta-catenin signaling pathway in carcinogenesis and cancer therapy. J Hematol Oncol. 2024;17(1):46. https://doi.org/10.1186/s13045-024-01563-4.
67. Zuo W, Yang H, Li N, Ouyang Y, Xu X, Hong J. *H. pylori* infection activates Wnt/beta-catenin pathway to promote the occurrence of gastritis by upregulating ASCL1 and AQP5. Cell Death Discov. 2022;8(1):257. https://doi.org/10.1038/s41420-022-01026-0.
68. Tian S, Peng P, Li J, Deng H, Zhan N, Zeng Z, et al. SERPINH1 regulates EMT and gastric cancer metastasis via the Wnt/beta-catenin signaling pathway. Aging (Albany NY). 2020;12(4):3574–93. https://doi.org/10.18632/aging.102831.
69. Tang Q, Chen J, Di Z, Yuan W, Zhou Z, Liu Z, et al. TM4SF1 promotes EMT and cancer stemness via the Wnt/beta-catenin/SOX2 pathway in colorectal cancer. J Exp Clin Cancer Res. 2020;39(1):232. https://doi.org/10.1186/s13046-020-01690-z.
70. Peng Y, Qin Y, Zhang X, Deng S, Yuan Y, Feng X, et al. MiRNA-20b/SUFU/Wnt axis accelerates gastric cancer cell proliferation, migration and EMT. Heliyon. 2021;7(4):e06695. https://doi.org/10.1016/j.heliyon.2021.e06695.
71. Katoh M, Katoh M. WNT signaling and cancer stemness. Essays Biochem. 2022;66(4):319–31. https://doi.org/10.1042/EBC20220016.
72. Sharafutdinov I, Tegtmeyer N, Linz B, Rohde M, Vieth M, Tay AC, et al. A single-nucleotide polymorphism in *H. pylori* promotes gastric cancer development. Cell Host Microbe. 2023;31(8):1345–58 e6. https://doi.org/10.1016/j.chom.2023.06.016.
73. Li N, Xu X, Zhan Y, Fei X, Ouyang Y, Zheng P, et al. YAP and beta-catenin cooperate to drive *H. pylori*-induced gastric tumorigenesis. Gut Microbes. 2023;15(1):2192501. https://doi.org/10.1080/19490976.2023.2192501.

74. Pan G, Wang X, Wang Y, Li R, Li G, He Y, et al. *H. pylori* promotes gastric cancer progression by upregulating semaphorin 5A expression via ERK/MMP9 signaling. Mol Ther Oncolytics. 2021;22:256–64. https://doi.org/10.1016/j.omto.2021.06.002.
75. Sharafutdinov I, Backert S, Tegtmeyer N. The *H. pylori* type IV secretion system upregulates epithelial cortactin expression by a CagA- and JNK-dependent pathway. Cell Microbiol. 2021;23(10):e13376. https://doi.org/10.1111/cmi.13376.
76. Peng Y, Lei X, Yang Q, Zhang G, He S, Wang M, et al. *H. pylori* CagA-mediated ether lipid biosynthesis promotes ferroptosis susceptibility in gastric cancer. Exp Mol Med. 2024;56(2):441–52. https://doi.org/10.1038/s12276-024-01167-5.
77. Xu W, Huang Y, Yang Z, Hu Y, Shu X, Xie C, et al. *H. pylori* promotes gastric epithelial cell survival through the PLK1/PI3K/Akt pathway. Onco Targets Ther. 2018;11:5703–13. https://doi.org/10.2147/OTT.S164749.
78. Meng Z, Moroishi T, Guan KL. Mechanisms of hippo pathway regulation. Genes Dev. 2016;30(1):1–17. https://doi.org/10.1101/gad.274027.115.
79. Zhao B, Wei X, Li W, Udan RS, Yang Q, Kim J, et al. Inactivation of YAP oncoprotein by the hippo pathway is involved in cell contact inhibition and tissue growth control. Genes Dev. 2007;21(21):2747–61. https://doi.org/10.1101/gad.1602907.
80. Liu Y, Zhang B, Zhou Y, Xing Y, Wang Y, Jia Y, et al. Targeting hippo pathway: a novel strategy for *H. pylori*-induced gastric cancer treatment. Biomed Pharmacother. 2023;161:114549. https://doi.org/10.1016/j.biopha.2023.114549.
81. Saadat I, Higashi H, Obuse C, Umeda M, Murata-Kamiya N, Saito Y, et al. *H. pylori* CagA targets PAR1/MARK kinase to disrupt epithelial cell polarity. Nature. 2007;447(7142):330–3. https://doi.org/10.1038/nature05765.
82. Hof P, Pluskey S, Dhe-Paganon S, Eck MJ, Shoelson SE. Crystal structure of the tyrosine phosphatase SHP-2. Cell. 1998;92(4):441–50. https://doi.org/10.1016/s0092-8674(00)80938-1.
83. Messina B, Lo Sardo F, Scalera S, Memeo L, Colarossi C, Mare M, et al. Hippo pathway dysregulation in gastric cancer: from *H. pylori* infection to tumor promotion and progression. Cell Death Dis. 2023;14(1):21. https://doi.org/10.1038/s41419-023-05568-8.
84. Coulombe G, Rivard N. New and unexpected biological functions for the Src-homology 2 domain-containing phosphatase SHP-2 in the gastrointestinal tract. Cell Mol Gastroenterol Hepatol. 2016;2(1):11–21. https://doi.org/10.1016/j.jcmgh.2015.11.001.
85. He XS, Ye WL, Zhang YJ, Yang XQ, Liu F, Wang JR, et al. Oncogenic potential of BEST4 in colorectal cancer via activation of PI3K/Akt signaling. Oncogene. 2022;41(8):1166–77. https://doi.org/10.1038/s41388-021-02160-2.
86. Peng Y, Wang Y, Zhou C, Mei W, Zeng C. PI3K/Akt/mTOR pathway and its role in cancer therapeutics: are we making headway? Front Oncol. 2022;12:819128. https://doi.org/10.3389/fonc.2022.819128.
87. Lee SD, Jeong H, Hwang BR, Yu BM, Cho Y, Nam KT, et al. *H. pylori* promotes epithelial-to-mesenchymal transition by downregulating CK2beta in gastric cancer cells. Biochim Biophys Acta Mol basis Dis. 2023;1869(1):166588. https://doi.org/10.1016/j.bbadis.2022.166588.
88. Zaryouh H, De Pauw I, Baysal H, Peeters M, Vermorken JB, Lardon F, et al. Recent insights in the PI3K/Akt pathway as a promising therapeutic target in combination with EGFR-targeting agents to treat head and neck squamous cell carcinoma. Med Res Rev. 2022;42(1):112–55. https://doi.org/10.1002/med.21806.
89. Morgos DT, Stefani C, Miricescu D, Greabu M, Stanciu S, Nica S, et al. Targeting PI3K/AKT/mTOR and MAPK signaling pathways in gastric cancer. Int J Mol Sci. 2024;25(3) https://doi.org/10.3390/ijms25031848.
90. Chen ZW, Dong ZB, Xiang HT, Chen SS, Yu WM, Liang C. *H. pylori* CagA protein induces gastric cancer stem cell-like properties through the Akt/FOXO3a axis. J Cell Biochem. 2024;125(3):e30527. https://doi.org/10.1002/jcb.30527.
91. Liu Q, Chen C, He Y, Mai W, Ruan S, Ning Y, et al. Notch signaling regulates the function and phenotype of dendritic cells in *H. pylori* infection. Microorganisms. 2023;11(11) https://doi.org/10.3390/microorganisms11112818.

92. Koh V, Chakrabarti J, Torvund M, Steele N, Hawkins JA, Ito Y, et al. Hedgehog transcriptional effector GLI mediates mTOR-induced PD-L1 expression in gastric cancer organoids. Cancer Lett. 2021;518:59–71. https://doi.org/10.1016/j.canlet.2021.06.007.
93. Jinling X, Guoan L, Chuxi C, Qiaoyuan L, Yinzhong C, Shihao C, et al. NOTCH1 is positively correlated with IL17F in *H. pylori* infection and a biomarker for mucosal injury. iScience. 2024;27(7):110323. https://doi.org/10.1016/j.isci.2024.110323.
94. Sakamoto N, Tsujimoto H, Takahata R, Cao B, Zhao P, Ito N, et al. MET4 expression predicts poor prognosis of gastric cancers with *H. pylori* infection. Cancer Sci. 2017;108(3):322–30. https://doi.org/10.1111/cas.13146.
95. Yong X, Tang B, Li BS, Xie R, Hu CJ, Luo G, et al. *H. pylori* virulence factor CagA promotes tumorigenesis of gastric cancer via multiple signaling pathways. Cell Commun Signal. 2015;13:30. https://doi.org/10.1186/s12964-015-0111-0.
96. Sah DK, Arjunan A, Lee B, Jung YD. Reactive oxygen species and *H. pylori* infection: a comprehensive review of their roles in gastric cancer development. Antioxidants (Basel). 2023;12(9) https://doi.org/10.3390/antiox12091712.
97. Wu S, Chen Y, Chen Z, Wei F, Zhou Q, Li P, et al. Reactive oxygen species and gastric carcinogenesis: the complex interaction between *H. pylori* and host. Helicobacter. 2023;28(6):e13024. https://doi.org/10.1111/hel.13024.
98. Butcher LD, den Hartog G, Ernst PB, Crowe SE. Oxidative stress resulting from *H. pylori* infection contributes to gastric carcinogenesis. Cell Mol Gastroenterol Hepatol. 2017;3(3):316–22. https://doi.org/10.1016/j.jcmgh.2017.02.002.
99. Bagchi D, McGinn TR, Ye X, Bagchi M, Krohn RL, Chatterjee A, et al. *H. pylori*-induced oxidative stress and DNA damage in a primary culture of human gastric mucosal cells. Dig Dis Sci. 2002;47(6):1405–12. https://doi.org/10.1023/a:1015399204069.
100. Ding SZ, Minohara Y, Fan XJ, Wang J, Reyes VE, Patel J, et al. *H. pylori* infection induces oxidative stress and programmed cell death in human gastric epithelial cells. Infect Immun. 2007;75(8):4030–9. https://doi.org/10.1128/IAI.00172-07.
101. Tsugawa H, Suzuki H, Saya H, Hatakeyama M, Hirayama T, Hirata K, et al. Reactive oxygen species-induced autophagic degradation of *H. pylori* CagA is specifically suppressed in cancer stem-like cells. Cell Host Microbe. 2012;12(6):764–77. https://doi.org/10.1016/j.chom.2012.10.014.
102. Hofseth LJ, Khan MA, Ambrose M, Nikolayeva O, Xu-Welliver M, Kartalou M, et al. The adaptive imbalance in base excision-repair enzymes generates microsatellite instability in chronic inflammation. J Clin Invest. 2003;112(12):1887–94. https://doi.org/10.1172/JCI19757.
103. den Hartog G, Chattopadhyay R, Ablack A, Hall EH, Butcher LD, Bhattacharyya A, et al. Regulation of Rac1 and reactive oxygen species production in response to infection of gastrointestinal epithelia. PLoS Pathog. 2016;12(1):e1005382. https://doi.org/10.1371/journal.ppat.1005382.
104. Kim JM, Kim JS, Lee JY, Kim YJ, Youn HJ, Kim IY, et al. Vacuolating cytotoxin in *H. pylori* water-soluble proteins upregulates chemokine expression in human eosinophils via Ca2+ influx, mitochondrial reactive oxygen intermediates, and NF-kappaB activation. Infect Immun. 2007;75(7):3373–81. https://doi.org/10.1128/IAI.01940-06.
105. Handa O, Naito Y, Yoshikawa T. Redox biology and gastric carcinogenesis: the role of *H. pylori*. Redox Rep. 2011;16(1):1–7. https://doi.org/10.1179/174329211X12968219310756.
106. Chaturvedi R, de Sablet T, Peek RM, Wilson KT. Spermine oxidase, a polyamine catabolic enzyme that links *H. pylori* CagA and gastric cancer risk. Gut Microbes. 2012;3(1):48–56. https://doi.org/10.4161/gmic.19345.
107. Srinivas US, Tan BWQ, Vellayappan BA, Jeyasekharan AD. ROS and the DNA damage response in cancer. Redox Biol. 2019;25:101084. https://doi.org/10.1016/j.redox.2018.101084.
108. Touati E. When bacteria become mutagenic and carcinogenic: lessons from *H. pylori*. Mutat Res. 2010;703(1):66–70. https://doi.org/10.1016/j.mrgentox.2010.07.014.

109. Koeppel M, Garcia-Alcalde F, Glowinski F, Schlaermann P, Meyer TF. *H. pylori* infection causes characteristic DNA damage patterns in human cells. Cell Rep. 2015;11(11):1703–13. https://doi.org/10.1016/j.celrep.2015.05.030.
110. Dorer MS, Sessler TH, Salama NR. Recombination and DNA repair in *H. pylori*. Ann Rev Microbiol. 2011;65:329–48. https://doi.org/10.1146/annurev-micro-090110-102931.
111. Farinati F, Cardin R, Bortolami M, Nitti D, Basso D, de Bernard M, et al. Oxidative DNA damage in gastric cancer: CagA status and OGG1 gene polymorphism. Int J Cancer. 2008;123(1):51–5. https://doi.org/10.1002/ijc.23473.
112. Chaturvedi R, Asim M, Romero-Gallo J, Barry DP, Hoge S, de Sablet T, et al. Spermine oxidase mediates the gastric cancer risk associated with *H. pylori* CagA. Gastroenterology. 2011;141(5):1696. https://doi.org/10.1053/j.gastro.2011.07.045.
113. McNamara KM, Sierra JC, Latour YL, Hawkins CV, Asim M, Williams KJ, et al. Spermine oxidase promotes *H. pylori*-mediated gastric carcinogenesis through acrolein production. Oncogene. 2025;44(5):296–306. https://doi.org/10.1038/s41388-024-03218-7.
114. Handa O, Naito Y, Yoshikawa T. *H. pylori*: a ROS-inducing bacterial species in the stomach. Inflamm Res. 2010;59(12):997–1003. https://doi.org/10.1007/s00011-010-0245-x.
115. Woodward M, Tunstall-Pedoe H, McColl K. *H. pylori* infection reduces systemic availability of dietary vitamin C. Eur J Gastroenterol Hepatol. 2001;13(3):233–7. https://doi.org/10.1097/00042737-200103000-00003.
116. Moloney JN, Cotter TG. ROS signalling in the biology of cancer. Semin Cell Dev Biol. 2018;80:50–64. https://doi.org/10.1016/j.semcdb.2017.05.023.
117. Naito Y, Yoshikawa T. Molecular and cellular mechanisms involved in *H. pylori*-induced inflammation and oxidative stress. Free Radic Biol Med. 2002;33(3):323–36. https://doi.org/10.1016/s0891-5849(02)00868-7.
118. Mimuro H, Suzuki T, Tanaka J, Asahi M, Haas R, Sasakawa C. Grb2 is a key mediator of *H. pylori* CagA protein activities. Mol Cell. 2002;10(4):745–55. https://doi.org/10.1016/s1097-2765(02)00681-0.
119. Bayraktar N, Guler I, Bayraktar M, Koyuncu I. Investigation the levels of endotoxin and 8-hydroxy-2′-deoxyguanosine in sera of patients with *H. pylori*-positive peptic ulcer. Int J Clin Pract. 2021;75(10):e14501. https://doi.org/10.1111/ijcp.14501.
120. Toller IM, Neelsen KJ, Steger M, Hartung ML, Hottiger MO, Stucki M, et al. Carcinogenic bacterial pathogen *H. pylori* triggers DNA double-strand breaks and a DNA damage response in its host cells. Proc Natl Acad Sci USA. 2011;108(36):14944–9. https://doi.org/10.1073/pnas.1100959108.
121. Zhang X, Liu F, Bao H, Wang A, Han M, Wu X, et al. Distinct genomic profile in *h. pylori*-associated gastric cancer. Cancer Med. 2021;10(7):2461–9. https://doi.org/10.1002/cam4.3765.
122. Figura N, Marano L, Moretti E, Ponzetto A. *H. pylori* infection and gastric carcinoma: not all the strains and patients are alike. World J Gastrointest Oncol. 2016;8(1):40–54. https://doi.org/10.4251/wjgo.v8.i1.40.
123. Imai S, Ooki T, Murata-Kamiya N, Komura D, Tahmina K, Wu W, et al. *H. pylori* CagA elicits BRCAness to induce genome instability that may underlie bacterial gastric carcinogenesis. Cell Host Microbe. 2021;29(6):941–58. https://doi.org/10.1016/j.chom.2021.04.006.
124. Kodaman N, Sobota RS, Mera R, Schneider BG, Williams SM. Disrupted human-pathogen co-evolution: a model for disease. Front Genet. 2014;5:290. https://doi.org/10.3389/fgene.2014.00290.
125. Perryman SV, Sylvester KG. Repair and regeneration: opportunities for carcinogenesis from tissue stem cells. J Cell Mol Med. 2006;10(2):292–308. https://doi.org/10.1111/j.1582-4934.2006.tb00400.x.
126. Noto JM, Piazuelo MB, Shah SC, Romero-Gallo J, Hart JL, Di C, et al. Iron deficiency linked to altered bile acid metabolism promotes *H. pylori*-induced inflammation-driven gastric carcinogenesis. J Clin Invest. 2022;132(10) https://doi.org/10.1172/JCI147822.

127. Viala J, Chaput C, Boneca IG, Cardona A, Girardin SE, Moran AP, et al. Nod1 responds to peptidoglycan delivered by the *H. pylori* cag pathogenicity Island. Nat Immunol. 2004;5(11):1166–74. https://doi.org/10.1038/ni1131.
128. Teng Y, Xie R, Xu J, Wang P, Chen W, Shan Z, et al. Tubulointerstitial nephritis antigen-like 1 is a novel matricellular protein that promotes gastric bacterial colonization and gastritis in the setting of *H. pylori* infection. Cell Mol Immunol. 2023;20(8):924–40. https://doi.org/10.1038/s41423-023-01055-4.
129. Salvatori S, Marafini I, Laudisi F, Monteleone G, Stolfi C. *H. pylori* and gastric cancer: pathogenetic mechanisms. Int J Mol Sci. 2023;24(3) https://doi.org/10.3390/ijms24032895.
130. Prichard A, Khuu L, Whitmore LC, Irimia D, Allen LH. *H. pylori*-infected human neutrophils exhibit impaired chemotaxis and a uropod retraction defect. Front Immunol. 2022;13:1038349. https://doi.org/10.3389/fimmu.2022.1038349.
131. Peng R, Xu C, Zhang L, Liu X, Peng D, Chen X, et al. M2 macrophages participate in ILC2 activation induced by *H. pylori* infection. Gut Microbes. 2024;16(1):2347025. https://doi.org/10.1080/19490976.2024.2347025.
132. Sit WY, Cheng ML, Chen TJ, Chen CJ, Chen BN, Huang DJ, et al. *H. pylori* PldA modulates TNFR1-mediated p38 signaling pathways to regulate macrophage responses for its survival. Gut Microbes. 2024;16(1):2409924. https://doi.org/10.1080/19490976.2024.2409924.
133. Neuper T, Frauenlob T, Dang HH, Krenn PW, Posselt G, Regl C, et al. ADP-heptose attenuates *H. pylori*-induced dendritic cell activation. Gut Microbes. 2024;16(1):2402543. https://doi.org/10.1080/19490976.2024.2402543.
134. Koch MRA, Gong R, Friedrich V, Engelsberger V, Kretschmer L, Wanisch A, et al. CagA-specific gastric CD8(+) tissue-resident T cells control *H. pylori* during the early infection phase. Gastroenterology. 2023;164(4):550–66. https://doi.org/10.1053/j.gastro.2022.12.016.
135. Li B, Rong Q, Du Y, Zhang R, Li J, Tong X, et al. Regulation of beta1-integrin in autophagy and apoptosis of gastric epithelial cells infected with *H. pylori*. World J Microbiol Biotechnol. 2021;38(1):12. https://doi.org/10.1007/s11274-021-03199-9.
136. Dianov GL, Hubscher U. Mammalian base excision repair: the forgotten archangel. Nucleic Acids Res. 2013;41(6):3483–90. https://doi.org/10.1093/nar/gkt076.
137. Tang PC, Chan MK, Chung JY, Chan AS, Zhang D, Li C, et al. Hematopoietic transcription factor RUNX1 is essential for promoting macrophage-Myofibroblast transition in non-small-cell lung carcinoma. Adv Sci (Weinh). 2024;11(1):e2302203. https://doi.org/10.1002/advs.202302203.
138. Lian S, Li S, Zhu J, Xia Y, Do Jung Y. Nicotine stimulates IL-8 expression via ROS/NF-kappaB and ROS/MAPK/AP-1 axis in human gastric cancer cells. Toxicology. 2022;466:153062. https://doi.org/10.1016/j.tox.2021.153062.
139. Zhao M, Wang Y, Li L, Liu S, Wang C, Yuan Y, et al. Mitochondrial ROS promote mitochondrial dysfunction and inflammation in ischemic acute kidney injury by disrupting TFAM-mediated mtDNA maintenance. Theranostics. 2021;11(4):1845–63. https://doi.org/10.7150/thno.50905.
140. Peek RM Jr, Blaser MJ, Mays DJ, Forsyth MH, Cover TL, Song SY, et al. *H. pylori* strain-specific genotypes and modulation of the gastric epithelial cell cycle. Cancer Res. 1999;59(24):6124–31.
141. Palrasu M, Zaika E, Paulrasu K, Caspa Gokulan R, Suarez G, Que J, et al. *H. pylori* pathogen inhibits cellular responses to oncogenic stress and apoptosis. PLoS Pathog. 2022;18(6):e1010628. https://doi.org/10.1371/journal.ppat.1010628.
142. Oldani A, Cormont M, Hofman V, Chiozzi V, Oregioni O, Canonici A, et al. *H. pylori* counteracts the apoptotic action of its VacA toxin by injecting the CagA protein into gastric epithelial cells. PLoS Pathog. 2009;5(10):e1000603. https://doi.org/10.1371/journal.ppat.1000603.
143. Gonciarz W, Krupa A, Hinc K, Obuchowski M, Moran AP, Gajewski A, et al. The effect of *H. pylori* infection and different *H. pylori* components on the proliferation and apoptosis of gastric epithelial cells and fibroblasts. PLoS One. 2019;14(8):e0220636. https://doi.org/10.1371/journal.pone.0220636.

144. Park M, Bang C, Yun WS, Jeong YM. Low-molecular-weight fucoidan inhibits the proliferation of melanoma via Bcl-2 phosphorylation and PTEN/AKT pathway. Oncol Res. 2023;32(2):273–82. https://doi.org/10.32604/or.2023.044362.
145. Zhang H, Fang DC, Wang RQ, Yang SM, Liu HF, Luo YH. Effect of *H. pylori* infection on expression of Bcl-2 family members in gastric adenocarcinoma. World J Gastroenterol. 2004;10(2):227–30. https://doi.org/10.3748/wjg.v10.i2.227.
146. Wei J, Noto J, Zaika E, Romero-Gallo J, Correa P, El-Rifai W, et al. Pathogenic bacterium *H. pylori* alters the expression profile of p53 protein isoforms and p53 response to cellular stresses. Proc Natl Acad Sci USA. 2012;109(38):E2543–50. https://doi.org/10.1073/pnas.1205664109.
147. Yanai A, Hirata Y, Mitsuno Y, Maeda S, Shibata W, Akanuma M, et al. *H. pylori* induces antiapoptosis through buclear factor-kappaB activation. J Infect Dis. 2003;188(11):1741–51. https://doi.org/10.1086/379629.
148. Nakajima A, Wada K, Miki H, Kubota N, Nakajima N, Terauchi Y, et al. Endogenous PPAR gamma mediates anti-inflammatory activity in murine ischemia-reperfusion injury. Gastroenterology. 2001;120(2):460–9. https://doi.org/10.1053/gast.2001.21191.
149. Yang H, Wei B, Hu B. Chronic inflammation and long-lasting changes in the gastric mucosa after *H. pylori* infection involved in gastric cancer. Inflamm Res. 2021;70(10–12):1015–26. https://doi.org/10.1007/s00011-021-01501-x.
150. Lim MCC, Jantaree P, Naumann M. The conundrum of *H. pylori*-associated apoptosis in gastric cancer. Trends Cancer. 2023;9(8):679–90. https://doi.org/10.1016/j.trecan.2023.04.012.
151. Shirin H, Sordillo EM, Oh SH, Yamamoto H, Delohery T, Weinstein IB, et al. *H. pylori* inhibits the G1 to S transition in AGS gastric epithelial cells. Cancer Res. 1999;59(10):2277–81.
152. Li SP, Chen XJ, Sun AH, Zhao JF, Yan J. CagA(+) *H. pylori* induces Akt1 phosphorylation and inhibits transcription of p21(WAF1/CIP1) and p27(KIP1) via PI3K/Akt1 pathway. Biomed Environ Sci. 2010;23(4):273–8. https://doi.org/10.1016/S0895-3988(10)60063-3.
153. Hardbower DM, Peek RM Jr, Wilson KT. At the bench: *H. pylori*, dysregulated host responses, DNA damage, and gastric cancer. J Leukoc Biol. 2014;96(2):201–12. https://doi.org/10.1189/jlb.4BT0214-099R.
154. van der Veen S, Tang CM. The BER necessities: the repair of DNA damage in human-adapted bacterial pathogens. Nat Rev Microbiol. 2015;13(2):83–94. https://doi.org/10.1038/nrmicro3391.
155. Ayub A, Hasan MK, Mahmud Z, Hossain MS, Kabir Y. Dissecting the multifaceted roles of autophagy in cancer initiation, growth, and metastasis: from molecular mechanisms to therapeutic applications. Med Oncol. 2024;41(7):183. https://doi.org/10.1007/s12032-024-02417-2.
156. Huang J, Brumell JH. Bacteria-autophagy interplay: a battle for survival. Nat Rev Microbiol. 2014;12(2):101–14. https://doi.org/10.1038/nrmicro3160.
157. Zhang L, Hu W, Cho CH, Chan FK, Yu J, Fitzgerald JR, et al. Reduced lysosomal clearance of autophagosomes promotes survival and colonization of *H. pylori*. J Pathol. 2018;244(4):432–44. https://doi.org/10.1002/path.5033.
158. Li B, Du Y, He J, Lv X, Liu S, Zhang X, et al. Chloroquine inhibited *H. pylori*-related gastric carcinogenesis by YAP-beta-catenin-autophagy axis. Microb Pathog. 2023;184:106388. https://doi.org/10.1016/j.micpath.2023.106388.
159. Tsugawa H, Mori H, Matsuzaki J, Sato A, Saito Y, Imoto M, et al. CAPZA1 determines the risk of gastric carcinogenesis by inhibiting *H. pylori* CagA-degraded autophagy. Autophagy. 2019;15(2):242–58. https://doi.org/10.1080/15548627.2018.1515530.
160. Capurro MI, Greenfield LK, Prashar A, Xia S, Abdullah M, Wong H, et al. VacA generates a protective intracellular reservoir for *H. pylori* that is eliminated by activation of the lysosomal calcium channel TRPML1. Nat Microbiol. 2019;4(8):1411–23. https://doi.org/10.1038/s41564-019-0441-6.
161. Greenfield LK, Jones NL. Modulation of autophagy by *H. pylori* and its role in gastric carcinogenesis. Trends Microbiol. 2013;21(11):602–12. https://doi.org/10.1016/j.tim.2013.09.004.

162. Li H, Lin J, Cheng S, Chi J, Luo J, Tang Y, et al. Comprehensive analysis of differences in N6-methyladenosine RNA methylomes in *H. pylori* infection. Front Cell Dev Biol. 2023;11:1136096. https://doi.org/10.3389/fcell.2023.1136096.
163. Xi Y, Zhang XL, Luo QX, Gan HN, Liu YS, Shao SH, et al. *H. pylori* regulates stomach diseases by activating cell pathways and DNA methylation of host cells. Front Cell Dev Biol. 2023;11:1187638. https://doi.org/10.3389/fcell.2023.1187638.
164. Ralser A, Dietl A, Jarosch S, Engelsberger V, Wanisch A, Janssen KP, et al. *H. pylori* promotes colorectal carcinogenesis by deregulating intestinal immunity and inducing a mucus-degrading microbiota signature. Gut. 2023;72(7):1258–70. https://doi.org/10.1136/gutjnl-2022-328075.
165. Cao C, Yue S, Lu A, Liang C. Host-gut microbiota metabolic interactions and their role in precision diagnosis and treatment of gastrointestinal cancers. Pharmacol Res. 2024;207:107321. https://doi.org/10.1016/j.phrs.2024.107321.
166. Chen CC, Liou JM, Lee YC, Hong TC, El-Omar EM, Wu MS. The interplay between *H. pylori* and gastrointestinal microbiota. Gut Microbes. 2021;13(1):1–22. https://doi.org/10.1080/19490976.2021.1909459.
167. Qiang L, Hu J, Tian M, Li Y, Ren C, Deng Y, et al. Extracellular vesicles from *H. pylori*-infected cells and *H. pylori* outer membrane vesicles in atherosclerosis. Helicobacter. 2022;27(2):e12877. https://doi.org/10.1111/hel.12877.
168. Ramesh P, Babu S, Ammankallu S, Codi JAK, Prasad TSK, Raju R. *H. pylori* regulated microRNA map of human gastric cells. Helicobacter. 2023;28(1):e12941. https://doi.org/10.1111/hel.12941.
169. Han T, Jing X, Bao J, Zhao L, Zhang A, Miao R, et al. *H. pylori* infection alters repair of DNA double-strand breaks via SNHG17. J Clin Invest. 2020;130(7):3901–18. https://doi.org/10.1172/JCI125581.

# Managing *H. pylori* in the Era of Antibiotic Resistance

# 6

Chuan Yin, Te Shi, and Jing Wang

*Helicobacter Pylori* is recognized as a definite carcinogen by both the World Health Organization and the US Department of Health and Human Services. Eradication of *H. pylori* infection is of critical public health importance for the prevention of gastric cancer and related gastroduodenal diseases. Current international consensus guidelines strongly recommend that all infected individuals without compelling competing factors should receive eradication therapy. Although antibiotics remain the cornerstone of eradication regimens, the growing global challenge of antibiotic resistance is fundamentally undermining this foundation, pushing the clinical management of *H. pylori* into an unprecedented dilemma—characterized by declining efficacy of standard regimens, increasing treatment failure rates, and a potential risk of clinicians facing a scenario of limited therapeutic options.

The relentless rise in antibiotic resistance has become the central challenge in *H. pylori* eradication, driving a profound transformation in clinical strategies. Since the publication of the *Maastricht V/Florence Consensus Report* in 2017, the global landscape of *H. pylori* resistance has become increasingly complex, leading to a continuous decline in the efficacy of traditional standard therapies [1]. In response, the European *Helicobacter* and Microbiota Study Group (EHSG) published the *Maastricht VI/Florence Consensus Report* in 2023 [2]. This consensus represents not merely a version update but a strategic shift in the antibiotic resistance era. It

C. Yin (✉)
Department of Gastroenterology, Changzheng Hospital, Naval Medical University, Shanghai, China
e-mail: chuanyinmd@smmu.edu.cn

Te Shi
Department of Gastroenterology, Chinese People's Liberation Army of Naval Medical Center, Shanghai, China

J. Wang
College of Life Science, Mudanjiang Medical University, Mudanjiang, China


J. Gao, Y. Du (eds.), *Non-Antibiotic Therapies for Helicobacter pylori*,
https://doi.org/10.1007/978-981-95-9598-3_6

involves a reevaluation of eradication indications, optimization of diagnostic pathways, and emphasizes novel strategies such as high-dose dual therapy, potassium-competitive acid blocker (P-CAB)-based regimens (e.g., vonoprazan), and susceptibility testing-guided personalized treatment, aiming to overcome the limitations of traditional therapies. Concurrently, the consensus highlights secondary issues like dysbiosis of the gut microbiota resulting from widespread antibiotic use. This section will delve into the current challenges and evolving directions of antibiotic therapy, incorporating recent research findings.

## 6.1 Evolution of Mainstream Antibiotic Treatment Regimens

The antibiotic treatment strategy for *H. pylori* has undergone significant evolution over the past decades, primarily driven by the increasing severity of antibiotic resistance. The therapeutic approach is shifting from broad, empirical combination therapy toward more precise and ecologically conscious individualized regimens.

### 6.1.1 Transition from Standard Triple Therapy to Quadruple Therapy

Initially, standard triple therapy consisting of a proton pump inhibitor (PPI) and two antibiotics (e.g., amoxicillin and clarithromycin) was the mainstream regimen, achieving eradication rates of approximately 80% during periods of low resistance [3]. However, with a significant increase in resistance to antibiotics like clarithromycin, the efficacy of this regimen has fallen to unacceptable levels (below 80%) in many regions and it is no longer recommended as first-line treatment. To combat resistance, bismuth-containing quadruple therapy has emerged as an important advancement. This regimen adds bismuth to a PPI and two antibiotics, typically administered over 14 days. Bismuth not only directly inhibits *H. pylori* but also forms a protective layer on the gastric mucosa. Studies show that bismuth-containing quadruple therapy, particularly when based on amoxicillin or tetracycline, can achieve eradication rates exceeding 90% in China [4].

### 6.1.2 Exploration of Non-Bismuth Quadruple Therapies

Due to the declining efficacy of standard triple therapy and challenges in bismuth availability in some countries, non-bismuth quadruple therapies have been proposed as alternatives. The Maastricht consensus reports recommend various dosing strategies, including concomitant, sequential, and hybrid therapies [2]. Particularly in areas with low clarithromycin resistance, if bismuth is unavailable, the *Maastricht VI/Florence Consensus Report* recommends concomitant therapy (simultaneous administration of a PPI, amoxicillin, clarithromycin, and a nitroimidazole) as the preferred non-bismuth quadruple therapy [2].

### 6.1.3 Exploration of High-Dose Dual Therapy (HDDT)

In recent years, heightened awareness of the potential drawbacks of quadruple therapy—such as polypharmacy, increased adverse effects, higher cost, and poor patient adherence—has renewed interest in high-dose PPI-amoxicillin dual therapy. Numerous studies over the past decade have shown that this regimen can achieve eradication rates up to 90% in rescue therapy [5]. The *Maastricht VI* consensus significantly elevated the status of HDDT, adding it as a recommended second-line option after failure of initial regimens (e.g., PPI-clarithromycin-amoxicillin triple therapy or non-bismuth quadruple therapy). HDDT can even be considered after failure of first- and second-line treatments in areas with low clarithromycin resistance [2]. These recommendations highlight HDDT's potential as a key strategy against resistance, leveraging potent acid suppression to enhance amoxicillin efficacy while avoiding antibiotics with high resistance rates like clarithromycin and fluoroquinolones. The simplicity of the HDDT regimen favors patient adherence, is more economical, and preliminary studies suggest promising efficacy, potentially positioning it as a new first-line option. However, further multicenter studies are needed to confirm its efficacy across different regions, populations, and conditions.

## 6.2 Major Challenges and Dilemmas in Antibiotic Therapy for *H. pylori*

### 6.2.1 Antibiotic Resistance

Despite the availability of multiple treatment regimens, the global rise in antibiotic resistance against *H. pylori* continues to significantly reduce the efficacy of most therapies. This is closely related to bacterial resistance gene mutations and host factors such as CYP2C19 gene polymorphisms [6].

Antibiotics commonly used for eradication, such as metronidazole and clarithromycin, exhibit high resistance rates in many regions, directly compromising their bactericidal effect. *H. pylori* can develop diverse resistance mechanisms through multi-locus mutations, making eradication particularly difficult. In recent years, dual resistance (e.g., metronidazole-clarithromycin, metronidazole-levofloxacin) and even multidrug resistance have become research hotspots [7]. This phenomenon directly leads to reduced eradication rates with bismuth-containing quadruple therapy, traditional quadruple therapy, and triple therapy containing clarithromycin or levofloxacin, resulting in persistent infection and increased recurrence rates. The spread of resistance is primarily associated with antibiotic misuse and non-standardized treatment, underscoring the critical importance of strict adherence to antibiotic stewardship principles in *H. pylori* management. Clinical management of

*H. pylori* infection typically follows international or national consensus guidelines. With ongoing research, the scope of antibiotic use, along with dosages and treatment duration, continues to be refined. Antibiotic resistance is an inherent bacterial trait, with mechanisms like enzyme production, altered membrane permeability, and target modification serving as bacterial self-defense strategies. Over the past three decades, consensus recommendations have evolved from standard triple therapy to bismuth-containing quadruple therapy, with treatment duration extending from 7 to 10 or even 14 days. Currently, increases in antibiotic dosage and duration are nearing their limits, accompanied by increasingly prominent adverse effects, creating a bottleneck in antibiotic selection. Therefore, the focal point of current *H. pylori* treatment is how to rationally use antibiotics to counteract resistance and genuinely improve eradication rates. From a precision medicine perspective, guiding antibiotic selection based on antibiotic susceptibility testing (AST) or resistance gene detection is the optimal strategy against resistance [8]. When AST is unavailable, empirical therapy should follow these principles: the first treatment should select the regimen with the best expected efficacy, minimal adverse effects, and suitability for the majority of patients; rescue therapy must involve changing antibiotic classes or appropriately extending duration (maximum 14 days); after repeated failures, treatment should adhere to the principles for "difficult-to-treat *H. pylori* infection," involving a treatment pause of 3–6 months to allow bacterial metabolic recovery before re-treatment. A comprehensive "individualized overall assessment" before treatment and the formulation of an "individualized overall treatment" plan are key to successful eradication in these patients [8].

Proton pump inhibitors play a crucial role in *H. pylori* eradication, but their efficacy is significantly influenced by the patient's CYP2C19 genotype, which has become an important factor in refractory infections. CYP2C19 is the primary enzyme metabolizing PPIs, and its polymorphisms lead to significant interindividual variation in PPI metabolism. The Chinese population is predominantly composed of rapid and intermediate metabolizers. In rapid metabolizers, accelerated hepatic metabolism of proton pump inhibitors (PPIs) results in decreased plasma concentrations and attenuated acid-suppressive effects, failing to raise the intragastric pH to the optimal level (>5.5). Inadequate acid suppression hinders *H. pylori* from entering an active growth phase, significantly reducing the antibacterial activity of antibiotics like amoxicillin. To address this, two clinical strategies can be employed: increasing the PPI dose or using agents less affected by CYP2C19 (e.g., rabeprazole, esomeprazole); or directly opting for novel potassium-competitive acid blockers (P-CABs, e.g., vonoprazan), which provide stronger acid inhibition independent of CYP2C19 genotype. For refractory infections, genotype-guided personalized dosing is recommended to improve eradication success.

### 6.2.2 Impact on Gut Microbiota

The gut microbial community, particularly the dominant Bacteroidetes and Firmicutes phyla, establishes a mutually beneficial symbiotic relationship with the host, primarily through three functional networks: digestive metabolism, immune regulation, and barrier defense [9–11]. Current antibiotic-based *H. pylori* eradication regimens, especially those involving broad-spectrum antibiotics, indiscriminately impact the host's normal gut microbiota while targeting the pathogen, leading to dysbiosis.

Studies show that *H. pylori* infection significantly alters the gastric microbiome, characterized by reduced gastric microbial diversity and dominance of *H. pylori*. Successful eradication has been shown to significantly increase gastric microbial alpha-diversity, potentially restoring it to levels comparable to uninfected controls, indicating that eradication therapy can partially reverse the ecological imbalance caused by *H. pylori* colonization [12, 13]. Regarding the short-term impact of eradication therapy on the gut microbiota, different regimens consistently cause disturbance. Multiple studies note a significant decline in gut microbial diversity shortly after treatment completion (e.g., at 2 weeks post-therapy). This effect is universal, observed with traditional triple therapy, bismuth-containing quadruple therapy, and concomitant therapy [14, 15]. However, the extent of disruption and the recovery rate vary among regimens. Some research found that gut microbial diversity showed signs of recovery at 2 months post-therapy with triple therapy, whereas no significant recovery was observed with bismuth-containing quadruple therapy or concomitant therapy at the same time point [15]. Further studies suggest that the number and intensity of antibiotics used are key factors [16, 17]. For instance, comparing vonoprazan-based triple therapy with dual therapy revealed that alpha- and beta-diversity were significantly reduced in the triple therapy group at 1 and 8 weeks posttreatment compared to baseline, while the dual therapy group (typically involving fewer antibiotic types or different dosing strategies) showed no statistically significant difference from baseline [17]. This indicates a positive correlation between the level of antibiotic exposure in a treatment regimen and the degree of short-term disruption to the gut microbiota—the more antibiotics used and the more complex the regimen, the greater the potential short-term negative impact on microbial diversity and the longer the recovery time. In summary, while the short-term impact of *H. pylori* eradication on the gut microbiota is clear, it varies by regimen. The general trend suggests that regimens with a lower antibiotic burden (e.g., dual therapy) cause relatively less disruption, potentially allowing for quicker microbial recovery. Future high-quality research is needed to explore the specific mechanisms by which different eradication regimens affect the gut microbiota, providing a more robust scientific basis for optimizing clinical protocols that maximize eradication efficacy while minimizing negative impacts on the host microbiome.

Current evidence suggests that the long-term impact of *H. pylori* eradication on the gut microbiota is generally manageable, but the effects differ among regimens, and the underlying mechanisms require further investigation. Most studies with follow-up periods of at least 6 months indicate that the overall diversity (alpha and beta diversity) of the gut microbial community can recover to baseline levels post-eradication. However, differences may persist in the relative abundance at the genus level, suggesting subtle alterations in the microbial community structure [14, 16, 18]. One study observed that at 6 months post-eradication, the relative abundance of

Firmicutes and Actinobacteria increased, while Bacteroidetes and Proteobacteria decreased, with the composition becoming similar to that of healthy controls [12]. The long-term effects vary by regimen. Studies suggest that triple therapy has a relatively smaller long-term impact on the gut microbiota, with diversity largely restored by 1 year posttreatment. In contrast, the gut microbiota diversity in concomitant therapy and bismuth-containing quadruple therapy groups may not fully recover at the same time point [15]. Dual therapy, which uses fewer antibiotics, also appears to cause less disruption. However, the long-term impact mechanisms of various regimens and their clinical significance require further elucidation through more high-quality studies.

It is important to note that for individuals with immature or unstable gastrointestinal microbiota, such as infants, the elderly, and immunocompromised patients, *H. pylori* eradication may pose additional risks. Therefore, a careful benefit–risk assessment and individualized management are crucial before initiating eradication therapy in these populations. Future research needs to further clarify the long-term impact mechanisms of various eradication regimens on the gut microbiota and provide more evidence for optimizing clinical treatment strategies.

### 6.2.3 Common Adverse Events in *H. pylori* Eradication Therapy

Patients undergoing *H. pylori* eradication may experience various drug-related adverse events. Different drug classes cause distinct types of adverse reactions. Antibiotic-related adverse events are most common, with gastrointestinal reactions being prominent. For example, metronidazole and clarithromycin often cause nausea and vomiting; amoxicillin can lead to diarrhea, sometimes severe enough to trigger pseudomembranous colitis; tetracycline commonly causes local irritant effects like glossitis and anal itching [19]. Some antibiotics cause specific symptoms; for instance, clarithromycin and metronidazole, secreted in saliva, can lead to dysgeusia (metallic taste) or taste disturbance [20]. Allergic reactions are another concern. Amoxicillin can cause rash, pruritus, or even anaphylactic shock in penicillin-allergic individuals. Metronidazole, clarithromycin, etc., may cause neurological symptoms like headache and dizziness. Bismuth is generally safe for short-term use but may cause black stools or tongue staining, which usually resolve after discontinuation [21]. PPIs are generally well-tolerated, with occasional headaches or diarrhea; however, long-term high-dose use may increase fracture risk and potentially interfere with vitamin B12 and iron absorption [22]. These adverse events are the primary reason for patient self-discontinuation or missed doses. The occurrence of diarrhea, nausea, or other significant discomfort drastically reduces willingness to continue medication. To manage these effects and improve adherence, strategies include rigorously inquiring about allergy history and clearly explaining medication risks; for patients with significant gastrointestinal symptoms, dietary adjustments or specific probiotic supplements under medical guidance may help. Additionally, selecting antibiotic combinations with low resistance rates and relatively minor side effects can help reduce adverse events while ensuring efficacy.

In summary, adverse events associated with *H. pylori* eradication are most commonly gastrointestinal reactions caused by antibiotics, alongside allergies, neurological reactions, and specific issues related to bismuth and PPIs. Rational drug selection, thorough pretreatment assessment, close monitoring during therapy, and necessary supportive interventions can help mitigate the impact of adverse events, improving treatment completion rates and overall eradication success.

## 6.3 Conclusion

The clinical management of *H. pylori* in the era of antibiotic resistance is a systematic endeavor. It requires clinicians to develop strategies that are effective, safe, and promote adherence, based on a deep understanding of local resistance patterns, the characteristics of different regimens, and individual patient factors. Actively embracing precision medicine and novel therapeutic approaches is essential to overcome the current challenges.

## References

1. Malfertheiner P, Megraud F, O'morain CA, et al. Management of *H. pylori* infection-the Maastricht V/Florence consensus report. Gut. 2017;66(1):6–30. https://doi.org/10.1136/gutjnl-2016-312288.
2. Malfertheiner P, Megraud F, Rokkas T, et al. Management of *H. pylori* infection: the Maastricht VI/Florence consensus report. Gut. 2022;71(9):1724–62. https://doi.org/10.1136/gutjnl-2022-327745.
3. Sugano K, Tack J, Kuipers EJ, et al. Kyoto global consensus report on *H. pylori* gastritis. Gut. 2015;64(9):1353–67. https://doi.org/10.1136/gutjnl-2015-309252.
4. Chinese Society of Gastroenterology, H. pylori Group. Chinese national clinical practice guideline on *H. pylori* eradication therapy. Chinese J Digestion. 2022;42(11):745–56. https://doi.org/10.3760/cma.j.cn311367-20220929-00479.
5. Yu L, Luo L, Long X, et al. High-dose PPI-amoxicillin dual therapy with or without bismuth for first-line *H. pylori* therapy: a randomized trial. Helicobacter. 2019;24(4):e12596. https://doi.org/10.1111/hel.12596.
6. Shao Y, Lin Y, Wang B, Miao M, et al. Antibiotic resistance status of *H. pylori* strains isolated from initial eradication patients in Ningbo, China, from 2017 to 2021. Helicobacter. 2022;27(5):e12920. https://doi.org/10.1111/hel.12920.
7. Yao SY, Wang F. New trends in the management of *H. pylori*: synchronously focusing on the new progress in digestive disease week 2023. Chinese J Dig Endosc. 2023;40(12):961–8.
8. Lu H. Individualized *H. pylori* treatment regimen based on antibiotic susceptibility testing. Chinese J Pract Internal Med. 2023;43(4):282–6.
9. Clemente JC, Ursell LK, Parfrey LW, et al. The impact of the gut microbiota on human health: an integrative view. Cell. 2012;148(6):1258–70. https://doi.org/10.1016/j.cell.2012.01.035.
10. Sun M, Wu W, Chen L, et al. Microbiota-derived short-chain fatty acids promote Th1 cell IL-10 production to maintain intestinal homeostasis. Nat Commun. 2018;9(1):3555. https://doi.org/10.1038/s41467-018-05901-2.
11. Sekirov I, Russell SL, Antunes LC, et al. Gut microbiota in health and disease. Physiol Rev. 2010;90(3):859–904. https://doi.org/10.1152/physrev.00045.2009.

12. He C, Peng C, Wang H, Ouyang Y, Zhu Z, et al. The eradication of *H. pylori* restores rather than disturbs the gastrointestinal microbiota in asymptomatic young adults. Helicobacter. 2019 Aug;24(4):e12590. https://doi.org/10.1111/hel.12590.
13. Li TH, Qin Y, Sham PC, Lau KS, Chu KM, et al. Alterations in gastric microbiota after *H. Pylori* eradication and in different histological stages of gastric carcinogenesis. Sci Rep. 2017;7:44935. https://doi.org/10.1038/srep44935.
14. Jakobsson HE, Jernberg C, Andersson AF, Sjölund-Karlsson M, Jansson JK, et al. Short-term antibiotic treatment has differing long-term impacts on the human throat and gut microbiome. PLoS One. 2010;5(3):e9836. https://doi.org/10.1371/journal.pone.0009836.
15. Liou JM, Chen CC, Chang CM, et al. Long-term changes of gut microbiota, antibiotic resistance, and metabolic parameters after *H. pylori* eradication: a multicentre, open-label, randomised trial. Lancet Infect Dis. 2019;19(10):1109–20. https://doi.org/10.1016/S1473-3099(19)30272-5.
16. Hsu PI, Pan CY, Kao JY, et al. Short-term and long-term impacts of *H. pylori* eradication with reverse hybrid therapy on the gut microbiota. J Gastroenterol Hepatol. 2019;34(11):1968–76. https://doi.org/10.1111/jgh.14736.
17. Horii T, Suzuki S, Takano C, et al. Lower impact of vonoprazan-amoxicillin dual therapy on gut microbiota for *H. pylori* eradication. J Gastroenterol Hepatol. 2021;36(12):3314–21. https://doi.org/10.1111/jgh.15572.
18. Yap TW, Gan HM, Lee YP, Leow AH, Azmi AN, et al. *H. pylori* eradication causes perturbation of the human gut microbiome in young adults. PLoS One. 2016;11(3):e0151893. https://doi.org/10.1371/journal.pone.0151893.
19. Malfertheiner P. Compliance, adverse events and antibiotic resistance in *H. pylori* treatment. Scand J Gastroenterol Suppl. 1993;196:34–7. https://doi.org/10.3109/00365529309098341.
20. Hafeez M, Qureshi ZA, Khattak AL, et al. *H. pylori* eradication therapy: still a challenge. Cureus. 2021;13(5):e14872. https://doi.org/10.7759/cureus.14872.
21. Graham DY, Lee SY. How to effectively use bismuth quadruple therapy: the good, the bad, and the ugly. Gastroenterol Clin N Am. 2015;44(3):537–63. https://doi.org/10.1016/j.gtc.2015.05.003.
22. Haastrup PF, Thompson W, Søndergaard J, et al. Side effects of long-term proton pump inhibitor use: a review. Basic Clin Pharmacol Toxicol. 2018;123(2):114–21. https://doi.org/10.1111/bcpt.13023.

# Part II

# Nanomaterials for Non-antibiotic Treatment

# Nonresponsive Nanomaterials

# 7

Min Lu, Huizhen Fan, Chenxi Wang, and Qiang Yu

## 7.1 Introduction

The increasing prevalence of antibiotic resistance in *H. pylori* has stimulated the search for alternative therapeutic strategies that do not rely solely on conventional antimicrobials. Among these, nanotechnology-based approaches have emerged as promising candidates for effective eradication [1]. Over the past two decades, nanomaterials have undergone rapid development, providing versatile platforms for drug delivery and antibacterial therapy [2]. Their nanoscale dimensions confer unique advantages, including enhanced stability of bioactive compounds, improved penetration through biological barriers, and the capacity for surface functionalization to achieve muco-adhesion or targeting [3]. In the context of *H. pylori* infection, nanomaterials offer distinct benefits by protecting therapeutic agents from gastric acid degradation, enabling sustained release at the infection site, and in some cases, exerting direct bactericidal effects [4]. Depending on whether they rely on environmental triggers to exert their activity, nanomaterials for *H. pylori* treatment can be broadly classified into four categories: nonresponsive nanomaterials, externally responsive nanomaterials, internally responsive nanocarriers, and multi-responsive nanocarriers.

Building on the above advances in nanodelivery and stimulus-responsive properties, researchers have proposed a nanomaterial-based "3R" design concept—Remove, Remodel, and Repair—to elevate material-based interventions from simple bactericidal action to integrated regulation and to guide the comprehensive management of *H. pylori* infection [5]. Within this functional framework, "Remove"

M. Lu (✉) · H. Fan · C. Wang · Q. Yu
Department of Orthopaedics, Shanghai Key Laboratory for Prevention and Treatment of Bone and Joint Diseases, Shanghai Institute of Traumatology and Orthopaedics, Ruijin Hospital, Shanghai Jiao Tong University School of Medicine, Shanghai, China
e-mail: lumin111@sjtu.edu.cn



J. Gao, Y. Du (eds.), *Non-Antibiotic Therapies for Helicobacter pylori*,
https://doi.org/10.1007/978-981-95-9598-3_7

focuses on directly eliminating *H. pylori* through mechanisms that bypass conventional antibiotic resistance (e.g., ion interference, reactive oxygen species generation); “Remodel” emphasizes modulation of the local immune response and gastric microenvironment to promote sustained pathogen clearance; and “Repair” targets reconstruction of the gastric mucosal barrier and preservation of the gut microbiota. Notably, the 3R concept serves as an integrative design paradigm for organizing the division of labor and synergistic application of different nanomaterial classes (including both nonresponsive and various stimulus-responsive carriers), rather than as a new material category per se [6]. Within this framework, nonresponsive nanomaterials, owing to their relative structural simplicity, high stability, and suitability for implementing “Remove” while partially contributing to “Repair,” have become one of the earliest and most intensively investigated branches in *H. pylori*-oriented nanotherapeutic strategies (Fig. 7.1).

Nonresponsive nanomaterials have attracted considerable interest due to their inherent physicochemical properties that enable antibacterial activity or efficient drug delivery without the need for external or internal triggers. Unlike stimulus-responsive systems, nonresponsive nanomaterials exert their effects primarily through intrinsic mechanisms, such as direct membrane disruption, generation of reactive oxygen species (ROS), ion release, or protection and sustained delivery of therapeutic agents in the harsh gastric environment. These features make them especially relevant for *H. pylori* therapy, where acidic pH, enzymatic degradation, and mucus barriers represent significant challenges for conventional treatments.

This chapter provides a comprehensive overview of nonresponsive nanomaterials investigated for nonantibiotic *H. pylori* eradication. To facilitate clarity, the discussion is organized according to material categories: lipid-based nanocarriers, polymer-based nanoparticles, metal and metal oxide nanoparticles. Within each section, representative systems, mechanisms of antibacterial action, and preclinical applications are summarized, followed by a discussion of their advantages, limitations, and future perspectives (Fig. 7.2).

## 7.2 Lipid-Based Nanocarriers

Lipid-based nanocarriers are among the most widely investigated nanosystems for antimicrobial therapy, owing to their compositional similarity to biological membranes and their favorable biocompatibility profile. Constructed from natural or synthetic lipids, these nanocarriers are capable of encapsulating both hydrophilic and hydrophobic bioactive compounds, thereby improving solubility, enhancing stability, and protecting labile drugs from gastric acid and enzymatic degradation. Their nanoscale dimensions and tunable surface chemistry allow for muco-adhesive designs that prolong gastric residence time or muco-penetrating modifications that facilitate access to *H. pylori* niches within the mucus layer. Importantly, certain lipid components themselves, such as phospholipids and fatty acids, may exert intrinsic antibacterial activity, further strengthening therapeutic efficacy. When combined with other functional materials, lipid nanocarriers can also provide

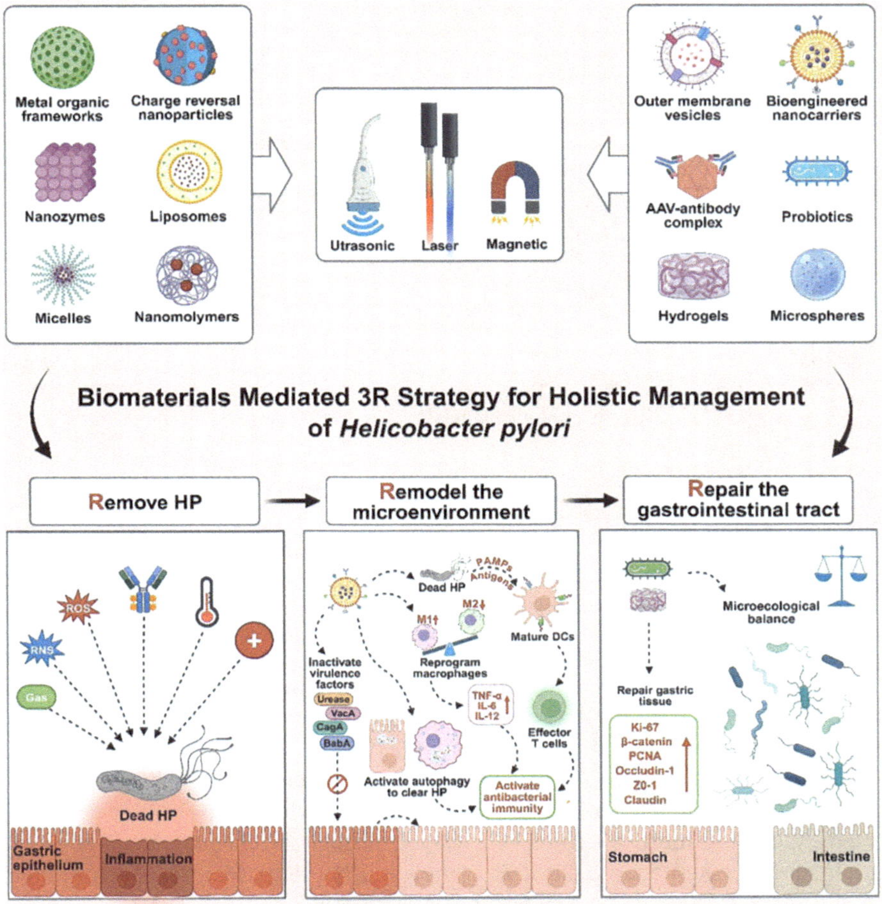


**Fig. 7.1** Biomaterials-mediated 3R strategy for holistic management of *H. pylori*. Various nanomaterials (e.g., metal-organic frameworks, charge-reversal nanoparticles, liposomes) are activated by noninvasive stimuli (ultrasound, laser, magnetic fields) to achieve "Remove" (direct *H. pylori* eradication), "Remodel" (immune microenvironment modulation), and "Repair" (gastric mucosal restoration and gut microbiota protection), providing a patient-centered and sustainable therapeutic paradigm. (Reprinted with permission from Ref. [6], Springer Nature)

site-specific delivery and controlled release, enabling efficient drug utilization while reducing the risk of resistance development.

Beyond passive delivery, lipid-based nanoparticles can actively interact with host and bacterial systems. For instance, liposomes and related carriers have demonstrated the ability to penetrate gastric epithelial cells, macrophages, and bacterial biofilms, thereby targeting both extracellular and intracellular *H. pylori*. This is particularly significant, as intracellular reservoirs and biofilm-associated colonies are major contributors to persistent infection and treatment failure. In this section,

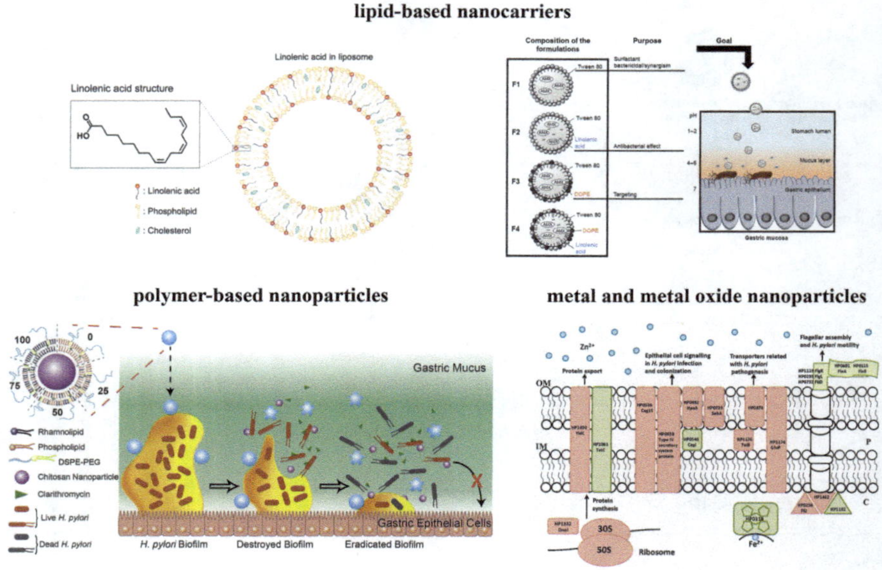


**Fig. 7.2** Design and mechanisms of nonresponsive nanomaterials: Lipid-based nanocarriers, polymer-based nanoparticles, and metal and metal oxide nanoparticles. (Reprinted with permission from Ref. [7], American Chemical Society. Reprinted with permission from Ref. [8], Dove Medical Press. Reprinted with permission from Ref. [9, 10], Elsevier)

lipid-based nanocarriers are categorized into three representative classes—liposomes, solid lipid nanoparticles (SLNs), and nanostructured lipid carriers (NLCs)—reflecting their structural evolution and functional diversification. For each category, design principles, mechanisms of antibacterial action, and preclinical applications in nonantibiotic strategies against *H. pylori* will be discussed, highlighting their stability, drug-transportation capacity, and antimicrobial performance.

## 7.2.1 Liposomes

Liposomes are nanoscale vesicles composed of one or more phospholipid bilayers surrounding an aqueous core. This unique architecture enables them to encapsulate both hydrophilic compounds within the internal aqueous phase and hydrophobic agents within the lipid bilayer, providing a flexible drug delivery platform [11]. Owing to their structural similarity to biological membranes, liposomes exhibit excellent biocompatibility and generally low toxicity, which has made them one of the most extensively investigated nanocarriers in anti-*H. pylori* therapy.

In addition to delivering conventional antibiotics such as amoxicillin, furazolidone, doxycycline, and metronidazole, liposomes have been designed to carry antimicrobial lipids, which are less likely to induce bacterial resistance [12]. A representative system is liposomal linolenic acid (LipoLLA), which incorporates a combination of unsaturated fatty acids to overcome the poor solubility of free linolenic acid. LipoLLA was effective in killing both spiral and coccoid forms of the bacteria via disrupting bacterial membranes. Using a metronidazole-resistant strain of *H. pylori* and seven clinically isolated strains, researchers further demonstrated that LipoLLA eradicated all strains of the bacteria regardless of their antibiotic resistance status [7]. This formulation was shown to eradicate more than 99.9% of *H. pylori* within minutes and to remain effective against dormant coccoid forms, while exhibiting a lower tendency to induce resistance compared with free fatty acids [13]. Other studies have modified liposomes with targeting moieties or stabilizing polymers to improve specificity and performance. For example, pectin-coated liposomes have been engineered to interact with mucins and bacterial adhesion proteins, thereby enhancing bactericidal activity [14], while liposome–polymer hybrids have demonstrated improved stability under acidic gastric conditions and even biofilm-disrupting functions when combined with rhamnolipids [15].

Despite their potential, several obstacles limit the translation of liposomes into clinical practice for *H. pylori* eradication. Conventional liposomes often display suboptimal encapsulation efficiencies for antibiotics, in part due to instability in the gastric environment [15]. Incorporating cholesterol into formulations can improve stability and strengthen bacterial membrane interactions, but excessive cholesterol content disrupts bilayer organization and compromises drug loading and release kinetics [16]. For instance, systematic optimization of cholesterol levels has been shown to markedly affect encapsulation efficiency and release behavior of furazolidone-loaded liposomes [17]. Moreover, the relatively complex and costly manufacturing processes further hinder large-scale production and routine application [18].

Taken together, liposomes combat *H. pylori* by protecting sensitive drugs from gastric degradation, enhancing solubility, facilitating mucosal adhesion or penetration, and in some cases, directly disrupting bacterial membranes. Future efforts should focus on optimizing formulation stability, reducing manufacturing complexity, and integrating multifunctional designs to fully exploit their therapeutic potential in nonantibiotic as well as antibiotic-based strategies.

### 7.2.2 Solid Lipid Nanoparticles

Solid lipid nanoparticles (SLNs) are colloidal carriers characterized by a solid lipid core stabilized by surfactants. Unlike liposomes, which rely on bilayered vesicles, SLNs feature a solid lipid matrix that enhances stability and allows for the encapsulation of hydrophobic compounds. Their use of Generally Recognized as Safe (GRAS) lipids confers high biocompatibility and low toxicity, making them promising candidates for oral delivery in *H. pylori* therapy.

In nonantibiotic strategies, SLNs have been widely employed to encapsulate phytochemicals and other bioactive molecules. Curcumin-loaded SLNs have been successfully prepared with appropriate size, loading capacity, and sustained release properties, making them suitable for oral formulations to enhance bioavailability [19]. Although direct antibacterial assays were not performed, previous evidence of curcumin's activity against *H. pylori* suggests that SLNs could improve its therapeutic potential and may even be combined with antibiotics for synergistic effects. Similarly, catechin- and resveratrol-loaded SLNs displayed enhanced stability in simulated gastric fluid and sustained antibacterial effects over several hours [20]. In addition to phytochemicals, SLNs have been developed for antibiotic encapsulation; for instance, amoxicillin-loaded lipid nanoparticles exhibited uniform particle size (~200 nm), high loading capacity, and long-term stability. They protected amoxicillin from acid degradation, maintained stability under harsh gastric conditions, and achieved localized release near the infection site by exploiting *H. pylori* lipolytic enzymes. Their interaction with mucins further prolonged gastric residence, which is critical for the time-dependent efficacy of amoxicillin. Moreover, the combined presence of amoxicillin, surfactants, and fatty acids reduced the likelihood of bacterial resistance, while cytotoxicity studies confirmed good compatibility with gastric cells [8].

Innovative applications have further expanded the scope of SLNs. For example, cationic SLNs incorporating the adjuvant monophosphoryl lipid A (MPL-A) were developed as carriers for a DNA vaccine encoding the *H. pylori* urease alpha (UreA) antigen. These SLN-based lipoplexes efficiently bound and protected DNA from nuclease degradation, promoted rapid cellular uptake, and localized within endosomes before enabling antigen expression in murine immune cells. Importantly, they also stimulated TNF-$\alpha$ secretion in THP-1 macrophage-like cells, confirming both immunogenicity and biocompatibility. Compared with conventional liposomal vectors, the solid lipid matrix offered higher stability and encapsulation efficiency, underscoring the potential of SLNs as versatile platforms not only for drug delivery but also for vaccine and immunotherapy applications [21].

Despite these promising outcomes, SLNs face inherent limitations. Their rigid crystalline structure can restrict drug loading and may lead to drug expulsion during long-term storage. Furthermore, their interaction with gastric mucus is weaker compared with muco-adhesive polymeric systems, potentially limiting residence time at infection sites. To address these drawbacks, nanostructured lipid carriers have been developed as the next generation of lipid-based nanocarriers.

### 7.2.3 Nanostructured Lipid Carriers

Nanostructured lipid carriers (NLCs) were designed to overcome the intrinsic limitations of SLNs. By combining solid and liquid lipids, NLCs form a less-ordered lipid matrix that increases drug-loading capacity, prevents drug expulsion during storage, and allows for more flexible and sustained drug release. This structural modification makes NLCs highly attractive for delivering bioactives in the harsh gastric environment encountered during *H. pylori* eradication.

Compared with SLNs, NLCs generally provide superior encapsulation efficiency, enhanced stability, and more controlled drug release. For instance, curcumin-loaded NLCs achieved encapsulation efficiencies of around 80% and maintained improved stability under physiological conditions, while resveratrol-loaded NLCs exhibited sustained release in simulated gastric fluid for up to 24 h, resulting in prolonged antibacterial activity in vitro [22, 23]. Co-delivery strategies have further demonstrated the therapeutic potential of NLCs. Hesperidin and clarithromycin co-loaded NLCs showed synergistic antibacterial effects against *H. pylori*. Mechanistic studies revealed that these NLCs adhered to bacterial membranes, disrupted membrane integrity, and caused cytoplasmic leakage, thereby enhancing bacterial eradication [24]. Similarly, docosahexaenoic acid (DHA)-loaded NLCs (~300 nm, ~66% drug loading) directly interacted with *H. pylori* membranes, inducing expansion of the periplasmic space and cytoplasmic leakage, while remaining nontoxic to gastric epithelial cells at bactericidal concentrations [25]. Notably, the nanoencapsulation of DHA allowed effective bacterial inhibition at much lower concentrations than free DHA, highlighting the efficiency of lipid-based carriers in improving bioactive molecule delivery. Taken together, these studies emphasize that NLCs not only enhance drug stability and prolong antibacterial activity but also enable targeted interaction with bacterial membranes. By combining antibiotics or natural bioactive compounds with the advantages of nanoscale delivery, NLCs present a versatile platform for more effective eradication of *H. pylori* while minimizing potential toxicity to host cells.

Surface engineering has further expanded the utility of NLCs. Coatings with muco-adhesive polymers such as chitosan improve gastric retention, while other modifications enhance mucus penetration, thereby ensuring closer contact with bacterial colonies. These properties, combined with controlled and prolonged drug release, position NLCs as versatile platforms to address the challenges of *H. pylori* eradication. Nevertheless, NLCs remain at the preclinical stage. Issues such as formulation complexity, variability in lipid composition, and the need for long-term biosafety evaluation continue to limit clinical translation. Despite these challenges, the superior drug-loading ability, stability, and multifunctional design potential of NLCs suggest they represent a promising next-generation lipid-based nanocarrier for nonantibiotic as well as antibiotic strategies against *H. pylori*.

## 7.3 Polymer-Based Nanoparticles

Polymer-based nanoparticles are a diverse class of nanocarriers constructed from natural or synthetic polymers, offering tunable physicochemical properties for drug delivery in the gastric environment. Natural polymers, such as chitosan, provide intrinsic antibacterial activity and strong muco-adhesive properties, enabling prolonged gastric residence and improved contact with *H. pylori* colonization sites. Synthetic polymers, including PLGA and PEGylated systems, can be engineered to achieve controlled drug release, enhanced stability, and targeted delivery. Polymer-based systems are particularly valuable for encapsulating nonantibiotic agents—such as antimicrobial peptides, phytochemicals, and nanozymes—protecting them from degradation while ensuring sustained therapeutic activity. This section will explore chitosan-based systems, synthetic polymer formulations, and muco-adhesive or muco-penetrating designs tailored for *H. pylori* eradication.

### 7.3.1 Chitosan-Based Systems

Chitosan is a naturally occurring cationic polysaccharide obtained by deacetylation of chitin, the second most abundant polysaccharide after cellulose and a major component of crustacean shells and fungal cell walls. It is composed of D-glucosamine and N-acetyl-D-glucosamine units, with its biological activity largely determined by molecular weight and degree of deacetylation (DD). Higher DD increases the number of free amine groups, enhancing chitosan's positive charge and facilitating chemical modification [26].

Chitosan has drawn considerable interest in *H. pylori* therapy due to two main properties: muco-adhesion and intrinsic antibacterial activity. In acidic gastric conditions, protonated amino groups strongly interact with negatively charged mucins, prolonging gastric residence and enabling closer contact with infection sites [27]. At the same time, electrostatic binding between chitosan and the anionic bacterial cell wall disrupts membrane integrity and inhibits bacterial proliferation. Notably, chitosan with higher DD (95%) showed stronger inhibition of *H. pylori* strains compared with chitosan of lower DD (88.5%), and in vivo administration of chitosan nanoparticles eradicated infection in up to 75% of mice, compared with 55% for chitosan solution [28].

As a versatile nanocarrier, chitosan has been employed to encapsulate antibiotics, natural polyphenols, and antimicrobial peptides. Such nanoparticles generally range from 100 to 200 nm in size and possess positive zeta potentials of +20 to +40 mV, properties that facilitate both muco-adhesion and muco-penetration. For instance, curcumin-loaded alginate beads have been designed as a gastric-targeted delivery system, exhibiting high encapsulation efficiency (~91%) and sustained release in simulated gastric fluid, with approximately 75% of curcumin released within 12 h. These beads also showed prolonged gastric residence and buoyancy for more than 8 h, indicating improved therapeutic potential against *H. pylori*-associated gastritis through localized and sustained drug delivery [29]. In another strategy, Aloe vera gel was incorporated into chitosan nanoparticles (CSNPs), which significantly enhanced its anti-*H. pylori* activity compared with Aloe vera extract alone. The CSNPs, with a mean size of ~36 nm, generated larger inhibition zones (28–30 mm) against both resistant and sensitive *H. pylori* strains and achieved lower MIC and MBC values [30]. Collectively, these studies demonstrate that natural product–based nanocarriers, including alginate beads and chitosan nanoparticles, can provide effective and sustained antibacterial activity under gastric conditions.

Surface modifications such as quaternized or thiolated derivatives further improve solubility and drug encapsulation efficiency. Targeted chitosan-based systems have also been developed. Ligand conjugation—such as lectins, fucose, or carbohydrate epitopes—can guide nanoparticles to bacterial adhesins. Fucose-modified chitosan nanoparticles carrying triple therapy drugs achieved eradication rates of 97% in vitro, outperforming both non-functionalized nanoparticles (91%) and free therapy (81%). In *H. pylori*-infected mice, this system completely eradicated infection, demonstrating the potential of glycan-targeted chitosan systems for precision treatment [31]. In addition, chitosan microspheres have been explored as

"*H. pylori*-binding systems." Cross-linked microspheres adhered strongly to bacteria at acidic pH, leading to 76–93% bacterial death depending on conditions. In vivo studies showed that these microspheres could bind to and remove adherent bacteria during mucus turnover, providing a nonantibiotic antibacterial strategy [32, 33]. Advanced platforms now leverage chitosan within complex, antibiotic-free architectures to achieve synergistic eradication. A representative example is the AP@CS@Lip@HKUST-1 platform, which integrates a chitosan shell with lipid layers and a copper-based metal–organic framework core [34]. The chitosan outer layer enables electrostatic targeting to inflammatory sites, where MMP-triggered release of nanoparticles disrupts bacterial membranes and urease activity. This multimodal system further employs rhamnolipids for biofilm dispersion and phosphatidic acid to promote autophagy for clearing intracellular infections. Notably, the delayed release of $Cu^{2+}$ facilitates mucosal repair while preserving the gut microbiota. Such integration highlights chitosan's evolution from a simple carrier to a crucial structural element in responsive, multi-target systems capable of overcoming antibiotic resistance and promoting gastric health.

Despite these advantages, several limitations remain. Native chitosan shows poor solubility at neutral pH, and excessive cationic charge may induce cytotoxicity at high concentrations. Batch-to-batch variability also affects reproducibility. Future directions should prioritize standardized preparation methods, rational structural modifications, and comprehensive evaluation of biosafety. Overall, chitosan-based nanocarriers represent a promising class of nonresponsive nanomaterials, combining inherent antibacterial properties, muco-adhesion, and flexible functionalization for *H. pylori* eradication.

### 7.3.2 Synthetic Polymers

Synthetic polymers have been extensively studied as nanocarrier platforms owing to their tunable physicochemical properties, biodegradability, and versatility in encapsulating both hydrophilic and hydrophobic drugs [35]. Among these, poly(lactic-co-glycolic acid) (PLGA) and polyethylene glycol (PEG) are the most widely used. PLGA nanoparticles provide controlled biodegradation and sustained release of encapsulated agents, while PEGylation improves hydrophilicity, mucus penetration, and systemic stability [36]. These features are especially valuable for gastric delivery, where drugs must be protected from rapid enzymatic and acidic degradation and maintained at effective concentrations at the infection site.

Encapsulation within synthetic polymers is particularly advantageous for poorly water-soluble bioactives. By incorporating such compounds into PLGA or PEG-based nanoparticles, their solubility, stability, and therapeutic performance against *H. pylori* can be significantly improved. For example, PLGA nanoparticles loaded with plant-derived polyphenols exhibited prolonged gastric retention and enhanced antibacterial effects compared with free compounds [36]. PEGylated polymeric nanoparticles, by contrast, demonstrated improved diffusion through the viscoelastic mucus layer, resulting in better drug distribution within the gastric mucosa and higher antibacterial efficacy compared with unmodified carriers. Typical formulations show particle sizes in the range of 100–200 nm, with encapsulation efficiencies >70% and sustained release profiles extending up to 24 h.

Beyond monofunctional systems, lipid–polymer hybrid nanoparticles (LPNs) have been developed to combine the advantages of polymeric cores with lipid shells. One representative design utilized a chitosan nanoparticle core, a rhamnolipid-based lipid shell, and DSPE-PEG2000 surface modification to encapsulate clarithromycin. These LPNs not only eradicated mature *H. pylori* biofilms by reducing biomass, destroying biofilm architecture, and eliminating extracellular polymeric substances, but also inhibited bacterial adhesion and new biofilm formation through the combined action of clarithromycin, chitosan, and rhamnolipids. Importantly, PEGylation enabled rapid penetration through the gastric mucus barrier without strong mucin interactions, ensuring drug delivery to bacteria residing beneath the mucus. Together, these features highlight LPNs as a promising strategy for overcoming both mucus obstruction and biofilm-associated resistance in persistent *H. pylori* infections [10]. Similarly, PEG–polycaprolactone (PEG–PCL)-based nanoclusters incorporating $Zn_{0.3}Fe_{2.7}O_4$ nanoparticles exhibited strong photothermal effects under external heating, which not only killed *H. pylori* but also reduced bacterial adhesion, vacuolization in host cells, and biofilm formation [37]. Interestingly, this treatment also reduced resistance to clarithromycin and levofloxacin, further underscoring the potential of multifunctional polymer-based platforms.

Despite these advances, challenges remain. Preparation methods often involve organic solvents, raising concerns over residual toxicity and safety. Additionally, large-scale production and cost-effectiveness remain obstacles for clinical translation. Future research should focus on solvent-free or green synthesis approaches, scalable manufacturing, and rigorous preclinical safety evaluation to accelerate the clinical application of synthetic polymer nanoparticles.

### 7.3.3 Muco-Adhesive and Muco-Penetrating Polymers

The gastric mucus layer presents a major biological barrier to the eradication of *H. pylori.* Polymer-based nanocarriers have therefore been designed with either muco-adhesive or muco-penetrating properties to optimize drug delivery at the infection site [38].

Muco-adhesive polymers—such as chitosan, carbopol, and their derivatives—bind strongly to gastric mucins through electrostatic interactions, hydrogen bonding, and chain interpenetration. This prolonged residence time allows sustained exposure of bioactive agents to colonization sites. For example, carbopol-loaded amoxicillin nanospheres adhered effectively to gastric mucosa and provided controlled release for up to 12 h, while maintaining long-term stability over 12 months [39]. Another study reported mannose-functionalized chitosan nanoparticles that not only adhered to mucus but also exhibited strong anti-adherence and biofilm-disruption effects against resistant *H. pylori* strains. Advanced imaging (CLSM, TEM, SEM) confirmed that mannose-decorated nanoparticles disrupted bacterial membranes, leading to amino acid leakage and bacterial death [40].

Muco-penetrating polymers are designed to overcome the viscoelastic barrier of mucus and deliver drugs deeper into infection niches. PEGylation and surface charge modulation are common strategies to reduce adhesive interactions with mucins, thereby improving nanoparticle diffusivity. PEGylated nanoparticles showed superior mucus permeability and higher antibacterial efficacy than non-PEGylated carriers, ensuring more uniform drug distribution within the gastric mucosa [36].

Both strategies have demonstrated distinct advantages. Muco-adhesive systems maximize gastric residence and localized concentration, while muco-penetrating carriers facilitate deeper access to bacteria protected within mucus layers and biofilms. In some cases, hybrid designs combining adhesion and penetration offer a balanced approach. For instance, fucose-conjugated chitosan nanoparticles encapsulating berberine effectively adhered to mucus while achieving controlled release in simulated gastrointestinal conditions. In vivo, they cleared *H. pylori* infection in mice more effectively than free berberine [38]. Similarly, lectin-conjugated gliadin nanoparticles loaded with acetohydroxamic acid were designed to exploit carbohydrate-binding interactions with *H. pylori* surface receptors. This targeted approach significantly enhanced bacterial adhesion and uptake, leading to nearly a twofold increase in inhibitory efficacy compared with non-functionalized gliadin nanoparticles. Such receptor-mediated strategies demonstrate the potential of surface engineering to improve both specificity and therapeutic outcomes in *H. pylori* eradication [41]. Gliadin nanoparticles co-loaded with clarithromycin and omeprazole also demonstrated improved eradication efficacy in vitro compared with single-drug formulations [42].

The main challenge lies in balancing adhesion and penetration. Excessive muco-adhesion may hinder diffusion through mucus, while strong penetration might reduce gastric residence. Rational combinations of both features, together with ligand functionalization for bacterial targeting, represent promising directions for the design of next-generation polymer-based nanocarriers for *H. pylori* eradication.

## 7.4 Metal and Metal Oxide Nanoparticles

Metal and metal oxide nanoparticles (NPs) have emerged as a promising class of nonresponsive nanomaterials for *H. pylori* eradication due to their intrinsic antimicrobial activity that does not rely on external or internal triggers. Their unique physicochemical and catalytic properties allow them to exert bactericidal effects directly within the gastric environment. Transition metals such as silver, zinc, copper, and gold, along with heavier elements like bismuth, are essential micronutrients at trace levels but display pronounced toxicity when delivered in higher concentrations, a phenomenon that can be exploited in nanomedicine [43–45]. Interestingly, this concept parallels the innate immune strategy of macrophages, which elevate intracellular concentrations of copper and zinc in phagosomes to eliminate engulfed pathogens [46, 47].

The antibacterial effects of metal- and metal oxide-based NPs are mediated primarily through: (1) Release of metal ions, which disrupt bacterial membranes, denature proteins, and interfere with nucleic acids; (2) Generation of ROS, which induce oxidative stress, protein dysfunction, and DNA fragmentation; (3) Direct structural disruption, including perturbation of lipid bilayers and inhibition of biofilm formation [48, 49]. These multifaceted mechanisms make metal and metal oxide NPs particularly attractive for addressing multidrug-resistant *H. pylori*, as their antibacterial action bypasses conventional resistance pathways.

### 7.4.1 Silver Nanoparticles

Silver nanoparticles (AgNPs) are among the most extensively investigated metallic nanomaterials due to their broad-spectrum antimicrobial properties and well-documented ability to disrupt bacterial physiology. Their antibacterial activity against *H. pylori* primarily involves multiple pathways, including the release of silver ions, direct membrane disruption, inhibition of respiratory enzymes, and induction of oxidative stress through ROS generation [50–52]. The multifaceted mode of action makes AgNPs highly effective, as they simultaneously target bacterial membranes, proteins, and nucleic acids, thereby reducing the likelihood of single-step resistance.

AgNPs can adhere to bacterial membranes, causing pits and permeability changes, and once internalized, they enhance oxidative stress [50]. Several formulations of AgNPs have been developed to improve their stability and enhance their therapeutic potential against *H. pylori*. For instance, Camargo et al. reported that incorporating a silver(I) complex into polymeric nanoparticles significantly potentiated its antibacterial effect, reducing the minimum inhibitory concentration (MIC) from 3.90 μg/mL to 0.781 μg/mL and the minimum bactericidal concentration (MBC) fivefold to 1.56 μg/mL. Importantly, this formulation also reduced cytotoxicity toward mammalian fibroblasts and inhibited biofilm formation more efficiently than the free silver complex, highlighting the benefits of nanocarrier-mediated delivery [51]. Similarly, green-synthesized AgNPs prepared using Artemisia princeps leaf extract exhibited size distributions between 10 and 40 nm and demonstrated potent dose-dependent inhibition of *H. pylori* growth and biofilms, accompanied by ROS generation and DNA fragmentation [52]. Interestingly, these biogenic AgNPs also induced apoptosis in human gastric carcinoma cells while sparing normal lung fibroblasts, suggesting potential dual antibacterial and anticancer properties.

Ultrasmall silver nanoclusters (SUNCs) with diameters below 5 nm have shown particular promise for overcoming biofilm-associated *H. pylori* infections. In vitro studies demonstrated MIC and MBC values in the sub-milligram per liter range, with MBEC assays confirming efficient biofilm eradication. Furthermore, SUNCs exhibited synergistic effects with conventional antibiotics such as clarithromycin and metronidazole, suggesting that they could be employed as adjuvants in combination therapy to restore antibiotic efficacy. Cytotoxicity assays revealed minimal toxicity toward gastric epithelial cells at bactericidal concentrations, further supporting their translational potential [53]. Despite these advances, safety concerns remain a critical barrier to clinical translation. Long-term exposure to AgNPs has been associated with the risk of bacterial adaptation. An experimental evolution study demonstrated that *Escherichia coli* could acquire resistance to citrate-coated AgNPs after only 200 generations of exposure, with genomic analyses identifying several mutations conferring improved fitness in silver-rich environments [54]. These findings caution that, despite the broad-spectrum activity of silver, resistance development is possible and should be carefully monitored. Moreover, prolonged exposure raises concerns of cytotoxicity, as silver accumulation in tissues can disrupt mitochondrial function and induce oxidative stress in host cells.

To mitigate these limitations, innovative strategies are being pursued. These include the development of ultrasmall clusters to enhance bacterial specificity while minimizing host toxicity, the use of polymeric or lipid coatings to control silver ion release, and green synthesis methods that yield biocompatible and stable formulations. Furthermore, combining AgNPs with antibiotics, biofilm-disrupting agents, or natural products may provide synergistic activity and reduce the selective pressure that drives resistance. Overall, while AgNPs remain one of the most promising classes of nonresponsive nanomaterials for *H. pylori* eradication, their long-term safety, potential for resistance, and environmental implications require rigorous preclinical and translational evaluation before clinical application.

### 7.4.2 Gold Nanoparticles

Gold nanoparticles (AuNPs) are generally considered biologically inert and have limited intrinsic antibacterial activity compared with other metal-based nanoparticles. However, their exceptional stability, tunable size and shape, and ease of surface functionalization make them highly attractive as nanoplatforms for biomedical use, including *H. pylori* eradication. A major focus has been the development of green-synthesized or phyto-derived AuNPs, which avoid toxic reducing agents and simultaneously incorporate bioactive phytochemicals, yielding safer and multifunctional formulations. For instance, Tribulus terrestris fruit extract was used to synthesize anisotropic AuNPs with average diameters of 7 nm and 55 nm, both of which exhibited size-dependent antibacterial effects against multidrug-resistant *H. pylori* strains while maintaining excellent biocompatibility in gastric epithelial cells [55]. Similarly, gold-coated iron nanoparticles (Fe@Au) prepared using olive oil and licorice root extracts demonstrated not only strong anti-*H. pylori* activity but also antioxidant and gastroprotective effects, suggesting that eco-friendly core–shell structures may provide dual therapeutic benefits in the gastric environment [56]. Phyto-synthesized AuNPs from *Pituranthos tortuosus* aqueous extracts also displayed broad-spectrum bioactivity, including antidiabetic and anticancer effects, and showed potent anti-*H. pylori* activity against resistant isolates, with MIC values as low as 15.63 μg/mL [57].

These studies underscore the versatility of AuNPs as therapeutic candidates, especially when synthesized via green chemistry approaches that integrate natural compounds with intrinsic antimicrobial or protective functions. In addition to direct antibacterial activity, AuNPs can serve as functional carriers or adjuvants: their surface can be easily conjugated with antibiotics, peptides, or plant-derived molecules to enhance antibacterial efficacy and biofilm disruption. Importantly, biosynthesized AuNPs typically exhibit low cytotoxicity toward mammalian cells at bactericidal concentrations, addressing one of the main concerns associated with heavy metal nanoparticles. Beyond direct therapeutic use, AuNPs have also been widely explored as plasmonic nanomaterials for biosensing and detection of *H. pylori* DNA, owing to their unique optical properties [58]. While such diagnostic applications fall outside the therapeutic scope of this chapter, they illustrate the broad biomedical potential of AuNPs and reinforce their safety and multifunctionality in gastric disease management. Overall, AuNPs represent a promising adjunct or alternative strategy for *H. pylori* eradication, especially when designed as green-synthesized, multifunctional nanocarriers that combine stability, biocompatibility, and the capacity for targeted therapeutic delivery.

Compared with silver nanoparticles, which exhibit strong bactericidal activity but may induce bacterial adaptation under prolonged exposure, AuNPs are relatively less potent in their intrinsic antibacterial effects. However, their excellent biocompatibility, chemical stability, and functionalization capacity make them highly suitable as delivery vehicles or synergistic agents in *H. pylori* therapy. In this regard, AgNPs can be viewed as direct antimicrobial agents, while AuNPs serve more effectively as biocompatible carriers and multifunctional scaffolds. The complementary characteristics of these two noble metal systems highlight the broader potential of metallic nanomaterials in designing safe, effective, and sustainable anti-*H. pylori* strategies.

### 7.4.3 Zinc Oxide Nanoparticles

Zinc oxide nanoparticles (ZnO NPs) have been extensively explored as promising nonresponsive nanomaterials against *H. pylori*, owing to their dual antibacterial mechanisms of zinc ion release and ROS generation. Upon contact with bacterial cells, $Zn^{2+}$ disrupts metabolic pathways and destabilizes membranes, while ROS production further exacerbates oxidative damage, leading to impaired viability. Transcriptomic profiling studies revealed that elevated zinc concentrations significantly repress genes associated with *H. pylori* motility, flagellar assembly, and type IV secretion systems, while simultaneously altering ribosomal biogenesis and metabolic functions, thereby exerting a broad-spectrum inhibitory effect at the genetic level [9]. These findings suggest that ZnO exposure not only induces immediate oxidative and membrane stress but also reprograms bacterial physiology in ways that weaken pathogenicity.

Surface-engineered ZnO NPs have demonstrated enhanced potency against drug-resistant *H. pylori*. For example, PEI-functionalized ZnO NPs (3–7 nm) were shown to internalize efficiently into *H. pylori* cytoplasm without agglomeration, where they caused severe membrane perturbation, rapid intracellular ROS accumulation, and extensive degradation of 16S and 23S rRNA. This resulted in strong bactericidal activity against metronidazole-resistant strains. Importantly, these PEI-ZnO NPs synergized with conventional antibiotics, enabling inhibition of bacterial growth at sub-MIC concentrations while remaining nontoxic to mammalian cells at doses up to 20 μg/mL [59]. Such synergy highlights the potential of ZnO NPs to resensitize resistant strains and extend the clinical lifespan of existing antibiotics.

Green-synthesized ZnO NPs have further advanced their biomedical relevance. Using Quercus infectoria (oak gall) extracts as reducing agents, ultrasmall ZnO NPs (~6–8 nm) were prepared with high stability and characterized by multiple analytical techniques. These biosynthesized nanoparticles exhibited moderate standalone antibacterial activity, but when combined with amoxicillin, they reduced the $MIC_{90}$ of resistant *H. pylori* strains up to fourfold, outperforming plant extract controls. The synergistic effect was attributed to both the phenolic-rich phytochemicals and the unique physicochemical properties of ZnO NPs, suggesting that eco-friendly synthesis routes can yield safe, low-cost formulations for practical therapy [60]. Safety is a crucial consideration for gastric nanomedicine. Biogenic ZnO NPs synthesized via *Bacillus megaterium* extracts showed selective toxicity against *H. pylori* strains while maintaining excellent compatibility with human mesenchymal stem cells (hMSCs), even at bactericidal concentrations. The nanoparticles induced significant bacterial killing without measurable cytotoxicity in mammalian cells, underscoring their potential as nano-antibiotics that balance efficacy and safety [61].

Taken together, ZnO NPs represent one of the most versatile metal oxide systems for *H. pylori* eradication. Their antibacterial activity arises from a combination of physicochemical stressors—ionic, oxidative, and structural—providing multiple simultaneous attack pathways that reduce the likelihood of resistance development. However, challenges remain regarding formulation stability in gastric conditions, optimization of surface coatings for mucus penetration, and long-term biosafety evaluation. Addressing these aspects through controlled synthesis, functionalization strategies, and rigorous preclinical studies will be critical to translating ZnO-based therapies into clinical applications.

### 7.4.4 Bismuth Nanoparticles

Bismuth-based compounds have been central to gastric therapies for decades, forming the cornerstone of quadruple regimens for *H. pylori* eradication. Their therapeutic success is attributed to the unique chemistry of trivalent bismuth ions ($Bi^{3+}$), which readily bind to sulfur-, nitrogen-, and oxygen-containing biomolecules, thereby disrupting key metabolic enzymes such as fumarase and interfering with urease activity—an essential virulence factor for *H. pylori* survival in acidic environments [62, 63]. Recent advances in nanotechnology have extended these applications into nanoparticulate formulations, offering enhanced bioavailability, stability, and antibacterial potency.

For example, biologically synthesized elemental bismuth nanoparticles (BiNPs) exhibited MICs ranging from 60 to 100 μg/mL against multiple clinical *H. pylori* isolates, demonstrating superior antibacterial activity compared with bismuth ions alone [64]. Metabolomic profiling further revealed that BiNP exposure disrupted central metabolic pathways, including the Krebs cycle, amino acid biosynthesis, and nucleotide metabolism, leading to the leakage of metabolites such as acetate, glutamate, and uracil into the extracellular milieu. In another study, bismuth subcarbonate nanoparticles ($BiO_2CO_3$, ~9.2 nm) synthesized *via* a microemulsion-assisted hydrothermal method achieved over 85% growth inhibition at 80 μg/mL, significantly outperforming bismuth nanotubes and showing comparable activity to the clinically used colloidal bismuth subcitrate (CBS) [65]. These findings highlight the potential of BiNPs to match or even exceed the performance of conventional bismuth formulations while reducing required doses.

Beyond simple nanoparticle formulations, innovative designs such as bismuth-based metal–organic frameworks (Bi-MOFs) are emerging as multifunctional platforms. Owing to their high porosity and tunable surface chemistry, Bi-MOFs hold promise for applications in catalysis, drug delivery, and biomedical imaging and could be tailored as next-generation anti-*H. pylori* therapeutics [62]. Mechanistic studies further reveal that bismuth drugs can disrupt multiple bacterial pathways simultaneously, reducing the likelihood of resistance development. For instance, proteomic and system pharmacology analyses have identified glutathione- and transporter-mediated pathways in human cells that allow selective toxicity toward *H. pylori*. Moreover, $Bi^{3+}$ can displace $Zn^{2+}$ cofactors from metallo-β-lactamases, conferring potential as broad-spectrum inhibitors that could be repurposed to combat antibiotic resistance [63].

Despite these advances, translation of bismuth-based nanoparticles into clinical settings remains challenging. Safety concerns include potential gastric mucosal irritation, long-term accumulation in tissues, and variability in nanoparticle dosing or stability. Furthermore, although *H. pylori* has not yet shown widespread adaptation to bismuth therapies, the possibility of resistance emerging with prolonged nanoparticle exposure cannot be excluded, as evidenced in other metal systems such as silver. To overcome these limitations, strategies such as green synthesis, polymeric or biomimetic surface coatings, and rational dose optimization are being investigated to improve both efficacy and safety profiles.

Overall, BiNPs represent a promising class of nonantibiotic nanotherapeutics that build upon the long-established clinical utility of bismuth. By integrating their inherent multi-targeted mechanisms with the advantages of nanoscale formulations, bismuth-based nanoparticles hold strong potential to enhance *H. pylori* eradication outcomes, provided that rigorous preclinical validation and careful safety assessments are performed. Like other metal- and metal oxide–based nanomaterials such as silver, gold, and zinc nanoparticles, BiNPs act through complementary multi-targeted mechanisms that impair bacterial survival and virulence. Their integration into therapeutic regimens highlights the broader promise of metallic nanostructures as effective nonantibiotic alternatives or adjuvants for *H. pylori* eradication.

## 7.5 Challenges and Future Perspectives

Although nonresponsive nanomaterials provide promising strategies for *H. pylori* eradication, several barriers must be addressed before their clinical translation can be realized. The stomach represents a highly hostile and dynamic environment, where strong acidity (pH 1–2 in the lumen), digestive enzymes such as pepsin, and a thick mucus layer with a steep pH gradient collectively compromise the stability of nanocarriers and hinder drug bioavailability [1, 66]. These barriers complicate eradication strategies, increase treatment costs, and significantly reduce patient compliance. Rational design of acid-stable formulations, protective coatings, and mucus-interactive nanoplatforms is therefore critical to improve delivery efficiency in this challenging microenvironment.

Safety remains another key concern. While many lipid- and polymer-based nanomaterials are biodegradable and metabolized into nontoxic byproducts, metal and metal oxide nanoparticles may persist in gastric tissues, releasing ions that induce oxidative stress or disrupt normal gastric microbiota [67]. Long-term toxicological studies, including assessments of systemic distribution and chronic exposure, are essential to ensure safe translation. Equally important is maintaining ecological balance in the gut while achieving selective toxicity against *H. pylori*.

Another major obstacle lies in antibiotic resistance and biofilm-associated infections. Bacteria within biofilms can display resistance levels up to 1000 times higher than planktonic cells [68], greatly undermining therapeutic efficacy. Nonresponsive nanomaterials such as silver, zinc oxide, or chitosan-based systems have shown potential in disrupting biofilm structures and altering bacterial membrane permeability [69]. Yet, ensuring consistent activity across resistant strains and diverse biofilm conditions remains a challenge. Designing nanomaterials that can effectively target intracellular and biofilm-embedded *H. pylori* will be crucial for raising cure rates.

Despite encouraging preclinical results, the clinical translation of nanomaterial-based systems for *H. pylori* remains limited. Barriers include high production costs, challenges in large-scale manufacturing, lack of standardized regulatory frameworks, and limited reproducibility across biological models and patient populations [67]. Standardized evaluation protocols addressing efficacy, safety, and pharmacokinetics are urgently needed to accelerate the transition from bench to bedside.

Looking forward, future research should focus on multifunctional and targeted nonresponsive nanoplatforms that integrate acid protection, muco-adhesion, controlled release, and biofilm disruption [70]. Targeted designs, such as ligand-modified carriers capable of binding directly to *H. pylori* surface receptors, may further enhance localization at colonization sites while reducing off-target toxicity [14]. Antibiotic-independent approaches, including antimicrobial lipids, phytochemicals, and metal nanoparticles, represent additional avenues to bypass resistance. Equally important will be the development of biodegradable, scalable, and cost-effective formulations that facilitate widespread clinical application.

In summary, nonresponsive nanomaterials hold strong potential to overcome the limitations of current *H. pylori* therapies, but achieving clinical translation requires careful optimization of stability, safety, efficacy against resistant and biofilm-associated strains, and scalability. With continued innovation, these platforms may evolve into viable nanomedicine strategies that complement or even replace conventional eradication regimens in the future.

## References


1. Zhang Q, Wu W, Zhang J, Xia X. Eradication of *H. pylori*: the power of nanosized formulations. Nanomedicine (Lond). 2020;15(5):527–42. https://doi.org/10.2217/nnm-2019-0329.
2. Geng Z, Cao Z, Liu J. Recent advances in targeted antibacterial therapy basing on nanomaterials. Exploration (Beijing). 2023;3(1):20210117. https://doi.org/10.1002/EXP.20210117.
3. Lamprecht A. IBD: selective nanoparticle adhesion can enhance colitis therapy. Nat Rev Gastroenterol Hepatol. 2010;7(6):311–2. https://doi.org/10.1038/nrgastro.2010.66.
4. Garg A, Karhana S, Khan MA. Nanomedicine for the eradication of *H. pylori*: recent advances, challenges and future perspective. Future Microbiol. 2024;19:431–47. https://doi.org/10.2217/fmb-2023-0189.
5. Zheng Y, Zhang T, Shao J, Du Y, Li Z, et al. Antibiotic-free responsive biomaterials for specific and targeted *H. pylori* eradication. J Control Release. 2025;379:708–29. https://doi.org/10.1016/j.jconrel.2025.01.054.
6. Zhang T, Zheng Y, Chen T, Gu Y, Gong Y, et al. Biomaterials mediated 3R (remove-remodel-repair) strategy: holistic management of *H. pylori* infection. J Nanobiotechnol. 2025;23(1):475. https://doi.org/10.1186/s12951-025-03455-2.
7. Obonyo M, Zhang L, Thamphiwatana S, Pornpattananangkul D, Fu V, et al. Antibacterial activities of liposomal linolenic acids against antibiotic-resistant *H. pylori*. Mol Pharm. 2012;9(9):2677–85. https://doi.org/10.1021/mp300243w.
8. Lopes-de-Campos D, Pinto RM, Lima SAC, Santos T, Sarmento B, et al. Delivering amoxicillin at the infection site – a rational design through lipid nanoparticles. Int J Nanomedicine. 2019;14:2781–95. https://doi.org/10.2147/IJN.S193992.
9. Fan D, Gong Y, Sun L, Zhang Y, Zhang J. Comparative transcriptome analysis to investigate the mechanism of anti-*H. pylori* activity of zinc. Microb Pathog. 2022;168:105611. https://doi.org/10.1016/j.micpath.2022.105611.
10. Li P, Chen X, Shen Y, Li H, Zou Y, et al. Mucus penetration enhanced lipid polymer nanoparticles improve the eradication rate of *H. pylori* biofilm. J Control Release. 2019;300:52–63. https://doi.org/10.1016/j.jconrel.2019.02.039.
11. Lai Y, Wei W, Du Y, Gao J, Li Z. Biomaterials for *H. pylori* therapy: therapeutic potential and future perspectives. Gut Microbes. 2022;14(1):2120747. https://doi.org/10.1080/19490976.2022.2120747.
12. Bardonnet PL, Faivre V, Boullanger P, Piffaretti JC, Falson F. Pre-formulation of liposomes against *H. pylori*: characterization and interaction with the bacteria. Eur J Pharm Biopharm. 2008;69(3):908–22. https://doi.org/10.1016/j.ejpb.2008.01.018.
13. Thamphiwatana S, Gao W, Obonyo M, Zhang L. In vivo treatment of *H. pylori* infection with liposomal linolenic acid reduces colonization and ameliorates inflammation. Proc Natl Acad Sci USA. 2014;111(49):17600–5. https://doi.org/10.1073/pnas.1418230111.
14. Gottesmann M, Goycoolea FM, Steinbacher T, Menogni T, Hensel A. Smart drug delivery against *H. pylori*: pectin-coated, mucoadhesive liposomes with antiadhesive activity and antibiotic cargo. Appl Microbiol Biotechnol. 2020;104(13):5943–57. https://doi.org/10.1007/s00253-020-10647-3.
15. Cai J, Huang H, Song W, Hu H, Chen J, et al. Preparation and evaluation of lipid polymer nanoparticles for eradicating *H. pylori* biofilm and impairing antibacterial resistance in vitro. Int J Pharm. 2015;495(2):728–37. https://doi.org/10.1016/j.ijpharm.2015.09.055.

16. Plaza-Oliver M, Santander-Ortega MJ, Lozano MV. Current approaches in lipid-based nanocarriers for oral drug delivery. Drug Deliv Transl Res. 2021;11(2):471–97. https://doi.org/10.1007/s13346-021-00908-7.
17. Alam MI, Paget T, Elkordy AA. Formulation and advantages of furazolidone in liposomal drug delivery systems. Eur J Pharm Sci. 2016;84:139–45. https://doi.org/10.1016/j.ejps.2016.01.017.
18. Shah S, Dhawan V, Holm R, Nagarsenker MS, Perrie Y. Liposomes: advancements and innovation in the manufacturing process. Adv Drug Deliv Rev. 2020;154-155:102–22. https://doi.org/10.1016/j.addr.2020.07.002.
19. Jourghanian P, Ghaffari S, Ardjmand M, Haghighat S, Mohammadnejad M. Sustained release curcumin loaded solid lipid nanoparticles. Adv Pharm Bull. 2016;6(1):17–21. https://doi.org/10.15171/apb.2016.04.
20. Shtay R, Keppler JK, Schrader K, Schwarz K. Encapsulation of (—)-epigallocatechin-3-gallate (EGCG) in solid lipid nanoparticles for food applications. J Food Eng. 2019;244:91–100. https://doi.org/10.1016/j.jfoodeng.2018.09.008.
21. Francis JE, Skakic I, Dekiwadia C, Shukla R, Taki AC, et al. Solid lipid nanoparticle carrier platform containing synthetic TLR4 agonist mediates non-viral DNA vaccine delivery. Vaccines (Basel). 2020;8(3) https://doi.org/10.3390/vaccines8030551.
22. Chanburee S, Tiyaboonchai W. Mucoadhesive nanostructured lipid carriers (NLCs) as potential carriers for improving oral delivery of curcumin. Drug Dev Ind Pharm. 2017;43(3):432–40. https://doi.org/10.1080/03639045.2016.1257020.
23. Manikandan S, Jose PA, Karuppaiah A, Rahman H. The effect of physical stability and modified gastrointestinal tract behaviour of resveratrol-loaded NLCs encapsulated alginate beads. Naunyn Schmiedeberg's Arch Pharmacol. 2024;397(11):9007–21. https://doi.org/10.1007/s00210-024-03223-3.
24. Sharaf M, Arif M, Khan S, Abdalla M, Shabana S, et al. Co-delivery of hesperidin and clarithromycin in a nanostructured lipid carrier for the eradication of *H. pylori* in vitro. Bioorg Chem. 2021;112:104896. https://doi.org/10.1016/j.bioorg.2021.104896.
25. Seabra CL, Nunes C, Gomez-Lazaro M, Correia M, Machado JC, et al. Docosahexaenoic acid loaded lipid nanoparticles with bactericidal activity against *H. pylori*. Int J Pharm. 2017;519(1–2):128–37. https://doi.org/10.1016/j.ijpharm.2017.01.014.
26. Dash M, Chiellini F, Ottenbrite RM, Chiellini E. Chitosan – a versatile semi-synthetic polymer in biomedical applications. Prog Polym Sci. 2011;36(8):981–1014. https://doi.org/10.1016/j.progpolymsci.2011.02.001.
27. Sogias IA, Williams AC, Khutoryanskiy VV. Why is chitosan mucoadhesive? Biomacromolecules. 2008;9(7):1837–42. https://doi.org/10.1021/bm800276d.
28. Luo D, Guo J, Wang F, Sun J, Li G, et al. Preparation and evaluation of anti-*H. pylori* efficacy of chitosan nanoparticles in vitro and in vivo. J Biomater Sci Polym Ed. 2009;20(11):1587–96. https://doi.org/10.1163/092050609X12464345137685.
29. Adil Abdelrahim Mohammed Y. Development of curcumin-loaded alginate beads for targeted gastric mucosa delivery in *H. pylori*-associated gastritis. Trends in Pharm Biotechnol. 2024;2(2):7–11. https://doi.org/10.57238/tpb.2024.153196.1018.
30. Yahya R, Al-Rajhi AMH, Alzaid SZ, Al Abboud MA, Almuhayawi MS, et al. Molecular docking and efficacy of Aloe vera gel based on chitosan nanoparticles against *H. pylori* and its antioxidant and anti-inflammatory activities. Polymers (Basel). 2022;14(15) https://doi.org/10.3390/polym14152994.
31. Ramteke S, Ganesh N, Bhattacharya S, Jain NK. Amoxicillin, clarithromycin, and omeprazole based targeted nanoparticles for the treatment of *H. pylori*. J Drug Target. 2009;17(3):225–34. https://doi.org/10.1080/10611860902718649.
32. Goncalves IC, Magalhaes A, Fernandes M, Rodrigues IV, Reis CA, et al. Bacterial-binding chitosan microspheres for gastric infection treatment and prevention. Acta Biomater. 2013;9(12):9370–8. https://doi.org/10.1016/j.actbio.2013.07.034.

33. Fernandes M, Goncalves IC, Nardecchia S, Amaral IF, Barbosa MA, et al. Modulation of stability and mucoadhesive properties of chitosan microspheres for therapeutic gastric application. Int J Pharm. 2013;454(1):116–24. https://doi.org/10.1016/j.ijpharm.2013.06.068.
34. Lai Y, Zhang T, Yin X, Zhu C, Du Y, et al. An antibiotic-free platform for eliminating persistent *H. pylori* infection without disrupting gut microbiota. Acta Pharm Sin B. 2024;14(7):3184–204. https://doi.org/10.1016/j.apsb.2024.03.014.
35. Yadav HK, Almokdad AA, Sumia IM, Debe MS. Polymer-based nanomaterials for drug-delivery carriers. In: Nanocarriers for drug delivery. Elsevier; 2019. p. 531–56.
36. Vrignaud S, Benoit JP, Saulnier P. Strategies for the nanoencapsulation of hydrophilic molecules in polymer-based nanoparticles. Biomaterials. 2011;32(33):8593–604. https://doi.org/10.1016/j.biomaterials.2011.07.057.
37. Meng F, Tao H, Mi Y, Yang T, Wang X, et al. Nanocluster-mediated photothermia improves eradication efficiency and antibiotic sensitivity of *H. pylori*. Cancer Nanotechnol. 2022;13(1) https://doi.org/10.1186/s12645-022-00121-2.
38. Lin YH, Lin JH, Chou SC, Chang SJ, Chung CC, et al. Berberine-loaded targeted nanoparticles as specific *H. pylori* eradication therapy: in vitro and in vivo study. Nanomedicine (Lond). 2015;10(1):57–71. https://doi.org/10.2217/nnm.14.76.
39. Harsha S. Dual drug delivery system for targeting *H. pylori* in the stomach: preparation and in vitro characterization of amoxicillin-loaded Carbopol(R) nanospheres. Int J Nanomedicine. 2012;7:4787–96. https://doi.org/10.2147/IJN.S34312.
40. Arif M, Ahmad R, Sharaf M, Samreen, Muhammad J, et al. Antibacterial and antibiofilm activity of mannose-modified chitosan/PMLA nanoparticles against multidrug-resistant *H. pylori*. Int J Biol Macromol. 2022;223(Pt A):418–32. https://doi.org/10.1016/j.ijbiomac.2022.10.265.
41. Umamaheshwari RB, Jain NK. Receptor mediated targeting of lectin conjugated gliadin nanoparticles in the treatment of *H. pylori*. J Drug Target. 2003;11(7):415–23. https://doi.org/10.1080/10611860310001647771.
42. Ramteke S, Jain NK. Clarithromycin- and omeprazole-containing gliadin nanoparticles for the treatment of *H. pylori*. J Drug Target. 2008;16(1):65–72. https://doi.org/10.1080/10611860701733278.
43. Mourino V, Cattalini JP, Boccaccini AR. Metallic ions as therapeutic agents in tissue engineering scaffolds: an overview of their biological applications and strategies for new developments. J R Soc Interface. 2012;9(68):401–19. https://doi.org/10.1098/rsif.2011.0611.
44. Begg SL. The role of metal ions in the virulence and viability of bacterial pathogens. Biochem Soc Trans. 2019;47(1):77–87. https://doi.org/10.1042/BST20180275.
45. Chandrangsu P, Rensing C, Helmann JD. Metal homeostasis and resistance in bacteria. Nat Rev Microbiol. 2017;15(6):338–50. https://doi.org/10.1038/nrmicro.2017.15.
46. Stafford SL, Bokil NJ, Achard ME, Kapetanovic R, Schembri MA, et al. Metal ions in macrophage antimicrobial pathways: emerging roles for zinc and copper. Biosci Rep. 2013;33(4) https://doi.org/10.1042/BSR20130014.
47. Djoko KY, Ong CL, Walker MJ, McEwan AG. The role of copper and zinc toxicity in innate immune defense against bacterial pathogens. J Biol Chem. 2015;290(31):18954–61. https://doi.org/10.1074/jbc.R115.647099.
48. Han D, Liu X, Wu S. Metal organic framework-based antibacterial agents and their underlying mechanisms. Chem Soc Rev. 2022;51(16):7138–69. https://doi.org/10.1039/d2cs00460g.
49. Yin X, Lai Y, Du Y, Zhang T, Gao J, et al. Metal-based nanoparticles: a prospective strategy for *H. pylori* treatment. Int J Nanomedicine. 2023;18:2413–29. https://doi.org/10.2147/IJN.S405052.
50. Jana S, Pal T. Synthesis, characterization and catalytic application of silver nanoshell coated functionalized polystyrene beads. J Nanosci Nanotechnol. 2007;7(6):2151–6. https://doi.org/10.1166/jnn.2007.785.
51. Furquim A, de Camargo B, Soares Silva DE, Noronha da Silva A, Campos DL, Machado Ribeiro TR, et al. New silver(I) coordination compound loaded into polymeric nanoparticles as a strategy to improve in vitro anti-*H. pylori* activity. Mol Pharm. 2020;17(7):2287–98. https://doi.org/10.1021/acs.molpharmaceut.9b01264.

52. Gurunathan S, Jeong JK, Han JW, Zhang XF, Park JH, et al. Multidimensional effects of biologically synthesized silver nanoparticles in *H. pylori*, helicobacter felis, and human lung (L132) and lung carcinoma A549 cells. Nanoscale Res Lett. 2015;10:35. https://doi.org/10.1186/s11671-015-0747-0.
53. Grande R, Sisto F, Puca V, Carradori S, Ronci M, et al. Antimicrobial and Antibiofilm activities of new synthesized silver ultra-NanoClusters (SUNCs) against *H. pylori*. Front Microbiol. 2020;11:1705. https://doi.org/10.3389/fmicb.2020.01705.
54. Graves JL Jr, Tajkarimi M, Cunningham Q, Campbell A, Nonga H, et al. Rapid evolution of silver nanoparticle resistance in Escherichia coli. Front Genet. 2015;6:42. https://doi.org/10.3389/fgene.2015.00042.
55. Gopinath V, Priyadarshini S, MubarakAli D, Loke MF, Thajuddin N, et al. Anti-*H. pylori*, cytotoxicity and catalytic activity of biosynthesized gold nanoparticles: multifaceted application. Arab J Chem. 2019;12(1):33–40. https://doi.org/10.1016/j.arabjc.2016.02.005.
56. Al-Radadi NS. Microwave assisted green synthesis of Fe@Au core–shell NPs magnetic to enhance olive oil efficiency on eradication of *H. pylori* (life preserver). Arab J Chem. 2022;15(5):103685. https://doi.org/10.1016/j.arabjc.2022.103685.
57. Abd El-Moaty HI, Soliman NA, Hamad RS, Ismail EH, Sabry DY, et al. Comparative therapeutic effects of Pituranthos tortuosus aqueous extract and phyto-synthesized gold nanoparticles on *H. pylori*, diabetic and cancer proliferation. S Afr J Bot. 2021;139:167–74. https://doi.org/10.1016/j.sajb.2021.02.009.
58. Lv MM, Fan SF, Wang QL, Lv QY, Song X, et al. An enzyme-free electrochemical sandwich DNA assay based on the use of hybridization chain reaction and gold nanoparticles: application to the determination of the DNA of *H. pylori*. Mikrochim Acta. 2019;187(1):73. https://doi.org/10.1007/s00604-019-3999-z.
59. Chakraborti S, Bhattacharya S, Chowdhury R, Chakrabarti P. The molecular basis of inactivation of metronidazole-resistant *H. pylori* using polyethyleneimine functionalized zinc oxide nanoparticles. PLoS One. 2013;8(8):e70776. https://doi.org/10.1371/journal.pone.0070776.
60. Attia HG, Albarqi HA, Said IG, Alqahtani O, Raey MAE. Synergistic effect between amoxicillin and zinc oxide nanoparticles reduced by oak gall extract against *H. pylori*. Molecules. 2022;27(14) https://doi.org/10.3390/molecules27144559.
61. Saravanan M, Gopinath V, Chaurasia MK, Syed A, Ameen F, et al. Green synthesis of anisotropic zinc oxide nanoparticles with antibacterial and cytofriendly properties. Microb Pathog. 2018;115:57–63. https://doi.org/10.1016/j.micpath.2017.12.039.
62. Wang Z, Zeng Z, Wang H, Zeng G, Xu P, et al. Bismuth-based metal–organic frameworks and their derivatives: opportunities and challenges. Coord Chem Rev. 2021;439:213902. https://doi.org/10.1016/j.ccr.2021.213902.
63. Li H, Wang R, Sun H. Systems approaches for unveiling the mechanism of action of bismuth drugs: new medicinal applications beyond *H. Pylori* infection. Acc Chem Res. 2019;52(1):216–27. https://doi.org/10.1021/acs.accounts.8b00439.
64. Nazari P, Dowlatabadi-Bazaz R, Mofid MR, Pourmand MR, Daryani NE, et al. The antimicrobial effects and metabolomic footprinting of carboxyl-capped bismuth nanoparticles against *H. pylori*. Appl Biochem Biotechnol. 2014;172(2):570–9. https://doi.org/10.1007/s12010-013-0571-x.
65. Chen R, Cheng G, So MH, Wu J, Lu Z, et al. Bismuth subcarbonate nanoparticles fabricated by water-in-oil microemulsion-assisted hydrothermal process exhibit anti-*H. pylori* properties. Mater Res Bull. 2010;45(5):654–8. https://doi.org/10.1016/j.materresbull.2009.12.035.
66. Narayanaswamy R, Torchilin VP. Hydrogels and their applications in targeted drug delivery. Molecules. 2019;24(3) https://doi.org/10.3390/molecules24030603.
67. Raju D, Hussey S, Ang M, Terebiznik MR, Sibony M, et al. Vacuolating cytotoxin and variants in Atg16L1 that disrupt autophagy promote *H. pylori* infection in humans. Gastroenterology. 2012;142(5):1160–71. https://doi.org/10.1053/j.gastro.2012.01.043.

68. Arif M, Sharaf M, Samreen, Khan S, Chi Z, et al. Chitosan-based nanoparticles as delivery-carrier for promising antimicrobial glycolipid biosurfactant to improve the eradication rate of *H. pylori* biofilm. J Biomater Sci Polym Ed. 2021;32(6):813–32. https://doi.org/10.1080/09205063.2020.1870323.
69. Moghadam MT, Chegini Z, Khoshbayan A, Farahani I, Shariati A. *H. pylori* biofilm and new strategies to combat it. Curr Mol Med. 2021;21(7):549–61. https://doi.org/10.2174/1566524020666201203165649.
70. Huang P, Wang X, Liang X, Yang J, Zhang C, et al. Nano-, micro-, and macroscale drug delivery systems for cancer immunotherapy. Acta Biomater. 2019;85:1–26. https://doi.org/10.1016/j.actbio.2018.12.028.

# Externally Responsive Nanomaterials

# 8

Min Lu, Huizhen Fan, Chenxi Wang, and Qiang Yu

## 8.1 Introduction

Externally stimuli-responsive nanomaterials are particularly attractive for *H. pylori* therapy because they address the unique challenges of the gastric environment. The stomach presents harsh and heterogeneous conditions, including fluctuating pH, a thick mucus barrier, and biofilm-protected bacterial colonies that limit the efficacy of conventional drugs. By responding to physical triggers such as light, ultrasound, magnetic fields, or heat, these nanomaterials allow antibacterial activity to be activated with precise spatial and temporal control. Such on-demand activation not only enhances drug localization at the infection site but also improves penetration through mucus and biofilms, reduces systemic exposure, and minimizes disturbance to the gut microbiota. These advantages are especially relevant for overcoming antibiotic resistance, where conventional therapies often fail to achieve sufficient local drug concentration or disrupt bacterial persistence niches.

Although challenges remain in stimulus delivery within the gastric environment, especially due to limited penetration depth and gastric motility, the development of minimally invasive techniques such as endoscopy-assisted irradiation or magnetic targeting provides feasible strategies for clinical application. These approaches highlight the potential of integrating responsive nanomaterials with nonantibiotic therapeutics for synergistic and antibiotic-sparing *H. pylori* therapy (Fig. 8.1).

M. Lu (✉) · H. Fan · C. Wang · Q. Yu
Department of Orthopaedics, Shanghai Key Laboratory for Prevention and Treatment of Bone and Joint Diseases, Shanghai Institute of Traumatology and Orthopaedics, Ruijin Hospital, Shanghai Jiao Tong University School of Medicine, Shanghai, China
e-mail: lumin111@sjtu.edu.cn



J. Gao, Y. Du (eds.), *Non-Antibiotic Therapies for Helicobacter pylori*,
https://doi.org/10.1007/978-981-95-9598-3_8

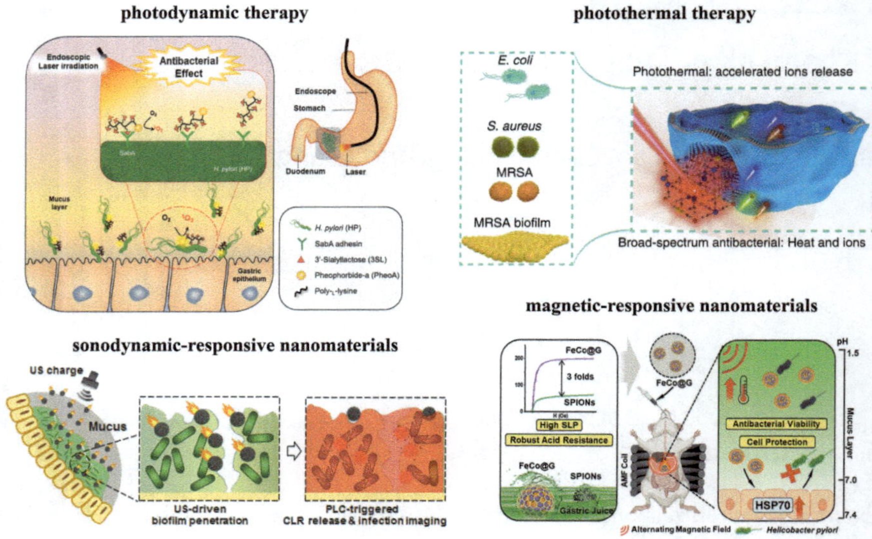


**Fig. 8.1** Design and mechanisms of externally responsive nanomaterials: Photodynamic therapy, photothermal therapy, sonodynamic-responsive nanomaterials, and magnetic-responsive nanomaterials. (Reprinted with permission from Ref. [1], Elsevier. Reprinted with permission from Ref. [2], Springer Nature. Reprinted with permission from Refs. [3, 4], American Chemical Society)

## 8.2 Light-Responsive Nanomaterials

Light-responsive nanomaterials exploit external light irradiation to trigger bactericidal activity primarily through two modalities: photodynamic therapy (PDT) and photothermal therapy (PTT). PDT involves the use of photosensitizers that, upon excitation by specific wavelengths of light, transfer energy to molecular oxygen and generate ROS. These ROS induce oxidative stress, damaging bacterial membranes, proteins, and DNA, thereby overcoming protective mechanisms such as biofilms [5]. In contrast, PTT relies on nanomaterials such as gold nanoparticles, graphene oxide, or transition metal sulfides that efficiently convert absorbed light into localized heat. This photothermal effect raises the local temperature to bactericidal levels, leading to disruption of bacterial membranes, denaturation of proteins, and eventual cell death [2]. Both PDT and PTT offer unique advantages in *H. pylori* therapy: they enable spatiotemporally controlled antibacterial effects, act independently of conventional antibiotics, and exhibit strong efficacy against drug-resistant strains or bacteria embedded within gastric biofilms. Moreover, the rapid and localized nature of these therapies reduces systemic toxicity and helps preserve the balance of commensal gut microbiota.

### 8.2.1 Photodynamic Therapy

Photodynamic therapy (PDT) relies on photosensitizers such as porphyrins, chlorin e6, methylene blue, or curcumin, which can be activated under light exposure to generate singlet oxygen and other ROS. These ROS cause oxidative damage to multiple bacterial targets, including lipid peroxidation of membranes, cross-linking or fragmentation of proteins, and irreversible DNA lesions. Importantly, *H. pylori* has only a poorly efficient *uvrA*-dependent DNA repair pathway and lacks many oxidative stress defense mechanisms, making it particularly vulnerable to ROS-mediated killing [6]. Early proof-of-concept work by Bedwell et al. first demonstrated the feasibility of PDT for *H. pylori* eradication [7]. Since then, several studies have validated its potency: ruthenium(II)-based complexes achieved >7-log10 reduction after 30 min of blue light irradiation [5], and methylene blue-mediated PDT induced >5-log10 reduction when combined with endoscopic light delivery [8].

Beyond monotherapy, synergistic strategies have been explored to enhance efficacy. For instance, methylene blue-PDT combined with chitosan demonstrated stronger bactericidal effects, attributed to improved photosensitizer uptake and increased membrane permeability [9]. A notable study developed a *H. pylori*-targeting PDT system using 3′-sialyllactose (3SL)-conjugated poly-L-lysine-based photosensitizers (p3SLP). This system takes advantage of the specific interaction between 3SL and the sialic acid-binding adhesin (SabA) on the *H. pylori* membrane, enabling targeted PDT. When administered orally to *H. pylori*-infected mice and irradiated with an endoscopic laser system, p3SLP demonstrated significant antibacterial effects against *H. pylori*, without affecting normal tissues. Furthermore, an anti-inflammatory response was observed at the site of infection following treatment, highlighting the potential of p3SLP as an effective alternative to antibiotic-based therapies [1]. Such molecular targeting strategies not only improve therapeutic selectivity but also reduce collateral damage to gastric epithelial cells.

Technological innovations are also addressing delivery limitations. Since light penetration in the gastric mucosa is shallow, specialized devices such as endoscopic light guides, fiber-optic catheters, or ingestible light capsules have been developed to ensure localized irradiation within the stomach. In parallel, research is focusing on optimizing photosensitizer biocompatibility to minimize phototoxicity to host tissues while maintaining strong ROS generation efficiency [10]. Taken together, PDT represents a promising antibiotic-free strategy for *H. pylori* eradication, with its dual advantages of overcoming bacterial resistance mechanisms and enabling targeted, spatiotemporally controllable therapy.

### 8.2.2 Photothermal Therapy

Photothermal therapy (PTT) employs nanomaterials with strong photothermal conversion efficiency, such as gold nanoparticles, graphene oxide, and transition-metal oxides to transform near-infrared (NIR) irradiation into localized heat. This localized hyperthermia induces multiple bactericidal mechanisms: disruption of bacterial membranes, protein denaturation, enzyme inactivation, and irreversible collapse of biofilm architecture [2]. In contrast to conventional antibiotics, PTT operates through a physical mechanism that bypasses genetic resistance, making it particularly attractive against multidrug-resistant *H. pylori*.

Representative studies have highlighted the therapeutic potential of PTT. Meng et al. engineered $Zn_{0.3}Fe_{2.7}O_4$ nanoclusters coated with PEG–PCL that elevated local gastric temperatures to ~41 °C within 240 s under NIR exposure, suppressing *H. pylori* growth to only 6.7% of the untreated control and simultaneously restoring antibiotic sensitivity in resistant strains [11, 12]. Gold nanostructures—including nanorods and nanoshells—have also been widely investigated due to their high extinction coefficients in the NIR region, enabling efficient bacterial eradication and biofilm disruption within minutes. Furthermore, graphene oxide nanosheets and their derivatives provide large surface areas for drug loading in addition to robust photothermal performance, making them promising dual-function platforms for combined PTT and drug delivery.

In addition to monotherapy, PTT can be integrated with antibiotics or other non-antibiotic therapeutics to achieve synergistic effects. Sublethal hyperthermia has been shown to increase bacterial membrane permeability, thereby enhancing antibiotic uptake and reducing minimum inhibitory concentrations. This combinatorial approach not only accelerates bacterial killing but also offers a strategy to reverse drug resistance and lower systemic antibiotic exposure.

Despite these advantages, several translational challenges remain. Precise thermal control is critical: although local temperatures above 50 °C achieve rapid bacterial killing, excessive hyperthermia risks collateral injury to gastric epithelial cells and disruption of mucosal integrity. Biodegradability and long-term biosafety of inorganic photothermal agents also warrant careful evaluation, particularly with repeated treatments. Advances in biodegradable or organic photothermal agents, such as polydopamine or porphyrin-based nanomaterials, may help address these concerns. Furthermore, the stomach's dynamic environment, including constant peristalsis and variable mucus thickness, complicates localized irradiation, highlighting the need for improved targeting strategies and integration with endoscopic or capsule-based light delivery devices.

Overall, PTT offers a rapid, controllable, and resistance-independent modality for *H. pylori* eradication, especially in the context of biofilm-associated infections. Continued innovation in photothermal agent design and delivery methods will be essential to translate this promising approach into safe and effective clinical therapies.

## 8.3 Sonodynamic-Responsive Nanomaterials

Sonodynamic therapy (SDT) utilizes ultrasound to activate sonosensitizers, resulting in the production of ROS and acoustic cavitation that disrupt bacterial membranes, proteins, and biofilms [13]. Unlike PDT, which is constrained by limited light penetration, ultrasound can penetrate several centimeters into soft tissues

(>10 cm), making SDT particularly attractive for noninvasive gastric applications. Furthermore, ultrasound waves can be precisely controlled in terms of frequency, intensity, and duration, offering spatiotemporal selectivity for antibacterial therapy.

The antibacterial effects of SDT arise from multiple mechanisms. First, activated sonosensitizers such as porphyrins, indocyanine green, and titanium dioxide derivatives generate singlet oxygen and hydroxyl radicals upon ultrasound irradiation, leading to oxidative damage of bacterial DNA and cell membranes. Second, acoustic cavitation—the formation, oscillation, and collapse of microbubbles under ultrasound—induces strong mechanical shear forces that physically disrupt biofilm structures and enhance penetration of therapeutic agents. Third, cavitation can generate secondary effects such as localized hyperthermia and shock waves, further amplifying bacterial killing. These multimodal actions make SDT particularly suited to overcome *H. pylori*'s protective biofilm barrier, which is often responsible for antibiotic resistance.

Several studies have demonstrated the promise of SDT for *H. pylori* eradication. Antibody-functionalized liposomes encapsulating indocyanine green achieved a 99.9% killing efficiency under ultrasound exposure, attributed to singlet oxygen generation and membrane disruption [14]. In another example, ultrasound was used to activate nanostructured $TiO_2$ particles, resulting in sustained ROS production and significant suppression of *H. pylori* viability in vitro. More recently, a targeted sonodynamic nanoplatform—chitosan-conjugated fucose loaded with indocyanine green (ICG@FCS)—was developed to address the limitations of conventional therapies [15]. This system leverages dual-targeting mechanisms: fucose-mediated molecular recognition of *H. pylori* and ultrasound-enhanced physical penetration of the gastric mucosa. Upon ultrasound activation, ICG@FCS generates singlet oxygen to eradicate planktonic bacteria, disrupt biofilms, and clear intracellular infections (including multidrug-resistant strains) by inducing autophagy. Notably, the platform minimizes disruption to gut microbiota and promotes gastric mucosa repair, embodying a "holistic integrative" approach that prioritizes both pathogen eradication and gastrointestinal health. This study underscores SDT's potential to transcend antibiotic resistance while preserving host tissue integrity, offering a paradigm shift in *H. pylori* management. In parallel, ultrasound-triggered mechanoluminescent nanomaterials have been proposed for theranostic applications, simultaneously enabling ROS-mediated bacterial killing and real-time imaging guidance [3]. These findings highlight the versatility of ultrasound as both a therapeutic and diagnostic tool.

Despite these advances, SDT for *H. pylori* remains at an early stage, with several hurdles to clinical translation. The acoustic environment of the stomach is complex, and the optimization of ultrasound parameters—such as frequency, intensity, duty cycle, and exposure time—requires systematic investigation to balance efficacy with safety. Additionally, the design of biocompatible and gastric-stable sonosensitizers is critical, as current agents may suffer from rapid degradation in acidic conditions or limited retention at the gastric mucosa. Another challenge is the lack of in vivo validation: most reported studies remain at the in vitro proof-of-concept level, and comprehensive animal studies are needed to assess long-term biosafety, biodistribution, and potential off-target effects on gastric tissues.

Overall, SDT offers a unique combination of deep tissue penetration, controllable activation, and multimodal antibacterial mechanisms, positioning it as a promising strategy for nonantibiotic *H. pylori* therapy. With advances in sensitizer design, ultrasound delivery systems, and in vivo validation, SDT could complement or even synergize with other nanotechnology-based approaches to achieve effective eradication of *H. pylori* in the future.

## 8.4 Magnetic-Responsive Nanomaterials

Magnetic-responsive systems employ external magnetic fields to guide, concentrate, or activate nanoparticles at infection sites, offering unique advantages for localized *H. pylori* therapy. Among these, iron oxide-based nanoparticles—particularly superparamagnetic iron oxide nanoparticles (SPIONs)—are the most widely investigated due to their biocompatibility, imaging capabilities, and responsiveness to alternating magnetic fields. Two primary therapeutic modes have been explored: magnetic hyperthermia and magnetic-targeted drug delivery.

In magnetic hyperthermia, nanoparticles dissipate thermal energy under an alternating magnetic field, generating localized heat that disrupts bacterial membranes, flagella, and biofilm structures while simultaneously sensitizing bacteria to antibiotics [16, 17]. Wu et al. developed $Mn_{0.3}Fe_{2.7}O_4$@$SiO_2$-amoxicillin nanocomposites that remained stable in acidic gastric fluid and achieved 93.8% growth inhibition under magnetic field exposure, nearly doubling the efficacy of free amoxicillin (53.6%) [17]. Similarly, FeCo@graphitic shell nanostructures were shown to withstand harsh gastric acid conditions and deliver effective magnetothermal stimulation, reducing bacterial load in infected mice by approximately 72.9% compared with controls [4]. Beyond direct killing, magnetothermal effects have been reported to restore host protective pathways, such as heat shock protein 70 expression in gastric epithelial cells, which may help preserve mucosal integrity during *H. pylori* infection [18].

Magnetic fields also provide a means for spatially controlled drug delivery. Superparamagnetic particles can be magnetically guided through the gastric mucus layer, prolonging drug retention and improving penetration to sites of bacterial colonization. For instance, Yang et al. engineered SPION-loaded chitosan/poly(acrylic acid) carriers co-delivering amoxicillin. Under external magnetic stimulation, these particles penetrated the mucus barrier more efficiently, achieved prolonged gastric retention (detectable up to 24 h post-administration), and increased bacterial eradication rates to 60%, compared with only 20% using free antibiotics and 10% using conventional triple therapy [19]. This strategy not only enhanced antibacterial efficacy but also protected antibiotics from rapid acid degradation.

Despite these promising results, several challenges remain for magnetic-responsive systems in *H. pylori* therapy. Safe heating thresholds must be maintained to avoid damage to gastric mucosa, and the potential accumulation or long-term retention of magnetic nanoparticles in tissues requires thorough toxicological assessment. Furthermore, magnetic stimulation cannot be applied to patients with certain biomedical implants such as pacemakers, limiting its clinical applicability. Delivery of a sufficiently strong yet safe magnetic field within the gastric environment also remains technically challenging.

Overall, magnetic-responsive nanomaterials combine diagnostic potential, targeted delivery, and controllable hyperthermia into a single platform, offering significant promise for nonantibiotic *H. pylori* eradication. Future efforts should focus on designing biodegradable magnetic nanocarriers, developing minimally invasive gastric field–delivery devices, and integrating magnetic therapy with other external stimuli (e.g., light or ultrasound) to achieve synergistic antibacterial outcomes.

## 8.5 Limitations and Future Perspectives

Despite encouraging progress, several hurdles hinder the clinical translation of external stimuli-responsive nanomaterials for *H. pylori* eradication. Technical barriers include the limited penetration depth of light and magnetic fields, which restricts activation to superficial gastric regions unless assisted by endoscopic devices [2, 20]. Ultrasound provides deeper tissue penetration but requires careful tuning of frequency and intensity to balance efficacy with safety. Both photothermal and magnetothermal therapies face challenges in maintaining therapeutic heating within a narrow safety window—excessive heat risks damaging gastric mucosa, whereas insufficient heat fails to eradicate bacteria. In addition, ROS generated during photodynamic or sonodynamic therapy may cause collateral damage to host epithelial cells. Furthermore, most of these systems rely on specialized external devices (e.g., lasers, magnetic coils, ultrasound probes), which complicates their clinical adoption.

From a biological perspective, the gastric environment remains a significant challenge. Gastric motility, mucus turnover, and variable acid secretion can reduce the residence time and stability of nanomaterials, resulting in inconsistent therapeutic outcomes. Long-term biosafety is another concern: repeated administration of metallic or carbon-based nanoparticles raises questions about tissue accumulation, immune reactions, and potential off-target effects. Current evidence is mostly derived from in vitro or small-animal models, with limited insight into chronic exposure, microbiota interactions, or large-animal validation.

Looking ahead, several future directions are particularly promising: (i) Developing multifunctional theranostic platforms that integrate antibacterial therapy with real-time monitoring, allowing precise feedback control of external stimulus application. (ii) Designing minimally invasive or ingestible devices, such as endoscopic light probes, magnetic-guided microrobots, or ultrasound capsules, to enable safe and practical delivery of external triggers inside the stomach. (iii) Conducting systematic in vivo studies in clinically relevant models to evaluate eradication efficacy, long-term safety, and host–pathogen–microbiota interactions under repeated treatments. (iv) Advancing the use of biodegradable and eco-friendly nanomaterials [21–23], which can reduce concerns over toxicity, cost, and environmental burden, thereby improving scalability and clinical acceptance. (v) Exploring combinational external modalities (e.g., light + ultrasound, magnetic + heat), which may enhance bacterial eradication while lowering the required energy dose for each stimulus, thus minimizing tissue damage.

In conclusion, externally responsive nanomaterials represent a promising next-generation approach for *H. pylori* therapy. By offering controlled, localized, and on-demand antibacterial activity, these systems could complement or potentially replace conventional antibiotic regimens. Continued interdisciplinary collaboration across nanotechnology, gastroenterology, and biomedical engineering will be crucial to overcoming current barriers and moving toward clinical translation.

## References


1. Im BN, Shin H, Lim B, Lee J, Kim KS, et al. *H. pylori*-targeting multiligand photosensitizer for effective antibacterial endoscopic photodynamic therapy. Biomaterials. 2021;271:120745. https://doi.org/10.1016/j.biomaterials.2021.120745.
2. Li J, Liu X, Tan L, Cui Z, Yang X, et al. Zinc-doped Prussian blue enhances photothermal clearance of Staphylococcus aureus and promotes tissue repair in infected wounds. Nat Commun. 2019;10(1):4490. https://doi.org/10.1038/s41467-019-12429-6.
3. Zhou D, Zhang Z, Qiu B, Zhang D, Xie S, et al. Ultrasound-activated persistent luminescence imaging and bacteria-triggered drug release for *H. pylori* infection theranostics. ACS Appl Mater Interfaces. 2022;14(23):26418–30. https://doi.org/10.1021/acsami.2c04683.
4. Xia X, Yin Z, Yang Y, Li S, Wang L, et al. In situ upregulating heat shock protein 70 via gastric nano-heaters for the interference of *H. pylori* infection. ACS Nano. 2022;16(9):14043–54. https://doi.org/10.1021/acsnano.2c03911.
5. Calvino-Fernandez M, Garcia-Fresnadillo D, Benito-Martinez S, McNicholl AG, Calvet X, et al. *H. pylori* inactivation and virulence gene damage using a supported sensitiser for photodynamic therapy. Eur J Med Chem. 2013;68:284–90. https://doi.org/10.1016/j.ejmech.2013.07.023.
6. Sinha RP, Hader DP. UV-induced DNA damage and repair: a review. Photochem Photobiol Sci. 2002;1(4):225–36. https://doi.org/10.1039/b201230h.
7. Bedwell J, Holton J, Vaira D, Macrobert AJ, Bown SG. Invitro killing of helicobacter-pylori with photodynamic therapy. Lancet. 1990;335(8700):1287. https://doi.org/10.1016/0140-6736(90)91361-D.
8. Choi SS, Lee HK, Chae HS. In vitro photodynamic antimicrobial activity of methylene blue and endoscopic white light against *H. pylori* 26695. J Photochem Photobiol B. 2010;101(3):206–9. https://doi.org/10.1016/j.jphotobiol.2010.07.004.
9. Kim EJ, Choi JH, Yang HJ, Choi SS, Lee HK, et al. Comparison of high and low molecular weight chitosan as in-vitro boosting agent for photodynamic therapy against *H. pylori* using methylene blue and endoscopic light. Photodiagn Photodyn Ther. 2019;26:111–5. https://doi.org/10.1016/j.pdpdt.2019.03.005.
10. Wang Y, Guo X, Zhou S, Wang L, Fang Y, et al. Selective photodynamic inactivation of *H. pylori* by a cationic benzylidene cyclopentanone photosensitizer - an in vitro and ex vivo study. J Photochem Photobiol B. 2021;223:112287. https://doi.org/10.1016/j.jphotobiol.2021.112287.
11. Meng F, Tao H, Mi Y, Yang T, Wang X, et al. Nanocluster-mediated photothermia improves eradication efficiency and antibiotic sensitivity of *H. pylori*. Cancer Nanotechnol. 2022;13(1) https://doi.org/10.1186/s12645-022-00121-2.
12. Giske CG, Turnidge J, Cantón R, Kahlmeter G, Comm ES. Update from the European Committee on Antimicrobial Susceptibility Testing (EUCAST). J Clin Microbiol. 2022;60(3):ARTN e00276-21. https://doi.org/10.1128/jcm.00276-21.

13. Wang D, Cheng DB, Ji L, Niu LJ, Zhang XH, et al. Precise magnetic resonance imaging-guided sonodynamic therapy for drug-resistant bacterial deep infection. Biomaterials. 2021;264:120386. https://doi.org/10.1016/j.biomaterials.2020.120386.
14. Wang R, Song C, Gao A, Liu Q, Guan W, et al. Antibody-conjugated liposomes loaded with indocyanine green for oral targeted photoacoustic imaging-guided sonodynamic therapy of *H. pylori* infection. Acta Biomater. 2022;143:418–27. https://doi.org/10.1016/j.actbio.2022.02.031.
15. Yin X, Lai Y, Zhang X, Zhang T, Tian J, et al. Targeted Sonodynamic therapy platform for holistic integrative *H. pylori* therapy. Adv Sci. 2025;12(2):e2408583. https://doi.org/10.1002/advs.202408583.
16. Huang J, Zhang X, Fu K, Wei G, Su Z. Stimulus-responsive nanomaterials under physical regulation for biomedical applications. J Mater Chem B. 2021;9(47):9642–57. https://doi.org/10.1039/d1tb02130c.
17. Wu T, Wang L, Gong M, Lin Y, Xu Y, et al. Synergistic effects of nanoparticle heating and amoxicillin on *H. pylori* inhibition. J Magn Magn Mater. 2019;485:95–104. https://doi.org/10.1016/j.jmmm.2019.04.076.
18. Abdelwahab SI. Protective mechanism of gallic acid and its novel derivative against ethanol-induced gastric ulcerogenesis: involvement of immunomodulation markers, Hsp70 and Bcl-2-associated X protein. Int Immunopharmacol. 2013;16(2):296–305. https://doi.org/10.1016/j.intimp.2013.04.005.
19. Yang SJ, Huang CH, Yang JC, Wang CH, Shieh MJ. Residence time-extended nanoparticles by magnetic field improve the eradication efficiency of *H. pylori*. ACS Appl Mater Interfaces. 2020;12(49):54316–27. https://doi.org/10.1021/acsami.0c13101.
20. He Z, Wang P, Ye X. Novel endoscopic optical diagnostic technologies in medical trial research: recent advancements and future prospects. Biomed Eng Online. 2021;20(1):5. https://doi.org/10.1186/s12938-020-00845-5.
21. Gopinath V, Priyadarshini S, MubarakAli D, Loke MF, Thajuddin N, et al. Anti-*H. pylori*, cytotoxicity and catalytic activity of biosynthesized gold nanoparticles: multifaceted application. Arab J Chem. 2019;12(1):33–40. https://doi.org/10.1016/j.arabjc.2016.02.005.
22. MubarakAli D, Akshaya T, Sathya R, Irfan N. Study on the interaction of algal peptides on virulence factors of *H. pylori*: in silico approach. Appl Biochem Biotechnol. 2022;194(1):37–53. https://doi.org/10.1007/s12010-021-03716-4.
23. Saravanakumar K, Chelliah R, MubarakAli D, Oh DH, Kathiresan K, et al. Unveiling the potentials of biocompatible silver nanoparticles on human lung carcinoma A549 cells and *H. pylori*. Sci Rep. 2019;9(1):5787. https://doi.org/10.1038/s41598-019-42112-1.

# Internally Environment-Responsive Nanomaterials

# 9

Min Lu, Chenxi Wang, Huizhen Fan, and Qiang Yu

## 9.1 Introduction

As mentioned earlier, the increasing antibiotic resistance presents significant challenges to the traditional antibiotic treatment of *Helicobacter pylori* infections. To overcome these challenges, researchers are gradually shifting toward nonantibiotic treatment strategies aimed at effectively inhibiting or eradicating *H. pylori* through various physical, chemical, or biological methods while minimizing the risk of resistance. These innovative strategies include vaccine development, immunotherapy, probiotic regulation, antimicrobial peptides, and more recently, the emergence of nanomedicine materials, which have shown immense potential in this field [1]. Among the applications of nanomedicine materials, environment-responsive nanocarriers have garnered significant attention [2]. Compared to traditional drug delivery systems, these nanocarriers are capable of responding to specific environmental signals at the stomach or infection sites, such as pH changes, elevated enzyme concentrations, and abnormal redox states, thereby enabling the selective release and precise targeting of drugs [3]. This "smart response" feature allows for the efficient release of drugs at the site of infection, enhancing therapeutic efficacy while minimizing damage to normal tissues.

Specifically, environment-responsive nanocarriers can be classified based on different environmental characteristics [4, 5]: (1) pH-responsive nanocarriers, which utilize the acidic environment of the stomach or local pH changes around *H. pylori* to trigger the release of drugs sensitive to acid; (2) enzyme-responsive nanocarriers, which use enzymes uniquely expressed by *H. pylori* or abnormally present at the

M. Lu (✉) · C. Wang · H. Fan · Q. Yu
Department of Orthopaedics, Shanghai Key Laboratory for Prevention and Treatment of Bone and Joint Diseases, Shanghai Institute of Traumatology and Orthopaedics, Ruijin Hospital, Shanghai Jiao Tong University School of Medicine, Shanghai, China
e-mail: lumin111@sjtu.edu.cn



J. Gao, Y. Du (eds.), *Non-Antibiotic Therapies for Helicobacter pylori*,
https://doi.org/10.1007/978-981-95-9598-3_9

infection site (e.g., urease, pepsin) as "switches" to trigger the degradation of nanomaterials and drug release; (3) redox-responsive nanocarriers, which adjust the stability of the carrier structure through oxidative stress or changes in glutathione (GSH) concentrations commonly found in inflamed areas; and (4) multi-responsive nanocarriers, which combine multiple environmental signals (e.g., pH + enzyme, pH + ROS) to achieve higher selectivity and stability. In the context of nonantibiotic therapies for *H. pylori*, the application of environment-responsive nanocarriers not only provides a more efficient drug delivery strategy but also has the potential to synergize with other nonantibiotic treatment methods such as probiotics, antimicrobial peptides, metal nanoparticles, and immune modulation, offering new therapeutic approaches for clinical practice.

In conclusion, environment-responsive nanocarriers, as an important branch of nanomedicine, combine the advantages of microenvironment recognition, intelligent release, and targeted therapy, making them a cutting-edge approach in the nonantibiotic treatment of *H. pylori* infections. This chapter will systematically introduce the pathological characteristics of the gastric internal environment, the types and design principles of responsive nanocarriers, relevant material systems, and the latest advancements in *H. pylori* treatment.

## 9.2 Characteristics and Pathological Basis of the Internal Environment for *H. pylori* Colonization

To effectively design and apply environment-responsive nanocarriers, it is essential to deeply understand the physicochemical characteristics of the *H. pylori* colonization site and its surrounding gastric environment. The ability of *H. pylori* to survive in the highly acidic gastric fluid over a long period and to induce chronic inflammation is fundamentally linked to its ability to cleverly utilize and modify the local environment [6]. This capability allows *H. pylori* to thrive and cause disease in the harsh conditions of the stomach [7]. Therefore, understanding the interaction between *H. pylori* and the gastric environment, as well as its pathological basis, is the prerequisite for developing effective therapeutic strategies, particularly for the design and application of environment-responsive nanocarriers. Below, we discuss in detail the characteristics of the gastric internal environment during *H. pylori* infection from multiple perspectives (Fig. 9.1).

### 9.2.1 Gastric Acidic Environment and pH Gradient

Under normal physiological conditions, the pH of human gastric fluid typically ranges from 1.0 to 3.0, creating an extremely acidic environment that is lethal to most microorganisms. This low pH is essential for digestion and serves as a defense mechanism against pathogens that may enter the stomach. However, *H. pylori* has evolved unique mechanisms that enable it to survive and proliferate in this hostile environment. The remarkable survival advantage of *H. pylori* in the stomach is

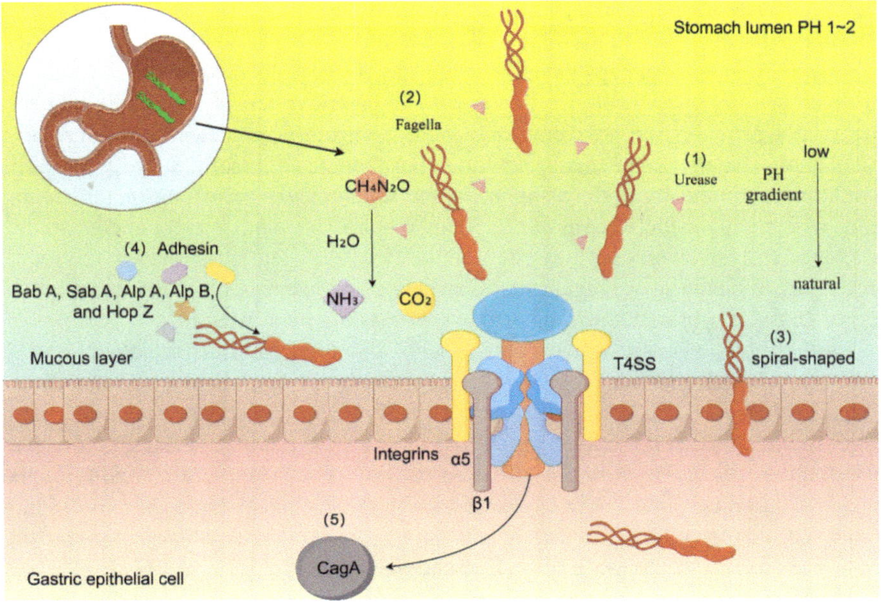


**Fig. 9.1** Characteristics of the internal environment and pathological basis of *Helicobacter pylori* colonization. (*H. pylori* infection process: from the molecular world to clinical treatment). (Reprinted with permission from Ref. [5], Frontiers)

primarily due to its ability to produce urease, an enzyme that catalyzes the hydrolysis of urea into ammonia and carbon dioxide [8]. The ammonia produced in this reaction forms a local alkaline microenvironment around the bacteria, effectively neutralizing the surrounding gastric acid. This neutralization allows *H. pylori* to maintain a stable internal pH, which is essential for its survival and persistence in the stomach, a site where most other microorganisms would not survive.

In addition to urease production, *H. pylori* has several other adaptations that contribute to its survival in the acidic gastric environment. The bacterium's unique spiral shape and flagellar motility enable it to navigate through the thick gastric mucus layer, which is composed of glycoproteins and acts as a physical barrier to pathogens [9]. Once it penetrates the mucus, *H. pylori* reaches regions near the gastric epithelial cells, where the pH is relatively higher (approximately 6–7), offering a less acidic and more buffered environment. This near-neutral pH zone allows *H. pylori* to colonize the stomach lining more effectively, where it can evade the acidic conditions of the gastric lumen and establish long-term infection. These adaptive strategies allow *H. pylori* to not only survive but also proliferate in the stomach, leading to persistent infections that can cause conditions such as chronic gastritis, peptic ulcers, and even gastric cancer.

The pH gradient between the highly acidic gastric lumen and the near-neutral gastric mucus layer provides a natural and highly specific environmental cue that can be leveraged for the design of pH-responsive nanocarriers. This gradient can be exploited to develop nanomaterials that are sensitive to pH changes [10]. For example, materials designed to dissolve or expand under low pH conditions can be used to selectively release encapsulated drugs in specific regions of the stomach, such as those close to *H. pylori* colonization sites. The gastric mucosal layer, with its near-neutral pH, provides a targeted environment for controlled drug release. Nanocarriers, such as pH-sensitive nanogels or nanomicelles, can be engineered to respond to the acidic environment by undergoing structural changes, like swelling or disintegration, which trigger the release of the drug. When these nanocarriers encounter the acidic gastric environment, they can release their therapeutic payload in the localized areas of bacterial aggregation or inflammation, ensuring more effective targeting of *H. pylori* without causing harm to surrounding healthy tissues.

Furthermore, pH-responsive nanocarriers offer the potential for selective drug delivery [3], which is particularly advantageous in the treatment of gastric infections like *H. pylori*. By optimizing the design of these nanocarriers, researchers can develop systems that not only respond to the gastric pH but also to the specific local pH changes associated with the infection process. For example, the acidity of the gastric environment may vary depending on the infection's progression, providing additional opportunities to design smart delivery systems that can adjust their drug release profiles in response to these fluctuations.

In addition to enhancing drug delivery efficiency, the use of pH-responsive materials in nanocarriers offers a promising strategy for overcoming some of the limitations associated with traditional drug delivery systems [11]. By ensuring that drugs are released at the precise location where they are needed most, pH-responsive nanocarriers can increase the bioavailability of drugs at infection sites, reduce systemic exposure, and minimize side effects, which are common issues with conventional oral drug treatments. This ability to deliver therapeutic agents specifically to the site of infection is particularly important in the context of *H. pylori* infection, as it helps to improve the therapeutic efficacy while reducing the risk of adverse effects on the surrounding healthy gastric tissue.

In conclusion, the gastric acidic environment and the pH gradient present a natural and unique challenge for therapeutic interventions targeting *H. pylori* infection. By exploiting this gradient and utilizing pH-responsive nanocarriers, researchers can develop more effective and targeted treatment strategies that enhance drug delivery, reduce side effects, and improve the overall outcome of treatment. The potential for pH-responsive nanomaterials to revolutionize the treatment of gastric infections is vast, and ongoing research is essential to optimize these systems for clinical application.

### 9.2.2 Enzyme Environment Associated with *H. pylori*

In addition to urease secretion, *H. pylori* infection is also linked to the abnormal expression or activation of various enzymes that contribute significantly to gastric mucosal damage. These enzymes not only play a critical role in the pathogenicity of *H. pylori* but also serve as potential targets for the development of enzyme-responsive nanocarriers designed for targeted therapy. Understanding the roles of these enzymes in the pathogenesis of *H. pylori* infection is crucial for the design of effective therapeutic strategies.

One of the key enzymes involved in *H. pylori* infection is pepsin, a digestive enzyme that is responsible for breaking down proteins in the stomach [12]. During *H. pylori* infection, pepsin activity is often elevated. Overactive pepsin can exacerbate the damage to the gastric mucosa by increasing the breakdown of proteins in the gastric lining. This heightened enzymatic activity not only worsens the mucosal injury but also plays a pivotal role in promoting the onset of inflammation. The inflammatory environment created by excessive pepsin activity creates a favorable niche for *H. pylori* to thrive, facilitating bacterial colonization and persistence. This interaction between pepsin and gastric mucosal injury is a critical factor in the development of peptic ulcers and chronic gastritis, conditions commonly associated with *H. pylori* infection.

In addition to pepsin, *H. pylori* secretes a range of other enzymes, such as mucopolysaccharidases and phospholipases, which further exacerbate gastric mucosal damage [13]. Mucopolysaccharidases break down glycoproteins in the gastric mucus barrier, while phospholipases degrade the phospholipids that make up the gastric epithelial cell membranes. These enzymes disrupt the structural integrity of the gastric mucosal layer, compromising its protective function. As a result, the gastric mucosa becomes more susceptible to damage from stomach acid and *H. pylori*-induced inflammation. The degradation of the gastric mucus barrier by these enzymes facilitates deeper bacterial infiltration, promoting persistent *H. pylori* colonization and increasing the severity of infection. This cascade of enzymatic activity not only enhances bacterial survival but also accelerates the progression of gastric diseases [14].

Moreover, during the chronic inflammation that characterizes *H. pylori* infection, several inflammatory enzymes are upregulated, including myeloperoxidase (MPO) and matrix metalloproteinases (MMPs) [15]. MPO, primarily produced by neutrophils, contributes to the production of reactive oxygen species (ROS) in response to infection [16]. ROS are known to damage cellular structures and play a role in the inflammatory response. MMPs, a group of enzymes that degrade extracellular matrix components, are also significantly elevated during *H. pylori* infection. The upregulation of MMPs leads to the breakdown of the extracellular matrix, further disrupting tissue integrity and promoting the infiltration of immune cells. The local accumulation of these enzymes in the gastric mucosa not only exacerbates the damage to the mucosal lining but also contributes to the persistence of inflammation. This prolonged inflammatory state is one of the key factors in the development of chronic gastric conditions, including ulcers and, in some cases, gastric cancer.

The overexpression and enhanced activity of these enzymes present a unique opportunity for the design of enzyme-responsive nanocarriers. By incorporating enzyme-sensitive peptide bonds or polymer chains into the nanocarrier structure, it is possible to create nanoparticles that undergo cleavage in the presence of specific enzymes. This cleavage triggers the release of the encapsulated drug directly at the site of infection, where the activity of these enzymes is highest. The ability to deliver drugs in response to specific enzymatic cues allows for highly targeted therapy, which not only increases the local concentration of the drug at the infection site but also minimizes off-target effects, thus reducing the potential for adverse effects on surrounding healthy tissues.

This enzyme-responsive drug delivery system has several advantages in treating *H. pylori* infections [17]. First, it enhances the efficacy of the drug by ensuring that it is released only at the site where it is needed most. Second, it reduces the systemic distribution of the drug, which can lower the risk of systemic side effects. Moreover, by targeting the enzymes that play a key role in *H. pylori* pathogenicity, this approach could potentially disrupt the bacterial colonization process and reduce the recurrence of infection. Furthermore, the use of enzyme-responsive nanocarriers offers the possibility of combining antibiotic therapy with other therapeutic modalities, such as anti-inflammatory agents or probiotics, creating a more comprehensive treatment strategy that addresses both the infection and the inflammation associated with *H. pylori* infection.

In conclusion, the enzymes associated with *H. pylori* infection, including pepsin, mucopolysaccharidases, phospholipases, MPO, and MMPs, play a critical role in the pathogenesis of gastric diseases. These enzymes present unique opportunities for the design of enzyme-responsive nanocarriers that can selectively release therapeutic agents at the site of infection, offering a promising approach to improve the efficacy and specificity of treatments for *H. pylori* infection. The continued development of enzyme-responsive nanocarriers holds great potential for advancing targeted therapies for gastric diseases caused by *H. pylori*.

### 9.2.3 Redox Environment

*H. pylori* infection is frequently associated with oxidative stress, a condition primarily induced by the infiltration of inflammatory cells and the subsequent generation of free radicals [18]. Inflammatory cells, such as neutrophils and macrophages, release reactive oxygen species (ROS) as part of the immune response to the bacterial infection. The resulting oxidative stress causes a significant increase in the local concentration of ROS, which leads to cellular damage, tissue inflammation, and further exacerbates the infection. ROS, including hydrogen peroxide ($H_2O_2$), superoxide anions ($O_2^-$), and hydroxyl radicals ($O\dot{H}$), are highly reactive and capable of damaging cellular components like lipids, proteins, and DNA. This elevated ROS level contributes to the destruction of the gastric mucosal barrier, impairing its protective function against gastric acid and promoting the pathogenesis of gastric diseases like chronic gastritis and peptic ulcers.

Simultaneously, within gastric epithelial cells, reductive molecules, such as GSH, exhibit abnormal fluctuations [6]. GSH is a major intracellular antioxidant that plays a critical role in maintaining the redox balance in cells by neutralizing ROS and protecting against oxidative damage. Under normal conditions, GSH maintains the redox homeostasis in gastric cells, but during *H. pylori* infection, the balance between oxidative and reductive molecules is disturbed. In the inflamed areas, the increased production of ROS overwhelms the cellular antioxidant defenses, leading to a state of oxidative stress. This imbalance between ROS and GSH creates a dynamic redox environment with both oxidative and reductive characteristics, further aggravating the infection and enhancing the persistence of *H. pylori*.

The dual nature of the oxidative/reductive environment provides a solid foundation for the design of redox-responsive nanocarriers [19, 20]. These nanocarriers can exploit the differences in the local redox environment to achieve precise, site-specific drug release. First, the high levels of ROS present in inflamed regions of the stomach serve as an effective triggering signal for the degradation of nanocarriers containing ROS-sensitive chemical bonds, such as thioethers or selenides. Upon exposure to ROS, these bonds undergo cleavage, leading to the controlled release of the encapsulated drug at the infection site. For example, certain nanocarriers have been designed to carry therapeutic agents that are protected by ROS-sensitive linkers. When the nanocarriers encounter the elevated ROS concentrations in the inflamed gastric tissue, these linkers are broken, allowing the drugs to be released precisely where they are needed most.

In addition to ROS, the glutathione (GSH) levels in inflamed tissues are significantly elevated [21]. GSH is particularly abundant in the cytoplasm of gastric epithelial cells, and it plays a pivotal role in maintaining cellular redox balance by reducing disulfide bonds, thereby restoring cellular function after oxidative damage. The ability of GSH to cleave disulfide bonds provides another mechanism for triggering drug release from redox-sensitive nanocarriers. When these carriers encounter high concentrations of GSH in the inflamed stomach lining, GSH can cleave the disulfide bonds within the nanocarrier, thereby triggering the release of the encapsulated therapeutic agents. This redox-triggered release mechanism is particularly advantageous because it ensures that drugs are released in response to the specific biochemical environment of the infection site, where oxidative stress is prevalent.

By exploiting these differences in the oxidative/reductive environment, redox-responsive nanocarriers enable precise control over drug release at the site of infection. This selective release not only enhances the therapeutic effects of the drugs by ensuring their high local concentration but also minimizes the potential side effects on healthy tissues. Unlike conventional drug delivery systems, which often suffer from low bioavailability and nonspecific distribution, redox-responsive nanocarriers can be designed to release their payload only in the presence of specific biomarkers of infection, such as elevated ROS or GSH levels [21]. This ability to target the infection site with high specificity offers a promising strategy for improving the safety and efficacy of *H. pylori* treatments.

Furthermore, the redox-responsive mechanism provides additional advantages over traditional drug delivery methods. Since the drug release is triggered by the unique oxidative/reductive characteristics of the infection site, this approach reduces systemic exposure to the drug, thereby lowering the risk of adverse side effects and enhancing the overall therapeutic outcome. In particular, this approach is highly effective in treating diseases associated with chronic inflammation [22], such as those caused by *H. pylori*, where sustained drug release is required to manage ongoing infections and inflammation. Additionally, redox-responsive nanocarriers can be used to co-deliver multiple therapeutic agents, such as antibiotics and anti-inflammatory drugs, creating a synergistic treatment strategy that addresses both the infection and the inflammation simultaneously.

In conclusion, the imbalance in the oxidative/reductive environment during *H. pylori* infection offers a unique opportunity for the development of redox-responsive nanocarriers. By leveraging the differences in ROS and GSH levels at the infection site, these nanocarriers can provide highly targeted drug delivery, significantly enhancing therapeutic efficacy while minimizing side effects on healthy tissues. The development of redox-responsive nanomaterials holds great promise for revolutionizing the treatment of gastric infections, and ongoing research into these systems will be crucial for optimizing their clinical application.

### 9.2.4 Gastric Mucus Barrier and Adhesion Environment

The gastric mucus layer plays a crucial role in protecting the stomach lining from the harsh acidic environment and serves as a critical habitat for *H. pylori*, the bacterium responsible for causing chronic gastric infections [23]. This mucus layer is primarily composed of glycoproteins and mucins, which form a dense, gel-like substance that acts as a physical barrier, shielding the gastric epithelium from the corrosive effects of gastric acid and enzymes. In addition to its protective role, the mucus layer also provides a suitable environment for microbial colonization, particularly for pathogens such as *H. pylori*.

For *H. pylori*, the mucus layer not only serves as a physical barrier but also as a medium for adhesion. The bacterium's ability to survive in the stomach is largely due to its ability to penetrate this mucus layer and adhere to the gastric epithelium [14]. *H. pylori* utilizes its unique spiral shape and flagellar motility to navigate through the thick mucus barrier. The bacterium's flagella enable it to move through the viscous mucus, and its chemotactic ability helps it to locate regions of the mucus with a more favorable microenvironment, closer to the gastric epithelium where the pH is slightly more neutral. Once *H. pylori* reaches the epithelial surface, it employs a range of adhesion molecules, known as adhesins, to bind firmly to the gastric mucosal cells. This binding is a critical step in the colonization process and is vital for the persistence of the infection. Through this mechanism, *H. pylori* is able to evade the harsh acidic conditions of the gastric lumen, which would otherwise destroy most other pathogens. The mucus barrier also exhibits adhesive and fluidic properties, which allow it to maintain its protective function while also enabling the movement of cells and microorganisms within the stomach. This dual functionality—acting as both a protective barrier and a substrate for microbial adhesion—creates a unique challenge for therapeutic interventions targeting *H. pylori*. The thick, sticky nature of the mucus layer prevents drugs from easily reaching the infected sites, making it difficult for conventional drugs to stay in the stomach long enough to exert their therapeutic effects. Most traditional drugs, including antibiotics, are rapidly cleared from the stomach by gastric motility, reducing their effectiveness in treating infections caused by *H. pylori*.

One of the primary challenges in the design of therapeutic strategies for *H. pylori* infection is the development of nanocarriers capable of overcoming this mucus barrier while ensuring sustained drug release at the infection site. To address this challenge, recent advances in nanomedicine have focused on designing responsive nanocarriers that can interact with the mucus layer and deliver drugs specifically to the site of infection. These nanocarriers are engineered with surface modifications that enhance their ability to adhere to the mucus layer or bacterial surfaces. For example, the use of chitosan, a biopolymer with mucoadhesive properties, has been shown to improve the adhesion of nanocarriers to the mucus layer, thereby increasing their residence time in the stomach [24]. Chitosan's positive charge enables it to interact electrostatically with the negatively charged components of the mucus, promoting adhesion and enhancing the stability of the nanocarriers within the mucus layer. Similarly, the incorporation of adhesion peptides or other molecular ligands that specifically bind to the surface of *H. pylori* can increase the targeting efficiency of nanocarriers, allowing them to adhere more effectively to the bacterial surface and release their therapeutic payload directly at the infection site.

Furthermore, the design of responsive nanocarriers that can release their drug payloads in response to internal environmental signals, such as changes in pH or the presence of specific enzymes, offers a promising approach for improving the efficacy of treatments for *H. pylori* infection. These nanocarriers can be designed to cross the mucus barrier and, once localized at the infection site, release their drug payloads in a controlled manner. For example, nanocarriers designed to respond to the acidic environment of the stomach can release their contents when exposed to the low pH of the gastric lumen, while those designed to respond to enzymes such as urease or phospholipase—enzymes that are overexpressed in *H. pylori* infection—can trigger drug release specifically at the site of bacterial colonization. This targeted approach allows for the controlled release of drugs at the site of infection, enhancing their therapeutic effect while minimizing systemic side effects. By ensuring that drugs are released only at the infection site, responsive nanocarriers can increase the concentration of therapeutic agents where they are most needed, thus improving treatment efficacy. Moreover, by overcoming the protective mucus barrier and ensuring localized drug release, these nanocarriers reduce the need for high systemic drug concentrations, which are often associated with side effects such as gastrointestinal disturbances or systemic toxicity.

In conclusion, the gastric mucus barrier and its adhesion properties present significant challenges for drug delivery in the treatment of *H. pylori* infections. However, by leveraging advances in nanotechnology, particularly through the use of mucoadhesive nanocarriers and responsive drug delivery systems, it is possible to overcome these barriers. The use of nanocarriers that can interact with the mucus layer, target *H. pylori*, and release drugs in a controlled, site-specific manner offers a promising strategy for improving the treatment of gastric infections and addressing the limitations of traditional therapeutic approaches.

### 9.2.5 Inflammatory Microenvironment Characteristics

*H. pylori* infection is known to induce chronic gastritis and initiate a cascade of immune responses, ultimately giving rise to a characteristic inflammatory microenvironment within the stomach. This inflammatory response is crucial to the pathogenesis of *H. pylori* infection, as it enables the host's immune system to attempt to control and eliminate the pathogen [25]. However, in the case of *H. pylori* infection, the prolonged and excessive immune reaction leads to ongoing tissue damage and inflammation, contributing to the persistence of the infection. Inflammatory cells, primarily neutrophils, macrophages, and lymphocytes, infiltrate the gastric mucosa in response to infection. These immune cells release a wide array of cytokines, chemokines, and other signaling molecules, such as interleukins (ILs), tumor necrosis factor-alpha (TNF-α), and interferons (IFNs), which recruit more immune cells to the site of infection. This cascade of immune activity not only intensifies the local inflammatory response but also contributes to tissue damage by inducing the release of proteases, reactive oxygen species (ROS), and other inflammatory mediators.

The persistent inflammatory state that develops during *H. pylori* infection significantly alters the local gastric microenvironment. As inflammation continues, several key changes occur, particularly at the infection site, which ultimately exacerbate the pathology of the disease. One of the most notable changes is the modification of the local pH. In areas of inflammation, the pH often becomes more acidic, particularly due to the release of acidic mediators such as protons from inflamed immune cells and the ongoing degradation of the gastric mucosal barrier. This acidification creates a progressively worsening environment for the host tissue, as well as for any therapeutic interventions aimed at treating the infection. The increasing acidity in the local environment not only facilitates the persistence of *H. pylori*, which thrives in such conditions, but also disrupts the normal function of gastric epithelial cells, leading to further tissue damage.

In addition to the changes in pH, other aspects of the gastric microenvironment are affected during *H. pylori* infection. Temperature, osmotic pressure, and blood flow can also experience significant abnormalities in inflamed areas [26]. As the immune response intensifies, the temperature of the tissue may increase as a result of localized inflammation and metabolic activity. This localized increase in temperature, coupled with alterations in osmotic pressure, contributes to a microenvironment that favors bacterial survival and persistence. Additionally, during chronic inflammation, the gastric mucosal barrier, which normally acts as a protective shield for the stomach lining, is compromised. This damage to the mucosal barrier increases vascular permeability, allowing pathogens like *H. pylori* to penetrate deeper into the gastric tissue. The increased permeability also facilitates the spread of inflammatory mediators, exacerbating the inflammation and causing a vicious cycle that further damages the gastric epithelium.

These pathological changes in the inflammatory microenvironment not only accelerate the progression of *H. pylori* infection but also present a unique opportunity for targeted therapeutic intervention. By recognizing specific inflammatory signals, such as the cytokines and chemokines secreted during the immune response, drug delivery systems can be designed to exploit these signals as "location coordinates." These inflammatory markers, which are often highly concentrated at the site of infection, can guide nanocarriers to precisely deliver therapeutic agents directly to the infected tissue. This targeted approach ensures that the drug is released in the areas where it is most needed, maximizing its therapeutic effect while minimizing systemic exposure and reducing the risk of off-target effects. The ability to target drugs specifically to the site of infection offers a significant advantage over traditional systemic therapies, which often expose the entire body to therapeutic agents and can lead to widespread side effects. Targeted drug delivery, particularly using environment-responsive nanocarriers, increases the concentration of drugs at the infection site, allowing for more effective treatment while sparing healthy tissues from unnecessary exposure. This approach not only enhances the therapeutic efficacy of drugs but also contributes to the development of precision medicine, where treatments are tailored to the specific needs of the patient, minimizing side effects and improving outcomes.

In conclusion, the inflammatory microenvironment induced by *H. pylori* infection provides valuable insights into how nanocarriers can be designed to respond to specific signals and deliver drugs with high precision. By targeting the inflammation-associated changes in pH, temperature, osmotic pressure, and cytokine secretion, nanocarriers can be optimized for controlled drug release, offering a promising strategy for more effective and selective treatment of *H. pylori* infections. The ability to deliver drugs precisely where they are needed most not only improves the efficacy of the treatment but also reduces the risk of adverse effects, demonstrating the tremendous potential of nanomedicine in the field of precision therapeutics.

## 9.3 Classification and Principles of Environment-Responsive Nanocarriers

The design of environment-responsive nanocarriers for the treatment of *H. pylori* infections relies on an in-depth understanding of the gastric pathological environment and the specific characteristics of *H. pylori* colonization. The primary goal of these carriers is to exploit microenvironmental signals to control the precise spatiotemporal release of therapeutic agents, ultimately increasing local drug concentrations while minimizing systemic side effects and damage to healthy tissues. Among the various environmental signals that can be exploited, pH, enzymes, and redox states are the most extensively studied, with each providing unique opportunities for targeted drug delivery [27]. Moreover, the development of multi-responsive systems that can simultaneously respond to multiple environmental cues has emerged as a promising strategy to further refine and enhance drug delivery precision (Fig. 9.2).

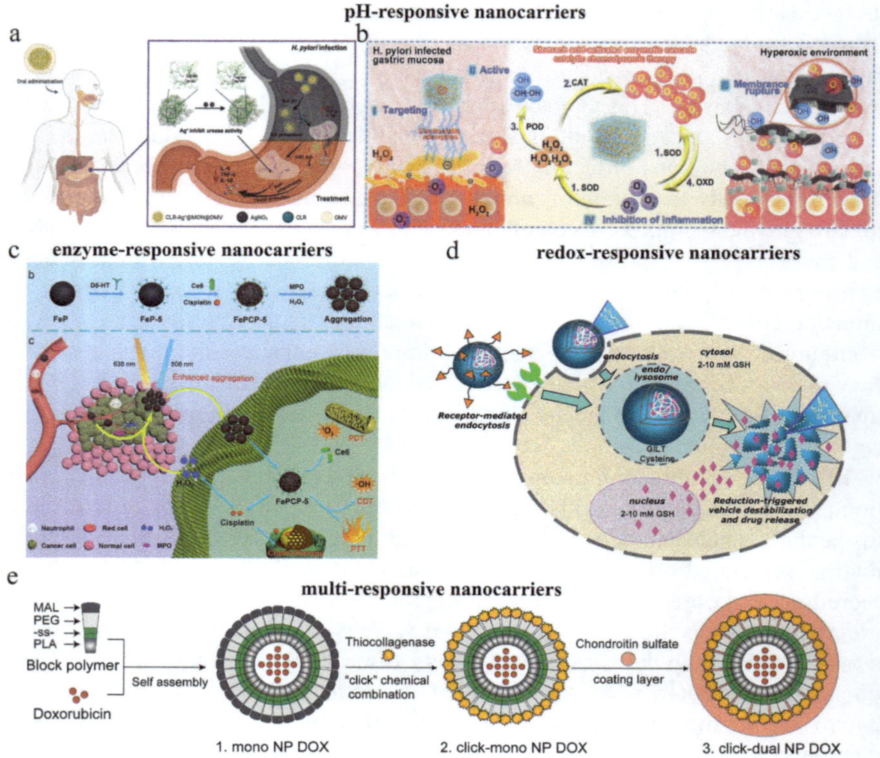


**Fig. 9.2** Design and mechanisms of environment-responsive nanocarriers for targeted drug delivery: pH, enzyme, redox, and multi-responsive systems. (**a**) Biodegradable acid-responsive nanocarrier for enhanced antibiotic therapy against drug-resistant *Helicobacter pylori* via urease Inhibition. (**b**) An acid-responsive nanozyme that activates cascade enzyme-mimicking reactions to selectively kill drug-resistant *H. pylori* and aid gastric mucosal repair. (**c**) Myeloperoxidase-targeted nanotheranostics for self-enhanced synergistic photo−/chemo−/chemodynamic therapy. (**d**) Glutathione-responsive nano-vehicles as a promising platform for targeted intracellular drug and gene delivery.e: Dual-responsive nanoparticles for enhanced drug delivery in breast cancer chemotherapy. (Reprinted with permission from Refs. [3, 28], Wiley. Reprinted with permission from Refs. [29–31], Elsevier)

## 9.3.1 pH-Responsive Nanocarriers for Targeted Treatment

The design of pH-responsive nanocarriers represents one of the most promising strategies for the targeted treatment of *H. pylori* infections, leveraging the unique pH gradient present within the stomach. The human stomach is characterized by a highly acidic gastric lumen with a pH typically ranging from 1.0 to 3.0, which serves not only as a crucial digestive mechanism but also as an innate defense system against pathogens. However, the regions where *H. pylori* colonizes, particularly the gastric mucus layer, exhibit a near-neutral pH (approximately 6–7). This stark contrast between the acidic gastric lumen and the relatively neutral mucus environment provides an ideal natural trigger for the targeted and controlled release of therapeutic agents encapsulated in nanocarriers. By exploiting this pH gradient, nanocarriers can be designed to release drugs selectively at the site of infection, enhancing the localized treatment of *H. pylori* while minimizing systemic exposure.

Nanocarriers designed to be pH-responsive typically incorporate materials that undergo conformational changes, dissolution, or degradation upon encountering an acidic environment, ensuring the efficient release of the encapsulated drug at the infection site. Such materials are often based on polymers or inorganic substances that are sensitive to changes in pH. For example, pH-sensitive polymers such as polyacrylic acid or poly(ethylene glycol)-co-poly(lactic acid) can undergo protonation or hydrolysis in response to low pH, leading to structural collapse and the release of the encapsulated drugs [32]. Similarly, inorganic materials, such as mesoporous silica nanoparticles or other inorganic nanoparticles, can be engineered to undergo dissolution under acidic conditions, releasing their payload in a controlled and localized manner. A common and widely used approach in the design of pH-responsive nanocarriers is the incorporation of pH-sensitive chemical linkages into the polymer backbone. These linkages can include hydrazone bonds, ketone linkages, or polyacrylic acid segments. These materials are designed to break, protonate, or undergo hydrolysis at low pH, resulting in the disassembly or degradation of the nanocarrier structure. For instance, nanomicelles—self-assembled nanoparticles formed from amphiphilic copolymers—often feature hydrophobic cores that encapsulate poorly soluble drugs. When exposed to the acidic environment in the stomach, the hydrophobic cores of these micelles can undergo a structural collapse, thereby releasing the encapsulated drugs. This mechanism is highly advantageous as it ensures rapid and localized release of the drug at the site of infection, without affecting the surrounding healthy tissues. Such nanocarriers can be engineered to maintain their stability in the neutral or slightly acidic environments of the stomach lining, only triggering drug release upon encountering the more acidic regions near the infection sites, thus enhancing the therapeutic efficacy.

Since *H. pylori* typically colonizes the near-neutral pH regions of the gastric mucus layer, such as those close to the gastric epithelial cells, pH-responsive nanocarriers offer a highly targeted treatment modality. The near-neutral pH of the mucus layer provides an ideal site for the localized release of drugs. Drugs released at these sites are more likely to act on bacterial aggregates, as the bacterial cells are concentrated in these regions. Recent research has introduced innovative pH-responsive nanocarriers like CAMO [3], which can effectively target *H. pylori* and alleviate associated gastric inflammation. CAMO uses a mesoporous organic silica nanoparticle core, which is redox-responsive and loaded with therapeutic agents, such as silver ions ($Ag^{+}$) and clarithromycin (CLR). Both $Ag^{+}$ and CLR are released in response to the acidic gastric environment, thereby providing effective antimicrobial treatment. The surface of the nanocarrier is encapsulated in hyaluronic acid, a biocompatible polymer that prevents premature drug leakage while ensuring controlled release in response to the acidic conditions found in the stomach. This innovative system not only delivers a dual-action therapeutic approach but also ensures that the drugs are only released when and where they are needed, minimizing side effects and improving the overall treatment efficacy. In addition to enhancing the precision of drug delivery, pH-responsive nanocarriers offer several other advantages in treating *H. pylori* infections. Traditional oral drug delivery systems often struggle to maintain therapeutic concentrations in the stomach for prolonged periods, leading to insufficient therapeutic effects and the risk of resistance. The use of pH-responsive nanocarriers, however, provides a sustained release profile that ensures drugs are available at the infection site for a longer period, thus improving treatment outcomes. Furthermore, this approach reduces the likelihood of drug degradation or absorption by nontarget tissues, increasing the bioavailability of the therapeutic agents at the infection site and reducing potential side effects on surrounding healthy tissues.

Moreover, pH-responsive nanocarriers can be further optimized to account for the variations in the pH within different regions of the stomach. For example, the pH within the gastric lumen may fluctuate in response to food intake or the progression of the infection [33]. By tailoring the release mechanism to respond to these dynamic changes in pH, nanocarriers can be designed to release their therapeutic payload precisely when it is needed most. This level of precision enhances the ability of the nanocarrier to treat *H. pylori* infections more effectively, while minimizing the systemic exposure that often results in adverse side effects.

The development of pH-responsive nanocarriers for the treatment of *H. pylori* infections represents a significant step forward in the field of nanomedicine. These carriers not only improve the therapeutic efficacy of existing drugs but also offer new avenues for the treatment of gastric diseases, particularly those associated with *H. pylori* infections. Ongoing research is focused on optimizing the design and functionality of these systems, with the ultimate goal of providing a safer, more effective, and targeted approach to treating gastric infections while minimizing the risks of side effects and resistance. In conclusion, the pH-responsive nanocarrier system holds great promise in revolutionizing the treatment of *H. pylori* infections. By exploiting the natural pH gradient in the stomach and using smart nanocarrier systems that respond to this gradient, researchers can develop highly effective, localized, and controlled drug delivery strategies that provide targeted treatments. These systems are not only designed to overcome the challenges posed by the gastric environment but also to provide precision in targeting bacterial colonies, thus improving both the therapeutic outcomes and patient safety.

### 9.3.2 Enzyme-Responsive Nanocarriers for Specific Targeting

In the quest for more precise and targeted treatments for *H. pylori* infections, enzyme-responsive nanocarriers have emerged as an innovative and highly promising approach. These nanocarriers take advantage of the upregulation of specific enzymes associated with the infection, enabling the controlled release of therapeutic agents directly at the infection site. Among the key enzymes involved in *H. pylori* infection, urease plays a pivotal role, and its activity offers a unique opportunity for targeted drug delivery. Other enzymes, such as matrix metalloproteinases (MMPs), which are upregulated during the inflammatory response, also present significant potential for the development of enzyme-responsive systems. Together, these enzymes provide both a natural and specific cue for the design of enzyme-sensitive nanocarriers [11].

Urease is a hallmark enzyme of *H. pylori* and is crucial to the bacterium's ability to survive in the acidic environment of the stomach. Upon infection, *H. pylori* secretes large amounts of urease, which catalyzes the hydrolysis of urea into ammonia and carbon dioxide. The ammonia produced in this reaction neutralizes the surrounding gastric acid, creating a localized alkaline microenvironment around the bacteria, which is essential for *H. pylori* survival and colonization. This ability to generate an alkaline microenvironment protects the bacterium from the harsh acidic conditions of the stomach, enabling it to persist and proliferate. The enzymatic activity of urease can be strategically exploited in the design of nanocarriers for targeted drug delivery. By incorporating urease-sensitive chemical bonds or peptide sequences into the nanocarrier's structure, researchers can design carriers that degrade or release their payload in response to the presence of urease. For example, urease-sensitive polymers or peptides can be embedded into the nanocarrier matrix, ensuring that the nanocarrier remains stable in the stomach's acidic environment but undergoes degradation when it encounters high concentrations of urease. This localized enzymatic degradation results in the controlled release of the encapsulated drugs at the site of infection, where urease is abundant. This highly targeted release improves the therapeutic efficacy of the drug while minimizing the exposure of healthy tissues to the active compound, thus reducing the potential for side effects. The primary advantage of this approach is its specificity. Since urease is exclusively associated with *H. pylori* infection, the targeted release mechanism ensures that drugs are delivered directly to the bacterial colonization sites, improving treatment outcomes. Additionally, urease-sensitive nanocarriers can be designed to respond to changes in the local pH that occur as a result of urease activity. By combining pH sensitivity with enzyme responsiveness, these nanocarriers can provide a dual-triggered release mechanism, ensuring that the drug is delivered precisely at the site of infection and only when the necessary conditions are met.

In addition to urease, the inflammatory environment in the stomach during *H. pylori* infection involves the upregulation of several other enzymes, notably matrix metalloproteinases (MMPs) and myeloperoxidase (MPO). MMPs are a group of enzymes involved in the degradation of extracellular matrix (ECM) proteins and are key players in tissue remodeling, particularly during inflammation. During *H. pylori* infection, MMPs are upregulated in the gastric tissue, where they facilitate the breakdown of the gastric mucosal barrier. This further exacerbates tissue damage and inflammation, contributing to the progression of chronic gastritis and other gastric diseases. Myeloperoxidase (MPO), an enzyme secreted by neutrophils, is also elevated in the stomach during *H. pylori* infection and plays a critical role in the inflammatory process. MPO generates reactive oxygen species (ROS) and contributes to oxidative stress, further damaging the gastric mucosa. Elevated levels of MPO can also enhance the activity of MMPs, creating a feedback loop that exacerbates tissue injury. Both MMPs and MPO present valuable targets for the development of enzyme-responsive nanocarriers [30]. By incorporating MMP- and MPO-sensitive peptide sequences, such as those containing proline or glycine-rich motifs, into the nanocarrier structure, researchers can design nanocarriers that are selectively degraded in the presence of elevated MMPs and MPO in the inflamed gastric tissue. Upon exposure to these enzymes, the peptide sequences in the nanocarrier undergo cleavage, triggering the release of the encapsulated drugs. This approach allows for highly specific drug delivery to the inflammatory regions, where the drug is most needed, while minimizing systemic exposure and the potential for side effects on healthy tissues.

A particularly innovative development in the field of enzyme-responsive drug delivery is the use of enzyme-driven bacterial outer membrane vesicle (OMV) nanorobots. OMVs are nanoscale vesicles naturally secreted by bacteria, including *H. pylori*. These vesicles play a critical role in bacterial communication and the transport of virulence factors. Researchers have harnessed the catalytic activity of urease on the surface of OMVs to create nanorobots capable of responding to the high urease concentrations found in the stomach during infection [34]. These nanorobots are designed to utilize urease catalysis to break down urea, generating the necessary energy to power the release of encapsulated drugs. The OMV nanorobots are specifically targeted to the high-concentration urease environment surrounding *H. pylori* colonies. By exploiting the catalytic properties of urease, these nanorobots can deliver therapeutic agents directly to the infection site, ensuring precise targeting and maximizing therapeutic efficacy. Moreover, the use of OMVs as nanocarriers offers several advantages, including their biocompatibility, ease of functionalization, and ability to interact with *H. pylori* in a natural and biologically relevant manner.

Enzyme-responsive nanocarriers offer several significant advantages in the treatment of *H. pylori* infections. Their ability to specifically target infection sites based on the presence of enzymes such as urease and MMPs ensures that drugs are delivered precisely where they are needed, improving therapeutic outcomes while minimizing off-target effects. Furthermore, the incorporation of multiple enzymatic triggers into the same nanocarrier system allows for a more nuanced and adaptable release mechanism, which can respond to changes in the local environment, such as shifts in enzyme levels or pH. However, there are also challenges associated with the development of enzyme-responsive nanocarriers. The stability of the nanocarrier in the gastric environment, especially in the presence of varying pH levels and digestive enzymes, is a key concern. Additionally, the development of enzyme-responsive systems requires careful design to ensure that the carrier's response to the target enzyme is highly specific, minimizing the risk of premature degradation or unspecific drug release. Despite these challenges, the potential for enzyme-responsive nanocarriers to provide a more targeted, efficient, and controlled drug delivery system for *H. pylori* infections is vast. In conclusion, enzyme-responsive nanocarriers represent an innovative approach to the treatment of *H. pylori* infections. By leveraging the enzymatic activity of key players like urease and MMPs, these nanocarriers provide a highly specific and localized delivery of drugs to infection sites, improving therapeutic efficacy while reducing systemic exposure.

Ongoing research in this field continues to refine and optimize these systems, with the potential to significantly improve the treatment outcomes for *H. pylori* infections and other gastric diseases.

### 9.3.3 Redox-Responsive Nanocarriers for Enhanced Drug Delivery

Redox-responsive nanocarriers represent another promising class of nanomaterials, especially in the context of *H. pylori* infection. During infection, gastric epithelial cells and infiltrating immune cells release large amounts of ROS, which contribute to local oxidative stress and tissue damage. Simultaneously, molecules like GSH [29], an important antioxidant, are elevated in inflammatory tissues. The imbalance between ROS and GSH in the infection site creates an ideal condition for the development of redox-responsive nanocarriers.

Polymers containing disulfide bonds or selenide structures are commonly used in redox-responsive nanocarriers. In the presence of high GSH concentrations, these nanomaterials undergo cleavage, triggering the release of the encapsulated drug. ROS-sensitive groups, such as thiol groups or selenium-based compounds, can also be incorporated into nanocarriers to achieve degradation under oxidative conditions. This approach ensures that drug release is highly localized in areas with elevated ROS, such as the sites of *H. pylori* infection, where oxidative stress is prevalent. Researchers at Sun Yat-sen University, for example, have developed dual-redox-responsive nanodrug delivery systems that can precisely target high-GSH environments to release drugs, offering a promising strategy for enhancing *H. pylori* clearance.

### 9.3.4 Multi-responsive Nanocarriers: The Future of Precision Drug Delivery

While pH, enzyme, and redox-responsive nanocarriers represent individual approaches for targeted drug delivery, the integration of multiple response mechanisms into a single nanocarrier system holds tremendous promise [27, 31]. Multi-responsive nanocarriers are designed to respond to a combination of environmental signals, such as pH, enzymes, and ROS, ensuring that drugs are released precisely when and where they are needed most. For example, a nanocarrier designed with both acid-sensitive and enzyme-sensitive bonds can initially respond to the acidic gastric environment, then further degrade in the presence of urease, providing dual protection and a more controlled drug release profile [3].

Other multi-responsive systems combine pH and ROS responses to ensure that drugs are released in areas of inflammation characterized by both low pH and high oxidative stress [21, 35]. These systems offer high stability and selectivity, allowing for more precise targeting and reducing the risk of nonspecific drug release. Multi-responsive nanocarriers represent the future of nanomedicine, where the combination of multiple environmental cues can be exploited to optimize drug delivery systems, achieving the highest therapeutic efficacy while minimizing side effects.

In summary, the development of environment-responsive nanocarriers for the treatment of *H. pylori* infections has revolutionized drug delivery strategies. By utilizing pH, enzyme, and redox-sensitive mechanisms, these carriers can be tailored to respond to specific microenvironmental cues, significantly enhancing the targeted delivery of therapeutic agents. The ongoing research into multi-responsive systems further holds the potential to create more sophisticated and precise drug delivery platforms. The integration of these technologies not only offers improved efficacy in treating *H. pylori* infections but also paves the way for more effective nonantibiotic therapeutic strategies, minimizing the risks of antibiotic resistance and improving patient outcomes in the treatment of gastric diseases.

## 9.4 Key Nanomaterial Systems

The choice of material systems plays a decisive role in the research and development of environment-responsive nanocarriers. Different materials not only determine the stability, degradability, and biocompatibility of the carriers but also determine their response modes to environmental signals. Currently, the most widely used material systems include polymeric nanomaterials, inorganic nanomaterials, biologically derived materials, and composite materials, each of which has its own unique advantages and limitations [36].

Polymeric nanomaterials have become a research hotspot due to their controllable structure and modifiability. Through molecular design, acid-sensitive, enzyme-sensitive, or redox-sensitive chemical bonds can be incorporated into the polymer backbone, imparting responsiveness to the carrier. Common forms include nanomicelles, nanogels, and polymeric vesicles. Nanomicelles are typically self-assembled from amphiphilic polymers, where the hydrophobic core can encapsulate poorly soluble drugs, while the hydrophilic shell provides stability. When the external environment changes, such as an increase in acidity or enzymatic cleavage, the micelle structure disassembles, releasing the drug rapidly. Nanogels, on the other hand, provide higher drug loading capacity through cross-linked networks and can expand or degrade under specific stimuli, thus modulating drug release kinetics. This flexibility allows for the precise control of drug release based on environmental changes [37].

Inorganic nanomaterials, due to their unique physicochemical properties, have also attracted significant attention [20, 38]. For example, iron oxide nanoparticles not only exhibit magnetic responsiveness but can also be surface-modified to achieve acid-sensitive or ROS-sensitive drug release. Metal-organic frameworks (MOFs) are a class of porous crystalline materials composed of metal ions and organic ligands, characterized by an extremely high surface area and tunable pore size. MOFs can gradually dissociate under acidic conditions, releasing the encapsulated drug, and their surfaces can be further functionalized to enable multi-responsive behavior. Compared to polymeric materials, inorganic materials often offer advantages in structural stability and functional diversity, but their biodegradability and long-term safety still require further validation.

Biological materials, such as exosomes and liposomes, have gained increasing attention in recent years [39]. Exosomes are small vesicles secreted by cells, naturally endowed with excellent biocompatibility and targeting properties, allowing them to cross the gastric mucus barrier and achieve effective delivery. Liposomes, formed by lipid bilayers, encapsulate drugs and mimic the natural properties of cell membranes. In the design of internal environment-responsive systems, researchers often modify the surface of liposomes with acid-sensitive or enzyme-sensitive molecules to enable smart drug release. The primary advantage of these biological materials is their excellent biological safety and in vivo metabolic properties, making them suitable for further clinical translation.

Composite nanomaterials aim to integrate the advantages of various material types [40]. For example, by coating a layer of inorganic materials on the surface of polymeric nanoparticles or embedding lipid structures into the pores of MOFs, multilayered response mechanisms can be achieved [11]. This composite strategy not only improves drug loading capacity and controllability but also enhances the carrier's stability in complex gastric environments. In recent years, some composite systems have been designed to remain stable in the gastric acidic environment and release drugs upon reaching bacteria-enriched areas, triggered by enzymatic or ROS signals, showing promising prospects for in vivo applications.

Overall, the diversity of key nanomaterial systems provides a wealth of options for the construction of environment-responsive nanocarriers. Through rational design and optimization, these materials can adapt to the complex gastric environment, offering a powerful tool for the precise treatment of *H. pylori* infections.

## 9.5 Drug/Active Molecule Loading and Release Mechanisms

The therapeutic efficacy of environment-responsive nanocarriers depends not only on the design of the material itself but also on the method of loading drugs or active molecules and their release mechanisms. Due to the specific nature of *H. pylori* infection, the therapeutic agents to be delivered may not only include traditional small molecule drugs but also antimicrobial peptides, probiotics, metal ions, and other bioactive molecules. Therefore, the carrier system must possess strong loading versatility and controllable release capabilities [41].

Drug loading methods are generally divided into three categories: physical adsorption, electrostatic interactions, and covalent modification [42]. Physical adsorption is the most common method, relying on hydrophobic interactions or van der Waals forces between the drug molecules and the carrier. This method is easy to operate but carries a higher risk of drug leakage. Electrostatic interactions are often used for loading charged drugs or peptide molecules. For example, cationic polymers can form stable complexes with negatively charged antimicrobial peptides. Covalent modification involves linking drugs to the carrier framework using degradable chemical bonds. This method ensures stable drug binding but requires the introduction of specific environment-sensitive bonds in the design to ensure drug release at the target site.

The release mechanism is the core of environment-responsive nanocarriers. For example, in acid-sensitive systems, chemical bonds that are hydrolyzable or dissociable under low pH conditions, such as hydrazone or acetal bonds, are often introduced. When the carrier reaches the acidic environment of the stomach, these chemical bonds break, leading to rapid drug release. Enzyme-responsive systems rely on enzyme-sensitive segments embedded in the carrier, which degrade when exposed to specific enzymes, such as urease or matrix metalloproteinases (MMPs), triggering drug release. Redox-responsive systems use ROS or high glutathione (GSH) concentrations at the inflammatory site to trigger drug release through the cleavage of ROS-sensitive groups or reduction of disulfide bonds. These mechanisms significantly enhance the local concentration of the drug at the target site, reducing the systemic exposure and associated side effects. The kinetics of drug release also affects the therapeutic outcome. A rapid release is suitable for clearing acute bacterial loads, while a sustained-release mode is ideal for maintaining drug concentrations over an extended period to prevent relapse. By controlling the degradation rate, cross-linking density, and response thresholds of the materials, researchers can achieve various release kinetics, from "burst release" to "sustained release." Furthermore, the introduction of multi-responsive systems enhances the precision of drug release. For instance, materials may partially dissociate and release the drug in an acidic environment, and upon subsequent exposure to urease, further accelerate drug release. This staged release strategy improves the therapeutic efficacy.

In the context of nonantibiotic therapies for *H. pylori*, the active molecules may include metal ions (such as silver and zinc ions), natural products (such as polyphenols), probiotics, or their metabolic products. For the delivery of probiotics, the carrier needs to provide protection in the acidic gastric environment to prevent bacterial death and ensure the release of viable bacteria when approaching the mucus layer. This special requirement has led to the development of more complex multilayer carrier designs.

Therefore, the loading and release mechanisms of drugs and active molecules are not only the technical core of environment-responsive nanocarrier design but also the key factors determining their clinical efficacy. By precisely controlling the loading strategies and release patterns, the full potential of nonantibiotic treatments can be realized.

## 9.6 Advantages, Challenges, and Future Directions of Environment-Responsive Nanocarriers

Environment-responsive nanocarriers offer significant advantages in the treatment of *H. pylori* infections. First, they can achieve selective drug release based on specific environmental signals at the stomach or infection site, such as pH, enzyme activity, and redox status. This allows for efficient treatment at the target site, significantly increasing the local concentration of the drug while minimizing its impact on normal tissues. Through this "smart response" mechanism, nanocarriers can enhance the therapeutic effect at the infection site and substantially reduce the common side effects associated with traditional drug treatments. Second, the design of environment-responsive nanocarriers effectively addresses the issue of antibiotic resistance. Traditional antibiotic treatments rely on the widespread use of antibiotics to suppress *H. pylori*, but prolonged use can lead to the emergence of resistant strains. In contrast, environment-responsive nanocarriers use a targeted drug release strategy that delivers therapeutic agents directly to the infection site, reducing systemic drug exposure and mitigating the risk of bacterial resistance. Additionally, environment-responsive nanocarriers can be combined with nonantibiotic therapeutic approaches, such as probiotics and antimicrobial peptides, to generate synergistic effects, offering new treatment strategies for clinical applications.

Despite their immense potential, the clinical application of environment-responsive nanocarriers still faces several challenges. First, the complexity of the gastric environment poses difficulties for the design and application of nanocarriers. The gradient difference between the highly acidic environment of the stomach and the near-neutral environment at the *H. pylori* colonization site, as well as the physical barrier of the gastric mucus layer, requires nanocarriers to be sensitive to environmental changes while being able to effectively penetrate these barriers and precisely release drugs at the infection site. This places higher demands on the stability, targeting ability, and drug release control of the nanocarriers. Furthermore, although a variety of material systems have been designed for environment-responsive nanocarriers, their production costs remain high, particularly due to the complexity of the synthesis process and the technical challenges of large-scale production. These factors continue to limit their widespread application. Moreover, some nanomaterials may pose biocompatibility issues or potential long-term toxicity, especially if they accumulate in the body, which requires further validation and optimization. Lastly, clinical translation remains a significant barrier. The clinical application of nanocarriers must meet stringent requirements for drug safety and efficacy, as well as resolve issues related to reproducibility in manufacturing, quality control, and regulatory approval. While many nanocarriers have entered preclinical stages, their progress toward market entry and widespread application remains slow, necessitating more clinical trial data and standardized production processes to support their commercialization.

Despite these challenges, the future of environment-responsive nanocarriers in *H. pylori* treatment remains promising. Future research should focus on several key areas: (1) Material innovation and multifunctional design are essential for improving drug loading capacity, release efficiency, and targeting specificity. Researchers should prioritize the development of novel nanomaterials, particularly composite nanocarriers capable of responding to multiple environmental signals. By combining pH, enzyme, and redox responses, more precise drug release can be achieved. (2) Intelligent and controllable release technologies will allow nanocarriers to respond to environmental changes at specific spatiotemporal conditions. Incorporating multi-signal-trigger systems will enable efficient drug release, improving therapeutic outcomes while reducing side effects. (3) Production process optimization and cost control are critical to overcoming the challenge of high production costs. Research should focus on simplifying synthesis methods, improving production efficiency, and identifying lower-cost raw materials to enable large-scale production and reduce costs, facilitating clinical applications. (4) The advancement of preclinical and clinical trials will lay the foundation for the clinical translation of nanocarriers. In-depth toxicological studies, pharmacokinetic analysis, and clinical efficacy assessments will ensure the safety and effectiveness of nanocarriers, while standardized production and quality control systems must be established to meet clinical demands.

In conclusion, with the continuous development of nanotechnology, environment-responsive nanocarriers hold great promise as an innovative therapeutic strategy for the treatment of *H. pylori* and other infectious diseases, showcasing immense potential for clinical application.

## References

1. Zhang Q, Zhou J, Yuan J, Sun W, Dong Q, et al. Novel therapeutic strategies for antibiotic resistant *H. pylori*. Chin J Microecol. 2020;32(7):842–8.
2. Shadvar N, Akrami S, Mousavi Sagharchi SM, Askandar RH, Merati A, et al. A review for non-antibiotic treatment of *H. pylori*: new insight. Front Microbiol. 2024;15:1379209. https://doi.org/10.3389/fmicb.2024.1379209.
3. Fan H, Wong KI, Ma Y, Li M, Li H, et al. Biodegradable acid-responsive Nanocarrier for enhanced antibiotic therapy against drug-resistant *H. pylori* via urease inhibition. Adv Funct Mater. 2025;35(2) https://doi.org/10.1002/adfm.202412893.
4. Chitas R, Fonseca DR, Parreira P, Martins MCL. Targeted nanotherapeutics for the treatment of *H. pylori* infection. J Biomed Sci. 2024;31(1) https://doi.org/10.1186/s12929-024-01068-9.
5. Yi M, Chen S, Yi X, Zhang F, Zhou X, et al. *H. pylori* infection process: from the molecular world to clinical treatment. Front Microbiol. 2025;16:1541140. https://doi.org/10.3389/fmicb.2025.1541140.
6. Ansari S, Yamaoka Y. Survival of *H. pylori* in gastric acidic territory. Helicobacter. 2017;22(4) https://doi.org/10.1111/hel.12386.
7. Hu W, Chen ZM, Wang Y, Yang C, Wu ZY, et al. Single-cell RNA sequencing dissects the immunosuppressive signatures in *H. pylori* -infected human gastric ecosystem. Nat Commun. 2025;16(1) https://doi.org/10.1038/s41467-025-59339-4.
8. McNulty R, Ulmschneider JP, Luecke H, Ulmschneider MB. Mechanisms of molecular transport through the urea channel of *H. pylori*. Nat Commun. 2013:4. https://doi.org/10.1038/ncomms3900.
9. Thorpe HA, Tourrette E, Yahara K, Vale FF, Liu S, et al. Repeated out-of-Africa expansions of *H. pylori* driven by replacement of deleterious mutations. Nat Commun. 2022;13(1) https://doi.org/10.1038/s41467-022-34475-3.
10. Aberg A, Gideonsson P, Bhat A, Ghosh P, Arnqvist A. Molecular insights into the fine-tuning of pH-dependent ArsR-mediated regulation of the SabA adhesin in *H. pylori*. Nucleic Acids Res. 2024;52(10):5572–95. https://doi.org/10.1093/nar/gkae188.
11. Zhang W, Zhou Y, Fan Y, Cao R, Xu Y, et al. Metal-organic-framework-based hydrogen-release platform for multieffective *H. pylori* targeting therapy and intestinal flora protective capabilities. Adv Mater. 2022;34(2) https://doi.org/10.1002/adma.202105738.

12. Zhu H, Hart CA, Sales D, Roberts NB. Bacterial killing in gastric juice -: effect of pH and pepsin on Escherichia coli and *H. pylori*. J Med Microbiol. 2006;55(9):1265–70. https://doi.org/10.1099/jmm.0.46611-0.
13. Kodama K, Sumii K, Kawano M, Kido T, Nojima K, et al. *H. pylori* infection increases serum nitrate and nitrite more prominently than serum pepsinogens. Helicobacter. 2002;7(1):9–13. https://doi.org/10.1046/j.1523-5378.2002.00050.x.
14. Zavros Y, Merchant JL. The immune microenvironment in gastric adenocarcinoma. Nat Rev Gastroenterol Hepatol. 2022;19(7):451–67. https://doi.org/10.1038/s41575-022-00591-0.
15. Sokolova O, Naumann M. Matrix metalloproteinases in *H. pylori* -associated gastritis and gastric cancer. Int J Mol Sci. 2022;23(3) https://doi.org/10.3390/ijms23031883.
16. Shaalan H, Azrad M, Peretz A. The effect of three urease inhibitors on *H. pylori* viability, urease activity and urease gene expression. Front Microbiol. 2024:15. https://doi.org/10.3389/fmicb.2024.1464484.
17. Chen SQ, Zhong Y, Fan WF, Xiang JJ, Wang GW, et al. Enhanced tumour penetration and prolonged circulation in blood of polyzwitterion-drug conjugates with cell-membrane affinity. Nat Biomed Eng. 2021;5(9):1019-+. https://doi.org/10.1038/s41551-021-00701-4.
18. Noszka M, Strzalka A, Muraszko J, Kolenda R, Meng C, et al. Profiling of the *H. pylori* redox switch HP1021 regulon using a multi-omics approach. Nat Commun. 2023;14(1) https://doi.org/10.1038/s41467-023-42364-6.
19. Daund V, Chalke S, Sherje AP, Kale PP. ROS responsive mesoporous silica nanoparticles for smart drug delivery: a review. J Drug Delivery Sci Technol 2021;64. DOI: https://doi.org/10.1016/j.jddst.2021.102599.
20. Dai LL, Yao MJ, Fu ZX, Li X, Zheng XM, et al. Multifunctional metal-organic framework-based nanoreactor for starvation/oxidation improved indoleamine 2,3-dioxygenase-blockade tumor immunotherapy. Nat Commun. 2022;13(1) https://doi.org/10.1038/s41467-022-30436-y.
21. Li YJ, Cao YJ, Ma KR, Ma R, Zhang MZ, et al. A triple-responsive polymeric prodrug nanoplatform with extracellular ROS consumption and intracellular H2O2 self-generation for imaging-guided tumor chemo-ferroptosis-immunotherapy. Adv Healthc Mater. 2024;13(16) https://doi.org/10.1002/adhm.202303568.
22. Ji Y, Qu S, Shi G, Fan L, Qian J, et al. Triggered cascade-activation nanoplatform to alleviate hypoxia for effective tumor immunotherapy guided by NIR-II imaging. ACS Nano. 2024;18(45):31421–34. https://doi.org/10.1021/acsnano.4c11334.
23. Hu W, Chen ZM, Wang Y, Yang C, Wu ZY, et al. Single-cell RNA sequencing dissects the immunosuppressive signatures in *H. pylori* -infected human gastric ecosystem. Nat Commun. 2025;16(1):3903. https://doi.org/10.1038/s41467-025-59339-4.
24. Ghaffarian R, Pérez-Herrero E, Oh H, Raghavan SR, Muro S. Chitosan–alginate microcapsules provide gastric protection and intestinal release of ICAM-1-targeting nanocarriers, enabling GI targeting in vivo. Adv Funct Mater. 2016;26(20):3382–93. https://doi.org/10.1002/adfm.201600084.
25. Coorens THH, Collord G, Jung HYC, Wang YC, Moore L, et al. The somatic mutation landscape of normal gastric epithelium. Nature. 2025;640(8058) https://doi.org/10.1038/s41586-025-08708-6.
26. Malfertheiner P, Camargo MC, El-Omar E, Liou JM, Peek R, et al. *H. pylori* infection. Nat Rev Dis Primers. 2023;9(1):19. https://doi.org/10.1038/s41572-023-00431-8.
27. Jiang Q, Zhang SF. Stimulus-responsive drug delivery nanoplatforms for osteoarthritis therapy. Small. 2023;19(23) https://doi.org/10.1002/smll.202206929.
28. Tong Y, Liu Q, Fu H, Han M, Zhu H, et al. Cascaded nanozyme based pH-responsive oxygenation for targeted eradication of resistant *H. pylori*. Small. 2024;20(36):2401059. https://doi.org/10.1002/smll.202401059.
29. Cheng R, Feng F, Meng FH, Deng C, Feijen J, et al. Glutathione-responsive nano-vehicles as a promising platform for targeted intracellular drug and gene delivery. J Control Release. 2011;152(1):2–12. https://doi.org/10.1016/j.jconrel.2011.01.030.

30. Liu N, Zuo W, Wu L, Chen L, Xiao Z, et al. Myeloperoxidase-targeted nanotheranostics for self-enhanced synergetic photo/chemo/chemodynamic therapy. Mater Today Chem. 2022:23. https://doi.org/10.1016/j.mtchem.2021.100740.
31. Ren Y, Li P, Xie Y, Xu J, Luo Q, et al. Dual-responsive nanoparticles for enhanced drug delivery in breast cancer chemotherapy. J Control Release. 2025;377:146–61. https://doi.org/10.1016/j.jconrel.2024.11.026.
32. Qin JY, Zhu YB, Zheng DS, Zhao QW. pH-sensitive polymeric nanocarriers for anti-tumor biotherapeutic molecules targeting delivery. Bio-Des Manuf. 2021;4(3):612–26. https://doi.org/10.1007/s42242-020-00105-4.
33. Dong HH, Guan AX, Yang SX, Wang LY, Zhao LJ, et al. An pH-responsive liposome for the targeted treatment of *H. pylori*. Mater Today Bio. 2025:34. https://doi.org/10.1016/j.mtbio.2025.102194.
34. Tang SS, Tang DT, Zhou HH, Li YY, Zhou DW, et al. Bacterial outer membrane vesicle nanorobot. Proc Natl Acad Sci USA. 2024;121(30) https://doi.org/10.1073/pnas.2403460121.
35. Dai FJ, Chen FJ, Zhang JY, Chen XW, Liang HZ, et al. Folate-modified pH and ROS dual-responsive polymeric Nanocarriers for targeted anticancer drug delivery. ACS Appl Nano Mater. 2024;7(7):7289–99. https://doi.org/10.1021/acsanm.4c00021.
36. Lamprecht A. IBD selective nanoparticle adhesion can enhance colitis therapy. Nat Rev Gastroenterol Hepatol. 2010;7(6):311–2. https://doi.org/10.1038/nrgastro.2010.66.
37. Gimenez C, de la Torre C, Gorbe M, Aznar E, Sancenon F, et al. Gated mesoporous silica nanoparticles for the controlled delivery of drugs in cancer cells. Langmuir. 2015;31(12):3753–62. https://doi.org/10.1021/acs.langmuir.5b00139.
38. Janjua TI, Cao YX, Yu CZ, Popat A. Clinical translation of silica nanoparticles. Nat Rev Mater. 2021;6(12):1072–4. https://doi.org/10.1038/s41578-021-00385-x.
39. Liu L, Guo Y, Wang B, Pan Y, Zhao C, et al. Yogurt-inspired hybrid membrane vesicles for the prevention and treatment of *H. pylori* infection. Cell Biomater. 2025;1(4):100072. https://doi.org/10.1016/j.celbio.2025.100072.
40. Tylawsky DE, Kiguchi H, Vaynshteyn J, Gerwin J, Shah JK, et al. P-selectin-targeted nanocarriers induce active crossing of the blood-brain barrier via caveolin-1-dependent transcytosis. Nat Mater. 2023;22(3):391-+. https://doi.org/10.1038/s41563-023-01481-9.
41. Lee JS, Han PT, Chaudhury R, Khan S, Bickerton S, et al. Metabolic and immunomodulatory control of type 1 diabetes via orally delivered bile-acid-polymer nanocarriers of insulin or rapamycin. Nat Biomed Eng. 2021;5(11):1403. https://doi.org/10.1038/s41551-021-00828-4.
42. Liu QY, Zou JH, Chen ZJ, He W, Wu W. Current research trends of nanomedicines. Acta Pharm Sin B. 2023;13(11):4391–416. https://doi.org/10.1016/j.apsb.2023.05.018.

# Multi-stimuli-Responsive Nanomaterials 10

Xin Ding, Xiangyu Luo, Li Yang, and Xinyi Zhong

## 10.1 Concept and Design of Multi-stimuli-Responsive Nanomaterials

The dynamic and complex gastric environment presents a significant challenge for the effective treatment of *Helicobacter pylori* infections [1–3]. The stomach exhibits steep pH gradients (ranging from 1.5 to 7.0), fluctuating temperatures, enzymatic activities, and mechanical motion, all of which complicate the delivery and efficacy of therapeutic agents [4, 5]. Single-stimulus responsive systems [6], which respond to only one external or internal stimulus, often fail to adapt to these multifaceted conditions, leading to limited therapeutic outcomes. Multi-stimuli-responsive nanomaterials, on the other hand, represent a breakthrough in overcoming these limitations by enabling simultaneous responses to multiple stimuli [7].

Multi-stimuli-responsive nanomaterials are intelligent systems engineered to respond to two or more stimuli, such as pH, temperature, light, magnetic fields, ultrasound, or enzymatic activity [8]. These materials are designed by integrating multiple functional components into a single nanostructure, allowing for synergistic interactions and versatile applications. For example, pH- and light-responsive nanomaterials can combine the benefits of pH-triggered activation in the acidic gastric

X. Ding (✉)
School of Medicine, Shenzhen Campus of Sun Yat-sen University, Shenzhen, China

State Key Laboratory of Anti-Infective Drug Discovery and Development, School of Pharmaceutical Sciences, Sun Yat-sen University, Guangzhou, China
e-mail: dingxin3@mail.sysu.edu.cn

X. Luo
School of Pharmaceutical Sciences (Shenzhen), Shenzhen Campus of Sun Yat-sen University, Shenzhen, China

L. Yang · X. Zhong
School of Medicine, Shenzhen Campus of Sun Yat-sen University, Shenzhen, China



J. Gao, Y. Du (eds.), *Non-Antibiotic Therapies for Helicobacter pylori*,
https://doi.org/10.1007/978-981-95-9598-3_10

environment with light-activated control for precise spatial and temporal drug delivery [9]. This dual functionality ensures that therapeutic agents are released in a controlled and targeted manner, enhancing efficacy while minimizing off-target effects [10].

In the context of *H. pylori* therapy, the ability of multi-stimuli-responsive nanomaterials to adapt to the dynamic gastric environment is particularly advantageous. The bacterium thrives in the acidic regions of the stomach, and its colonization often leads to biofilm formation [11], which complicates treatment. Multi-stimuli-responsive systems can address these challenges by:

1. Enhancing Drug Delivery: By responding to multiple stimuli, such as pH and ultrasound [12], nanocarriers can ensure that therapeutic agents are released specifically in the infection site. For instance, pH-responsive nanomaterials can encapsulate drugs and remain stable in the neutral environment, releasing their payload only in the acidic stomach where *H. pylori* resides [13].
2. Increasing Selectivity: Multi-stimuli-responsive systems can reduce systemic toxicity by targeting drug release exclusively to the infection site. For example, light-responsive nanomaterials can be activated by external light sources, enabling controlled drug release in the exact location of the infection [9, 14].
3. Combating Biofilm Formation: *H. pylori* often form biofilms to resist conventional therapies [11]. Multi-stimuli-responsive nanomaterials can disrupt biofilm formation by combining stimuli such as pH and enzyme responsiveness, ensuring more effective penetration and delivery of therapeutic agents [15].
4. Improving Therapeutic Efficacy: By integrating multiple functions into a single platform, these nanomaterials can enhance the overall effectiveness of the therapy. For instance, magnetic and pH-responsive nanoparticles can be guided to the infection site using an external magnetic field, while pH-triggered release ensures drug delivery in the acidic environment [16, 17].

The design of multi-stimuli-responsive nanomaterials involves careful consideration of their composition, structure, and functional groups [18]. Polymers, such as poly(N-isopropylacrylamide) (PNIPAM) and poly (acrylic acid) (PAA), are commonly used due to their temperature and pH responsiveness, respectively [19, 20]. Inorganic nanoparticles, such as gold, silver, or iron oxide, can be incorporated to add functionalities like photothermal or magnetic responsiveness [21]. Additionally, hybrid materials, which combine organic and inorganic components, offer enhanced versatility and performance.

Despite their immense potential, the development of multi-stimuli-responsive nanomaterials for *H. pylori* therapy is not without challenges. Issues such as biocompatibility, stability under physiological conditions, scalability of synthesis, and regulatory approval must be addressed. However, ongoing research is focused on optimizing these materials to overcome these limitations and translate their promising capabilities into clinical applications.

## 10.2 Applications of Multi-stimuli-Responsive Nanomaterials in *H. pylori* Therapy

### 10.2.1 Drug Delivery

Effective drug delivery to eradicate *H. pylori* is fraught with challenges. The harsh acidic environment of the stomach can degrade antibiotics before they reach their target [22]. Furthermore, the bacterium's niche beneath the gastric mucus layer acts as a physical barrier, limiting drug penetration and retention [23]. Consequently, high and frequent dosing is often required, leading to side effects and contributing to the development of resistance [24]. Multi-stimuli-responsive nanomaterials address these issues by protecting their payload in transit, leveraging the pathological triggers of the infection site (e.g., acidity, bacterial enzymes) for targeted release [25, 26], and enhancing localization and retention through functionalized surfaces, thereby drastically improving bioavailability and therapeutic efficacy [27, 28].

#### 10.2.1.1 Case Study 1: pH and ROS-Responsive Delivery of Antibiotics [29]

This innovative study presents a breakthrough approach to *H. pylori* treatment by developing a dual-stimuli-responsive nanoplatform that utilizes both gastric acidity and reactive oxygen species (ROS) to enhance antibiotic delivery and therapeutic efficacy. The research team engineered a sophisticated tannic acid-coated iron-engineered hollow mesoporous silica nanoparticle system (TA-FeHMSN@Amox) that responds to both the physiological stimulus of gastric pH and ROS generated upon chemodynamic therapy, creating a powerful synergistic antibacterial effect against drug-resistant *H. pylori* while minimizing systemic antibiotic exposure and preserving gut microbiota homeostasis (Fig. 10.1).

***Material Design and Stimuli-Responsiveness*** The nanoplatform was meticulously designed with a core-shell structure featuring iron-engineered hollow mesoporous silica nanoparticles (FeHMSN) as the core carrier, loaded with amoxicillin (Amox), and encapsulated with a pH-responsive metal-polyphenol network composed of tannic acid (TA) and $Fe^{3+}$. The FeHMSN core was synthesized via hydrothermal reaction, creating a hollow mesoporous structure with a surface area of 33.056 $m^2\ g^{-1}$ and pore size of 4.9 nm, ideal for high antibiotic loading capacity. The system was engineered to respond to two distinct stimuli: (1) Gastric acidity(pH 2.2–6.0) triggers the degradation of both the silica framework and TA-$Fe^{3+}$ shell, releasing Fe ions and amoxicillin while initiating Fenton reactions; and (2) $H_2O_2$ induces Fenton reaction, generating ROS to damage bacterial cell wall. The synergistic effect of ROS and released antibiotic leads to eradication of *H. pylori*.

***Characteristics and Performance*** Comprehensive material characterization revealed the successful integration of both responsive elements. TEM imaging showed the hollow mesoporous structure with a distinct ~10 nm TA-Fe shell, while

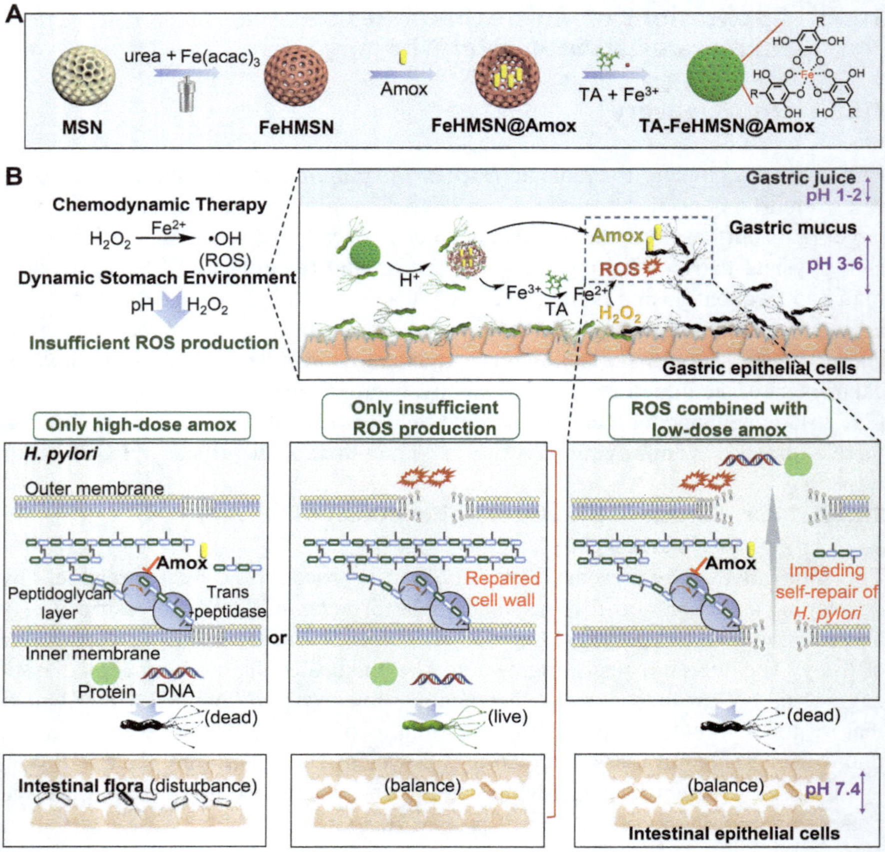


**Fig. 10.1** Preparation and anti-*Helicobacter pylori* mechanism of TA-FeHMSN@Amox. (Reproduced with permission from Ref. [29], American Chemical Society)

element mapping confirmed uniform distribution of Fe throughout the silica matrix and successful amoxicillin loading (8% loading content). The responsive behavior was demonstrated through controlled release studies: under acidic conditions (pH 2.2) alone, the system achieved 46.50% amoxicillin release within 12 h, but when combined with ultrasound stimulation, the release efficiency dramatically increased to 86.46%. The pH-responsive component ensured targeted activation specifically in the gastric environment.

The combination of pH and ultrasound responsiveness enabled multiple antibacterial mechanisms: (1) Acid-triggered release of Fe ions catalyzed Fenton reactions with endogenous $H_2O_2$ at the infection site, generating bactericidal hydroxyl radicals (·OH) that damaged bacterial membranes; (2) Ultrasound-enhanced antibiotic penetration ensured effective drug delivery to *H. pylori* colonies within biofilms and mucosal layers; (3) The sequential activation—first by gastric pH initiating the process, then by ultrasound enhancing it—created a powerful synergistic effect where OH-mediated membrane damage facilitated improved antibiotic uptake, while antibiotic action prevented bacterial repair of ROS-induced damage. This approach demonstrated exceptional efficacy against clinical drug-resistant strains, achieving >99% bactericidal activity against multidrug-resistant *H. pylori* isolates, including those resistant to clarithromycin, metronidazole, and levofloxacin.

In a BALB/c mouse infection model, the dual-responsive system showed superior therapeutic outcomes compared to conventional treatments. The pH responsiveness ensured gastric-specific activation, with nanoparticles maintaining integrity during transit but rapidly releasing their payload upon reaching the acidic stomach environment. The combination with chemodynamic therapy achieved a remarkable 2.42-log reduction in bacterial burden, significantly outperforming equivalent doses of free amoxicillin (0.75-log reduction) and demonstrating comparable efficacy to high-dose dual therapy but with reduced antibiotic quantity and minimal impact on intestinal flora. Histopathological analysis revealed complete eradication of *H. pylori* from gastric tissues with no significant inflammatory response or tissue damage. A crucial advantage of this dual-responsive system was its excellent safety profile and minimal disruption to gut microbiota. The pH-dependent activation preserves the natural gut microbiome composition—a significant improvement over conventional antibiotic therapies that cause substantial microbiota disruption. α-Diversity analysis (Chao1, Shannon, Simpson, and Pielou indexes) confirmed no significant changes in microbial diversity following treatment, and histological examination of major organs showed no evidence of toxicity or tissue damage (Fig. 10.2).

***Future Perspectives and Clinical Translation*** This dual-stimuli-responsive platform represents a significant advancement in targeted antibiotic therapy for gastrointestinal infections. The combination of intrinsic responsiveness offers precise spatiotemporal control over drug release, enabling highly targeted therapy while minimizing systemic exposure and collateral damage to beneficial microbiota. The technology platform could also be adapted for delivery of other antibiotics or combination therapies, providing a versatile approach for treating various gastrointestinal infections beyond *H. pylori*. This innovative strategy addresses the critical challenges of antibiotic resistance and microbiota preservation while leveraging the physiological features of the gastrointestinal tract for enhanced therapeutic outcomes.

#### 10.2.1.2 Case Study 2: pH- and Light-Responsive Nanomaterials Delivering Chinese Medicine Compound [30]

***Material Design and Stimuli-Responsiveness*** The nanoplatform was engineered through a sophisticated "bottom-up and top-down" approach. Black phosphorus nanosheets (BP) were first exfoliated from bulk black phosphorus using an intermittent ultrasonic probe method in DMF, creating a stable BP dispersion with excellent drug-loading capacity. The ISL was then loaded onto the BP nanosheets with

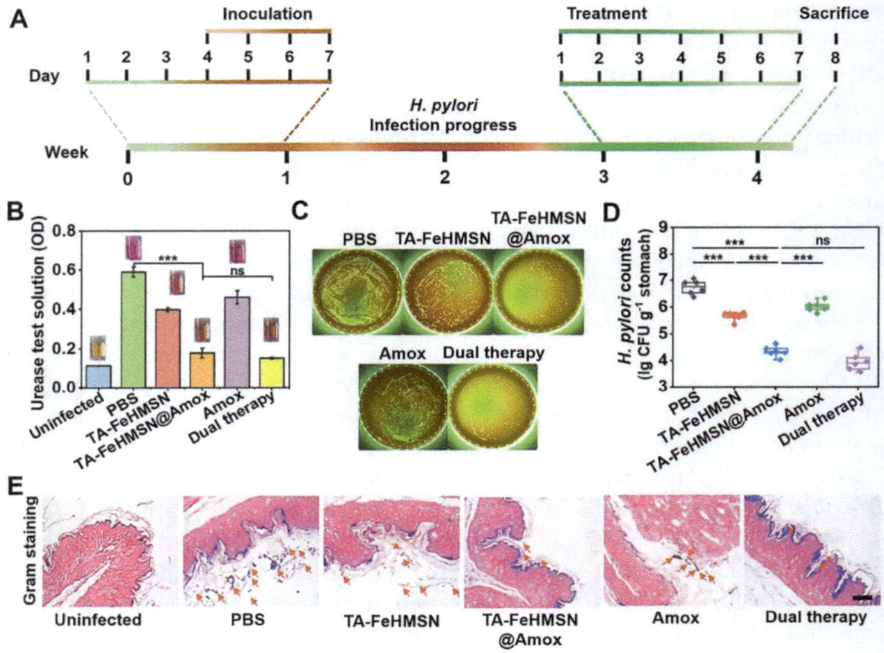


**Fig. 10.2** Therapeutic efficacy of TA-FeHMSN@Amox. (Reproduced with permission from Ref. [29], American Chemical Society)

remarkable efficiency (98.5% loading efficiency at 15% feeding ratio). The critical innovation came through the encapsulation of BP/ISL with rhamnolipid (RHL), a biosurfactant that formed protective micelles around the nanocomposite through hydrophobic interactions.

The system was designed to respond to two key stimuli: (1) Acidic pH of the gastric environment, which promotes swelling of the RHL layer and accelerates BP degradation, facilitating ISL release; and (2) Near-Infrared (NIR) Laser irradiation (808 nm), where the BP core generates intense local heat through photothermal conversion, further promoting drug release and enabling photothermal therapy.

***Characteristics and Performance*** The RHL coating conferred exceptional stability to the otherwise fragile BP nanosheets. While uncoated BP/ISL degraded within 24 h, RHL@BP/ISL maintained structural integrity for over 3 days, with only minimal reduction in photothermal efficiency (from 45.1 °C to 43 °C after 10 min NIR irradiation). The stimuli-responsive release profile was remarkable—under acidic conditions (pH 3.0) combined with NIR irradiation, 86.46% of ISL was released within 12 h, compared to only 5.50% at neutral pH without NIR. The nanocomposite demonstrated excellent biocompatibility with no significant cytotoxicity observed on gastric epithelial cells even at high concentrations. The RHL@BP/ISL platform operated through multiple antibacterial mechanisms: (1) RHL itself inhibited and disrupted *H. pylori* biofilm formation; (2) BP-mediated photothermal and photodynamic effects under NIR irradiation generated reactive oxygen species (ROS) and directly damaged bacterial membranes; (3) Acid-triggered release of ISL provided potent direct antibacterial activity against drug-resistant strains.

The combination therapy demonstrated exceptional efficacy against clinical drug-resistant *H. pylori* strains, with RHL@BP/ISL (0.5 μg/mL ISL) eliminating over 90% of bacteria within 1 h at acidic pH without requiring NIR irradiation. In a C57BL/6 mouse infection model, the nanocomposite showed excellent gastric retention and distribution, with fluorescence signals maintained in the stomach for over 4 h post-administration.

Most impressively, RHL@BP/ISL combined with NIR irradiation achieved a remarkable 72.9% reduction in bacterial burden, significantly outperforming conventional triple therapy ($P < 0.001$). The treatment also significantly reduced expression of inflammatory cytokines IL-6 and TNF-α, indicating both antibacterial and anti-inflammatory effects. The platform demonstrated excellent safety profile with no apparent organ damage observed even at high doses (100 mg/kg).

***Future Perspectives and Clinical Translation*** This multifunctional nanoplatform represents a significant advancement in *H. pylori* therapy, particularly for drug-resistant infections. The use of RHL as both a stabilizer and bioactive component, combined with BP's excellent photothermal properties and ISL's potent antibacterial activity, creates a comprehensive therapeutic strategy that addresses multiple challenges simultaneously: gastric acid stability, mucus penetration, biofilm disruption, and drug resistance. The platform's ability to function without proton pump inhibitors (antacids) eliminates gastrointestinal side effects associated with conventional triple therapy. Future development should focus on scaling up production, optimizing NIR irradiation protocols for endoscopic compatibility, and exploring combination therapies with other antibacterial agents for enhanced eradication rates. This technology bridges traditional Chinese medicine with modern nanotechnology, offering a promising solution for the growing challenge of antibiotic-resistant *H. pylori* infections.

### 10.2.2 Innovative Therapeutic Approaches

Traditional antibiotic therapies face limitations due to poor bioavailability, nonspecific targeting, and disruption of beneficial gut microbiota [31–34]. Photodynamic therapy (PDT) [35], sonodynamic therapy (SDT) [36], and chemodynamic therapy (CDT) [37] have emerged as promising alternatives, leveraging reactive oxygen species (ROS) to target diverse bacterial states while minimizing collateral damage [38]. These approaches utilize nanoplatforms designed to respond to specific stimuli, enabling precise spatiotemporal control over therapeutic action. The following case studies highlight innovative strategies that integrate multiple antimicrobial mechanisms for enhanced efficacy against *H. pylori*.

#### 10.2.2.1 Case Study 1: Stimuli-Responsive Sonodynamic Therapy [12]

Sonodynamic therapy (SDT) is an emerging noninvasive treatment modality that utilizes low-intensity ultrasound to activate sonosensitizers, generating ROS such as singlet oxygen ($^1O_2$) for targeted therapeutic effects. Unlike conventional antibiotics, SDT offers deep tissue penetration and precise spatiotemporal control through external ultrasound application, making it particularly suitable for treating gastric infections like *H. pylori*.

***Material Design and Stimuli-Responsiveness*** The Fe-HMME@DHA@MPN nanogenerator represents an innovative approach to combining SDT and CDT for enhanced ROS production. The platform was constructed through a meticulous three-step procedure: first, creating a mesoporous metal-organic nanostructure (Fe-HMME) via coordination between hematoporphyrin monomethyl ether (HMME) and ferric iron; second, loading the antimalarial drug dihydroartemisinin (DHA) as a hydroperoxide source; and finally, coating with a pH-responsive metal-phenolic network (MPN) shell using tannic acid (TA). This sophisticated design enables multiple therapeutic mechanisms working in concert. The MPN shell provides pH-responsive behavior, dissociating under acidic gastric conditions to release therapeutic components while remaining stable at neutral intestinal pH. This ensures targeted activation specifically in the stomach where *H. pylori* reside. The released $Fe^{2+}$ ions catalyze Fenton reactions with endogenous $H_2O_2$ at the infection site, generating hydroxyl radicals (•OH), while simultaneously producing oxygen that alleviates hypoxia and enhances SDT efficacy. The incorporated DHA serves as an alternative hydroperoxide source in microenvironments with insufficient $H_2O_2$, ensuring consistent •OH generation regardless of local conditions (Fig. 10.3).

***Characteristics and Performance*** Comprehensive material characterization revealed excellent properties for gastric applications. The nanogenerator exhibited a spherical morphology with an average diameter of approximately 80 nm and a negative surface potential of −11.03 mV. The mesoporous structure facilitated high DHA loading capacity (0.302 mg mg$^{-1}$). X-ray photoelectron spectroscopy (XPS) analysis confirmed the presence of ferrous Fe(II) (19.5% peak area), crucial for Fenton reaction activity. The pH-responsive degradation was demonstrated through TEM imaging, showing complete dissociation at pH 2.2 within 30 min while maintaining stability at pH 7.4. This selective activation ensured controlled DHA release, with 86.3% released at gastric pH compared to only 8.5% at intestinal pH. ROS generation studies confirmed production of both $^1O_2$ and •OH, with the latter being particularly effective against bacterial membranes. The platform also exhibited peroxidase-like activity that was strongly pH-dependent, minimizing off-target effects in neutral environments.

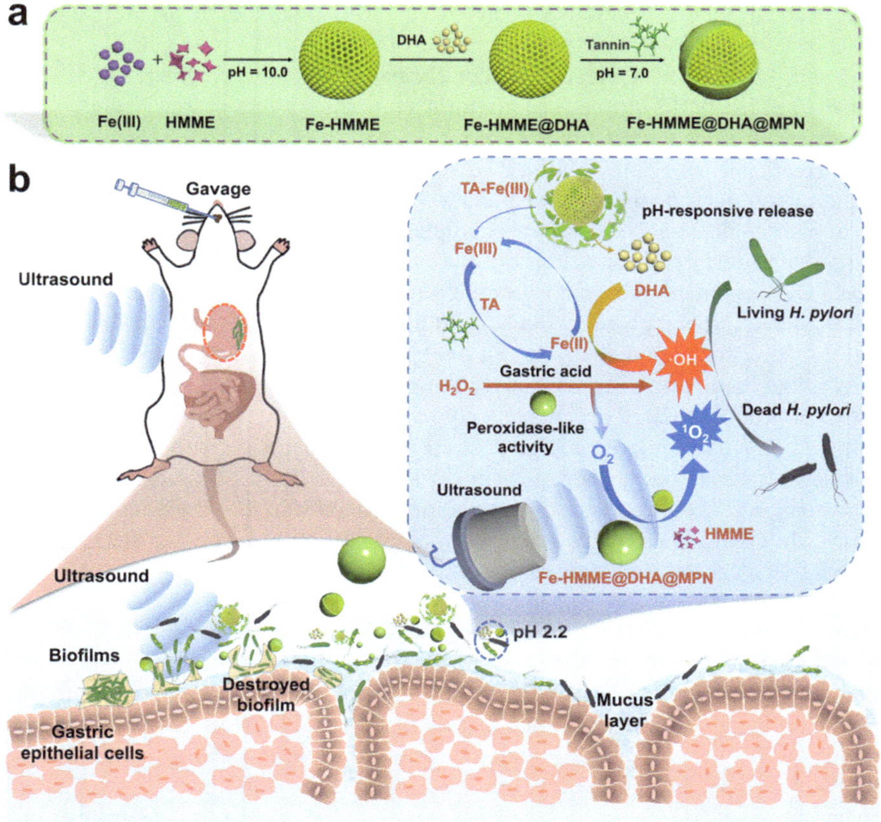


**Fig. 10.3** Preparation and anti-*Helicobacter pylori* mechanism of Fe-HMME@DHA@MPN. (Reproduced with permission from Ref. [12], Wiley)

The antimicrobial efficacy was demonstrated through multiple mechanisms. The nanogenerator showed excellent binding to *H. pylori* membranes through TA-mediated interactions, localizing ROS production precisely at bacterial surfaces. Against planktonic bacteria, the combination of CDT and SDT achieved complete eradication within 20 min against various strains including multidrug-resistant isolates. Biofilm disruption was particularly impressive, with 94.5% biomass removal achieved through combined physical disruption by ultrasound and chemical degradation by ROS. Live/dead staining and SEM imaging confirmed extensive membrane damage and protein leakage in treated bacteria. The platform's ability to function in hypoxic conditions commonly found in biofilms was particularly valuable, as oxygen generated through Fenton reactions enhanced SDT efficacy where conventional therapies often fail. The multiple antibacterial mechanisms working simultaneously—membrane disruption by •OH, oxidative damage by $^1O_2$, and enhanced antibiotic penetration—prevented development of resistance and provided robust activity against persistent infections..

In a BALB/c mouse infection model, the nanogenerator demonstrated outstanding therapeutic performance. The platform showed excellent gastric retention, with fluorescence signals persisting for over 24 h due to TA-mediated adhesion properties. Treatment with Fe-HMME@DHA@MPN + US achieved a 99% reduction in bacterial burden, matching the efficacy of conventional triple therapy (OAC). Histopathological analysis revealed complete eradication of *H. pylori* from gastric tissues with restored mucosal architecture and minimal inflammatory cell infiltration. The treatment preserved gut microbiota homeostasis, with no significant changes in microbial diversity indices and maintenance of beneficial bacteria populations. This contrasted sharply with antibiotic therapy, which caused substantial microbiota disruption and increased pathogenic bacteria abundance. The nanogenerator also demonstrated excellent safety profile, with no evidence of systemic toxicity or organ damage even at high doses. Blood biochemistry parameters remained within normal ranges, and H&E staining showed intact tissue architecture in all major organs. The pH-dependent activation ensured minimal impact on intestinal flora while providing potent antibacterial effects specifically in the stomach, addressing a critical limitation of conventional antibiotics (Fig. 10.4).

***Future Perspectives and Clinical Translation*** The development of SDT- and CDT-based nanoplatforms for *H. pylori* treatment opens several promising directions for future research and clinical application. Additionally, exploration of these platforms for other gastrointestinal infections beyond *H. pylori* could expand their clinical utility. The successful implementation of these technologies would represent a significant advancement in infectious disease treatment, moving from broad-spectrum antibiotic approaches to precisely targeted, mechanism-based therapies that preserve the body's natural defenses and microbial ecosystems.

#### 10.2.2.2 Case Study 2: Multi-stimuli-Responsive Photothermal/Photodynamic Therapy [39]

PTT and PDT represent advanced light-activated therapeutic approaches that utilize near-infrared light to generate localized heat and ROS for targeted antibacterial effects. These modalities offer significant advantages over conventional antibiotics through their precise spatiotemporal control and ability to overcome multidrug resistance, making them particularly suitable for treating challenging infections such as *H. pylori*. The multi-stimuli-responsive nanoplatform discussed herein demonstrates how sophisticated material design can synergistically combine PTT and PDT while overcoming biological barriers..

***Material Design and Stimuli-Responsiveness*** The RLs@T780TG nanoplatform exemplifies advanced multi-stimuli-responsive design through its sophisticated assembly and dual light-activation mechanisms. The system was constructed through precise engineering of a near-infrared photosensitizer T780T-Gu with anionic biosurfactant rhamnolipids (RLs), forming an optimal nanostructure for

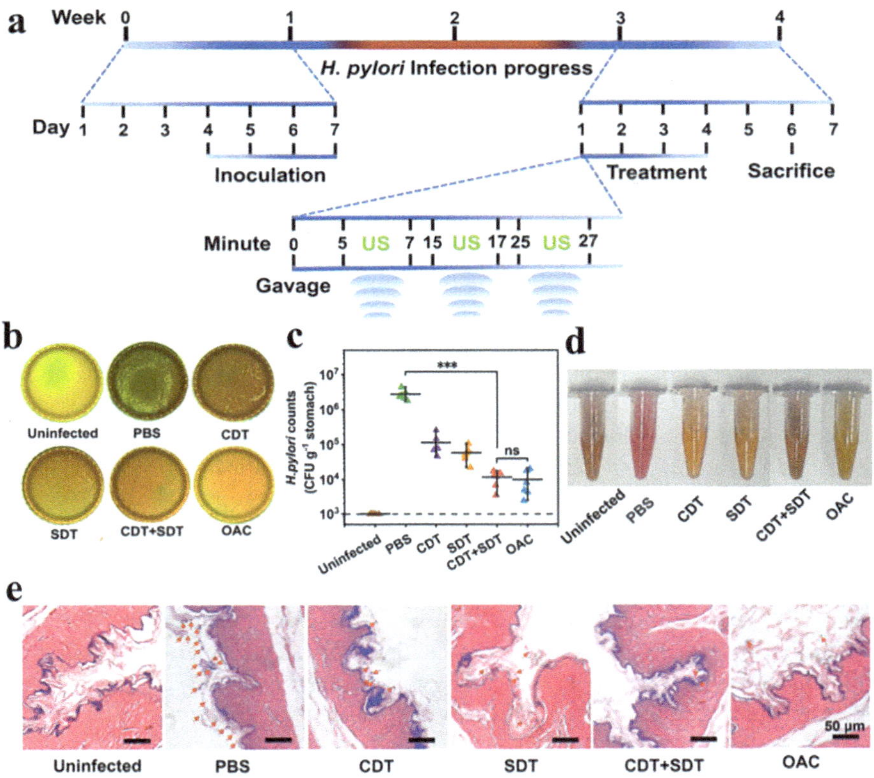


**Fig. 10.4** Therapeutic effect of Fe-HMME@DHA@MPN against *Helicobacter pylori* infection. (Reproduced with permission from Ref. [12], Wiley)

gastric applications. The platform exhibits intelligent responsiveness to both the gastric environment and external light stimulation: first, maintaining stability in acidic conditions while facilitating deep mucus penetration, and second, rapidly disassembling under near-infrared light irradiation (808 nm) to activate therapeutic functions. Molecular dynamics simulations revealed spontaneous self-assembly through van der Waals forces and electrostatic interactions, with the system reaching equilibrium within 100 ns. This sophisticated design ensures the platform remains intact during gastric transit but immediately activates upon light exposure at the infection site.

***Characteristics and Performance*** Comprehensive characterization demonstrated exceptional photophysical properties optimized for dual-mode therapy. The nanoplatform exhibited a remarkable 20 nm redshift in UV absorbance spectra due to intermolecular π-π stacking, significantly enhancing light penetration depth and therapeutic efficacy. The photothermal performance was outstanding, with temperature increases proportional to both laser power intensity (0.3–1.0 W $cm^{-2}$) and photosensitizer concentration (2–20 μM), achieving temperatures sufficient for bacterial ablation while maintaining tissue safety. The system demonstrated excellent photostability through multiple heating-cooling cycles without significant efficiency loss. Simultaneously, the platform generated substantial singlet oxygen under light irradiation, as confirmed by SOSG fluorescence assays, enabling potent combined photothermal and photodynamic effects. The mesoporous structure facilitated high drug loading capacity (≈60% for RLs) while maintaining ideal particle size (≈80 nm) for biological applications.

The dual-mode phototherapeutic approach demonstrated exceptional efficacy against both planktonic and biofilm-associated *H. pylori*. The platform achieved complete bacterial eradication within 20 min of treatment against multiple drug-resistant strains, including clinical isolates resistant to first-line antibiotics. The antimicrobial mechanism involved three synergistic pathways: first, enhanced mucus penetration through optimized size and surface properties allowing deep access to infection sites; second, specific targeting of inflammatory regions through charge-based interactions; third, light-activated generation of both localized heat and ROS that caused comprehensive bacterial membrane disruption and protein leakage. Against biofilms, the platform achieved 94.5% biomass removal through combined physical disruption and chemical degradation, effectively addressing one of the most challenging aspects of *H. pylori* treatment. Live/dead staining and TEM imaging confirmed extensive membrane damage and structural breakdown in treated bacteria.

In a BALB/c mouse infection model, the phototherapeutic platform demonstrated outstanding treatment outcomes that significantly surpassed conventional approaches. The system showed excellent gastric retention with fluorescence signals persisting for over 24 h, ensuring prolonged therapeutic exposure at the infection site. Treatment with RLs@T780TG + NIR irradiation achieved a remarkable 99.9% reduction in bacterial burden, matching the efficacy of conventional triple therapy while using significantly lower drug concentrations. Histopathological analysis revealed complete eradication of *H. pylori* from gastric tissues with restored mucosal architecture and minimal inflammatory cell infiltration. Crucially, the treatment preserved gut microbiota homeostasis, with no significant changes in microbial diversity indices, contrasting sharply with antibiotic therapy that caused substantial microbiota disruption. The safety profile was excellent, with no evidence of systemic toxicity or organ damage even at high doses.

***Future Perspectives and Clinical Translation*** This multi-stimuli-responsive phototherapeutic platform represents a paradigm shift in antimicrobial treatment, moving from broad-spectrum approaches to precisely targeted, light-activated therapies. The technology offers several translational advantages: the use of biocompatible components with FDA-approved precedents, wavelength-specific activation that minimizes off-target effects, and compatibility with endoscopic light delivery systems for clinical application. Future development should focus on optimizing light parameters for human applications, developing portable light sources for patient self-administration, and exploring combination therapies with other treatment modalities. The platform's modular design also allows for adaptation to various gastrointestinal infections, potentially expanding its clinical utility beyond *H. pylori* treatment. This approach effectively addresses the critical challenges of antibiotic resistance while preserving the body's natural microbial ecosystems, representing a significant advancement in infectious disease management.

### 10.2.3 Theranostics

Traditional antibiotic therapies face significant limitations, including poor bioavailability, nonspecific targeting, and disruption of beneficial gut microbiota [31]. Theranostic nanoplatforms, which integrate diagnostic and therapeutic functions into a single system, offer a promising strategy to overcome these hurdles [40]. By enabling real-time imaging of the infection site and spatially controlled, stimulus-responsive drug release, these systems facilitate precise diagnosis and highly efficient eradication of *H. pylori* [40–42]. The following case studies highlight two innovative theranostic approaches that utilize external energy sources (ultrasound and near-infrared light) for targeted diagnosis and treatment.

#### 10.2.3.1 Case Study 1: Stimuli-Responsive Imaging and Bacteria-Triggered Therapy [25]

The challenge of eradicating *H. pylori* is compounded by its strategic colonization beneath the protective gastric mucus layer and its tendency to form dense biofilms. These biofilms act as a formidable barrier, preventing imaging probes and antimicrobial agents from reaching the bacteria, leading to diagnostic uncertainties and therapeutic failures. Standard optical imaging is hampered by tissue autofluorescence and shallow light penetration. Furthermore, achieving sufficient antibiotic penetration into these biofilms with systemic administration is difficult, often requiring high doses that contribute to resistance and collateral damage to the gut flora. To address these interconnected diagnostic and therapeutic challenges, a theranostic nanoparticle (mSZ-Clr@ePDA NP) was developed. This platform was specifically designed to actively penetrate biological barriers, provide autofluorescence-free imaging from deep tissues, and release its antibiotic payload only upon encountering the specific enzymatic signature of the target bacteria, thereby maximizing efficacy and specificity.

***Material Design and Stimuli-Responsiveness*** This platform was constructed by depositing mechanoluminescent $SrAl_2O_4:Eu^{2+}$ (SAOE) nanodots and persistent luminescent $ZnGa_2O_4:Cr^{3+}$ (ZGC) nanodots onto mesoporous silica nanoparticles (MSNs) to form a luminescent core (mSZ NPs). The antibiotic clarithromycin (CLR) was adsorbed into the pores, and the construct was coated with an enzyme-responsive polydopamine shell (ePDA) containing phosphate ester linkages. The theranostic mechanism is multifaceted: Firstly, US serves a dual purpose—it drives the nanomotors through viscous mucus and biofilms, and it mechanically activates the SAOE nanodots to emit green light, which in turn excites the ZGC nanodots to emit a near-infrared (NIR) afterglow for deep-tissue imaging. Secondly, the ePDA shell acts as a "gatekeeper," quenching the luminescence of the core and protecting the CLR payload during transit. Upon reaching the infection site, phospholipase C (PLC) secreted by *H. pylori* specifically cleaves the phosphate bonds in the ePDA layer, causing its disintegration. This collapse simultaneously restores the NIR luminescence for precise bacterial imaging and triggers the localized release of CLR.

***Characteristics and Performance*** Characterization confirmed the successful synthesis of the core-shell structure with a hydrodynamic size of ~147 nm. The ePDA coating provided excellent stability in simulated gastric environments (pH 1.2 and 4.5). US activation at 2.0 W/cm$^2$ for 5 min generated a strong NIR afterglow from the mSZ core that lasted for hours, and this could be repeatedly recharged over multiple cycles. Motion studies in agarose gels (simulating mucus) demonstrated that US propulsion significantly enhanced NP penetration depth in a power-dependent manner, reaching 290 μm at 2.0 W/cm$^2$. The drug release profile was highly specific: the ePDA shell minimized premature CLR release in acidic conditions (only ~7% in pH 4.5 buffer over 2 h), but rapid release occurred (~82% over 8 h) in the presence of *H. pylori* due to enzymatic degradation of the shell.

The US-driven motion dramatically improved the penetration of NPs into *H. pylori* biofilms, with fluorescence signals detected throughout the biofilm depth, a feat not achieved by passive diffusion. The combination of US-propelled delivery and bacteria-triggered CLR release led to near-complete destruction of biofilms and eradication of embedded bacteria. In an *H. pylori*-infected mouse model, oral administration of mSZ-Clr@ePDA NPs followed by US application resulted in significantly higher retention of NPs in the stomach and accumulation on the epithelial layer compared to non-US treated groups. The US-activated persistent luminescence allowed for real-time, autofluorescence-free imaging of the infection. Most importantly, the mSZ-Clr@ePDA/US treatment achieved a significant reduction in bacterial burden at a lower antibiotic dose, comparable to conventional triple therapy (OAC), but without causing damage to the gastrointestinal tract or major organs.

***Future Perspectives and Clinical Translation*** The path to clinical translation mSZ-Clr@ePDA nanomotor platform involves overcoming technical and regulatory hurdles. A critical next step is a comprehensive investigation into the long-term biodistribution and biodegradation of its multicomponent inorganic core (SAOE and ZGC). While the study confirmed short-term excretion, detailed pharmacokinetic and toxicological studies over extended periods are essential to ensure the complete clearance of metal ions (e.g., $Sr^{2+}$, $Zn^{2+}$, $Cr^{3+}$) and long-term biological safety. Furthermore, the scalability and reproducibility of the complex, multistep synthesis process—including the in situ deposition of nanodots and the coating of the enzyme-responsive polymer—need to be rigorously optimized for Good

Manufacturing Practice (GMP) standards. From a clinical application perspective, future work should focus on engineering integrated, minimally invasive US devices. Ideally, this would involve the development of a dedicated endoscopic US probe capable of simultaneously performing imaging (to locate infected areas) and providing localized, therapeutic US energy to activate the nanomotors and enhance their penetration directly at the gastric wall. This would maximize efficacy while minimizing off-target effects. The platform's modular design also allows for adaptation beyond clarithromycin. Future studies could explore loading with other antibiotics, or even novel antimicrobial agents, to combat specific resistance patterns, potentially creating a versatile "plug-and-play" theranostic platform for various gastrointestinal infections.

#### 10.2.3.2 Case Study 2: Stimuli-Responsive Photoacoustic Imaging and Targeted Photothermal Therapy [14]

The rising global prevalence of antibiotic-resistant *H. pylori* strains renders standard triple therapies increasingly ineffective, creating an urgent need for treatment strategies that are independent of traditional antibacterial mechanisms [43]. Physical methods, such as photothermal therapy (PTT), which kills bacteria through localized heat generation, offer a promising solution to overcome existing and future resistance mechanisms. However, a major challenge for any orally administered agent is achieving specific targeting of *H. pylori* in the complex and dynamic environment of the stomach, while avoiding unintended harm to gastric tissues and the delicate balance of the intestinal microbiome. Without specific targeting, PTT would require high, potentially damaging laser power. This case study addresses the critical need for a highly selective antimicrobial approach by developing a nanoprobe that not only provides a noninvasive diagnostic signal through photoacoustic imaging but also ensures that the potent photothermal effect is confined exclusively to the pathogen, thereby preserving the host's microbiome and preventing damage to healthy tissue..

***Material Design and Stimuli-Responsiveness*** The design of this theranostic platform is a sophisticated response to the challenges of the gastric environment. The gold nanostars (GNS) were synthesized via a surfactant-free method, ensuring high biocompatibility and a clean surface for subsequent functionalization. The unique structure of GNS, with their multiple sharp branches, is critical as it creates intense localized surface plasmon resonance (LSPR) effects, resulting in a strong absorption peak in the near-infrared (NIR) region. This NIR window is ideal for biomedical applications because biological tissues have minimal absorption and scattering in this range, allowing for deeper light penetration. The GNS were first modified with carboxyl-terminated polyethylene glycol (HS-PEG-COOH), which confers colloidal stability in biological fluids and provides functional groups for conjugation. The key innovation lies in the conjugation of polyclonal antibodies against *H. pylori* using a pH-sensitive linker, cis-aconitic anhydride. This linker is essential for preserving the targeting capability in the stomach. In the acidic environment of the gastric juice (pH ~1.5–3.5), the linker remains stable, preventing the premature detachment of antibodies and ensuring that the GNS@Ab nanoprobes reach the gastric mucosa with their targeting ligands intact. Upon reaching the neutral pH microenvironment close to the epithelial layer where *H. pylori* colonizes, the cis-aconitic amide bond undergoes hydrolysis, restoring the antibody's native charge and conformation, thereby facilitating optimal binding to bacterial surface antigens. This pH-triggered "activation" mechanism ensures highly specific targeting where it is needed most.

***Characteristics and Performance*** Comprehensive characterization validated the successful synthesis and excellent properties of the GNS@Ab nanoprobes. High-resolution transmission electron microscopy (HR-TEM) confirmed the star-like morphology with an average core size of 80 ± 7 nm and sharp, spike-like tips approximately 20 ± 5 nm in length, which are responsible for the enhanced photothermal efficiency. The conjugation of antibodies was verified through multiple techniques: UV-Vis spectroscopy showed a distinct redshift of the LSPR peak from 808 nm to 826 nm, indicating a change in the local refractive index due to the antibody coating. Fourier-Transform Infrared Spectroscopy (FT-IR) further confirmed the presence of amide bonds and Au-S bonds, proving successful linkage. The nanoprobes demonstrated exceptional specificity; scanning electron microscopy (SEM) images clearly showed dense binding of GNS@Ab to the surface and flagella of *H. pylori*, while no such binding was observed with control bacteria like *Escherichia coli* DH5α. This high specificity is the foundation for targeted therapy. The photothermal performance was rigorously quantified: under irradiation with a 790 nm NIR laser at a power density of 3 W/cm$^2$, the GNS@Ab suspension experienced a rapid and concentration-dependent temperature increase, capable of reaching the hyperthermia threshold (>50 °C) required for effective bacterial ablation within minutes. Furthermore, the platform exhibited outstanding biocompatibility. Cytotoxicity assays on human gastric epithelial cells (GES-1) showed no significant reduction in cell viability even at high concentrations up to 80 μg/mL, and hemolysis assays confirmed excellent blood compatibility with a hemolysis rate below 5%, underscoring its safety profile for in vivo application.

***Future Perspectives and Clinical Translation*** The GNS@Ab nanoprobe holds significant clinical potential due to the well-established safety profile of gold in biomedical applications. The immediate research priority is to transition from using polyclonal antibodies to more specific and scalable targeting ligands. Replacing polyclonal antibodies with humanized monoclonal antibodies or smaller biomolecules (e.g., affibodies, peptides) would reduce potential immunogenicity and improve batch-to-batch consistency, which is critical for regulatory approval. Additionally, detailed studies on the potential for bacterial photothermal resistance, although theoretically low, are necessary to rule out any adaptive mechanisms in response to hyperthermia. Clinically, the translation of GNS@Ab hinges on the development of a clinically viable light delivery system. The most promising route is its integration with standard gastrointestinal endoscopy. A forward-looking application could involve a theranostic endoscope that combines photoacoustic imaging to confirm the presence and location of the targeted nanoprobes with a built-in, low-power NIR laser for subsequent photothermal ablation. This would enable a seamless "see-and-treat" paradigm during a single endoscopic procedure, allowing physicians to visually confirm targeting before activating treatment. This approach could be particularly valuable for treating focal infections or antibiotic-resistant strains identified during routine screening. Finally, despite the favorable excretion profile, formal toxicity studies following Good Laboratory Practice (GLP) guidelines are the essential final step to initiate clinical trials.

## 10.3 Design Considerations and Challenges

The development of multi-stimuli-responsive nanomaterials for *H. pylori* therapy, while promising, is fraught with significant challenges that must be addressed to translate these sophisticated systems from laboratory research to clinical practice [44]. The intricate design that enables their smart functionality also introduces complexities related to their safety, stability, and scalable production. This section critically examines the primary design considerations and hurdles across four key domains: biocompatibility, stability, large-scale manufacturing, and the clinical translation pathway.

### 10.3.1 Biocompatibility and Toxicity

The foremost concern for any biomedical nanomaterial is its biocompatibility [45]. For multi-stimuli-responsive systems, this challenge is amplified due to their complex composition, which often includes inorganic nanoparticles (e.g., gold, iron oxide, silica), organic polymers, and synthetic linkers [46, 47]. The long-term fate of these components within the human body must be rigorously evaluated [48]. While materials like gold nanostars are generally considered biocompatible, their unique nanostructure (e.g., sharp tips) could potentially induce unforeseen cellular stress or immune responses [49, 50]. Similarly, the degradation products of materials like mesoporous silica or metal-organic frameworks need careful scrutiny [51]. For instance, the gradual degradation of SrA12O4:Eu2+ and ZnGa2O4:Cr3+ in the mSZ-Clr@ePDA platform raises questions about the chronic toxicity of released metal ions ($Sr^{2+}$, $Zn^{2+}$, $Cr^{3+}$) [25]. Comprehensive pharmacokinetic studies are required to track the absorption, distribution, metabolism, excretion, and potential accumulation of these materials and their breakdown products in organs beyond the gastrointestinal tract. Furthermore, the immune response triggered by these foreign nanomaterials must be characterized, assessing for unintended inflammation or immunogenicity that could negate their therapeutic benefits. Ensuring that the "off" state of these responsive materials is truly inert and that their activation is confined strictly to the target site is critical for minimizing off-target toxicity.

### 10.3.2 Stability and Performance In Vivo

A significant gap exists between demonstrating performance in controlled laboratory settings and functioning reliably within the dynamic, harsh environment of the human stomach. A primary challenge is the in vivo stability of the nanomaterial itself. For example, the enzyme-responsive ePDA shell in the mSZ-Clr@ePDA NPs must remain intact during transit through the highly acidic and enzyme-rich gastric juice until it encounters the specific phospholipase C produced by *H. pylori*. Premature degradation could lead to unintended drug release and diagnostic signal loss. Similarly, the stability of sensitive components like black phosphorus nanosheets is a major hurdle; without proper coating (e.g., with rhamnolipids), they can rapidly oxidize and lose their functionality [30]. Beyond material stability, the consistency of the stimuli-responsive behavior in vivo is another concern. Factors such as inter-patient variability in gastric pH, mucus thickness, and bacterial load can lead to inconsistent activation of pH- or enzyme-responsive systems. The penetration efficiency of external stimuli like ultrasound or near-infrared light can also vary significantly based on an individual's anatomy, potentially leading to subtherapeutic outcomes in some patients [52–54]. Ensuring that these smart systems perform reliably across a diverse patient population is a major hurdle that requires extensive validation in physiologically relevant models.

### 10.3.3 Large-Scale Manufacturing and Quality Control

The transition from a successful laboratory-scale synthesis to a robust, reproducible, and cost-effective industrial manufacturing process presents a formidable challenge [55]. Many of the described nanomaterials require complex, multistep synthesis protocols. The preparation of mSZ-Clr@ePDA NPs, for instance, involves the in situ deposition of two types of luminescent nanodots onto MSNs, followed by drug loading and a precise polymerization process to form the ePDA shell [25]. Each step introduces potential batch-to-batch variations in critical parameters such as particle size, polydispersity, drug loading efficiency, and responsiveness. Reproducibly achieving a high degree of monodispersity and consistent functional performance on a large scale is exceptionally difficult. Furthermore, stringent quality control measures must be developed to characterize these complex products. This goes beyond standard metrics like size and zeta potential to include functional assessments, such as quantifying the responsiveness to specific stimuli (e.g., pH-dependent release profiles, photothermal conversion efficiency) for every batch. The high cost of some raw materials, such as gold salts for GNS or specialized polymers, could also impede the economic viability of these therapies, making scalable and affordable synthesis routes a critical area for future research and development [14].

### 10.3.4 Clinical Translation and Regulatory Pathways

The path to clinical approval for these multifunctional nanomaterials is complex and uncharted. A significant challenge is the regulatory classification of these entities, as they combine aspects of drugs, medical devices (if they include an activation device), and potentially even diagnostics, falling into a challenging "combination product" category [56, 57]. Regulatory bodies like the FDA and EMA have limited experience with such complex, stimuli-responsive systems, and establishing a clear regulatory pathway will require extensive dialogue and the development of new evaluation criteria. The preclinical data required for an Investigational New Drug (IND) application will be substantially more comprehensive than for a conventional drug [58]. It must not only demonstrate efficacy but also rigorously prove the safety of all components, the reliability of the activation mechanism, and the safety of the external stimuli (e.g., long-term effects of repeated ultrasound or laser application on gastric tissue). Designing and executing clinical trials also presents unique challenges. For example, blinding will be difficult when the treatment involves a physical intervention like light or ultrasound [59]. Identifying the optimal patient population, likely those with antibiotic-resistant infections, and defining clinically meaningful endpoints beyond bacterial eradication (e.g., microbiome preservation, long-term recurrence rates) will be crucial. Successfully navigating these regulatory and clinical hurdles is the ultimate challenge in bringing these innovative therapies to patients.

## References


1. Vakil N, Megraud F. Eradication therapy for *H. pylori*. Gastroenterology. 2007;133(3):985–1001. https://doi.org/10.1053/j.gastro.2007.07.008.
2. Bloom SR, Polak JM. Physiology of gastrointestinal hormones. Biochem Soc Trans. 1980;8(1):15–7. https://doi.org/10.1042/bst0080015.
3. Wu W, Yang Y, Sun G. Recent insights into antibiotic resistance in *H. pylori* eradication. Gastroenterol Res Pract. 2012;2012:723183. https://doi.org/10.1155/2012/723183.
4. Abuhelwa AY, Williams DB, Upton RN, Foster DJ. Food, gastrointestinal pH, and models of oral drug absorption. Eur J Pharm Biopharm. 2017;112:234–48. https://doi.org/10.1016/j.ejpb.2016.11.034.
5. Goelen N, Morales JF, Tack J, Janssen P. Repeated gastric motility measurement affects gastric motility and epigastric symptom sensation. Front Med (Lausanne). 2020;7:172. https://doi.org/10.3389/fmed.2020.00172.
6. Hu Y, Li S, Dong H, Weng L, Yuwen L, et al. Environment-responsive therapeutic platforms for the treatment of implant infection. Adv Healthc Mater. 2023;12(26):e2300985. https://doi.org/10.1002/adhm.202300985.
7. Singh R, Sharma A, Saji J, Umapathi A, Kumar S, et al. Smart nanomaterials for cancer diagnosis and treatment. Nano Converg. 2022;9(1):21. https://doi.org/10.1186/s40580-022-00313-x.
8. Zheng Y, Zhang T, Shao J, Du Y, Li Z, et al. Antibiotic-free responsive biomaterials for specific and targeted *H. pylori* eradication. J Control Release. 2025;379:708–29. https://doi.org/10.1016/j.jconrel.2025.01.054.

9. Wong KI, Wang S, Li M, Zhao G, Wang C, et al. Combating drug-resistant *H. pylori* infection with zinc peroxide-based nanoparticles: a ROS reservoir via photochemical reaction. Chem Eng J. 2024;483:149287. https://doi.org/10.1016/j.cej.2024.149287.
10. Kantaros A, Ganetsos T. From static to dynamic: smart materials pioneering additive manufacturing in regenerative medicine. Int J Mol Sci. 2023;24(21) https://doi.org/10.3390/ijms242115748.
11. Gupta A, Shetty S, Mutalik S, Chandrashekar HR, Nandakumar K, et al. Treatment of *H. pylori* infection and gastric ulcer: need for novel pharmaceutical formulation. Heliyon. 2023;9(10):e20406. https://doi.org/10.1016/j.heliyon.2023.e20406.
12. Yu J, Guo Z, Yan J, Bu C, Peng C, et al. Gastric acid-responsive ROS nanogenerators for effective treatment of *H. pylori* infection without disrupting homeostasis of intestinal flora. Adv Sci. 2023;10(20):2206957. https://doi.org/10.1002/advs.202206957.
13. Yan L-X, Wang B-B, Zhao X, Chen L-J, Yan X-P. A pH-responsive persistent luminescence nanozyme for selective imaging and killing of *H. pylori* and common resistant bacteria. ACS Appl Mater Interfaces. 2021;13(51):60955–65. https://doi.org/10.1021/acsami.1c21318.
14. Zhi X, Liu Y, Lin L, Yang M, Zhang L, et al. Oral pH sensitive GNS@ab nanoprobes for targeted therapy of *H. pylori* without disturbance gut microbiome. Nanomedicine. 2019;20:102019. https://doi.org/10.1016/j.nano.2019.102019.
15. Lai Y, Zhang T, Yin X, Zhu C, Du Y, et al. An antibiotic-free platform for eliminating persistent *H. pylori* infection without disrupting gut microbiota. Acta Pharm Sin B. 2024;14(7):3184–204. https://doi.org/10.1016/j.apsb.2024.03.014.
16. Xia X, Yin Z, Yang Y, Li S, Wang L, et al. In situ upregulating heat shock protein 70 via gastric nano-heaters for the interference of *H. pylori* infection. ACS Nano. 2022;16(9):14043–54. https://doi.org/10.1021/acsnano.2c03911.
17. Cai X, Li Z, Zhou WJ, Deng H, Cao X, et al. Single-atom-anchored microsweepers for *H. pylori* inhibition through dynamically navigated reciprocating locomotion. Chem Commun (Camb). 2023;59(36):5455–8. https://doi.org/10.1039/d3cc00933e.
18. Venditti I. Engineered gold-based nanomaterials: morphologies and functionalities in biomedical applications. A mini review. Bioengineering (Basel). 2019;6(2) https://doi.org/10.3390/bioengineering6020053.
19. Ahmadkhani L, Abbasian M, Akbarzadeh A. Synthesis of sharply thermo and PH responsive PMA-b-PNIPAM-b-PEG-b-PNIPAM-b-PMA by RAFT radical polymerization and its schizophrenic micellization in aqueous solutions. Des Monomers Polym. 2017;20(1):406–18. https://doi.org/10.1080/15685551.2017.1314654.
20. Nigro V, Angelini R, Bertoldo M, Buratti E, Franco S, et al. Chemical-physical behaviour of microgels made of interpenetrating polymer networks of PNIPAM and poly(acrylic acid). Polymers (Basel). 2021;13(9) https://doi.org/10.3390/polym13091353.
21. Banerjee D, Harfouche R, Sengupta S. Nanotechnology-mediated targeting of tumor angiogenesis. Vasc Cell. 2011;3(1):3. https://doi.org/10.1186/2045-824x-3-3.
22. Tsuji A, Nakashima E, Hamano S, Yamana T. Physicochemical properties of amphoteric β-lactam antibiotics I: stability, solubility, and dissolution behavior of amino penicillins as a function of pH. J Pharm Sci. 1978;67(8):1059–66. https://doi.org/10.1002/jps.2600670810.
23. Miyazaki K, Sasaki A, Mizuuchi H. Advances in the evaluation of gastrointestinal absorption considering the mucus layer. Pharmaceutics. 2023;15(12) https://doi.org/10.3390/pharmaceutics15122714.
24. Lopes D, Nunes C, Martins MC, Sarmento B, Reis S. Eradication of *H. pylori*: past, present and future. J Control Release. 2014;189:169–86. https://doi.org/10.1016/j.jconrel.2014.06.020.
25. Zhou D, Zhang Z, Qiu B, Zhang D, Xie S, et al. Ultrasound-activated persistent luminescence imaging and bacteria-triggered drug release for *H. pylori* infection theranostics. ACS Appl Mater Interfaces. 2022;14(23):26418–30. https://doi.org/10.1021/acsami.2c04683.
26. Zhang W, Zhou Y, Fan Y, Cao R, Xu Y, et al. Metal-organic-framework-based hydrogen-release platform for multieffective *H. pylori* targeting therapy and intestinal flora protective capabilities. Adv Mater. 2022;34(2):e2105738. https://doi.org/10.1002/adma.202105738.

27. Anwar A, Khalid S, Perveen S, Ahmed S, Siddiqui R, et al. Synthesis of 4-(dimethylamino) pyridine propylthioacetate coated gold nanoparticles and their antibacterial and photophysical activity. J Nanobiotechnol. 2018;16:1–8.
28. Wang S, Yao J, Zhou B, Yang J, Chaudry MT, et al. Bacteriostatic effect of quercetin as an antibiotic alternative in vivo and its antibacterial mechanism in vitro. J Food Prot. 2018;81(1):68–78. https://doi.org/10.4315/0362-028X.JFP-17-214.
29. Yan J, Yu J, Bu C, Yang L, Chen J, et al. Antibiotic-augmented chemodynamic therapy for treatment of *H. pylori* infection in the dynamic stomach environment. Nano Lett. 2024;24(47):14983–92. https://doi.org/10.1021/acs.nanolett.4c03692.
30. Huang Y, Li R, Dai Y, Qin C, He J, et al. Rhamnolipid-assisted black phosphorus nanosheets with efficient isolinderalactone loading against drug resistant *H. pylori*. Mater Des. 2022;216:110536. https://doi.org/10.1016/j.matdes.2022.110536.
31. Umarje SC, Banerjee SK. Non-traditional approaches for control of antibiotic resistance. Expert Opin Biol Ther. 2023;23(11):1113–35. https://doi.org/10.1080/14712598.2023.2279644.
32. Yuan J, Hu YJ, Zheng J, Kim JH, Sumerlin T, et al. Long-term use of antibiotics and risk of type 2 diabetes in women: a prospective cohort study. Int J Epidemiol. 2020;49(5):1572–81. https://doi.org/10.1093/ije/dyaa122.
33. Nguyen LH, Örtqvist AK, Cao Y, Simon TG, Roelstraete B, et al. Antibiotic use and the development of inflammatory bowel disease: a national case-control study in Sweden. Lancet Gastroenterol Hepatol. 2020;5(11):986–95. https://doi.org/10.1016/s2468-1253(20)30267-3.
34. Pennycook JH, Scanlan PD. Ecological and evolutionary responses to antibiotic treatment in the human gut microbiota. FEMS Microbiol Rev. 2021;45(5) https://doi.org/10.1093/femsre/fuab018.
35. Fan L, Xu X, Zhu C, Han J, Gao L, et al. Tumor catalytic-Photothermal therapy with yolk-Shell gold@carbon nanozymes. ACS Appl Mater Interfaces. 2018;10(5):4502–11. https://doi.org/10.1021/acsami.7b17916.
36. Sun D, Pang X, Cheng Y, Ming J, Xiang S, et al. Ultrasound-switchable nanozyme augments sonodynamic therapy against multidrug-resistant bacterial infection. ACS Nano. 2020;14(2):2063–76. https://doi.org/10.1021/acsnano.9b08667.
37. Liu Z, Zhao X, Yu B, Zhao N, Zhang C, et al. Rough carbon-iron oxide nanohybrids for near-infrared-II light-responsive synergistic antibacterial therapy. ACS Nano. 2021;15(4):7482–90. https://doi.org/10.1021/acsnano.1c00894.
38. Li H, Ren Y, Zhan Y, Yu X, Zhang X, et al. High-entropy nanozyme-driven "Chinese acupuncture" hydrogel microneedles for combined therapy of infected scalds. Mater Today Bio. 2025;33:101953. https://doi.org/10.1016/j.mtbio.2025.101953.
39. Qiao Y, Han M, Fu H, Xu Y, Bai Y, et al. Mucus-penetrating Nanoassembly as potential Oral phototherapeutic formulation against multi-drug resistant *H. pylori* infection. Small. 2024;20(20):e2306909. https://doi.org/10.1002/smll.202306909.
40. Wang R, Song C, Gao A, Liu Q, Guan W, et al. Antibody-conjugated liposomes loaded with indocyanine green for oral targeted photoacoustic imaging-guided sonodynamic therapy of *H. pylori* infection. Acta Biomater. 2022;143:418–27. https://doi.org/10.1016/j.actbio.2022.02.031.
41. Canaparo R, Foglietta F, Giuntini F, Della Pepa C, Dosio F, et al. Recent developments in antibacterial therapy: focus on stimuli-responsive drug-delivery systems and therapeutic nanoparticles. Molecules. 2019;24(10) https://doi.org/10.3390/molecules24101991.
42. Li Y, Hu X, Ding D, Zou Y, Xu Y, et al. In situ targeted MRI detection of *H. pylori* with stable magnetic graphitic nanocapsules. Nat Commun. 2017;8:15653. https://doi.org/10.1038/ncomms15653.
43. Safina DD, Abdulkhakov SR, Abdulkhakov RA. Eradication therapy of *H. pylori*: Present and future. Eksp Klin Gastroenterol. 2016;11:84–93.
44. Si Z, Chan-Park MB. Chemical innovations of antimicrobial polymers for combating antimicrobial resistance. ACS Biomater Sci Eng. 2025;11(5):2470–80. https://doi.org/10.1021/acsbiomaterials.4c02147.

45. Gedda G, Park YJ, Pang MG. Recent development of nanotechnology-based approaches for gynecologic cancer therapy. Obstet Gynecol Sci. 2025;68(1):18–29. https://doi.org/10.5468/ogs.24180.
46. He C, Lu J, Lin W. Hybrid nanoparticles for combination therapy of cancer. J Control Release. 2015;219:224–36. https://doi.org/10.1016/j.jconrel.2015.09.029.
47. Li Z, Tan S, Li S, Shen Q, Wang K. Cancer drug delivery in the nano era: an overview and perspectives (review). Oncol Rep. 2017;38(2):611–24. https://doi.org/10.3892/or.2017.5718.
48. Zheng Z, Cui J, Wu S, Cao Z, Cao P. Engineering metal-organic frameworks for enhanced antimicrobial efficacy: synthesis methodologies, mechanistic perspectives, and versatile applications. J Funct Biomater. 2025;16(9) https://doi.org/10.3390/jfb16090353.
49. Dey AK, Gonon A, Pécheur EI, Pezet M, Villiers C, et al. Impact of gold nanoparticles on the functions of macrophages and dendritic cells. Cells. 2021;10(1) https://doi.org/10.3390/cells10010096.
50. Koushki K, Biswal P, Vijay GV, Sadeghi M, Dehnavi S, et al. Immunomodulatory effects of gold nanoparticles: impacts on immune cells and mechanisms of action. Nanomaterials (Basel). 2025;15(15) https://doi.org/10.3390/nano15151201.
51. Song J-X, Zhang H-M, Shen X-F, Pang Y-H. Defect-engineered Fe-based metal-organic framework for synergistic photo-Fenton degradation and solid-phase microextraction analysis of triclosan and its degradation products. J Colloid Interface Sci. 2026;702:138993. https://doi.org/10.1016/j.jcis.2025.138993.
52. Salehpour F, Cassano P, Rouhi N, Hamblin MR, De Taboada L, et al. Penetration profiles of visible and near-infrared lasers and light-emitting diode light through the head tissues in animal and human species: a review of literature. Photobiomodul Photomed Laser Surg. 2019;37(10):581–95. https://doi.org/10.1089/photob.2019.4676.
53. Maruani A, Vierron E, Machet L, Giraudeau B, Boucaud A. Efficiency of low-frequency ultrasound sonophoresis in skin penetration of histamine: a randomized study in humans. Int J Pharm. 2010;385(1–2):37–41. https://doi.org/10.1016/j.ijpharm.2009.10.024.
54. Guo S, Gu D, Yang Y, Tian J, Chen X. Near-infrared photodynamic and photothermal co-therapy based on organic small molecular dyes. J Nanobiotechnol. 2023;21(1):348. https://doi.org/10.1186/s12951-023-02111-x.
55. Es-Haghi A, Amiri MS, Taghavizadeh Yazdi ME. Ferula latisecta gels for synthesis of zinc/silver binary nanoparticles: antibacterial effects against gram-negative and gram-positive bacteria and physicochemical characteristics. BMC Biotechnol. 2024;24(1):51. https://doi.org/10.1186/s12896-024-00878-x.
56. Reis ME, Bettencourt A, Ribeiro HM. The regulatory challenges of innovative customized combination products. Front Med (Lausanne). 2022;9:821094. https://doi.org/10.3389/fmed.2022.821094.
57. Zarepour A, Khosravi A, Iravani S, Zarrabi A. Biohybrid micro/nanorobots: pioneering the next generation of medical technology. Adv Healthc Mater. 2024;13(31):e2402102. https://doi.org/10.1002/adhm.202402102.
58. Jarab AS, Abu Heshmeh SR, Al Meslamani AZ. Bridging the gap: the future of biosimilars regulations. Hum Vaccin Immunother. 2024;20(1):2362450. https://doi.org/10.1080/21645515.2024.2362450.
59. Moraska AF, Stenerson L, Butryn N, Krutsch JP, Schmiege SJ, et al. Myofascial trigger point-focused head and neck massage for recurrent tension-type headache: a randomized, placebo-controlled clinical trial. Clin J Pain. 2015;31(2):159–68. https://doi.org/10.1097/ajp.0000000000000091.

# 11 Antimicrobial Mechanisms of Nanomaterials

Xin Ding, Xiangyu Luo, Li Yang, and Xinyi Zhong

## 11.1 Introduction

*Helicobacter pylori* infection remains a formidable global health challenge, intricately linked to chronic gastritis, peptic ulcer disease, and gastric adenocarcinoma [1]. The escalating prevalence of antibiotic-resistant strains, coupled with the inherent limitations of conventional therapies—such as poor drug bioavailability in the harsh gastric milieu and the bacterium's capacity for biofilm formation and intracellular persistence—has necessitated the exploration of innovative treatment modalities [2–4]. Nanomaterials have emerged as a transformative platform in this endeavor, offering a multifaceted arsenal of antimicrobial mechanisms that effectively circumvent traditional resistance pathways. These mechanisms, which include mechanical membrane disruption, targeted cytoskeletal interference, and immunomodulatory activation, provide a strategic advantage against *H. pylori*'s complex pathogenesis [5].

The distinct characteristic of nanomaterial-based strategies lies in their ability to deploy multi-mechanistic attacks. For instance, metal nanoparticles can inflict direct physical damage to bacterial membranes, while functionalized polymeric carriers enable the pH-responsive release of antimicrobial agents precisely at the

X. Ding (✉)
School of Medicine, Shenzhen Campus of Sun Yat-sen University, Shenzhen, China

State Key Laboratory of Anti-Infective Drug Discovery and Development, School of Pharmaceutical Sciences, Sun Yat-sen University, Guangzhou, China
e-mail: dingxin3@mail.sysu.edu.cn

X. Luo
School of Pharmaceutical Sciences (Shenzhen), Shenzhen Campus of Sun Yat-sen University, Shenzhen, China

L. Yang · X. Zhong
School of Medicine, Shenzhen Campus of Sun Yat-sen University, Shenzhen, China


J. Gao, Y. Du (eds.), *Non-Antibiotic Therapies for Helicobacter pylori*,
https://doi.org/10.1007/978-981-95-9598-3_11

colonization site [6]. This targeted approach, often enhanced by surface modifications with specific ligands, minimizes off-target effects and preserves the host's commensal microbiota. Furthermore, the ability of certain nanomaterials to generate reactive oxygen species (ROS) [7] or disrupt essential bacterial division proteins like FtsZ [8] introduces a level of bactericidal action that is difficult for *H. pylori* to evade through classical mutation-based resistance.

The principal advantages of applying these nanomaterial mechanisms to *H. pylori* infection are profound. Their nonspecific, physical-mode-of-action mechanisms significantly reduce the propensity for resistance development compared to molecular antibiotics [9]. The enhanced targeting and controlled release profiles improve therapeutic efficacy while lowering required drug doses, thereby mitigating systemic side effects [10]. By leveraging these advanced material properties, nanotechnology presents a promising avenue for achieving effective and sustainable eradication of *H. pylori*, potentially revolutionizing the management of this pervasive infection.

## 11.2 Physical Disruption Mechanisms

### 11.2.1 Mechanical Damage

Mechanical damage inflicted by nanomaterials involves physical interactions that compromise bacterial membrane integrity, leading to cell lysis and death [11]. This mechanism is particularly effective against *H. pylori* due to its Gram-negative outer membrane, which is rich in lipopolysaccharides and prone to disruption by nanostructured surfaces [12, 13]. Nanomaterials, such as metal nanoparticles (e.g., silver, gold, copper) and cationic polymers, exploit their high surface energy, sharp edges, and electrostatic properties to induce mechanical stress.

***Nanostructured Surfaces of Metal Nanoparticles*** Metal nanoparticles, such as silver (AgNPs) [14], copper (CuNPs) [15], and zinc oxide (ZnO NPs) [16], possess high surface energy and anisotropic shapes that enable them to physically puncture bacterial membranes. For instance, AgNPs adsorb onto the negatively charged surfaces of *H. pylori* via electrostatic and van der Waals forces [17], creating pores in the lipid bilayer [18]. This leads to osmotic imbalance, cytoplasmic leakage, and eventual cell lysis. Studies show that AgNPs with sizes below 10 nm exhibit enhanced bactericidal activity due to their increased surface-area-to-volume ratio, allowing deeper penetration into the membrane. In vitro experiments demonstrate that citrate-functionalized AgNPs achieve a 99.9% reduction in *H. pylori* viability within 30 min by causing membrane integrity loss, as confirmed by transmission electron microscopy (TEM) images showing disrupted cellular structures. The mechanical action is energy-independent and broad-spectrum, making it less susceptible to resistance development. Similarly, CuNPs generate mechanical stress through ion release and surface interactions, disrupting membrane potential and inducing oxidative stress synergistically. This approach is particularly effective against *H. pylori* due to its Gram-negative outer membrane, which is rich in lipopolysaccharides and prone to physical disruption [13].

***Sharp Edges and High Surface Energy*** Nanoparticles with sharp edges, such as gold nanostars (AuNSs), graphene oxide nanosheets, or carbon nanotubes, act as "nanoknives" that slice through bacterial membranes [18]. The high surface energy of these materials promotes strong adhesion to bacterial surfaces, amplifying mechanical stress. For example, AuNSs with sharp tips create microtears in *H. pylori* membranes, leading to rapid ion efflux and cell death. Carbon-based nanomaterials like graphene oxide exhibit similar effects; their two-dimensional structure and edges physically rupture cells, as observed in scanning electron microscopy (SEM) studies where bacterial membranes appeared shredded post-exposure [19]. This mechanism is advantageous for *H. pylori* therapy because it targets the bacterium's structural integrity without relying on specific receptor interactions, thereby reducing the risk of resistance.

***Cationic Polymers*** Cationic polymers, such as chitosan, polyethylenimine (PEI), and their derivatives, induce mechanical damage through electrostatic interactions with negatively charged bacterial membranes [20–22]. Chitosan nanoparticles, with their high density of amine groups, bind to lipopolysaccharides on *H. pylori* surfaces, disrupting membrane fluidity and causing lysis. This is enhanced by chitosan's mucoadhesive properties, which prolong gastric retention and facilitate targeted delivery to infection sites. Studies show that chitosan-based NPs reduce *H. pylori* adhesion to gastric epithelial cells by competing with host receptors like BabA and SabA [23]. In murine models, chitosan microparticles functionalized with fucose demonstrated a 76% reduction in bacterial load by mechanically disrupting membranes without inducing resistance. Similarly, PEI-based polymers form pores in membranes through a "proton sponge" effect, where their buffering capacity leads to osmotic swelling and rupture. The advantage of cationic polymers lies in their ability to cause physical damage while also acting as drug carriers, providing a dual antimicrobial approach. For *H. pylori*, which resides in the mucus layer, these polymers can penetrate biofilms and disrupt embedded cells, addressing a key challenge in eradication.

The application of mechanical damage mechanisms in *H. pylori* therapy is enhanced by nanomaterial functionalization. For instance, pH-responsive polymers like Eudragit ensure that mechanical disruption occurs specifically in the neutral pH environment of *H. pylori* colonization sites (pH 6–7), minimizing off-target effects in the acidic stomach [24]. This targeted approach reduces collateral damage to gastric microbiota and improves treatment outcomes.

### 11.2.2 Cytoskeletal Disruption by Nanomaterials

Cytoskeletal disruption involves nanomaterials interfering with bacterial proteins essential for cell division, morphology, and intracellular transport. In *H. pylori*, key targets include FtsZ (a homolog of tubulin) and MreB (an actin-like protein), which regulate septum formation and cell shape [25–27]. Nanomaterials can inhibit these proteins through direct binding or by inducing oxidative stress, leading to filamentation, aberrant division, and apoptosis-like death.

***Interference with FtsZ Polymerization*** Nanomaterials like gold nanoparticles (AuNPs) functionalized with FtsZ inhibitors (e.g., berberine or curcumin) bind to the GTPase domain of FtsZ, preventing its polymerization and Z-ring assembly. This leads to elongated cell morphologies, filamentation, and eventual lysis. For instance, AuNPs conjugated with FtsZ inhibitors caused division defects in *H. pylori* cells, as visualized by fluorescence microscopy, where cells failed to separate post-division [28]. This approach is highly specific to bacterial cells due to the conservation of FtsZ in prokaryotes, minimizing host toxicity. The mechanism is particularly effective against *H. pylori* because the bacterium's spiral shape and motility rely on precise cytoskeletal regulation; disruption impairs colonization and virulence. Studies demonstrate that FtsZ-targeting nanomaterials reduce *H. pylori* viability by over 80% in vitro, with synergistic effects when combined with mechanical damage agents like AgNPs.

***Reactive Oxygen Species (ROS)-Mediated Disruption*** Certain nanomaterials, such as iron oxide ($Fe_3O_4$) or titanium dioxide ($TiO_2$) nanoparticles, generate ROS under specific conditions (e.g., light irradiation or acidic environments) [29], which oxidize cytoskeletal proteins like FtsZ and MreB. In *H. pylori*, which is microaerophilic and susceptible to oxidative stress, ROS production disrupts cell rigidity and division. This mechanism synergizes with mechanical damage by weakening bacterial defenses, allowing enhanced nanomaterial penetration. The combination of physical and chemical disruption ensures comprehensive targeting of *H. pylori*, reducing the likelihood of escape mutants.

***Polymer-Based Nanosystems*** Cationic polymers, such as poly(lactic-co-glycolic acid) (PLGA) nanoparticles can penetrate *H. pylori* cells and release payloads that disrupt protein assembly. These systems often incorporate targeting ligands (e.g., antibodies against *H. pylori* urease) to enhance specificity, reducing host cell toxicity.

The combination of cytoskeletal disruption with other mechanisms, such as drug delivery, provides a multimodal attack. Nanomaterials like mesoporous silica nanoparticles (MSNs) can co-deliver antibiotics and cytoskeletal inhibitors, ensuring comprehensive eradication [30].

## 11.3 Chemical Interference Mechanisms

Nanomaterials offer promising solutions through chemical interference mechanisms, which involve targeted antimicrobial agent release, reactive oxygen species (ROS) generation, and disruption of bacterial membrane potential. These approaches leverage the unique physicochemical properties of nanomaterials to achieve precise, efficient, and resistance-resistant eradication of *H. pylori*.

### 11.3.1 Antimicrobial Agent Release

Nanocarriers, such as mesoporous silica nanoparticles (MSNs) and polymeric nanoparticles (e.g., poly(lactic-co-glycolic acid) (PLGA) NPs), are designed to encapsulate and release antimicrobial agents—including essential oils, polyphenols, and antimicrobial peptides—with high precision [31]. These systems employ controlled release mechanisms, such as pH-responsive or ultrasound-activated delivery [32], to ensure localized and sustained antimicrobial activity at *H. pylori* colonization sites.

***Mesoporous Silica Nanoparticles (MSNs)*** MSNs possess a high surface area and tunable pore sizes, allowing efficient loading of hydrophobic antimicrobial agents like curcumin or thymol. For instance, MSNs functionalized with pH-responsive polymers release amoxicillin specifically in the neutral pH environment of *H. pylori* niches, minimizing premature degradation in gastric acid [32]. The targeted release enhances drug bioavailability and reduces systemic side effects, addressing a key challenge in *H. pylori* therapy.

***Polymeric Nanoparticles*** Chitosan nanoparticles are biodegradable and biocompatible, making them ideal for encapsulating peptides or polyphenols [33]. For example, fucose-chitosan/gelatin NPs loaded with epigallocatechin-3-gallate (EGCG)—a green tea polyphenol—exhibit pH-dependent release, with minimal release at pH 1.2 and rapid release at pH 7.4 [34]. The controlled release paradigm prevents sublethal antibiotic exposure, which is a primary driver of resistance development.

***Light-Activated Delivery Systems*** Nanomaterials conjugated with photosensitizers enable light-triggered release of antimicrobials. Under near-infrared (NIR) light irradiation, these systems generate heat or ROS, which disrupt the nanocarrier structure and release encapsulated agents [35, 36]. This approach is particularly useful for eradicating biofilm-embedded *H. pylori*, as light penetration can be tuned to target deep tissue layers. In vitro studies show that light-activated nanocarriers achieve biofilm disruption, synergizing with mechanical damage mechanisms.

The advantage of antimicrobial agent release lies in its ability to maintain therapeutic drug levels at the infection site, reducing dosing frequency and minimizing resistance risks.

### 11.3.2 Generation of Reactive Oxygen Species (ROS)

Certain nanomaterials, such as titanium dioxide ($TiO_2$) and zinc oxide (ZnO) nanoparticles, generate ROS—including hydroxyl radicals (•OH), superoxide ions ($O_2^-$), and hydrogen peroxide ($H_2O_2$)—under specific conditions (e.g., UV light irradiation or acidic environments) [37–39]. ROS induce oxidative stress by damaging bacterial DNA, proteins, and lipid membranes, leading to cell death. The ROS-mediated damage is nonspecific, targeting multiple cellular components simultaneously, which makes it difficult for *H. pylori* to develop resistance.

***Zinc Oxide (ZnO) Nanoparticles*** ZnO NPs generate ROS through both photocatalytic and ion-release mechanisms [40–42]. In the acidic gastric environment, $Zn^{2+}$ ions leach from ZnO NPs and catalyze Fenton-like reactions, producing •OH radicals that oxidize bacterial proteins and DNA. Studies show that ZnO NPs reduce *H. pylori* biofilm biomass in vitro, with confocal microscopy confirming ROS-induced membrane damage. The combination of ROS generation and ionic effects provides a dual antimicrobial action, enhancing efficacy against resistant strains.

ROS-generating nanomaterials penetrate biofilms by degrading extracellular polymeric substances (EPS) through oxidative cleavage [43]. This is particularly advantageous for chronic infections, where biofilms confer tolerance to antibiotics. The key advantage of ROS generation is its broad-spectrum, multi-target action, which minimizes resistance development. Additionally, ROS-based therapies can be synergized with other mechanisms, such as drug delivery, for enhanced outcomes. However, careful design is needed to avoid off-target oxidative damage to host tissues, achieved through targeting ligands and controlled activation.

### 11.3.3 Disruption of Membrane Potential

Some nanomaterials, particularly cationic polymers and metal nanoparticles, insert into bacterial membranes and alter their electrostatic properties, disrupting membrane potential and ion transport [43, 44]. This leads to loss of proton motive force, impaired ATP synthesis, and eventual cell death.

***Cationic Polymers*** Polymers like chitosan and polyethylenimine (PEI) possess positively charged groups that interact with negatively charged phospholipids in *H. pylori* membranes. This interaction causes membrane depolarization, pore formation, and ion leakage. For example, chitosan nanoparticles increase membrane permeability, leading to cytoplasmic efflux and cell lysis, as observed in transmission electron microscopy (TEM) studies [45]. The mucoadhesive properties of chitosan enhance gastric retention, allowing sustained membrane disruption.

***Metal Nanoparticles*** Silver nanoparticles (AgNPs) and gold nanoparticles (AuNPs) disrupt membrane potential through physical insertion and ion release [46]. AgNPs, with their high surface energy, embed into lipid bilayers, causing thinning and increased permeability. This is accompanied by $Ag^+$ ion release, which further disrupts ion channels and potentiates oxidative stress [47]. The combination of physical and chemical disruption ensures rapid bactericidal activity.

***Synergy with Other Mechanisms*** Membrane disruption often synergizes with ROS generation or drug release. For instance, ZnO NPs simultaneously generate ROS and disrupt membrane integrity, leading to enhanced killing.

The advantage of membrane potential disruption is its rapid, nonspecific action, which is less prone to resistance compared to receptor-targeted therapies. It is particularly effective against *H. pylori* due to the bacterium's Gram-negative outer membrane, which is rich in lipopolysaccharides and susceptible to electrostatic interactions. Moreover, this mechanism can be tailored through surface functionalization to improve specificity and reduce host cell toxicity.

## 11.4 Immunomodulatory Mechanisms

Nanomaterials offer promising solutions through immunomodulatory mechanisms, which involve stimulating innate immunity, modulating biofilm formation, and reducing bacterial virulence. These approaches leverage the unique physicochemical properties of nanomaterials to enhance host defenses while directly targeting bacterial resilience.

### 11.4.1 Stimulation of Innate Immunity

Nanomaterials can activate innate immune cells [48], such as macrophages and neutrophils, to enhance their phagocytic activity and production of antimicrobial factors, including cytokines and reactive nitrogen species (RNS) [49, 50]. This activation occurs through pattern recognition receptors (e.g., Toll-like receptors, TLRs), leading to improved bacterial clearance and reduced immune evasion by *H. pylori*.

***Activation of Macrophages and Neutrophils*** Metal-based nanoparticles, such as silver nanoparticles (AgNPs) and zinc oxide nanoparticles (ZnO NPs), serve as potent immune stimulants. For instance, AgNPs upregulate TLR4 signaling in macrophages, prompting the release of pro-inflammatory cytokines (e.g., TNF-α, IL-6, and IL-1β) and nitric oxide (NO), which create a hostile microenvironment for

*H. pylori* [51]. Studies demonstrate that AgNPs with sizes below 20 nm induce macrophage polarization to the M1 phenotype, enhancing phagocytosis and bacterial killing in vitro. The key advantage here is the nonspecific immune activation, which reduces the likelihood of resistance development compared to targeted antibiotic therapies.

***Polymeric Nanomaterials as Adjuvants*** Cationic polymers, such as chitosan and polyethylenimine (PEI), act as immune adjuvants by enhancing antigen presentation and cytokine production [52]. Chitosan nanoparticles, with their mucoadhesive properties, prolong gastric retention and stimulate local immune responses. Chitosan-based NPs loaded with *H. pylori* antigens (e.g., urease subunit B) promote dendritic cell maturation and Th1 immune responses, leading to increased IgA and IgG production in mucosal tissues [53]. In preclinical studies, oral administration of chitosan-antigen NPs reduced *H. pylori* load in mice by enhancing mucosal immunity without causing excessive inflammation [54]. This approach is particularly beneficial for *H. pylori* infections, where the bacterium often suppresses host immune responses through virulence factors like CagA.

***Synergy with Antimicrobial Factors*** Nanomaterials can synergize with endogenous antimicrobial peptides (e.g., defensins) to amplify innate immunity [55, 56]. The combination of nanomaterial-induced immune stimulation and direct antimicrobial action provides a multifaceted attack, minimizing the risk of bacterial escape mutants.

## 11.4.2 Modulation of Biofilm Formation

*H. pylori* frequently forms biofilms, structured communities embedded in extracellular polymeric substances (EPS), that confer resistance to host immune responses and conventional therapies [57]. Nanomaterials can disrupt biofilm formation by inhibiting bacterial adhesion, degrading EPS, and preventing biofilm maturation through physical and chemical interactions.

***Inhibition of Bacterial Adhesion*** Nanomaterials functionalized with anti-adhesion ligands (e.g., fucose or lectins) compete with *H. pylori* for binding sites on gastric epithelial cells. For example, fucose-conjugated polymeric nanoparticles block the interaction between *H. pylori* BabA adhesin and host Lewis b antigens, reducing bacterial adhesion in vitro [58]. This prevention of initial attachment disrupts biofilm initiation and enhances the efficacy of immune clearance. Additionally, cationic polymers like chitosan form a protective barrier on mucosal surfaces, preventing *H. pylori* colonization and biofilm formation.

***Degradation of Extracellular Polymeric Substances (EPS)*** Enzymatic nanomaterials, such as liposomes loaded with DNase I or protease, break down EPS components (e.g., DNA, proteins, and polysaccharides), destabilizing biofilm integrity

[58]. pH-responsive systems, like Eudragit-coated nanoparticles, release EPS-degrading enzymes specifically in the neutral pH environment of *H. pylori* biofilms. Studies show that DNase I-loaded mesoporous silica nanoparticles (MSNs) reduce biofilm biomass in vitro, as quantified by crystal violet staining, allowing immune cells to penetrate and eradicate embedded bacteria [59]. This approach is crucial for chronic infections, where biofilms act as reservoirs for persistent *H. pylori*.

***Disruption of Biofilm Maturation*** Nanomaterials can interfere with quorum sensing, a key process in biofilm development [60–62]. The advantage of biofilm modulation lies in its ability to resensitize *H. pylori* to host defenses and antibiotics, addressing a major challenge in treatment failure.

### 11.4.3 Reduction of Bacterial Virulence

Nanomaterials can directly target and neutralize *H. pylori* virulence factors, such as flagella, adhesion proteins (e.g., BabA and SabA), and toxins (e.g., VacA and CagA), thereby attenuating pathogenicity and enhancing host immune clearance [63–65].

***Neutralization of Adhesion Proteins*** Gottesmann and colleagues, for instance, developed pectin-coated liposomes encapsulating amoxicillin, which leverage a specific interaction between pectin and mucins (BabA) as well as surface structures of *H. pylori* [66]. Their findings demonstrated that the pectin-coated liposomes had stronger bactericidal activity compared to their non-coated counterparts. This not only prevents adhesion but also opsonizes the bacteria for enhanced immune recognition. This targeted approach minimizes host tissue damage and reduces inflammation.

***Detoxification of Virulence Factors*** Some nanomaterials possess catalytic properties that degrade bacterial toxins. Cerium oxide nanoparticles ($CeO_2$ NPs) exhibit oxidase-like activity that neutralizes VacA toxin, mitigating its cytotoxic effects on gastric epithelial cells [67, 68]. $CeO_2$ NPs reduce VacA-induced vacuolation, preserving epithelial integrity and allowing immune cells to function effectively. Additionally, nanomaterials can deliver siRNA or antisense oligonucleotides to silence virulence genes, further reducing *H. pylori* pathogenicity.

The reduction of virulence complements innate immunity stimulation by making *H. pylori* less capable of evading host defenses, thereby enhancing the overall efficacy of nanomaterial-based therapies.

The integration of immunomodulatory mechanisms into nanomaterial design has shown promising results in preclinical models of *H. pylori* infection. For example, multifunctional nanoparticles combining innate immunity stimulation (e.g., TLR agonists) and virulence reduction (e.g., anti-adhesion agents) have achieved synergistic effects, reducing bacterial load in murine gastritis models. pH-sensitive liposomes co-delivering biofilm-disrupting enzymes and immunomodulators (e.g., cytokines) enhance bacterial clearance while minimizing gastric damage.

## 11.5 Synergistic Effects

Synergistic effects refer to the phenomenon where the combined action of nanomaterials and other therapeutics results in enhanced antimicrobial efficacy that is greater than the sum of their individual effects. This mechanism leverages multi-targeted approaches to overcome resistance, improve targeting, and reduce treatment duration.

### 11.5.1 Synergy with Natural Antimicrobial Agents

Natural antimicrobial agents, such as plant-derived polyphenols, essential oils, and antimicrobial peptides, exhibit potent activity against *H. pylori* but often face challenges like low stability, poor solubility, and rapid degradation in the gastric environment [69]. Nanomaterials can overcome these limitations by serving as delivery vehicles or enhancers, leading to synergistic effects that amplify bactericidal action. AMPs, such as defensins or LL-37, are prone to proteolytic degradation in the stomach [69]. Nanomaterials like chitosan nanoparticles protect AMPs and enhance their delivery [70]. In combination, AMPs (LL-37) and chitosan disrupt bacterial membranes and modulate immune responses synergistically. This approach leverages the innate immune stimulation by AMPs and the physical damage by nanomaterials, providing a multifaceted attack.

### 11.5.2 Synergy with Probiotics

Probiotics, such as *Lactobacillus* or *Bifidobacterium* species, compete with *H. pylori* for adhesion sites and produce antimicrobial substances like bacteriocins [71–73]. However, their efficacy is limited by low survival in gastric acid and poor colonization. Nanomaterials can enhance probiotic activity through protection, targeted delivery, and synergistic actions. Nanomaterials like alginate or chitosan-based particles encapsulate probiotics, shielding them from gastric acid and ensuring delivery to the infection site [74]. The nanomaterials disrupt *H. pylori* membranes, while probiotics restore microbial balance, reducing inflammation and recurrence.

### 11.5.3 Synergy with Phototherapy

Phototherapy, including photodynamic therapy (PDT) and photothermal therapy (PTT), uses light-activated agents to generate ROS or heat, selectively killing bacteria. Nanomaterials can enhance phototherapy by improving targeting, stability, and efficacy through synergistic effects.

***Photodynamic Therapy (PDT) Synergy*** In PDT, photosensitizers (e.g., porphyrins or methylene blue) produce ROS under light irradiation. Nanomaterials like $ZnO_2$ nanoparticles can serve as photosensitizer carriers or enhance ROS generation. For example, Liposomes loaded with $ZnO_2$-Ce6 under blue light produce ROS [75]. This combination achieves near-complete eradication of *H. pylori* biofilms in vitro. The nanomaterials penetrate biofilms, while ROS cause oxidative damage to bacterial DNA and proteins.

***Photothermal Therapy (PTT) Synergy*** PTT involves nanomaterials that convert light energy into heat, causing thermal ablation of bacteria. Gold nanorods or graphene oxide sheets under near-infrared (NIR) light generate localized heat, disrupting bacterial membranes [76, 77]. This approach is particularly useful for targeting deep-seated infections, as light penetration can be controlled.

The synergistic effects of nanomaterials have significant clinical implications for *H. pylori* management. Nano-formulations that combine antimicrobial agents with targeting ligands (e.g., fucose or lectins) enable precise delivery to *H. pylori*-infected regions, enhancing localized synergy while reducing systemic side effects. In preclinical studies, pH-responsive nanocarriers co-loaded with antibiotics and natural compounds have shown superior efficacy in eradicating resistant *H. pylori* strains in animal models. These systems also address biofilm-related challenges by disrupting EPS and resensitizing bacteria to treatments. Furthermore, synergistic approaches reduce treatment duration and antibiotic doses, aligning with antimicrobial stewardship goals. Future translation requires optimizing biocompatibility and scalability, with ongoing research focusing on multifunctional nanosystems that integrate immune modulation and microbiome conservation for comprehensive *H. pylori* therapy.

## References

1. Fock KM, Graham DY, Malfertheiner P. *H. pylori* research: historical insights and future directions. Nat Rev Gastroenterol Hepatol. 2013;10(8):495–500. https://doi.org/10.1038/nrgastro.2013.96.
2. Gupta A, Shetty S, Mutalik S, Chandrashekar HR, Nandakumar K, et al. Treatment of *H. pylori* infection and gastric ulcer: need for novel pharmaceutical formulation. Heliyon. 2023;9(10):e20406. https://doi.org/10.1016/j.heliyon.2023.e20406.
3. Lopes D, Nunes C, Martins MC, Sarmento B, Reis S. Eradication of *H. pylori*: past, present and future. J Control Release. 2014;189:169–86. https://doi.org/10.1016/j.jconrel.2014.06.020.
4. Nguyen LH, Örtqvist AK, Cao Y, Simon TG, Roelstraete B, et al. Antibiotic use and the development of inflammatory bowel disease: a national case-control study in Sweden. Lancet Gastroenterol Hepatol. 2020;5(11):986–95. https://doi.org/10.1016/s2468-1253(20)30267-3.
5. Xie M, Gao M, Yun Y, Malmsten M, Rotello VM, et al. Antibacterial nanomaterials: mechanisms, impacts on antimicrobial resistance and design principles. Angew Chem Int Ed. 2023;62(17):e202217345. https://doi.org/10.1002/anie.202217345.

6. Zhu C, Diao Z, Yang Y, Liao J, Wang C, et al. Recent advances and challenges in metal-based antimicrobial materials: a review of strategies to combat antibiotic resistance. J Nanobiotechnol. 2025;23(1):193. https://doi.org/10.1186/s12951-025-03249-6.
7. Yu J, Guo Z, Yan J, Bu C, Peng C, et al. Gastric acid-responsive ROS nanogenerators for effective treatment of *H. pylori* infection without disrupting homeostasis of intestinal flora. Adv Sci. 2023;10(20):2206957. https://doi.org/10.1002/advs.202206957.
8. Mistry TL, Truong L, Ghosh AK, Johnson ME, Mehboob S. Benzimidazole-based FabI inhibitors: a promising novel scaffold for anti-staphylococcal drug development. ACS Infect Dis. 2017;3(1):54–61. https://doi.org/10.1021/acsinfecdis.6b00123.
9. Zhang Q, Zhou H, Jiang P, Xiao X. Metal-based nanomaterials as antimicrobial agents: a novel driveway to accelerate the aggravation of antibiotic resistance. J Hazard Mater. 2023;455:131658. https://doi.org/10.1016/j.jhazmat.2023.131658.
10. Zhang J, Tang W, Zhang X, Song Z, Tong T. An overview of stimuli-responsive intelligent antibacterial nanomaterials. Pharmaceutics. 2023;15(8) https://doi.org/10.3390/pharmaceutics15082113.
11. Li Z, Ren J, Sui X, Yang N, Li S, et al. A win-win platform: stabilized black phosphorous nanosheets loading gallium ions for enhancing the healing of bacterial-infected wounds through synergistic antibacterial approaches. Int Wound J. 2024;21(6):e14940. https://doi.org/10.1111/iwj.14940.
12. Silva LM, Correia VG, Moreira ASP, Domingues MRM, Ferreira RM, et al. *H. pylori* lipopolysaccharide structural domains and their recognition by immune proteins revealed with carbohydrate microarrays. Carbohydr Polym. 2021;253:117350. https://doi.org/10.1016/j.carbpol.2020.117350.
13. Li H, Liao T, Debowski AW, Tang H, Nilsson HO, et al. Lipopolysaccharide structure and biosynthesis in *H. pylori*. Helicobacter. 2016;21(6):445–61. https://doi.org/10.1111/hel.12301.
14. Jyoti K, Singh A. Evaluation of antibacterial activity from phytosynthesized silver nanoparticles against medical devices infected with Staphylococcus spp. J Taibah Univ Med Sci. 2017;12(1):47–54. https://doi.org/10.1016/j.jtumed.2016.08.006.
15. Anyaogu KC, Fedorov AV, Neckers DC. Synthesis, characterization, and antifouling potential of functionalized copper nanoparticles. Langmuir. 2008;24(8):4340–6. https://doi.org/10.1021/la800102f.
16. McGuffie MJ, Hong J, Bahng JH, Glynos E, Green PF, et al. Zinc oxide nanoparticle suspensions and layer-by-layer coatings inhibit staphylococcal growth. Nanomedicine. 2016;12(1):33–42. https://doi.org/10.1016/j.nano.2015.10.002.
17. Fernando I, Tay YY, Karunasekera H, Zhou Y. Observation of the interactions of silver nanoparticles (AgNPs) mediated by acid in the aquatic matrices using in-situ liquid cell transmission electron microscopy. Anal Chim Acta. 2020;1104:47–52. https://doi.org/10.1016/j.aca.2019.12.072.
18. Tu Y, Lv M, Xiu P, Huynh T, Zhang M, et al. Destructive extraction of phospholipids from Escherichia coli membranes by graphene nanosheets. Nat Nanotechnol. 2013;8(8):594–601. https://doi.org/10.1038/nnano.2013.125.
19. Zheng H, Ma R, Gao M, Tian X, Li YQ, et al. Antibacterial applications of graphene oxides: structure-activity relationships, molecular initiating events and biosafety. Sci Bull (Beijing). 2018;63(2):133–42. https://doi.org/10.1016/j.scib.2017.12.012.
20. Carmona-Ribeiro AM, de Melo Carrasco LD. Cationic antimicrobial polymers and their assemblies. Int J Mol Sci. 2013;14(5):9906–46. https://doi.org/10.3390/ijms14059906.
21. Pachla J, Kopiasz RJ, Marek G, Tomaszewski W, Głogowska A, et al. Polytrimethylenimines: highly potent antibacterial agents with activity and toxicity modulated by the polymer molecular weight. Biomacromolecules. 2023;24(5):2237–49. https://doi.org/10.1021/acs.biomac.3c00139.
22. Lin YH, Tsai SC, Lai CH, Lee CH, He ZS, et al. Genipin-cross-linked fucose-chitosan/heparin nanoparticles for the eradication of *H. pylori*. Biomaterials. 2013;34(18):4466–79. https://doi.org/10.1016/j.biomaterials.2013.02.028.

23. Luo D, Guo J, Wang F, Sun J, Li G, et al. Preparation and evaluation of anti-*H. pylori* efficacy of chitosan nanoparticles in vitro and in vivo. J Biomater Sci Polym Ed. 2009;20(11):1587–96. https://doi.org/10.1163/092050609X12464345137685.
24. Kamankesh M, Yadegar A, Llopis-Lorente A, Liu C, Haririan I, et al. Future nanotechnology-based strategies for improved management of *H. pylori* infection. Small. 2024;20(3):2302532. https://doi.org/10.1002/smll.202302532.
25. Margolin W. FtsZ and the division of prokaryotic cells and organelles. Nat Rev Mol Cell Biol. 2005;6(11):862–71. https://doi.org/10.1038/nrm1745.
26. Taylor JA, Bratton BP, Sichel SR, Blair KM, Jacobs HM, et al. Distinct cytoskeletal proteins define zones of enhanced cell wall synthesis in *H. pylori*. elife. 2020:9. https://doi.org/10.7554/eLife.52482.
27. Scheffers D, Driessen AJ. The polymerization mechanism of the bacterial cell division protein FtsZ. FEBS Lett. 2001;506(1):6–10. https://doi.org/10.1016/s0014-5793(01)02855-1.
28. Sanyasi S, Majhi RK, Kumar S, Mishra M, Ghosh A, et al. Polysaccharide-capped silver nanoparticles inhibit biofilm formation and eliminate multi-drug-resistant bacteria by disrupting bacterial cytoskeleton with reduced cytotoxicity towards mammalian cells. Sci Rep. 2016;6:24929. https://doi.org/10.1038/srep24929.
29. Mandal TK. Nanomaterial-enhanced hybrid disinfection: a solution to combat multidrug-resistant bacteria and antibiotic resistance genes in wastewater. Nanomater Internet. 2024;14(22):1847.
30. Duan F, Feng X, Jin Y, Liu D, Yang X, et al. Metal-carbenicillin framework-based nanoantibiotics with enhanced penetration and highly efficient inhibition of MRSA. Biomaterials. 2017;144:155–65. https://doi.org/10.1016/j.biomaterials.2017.08.024.
31. de Ávila BE-F, Angsantikul P, Li J, Angel Lopez-Ramirez M, Ramírez-Herrera DE, et al. Micromotor-enabled active drug delivery for in vivo treatment of stomach infection. Nat Commun. 2017;8(1):272. https://doi.org/10.1038/s41467-017-00309-w.
32. Yan J, Yu J, Bu C, Yang L, Chen J, et al. Antibiotic-augmented chemodynamic therapy for treatment of *H. pylori* infection in the dynamic stomach environment. Nano Lett. 2024;24(47):14983–92. https://doi.org/10.1021/acs.nanolett.4c03692.
33. Refate A, Mohamed Y, Mohamed M, Sobhy M, Samhy K, et al. Influence of electrospinning parameters on biopolymers nanofibers, with emphasis on cellulose & chitosan. Heliyon. 2023;9(6):e17051. https://doi.org/10.1016/j.heliyon.2023.e17051.
34. Lin YH, Chen ZR, Lai CH, Hsieh CH, Feng CL. Active targeted nanoparticles for oral administration of gastric cancer therapy. Biomacromolecules. 2015;16(9):3021–32. https://doi.org/10.1021/acs.biomac.5b00907.
35. Xu P, Wen C, Gao C, Liu H, Li Y, et al. Near-infrared-II-activatable self-assembled manganese porphyrin-gold heterostructures for photoacoustic imaging-guided sonodynamic-augmented photothermal/photodynamic therapy. ACS Nano. 2024;18(1):713–27. https://doi.org/10.1021/acsnano.3c09011.
36. Qiao Y, Ma Y, Tong Y, Liu W, Wang S, et al. Phototherapy and mechanism exploration of biofilm and multidrug-resistant *H. pylori* by bacteria-targeted NIR photosensitizer. Small. 2023;19(4):e2205248. https://doi.org/10.1002/smll.202205248.
37. Sodeinde KO, Olusanya SO, Lawal OS, Sriariyanun M, Adediran AA. Enhanced adsorptional-photocatalytic degradation of chloramphenicol by reduced graphene oxide-zinc oxide nanocomposite. Sci Rep. 2022;12(1):17054. https://doi.org/10.1038/s41598-022-21266-5.
38. Rafique A, Amjad F, Janjua M, Naqvi SAR, Hassan SU, et al. Chia seed-mediated fabrication of ZnO/Ag/Ag(2)O nanocomposites: structural, antioxidant, anticancer, and wound healing studies. Front Chem. 2024;12:1405385. https://doi.org/10.3389/fchem.2024.1405385.
39. Martínez-Carmona M, Gun'ko Y, Vallet-Regí M. ZnO nanostructures for drug delivery and theranostic applications. Nanomaterials (Basel). 2018;8(4) https://doi.org/10.3390/nano8040268.
40. Mutukwa D, Taziwa RT, Khotseng L. A review of plant-mediated ZnO nanoparticles for photodegradation and antibacterial applications. Nanomaterials (Basel). 2024;14(14) https://doi.org/10.3390/nano14141182.

41. Spoială A, Ilie CI, Trușcă RD, Oprea OC, Surdu VA, et al. Zinc oxide nanoparticles for water purification. Materials (Basel). 2021;14(16) https://doi.org/10.3390/ma14164747.
42. Liu C, Chen S, Sun C, Zuo W, Wu P, et al. Protonated charge reversal nanodrugs for active targeting clearance of *H. pylori* accompanied by gut microbiota protection. Adv Funct Mater. 2023;33(35):2300682. https://doi.org/10.1002/adfm.202300682.
43. Astuty SD, Suhariningsih BA, Astuti SD. The efficacy of photodynamic inactivation of the diode laser in inactivation of the Candida albicans biofilms with exogenous photosensitizer of papaya leaf chlorophyll. J Lasers Med Sci. 2019;10(3):215–24. https://doi.org/10.15171/jlms.2019.35.
44. Durán N, Durán M, de Jesus MB, Seabra AB, Fávaro WJ, et al. Silver nanoparticles: a new view on mechanistic aspects on antimicrobial activity. Nanomedicine. 2016;12(3):789–99. https://doi.org/10.1016/j.nano.2015.11.016.
45. Lou MM, Zhu B, Muhammad I, Li B, Xie GL, et al. Antibacterial activity and mechanism of action of chitosan solutions against apricot fruit rot pathogen Burkholderia seminalis. Carbohydr Res. 2011;346(11):1294–301. https://doi.org/10.1016/j.carres.2011.04.042.
46. Kanwal Z, Raza MA, Riaz S, Manzoor S, Tayyeb A, et al. Synthesis and characterization of silver nanoparticle-decorated cobalt nanocomposites (Co@AgNPs) and their density-dependent antibacterial activity. R Soc Open Sci. 2019;6(5):182135. https://doi.org/10.1098/rsos.182135.
47. Amin M, Hameed S, Ali A, Anwar F, Shahid SA, et al. Green synthesis of silver nanoparticles: structural features and in vivo and in vitro therapeutic effects against *H pylori* induced gastritis. Bioinorg Chem Appl. 2014;2014:135824. https://doi.org/10.1155/2014/135824.
48. Cronin JG, Jones N, Thornton CA, Jenkins GJS, Doak SH, et al. Nanomaterials and innate immunity: a perspective of the current status in nanosafety. Chem Res Toxicol. 2020;33(5):1061–73. https://doi.org/10.1021/acs.chemrestox.0c00051.
49. Rubey KM, Mukhitov AR, Nong J, Wu J, Krymskaya VP, et al. Nanoparticle-induced augmentation of neutrophils' phagocytosis of bacteria. Front Pharmacol. 2022;13:923814. https://doi.org/10.3389/fphar.2022.923814.
50. Taratummarat S, Sangphech N, Vu CTB, Palaga T, Ondee T, et al. Gold nanoparticles attenuates bacterial sepsis in cecal ligation and puncture mouse model through the induction of M2 macrophage polarization. BMC Microbiol. 2018;18(1):85. https://doi.org/10.1186/s12866-018-1227-3.
51. Park J, Lim DH, Lim HJ, Kwon T, Choi JS, et al. Size dependent macrophage responses and toxicological effects of Ag nanoparticles. Chem Commun (Camb). 2011;47(15):4382–4. https://doi.org/10.1039/c1cc10357a.
52. Zhang Y, Gu P, Wusiman A, Xu S, Ni H, et al. The immunoenhancement effects of polyethylenimine-modified Chinese yam polysaccharide-encapsulated PLGA nanoparticles as an adjuvant. Int J Nanomedicine. 2020;15:5527–43. https://doi.org/10.2147/ijn.S252515.
53. Guo T, Li X, Lin M, Zhang L. Mucosal adjuvant activity of chitosan encapsulated nanoparticles as *H. pylori* epitope vaccine carrier. Nanosci Nanotechnol Lett. 2016;8(12):1106–11.
54. Chan YL, Chao YP, Li TL, Liao KW. Chitosan-based Oral DNA vaccine elicits mucosal immunity and reduces *H. pylori* colonization in mice. Vaccine. 2025;64:127692. https://doi.org/10.1016/j.vaccine.2025.127692.
55. Zheng Y, Liu W, Chen Y, Li C, Jiang H, et al. Conjugating gold nanoclusters and antimicrobial peptides: from aggregation-induced emission to antibacterial synergy. J Colloid Interface Sci. 2019;546:1–10. https://doi.org/10.1016/j.jcis.2019.03.052.
56. Ganz T. Defensins: antimicrobial peptides of innate immunity. Nat Rev Immunol. 2003;3(9):710–20. https://doi.org/10.1038/nri1180.
57. Elshenawi Y, Hu S, Hathroubi S. Biofilm of *H. pylori*: life cycle, features, and treatment options. Antibiot Internet. 2023;12(8)
58. Zou Y, Chen X, Sun Y, Li P, Xu M, et al. Antibiotics-free nanoparticles eradicate *H. pylori* biofilms and intracellular bacteria. J Control Release. 2022;348:370–85. https://doi.org/10.1016/j.jconrel.2022.05.044.

59. Tasia W, Lei C, Cao Y, Ye Q, He Y, et al. Enhanced eradication of bacterial biofilms with DNase I-loaded silver-doped mesoporous silica nanoparticles. Nanoscale. 2020;12(4):2328–32. https://doi.org/10.1039/c9nr08467c.
60. Muras A, Parga A, Mayer C, Otero A. Use of quorum sensing inhibition strategies to control microfouling. Mar Drugs Internet. 2021;19(2)
61. Nasra S, Pramanik S, Oza V, Kansara K, Kumar A. Advancements in wound management: integrating nanotechnology and smart materials for enhanced therapeutic interventions. Discov Nano. 2024;19(1):159. https://doi.org/10.1186/s11671-024-04116-3.
62. García-Lara B, Saucedo-Mora MÁ, Roldán-Sánchez JA, Pérez-Eretza B, Ramasamy M, et al. Inhibition of quorum-sensing-dependent virulence factors and biofilm formation of clinical and environmental Pseudomonas aeruginosa strains by ZnO nanoparticles. Lett Appl Microbiol. 2015;61(3):299–305. https://doi.org/10.1111/lam.12456.
63. Liu T, Chai S, Li M, Chen X, Xie Y, et al. A nanoparticle-based sonodynamic therapy reduces *H. pylori* infection in mouse without disrupting gut microbiota. Nat Commun. 2024;15(1):844. https://doi.org/10.1038/s41467-024-45156-8.
64. Yamaoka Y, Graham DY. *H. pylori* virulence and cancer pathogenesis. Future Oncol. 2014;10(8):1487–500. https://doi.org/10.2217/fon.14.29.
65. Saeidi Y, Pournajaf A, Gholami M, Hasannejad-Bibalan M, Yaghoubi S, et al. Determination of *H. pylori* virulence-associated genes in duodenal ulcer and gastric biopsies. Med J Islam Repub Iran. 2017;31:95.
66. Gottesmann M, Goycoolea FM, Steinbacher T, Menogni T, Hensel A. Smart drug delivery against *H. pylori*: pectin-coated, mucoadhesive liposomes with antiadhesive activity and antibiotic cargo. Appl Microbiol Biotechnol. 2020;104(13):5943–57. https://doi.org/10.1007/s00253-020-10647-3.
67. Filippova AD, Sozarukova MM, Baranchikov AE, Kottsov SY, Cherednichenko KA, et al. Peroxidase-like activity of CeO(2) nanozymes: particle size and chemical environment matter. Molecules. 2023;28(9) https://doi.org/10.3390/molecules28093811.
68. Su B, Xu Z, Shen M, Wang H, Tang Y, et al. Enzyme-like CeO2 nanomaterials with different structure–activity for ulcerative colitis therapy. ACS Appl Nano Mater. 2025;8(6):2847–60. https://doi.org/10.1021/acsanm.4c06461.
69. Li CH, Chen X, Landis RF, Geng Y, Makabenta JM, et al. Phytochemical-based nanocomposites for the treatment of bacterial biofilms. ACS Infect Dis. 2019;5(9):1590–6. https://doi.org/10.1021/acsinfecdis.9b00134.
70. Ergün FC, Kars MD, Kars G. Development and characterization of LL37 antimicrobial-peptide-loaded chitosan nanoparticles: an antimicrobial sustained release system. Polymers (Basel). 2025;17(13) https://doi.org/10.3390/polym17131884.
71. Reid G. Probiotics: definition, scope and mechanisms of action. Best Pract Res Clin Gastroenterol. 2016;30(1):17–25. https://doi.org/10.1016/j.bpg.2015.12.001.
72. Erickson KL, Hubbard NE. Probiotic immunomodulation in health and disease. J Nutr. 2000;130(2S Suppl):403s–9s. https://doi.org/10.1093/jn/130.2.403S.
73. Marco ML, Pavan S, Kleerebezem M. Towards understanding molecular modes of probiotic action. Curr Opin Biotechnol. 2006;17(2):204–10. https://doi.org/10.1016/j.copbio.2006.02.005.
74. Hao L, Liu Y, Szeto IM, Hao H, Zhang T, et al. Milk exosome-based delivery system for probiotic encapsulation that enhances the gastrointestinal resistance and adhesion of probiotics. Nutrients. 2025;17(5) https://doi.org/10.3390/nu17050923.
75. Wong KI, Wang S, Li M, Zhao G, Wang C, et al. Combating drug-resistant *H. pylori* infection with zinc peroxide-based nanoparticles: a ROS reservoir via photochemical reaction. Chem Eng J. 2024;483:149287. https://doi.org/10.1016/j.cej.2024.149287.
76. Singh R, Torti SV. Carbon nanotubes in hyperthermia therapy. Adv Drug Deliv Rev. 2013;65(15):2045–60. https://doi.org/10.1016/j.addr.2013.08.001.
77. Lv S, Zheng D, Chen Z, Jia B, Zhang P, et al. Near-infrared light-responsive size-selective lateral flow chip for single-cell manipulation of circulating tumor cells. Anal Chem. 2023;95(2):1201–9. https://doi.org/10.1021/acs.analchem.2c03947.

# Part III

# Immune-Based Strategies Against *H. pylori*

# Host Immune Response to *H. pylori* Infection

# 12

Yuan Zhuang

## 12.1 Innate Immunity

### 12.1.1 Granulocyte Response

Neutrophils are the most abundant white blood cells in the blood and constitute a crucial part of innate immunity. As the first responders and key components of the innate immune response, neutrophils can function rapidly and diversely to combat invading pathogens and maintain tissue homeostasis [1, 2]. They destroy pathogens through phagocytosis, intracellular degradation, release of granules, and the formation of extracellular traps. In fact, neutrophils possess an impressive arsenal of antimicrobial peptides and defensins and deploy a range of oxidative mechanisms to eliminate pathogens [3, 4]. The activation of the nicotinamide adenine dinucleotide phosphate (NADPH) oxidase complex within neutrophils produces reactive oxygen species (ROS) with bactericidal properties, such as superoxide anion and hydrogen peroxide [5, 6]. Furthermore, neutrophils form neutrophil extracellular traps (NETs), which are web-like structures composed of chromatin and antimicrobial proteins used to capture and neutralize extracellular pathogens [7, 8].

The *Helicobacter pylori* virulence factor, neutrophil-activating protein (HP-NAP), plays a significant role in neutrophil-associated gastritis. On one hand, HP-NAP can prolong the lifespan of monocytes and induce the expression of anti-apoptotic proteins A1, Mcl-1, Bcl-2, and Bcl-X(L) [9]. It can activate neutrophils in vitro and promote the adhesion of neutrophils to endothelial cells in vivo, enabling their efficient transendothelial migration. This HP-NAP-induced adhesion relies on the acquisition of a high-affinity state of β2 integrins on the neutrophil plasma

Y. Zhuang (✉)
National Engineering Research Center of Immunological Products, Department of Microbiology and Biochemical Pharmacy, College of Pharmacy and Laboratory Medicine, Third Military Medical University, Chongqing, China


J. Gao, Y. Du (eds.), *Non-Antibiotic Therapies for Helicobacter pylori*, https://doi.org/10.1007/978-981-95-9598-3_12

membrane, a conformational change that requires functional p38 MAPK activation [10]. HP-NAP can also stimulate neutrophils to synthesize and release various chemokines, including CXCL8, CCL3, and CCL4 [10]. Regarding the activation mechanism, HP-NAP activates neutrophils through its C-terminal region, and this activation is stimulated by structural elements located in the C-terminal region of both mutant and wild-type dodecameric HP-NAP; notably, the dodecamer conformation does not appear to be essential for activation, whereas helices H3 (Leu69-Leu75) and H4 (Lys89-Leu114) or the connecting loops (His63-Thr68 and Thr76-Ser88) are crucial for stimulating neutrophil activation [11].

Another mechanism mediating the successful extravasation of neutrophils through the perivascular basement membrane into inflamed tissue found that MST1 is a key regulator of neutrophil extravasation during inflammation; in an *H. pylori*-infected mouse model, Mst1−/− neutrophils also failed to extravasate from submucosal gastric vessels; the mechanism involves defective translocation of VLA-3, VLA-6, and neutrophil elastase from intracellular vesicles to the surface of Mst1−/− neutrophils. MST1 associates with the Rab27 effector protein synaptotagmin-like protein 1 (JFC1, encoded by Sytl1 in mice), but not with Munc13-4, thereby regulating the trafficking of Rab27-positive vesicles to the cell membrane [12]. Further research found that *H. pylori* infection can trigger a TNF-α/HDGF/COX-2 inflammatory cascade in the gastric microenvironment. HDGF secretion induced by *H. pylori* infection promotes neutrophil infiltration and participates in *H. pylori*-induced gastric cancer progression [13]. Additionally, increased expression of the adhesion molecule E-selectin, induced by the cag pathogenicity island (cagPAI)-encoded *H. pylori* type IV secretion system (T4SS), is also one of the reasons for increased neutrophil infiltration at gastritis sites [14]. However, in *H. pylori*-associated gastric cancer, neutrophil density correlates positively with cell proliferation and negatively with E-cadherin intensity [15]; suggesting that the *H. pylori*-induced neutrophil response is differentially regulated at different stages of gastric disease.

The *H. pylori* virulence factor CagA, encoded by the cag pathogenicity island (cagPAI), plays an important role in the neutrophil response. In vitro, oxidative burst and phagocytosis of *H. pylori* are strongly enhanced after neutrophils contact the bacteria, but expression of hCEACAM3/6 on neutrophils allows *H. pylori* to prolong its survival inside neutrophils in a HopQ-dependent manner. In a chronic mouse infection model, *H. pylori* can systemically downregulate the expression of hCEACAM1 and hCEACAM6 receptors on neutrophils, which may limit CagA translocation efficiency and gastritis progression [16]. The neutrophil response induced by *H. pylori* via CagA also involves different regulatory mechanisms at different stages of gastritis. *H. pylori* infection, dependent on the virulence factor CagA, can induce gastric epithelial cells to express ARRDC3 in the early stage of infection, promoting CXCL2 secretion via the ARRDC3-PAR1 axis, mediating neutrophil infiltration involved in gastritis development [17]. In the later stage of infection, *H. pylori* infection, dependent on the virulence factor CagA, can induce gastric epithelial cells to express ANGPTL4, which suppresses CXCL1 secretion via the ANGPTL4-ITGAV axis, thereby reducing neutrophil infiltration [18].

Simultaneously, *H. pylori* can also depend on the virulence factor CagA to suppress the expression of cathepsin C in gastric epithelial cells, thereby inhibiting neutrophil activation [19]; together, these mechanisms pathologically suppress the immune clearance effect of neutrophil phagocytosis and killing of *H. pylori*. Recently, a novel in vivo imaging model of the mouse gastric mucosa, for the first time, showed the in situ recruitment of neutrophils and direct interaction between *H. pylori* and neutrophils/macrophages during infection; using multiphoton intravital microscopy with real-time drift correction software (VivoSfollow) on LysM-eGFP and CX3CR1-eGFP reporter mouse strains, specific immune cell subsets were fluorescently labeled, demonstrating that multiphoton microscopy is an excellent tool for characterizing interactions between immune cells and pathogens in vivo [20].

Neutrophils are short-lived immune cells continuously released into the bloodstream as inactive cells [21]. They are generated in the bone marrow, enter the circulation under homeostatic conditions, patrol tissues, and seek potential pathogens, rapidly destroying them through phagocytosis, intracellular degradation, granule release, and extracellular trap formation [22]. During inflammation, external signals like infection can lead to the rapid release of neutrophils from the bone marrow to the site of infection, where they are activated to enhance cytotoxicity and lifespan, ultimately undergoing apoptosis [23]. First, granulopoiesis is a tightly regulated process in the bone marrow microenvironment involving multiple stages of differentiation and maturation [24, 25]. Second, molecules like PECAM and CD99 have traditionally been shown to play additional regulatory roles in key steps of neutrophil migration across endothelial cells during inflammation [26, 27]. Neutrophils arriving at the reaction site release large amounts of pro-inflammatory cytokines, chemokines, and lipid mediators, which recruit and activate other immune cells to the site of infection or tissue damage [28]. Notably, neutrophils interact with dendritic cells, macrophages, and lymphocytes, thereby influencing the initiation and resolution of immune responses [29, 30]. Finally, neutrophils can engage in different death pathways depending on environmental signals, including apoptosis, NETosis, ferroptosis, necroptosis, and pyroptosis [31].

Studies have found that *H. pylori*-infected neutrophils exhibit lower levels of apoptosis; *H. pylori* infection can inhibit neutrophil apoptosis and induce IL-1β secretion via autophagy; the mechanism involves *H. pylori* stimulation activating the NACHT/LRR/PYD domain-containing protein 3 (NLRP3)-gasdermin-D (GSDMD) pathway to secrete IL-1β; however, IL-1β secretion is not entirely dependent on GSDMD, as inhibiting autophagy significantly reduces IL-1β release, and autophagy-related molecules are significantly upregulated in *H. pylori*-infected neutrophils [32]. *H. pylori* infection can disrupt the targeting of NADPH oxidase in human neutrophils, leading to the release of superoxide anion into the extracellular environment without accumulation inside the *H. pylori* phagosome. This disruption of NADPH oxidase targeting allows unopsonized *H. pylori* to evade immune phagocytic killing, supporting the hypothesis that bacteria and neutrophils act synergistically to damage the gastric mucosa [33]. *H. pylori* infection can also activate the NLRP3 inflammasome in human neutrophils [34] and regulate human neutrophils at the epigenetic level, such as downregulating miR-223 [35]. Among these, the

*H. pylori* virulence factor CagA plays a significant role; studies show that ROS released upon CagA activation in neutrophils participates in mediating inflammatory damage [36]. In summary, neutrophils, as a key component of innate immunity, play an important effector role in the innate immune response induced by *H. pylori* infection.

### 12.1.2 Macrophage Response

Macrophages are professional phagocytes specialized in detecting, phagocytosing, and destroying harmful organisms, apoptotic cells, damage-associated debris, and metabolic by-products, providing immediate defense [37]. After ingesting pathogens via phagocytosis, macrophages can directly present peptide antigens via Major Histocompatibility Complex Class II (MHC II) to activate helper T cells (CD4+ T cells) [38]. In contrast to dendritic cells (DCs), which present antigens in lymph nodes and activate naive T cells, macrophages present antigens within tissues and cannot induce naive T cell activation [39]. Macrophages detect Pathogen-Associated Molecular Patterns (PAMPs), such as bacterial products, and Damage-Associated Molecular Patterns (DAMPs) generated in response to trauma, ischemia, or tissue damage. In this process, they use a system of Pattern Recognition Receptors (PRRs), like Toll-like Receptors (TLRs), which specifically bind pathogen components like bacterial lipopolysaccharide (LPS), RNA, DNA, or extracellular proteins, leading to the activation of signaling cascades and production of inflammatory mediators. Consequently, macrophages release soluble factors such as cytokines, enzymes, or metabolites that affect other immune cell types [40].

*H. pylori* can utilize various molecular mechanisms to evade immune clearance by macrophages, including impairing pathogen recognition and phagocytosis, disrupting phagosome maturation, and reducing clearance capacity; furthermore, *H. pylori* infection skews macrophage polarization to promote chronic inflammatory damage, suppresses antigen processing and presentation to evade adaptive immune responses, and induces macrophage apoptosis by activating apoptotic signaling pathways [41]. Multiple *H. pylori* virulence factors participate in these pathological activities.

Regarding macrophage phagocytosis, *H. pylori* phospholipase A (PldA) is a key factor modulating host macrophage responses, promoting *H. pylori* immune evasion and persistence. PldA alters membrane lipids through reversible acylation and deacylation, affecting their structure and function. PldA incorporates lysophosphatidylethanolamine into the macrophage membrane, disrupting its bilayer structure and impairing TNFR1-mediated p38-MK2 signaling; this disruption leads to reduced macrophage autophagy and increased RIP1-dependent apoptosis, thereby enhancing *H. pylori* survival; pharmacological inhibition of PldA significantly reduces *H. pylori* survival and increases macrophage survival [42].

Regarding macrophage activation, early activation and maturation of macrophages by *H. pylori* strongly depend on an intermediate metabolite from the *H. pylori* lipopolysaccharide (LPS) biosynthesis pathway, ADP-heptose [43, 44].

The *H. pylori* type IV secretion system (CagT4SS) can be recognized by macrophages and participate in macrophage activation independently of Toll-like receptor signaling [44]. Additionally, ADP-heptose can upregulate the expression of miR146b in macrophages, which downregulates the master regulator of HLA-II genes, CIITA, reducing the expression of the presentation molecule HLA-II and thereby impairing the antigen-presenting capacity of macrophages [45]. *H. pylori* can also more potently suppress HLA-II expression in macrophages by inducing the upregulation of let-7f-5p, let-7i-5p, miR-146b-5p, and – 185-5p, which target CIITA, leading to downregulation of CIITA expression [46].

Regarding macrophage cytokine release, *H. pylori* stimulates macrophages to secrete CCL3 via the JAK1-STAT1 pathway, contributing to inflammatory damage [47]. The *H. pylori* secreted protein HP1173 causes macrophages to produce and release the pro-inflammatory cytokines TNF and IL-1β, and the chemokine CXCL8, in a dose- and time-dependent manner [48]. *H. pylori* also induces increased levels of reactive oxygen species (ROS) and myeloperoxidase (MPO) in macrophages via NLRP3 inflammasome activation, participating in the regulation of macrophage differentiation [49].

Regarding macrophage differentiation, *H. pylori* activates NADPH oxidase activity to produce ROS via TRPM2, which acts as an oxidative stress sensor molecule in macrophages. After *H. pylori* stimulation, TRPM2-deficient macrophages fail to control intracellular calcium levels, leading to calcium overload; the increase in intracellular calcium enhances mitogen-activated protein kinase and NADPH oxidase activities; TRPM2 knockout mice exhibit increased gastric inflammation, reduced bacterial colonization upon long-term *H. pylori* infection, and enhanced polarization toward classically activated M1 macrophages [50]. *H. pylori* infection can promote both M1 and M2 macrophage polarization; infection can promote M1 macrophage polarization and gastric inflammation via TNF/TNFR1 axis activation of the NLRP3 inflammasome [51]; on the other hand, *H. pylori* infection can upregulate indoleamine 2,3-dioxygenase (IDO) expression in macrophages to induce M2 macrophage polarization, which in turn secretes thymic stromal lymphopoietin (TSLP), further inducing Group 2 innate lymphoid cell (ILC2) responses [52].

In adult hosts, macrophages primarily originate from blood monocytes derived from bone marrow (BM) myeloid progenitors, which leave the circulation and differentiate into macrophages within tissues [53]. Macrophages generated during early ontogeny, primarily from the yolk sac or fetal liver, are also present in various tissues. These embryo-derived macrophages persist throughout life as tissue-resident macrophages (TRMs) and participate in cancer evolution and metastasis [54, 55]. Adult bone marrow produces Ly6C- (nonclassical) circulating monocytes, which patrol for pathogens and maintain vascular integrity, and Ly6C+ (classical) inflammatory monocytes, which are recruited to sites of infection, tissue damage, and tumors. Ly6C+ monocytes continuously replenish DCs in an FLT3-controlled process, giving rise to monocyte-derived DCs (moDCs) [56]. Despite their different origins, the differentiation and expansion of all monocyte and macrophage lineages are regulated by CSF1R (the receptor for Colony-Stimulating Factor 1) and its ligands IL-34 and CSF1 [57, 58]. In some organs, such as the kidney, liver, brain, and lungs, BM-derived monocyte-originated macrophages coexist with embryo-derived TRMs [59, 60].

First, macrophages are the main effector cells in acute *H. pylori* infection-associated gastritis [61] and can rely on Myd88 to mediate Toll-like receptor (TLR) activation [62]. In chronic inflammation induced by *H. pylori* infection, macrophages increase in gastric mucosal inflammatory infiltrates, accompanied by increased expression of inducible NO synthase (iNOS) in the tissue and excess free radicals, exacerbating inflammation [63]. Second, the activation of macrophage surface and intracellular signaling pathways also participates in the *H. pylori*-related immune response; *H. pylori*-induced phosphorylation of macrophage surface EGFR [64] and intracellular Notch signaling activation [65] are crucial for the subsequent inflammatory Th1 and Th17 adaptive immune responses. Third, macrophage metabolic reprogramming is also involved in the *H. pylori*-related immune response; in *H. pylori* infection, cystathionine γ-lyase (CTH) is a key enhancer of macrophage activation. After CTH deletion, macrophages exhibit altered proteomes, reduced NF-κB activation, decreased expression of macrophage activation markers, and impaired oxidative phosphorylation and glycolysis [66]. Bromodomain-containing protein 4 (BRD4) is another key molecule in macrophages. During *H. pylori* infection, BRD4 is recruited to the promoters of Slc2a1 and Hk2 via Hypoxia-Inducible Factor-1α, promoting their expression. BRD4-mediated glycolysis stabilizes *H. pylori*-induced iNOS production [67]; mice with macrophage-specific BRD4 knockout exhibit reduced gastric inflammation, increased *H. pylori* colonization, and decreased iNOS expression in gastric macrophages. In summary, macrophages, as a key component of innate immunity, play an important effector role in the innate immune response induced by *H. pylori* infection.

## 12.2 Adaptive Immunity

### 12.2.1 T Cell Response

T cells are core components of the human immune system, primarily responsible for recognizing and clearing abnormal cells (such as virus-infected cells or cancer cells), coordinating immune responses, and playing a key role in immunological memory [68, 69]. T cells are mainly divided into cytotoxic T cells (CD8+ T cells), which directly recognize and kill virus-infected cells or cancer cells by releasing perforin and granzymes to disrupt target cell membranes, and helper T cells (CD4+ T cells), which secrete cytokines to activate other immune cells (such as B cells, macrophages), combat extracellular pathogens, and enhance the overall immune response [70].

*H. pylori* infection primarily induces a CD4+ T cell response. Different types of CD4+ T cells differentiate into distinct effector cells and provide different functional help to B cells [71]. Specifically:

Type I immune responses are primarily triggered by intracellular infections, inducing upregulation of the transcription factor T-bet in CD4+ T cells, leading to the differentiation of IFN-γ-secreting Th1 cells via IL-12 and IFN-γ-mediated STAT1 and STAT4 signaling [72, 73].

Type II immune responses are primarily triggered by extracellular parasites, certain allergens, or weak immunogens, promoting the differentiation of GATA3-high Th2 cells, characterized by the synthesis of IL-4, IL-5, and IL-13 induced via IL-4-mediated STAT6 signaling [74, 75].

Type III immune responses are primarily triggered by fungal and extracellular bacterial infections, driven by the induction of the transcription factor RORγt, leading to Th17 cell differentiation. Th17 cells secrete IL-17 via IL-6, IL-23, IL-1β, or TGF-β-mediated STAT3 signaling [76, 77].

Type I immune response cytokines promote CD8+ T cell and macrophage responses, type II immune response cytokines attract mast cells, basophils, and eosinophils, and type III immune responses preferentially induce neutrophil influx to inflammatory sites [78, 79].

Regarding the Th1 cell immune response induced by *H. pylori* infection, studies have found that infection can induce a strong Th1 response, which is one of the most prominent components of the inflammatory response [80]. Regarding the regulatory mechanisms inducing the Th1 cell immune response, *H. pylori* neutrophil-activating protein (HP-NAP) can induce upregulation of maturation markers CD83, CD80, CD86, HLA-DR, CD40, and CCR7 on human dendritic cells, as well as increased IL-12 secretion [81]. Mature dendritic cells subsequently induce a Th1 cell immune response [82]. HP-NAP can also induce IL-12 production in neutrophils, thereby inducing a Th1 cell immune response [83]. IL-12 then activates the T cell transcription factors STAT4 and T-bet, inducing Th1 cell differentiation [84]. Besides the IL-12-STAT4/T-bet pathway, the exogenous cytokine IL-21 [85] and activation of the intrinsic Notch signaling pathway in T cells [86] are also involved in *H. pylori* infection-related Th1 cell differentiation and the Th1 cell immune response. The strong Th1 cell immune response induced by *H. pylori* infection also has negative regulatory mechanisms; various *H. pylori*-induced effector molecules in the gastric mucosal inflammatory microenvironment, such as Tubulointerstitial nephritis antigen-like 1 (TINAGL1) [87], Rev-erbα [88], Protease-activated receptor 1 (PAR1) [89], etc., as well as mucosal dendritic cells [90] and eosinophils [91], participate in limiting the Th1 cell immune response to varying degrees.

Regarding the Th2 cell immune response induced by *H. pylori* infection, studies have found that infection can induce a relatively weaker Th2 response. IL-33 [92] and thymic stromal lymphopoietin (TSLP) [93] are involved in *H. pylori* infection-related Th2 cell differentiation and the Th2 cell immune response. The interaction between Lewis (Le) antigens on *H. pylori* lipopolysaccharide (LPS) and the dendritic cell C-type lectin DC-SIGN can modulate the balance between Th1 and Th2 cell immune responses induced by *H. pylori* infection [94].

Regarding the Th17 cell immune response induced by *H. pylori* infection, studies have found that infection can induce a relatively strong Th17 cell immune response, which primarily exerts pro-inflammatory effects [95]. Clinically, cytokines associated with the Th17 cell immune response, such as IL-17A, IL-23, IL-21, and TGF-β, are significantly elevated in the gastric mucosa of *H. pylori*-infected patients [96]. Regarding the regulatory mechanisms inducing the Th17 cell immune response, the *H. pylori* virulence factor urease B (UreB) can induce increased secretion of IL-6 and IL-23 from macrophages, thereby inducing Th17 cell differentiation [97, 98]. Furthermore, macrophage-derived BAFF can directly promote Th17 cell differentiation [99]. Simultaneously, *H. pylori* can also induce dendritic cells to secrete IL-23, thereby inducing Th17 cell differentiation [100]. In mouse models, *H. pylori* infection can induce REDD1 to regulate the Th17 cell immune response [101], which in turn regulates the Th1 cell immune response involved in gastritis progression [102]. The strong Th17 cell immune response induced by *H. pylori* infection also has negative regulatory mechanisms; various *H. pylori*-induced effector molecules in the gastric mucosal inflammatory microenvironment, such as B7-H2 [103], and mucosal dendritic cell-induced CD4+ regulatory T cells (Treg cells) [104] participate in limiting the Th17 cell immune response to varying degrees.

Additionally, *H. pylori* infection can induce a Th22 cell response. Clinically, the Th22 cell immune response-related cytokine IL-22 is significantly elevated in the gastric mucosa of patients with *H. pylori*-induced gastritis and peptic ulcers [105]. In mouse models, *H. pylori* infection can induce a Th22 cell immune response via the virulence factor CagA, which subsequently regulates myeloid-derived suppressor cell (MDSC) infiltration involved in gastritis progression [106].

Furthermore, Type I, II, and III responses are accompanied by the parallel differentiation of Bcl6-high T follicular helper (Tfh) cells [107]. Tfh cells are induced in a two-step differentiation process and provide help to B cells through IL-21 secretion and cell contact-dependent costimulation via CD40L-CD40 interaction [108]. Tfh cells within the Tfh1, Tfh2, or Tfh17 subsets promote antibody production such as IgG, Tfh2 cells can induce IgE class switching, and Tfh17 cells are associated with IgA class switching [109]. Most effector CD4+ T cells die, with about 5–10% of antigen-specific cells surviving to become memory T cells [110, 111]. Some studies suggest that memory CD4+ T cell numbers are less stable compared to memory CD8+ T cells, or that certain subsets within the CD4+ T memory compartment are less stable, but this may also depend on the nature of the pathogen studied [112, 113]. It was recently discovered that *H. pylori*-specific tissue-resident memory CD4+ and CD8+ T cells persist long term in the stomach and provide host protection against reinfection with the help of neutrophils [114].

Regarding the Tfh cell immune response induced by *H. pylori* infection, studies have found that signals related to TGFB-Induced Factor Homeobox 1 (Tgif1), retinoid X receptor (RXR), and glucocorticoid-induced tumor necrosis factor receptor (GITR) promote the polarization of Tfh cells and the Tfh cell immune response in *H. pylori*-positive gastritis [115, 116].

Besides the CD4+ T cell response, *H. pylori* infection also induces a CD8+ T cell response. In a pig model of *H. pylori* infection, the CD8+ T cell response was significantly increased [117]. In mouse models, *H. pylori* infection can induce MMP-10 via the virulence factor CagA, which subsequently regulates CD8+ T cell infiltration involved in gastritis progression [118]. In the absence of CD4+ T cells, CD8+ T cells may mediate more severe disease [119]. Recently, a population of CD8+ T cells with a tissue-resident memory (TRM) phenotype was found to infiltrate the gastric mucosa shortly after *H. pylori* infection and mediate pathogen control by executing antigen-specific effector properties; these induced CD8+ tissue-resident memory T cells (CD8+ TRM cells) display a unique T cell receptor β chain repertoire and are primarily reactive to CagA, suggesting a potential effector role for the CD8+ T cell response in *H. pylori* infection [120]. In summary, T cells, as important lymphocyte types, play a significant effector role in the adaptive immune response induced by *H. pylori* infection.

### 12.2.2 B Cell Response

B cells are a major type of lymphocyte. Belonging to the humoral immunity of the adaptive immune system, their primary function is to secrete antibodies [121, 122]. Additionally, B cells can present antigens (also functioning as antigen-presenting cells) and secrete cytokines. In mammals, B cells mature in the bone marrow, located at the core of most bones. They originate from hematopoietic stem cells in the bone marrow, mature there, and play an important role in humoral immunity by producing antibodies [123, 124]. Upon encountering an antigen, they differentiate into plasma cells with a larger nuclear-to-cytoplasmic ratio. The cytoplasm of plasma cells forms granules that stain readily with azure dyes such as methylene blue. These cells then produce antibodies, which are either displayed on the cell membrane or secreted. Another subset of B cells, upon antigen activation, does not become plasma cells but instead becomes memory B cells. Upon re-encountering the same antigen, memory B cells can respond rapidly, undergoing extensive differentiation and proliferation [125, 126].

*H. pylori* infection can cause infiltration of B cells into the human gastric mucosa, particularly with abundant B cell infiltration observed in the gastric mucosa of *H. pylori*-positive peptic ulcer patients [127]. In the tissues of *H. pylori*-infected gastritis patients, IgA+ memory B cells are recruited via CCL28-mediated migration, while IgA+ plasma cells are recruited via CXCL12-mediated migration; this explains the massive influx of IgA-secreting cells into the gastric mucosa caused by increased CCL28 production during human *H. pylori* infection [128]. Regarding the phenotype of infiltrating B cells, research found that mucosal IgA+ B cells in *H. pylori*-infected patients express iNOS, and the degree of induction correlates with disease progression [129]. Compared to uninfected individuals, studies also found increased infiltration of PDCA-1+ B cells, PDCA-1 + IgM+ B cells, CD93 + PDCA-1+ B cells, and CD93 + PDCA-1 + IgM+ B cells in *H. pylori*-infected patients; compared to *H. pylori*-positive chronic non-atrophic gastritis patients, higher proportions of PDCA-1+ B cells, CD93 + PDCA-1+ B cells, and CD137 + PDCA-1+ B cells were observed in *H. pylori*-positive atrophic gastritis or peptic ulcer patients [130].

Regulatory B cells (Breg cells) are a recently identified functional subset of B cells that primarily secrete effector factors like IL-10; studies have found that Breg cells mediate pathological roles in digestive tract diseases and conditions, particularly in *H. pylori* infection, where their presence and infiltration correlate with the severity and progression of related diseases [131]. In clinical patients, the proportion of peripheral blood Breg cells is significantly higher in *H. pylori*-infected patients compared to uninfected patients [132]. The number of CD19 + IL-10+ B cells in the gastric mucosa is significantly higher in *H. pylori*-infected patients than in uninfected patients. Furthermore, the number of CD19 + IL-10+ B cells is higher in *H. pylori*-positive gastritis patients compared to *H. pylori*-positive peptic ulcer patients [133]. In a mouse model of dextran sulfate sodium (DSS)-induced acute and chronic colitis, compared to the DSS-alone group, the *H. pylori*/DSS co-induced group showed significantly lower colon histology scores, reduced colon shortening, and a higher percentage of CD19 + IL-10+ Breg cells in the colonic mucosa [134].

B cells originate and mature in the bone marrow, where they undergo V(D)J recombination, leading to the surface expression of an IgM isotype B cell receptor (BCR). Up to 75% of immature B cells are self-reactive [135] and undergo clonal deletion or receptor editing, where a second immunoglobulin light chain is produced, providing a second chance to reduce self-reactivity. Nonetheless, a significant proportion of self-reactive B cells exit to the periphery in healthy individuals, accounting for about 20% of mature B cells [136]. This self-reactive B cell repertoire is considered crucial to avoid large "gaps" in the B cell repertoire that could otherwise become sanctuaries for pathogens expressing antigens mimicking host proteins [137]. Indeed, self-reactive B cells can participate in responses to infectious agents through a process called clonal redemption [138, 139]. Peripheral B cell tolerance is maintained by several mechanisms, primarily the balance between the suppressive action of CD4+ regulatory T cells (Tregs) and the stimulatory action of CD4+ T follicular helper (Tfh) cells and other T cell subsets. In fact, either a decrease in Tregs or an increase in Tfh cells can be sufficient to give rise to self-reactive B cells [140]. Many studies show that in other inflammatory contexts, such as acute and chronic infections, T cell–B cell interactions and peripheral tolerance mechanisms can indeed be pathologically disrupted [141, 142].

Research has found that *H. pylori* infection induces Treg cell production in vitro and in vivo via TLR-2 ligand-activated B cells. The conversion of Tregs depends on TCR signaling and direct T cell-B cell interaction via CD40/CD40L and CD80/CD28. B cell-induced Tregs acquire suppressive activity in vitro and suppress excessive gastric mucosal immunopathology induced by *H. pylori* infection in vivo. This study reveals a novel mechanism for the generation and function of B cell-dependent Tregs in a clinically relevant disease model, demonstrating that the B cell/Treg axis is crucial for balancing the control of *H. pylori* infection with the prevention of excessive Th1-driven gastric immunopathology, thereby promoting gastric mucosal homeostasis on one hand and facilitating *H. pylori* persistence on the other [143]. *H. pylori* infection can activate NF-κB via the alternative pathway in B lymphocytes [144] and can also utilize its outer membrane proteins HomA and HomB to downregulate activation-induced cytidine deaminase (AID) and Ig switch germline transcription, thereby affecting class switch recombination (CSR) of Ig genes in human B cells [145]. *H. pylori* infection, dependent on the virulence factor CagA, also participates in B cells activation. After infection of B cells, CagA is rapidly injected and undergoes tyrosine phosphorylation, and is cleaved into a 100 kDa N-terminal fragment and a 40 kDa C-terminal fragment. CagA phosphorylation is mediated by non-receptor tyrosine kinases of the Src and Abl families [146]. Transcriptome analysis of CagA-positive *H. pylori*-infected B cells found that 889 differentially expressed genes (DEGs) were upregulated, including genes encoding JUN, FOSL2, HSPA1B, SRC, CXCR3, TLR-4, TNF-α, CXCL8, CCL2, CCL4, MHC class I, and MHC class II molecules; overall, it revealed the downregulation of inflammation- and migration-related genes, while central signal transduction regulators of cell survival and death were upregulated [147]. In summary, B cells, as important lymphocyte types, play a significant effector role in the adaptive immune response induced by *H. pylori* infection.

## References

1. Khoyratty TE, et al. Distinct transcription factor networks control neutrophil-driven inflammation. Nat Immunol. 2021;22:1093–106.
2. Mantovani A, Cassatella MA, Costantini C, Jaillon S. Neutrophilsin the activation and regulation of innate and adaptive immunity. Nat Rev Immunol. 2011;11:519–31.
3. de Oliveira S, Rosowski EE, Huttenlocher A. Neutrophil migration in infection and wound repair: going forward in reverse. Nat Rev Immunol. 2016;16:378–91.
4. Amulic B, Cazalet C, Hayes GL, Metzler KD, Zychlinsky A. Neutrophil function: from mechanisms to disease. Annu Rev Immunol. 2012;30:459–89.
5. Clark SR, Ma AC, Tavener SA, McDonald B, Goodarzi Z, Kelly MM, Patel KD, Chakrabarti S, McAvoy E, Sinclair GD, et al. Platelet TLR4 activates neutrophil extracellular traps to ensnare bacteria in septicblood. Nat Med. 2007;13:463–9.
6. Deshmukh HS, Liu Y, Menkiti OR, Mei J, Dai N, O'Leary CE, Oliver PM, Kolls JK, Weiser JN, Worthen GS. The microbiota regulatesneutrophil homeostasis and host resistance to Escherichia coli K1 sepsisin neonatal mice. Nat Med. 2014;20:524–30.
7. Ackermann M, Anders HJ, Bilyy R, Bowlin GL, Daniel C, De Lorenzo R, Egeblad M, Henneck T, Hidalgo A, Hoffmann M, et al. Patientswith COVID-19: in the dark-NETs of neutrophils. Cell Death Differ. 2021;28:3125–39.
8. Brinkmann V, Reichard U, Goosmann C, Fauler B, Uhlemann Y, Weiss DS, Weinrauch Y, Zychlinsky A. Neutrophil extracellular traps kill bacteria. Science. 2004;303:1532.
9. Cappon A, Babolin C, Segat D, Cancian L, Amedei A, Calzetti F, Cassatella MA, D'Elios MM, de Bernard M. *H. pylori*-derived neutrophil-activating protein increases the lifespan of monocytes and neutrophils. Cell Microbiol. 2010;12(6):754–64.
10. Polenghi A, Bossi F, Fischetti F, Durigutto P, Cabrelle A, Tamassia N, Cassatella MA, Montecucco C, Tedesco F, de Bernard M. The neutrophil-activating protein of *H. pylori* crosses endothelia to promote neutrophil adhesion in vivo. J Immunol. 2007;178(3):1312–20.
11. Kottakis F, Papadopoulos G, Pappa EV, Cordopatis P, Pentas S, Choli-Papadopoulou T. *H. pylori* neutrophil-activating protein activates neutrophils by its C-terminal region even without dodecamer formation, which is a prerequisite for DNA protection – novel approaches against *H. pylori* inflammation. FEBS J. 2008;275(2):302–17.
12. Kurz ARM, et al. MST1-dependent vesicle trafficking regulates neutrophil transmigration through the vascular basement membrane. J Clin Invest. 2016;126(11):4125–39.
13. Chu T-H, et al. Hepatoma-derived growth factor participates in *H. pylori*-induced neutrophils recruitment, gastritis and gastric carcinogenesis. Oncogene. 2019;38(37):6461–77.
14. Svensson H, Hansson M, Kilhamn J, Backert S, Quiding-Järbrink M. Selective upregulation of endothelial E-selectin in response to *H. pylori*-induced gastritis. Infect Immun. 2009;77(7):3109–16.

15. Fu H, Ma Y, Yang M, Zhang C, Huang H, Xia Y, Lu L, Jin W, Cui D. Persisting and increasing neutrophil infiltration associates with gastric carcinogenesis and E-cadherin downregulation. Sci Rep. 2016;6:29762.
16. Behrens I-K, et al. The HopQ-CEACAM interaction controls CagA translocation, phosphorylation, and phagocytosis of *H. pylori* in Neutrophils. mBio. 2020;11(1):e03256–19.
17. Liu Y-G, et al. Arrestin domain containing 3 promotes *H. pylori*-associated gastritis by regulating protease-activated receptor 1. JCI Insight. 2020;5(15):e135849.
18. Xie R, et al. *H. Pylori*-induced angiopoietin-like 4 promotes gastric bacterial colonization and gastritis. Research (Wash D C). 2024;7:0409.
19. Liu Y-G, et al. Abrogation of cathepsin C by *H. pylori* impairs neutrophil activation to promote gastric infection. FASEB J. 2019;33(4):5018–33.
20. Ishikawa-Ankerhold H, et al. Novel multiphoton intravital imaging enables real-time study of *H. pylori* interaction with neutrophils and macrophages in the mouse stomach. PLoS Pathog. 2024;20(9):e1012580.
21. Evrard M, et al. Developmental analysis of bone marrow neutrophils reveals populations specialized in expansion, trafficking, and effector functions. Immunity. 2018;48:364–79.
22. Adrover JM, et al. A neutrophil timer coordinates immune defense and vascular protection. Immunity. 2019;51:966–7.
23. Furze RC, Rankin SM. The role of the bone marrow in neutrophil clearance under homeostatic conditions in the mouse. FASEB J. 2008;22:3111–9.
24. Giladi A, et al. Single-cell characterization of haematopoietic progenitors and their trajectories in homeostasis and perturbed haematopoiesis. Nat Cell Biol. 2018;20:836–46.
25. Mitroulis I, et al. Secreted protein Del-1 regulates myelopoiesis in the hematopoietic stem cell niche. J Clin Invest. 2017;127:3624–39.
26. Lou O, Alcaide P, Luscinskas FW, Muller WA. CD99 is a key mediator of the transendothelial migration of neutrophils. J Immunol. 2007;178:1136–43.
27. Sullivan DP, Watson RL, Muller WA. 4D intravital microscopy uncovers critical strain differences for the roles of PECAM and CD99 in leukocyte diapedesis. Am J Physiol Heart Circ Physiol. 2016;311:H621–32.
28. Liew PX, Kubes P. The neutrophil's role during health and disease. Physiol Rev. 2019;99:1223–48.
29. Guilliams M, et al. Alveolar macrophages develop from fetal monocytes that differentiate into long-lived cells in the first week of life via GM-CSF. J Exp Med. 2013;210:1977–92.
30. Hashimoto D, et al. Tissue-resident macrophages self-maintain locally throughout adult life with minimal contribution from circulating monocytes. Immunity. 2013;38:792–804.
31. Dejas L, Santoni K, Meunier E, Lamkanfi M. Regulated cell death in neutrophils: from apoptosis to NETosis and pyroptosis. Semin Immunol. 2023;70:101849.
32. Song Y, et al. *H. pylori* infection delays neutrophil apoptosis and exacerbates inflammatory response. Future Microbiol. 2024;19(13):1145–56.
33. Allen L-AH, Beecher BR, Lynch JT, Rohner OV, Wittine LM. *H. pylori* disrupts NADPH oxidase targeting in human neutrophils to induce extracellular superoxide release. J Immunol. 2005;174(6):3658–67.
34. Pérez-Figueroa E, Torres J, Sánchez-Zauco N, Contreras-Ramos A, Alvarez-Arellano L, Maldonado-Bernal C. Activation of NLRP3 inflammasome in human neutrophils by *H. pylori* infection. Innate Immun. 2016;22(2):103–12.
35. Zhang Y, Xu C, Zhong H-M, Song Y, Luo H, Liu P. *H. pylori* infection downregulates the expression and release of miR- 223 in neutrophils. Int Microbiol. 2025; https://doi.org/10.1007/s10123-025-00660-9.
36. Thai TD, et al. *H. pylori* extract induces purified neutrophils to produce reactive oxygen species only in the presence of plasma. Biomed Rep. 2023;19(6):89.
37. Watanabe S, Alexander M, Misharin AV, Budinger GRS. The role of macrophages in the resolution of inflammation. J Clin Invest. 2019;129:2619–28.
38. Hirayama D, Iida T, Nakase H. The phagocytic function of macrophage-enforcing innate immunity and tissue homeostasis. Int J Mol Sci. 2017;19:92.

39. Martinez FO, Sica A, Mantovani A, Locati M. Macrophage activation and polarization. Front Biosci. 2008;13:453–61.
40. Wynn TA, Chawla A, Pollard JW. Macrophage biology in development, homeostasis and disease. Nature. 2013;496:445–55.
41. Xiang M, et al. Dysregulated macrophage immunity in *H. pylori* infection: unveiling mechanistic insights and therapeutic implications. Front Immunol. 2025;16:1636768.
42. Sit WY, et al. *H. pylori* PldA modulates TNFR1-mediated p38 signaling pathways to regulate macrophage responses for its survival. Gut Microbes. 2024;16(1):2409924.
43. Faass L, et al. Contribution of Heptose metabolites and the cag Pathogenicity Island to the activation of monocytes/macrophages by *H. pylori*. Front Immunol. 2021;12:632154.
44. Koch M, et al. Macrophages recognize the *H. pylori* type IV secretion system in the absence of toll-like receptor signalling. Cell Microbiol. 2016;18(1):137–47.
45. Coletta S, et al. ADP-heptose enables *H. pylori* to exploit macrophages as a survival niche by suppressing antigen-presenting HLA-II expression. FEBS Lett. 2021;595(16):2160–8.
46. Codolo G, et al. *H. pylori* dampens HLA-II expression on macrophages via the up-regulation of miRNAs targeting CIITA. Front Immunol. 2020;10:2923.
47. Wei Y-F, et al. *H. pylori* disrupts gastric mucosal homeostasis by stimulating macrophages to secrete CCL3. Cell Commun Signal. 2024;22(1):263.
48. Tavares R, Pathak SK. Induction of TNF, CXCL8 and IL-1β in macrophages by *H. pylori* secreted protein HP1173 occurs via MAP-kinases, NF-κB and AP-1 signaling pathways. Microb Pathog. 2018;125:295–305.
49. Kumar S, Dhiman M. *H. pylori* secretary proteins-induced oxidative stress and its role in NLRP3 inflammasome activation. Cell Immunol. 2024;399–400:104811.
50. Beceiro S, et al. TRPM2 ion channels regulate macrophage polarization and gastric inflammation during *H. pylori* infection. Mucosal Immunol. 2017;10(2):493–507.
51. Xiao F, et al. *H. pylori* infection promotes M1 macrophage polarization and gastric inflammation by activation of NLRP3 inflammasome via TNF/TNFR1 axis. Cell Commun Signal. 2025;23(1):6.
52. Peng R, et al. M2 macrophages participate in ILC2 activation induced by *H. pylori* infection. Gut Microbes. 2024;16(1):2347025.
53. Ginhoux F, Guilliams M. Tissue-resident macrophage ontogeny and homeostasis. Immunity. 2016;44:439–49.
54. Carlin LM, et al. Nr4a1-dependent $Ly6C_{low}$ monocytes monitor endothelial cells and orchestrate their disposal. Cell. 2013;153:362–75.
55. Auffray C, et al. Monitoring of blood vessels and tissues by a population of monocytes with patrolling behavior. Science. 2007;317:666–70.
56. Geissmann F, et al. Development of monocytes, macrophages, and dendritic cells. Science. 2010;327:656–61.
57. Wang Y, et al. IL-34 is a tissue-restricted ligand of CSF1R required for the development of Langerhans cells and microglia. Nat Immunol. 2012;13:753–60.
58. Wei S, et al. Functional overlap but differential expression of CSF-1 and IL-34 in their CSF-1 receptor-mediated regulation of myeloid cells. J Leukoc Biol. 2010;88:495–505.
59. Ginhoux F, et al. Fate mapping analysis reveals that adult microglia derive from primitive macrophages. Science. 2010;330:841–5.
60. Ginhoux F, Merad M. Ontogeny and homeostasis of Langerhans cells. Immunol Cell Biol. 2010;88:387–92.
61. Kaparakis M, et al. Macrophages are mediators of gastritis in acute *H. pylori* infection in C57BL/6 mice. Infect Immun. 2008;76(5):2235–9.
62. Rad R, et al. Toll-like receptor-dependent activation of antigen-presenting cells affects adaptive immunity to *H. pylori*. Gastroenterology. 2007;133(1):150–63.
63. Cherdantseva LA, et al. Association of *H. pylori* and iNOS production by macrophages and lymphocytes in the gastric mucosa in chronic gastritis. J Immunol Res. 2014;2014:762514.
64. Hardbower DM, et al. EGFR regulates macrophage activation and function in bacterial infection. J Clin Invest. 2016;126(9):3296–312.

65. He Y, et al. Notch signaling aggravates *H. pylori*-induced inflammation by promoting macrophage activation and Proinflammatory Th1/Th17 responses. Am J Pathol. 2025;195(8):1428–42.
66. Latour YL, et al. Cystathionine γ-lyase exacerbates *H. pylori* immunopathogenesis by promoting macrophage metabolic remodeling and activation. JCI Insight. 2022;7(12):e155338.
67. Modi N, et al. BRD4 regulates glycolysis-dependent Nos2 expression in macrophages upon H pylori infection. Cell Mol Gastroenterol Hepatol. 2024;17(2):292–308.
68. Sallusto F, Lenig D, Forster R, Lipp M, Lanzavecchia A. Two subsets of memory T lymphocytes with distinct homing potentials and effector functions. Nature. 1999;401:708–12.
69. Mosmann TR, Coffman RL. TH1 and TH2 cells: different patterns of lymphokine secretion lead to different functional properties. Annu Rev Immunol. 1989;7:145–73.
70. Laidlaw BJ, Craft JE, Kaech SM. The multifaceted role of $CD4^+$ T cells in $CD8^+$ T cell memory. Nat Rev Immunol. 2016;16:102–11.
71. Tubo NJ, et al. Single naive $CD4^+$ T cells from a diverse repertoire produce different effector cell types during infection. Cell. 2013;153:785–96.
72. Szabo SJ, Kim ST, Costa GL, Zhang X, Fathman CG, Glimcher LH. A novel transcription factor, T-bet, directs Th1 lineage commitment. Cell. 2000;100:655–69.
73. Szabo SJ, Sullivan BM, Stemmann C, Satoskar AR, Sleckman BP, Glimcher LH. Distinct effects of T-bet in TH1 lineage commitment and IFN-γ production in CD4 and CD8 T cells. Science. 2002;295:338–42.
74. Lee GR, Fields PE, Flavell RA. Regulation of IL-4 gene expression by distal regulatory elements and GATA-3 at the chromatin level. Immunity. 2001;14:447–59.
75. Lee GR, Spilianakis CG, Flavell RA. Hypersensitive site 7 of the TH2 locus control region is essential for expressing TH2 cytokine genes and for long-range intrachromosomal interactions. Nat Immunol. 2005;6:42–8.
76. Harrington LE, Hatton RD, Mangan PR, Turner H, Murphy TL, et al. Interleukin 17-producing CD4+ effector T cells develop via a lineage distinct from the T helper type 1 and 2 lineages. Nat Immunol. 2005;6:1123–32.
77. Park H, Li Z, Yang XO, Chang SH, Nurieva R, et al. A distinct lineage of CD4 T cells regulates tissue inflammation by producing interleukin 17. Nat Immunol. 2005;6:1133–41.
78. Annunziato F, Romagnani C, Romagnani S. The 3 major types of innate and adaptive cell-mediated effector immunity. J Allergy Clin Immunol. 2015;135:626–35.
79. Medzhitov R. Recognition of microorganisms and activation of the immune response. Nature. 2007;449:819–26.
80. Bagheri N, et al. The role of T helper 1-cell response in *H. pylori*-infection. Microb Pathog. 2018;123:1–8.
81. Ramachandran M, et al. Vector-encoded *H. pylori* neutrophil-activating protein promotes maturation of dendritic cells with Th1 polarization and improved migration. J Immunol. 2014;193(5):2287–96.
82. Bimczok D, et al. Human primary gastric dendritic cells induce a Th1 response to *H. pylori*. Mucosal Immunol. 2010;3(3):260–9.
83. Amedei A, et al. The neutrophil-activating protein of *H. pylori* promotes Th1 immune responses. J Clin Invest. 2006;116(4):1092–101.
84. Pellicanò A, et al. Interleukin-12 drives the Th1 signaling pathway in *H. pylori*-infected human gastric mucosa. Infect Immun. 2007;75(4):1738–44.
85. Carbo A, et al. Systems modeling of the role of interleukin-21 in the maintenance of effector CD4+ T cell responses during chronic *H pylori* infection. MBio. 2014;5(4):e01243-14.
86. Xie J, et al. Notch 1 is involved in CD4+ T cell differentiation into Th1 subtype during *H. pylori* infection. Front Cell Infect Microbiol. 2020;10:575271.
87. Teng Y, et al. Tubulointerstitial nephritis antigen-like 1 is a novel matricellular protein that promotes gastric bacterial colonization and gastritis in the setting of *H. pylori* infection. Cell Mol Immunol. 2023;20(8):924–40.
88. Mao F-Y, et al. *H. pylori*-induced rev-erbα fosters gastric bacteria colonization by impairing host innate and adaptive defense. Cell Mol Gastroenterol Hepatol. 2021;12(2):395–425.

89. Chionh Y-T, et al. Protease-activated receptor 1 suppresses *H. pylori* gastritis via the inhibition of macrophage cytokine secretion and interferon regulatory factor 5. Mucosal Immunol. 2015;8(1):68–79.
90. Bimczok D, et al. Stromal regulation of human gastric dendritic cells restricts the Th1 response to *H. pylori*. Gastroenterology. 2011;141(3):929–38.
91. Arnold IC, et al. Eosinophils suppress Th1 responses and restrict bacterially induced gastrointestinal inflammation. J Exp Med. 2018;215(8):2055–72.
92. Buzzelli JN, et al. IL33 is a stomach Alarmin that initiates a skewed Th2 response to injury and infection. Cell Mol Gastroenterol Hepatol. 2015;1(2):203–221.e3.
93. Kido M, et al. *H. pylori* promotes the production of thymic stromal lymphopoietin by gastric epithelial cells and induces dendritic cell-mediated inflammatory Th2 responses. Infect Immun. 2010;78(1):108–14.
94. Bergman MP, et al. *H. pylori* modulates the T helper cell 1/T helper cell 2 balance through phase-variable interaction between lipopolysaccharide and DC-SIGN. J Exp Med. 2004;200(8):979–90.
95. Shamsdin SA, et al. Th17 cells in *H. pylori* infection: a dichotomy of help and harm. Infect Immun. 2019;87(11):e00363–19.
96. Shamsdin SA, et al. Alterations in Th17 and the respective cytokine levels in *H. pylori*-induced stomach diseases. Helicobacter. 2015;20(6):460–75.
97. Zhang J-Y, et al. Induction of a Th17 cell response by *H. pylori* urease subunit B. Immunobiology. 2011;216(7):803–10.
98. Zhuang Y, et al. *H. pylori*-infected macrophages induce Th17 cell differentiation. Immunobiology. 2011;216(1–2):200–7.
99. Munari F, et al. Cytokine BAFF released by *H. Pylori*-infected macrophages triggers the Th17 response in human chronic gastritis. J Immunol. 2014;193(11):5584–94.
100. Khamri W, et al. *H. pylori* stimulates dendritic cells to induce interleukin-17 expression from CD4+ T lymphocytes. Infect Immun. 2010;78(2):845–53.
101. Yan Z-B, et al. *H. pylori*-induced REDD1 modulates Th17 cell responses that contribute to gastritis. Clin Sci (Lond). 2021;135(22):2541–58.
102. Shi Y, et al. *H. pylori*-induced Th17 responses modulate Th1 cell responses, benefit bacterial growth, and contribute to pathology in mice. J Immunol. 2010;184(9):5121–9.
103. Lina TT, et al. CagA-dependent downregulation of B7-H2 expression on gastric mucosa and inhibition of Th17 responses during *H. pylori* infection. J Immunol. 2013;191(7):3838–46.
104. Kao JY, et al. *H. pylori* immune escape is mediated by dendritic cell-induced Treg skewing and Th17 suppression in mice. Gastroenterology. 2010;138(3):1046–54.
105. Shamsdin SA, et al. The importance of TH22 and TC22 cells in the pathogenesis of *H. pylori*-associated gastric diseases. Helicobacter. 2017;22(3)
106. Zhuang Y, et al. A pro-inflammatory role for Th22 cells in *H. pylori*-associated gastritis. Gut. 2015;64(9):1368–78.
107. Crotty S. T follicular helper cell biology: a decade of discovery and diseases. Immunity. 2019;50:1132–48.
108. Crotty S. Follicular helper CD4 T cells ($T_{FH}$). Annu Rev Immunol. 2011;29:621–63.
109. Eisenbarth SC, et al. $CD4^{+}$ T cells that help B cells – a proposal for uniform nomenclature. Trends Immunol. 2021;42:658–69.
110. Ahmed R, Gray D. Immunological memory and protective immunity: understanding their relation. Science. 1996;272:54–60.
111. Swain SL, Hu H, Huston G. Class II-independent generation of CD4 memory T cells from effectors. Science. 1999;286:1381–3.
112. Pepper M, Pagan AJ, Igyarto BZ, Taylor JJ, Jenkins MK. Opposing signals from the Bcl6 transcription factor and the interleukin-2 receptor generate T helper 1 central and effector memory cells. Immunity. 2011;35:583–95.
113. Pepper M, et al. Different routes of bacterial infection induce long-lived $T_H1$ memory cells and short-lived $T_H17$ cells. Nat Immunol. 2010;11:83–9.

114. Gong R, Huang B, Ralser A, Friedrich V, Mibus C, Engelsberger V, Koch MRA, Skerhut M, Giese T, Andrä I, Vieth M, van Gisbergen KPJM, Semper RP, Gerhard M, Mejías-Luque R. CagA-dependent Hobit(+) gastric tissue-resident memory T cells confer full protection from *H. pylori* reinfection. Gut. 2025;
115. Leber A, et al. Bistability analyses of CD4+ T follicular helper and regulatory cells during *H. pylori* infection. J Theor Biol. 2016;398:74–84.
116. Ming S, et al. GITR promotes the polarization of TFH-like cells in *H. pylori*-positive gastritis. Front Immunol. 2021;12:736269.
117. Kronsteiner B, et al. Novel insights on the role of CD8+ T cells and cytotoxic responses during *H. pylori* infection. Gut Microbes. 2014;5(3):357–62.
118. Lv Y-P, et al. *H. pylori*-induced matrix metallopeptidase-10 promotes gastric bacterial colonization and gastritis. Sci Adv. 2019;5(4):eaau6547.
119. Tan MP, et al. CD8+ T cells are associated with severe gastritis in *H. pylori*-infected mice in the absence of CD4+ T cells. Infect Immun. 2008;76(3):1289–97.
120. Koch MRA, et al. CagA-specific gastric CD8+ tissue-resident T cells control *H. pylori* during the early infection phase. Gastroenterology. 2023;164(4):550–66.
121. McHeyzer-Williams M, et al. Molecular programming of B cell memory. Nat Rev Immunol. 2011;12(1):24–34.
122. Ramadoss NS, et al. Characterizing the BCR repertoire in immune-mediated diseases. Nat Rev Rheumatol. 2020;16(1):7–8.
123. Clark MR, et al. Orchestrating B cell lymphopoiesis through interplay of IL-7 receptor and pre-B cell receptor signalling. Nat Rev Immunol. 2014;14(2):69–80.
124. Nemazee D. Mechanisms of central tolerance for B cells. Nat Rev Immunol. 2017;17(5):281–94.
125. Lund FE, Randall TD. Effector and regulatory B cells: modulators of CD4+ T cell immunity. Nat Rev Immunol. 2010;10(4):236–47.
126. Cancro MP. Signalling crosstalk in B cells: managing worth and need. Nat Rev Immunol. 2009;9(9):657–61.
127. Goll R, et al. Increased frequency of antral CD4 T and CD19 B cells in patients with *H. pylori*-related peptic ulcer disease. Scand J Immunol. 2005;61(1):92–7.
128. Hansson M, et al. CCL28 is increased in human *H. pylori*-induced gastritis and mediates recruitment of gastric immunoglobulin A-secreting cells. Infect Immun. 2008;76(7):3304–11.
129. Neumann L, et al. Mucosal inducible NO synthase-producing IgA+ plasma cells in *H. pylori*-infected patients. J Immunol. 2016;197(5):1801–8.
130. Ma Z, et al. *H. pylori* infection increases frequency of PDCA-1(+) (CD317(+)) B-cell subsets. Arch Med Res. 2016;47(2):96–104.
131. Moreira H, et al. Unraveling the role of Breg cells in digestive tract cancer and infectious immunity. Front Immunol. 2022;13:981847.
132. Hassuna NA, et al. Regulatory B cells (Bregs) in *H. pylori* chronic infection. Helicobacter. 2023;28(2):e12951.
133. Nahid-Samiei M, et al. Enhanced frequency of CD19+IL-10+B cells in human gastric mucosa infected by *H. pylori*. Am J Med Sci. 2020;359(6):347–53.
134. Li X, et al. H. pylori infection alleviates acute and chronic colitis with the expansion of regulatory B cells in mice. Inflammation. 2019;42(5):1611–21.
135. Wardemann H, et al. Predominant autoantibody production by early human B cell precursors. Science. 2003;301:1374–7.
136. Sng J, et al. AIRE expression controls the peripheral selection of autoreactive B cells. Sci Immunol. 2019;4:eaav6778.
137. Finney J, Watanabe A, Kelsoe G, Kuraoka M. Minding the gap: the impact of B-cell tolerance on the microbial antibody repertoire. Immunol Rev. 2019;292:24–36.
138. Burnett DL, Reed JH, Christ D, Goodnow CC. Clonal redemption and clonal anergy as mechanisms to balance B cell tolerance and immunity. Immunol Rev. 2019;292:61–75.
139. Brink R, Phan TG. Self-reactive B cells in the germinal center reaction. Annu Rev Immunol. 2018;36:339–60.

140. Meffre E, O'Connor KC. Impaired B-cell tolerance checkpoints promote the development of autoimmune diseases and pathogenic autoantibodies. Immunol Rev. 2019;292:90–101.
141. Wang EY, et al. Diverse functional autoantibodies in patients with COVID-19. Nature. 2021;595:283–8.
142. Qiu CC, Caricchio R, Gallucci S. Triggers of autoimmunity: the role of bacterial infections in the extracellular exposure of lupus nuclear autoantigens. Front Immunol. 2019;10:1–15.
143. Sayi A, et al. TLR-2-activated B cells suppress helicobacter-induced preneoplastic gastric immunopathology by inducing T regulatory-1 cells. J Immunol. 2011;186(2):878–90.
144. Ohmae T, et al. *H. pylori* activates NF-kappaB via the alternative pathway in B lymphocytes. J Immunol. 2005;175(11):7162–9.
145. Tamrakar A, Kodgire P. HomA and HomB, outer membrane proteins of *H. pylori* down-regulate activation-induced cytidine deaminase (AID) and Ig switch germline transcription and thereby affect class switch recombination (CSR) of Ig genes in human B-cells. Mol Immunol. 2022;142:37–49.
146. Krisch LM, et al. CagA phosphorylation in *H. pylori*-infected B cells is mediated by the nonreceptor tyrosine kinases of the Src and Abl families. Infect Immun. 2016;84(9):2671–80.
147. Chichirau BE, et al. Dissecting the *H. pylori*-regulated transcriptome of B cells. Pathog Dis. 2020;78(7):ftaa049.

# Immune Evasion Mechanisms of *H. pylori* 13

Yuan Zhuang

## 13.1 Gastric Epithelial Barrier Damage

The gastric epithelial barrier, composed of physical, chemical, cellular barriers, and immune auxiliary components, constitutes the core multilayered defense system of the gastric mucosa against gastric acid, pepsin, and external harmful substances. Its main components and functions are as follows: (1) The mucus-bicarbonate-phospholipid barrier: The gastric mucosal surface is covered by a thick mucus layer composed of glycoproteins, which lubricates food and slows gastric acid penetration [1, 2]. Bicarbonate embedded in the mucus is actively secreted by gastric mucosal epithelial cells, forming a pH gradient (near the epithelium pH ~7.0, away from the epithelium pH ~1.0), neutralizing gastric acid and inhibiting pepsin activity [3]; furthermore, phospholipid molecules enhance the barrier's acid resistance through hydrophobic interactions [4]. (2) Epithelial cells and tight junctions: The gastric mucosal epithelium consists of a single layer of columnar cells, forming a seamless physical barrier through tight junction proteins, preventing $H^+$ back-diffusion into the submucosa [5]. These cells renew rapidly (renewal cycle ~3–5 days), enabling timely repair of damaged areas [6]. (3) Ion channels and immune regulation: The chloride ion channel Slc26a9 is highly expressed in gastric epithelial cells, reducing epithelial cell death and maintaining barrier stability by regulating endoplasmic reticulum stress and mitochondrial-dependent apoptotic signaling [7]. Additionally, alarmins secreted by epithelial cells can activate immune cells [8], and secreted defensins can directly kill bacteria and viruses, reducing microbial damage to the barrier [9]. The gastric epithelial barrier is the host's first line of defense against *Helicobacter pylori*. By disrupting this barrier, *H. pylori* can not only breach the

Y. Zhuang (✉)
National Engineering Research Center of Immunological Products, Department of Microbiology and Biochemical Pharmacy, College of Pharmacy and Laboratory Medicine, Third Military Medical University, Chongqing, China



J. Gao, Y. Du (eds.), *Non-Antibiotic Therapies for Helicobacter pylori*,
https://doi.org/10.1007/978-981-95-9598-3_13

host's initial defense and achieve colonization but also indirectly interfere with subsequent immune responses, creating a safe colonization microenvironment for itself.

*H. pylori* primarily disrupts the mucus-bicarbonate-phospholipid barrier through two mechanisms, creating conditions for its subsequent colonization and pathogenicity. First, *H. pylori* secretes large amounts of urease, which specifically decomposes urea in the mucus layer, generating abundant ammonia ($NH_3$) [10]: (1) Directly disrupts the structural integrity of the mucus layer—the core component of the mucus layer is mucin, and ammonia hydrolyzes the sugar chains of mucin, leading to thinning of the mucus layer, decreased adhesiveness, inability to tightly cover the epithelial surface, creating "gaps" through which *H. pylori* can directly contact and damage epithelial cells [11, 12]. (2) Interferes with the pH gradient balance of the mucus layer—normally, a gradient exists from "near-epithelial side pH≈7.0 (neutral)" to "near-gastric lumen side pH≈1.0 (strongly acidic)," which is a key chemical defense inhibiting pathogen survival. Ammonia can neutralize local gastric acid within the mucus layer, raising the pH around *H. pylori*, not only weakening the antimicrobial effect of the chemical barrier but also creating a suitable neutral microenvironment for *H. pylori* itself [13]. (3) Inhibits the activity of Secretory Immunoglobulin A (SlgA)—SlgA is an important immune defense molecule in the mucus layer, which can specifically bind *H. pylori* surface adhesins (like BabA), inhibiting *H. pylori* colonization by blocking adhesin binding to epithelial cells [14, 15]. Ammonia degrades the antibody structure of SlgA (e.g., destroying its antigen-binding site) by regulating acidity/alkalinity, causing it to lose its ability to bind adhesins, further reducing the immune defense efficiency of the mucus layer [16]. Second, *H. pylori*'s flagella (mainly composed of FlaA, FlaB proteins) can rotate rapidly, providing sufficient motility, helping it overcome resistance in the viscous mucus layer, and move directionally toward epithelial cells, achieving "active penetration" of the mucus layer [17].

Simultaneously, when *H. pylori* approaches epithelial cells, its surface adhesins exert specific binding effects—e.g., BabA can bind Lewis b antigen on epithelial cell surfaces [18], SabA can bind sialylated glycoproteins on epithelial cell surfaces [19]. Once bound, *H. pylori* firmly anchors to the epithelial cell surface, effectively avoiding clearance due to mechanical flushing from gastric peristalsis. This synergistic effect of "active penetration of mucus layer + tight adhesion to epithelial cells" further weakens the physical barrier function of the mucus layer, preventing the barrier from effectively blocking direct contact between *H. pylori* and epithelial cells.

Tight junctions between gastric epithelial cells are core to maintaining epithelial barrier integrity; however, *H. pylori* can directly disrupt this structure through its core virulence factor Cytotoxin-associated gene A (CagA), leading to a significant increase in epithelial barrier permeability. First, *H. pylori* actively injects CagA protein into gastric epithelial cells via its own Type IV Secretion System (T4SS, a protein transport apparatus across the bacterial cell membrane). After entering the cell, CagA is not autonomously active but requires phosphorylation by host cell tyrosine kinases (like c-Src, Abl family kinases) to become active [20]; Activated CagA specifically binds tight junction-associated proteins, with the interaction with

Zonula Occludens-1 (ZO-1) being most critical, primarily disrupting tight junction function in two ways: (1) Disrupts the localization and distribution of tight junction proteins—ZO-1, as the core scaffolding protein of tight junctions, normally highly enriches at the tight junction region in the intercellular space, cooperatively assembling a continuous seal structure with proteins like occludin and claudin [21–23]. When CagA binds ZO-1, it disrupts the interaction between ZO-1 and other tight junction proteins, causing ZO-1 to detach from the tight junction region and diffuse into the cytoplasm [24]. This mislocalization directly leads to intermittent gaps in the tight junction "seal belt," breaking the physical barrier's integrity. (2) Interferes with the functional activity of tight junction proteins—besides affecting localization, CagA can also interfere with normal phosphorylation modifications of tight junction proteins by regulating signaling pathways. For example, it can inhibit the physiological phosphorylation of ZO-1 and occludin, which is necessary for correct protein polymerization and formation of a stable seal structure [25, 26]. Abnormal phosphorylation prevents normal assembly of tight junction proteins, further weakening their "sealing ability," ultimately greatly increasing epithelial barrier permeability. After increased permeability, *H. pylori* itself and its metabolites like urease, vacuolating toxin (VacA), etc., can freely diffuse through the tight junction gaps to adjacent epithelial cell surfaces or even the submucosa: on one hand directly damaging surrounding healthy epithelial cells, on the other hand creating favorable conditions for *H. pylori* to breach the epithelial layer and invade deeper into the mucosa for colonization [27].

*H. pylori* not only directly damages epithelial cells but also inhibits epithelial barrier repair and weakens its immune activation function. On one hand, the vacuolating toxin (VacA) secreted by *H. pylori* is a core factor causing direct epithelial cell damage, inducing epithelial cell vacuolation and apoptosis—VacA can actively insert into the gastric epithelial cell membrane, assembling to form transmembrane channels approximately 2–3 nm in diameter. These channels disrupt the ion balance of the cell membrane, causing massive efflux of intracellular potassium ($K^+$) and chloride ($Cl^-$) ions, while promoting calcium ($Ca^{2+}$) influx [28, 29]. Ion imbalance further triggers fusion of intracellular membrane structures like the endoplasmic reticulum and Golgi apparatus, forming numerous vacuoles (i.e., "cell vacuolation") [28]. Vacuolated cells activates caspase apoptotic pathways (e.g., caspase-3, caspase-9), ultimately leading to programmed cell death of epithelial cells, directly reducing barrier cell numbers [30, 31]. On the other hand, the virulence factor CagA can also weaken the chemical defense function of epithelial cells—CagA activates the Src kinase pathway within host cells (a key pathway regulating cytoskeleton and polarity) [32]. Activated Src kinase induces cytoskeletal rearrangement (e.g., actin), leading to disruption of epithelial cell polarity [33]. This polarity disruption directly weakens the ability of epithelial cells to secrete mucus and bicarbonate, causing a simultaneous decline in the chemical defense function of the mucosa (mucus-bicarbonate barrier) [34]. Furthermore, *H. pylori*'s lipopolysaccharide (LPS, a bacterial cell wall component) has unique structural modifications—through dephosphorylation or acylation modifications of lipid A (the active core of LPS), it significantly reduces its recognition efficiency by host cell surface Toll-like receptor

4 (TLR4), avoiding rapid clearance by the immune system [35]. Nevertheless, when the epithelial barrier develops gaps due to tight junction disruption, the modified LPS can still contact epithelial cells through these gaps, bind to cell surface TLR4, activate the TLR4-MyD88 signaling pathway, and trigger a mild inflammatory response [36, 37]. This mild inflammation is characterized by "insufficient intensity and limited effect": on one hand, the degree of inflammation is insufficient to recruit enough immune cells to clear *H. pylori* [38]; on the other hand, small quantities of inflammatory factors (e.g., IL-6) released during inflammation can promote epithelial cell apoptosis, exacerbating barrier damage [39], ultimately forming a vicious cycle of "inflammation fails to eliminate bacteria but instead aids bacterial colonization," providing a suitable microenvironment for long-term *H. pylori* survival.

Damage to the gastric epithelial barrier is the foundation of *H. pylori* immune evasion—only by first breaking down this defense line can *H. pylori*'s subsequent mechanisms, such as interfering with immune cells, inducing immune tolerance, and undergoing antigenic variation, take effect, ultimately achieving chronic colonization. This also explains why, in clinical treatment of *H. pylori* infection, besides antibiotic bactericidal therapy, gastric mucosal protectants are combined—by repairing the gastric epithelial barrier, restoring the host's initial defense, assisting in *H. pylori* clearance, and reducing recurrence.

## 13.2 Phagocytic Effector Impairment

Gastric mucosal phagocytosis is a key defense mechanism of the gastric mucosal immune system against pathogen invasion, clearing foreign substances and senescent cells. It is primarily mediated by professional phagocytes in the lamina propria (like macrophages, neutrophils) and dendritic cells that possess both phagocytic and antigen-presenting functions. This process, through the complete sequence of chemotaxis → recognition and adhesion → engulfment (forming phagosomes) → phagosome-lysosome fusion → killing and clearance, achieves defense against pathogens, regulation of gastric mucosal inflammatory balance, and assistance in gastric mucosal tissue repair [40–42]. *H. pylori* impairs phagocyte function or even triggers phagocyte apoptosis by disrupting multiple steps of the chemotaxis-recognition-phagocytosis-killing cascade, enabling the pathogen to evade immune clearance.

Phagocytes can directionally migrate to the *H. pylori* infection site under the induction of chemokines [25]. Subsequently, phagocytes recognize *H. pylori* surface Pathogen-Associated Molecular Patterns (PAMPs) specifically through their cell surface Pattern Recognition Receptors (PRRs, like TLR4, TLR5), thereby initiating the phagocytosis process of *H. pylori* [43]. *H. pylori* interferes with this process through multiple mechanisms: First, *H. pylori* infection induces host gastric epithelial cells to express TINAGL1 protein, which directly inhibits the production of the chemokine CCL21—CCL21 is a key chemokine mediating the CCR7 receptor-dependent migration of dendritic cells and other phagocytes. Downregulation of CCL21 leads to ineffective infiltration of phagocytes into the gastric infection site, thereby weakening the bacterial clearance capacity of *H. pylori*-specific Th1 cells [44, 45]. Second, *H. pylori* can modify its own PAMPs (e.g., its cell wall component LPS), significantly reducing the binding affinity of LPS to macrophage surface TLR4, thus failing to effectively activate the TLR4-mediated signaling pathway, making macrophages unable to effectively recognize *H. pylori,* and it is also difficult to initiate pre-phagocytosis preparatory reactions [35]. Third, *H. pylori*'s flagellin (FlaA/FlaB) has mutations at key TLR5 recognition sites, and TLR5 expressed on neutrophil surfaces cannot bind the mutated flagellin, further reducing the recognition efficiency of phagocytes for *H. pylori* [46].

Even if *H. pylori* is phagocytosed by professional phagocytes like macrophages and forms phagosomes, it can still interfere with the key bactericidal step of phagosome–lysosome fusion through multiple pathways, thereby evading degradation and killing by hydrolases like proteases and nucleases in lysosomes. First, *H. pylori*'s outer membrane protein OmpE can be released onto the phagosome membrane surface after phagocytosis, specifically binding to the key GTPase Rab5, which regulates early phagosome maturation in macrophages. This binding inhibits Rab5's GTPase activity, preventing the switch from Rab5 on the phagosome membrane to Rab7, which regulates late phagosome formation, causing the phagosome to arrest in an immature early state and be unable to fuse with lysosomes, ultimately allowing *H. pylori* to achieve intracellular safe survival within immature phagosomes [47, 48]. Second, specific *H. pylori* metabolites, like cholesteryl-α-glucoside (CAG), hijack the host autophagy pathway, reducing lysosome biogenesis and interfering with the autophagy process, creating an intracellular environment favorable for bacterial survival [49]. Additionally, during *H. pylori* infection, increased lysosome membrane permeabilization leads to structural damage and functional impairment of lysosomes, thus indirectly preventing effective fusion of phagosomes with lysosomes [50].

Even if a few phagosomes containing *H. pylori* successfully fuse with lysosomes, *H. pylori* can still reduce its probability of being degraded and by inhibiting phagocytes' generation of core oxidative bactericidal substances such as Reactive Oxygen Species (ROS) and nitric oxide and by clearing already generated bactericidal molecules. First, *H. pylori*'s vacuolating toxin VacA inhibits ROS generation through two pathways: (1) The transmembrane channels formed by VacA previously inserted into the phagosome membrane interfere with the structural integrity of the phagosome membrane, thereby preventing the normal assembly of NADPH oxidase—the core enzyme complex for ROS generation—on the phagosome membrane (this complex needs precise localization and assembly on the phagosome membrane to be activated) [51, 52] and (2) VacA can downregulate the concentration of free calcium ions in the macrophage cytoplasm, and free calcium ions are key signaling molecules for NADPH oxidase activation. The lack of signal ultimately leads to a significant reduction in ROS production [53, 54]. Second, *H. pylori* can secrete arginine deiminase, which specifically degrades arginine in the macrophage cytoplasm—arginine is the key substrate for inducible nitric oxide synthase (iNOS)—the key enzyme for NO synthesis. Through a "substrate depletion" mechanism, *H. pylori* further blocks the NO synthesis pathway [55–57]. Third, *H. pylori* itself can express two antioxidant enzymes: catalase and superoxide dismutase: the former directly decomposes hydrogen peroxide ($H_2O_2$) already produced by macrophages. The latter specifically scavenges superoxide anion ($O_2^-$), weakening the oxidative bactericidal capacity of phagocytes by directly neutralizing already generated ROS [58–60].

*H. pylori* can not only evade killing by immune cells but also actively induce phagocyte apoptosis or functional abnormalities through multiple virulence factors, further undermining the host's innate immune defense line. On one hand, *H. pylori*'s vacuolating toxin VacA can weaken immune clearance capacity by inducing macrophage apoptosis—VacA can form transmembrane channels on the macrophage membrane, leading to disruption of intracellular ion homeostasis (e.g., massive potassium efflux, abnormal calcium influx). This imbalance activates the caspase-9/caspase-3 cascade-mediated mitochondrial apoptotic pathway, ultimately triggering macrophage apoptosis. Apoptotic macrophages lose their phagocytic function and can no longer clear other *H. pylori* [28, 61, 62]. More critically, if apoptotic bodies released by apoptotic macrophages parcel *H. pylori*, they might be phagocytosed by adjacent macrophages, allowing *H. pylori* to achieve intercellular spread, creating conditions for persistent infection [63]. On the other hand, the virulence factor CagA (injected into macrophage cytoplasm via the Type IV secretion system) induces the host cell to produce the chemokine CXCL12a, activating CDC42 GTPase. The activation of the latter leads to F-actin polymerization and cytoskeletal rearrangement, disrupting the normal phagocytic and bactericidal function of macrophage [64]. Furthermore, *H. pylori*'s outer membrane protein Omp18 can inhibit macrophage function by interfering with energy metabolism—Omp18 can specifically bind the CD40 molecule on the macrophage surface, thereby blocking the CD40-CD40L (CD40 ligand, also known as CD154) signaling pathway. This pathway is key for regulating macrophage energy metabolism processes (like the glycolytic pathway). Inhibition of this pathway by Omp18 leads to insufficient energy production in macrophages, inability to maintain normal phagocytic activity and bactericidal function, ultimately rendering them functionally inactive [65–68].

*H. pylori*'s immune evasion targeting phagocytosis first reduces the probability of being phagocytosed by inhibiting chemokine production and modifying its own PAMPs, making it difficult for phagocytes to locate the infection site and effectively recognize *H. pylori*. If *H. pylori* is unfortunately phagocytosed, it then acts on the phagosome maturation process through collaborate actions of outer membrane proteins like OmpE and specific metabolites, preventing phagosome-lysosome fusion, thus achieving intracellular "refuge" survival within phagocytes. Simultaneously, by inhibiting the generation of ROS and NO, and clearing already produced ROS (e.g., via catalase, superoxide dismutase), it weakens the phagocytes' bactericidal ability, avoiding self-degradation. Additionally, by inducing phagocyte apoptosis or functional abnormalities, it further destructively damages the host's innate immune defense line, creating favorable conditions for its own long-term colonization in gastric mucosal tissue.

## 13.3 Pathological Suppression of Antimicrobial Molecules

Gastric mucosal antimicrobial molecules, primarily secreted by gastric epithelial cells and mucosal immune cells, can directly act on invading bacteria by disrupting bacterial cell membrane structure, inhibiting bacterial metabolic processes, or interfering with bacterial colonization on the mucosal surface, constituting an important defensive component of gastric mucosal innate immunity. Targeting this defense mechanism, *H. pylori* interferes with the generation, activity, or function of antimicrobial molecules that directly kill pathogens in host innate immunity (such as reactive oxygen species, nitric oxide, antimicrobial peptides, lysozyme.), thereby weakening the host's direct bactericidal ability against itself and avoiding clearance by these bactericidal substances, creating conditions for its colonization in the gastric mucosa.

*H.pylori* can inhibit host Reactive Oxygen Species (ROS) production or remove already generated ROS through various mechanisms. On one hand, the vacuolating toxin (VacA) secreted by *H. pylori* can insert into the phagosome membrane of phagocytes (like macrophages), forming transmembrane channels, leading to decreased stability of the phagosome membrane—ROS are primarily produced by the NADPH oxidase complex in neutrophils and macrophages, and this complex needs to localize to the phagosome membrane to complete assembly and activation—when the phagosome membrane structure is disrupted by VacA, this complex cannot assemble normally, ultimately leading to a significant reduction in ROS production [51, 52]. On the other hand, VacA can downregulate the concentration of free calcium ions in the macrophage cytoplasm, and free calcium ions are key signaling molecules for NADPH oxidase activation. The lack of signal ultimately leads to a significant reduction in ROS generation [53, 54]. Furthermore, *H. pylori* can actively degrade host-produced ROS through its own secreted antioxidant enzyme system, specifically including three key enzymes:

Catalase, efficiently decomposes host-produced hydrogen peroxide ($H_2O_2$) into water and oxygen, preventing direct oxidative damage to *H. pylori* by $H_2O_2$ [69, 70].

Superoxide dismutase (SOD): Converts superoxide anion ($O_2^-$) into $H_2O_2$, which is then further cleared by the aforementioned catalase [71, 72].

Glutathione reductase: Maintains the reduced state of glutathione (GSH, a key antioxidant molecule) inside *H. pylori*, thereby enhancing *H. pylori*'s own resistance to ROS [73, 74].

Nitric oxide (NO) is primarily generated by inducible nitric oxide synthase (iNOS) catalyzing L-arginine in macrophages and epithelial cells. *H. pylori* can block NO generation and function through substrate depletion and synthesis restriction aspects [40, 75]. On one hand, *H. pylori* can secrete arginine deiminase (ADI), which is released into the host cell microenvironment, efficiently decomposing L-arginine within host cells, converting it to citrulline and ammonia; Since iNOS-catalyzed NO generation requires L-arginine as an essential substrate, even if iNOS is normally expressed, its substrate being consumed by ADI leads to a significant decrease in NO production levels [76, 77]. On the other hand, *H. pylori* can inhibit iNOS synthesis, reducing the source of NO generation. Although *H. pylori*'s lipopolysaccharide (LPS) is structurally modified (reducing its ability to activate TLR4), it can still mildly activate the "TLR4-MyD88 signaling pathway," inducing epithelial cells to secrete interleukin-10 (IL-10, a key anti-inflammatory factor). IL-10 further inhibits iNOS expression, forming an indirect inhibitory loop of "*H. pylori*-LPS-IL-10-iNOS" [35, 78, 79].

Antimicrobial peptides are the first chemical barrier of the gastric mucosa, exerting killing effects by directly disrupting *H. pylori*'s cell membrane. *H. pylori* resists antimicrobial peptide attacks through the following two mechanisms. On one hand, *H. pylori* reduces the generation of antimicrobial peptides from the "source" by regulating host cell signaling pathways: After *H. pylori* infection, its virulence factor CagA upregulates the expression of the host transcriptional repressor Rev-erbα by activating the ERK-SP1 signaling pathway. Rev-erbα directly binds to the promoter regions of defensin genes (like β-defensin-1), inhibiting their transcription, thereby reducing defensin synthesis. Besides directly inhibiting defensins, Rev-erbα also downregulates other antimicrobial proteins (like Reg3b) and chemokines (like CCL21), weakening the overall antimicrobial capacity of gastric epithelial cells [80]. On the other hand, *H. pylori* reduces the binding sites for antimicrobial peptides and hinders their access to the cell membrane by altering the physical and chemical characteristics of its own outer membrane: (1) Spatial barrier effect of LPS: LPS in the outer membrane significantly increases the length of the O-antigen side chains, forming a "physical spatial barrier" on the cell membrane surface, directly preventing antimicrobial peptides from accessing the outer membrane surface [81, 82]. (2) Electrostatic repulsion effect of outer membrane proteins: Outer membrane proteins (like OmpA, OmpC) undergo surface charge modifications (significantly increasing negative charge density), while antimicrobial peptides are usually positively charged. Electrostatic repulsion between them greatly reduces the probability of antimicrobial peptides inserting into the *H. pylori* cell membrane [83, 84].

Lysozyme (core function: hydrolyzing bacterial cell wall peptidoglycan) and lactoferrin (core function: competitively binding iron ions) are key auxiliary antimicrobial molecules for the host against *H. pylori*. *H. pylori* weakens their antimicrobial functions through targeted mechanisms, respectively. On one hand, *H. pylori* blocks lysozyme's hydrolytic action by altering the chemical structure of its own cell wall peptidoglycan—it specifically adds O-acetyl groups to the N-acetylmuramic acid (MurNAc) residues of peptidoglycan [85]. The core action site of lysozyme is the β-1,4 glycosidic bond between N-acetylglucosamine (GlcNAc) and N-acetylmuramic acid in peptidoglycan, and the O-acetyl group hinders the binding of lysozyme to this glycosidic bond through a steric hindrance effect, preventing lysozyme from hydrolyzing the peptidoglycan backbone [86–89]. Peptidoglycan is the core structure of the *H. pylori* cell wall; its integrity being protected means lysozyme cannot destroy *H. pylori*'s cellular structure [88, 90]. On the other hand, lactoferrin achieves "nutritional deprivation" by binding iron ions (essential trace elements for *H. pylori* growth) in the host environment. *H. pylori* counteracts this effect through a collaborative strategy of "siderophore capture + transporter uptake" to compete for iron ions: (1) Priority capture by siderophores: *H. pylori* can secrete high-affinity siderophores, whose dissociation constant for iron binding is much lower than that of lactoferrin, allowing them to preferentially bind iron ions in the host microenvironment, forming "siderophore-iron" complexes [91–93]. (2) Efficient uptake by transporters: Specific iron transporters on the *H. pylori* cell membrane can recognize and efficiently uptake the "siderophore-iron" complexes, transporting iron ions into the cell [94]. Iron is an essential cofactor for key enzymes in the *H. pylori* respiratory chain (like cytochrome oxidase) and nucleic acid synthesis enzymes. This uptake process ensures *H. pylori*'s iron requirements are met, thereby resisting lactoferrin's "iron deprivation" effect [95, 96].

Through multiple strategies including virulence factor regulation (like VacA, arginine deiminase ADI.) and self-structural modifications (like modifications to LPS, cell wall peptidoglycan), *H. pylori* comprehensively blocks the bactericidal effects of host antimicrobial molecules (like ROS, NO, antimicrobial peptides, lysozyme.) from the generation stage (e.g., inhibiting the expression of antimicrobial-related enzymes, consuming their synthesis substrates) to the functional stage (e.g., directly clearing already generated antimicrobial molecules, resisting their damage to its own structure). This enables it to both colonize the gastric mucosal surface and survive within phagocytes, effectively evading clearance by the host's chemical defense mechanisms, ultimately achieving long-term colonization.

## 13.4 Pathological Suppression of Anti-*H. pylori* Host Immune Response

The host's anti-*H. pylori* immune response is a protective defense reaction initiated by the human immune system to resist *H. pylori* infection, mainly including two core components: innate immunity and adaptive immunity. The former, centered on macrophages, neutrophils, dendritic cells, and antimicrobial molecules (like ROS, antimicrobial peptides), is responsible for early rapid recognition and clearance of *H. pylori* [97, 98]. The latter relies on specific immune responses mediated by T lymphocytes and B lymphocytes, forming long-term immune memory to curb *H. pylori* spread [99, 100]. However, as a pathogenic bacterium capable of achieving persistent colonization in the gastric mucosa, *H. pylori* employs various strategies to suppress the effective activation of both host innate and adaptive immunity, while simultaneously inducing ineffective chronic inflammation—this inflammation cannot effectively clear *H. pylori* but continuously damages the gastric mucosal barrier, leading to the long-term maintenance of a chronic inflammatory state in the gastric mucosa [25, 101–103]. Ultimately, the "clearance failure" of the immune response and the "inflammatory damage" form a vicious cycle, gradually causing chronic gastric mucosal inflammation, which may progress to chronic gastritis, peptic ulcers, and in severe cases, even induce malignant transformation of gastric mucosal epithelial cells, increasing the risk of gastric cancer.

Innate immunity is the host's core first line of defense against *H. pylori.* Its defense system includes both preexisting defense structures like gastric acid and the gastric mucosal physical barrier, both immune cells like neutrophils and macrophages, and immune molecules like cytokines [43, 83]. However, *H. pylori* can disrupt the effective activation of the innate immune defense line through various precision mechanisms, even blocking innate immune signaling pathways to evade immune clearance. First, *H. pylori*'s lipopolysaccharide (LPS) exhibits more significant structural modifications compared to typical Gram-negative bacteria (e.g., elongated O-antigen side chains, lipid A modifications), which can significantly reduce chemotactic activity toward neutrophils, decreasing neutrophil aggregation at the *H. pylori* infection site [35, 104, 105]. Second, *H. pylori* can induce host cells to express immunosuppressive molecules, particularly programmed death-ligand 1 (PD-L1). This molecule, upon binding to the PD-1 receptor, inhibits T cell function, allowing bacterial immune evasion and maintaining persistent infection [106]. Furthermore, *H. pylori* infection can induce macrophages and dendritic cells to express the cytokine signaling suppressor SOCS, inhibiting cytokine signaling (like TNF-α or IL-6), thereby reducing inflammation and protecting the bacteria from clearance [107]. Additionally, *H. pylori* metabolites and specific effectors can directly affect immune cells, suppressing innate immune responses: (1) Metabolites derived from *H. pylori*-modified host cholesterol can exacerbate gastritis and inhibit innate immune detection through immune interactions, thereby promoting inflammation while evading clearance [108, 109]. (2) *H. pylori* expresses various effectors (like VacA and CagA). These proteins, different from other mucosal pathogens, can specifically inhibit IL-2 secretion and T cell proliferation, while simultaneously interfering with DNA repair and immune activation through signaling pathways like ALPK1/TIFA, indirectly suppressing innate immunity [42, 110].

Adaptive immunity is the specific immune response mediated by T cells and B cells. Its core function should be to initiate targeted killing and clearance mechanisms by precisely recognizing *H. pylori* antigens. However, *H. pylori* not only suppresses the effective killing function of adaptive immunity but also skews it toward a damaging inflammatory direction through multidimensional strategies including inducing immune tolerance, interfering with T cell differentiation, and disrupting B cell responses, ultimately achieving immune evasion. First, dendritic cells (DCs) are the "bridge cells" connecting innate and adaptive immunity. Their core function is to mature after uptaking *H. pylori* antigens and then present the antigens to naive T cells to initiate specific immunity [111, 112]. But *H. pylori* can cut this key link by targeting and inhibiting DC function: (1) Secreted by *H. pylori,* the metabolite ADP-heptose acts as a bacterial pathogen-associated molecular pattern that exerts dual inhibitory effects on DC maturation: it not only impairs the cells' ability to activate T cells, but also directly suppresses DC maturation through the downregulation of molecules related to the Toll-like receptor 2 (TLR2) signaling pathway. Immature DCs, although capable of normally phagocytosing *H. pylori* antigens, cannot upregulate the expression of co-stimulatory molecules (CD80,

CD86) and key antigen presentation molecules (MHC class II molecules) [113]. T cell activation requires" dual signals" (the "antigen signal" provided by MHC II—antigen peptide; the "co-stimulatory signal" from CD80/CD86 binding to T cell CD28). DCs lacking co-stimulatory signals cannot effectively activate naive T cells, preventing T cells from differentiating into effector T cells with killing function [114, 115]. (2) Induces T cells into an "anergic state"—Under conditions lacking co-stimulatory signals, immature DCs present *H. pylori* antigens and instead induce naive T cells to differentiate into an immune anergy state [116]. T cells in an anergic state, even upon subsequent re-exposure to *H. pylori* antigens, cannot initiate proliferation and cytokine secretion, completely losing their specific response capacity to *H. pylori* [117, 118]. Second, *H. pylori* can actively regulate the balance of T cell subsets in the local gastric mucosa, establishing immune tolerance to itself by inducing the proliferation and activation of regulatory T cells (Tregs): (1) Tregs suppress effector T cell function via anti-inflammatory factors—Activated Tregs can secrete anti-inflammatory factors like interleukin-10 (IL-10) and transforming growth factor-beta (TGF-β), directly inhibiting the proliferation and function of effector T cells (like T helper 1 (Th1) cells and T helper 17 (Th17) cells) [69, 70, 119]. Among them, Th1 cells secrete gamma-interferon (IFN-γ), which can mediate macrophage activation and enhance cytotoxic activity [71]. Th17 cells secrete IL-17, which can recruit neutrophils to the infection site via chemokines produced by gastric epithelial cells [72]. Both are key effector cells for clearing *H. pylori*. After their functions are suppressed by Tregs, the clearance efficiency of *H. pylori* is greatly reduced. (2) Clinical correlation between Tregs and *H. pylori* colonization: Clinical studies have found that the proportion of Treg cells in the gastric mucosal tissue of patients with chronic *H. pylori* infection is significantly higher than in noninfected individuals, and this proportion is positively correlated with the density of *H. pylori* gastric mucosal colonization [81]. This phenomenon suggests that Treg-mediated immune tolerance is not accidental but a key mechanism actively induced by *H. pylori* to maintain its own long-term colonization. Third, although *H. pylori* can suppress the clearance function of adaptive immunity, it does not suppress its inflammatory effects, but rather skews the immune response toward ineffective inflammation, which neither clears the bacteria nor continuously damages the gastric mucosa: (1) Abnormally induces Th17 cells, exacerbating mucosal damage—*H. pylori* infection can activate MHC class $II^+$ monocytes, causing them to secrete IL-23 (a cytokine), which induces $CD4^+$ T cells to polarize toward IL-17A-producing cells (i.e., Th17 cells), thereby increasing IL-17A secretion [82]. Although IL-17 can recruit neutrophils to the infection site, *H. pylori* has already defended against neutrophil killing through the aforementioned mechanisms (like secreting SOD to remove ROS, modifying the outer membrane to resist antimicrobial peptides) [38, 120–123][40]. Neutrophils that are not cleared will continuously release proteases (like elastase) and ROS, directly destruct the structural integrity of gastric mucosal epithelial cells, leading to mucosal erosion and glandular atrophy, which are typical pathological features of chronic atrophic gastritis and may further progress to intestinal metaplasia (a key step in gastric precancerous lesions)[41–43]. (2) Induces B cells to produce "ineffective antibody responses"—Although *H. pylori* can activate B cells to produce anti-*H. pylori* antibodies (like IgG in serum, IgA in the local mucosa), these antibodies have significant functional defects: on one hand, their affinity is low, unable to effectively neutralize *H. pylori* virulence factors (like CagA, VacA)[44]; on the other hand, they cannot block the binding of *H. pylori* to gastric mucosal epithelial cells via adhesins (like *H. pylori* AlpA, BabA)[45,46]. This ineffective "non-protective antibody response" not only fails to eliminate *H. pylori*, but may also indirectly exacerbate gastric mucosal inflammation through the deposition of antibody-antigen complexes.

Through the above mechanisms, *H. pylori* achieves pathological suppression of the host immune response, ultimately leading to two key outcomes: (1) Immune clearance failure: Innate immunity cannot clear *H. pylori* through early rapid killing effects, and adaptive immunity cannot initiate specific killing and long-term memory against *H. pylori*, ultimately allowing *H. pylori* to break through the immune defense lines and achieve long-term stable colonization on the gastric mucosal surface and within the mucosal layer. (2) Persistent chronic inflammation: The immune response regulated by *H. pylori* is not completely silent but shifts toward a low-intensity persistent inflammatory state that "has no clearance effect but only causes damage," chronically injuring the gastric mucosa, inducing chronic superficial gastritis → atrophic gastritis → intestinal metaplasia → dysplasia, ultimately increasing the risk of gastric cancer.

## References

1. McCoy R, Wang K, Treiber J, Fu Y, Malliaras GG, Salleo A, Owens RM. Mucus-on-a-chip: investigating the barrier properties of mucus with organic bioelectronics. J Mater Chem B. 2025;13(2):577–87.
2. Kavishvar D, Ramachandran A. The yielding behaviour of human mucus. Adv Colloid Interf Sci. 2023;322:103049.
3. Rampert DC, van de Graaf SFJ, Jongejan A, Oude Elferink RPJ, Beuers U. Hepatobiliary acid-base homeostasis: insights from analogous secretory epithelia. J Hepatol. 2021;74(2):428–41.
4. Lundstedt E, Kahne D, Ruiz N. Assembly and maintenance of lipids at the bacterial outer membrane. Chem Rev. 2021;121(9):5098–123.
5. Liebing E, Krug SM, Neurath MF, Siegmund B, Becker C. Wall of resilience: how the intestinal epithelium prevents inflammatory onslaught in the gut. Cell Mol Gastroenterol Hepatol. 2025;19(2):101423.
6. Lee SH, Won Y, Gibbs D, Caldwell B, Goldstein A, Choi E, Goldenring JR. Amphiregulin switches progenitor cell fate for lineage commitment during gastric mucosal regeneration. Gastroenterology. 2024;167(3):469–84.
7. Zhang M, Ma Z, Yi Z, Wang H, Zhu J, Wen G, Jin H, An J, Deng Z, Tuo B, Li T, Liu X. SLC26A9 promotes colorectal tumorigenesis by modulating Wnt/β-catenin signaling. Cell Death Discov. 2024;10(1):123.
8. Rizzi A, Di Gioacchino M, Gammeri L, Inchingolo R, Chini R, Santilli F, Nucera E, Gangemi S. The emerging role of innate lymphoid cells (ILCs) and alarmins in celiac disease: an update on pathophysiological insights, potential use as disease biomarkers, and therapeutic implications. Cells. 2023;12(14):1910.
9. Vllahu M, Voli A, Licursi V, Zagami C, D'Amore A, Traulsen J, Woelffling S, Schmid M, Crickley R, Lisle R, Link A, Tosco A, Meyer TF, Boccellato F. Inflammation promotes stom-

ach epithelial defense by stimulating the secretion of antimicrobial peptides in the mucus. Gut Microbes. 2024;16(1):2390680.

10. Khan Y, Solangi M, Khan KM, Ullah N, Iqbal J, Hussain Z, Khan IA, Salar U, Taha M. Exploration of thiazine Schiff bases as promising urease inhibitors: design, synthesis, enzyme inhibition, kinetic analysis, ADME/T evaluation, and molecular docking studies. Int J Biol Macromol. 2024;281(Pt 2):136361.
11. Reznik N, Gallo AD, Rush KW, Javitt G, Fridmann-Sirkis Y, Ilani T, Nairner NA, Fishilevich S, Gokhman D, Chacón KN, Franz KJ, Fass D. Intestinal mucin is a chaperone of multivalent copper. Cell. 2022;185(22):4206–4215.e11.
12. Li X, Zhang J, Liu X, Wu Z, Yu Z, Zhang W, Wang Y, Li Z, Cohen Stuart MA, Sagis LMC, Lv M, Xiao J, Li Y. Gastric-mucus penetrating and responsive microgels for alleviating H. pylori-induced gastritis. J Control Release. 2025;383:113741.
13. Jiang X, Wang W, Wang Z, Wang Z, Shi H, Meng L, Pang S, Fan M, Lin R. Gamma-glutamyl transferase secreted by H. pylori promotes the development of gastric cancer by affecting the energy metabolism and histone methylation status of gastric epithelial cells. Cell Commun Signal. 2024;22(1):402.
14. Raskova Kafkova L, Brokesova D, Krupka M, Stehlikova Z, Dvorak J, Coufal S, Fajstova A, Srutkova D, Stepanova K, Hermanova P, Stepankova R, Uberall I, Skarda J, Novak Z, Vannucci L, Tlaskalova-Hogenova H, Jiraskova Zakostelska Z, Sinkora M, Mestecky J, Raska M. Secretory IgA N-glycans contribute to the protection against E. coli O55 infection of germ-free piglets. Mucosal Immunol. 2021;14(2):511–22.
15. Zhang F, Chen J, Zhang Z, Wu J, Qu Y, Ni L, Zhang G, Liu K, Guo L. M cells targeted H. pylori antigen SAM-FAdE displayed on bacterium-like particles induce protective immunity. J Nanobiotechnol. 2025;23(1):23.
16. Xia N, Wang C, Zhu S. Interaction between pH-shifted ovalbumin and insoluble neohesperidin: experimental and binding mechanism studies. Food Chem. 2022;390:133104.
17. Tachiyama S, Rosinke K, Khan MF, Zhou X, Xin Y, Botting JM, Yue J, Roujeinikova A, Hoover TR, Liu J. Flg Y, PflA, and PflB form a spoke-ring network in the high-torque flagellar motor of *H. pylori*. Proc Natl Acad Sci USA. 2025;122(17):e2421632122.
18. Chahal G, Padra M, Erhardsson M, Jin C, Quintana-Hayashi M, Venkatakrishnan V, Padra JT, Stenbäck H, Thorell A, Karlsson NG, Lindén SK. A complex connection between the diversity of human gastric mucin O-glycans, H. pylori binding, helicobacter infection and fucosylation. Mol Cell Proteomics. 2022;21(11):100421.
19. Chitas R, Fonseca DR, Parreira P, Martins MCL. Targeted nanotherapeutics for the treatment of H. pylori infection. J Biomed Sci. 2024;31(1):78.
20. Blanc M, Lettl C, Guérin J, Vieille A, Furler S, Briand-Schumacher S, Dreier B, Bergé C, Plückthun A, Vadon-Le Goff S, Fronzes R, Rousselle P, Fischer W, Terradot L. Designed ankyrin repeat proteins provide insights into the structure and function of CagI and are potent inhibitors of CagA translocation by the H. pylori type IV secretion system. PLoS Pathog. 2023;19(5):e1011368.
21. Yu S, He J, Xie K. Zonula occludens proteins signaling in inflammation and tumorigenesis. Int J Biol Sci. 2023;19(12):3804–15.
22. Zhang Z, Xie Y, Yi Q, Liu J, Yang L, Wang R, Cai J, Li X, Feng X, Yao S, Pan Z, Paolino M, Zhou Q. PEAK1 maintains tight junctions in intestinal epithelial cells and resists colitis by inhibiting autophagy-mediated ZO-1 degradation. Nat Commun. 2025;16(1):6777.
23. Phillips CM, Johnson AM, Stamatovic SM, Keep RF, Andjelkovic AV. 20 kDa isoform of connexin-43 augments spatial reorganization of the brain endothelial junctional complex and lesion leakage in cerebral cavernous malformation type-3. Neurobiol Dis. 2023;186:106277.
24. Haas AJ, Zihni C, Krug SM, Maraspini R, Otani T, Furuse M, Honigmann A, Balda MS, Matter K. ZO-1 guides tight junction assembly and epithelial morphogenesis via cytoskeletal tension-dependent and -independent functions. Cells. 2022;11(23):3775.
25. Wei YF, Li X, Zhao MR, Liu S, Min L, Zhu ST, Zhang ST, Xie SA. H. pylori disrupts gastric mucosal homeostasis by stimulating macrophages to secrete CCL3. Cell Commun Signal. 2024;22(1):263.

26. Srivastava AK, Venkata BS, Sweat YY, Rizzo HR, Jean-François L, Zuo L, Kurgan KW, Moore P, Shashikanth N, Smok I, Sachleben JR, Turner JR, Meredith SC. Serine 408 phosphorylation is a molecular switch that regulates structure and function of the occludin α-helical bundle. Proc Natl Acad Sci USA. 2022;119(34):e2204618119.
27. Bernegger S, Vidmar R, Fonovic M, Posselt G, Turk B, Wessler S. Identification of desmoglein-2 as a novel target of H. pylori HtrA in epithelial cells. Cell Commun Signal. 2021;19(1):108.
28. Son YS, Kwon YH, Lee MS, Kwon O, Jeong YJ, Mun SJ, Jeon S, Park JH, Han MH, Bae JS, Hur K, Jang AR, Park JH, Cho HS, Jung CR, Ryu CM, Son MJ, Park DS, Son MY. H. pylori VacA-induced mitochondrial damage in the gastric pit cells of the antrum and therapeutic rescue. Biomaterials. 2025;314:122842.
29. Gandhi A, Kariyat R, Harikishore A, Ayati M, Bhunia A, Sahoo N. Deciphering the role of ion channels in early defense signaling against herbivorous insects. Cells. 2021;10(9):2219.
30. Guo H, Hu R, Huang G, Pu W, Chu X, Xing C, Zhang C. Molybdenum and cadmium co-exposure induces endoplasmic reticulum stress-mediated apoptosis by Th1 polarization in Shaoxing duck (Anas platyrhyncha) spleens. Chemosphere. 2022;298:134275.
31. Dörflinger B, Badr MT, Haimovici A, Fischer L, Vier J, Metz A, Eisele B, Bronsert P, Aumann K, Höppner J, Waguia Kontchou C, Parui I, Weber A, Kirschnek S, Häcker G. Mitochondria supply sub-lethal signals for cytokine secretion and DNA-damage in H. pylori infection. Cell Death Differ. 2022;29(11):2218–32.
32. Burikhanov R, Ganguly S, Ellingson S, Sviripa VM, Araujo N, Li S, Venkatraman P, Rao M, Choughule A, Brainson CF, Zhan CG, Spielmann HP, Watt DS, Govindan R, Rangnekar VM. Crizotinib induces Par-4 secretion from normal cells and GRP78 expression on the cancer cell surface for selective tumor growth inhibition. Am J Cancer Res. 2023;13(3):976–91.
33. Tamada M, Shi J, Bourdot KS, Supriyatno S, Palmquist KH, Gutierrez-Ruiz OL, Zallen JA. Toll receptors remodel epithelia by directing planar-polarized Src and PI3K activity. Dev Cell. 2021;56(11):1589–1602.e9.
34. Rehman T, Karp PH, Thurman AL, Mather SE, Jain A, Cooney AL, Sinn PL, Pezzulo AA, Duffey ME, Welsh MJ. WNK inhibition increases surface liquid pH and host defense in cystic fibrosis airway epithelia. Am J Respir Cell Mol Biol. 2022;67(4):491–502.
35. Pachathundikandi SK, Tegtmeyer N, Backert S. Masking of typical TLR4 and TLR5 ligands modulates inflammation and resolution by H. pylori. Trends Microbiol. 2023;31(9):903–15.
36. Luo D, Huang Z, Jia G, Zhao H, Liu G, Chen X. Naringin mitigates LPS-induced intestinal barrier injury in mice. Food Funct. 2023;14(3):1617–26.
37. Xie Y, Li X, Xu D, He D, Wang J, Bi J, Liu J, Fu S. Hordenine alleviates lipopolysaccharide-induced mastitis by suppressing inflammation and oxidative stress, modulating intestinal microbiota, and preserving the blood-milk barrier. J Agric Food Chem. 2024;72(39):21503–19.
38. Ishikawa-Ankerhold H, et al. Novel multiphoton intravital imaging enables real-time study of H. pylori interaction with neutrophils and macrophages in the mouse stomach. PLoS Pathog. 2024;20(9):e1012580.
39. Yu B, de Vos D, Guo X, Peng S, Xie W, Peppelenbosch MP, Fu Y, Fuhler GM. IL-6 facilitates cross-talk between epithelial cells and tumor-associated macrophages in H. pylori-linked gastric carcinogenesis. Neoplasia. 2024;50:100981.
40. Fei X, Li N, Xu X, Zhu Y. Macrophage biology in the pathogenesis of H. pylori infection. Crit Rev Microbiol. 2025;51(3):399–416.
41. Ishikawa-Ankerhold H, et al. Novel multiphoton intravital imaging enables real-time study of sinteraction with neutrophils and macrophages in the mouse stomach. PLoS Pathog. 2024;20(9):e1012580.
42. Neuper T, Frauenlob T, Dang HH, Krenn PW, Posselt G, Regl C, Fortelny N, Schäpertöns V, Unger MS, Üblagger G, Neureiter D, Mühlbacher I, Weitzendorfer M, Singhartinger F, Emmanuel K, Huber CG, Wessler S, Aberger F, Horejs-Hoeck J. ADP-heptose attenuates H. pylori-induced dendritic cell activation. Gut Microbes. 2024;16(1):2402543.
43. Faass L, Hauke M, Stein SC, Josenhans C. Innate immune activation and modulatory factors of H. pylori towards phagocytic and nonphagocytic cells. Curr Opin Immunol. 2023;82:102301.

44. Mao FY, Lv YP, Hao CJ, Teng YS, Liu YG, Cheng P, Yang SM, Chen W, Liu T, Zou QM, Xie R, Xu JY, Zhuang Y. H. pylori-induced rev-erbα fosters gastric bacteria colonization by impairing host innate and adaptive defense. Cell Mol Gastroenterol Hepatol. 2021;12(2):395–425.
45. Teng Y, Xie R, Xu J, Wang P, Chen W, Shan Z, Yan Z, Mao F, Cheng P, Peng L, Zhang J, Tian W, Yang S, Zhao Y, Chen W, Zou Q, Zhuang Y. Tubulointerstitial nephritis antigen-like 1 is a novel matricellular protein that promotes gastric bacterial colonization and gastritis in the setting of H. pylori infection. Cell Mol Immunol. 2023;20(8):924–40.
46. Clasen SJ, Bell MEW, Borbón A, Lee DH, Henseler ZM, de la Cuesta-Zuluaga J, Parys K, Zou J, Wang Y, Altmannova V, Youngblut ND, Weir JR, Gewirtz AT, Belkhadir Y, Ley RE. Silent recognition of flagellins from human gut commensal bacteria by toll-like receptor 5. Sci Immunol. 2023;8(79):eabq7001.
47. Xie J, Cools L, Van Imschoot G, Van Wonterghem E, Pauwels MJ, Vlaeminck I, De Witte C, El Andaloussi S, Wierda K, De Groef L, Haesebrouck F, Van Hoecke L, Vandenbroucke RE. H. pylori-derived outer membrane vesicles contribute to Alzheimer's disease pathogenesis via C3-C3aR signalling. J Extracell Vesicles. 2023;12(2):e12306.
48. Borchers AC, Janz M, Schäfer JH, Moeller A, Kümmel D, Paululat A, Ungermann C, Langemeyer L. Regulatory sites in the Mon1-Ccz1 complex control Rab5 to Rab7 transition and endosome maturation. Proc Natl Acad Sci USA. 2023;120(30):e2303750120.
49. Muthusamy S, Jan HM, Hsieh MY, Mondal S, Liu WC, Ko YA, Yang WY, MongKT CGC, Lin CH. Enhanced enzymatic production of cholesteryl 6′-acylglucoside impairs lysosomal degradation for the intracellular survival of H. pylori. J Biomed Sci. 2021;28(1):72.
50. Khan U, Karmakar BC, Basak P, Paul S, Gope A, Sarkar D, Mukhopadhyay AK, Dutta S, Bhattacharya S. Glycyrrhizin, an inhibitor of HMGB1 induces autolysosomal degradation function and inhibits H. pylori infection. Mol Med. 2023;29(1):51.
51. Huo C, Liu Y, Li X, Xu R, Jia X, Hou L, Wang X. LRRC8A contributes to angiotensin II-induced cardiac hypertrophy by interacting with NADPH oxidases via the C-terminal leucine-rich repeat domain. Free Radic Biol Med. 2021;165:191–202.
52. Fang S, Cheng Y, Deng F, Zhang B. RNF34 ablation promotes cerebrovascular remodeling and hypertension by increasing NADPH-derived ROS generation. Neurobiol Dis. 2021;156:105396.
53. Nakagawa K, Itoya M, Takemoto N, Matsuura Y, Tawa M, Matsumura Y, Ohkita M. Indoxyl sulfate induces ROS production via the aryl hydrocarbon receptor-NADPH oxidase pathway and inactivates NO in vascular tissues. Life Sci. 2021;265:118807.
54. Cui C, Jiang M, Jain N, Das S, Lo YH, Kermani AA, Pipatpolkai T, Sun J. Structural basis of human NOX5 activation. Nat Commun. 2024;15(1):3994.
55. Hernández VM, Arteaga A, Dunn MF. Diversity, properties and functions of bacterial arginases. FEMS Microbiol Rev. 2021;45(6):fuab034.
56. Zhang Y, Chung SF, Tam SY, Leung YC, Guan X. Arginine deprivation as a strategy for cancer therapy: an insight into drug design and drug combination. Cancer Lett. 2021;502:58–70.
57. Abu-Soud HM, Camp OG, Ramadoss J, Chatzicharalampous C, Kofinas G, Kofinas JD. Regulation of nitric oxide generation and consumption. Int J Biol Sci. 2025;21(3):1097–109.
58. da Mata AMOF, Paz MFCJ, de Menezes APM, Dos Reis AC, da Silva Souza B, de Carvalho Sousa CD, Machado SA, Medeiros TSG, Sarkar C, Islam MT, Sharifi-Rad J, Daştan SD, Alshehri MM, de Castro E Sousa JM, de Carvalho Melo Cavalcante AA. Evaluation of mutagenesis, necrosis and apoptosis induced by omeprazole in stomach cells of patients with gastritis. Cancer Cell Int. 2022;22(1):154.
59. Drozdz-Afelt JM, Koim-Puchowska BB, Kaminski P. Analysis of oxidative stress indicators in polish patients with prostate cancer. Environ Sci Pollut Res Int. 2022;29(3):4632–40.
60. Han X, Ju L, Saengow C, Ren W, Ewoldt R, Fan T, Irudayaraj J. Nano oxygen chamber by cascade reaction for hypoxia mitigation and reactive oxygen species scavenging in wound healing. Bioact Mater. 2024;35:67–81.
61. Jenner A, Garcia-Saez AJ. The regulation of the apoptotic pore – an immunological tightrope walk. Adv Immunol. 2024;162:59–108.

62. Vringer E, Tait SWG. Mitochondria and cell death-associated inflammation. Cell Death Differ. 2023;30(2):304–12.
63. Lim MCC, Jantaree P, Naumann M. The conundrum of H. pylori-associated apoptosis in gastric cancer. Trends Cancer. 2023;9(8):679–90.
64. Zhang Y, Xia N, Hu Y, Zhu W, Yang C, Su J. Bactericidal ability of target acidic phospholipids and phagocytosis of CDC42 GTPase-mediated cytoskeletal rearrangement underlie functional conservation of CXCL12 in vertebrates. Sci China Life Sci. 2024;67(12):2713–29.
65. Song L, Song M, Rabkin CS, Chung Y, Williams S, Torres J, Corvalan AH, Gonzalez R, Bellolio E, Shome M, LaBaer J, Qiu J, Camargo MC. Identification of anti-H. pylori antibody signatures in gastric intestinal metaplasia. J Gastroenterol. 2023;58(2):112–24.
66. Zhou A, Shi C, Fan Y, Zheng Y, Wang J, Liu Z, Xie H, Liu J, Jiao Q. Involvement of CD40-CD40L and ICOS-ICOSL in the development of chronic rhinosinusitis by targeting eosinophils. Front Immunol. 2023;14:1171308.
67. Liu PS, Chen YT, Li X, Hsueh PC, Tzeng SF, Chen H, Shi PZ, Xie X, Parik S, Planque M, Fendt SM, Ho PC. CD40 signal rewires fatty acid and glutamine metabolism for stimulating macrophage anti-tumorigenic functions. Nat Immunol. 2023;24(3):452–62.
68. Latour YL, Sierra JC, Finley JL, Asim M, Barry DP, Allaman MM, Smith TM, McNamara KM, Luis PB, Schneider C, Jacobse J, Goettel JA, Calcutt MW, Rose KL, Schey KL, Milne GL, Delgado AG, Piazuelo MB, Paul BD, Snyder SH, Wilson KT. Cystathionine γ-lyase exacerbates H. pylori immunopathogenesis by promoting macrophage metabolic remodeling and activation. JCI Insight. 2022;7(12):e155338.
69. Szczepanowski P, Noszka M, Żyła-Uklejewicz D, Pikuła F, Nowaczyk-Cieszewska M, Krężel A, Stingl K, Zawilak-Pawlik A. H. pylori1021 is a redox switch protein identified in H. pylori. Nucleic Acids Res. 2021;49(12):6863–79.
70. Noszka M, Strzałka A, Muraszko J, Kolenda R, Meng C, Ludwig C, Stingl K, Zawilak-Pawlik A. Profiling of the H. pylori redox switch H. pylori1021 regulon using a multi-omics approach. Nat Commun. 2023;14(1):6715.
71. Iskandar K, Foo J, Liew AQX, Zhu H, Raman D, Hirpara JL, Leong YY, Babak MV, Kirsanova AA, Armand AS, Oury F, Bellot G, Pervaiz S. A novel MTORC2-AKT-ROS axis triggers mitofission and mitophagy-associated execution of colorectal cancer cells upon drug-induced activation of mutant KRAS. Autophagy. 2024;20(6):1418–41.
72. Zhou J, Zhang L, Zeng L, Yu L, Duan Y, Shen S, Hu J, Zhang P, Song W, Ruan X, Jiang J, Zhang Y, Zhou L, Jia J, Hang X, Tian C, Lin H, Chen HZ, Cronan JE, Bi H, Zhang L. H. pylori FabX contains a 4Fe-4S cluster essential for unsaturated fatty acid synthesis. Nat Commun. 2021;12(1):6932.
73. Baskerville MJ, Kovalyova Y, Mejías-Luque R, Gerhard M, Hatzios SK. Isotope tracing reveals bacterial catabolism of host-derived glutathione during H. pylori infection. PLoS Pathog. 2023;19(7):e1011526.
74. Dumitrescu DG, Gordon EM, Kovalyova Y, Seminara AB, Duncan-Lowey B, Forster ER, Zhou W, Booth CJ, Shen A, Kranzusch PJ, Hatzios SK. A microbial transporter of the dietary antioxidant ergothioneine. Cell. 2022;185(24):4526–40.
75. Grubwieser P, Böck N, Soto EK, Hilbe R, Moser P, Seifert M, Dichtl S, Govrins MA, Posch W, Sonnweber T, Nairz M, Theurl I, Trajanoski Z, Weiss G. Human airway epithelium controls Pseudomonas aeruginosa infection via inducible nitric oxide synthase. Front Immunol. 2024;15:1508727.
76. Kawatra A, Dhankhar R, Gulati P. Microbial arginine deiminase: a multifaceted green catalyst in biomedical sciences. Int J Biol Macromol. 2022;196:151–62.
77. Peng Y, Liu Y, Wu J, Zhang Y, Wang Q, Shao S. ArgR-dependent bacterial resistance to host lipid droplets in Edwardsiella piscicida. Commun Biol. 2025;8(1):333.
78. Díaz-Dinamarca DA, Salazar ML, Escobar DF, Castillo BN, Valdebenito B, Díaz P, Manubens A, Salazar F, Troncoso MF, Lavandero S, Díaz J, Becker MI, Vásquez AE. Surface immunogenic protein from Streptococcus agalactiae and Fissurella latimarginata hemocyanin are TLR4 ligands and activate MyD88- and TRIF dependent signaling pathways. Front Immunol. 2023;14:1186188.

79. Cao M, Yang J, Wang X, Hu W, Xie X, Zhao Y, Liu M, Wei Y, Yu M, Hu T. Sophora subprostrate polysaccharide regulates histone acetylation to inhibit inflammation in PCV2-infected murine splenic lymphocytes in vitro and in vivo. Int J Biol Macromol. 2021;191:668–78.
80. Mao FY, Lv YP, Hao CJ, Teng YS, Liu YG, Cheng P, Yang SM, Chen W, Liu T, Zou QM, Xie R, Xu JY, Zhuang Y. H. pylori-induced rev-erbα fosters gastric bacterial colonization by impairing host innate and adaptive defense. Cell Mol Gastroenterol Hepatol. 2021;12(2):395–425.
81. Pernitzsch SR, Alzheimer M, Bremer BU, Robbe-Saule M, De Reuse H, Sharma CM. Small RNA mediated gradual control of lipopolysaccharide biosynthesis affects antibiotic resistance in H. pylori. Nat Commun. 2021;12(1):4433.
82. Tsai CE, Wang FQ, Yang CW, Yang LL, Nguyen TV, Chen YC, Chen PY, Hwang IS, Ting SY. Surface-mediated bacteriophage defense incurs fitness tradeoffs for interbacterial antagonism. EMBO J. 2025;44(9):2473–500.
83. Tang X, Tay A, Benghezal M, Marshall BJ, Tang H, Li H. Advances in H. pylori lipopolysaccharide structure and function. FEMS Microbiol Rev. 2025;49:fuaf034.
84. Gong H, Hu X, Liao M, Fa K, Ciumac D, Clifton LA, Sani MA, King SM, Maestro A, Separovic F, Waigh TA, Xu H, McBain AJ, Lu JR. Structural disruptions of the outer membranes of gram-negative bacteria by rationally designed amphiphilic antimicrobial peptides. ACS Appl Mater Interfaces. 2021;13(14):16062–74.
85. Jones CS, Anderson AC, Clarke AJ. Mechanism of Staphylococcus aureus peptidoglycan O-acetyltransferase a as an O-acyltransferase. Proc Natl Acad Sci USA. 2021;118(36):e2103602118.
86. Ghamarypour A, Aminzadeh S, Majd A, Movahedi M. A novel thermophilic lysozyme 4356 from Cohnella sp. A01: cloning, heterologous expression, biochemical and kinetic characterization. Int J Biol Macromol. 2024;279(Pt 4):135341.
87. van Lente JJ, Baig MI, de Vos WM, Lindhoud S. Biocatalytic membranes through aqueous phase separation. J Colloid Interface Sci. 2022;616:903–10.
88. Hellmann MJ, Marongiu GL, Gorzelanny C, Moerschbacher BM, Cord-Landwehr S. Hydrolysis of chitin and chitosans by the human chitinolytic enzymes: chitotriosidase, acidic mammalian chitinase, and lysozyme. Int J Biol Macromol. 2025;297:139789.
89. Coullon H, Candela T. Clostridioides difficile peptidoglycan modifications. Curr Opin Microbiol. 2022;65:156–61.
90. Klebba PE, Newton SMC, Six DA, Kumar A, Yang T, Nairn BL, Munger C, Chakravorty S. Iron acquisition systems of gram-negative bacterial pathogens define TonB-dependent pathways to novel antibiotics. Chem Rev. 2021;121(9):5193–239.
91. Guo C, Nolan EM. Exploring the antibacterial activity and cellular fates of Enterobactin-drug conjugates that target gram-negative bacterial pathogens. Acc Chem Res. 2024;57(7):1046–56.
92. Saha P, Yeoh BS, Xiao X, Golonka RM, Abokor AA, Wenceslau CF, Shah YM, Joe B, Vijay-Kumar M. Enterobactin induces the chemokine, interleukin-8, from intestinal epithelia by chelating intracellular iron. Gut Microbes. 2020;12(1):1–18.
93. Yang X, Yuan L, Sun X, Lan W. Inhibitory mechanism of chlorogenic acid-grafted chitosan against Pseudomonas fluorescens: interruption of iron uptake. Int J Biol Macromol. 2025;305(Pt 2):141210.
94. Lettl C, Schindele F, Mehdipour AR, Steiner T, Ring D, Brack-Werner R, Stecher B, Eisenreich W, Bilitewski U, Hummer G, Witschel M, Fischer W, Haas R. Selective killing of the human gastric pathogen H. pylori by mitochondrial respiratory complex I inhibitors. Cell Chem Biol. 2023;30(5):499–512.
95. Huang TT, Liu YN, Huang JX, Yan PP, Wang JJ, Cao YX, Cao L. Sodium sulfite-driven H. pylori eradication: Unraveling oxygen dynamics through multi-omics investigation. Biochem Pharmacol. 2024;222:116055.
96. Ni Q, Yang H, Rao H, Zhang L, Xiong M, Han X, Deng B, Wang L, Chen J, Shi Y. The role of the C5a-C5aR pathway in iron metabolism and gastric cancer progression. Front Immunol. 2025;15:1522181.
97. Fei X, Chen S, Li L, Xu X, Wang H, Ke H, He C, Xie C, Wu X, Liu J, Xie Y, Lu N, Zhu Y, Li N. H. pylori infection promotes M1 macrophage polarization and gastric inflamma-

tion by activation of NLRP3 inflammasome via TNF/TNFR1 axis. Cell Commun Signal. 2025;23(1):6.

98. Zhang X, He Y, Zhang X, Fu B, Song Z, Wang L, Fu R, Lu X, Xing J, Lv J, Guo M, Huo X, Liu X, Lu J, Du X, Ge Z, Chen Z, Li C. Sustained exposure to H. pylori induces immune tolerance by desensitizing TLR6. Gastric Cancer. 2024;27(2):324–42.
99. Koch MRA, Gong R, Friedrich V, Engelsberger V, Kretschmer L, Wanisch A, Jarosch S, Ralser A, Lugen B, Quante M, Vieth M, Vasapolli R, Schulz C, Buchholz VR, Busch DH, Mejías-Luque R, Gerhard M. CagA-specific gastric $CD8^+$ tissue-resident T cells control H. pylori during the early infection phase. Gastroenterology. 2023;164(4):550–66.
100. Tong J, Chen S, Gu X, Zhang X, Wei F, Xing Y. CD38 and extracellular $NAD^+$ regulate the development and maintenance of H. pylori vaccine-induced $CD4^+$ $T_{RM}$ in the gastric epithelium. Mucosal Immunol. 2024;17(5):990–1004.
101. Tegtmeyer N, Neddermann M, Lind J, Pachathundikandi SK, Sharafutdinov I, Gutiérrez-Escobar AJ, Brönstrup M, Tegge W, Hong M, Rohde M, Delahay RM, Vieth M, Sticht H, Backert S. Toll-like receptor 5 activation by the CagY repeat domains of H. pylori. Cell Rep. 2020;32(11):108159.
102. Hu W, Chen ZM, Wang Y, Yang C, Wu ZY, You LJ, Zhai ZY, Huang ZY, Zhou P, Huang SL, Li XX, Yang GH, Bao CJ, Cui XB, Xia GL, Ou Yang MP, Zhang L, Wu WKK, Li LF, Tan LK, Gong W. Single-cell RNA sequencing dissects the immunosuppressive signatures in H pylori-infected human gastric ecosystem. Nat Commun. 2025;16(1):3903.
103. Go DM, Lee SH, Lee SH, Woo SH, Kim K, Kim K, Park KS, Park JH, Ha SJ, Kim WH, Choi JH, Kim DY. Programmed death ligand 1-expressing classical dendritic cells mitigate helicobacter-induced gastritis. Cell Mol Gastroenterol Hepatol. 2021;12(2):715–39.
104. Li J, Song X, Huang S, Hu Y, Wang H, Wang R, Zhang Y, Zhang Y, Zhang X. A photo-driven immune modulator combined with chemo-photothermal therapy blocks immune checkpoint for H. pylori infection. Adv Healthcare Mater. 2025;14(18):e2501197.
105. Sarajlic M, Neuper T, Vetter J, Schaller S, Klicznik MM, Gratz IK, Wessler S, Posselt G, Horejs-Hoeck J. H. pylori modulates DC functions via T4SS/TNFα/p38-dependent SOCS3 expression. Cell Commun Signal. 2020;18(1):160.
106. Nagata M, Toyonaga K, Ishikawa E, Haji S, Okahashi N, Takahashi M, Izumi Y, Imamura A, Takato K, Ishida H, Nagai S, Illarionov P, Stocker BL, Timmer MSM, Smith DGM, Williams SJ, Bamba T, Miyamoto T, Arita M, Appelmelk BJ, Yamasaki S. H. pylori metabolites exacerbate gastritis through C-type lectin receptors. J Exp Med. 2021;218(1):e20200815.
107. Sirit IS, Peek RM Jr. Decoding the ability of H. pylori to evade immune recognition and cause disease. Cell Mol Gastroenterol Hepatol. 2025;19(5):101470.
108. Bauer M, Nascakova Z, Mihai AI, Cheng PF, Levesque MP, Lampart S, Hurwitz R, Pfannkuch L, Dobrovolna J, Jacobs M, Bartfeld S, Dohlman A, Shen X, Gall AA, Salama NR, Töpfer A, Weber A, Meyer TF, Janscak P, Müller A. The ALPK1/TIFA/NF-κB axis links a bacterial carcinogen to R-loop-induced replication stress. Nat Commun. 2020;11(1):5117.
109. van der Sluis RM, García-Rodríguez JL, Nielsen IH, Gris-Oliver A, Becker J, Costa B, Chaudhry MZ, Werner M, Laustsen A, Pedersen JG, Gammelgaard KR, Mogensen TH, Kalinke U, Cicin-Sain L, Bak RO, Kristensen LS, Jakobsen MR. Distinctive $CD8^+$ T cell activation by antigen-presenting plasmacytoid dendritic cells compared to conventional dendritic cells. Cell Rep. 2025;44(3):115413.
110. Lim KH, Wang L, Dotse E, Wang M, Tiu CY, Wijanarko KJ, Wang X, Chow KT. TLR4 sensitizes plasmacytoid dendritic cells for antiviral response against SARS-CoV-2 coronavirus. J Leukoc Biol. 2024;115(1):190–200.
111. You Y, Jin F, Du Y, Zhu L, Liu D, Zhu M, Du Y, Lang J, Li W, Ji JS, Du YZ. A photo-activable nano-agonist for the two-signal model of T cell in vivo activation. J Control Release. 2023;361:681–93.
112. Dadwal N, Mix C, Reinhold A, Witte A, Freund C, Schraven B, Kliche S. The multiple roles of the cytosolic adapter proteins ADAP, SKAP1 and SKAP2 for TCR/CD3-mediated signaling events. Front Immunol. 2021;12:703534.

113. ElTanbouly MA, Noelle RJ. Rethinking peripheral T cell tolerance: checkpoints across a T cell's journey. Nat Rev Immunol. 2021;21(4):257–67.
114. Jiang Q, Chen Z, Meng F, Zhang H, Chen H, Xue J, Shen X, Liu T, Dong L, Zhang S, Xue R. CD36-BATF2/MYB axis predicts anti-PD-1 immunotherapy response in gastric cancer. Int J Biol Sci. 2023;19(14):4476–92.
115. Artola-Borán M, Fallegger A, Priola M, Jeske R, Waterboer T, Dohlman AB, Shen X, Wild S, He J, Levesque MP, Yousefi S, Simon HU, Cheng PF, Müller A. Mycobacterial infection aggravates H. pylori-induced gastric preneoplastic pathology by redirection of de novo induced Treg cells. Cell Rep. 2022;38(6):110359.
116. Cardenas MA, Prokhnevska N, Sobierajska E, Gregorova P, Medina CB, Valanparambil RM, Greenwald R, DelBalzo L, Bilen MA, Joshi SS, Narayan VM, Master VA, Sanda MG, Kissick HT. Differentiation fate of a stem-like CD4 T cell controls immunity to cancer. Nature. 2024;636(8041):224–32.
117. Yan ZB, Zhang JY, Lv YP, Tian WQ, Shan ZG, Mao FY, Liu YG, Chen WY, Wang P, Yang Y, Cheng P, Peng LS, Liao YL, Yue GY, Xu XL, Zhao YL, Lü MH, Zhuang Y. H. pylori-induced REDD1 modulates Th17 cell responses that contribute to gastritis. Clin Sci. 2021;135(22):2541–58.
118. Casanova JL, MacMicking JD, Nathan CF. Interferon-γ and infectious diseases: lessons and prospects. Science. 2024;384(6693):eadl2016.
119. Mills KHG. IL-17 and IL-17-producing cells in protection versus pathology. Nat Rev Immunol. 2023;23(1):38–54.
120. Prichard A, Khuu L, Whitmore LC, Irimia D, Allen LH. H. pylori-infected human neutrophils exhibit impaired chemotaxis and a uropod retraction defect. Front Immunol. 2022;13:1038349.
121. Xiao S, Shen Y, Zhang M, Liu X, Cai T, Wang F. VacA promotes pyroptosis via TNFAIP3/TRAF1 signaling to induce onset of atrophic gastritis. Microbiol Res. 2025;296:128142.
122. Ji B, Ma Z, Liu S, Yao S, Mizuno K, Terai S, Tuo B, Li T, Liu X. Pyroptosis in gastric mucosal injury-related diseases (review). Int J Mol Med. 2025;56(4):165.
123. Wang G, Wang M, Pang J, Sun Q, Fan T, Li Z, You X. Effect of daphnetin, the coumarin derivative isolated from Daphne genus, on H. pylori adhesion to gastric epithelial cells. Phytomedicine. 2025;137:156357.

# Immunotherapeutic Strategies Against *H. pylori*

14

Xiaobo He

## 14.1 Introduction

*Helicobacter pylori* infection is a major causative factor in a spectrum of gastrointestinal diseases worldwide, including chronic gastritis, peptic ulcers, and gastric cancer. For decades, standard triple or quadruple antibiotic-based regimens have been the primary approach for eradicating *H. pylori* in clinical practice. Nevertheless, the relentless increase in antibiotic resistance has progressively undermined their therapeutic efficacy, evolving into a critical global health concern. Surveillance data from epidemiological studies indicate high resistance rates to key antibiotics such as metronidazole and clarithromycin, which can exceed 45.5% and 26.5%, respectively, in some regions [1]. This rise in antibiotic resistance directly contributes to higher failure rates of conventional therapies, presenting a major challenge in clinical management and underscoring the urgent need to develop alternative treatment strategies.

The pathogenesis of *H. pylori* infection involves complex and persistent interactions between the bacterium and the host gastric mucosal immune system. On one hand, *H. pylori* employs virulence factors such as cytotoxin-associated gene A (CagA) and vacuolating cytotoxin A (VacA) to act on host epithelial cells. This interaction activates pattern recognition receptor (PRR) signaling pathways, driving the robust release of pro-inflammatory cytokines (e.g., IL-8, TNF-α). These cytokines recruit and activate inflammatory cells, including neutrophils and macrophages, initiating and sustaining a state of chronic inflammation in the gastric mucosa [2].On the other hand, *H. pylori* has evolved multiple immune evasion mechanisms to facilitate persistent colonization. These strategies include structural

X. He (✉)
National Key Laboratory of Immunity and Inflammation & Institute of Immunology, College of Basic Medical Sciences, Naval Medical University, Shanghai, China
e-mail: hexiaobo@immunol.org





J. Gao, Y. Du (eds.), *Non-Antibiotic Therapies for Helicobacter pylori*,
https://doi.org/10.1007/978-981-95-9598-3_14

modifications of components such as bacterial RNA and lipopolysaccharide (LPS) to evade recognition and activation by PRRs. VacA also contributes by inhibiting macrophage phagocytic function and inducing their apoptosis through various pathways. Furthermore, the bacterium subverts adaptive immunity by inducing a tolerogenic dendritic cell (DC) phenotype, suppressing T-cell proliferation, and promoting the expansion of regulatory T cells (Tregs) [3, 4]. Consequently, the long-term persistence of *H. pylori* is closely linked to disrupted immune surveillance and clearance. This intricate host–pathogen interplay underscores the significant potential of immune-based strategies as promising alternative therapeutic approaches.

Immunotherapy, as a revolutionary treatment paradigm, has achieved milestone successes in areas such as cancer and autoimmune diseases. Its core concept—the precise modulation of the host's immune system to restore homeostatic function or enhance specific defensive capabilities—provides a novel and powerful perspective for treating persistent infections like those caused by *H. pylori.* Unlike the direct, broad-spectrum bactericidal mechanism of antibiotics, which exerts immense selective pressure for the development of resistanc, immunotherapy against *H. pylori* adopts a more nuanced and host-centric approach. It aims to correct the profound immune dysregulation instigated by the pathogen across multiple levels, from reversing local immunosuppression to reinvigorating ineffective adaptive responses, thereby reshaping a robust and effective anti-infection immunity. Its mechanisms primarily include: inducing effective adaptive immune protection in the body through active immunization and reshaping and optimizing the suppressive immune microenvironment through immunomodulatory strategies to enhance the body's innate and adaptive immune defenses against the bacteria. Based on these unique mechanisms of action, immunotherapeutic strategies are expected to become effective alternatives to antibiotic therapy. By reversing the chronic inflammation and immunosuppressive state established by *H. pylori* infection, they can block the malignant progression of related gastric diseases. Therefore, developing corresponding immune interventions based on the immunopathogenic mechanisms of *H. pylori* will provide new ideas and directions for overcoming antibiotic resistance and achieving disease prevention and radical cure.

## 14.2 Active Immunization Strategies

Vaccination is a classic immunotherapeutic strategy that enhances adaptive immunity through active immunization. It represents the most extensively studied and clinically promising immunotherapeutic approach against *H. pylori.* The fundamental principle of vaccination is to preemptively prime the immune system for rapid pathogen recognition and elimination upon subsequent exposure, offering a strategy for both prophylaxis and treatment. Vaccines function by delivering *H. pylori* antigens, often in combination with adjuvants, to induce robust humoral (antibody-mediated) and cellular (T cell-mediated) immune responses against the pathogen. This coordinated adaptive immunity can rapidly achieve bacterial clearance or prevent initial colonization. A successful *H. pylori* vaccine should overcome the bacterium's sophisticated evasion tactics and generate long-lasting immunity, particularly at the mucosal surface of the stomach, which represents the primary site of infection. Some studies indicate that vaccines can significantly induce mucosal immune responses and stimulate secretory immunoglobulin A (IgA) production, which is crucial for inhibiting *H. pylori* colonization.

To date, numerous *H. pylori*-associated proteins have been identified as possessing favorable immunogenic properties and have been investigated as vaccine antigens. Relevant targets include virulence factors (e.g., CagA and VacA), outer membrane proteins (e.g., Lpp20 and Omp18), adhesion factors, and flagellin. These antigens play important roles in the colonization, survival, and pathogenic mechanisms of *H. pylori*. Both single antigens and antigen combinations have demonstrated the potential to induce potent, specific immune responses in studies of various vaccine platforms. Looking forward, antigen screening will increasingly focus on antigens that combine strong immunogenicity with high conservation across different bacterial strains, addressing the genetic diversity and antigenic variability of *H. pylori* to provide broad and long-lasting immune protection.

### 14.2.1 Recombinant Subunit Vaccines

Subunit vaccines utilize recombinant antigenic proteins to induce protective immune responses against pathogens. They represent a major research focus due to their well-defined composition, high safety profile, and strong immunogenicity. Extensive studies have demonstrated that oral subunit vaccines can effectively stimulate Th1/Th17 responses and induce IgA production, thereby promoting the clearance of *H. pylori*. Notably, a phase III clinical trial by Zeng et al. confirmed that a urease-based subunit vaccine exhibited favorable safety and immunogenicity, effectively reducing the *H. pylori* infection rate in children [5]. In animal models, a vaccine combining recombinant CagA antigen with CpG adjuvant significantly elevated IgG levels and induced a robust Th1-biased immune response [6]. Furthermore, a multivalent subunit vaccine based on three antigens—neutrophil-activating protein (NAP), urease subunit A (UreA), and urease subunit B (UreB)—significantly reduced bacterial colonization in infected mice and overcame immune evasion by restoring the suppression of the Th2 immune response [7]. However, in a phase I/II clinical trial, a trivalent vaccine incorporating VacA, CagA, and NAP, while increasing specific antibody titers in humans, failed to provide additional protection against challenge with CagA-positive strains [8]. These findings collectively indicate that multivalent vaccines based on antigen combinations represent a rational and promising strategy for enhancing immunogenicity and optimizing immune efficacy. However, their clinical effectiveness remains influenced by a complex interplay of multiple factors. Key considerations include the selection of appropriate and potent adjuvants to steer the immune response toward a stronger and protective phenotype, and the inherent genetic diversity and antigenic variation of *H. pylori* strains across different geographical populations, which may affect vaccine coverage. Furthermore, the failure of some multivalent candidates in clinical trials, despite promising preclinical data, highlights the imperfect correlation between animal models and human immune responses to *H. pylori*, underscoring the need for more predictive models and careful patient stratification in future clinical studies.

### 14.2.2 Multi-Epitope Vaccines

Vaccines based on single protein antigens often elicit suboptimal or insufficient immune responses due to their limited repertoire of antigenic epitopes [9]. Although strategies combining different antigenic proteins can partially compensate for this limitation, their effectiveness remains constrained. Multi-epitope vaccines address this gap by directly integrating B-cell and T-cell epitopes from *H. pylori*, thereby inducing a more comprehensive and potent synergistic immune response. The design of multi-epitope vaccines typically relies on immunoinformatics approaches. Key antigenic proteins of *H. pylori* are screened to identify highly immunogenic B-cell and T-cell epitopes, which are then linked via spacers to construct the vaccine. For example, one study designed a multi-epitope vaccine incorporating epitopes from UreB, CagA, and HopE, and validated its binding capacity to Toll-like receptors through molecular dynamics simulations, demonstrating its potential to induce robust immune responses [10]. This strategy not only enhances immunogenicity but also reduces the likelihood of adverse effects [11, 12]. Studies in animal models have shown that multi-epitope vaccines significantly reduce gastric colonization by *H. pylori* in mice. They promote the production of specific IgG and secretory IgA (sIgA) antibodies, modulate Th1/Th17 immune responses, alleviate gastritis-related symptoms, and enhance protective immunity [13, 14]. Another study reported that a tetravalent epitope vaccine combined with a Th1-polarizing adjuvant induced a broader array of antibody subtypes and exerted stronger bactericidal effects in a Mongolian gerbil model of *H. pylori* infection [15]. By leveraging immunoinformatics, multi-epitope vaccines enable precise direction of immune responses, avoid potential adverse effects associated with noncritical components of full-length proteins, and improve vaccine safety. Furthermore, the combination of multiple epitopes can target diverse antigenic determinants, inducing broader cross-protective immunity. This approach effectively counteracts immune evasion caused by the genetic diversity and antigenic variation of *H. pylori.* Despite these notable advantages, the development of multi-epitope vaccines faces several challenges. The computational prediction of immunodominant and conserved epitopes is not entirely reliable, and due to HLA polymorphism, the selected epitopes may not always be efficiently processed and presented in all individuals. Furthermore, the immunogenicity of these engineered constructs can sometimes be lower than that of full-length proteins, necessitating the use of complex delivery systems or potent adjuvants. Therefore, while multi-epitope vaccines represent a highly rational and targeted approach, their ultimate success depends on continuous improvements in prediction algorithms, epitope selection, and formulation strategies.

### 14.2.3 DNA Vaccines

DNA vaccines are prepared using genetic engineering techniques to insert gene fragments encoding pathogen antigens into a DNA vector. Compared to traditional vaccines, DNA vaccines offer distinct advantages. On one hand, plasmid DNA can efficiently synthesize the encoded antigenic proteins and activate immune responses after entering host cells. On the other hand, DNA vaccines are characterized by flexible design, simple production, and stable properties. In mouse models, DNA vaccines based on single antigens such as flagellin and CagW have been shown to effectively induce immune responses against *H. pylori* [16, 17]. One study reported that a multi-epitope DNA vaccine targeting four antigens (outer membrane beta-barrel, outer membrane beta, HofA, and Hcp beta-lactamase-like protein) induced significant IgG levels and a Th1-type cellular response in mice. When combined with a CpG adjuvant, the treatment even completely blocked gastric colonization by *H. pylori* [18]. Furthermore, DNA vaccines utilizing novel delivery systems, such as nanomaterials, have demonstrated enhanced efficacy in inducing immune protection. For example, an alginate-based DNA vaccine targeting UreH induced significant intestinal and systemic Th1/Th17 responses in mice and promoted IL-17-mediated mucosal antigen-specific IgA production [19]. In this study, the alginate-based nanovaccine strategy effectively improved both the delivery efficiency of the DNA vaccine and the resulting immune response. Similarly, a chitosan nanoparticle-encapsulated DNA vaccine targeting CagW induced stronger specific Th1/Th2 immune responses and enhanced target antigen expression by delaying DNA degradation [17]. These studies highlight the potential of DNA vaccines for *H. pylori* applications. However, challenges such as the risk of genomic integration and insufficient immunogenicity require further attention and optimization.

### 14.2.4 Live-Attenuated Vaccine

Live-attenuated vaccines are developed by using genetic engineering to delete virulence-associated genes from a pathogen and inserting genes encoding target antigens. This approach eliminates the pathogen's pathogenicity while retaining its ability to infect and express vaccine antigens. Since most attenuated pathogens inherently possess characteristics that activate both innate and adaptive immunity, these engineered vectors can not only deliver *H. pylori*-related antigens but often synergistically induce more potent immune responses. *Bacillus subtilis* spores have been widely utilized as mucosal vaccine vectors and can effectively induce mucosal immune responses. Katsande et al. engineered *B. subtilis* spores to express the *H. pylori* antigens UreA and UreB. Oral immunization with the modified spores induced antigen-specific mucosal responses (fecal sIgA), including seroconversion and hyperimmunity in mice, and reduced *H. pylori* colonization upon challenge [20]. *Listeria monocytogenes* is another common bacterial vaccine vector known to enhance adaptive immune responses by promoting antigen presentation. Liu et al. constructed an attenuated *Listeria* vector vaccine that was shown to induce high levels of serum IgG and gastric sIgA and to reduce *H. pylori* colonization [21]. However, the study also revealed that the attenuated *Listeria* vector without the antigen induced similar immune protection, suggesting the potential for the vector itself to elicit cross-reactive immune responses. A recent study developed an intranasal influenza A virus (IAV) vector vaccine encoding *H. pylori* neutrophil-activating protein (NapA). In a mouse model, vaccination induced robust humoral, mucosal, and Th1/Th17 cellular immune responses. Both prophylactic and therapeutic vaccination reduced *H. pylori* colonization and gastric inflammation [22]. Additional studies have utilized other attenuated live vectors such as *Salmonella*, *Lactic acid bacteria*, *Saccharomyces cerevisiae*, and Bacillus Calmette-Guérin (BCG) to construct vaccines, with preliminary evidence supporting their efficacy against *H. pylori* infection [23–26]. Various types of live-attenuated vaccines have demonstrated potential for preventing or treating *H. pylori* in research settings. However, further optimization is required to address key challenges, including potential dose-limiting toxicity and the presence of preexisting neutralizing antibodies against the vector due to natural exposure.

### 14.2.5 OMV Vaccines

Bacterial outer membrane vesicles (OMVs) are bilayer nanostructures, 20–250 nm in diameter, naturally released from the outer membrane of Gram-negative bacteria. The OMV membrane contains components such as lipopolysaccharides and lipoproteins, while the lumen can encapsulate periplasmic proteins, virulence factors, and other cargo [27]. This structural composition enables OMVs to both activate innate immune responses via their carried pathogen-associated molecular patterns (PAMPs) and function as platforms for delivering pathogen-derived antigens to induce adaptive immunity. For instance, LPS present on the OMV surface can be recognized by the TLR4/MD2 complex, activating downstream signaling pathways mediated by NF-κB and IRF3, thereby effectively stimulating innate immune cells such as macrophages. Additionally, the abundant lipoprotein components contained within OMVs can trigger innate immune activation via TLR2 signaling [28, 29]. However, the diverse PAMPs carried by OMVs may exert complex and sometimes opposing immunomodulatory effects. For example, Opa proteins present in the outer membrane of pathogenic Neisseria, upon binding to CEACAM receptors, have been shown to mediate immunosuppressive effects [30]. This underscores the necessity to carefully consider the bacterial origin and specific composition of OMVs when developing them as vaccine platforms, as these factors critically influence the net immunostimulatory outcome. Collectively, the intrinsic immune-activating capacity of OMVs positions them as promising vaccine adjuvants. Purified *H. pylori* OMVs, when employed as an adjuvant in mice, induced vaccine-mediated immune protection that persisted for 12 weeks. Compared to the cholera toxin (CT) adjuvant, OMVs not only promoted a Th1 immune response but also preferentially enhanced Th2 and Th17 immune responses. This finding highlights the potential benefits of OMVs as adjuvants [31]. Liu et al. developed a recombinant

OMV vaccine by engineering *H. pylori* OMVs to incorporate the antigens UreB, VacA, and CagA [32]. This recombinant vaccine elicited robust humoral, mucosal, and mixed Th1/Th2/Th17 cellular immune responses in mice, effectively reducing *H. pylori* colonization and urease activity. Another study utilized OMVs derived from *Salmonella enterica* serovar Typhimurium mutant strains to deliver UreB and CagA antigens, observing similar protective immune responses in mice [33]. Currently, owing to their inherent immunostimulatory properties and convenient capacity for antigen display, OMVs are emerging as a promising technological platform for *H. pylori* vaccine development.

In summary, various vaccine strategies demonstrate considerable potential for managing *H. pylori* infection and represent the most promising approaches likely to reach clinical application (Fig. 14.1). Vaccines could serve not only as preventive measures to reduce infection incidence but also as therapeutic interventions to eradicate established colonization. However, despite significant progress in preclinical studies and advancements of several candidates into clinical trials, the majority remain in the preclinical stage, and no *H. pylori* vaccine has yet been approved for clinical use worldwide. This situation underscores the persistent challenges in achieving durable and broad immune protection. Future efforts, informed by a deeper understanding of *H. pylori*–host immune interactions and focused on the rational optimization and design of vaccine antigens, novel adjuvants, and delivery systems, hold promise for overcoming the current bottlenecks in vaccine

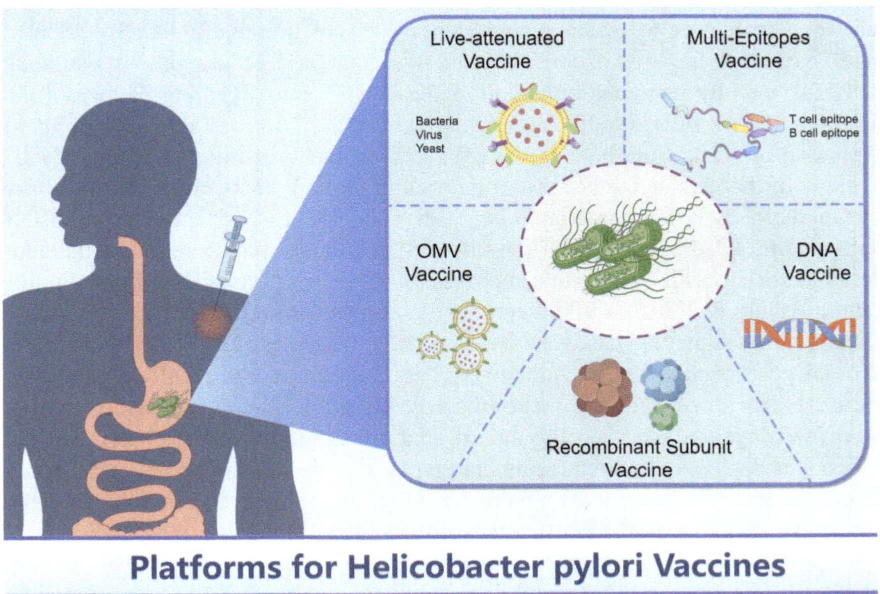


**Fig. 14.1** Platforms for *H. pylori* vaccines. (Image created by Figdraw with permission)

development and ultimately advancing *H. pylori* vaccines toward clinical translation and application.

## 14.3 Targeting Pattern Recognition Receptor Signaling Pathways

*H. pylori* infection initially triggers the host innate immune response. This process is initiated when pattern recognition receptors (PRRs) on innate immune cells—such as macrophages and dendritic cells—as well as on non-immune cells like gastric mucosal epithelial cells, recognize PAMPs. The PRR family primarily includes Toll-like receptors (TLRs), C-type lectin receptors (CLRs), RIG-I-like receptors (RLRs), and Nod-like receptors (NLRs). Typically, TLRs and CLRs recognize components of *H. pylori* such as LPS, neutrophil-activating protein (NAP), flagellin, DNA, and RNA. This recognition activates transcription factors including NF-κB, AP-1, and IRFs, leading to the secretion of pro-inflammatory cytokines and initiation of inflammatory responses. Meanwhile, NLRs, functioning as intracellular sensors, identify bacterial virulence factors and peptidoglycan fragments. This triggers inflammasome assembly and activates caspase-1 (CASP1), resulting in the release of IL-1β and IL-18 [34, 35].

However, *H. pylori* can attenuate PRR-induced immune responses through various mechanisms [36]. For instance, *H. pylori* antigens can modulate the recognition by multiple Toll-like receptors, including TLR4 and TLR5, thereby facilitating bacterial evasion of immune clearance. The LPS of *H. pylori* possesses a unique structure that enables it to evade recognition by innate immune cells. Specifically, modifications to the lipid A core structure of its surface LPS impair stable binding to TLR4, thereby reducing activation of the downstream signaling pathway [37]. Additionally, *H. pylori* modifies the O-antigen of its LPS to mimic Lewis antigens expressed on gastric epithelial cells [38]. This strategy of mimicking host "self" markers further assists the bacterium in evading immune recognition. Studies have shown that Lewis antigen-containing LPS can bind to and activate DC-SIGN (DC-specific ICAM-3 grabbing non-integrin), leading to the secretion of the anti-inflammatory cytokine IL-10 and suppression of Th1 responses [39]. Flagellin, the natural ligand for TLR5, would theoretically bind and activate TLR5-mediated pro-inflammatory signaling cascades. However, *H. pylori* evades TLR5 recognition through posttranslational modifications of its flagellin, consequently suppressing the mediated innate immune response [40]. Mutations in conserved domains of *H. pylori* flagellin—specifically, amino acid substitutions in the N-terminal D1 region (residues 89–96)—reduce its binding affinity for TLR5. Notably, analogous mutations have been shown to facilitate TLR5 evasion in Salmonella flagellin, underscoring a convergent evolutionary strategy for immune evasion [41]. TLR2 exhibits a dual role: it can bind *H. pylori* neutrophil-activating protein (HP-NAP) to mediate IL-8 secretion by neutrophils and synergize with TLR4 to enhance inducible nitric oxide synthase (iNOS) production, exerting pro-inflammatory effects; conversely, it can also mediate IL-10 secretion, exerting anti-inflammatory effects

[42, 43]. Studies have shown that TLR2 deficiency leads to reduced *H. pylori* colonization and more severe gastric mucosal inflammation, suggesting that *H. pylori* utilizes TLR2-mediated complex regulation to promote colonization [44]. Furthermore, *H. pylori* can suppress the STING and RIG-I signaling pathways by downregulating IRF3 activation, thereby achieving immune suppression [45]. Deacetylated *H. pylori* peptidoglycan can also activate a NOD1-mediated negative feedback pathway, assisting the bacteria in evading host clearance [46].

Currently, immune agonists targeting various PRRs have been developed for treating microbial infections and are often used as adjuvant therapies, such as vaccine adjuvants. The intranasal immunization of mice with TLR3 agonist poly(I:C) and TLR9 agonist CpG ODN, combined with *H. pylori* recombinant proteins LpoB and UreA, effectively induced immune protection and significantly reduced gastritis. The application of poly(I:C) enhanced the specific secretory IgA (sIgA) antibody response and reduced gastric bacterial colonization. Furthermore, the combined use of TLR agonists effectively promoted a mixed Th1/Th2/Th17 immune response, demonstrating a synergistic effect [47]. For example, the TLR9 agonist CpG-7909, a synthetic oligodeoxynucleotide, effectively activates NF-κB and MAPK signaling pathways and induces pro-inflammatory cytokine secretion. As an adjuvant, it effectively enhances the efficacy of HBV vaccines [48, 49]. In hepatitis B vaccine research, CpG ODN, as a TLR9 agonist, has been demonstrated to effectively promote the proliferation and activation of peripheral blood B cells in mice. It concurrently upregulates the expression of CD40, CD80, CD86, and MHC II, thereby enhancing the vaccine-induced humoral immune response [50]. The TLR4 agonist MPL, a derivative of LPS, is the first TLR agonist approved as a vaccine adjuvant and has been used in shingles vaccines [51]. Additionally, a TLR6 agonist was demonstrated in an animal model to restore IL-1β and IL-8 expression and reduce *H. pylori* colonization [52]. These PRR-targeting biologics, functioning as immunomodulators, are utilized in the treatment of various diseases to enhance antimicrobial immune responses, thereby providing a valuable reference for managing *H. pylori* infection. However, since PRRs recognize a broad spectrum of PAMPs, their activation lacks specificity. Over-activation may entail potential risks—such as severe systemic inflammatory responses or cytokine storms—which can cause immune-mediated damage to healthy tissues. Furthermore, PRR-mediated signaling pathways exhibit complex functional diversity; for instance, TLR2 demonstrates dual pro-inflammatory and anti-inflammatory roles. Consequently, PRR-targeted immunomodulation represents a promising yet complex therapeutic strategy for *H. pylori* infection. The primary challenge lies in achieving localized and controlled immune activation sufficient to promote bacterial clearance without triggering excessive systemic inflammation. The risk of cytokine release syndrome (CRS) or other immunopathologies cannot be overlooked, particularly with systemic administration of nonspecific PRR agonists. Furthermore, the functional pleiotropy and cross talk among different PRR pathways—as exemplified by the dual role of TLR2 described earlier—introduce additional complexity that is difficult to predict and control in clinical applications. Future research should focus on developing delivery systems that specifically target PRR agonists to the infection site, such as the gastric

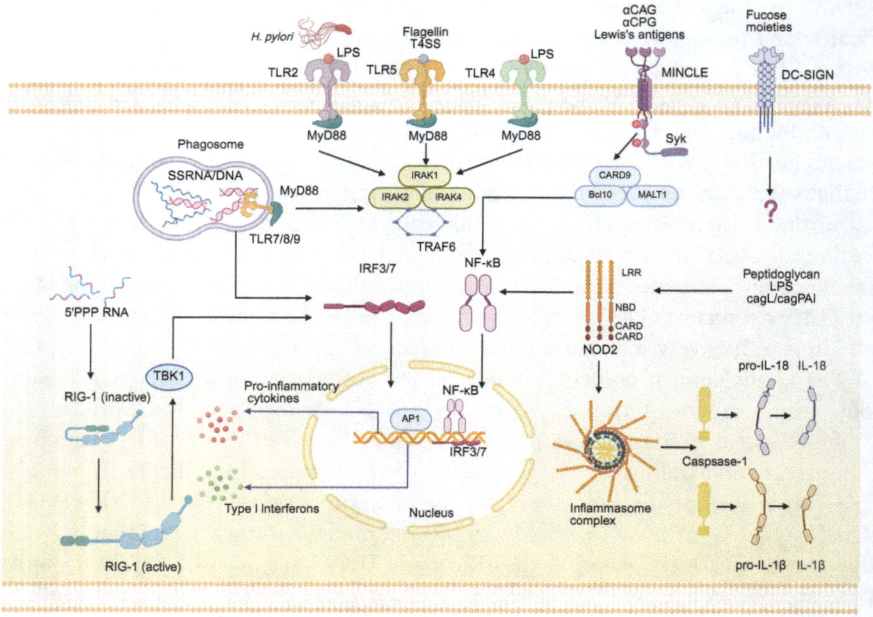


**Fig. 14.2** Recognition of *H. pylori* PAMPs by PRRs. (Image created with BioRender.com,with permission)

mucosa, to enable more effective local activation. Concurrently, exploring combination therapies with multiple agonists may achieve synergistic effects with reduced toxicity (Fig. 14.2).

## 14.4 Host Defense Peptides

Host defense peptides (HDPs) are short, positively charged peptides widely present across all living organisms and constitute an essential component of the innate immune defense against pathogenic infections in humans. The direct antimicrobial mechanisms of HDPs include: targeting and binding to the cell wall or its precursor molecules, leading to cell wall lysis or impaired synthesis and acting on specific intracellular bacterial targets to block processes such as nucleic acid replication and protein synthesis. The mechanism by which HDPs target the bacterial cell wall relies on their physicochemical properties: their positive charge facilitates binding to the negatively charged bacterial cell surface, while their amphipathic helical structure enables insertion into the lipid bilayer via hydrophobic interactions [53]. Furthermore, recent studies have highlighted the important role of their anti-biofilm activity in the antimicrobial process [54]. Windham et al. demonstrated that the antimicrobial peptides DJK-5 and IDR-1018 significantly reduce the maturity of *H. pylori* biofilms, thereby inhibiting bacterial colonization. Specifically, DJK-5 primarily acts by blocking the biofilm formation process, whereas IDR-1018 disrupts established biofilms and suppresses their further development [55]. Additionally, HDPs can directly or indirectly recruit and activate immune cells such as macrophages, neutrophils, and dendritic cells, and exert complex immunomodulatory effects on both innate and adaptive immunity by regulating pro- and anti-inflammatory cytokines including IL-1β, IL-10, and IL-6 [56, 57]. Owing to their dual antimicrobial and immunomodulatory properties, HDPs are extensively investigated for the treatment of multidrug-resistant bacterial infections and represent highly promising antibiotic alternatives. Their ability to target bacterial membranes physically and disrupt multiple intracellular biosynthetic processes makes them less susceptible to conventional resistance mechanisms compared to single-target antibiotics. Furthermore, their role as immunomodulators can ameliorate the chronic inflammation associated with *H. pylori* infection, thereby providing dual therapeutic benefits.

In humans, multiple HDPs participate in innate and adaptive immune responses, primarily categorized into the cathelicidin and defensin families. Defensins are arginine-rich HDPs mainly classified as α-defensins and β-defensins, exhibiting broad-spectrum antimicrobial activity against both Gram-positive and Gram-negative bacteria [58]. LL-37, belonging to the cathelicidin family, is a C-terminal peptide derived from human cationic antimicrobial peptide 18 (hCAP18). Studies have shown that LL-37 is secreted by gastric mucosal epithelial cells during *H. pylori* infection and exhibits potent bactericidal activity against various *H. pylori* strains in vitro, either alone or in synergy with human β-defensin 1 (hβD1) [59]. Concurrently, *H. pylori* employs various mechanisms to evade human HDPs, facilitating persistent colonization [60]. For example, cagPAI+ wild-type *H. pylori* suppresses hβD1 expression by downregulating NF-κB signaling [61]. Additionally, during early *H. pylori* infection, hβD3 secretion is rapidly induced via EGFR-dependent activation of MAP kinase and JAK/STAT signaling; however, its expression is subsequently suppressed due to CagA-mediated EGFR inactivation [62]. Therefore, restoring or enhancing HDP expression holds potential therapeutic significance for eradicating *H. pylori*. Numerous studies have validated the potential of HDPs in treating *H. pylori* infection. For instance, in a mouse model, vitamin D3 supplementation inhibited *H. pylori* infection by upregulating vitamin D receptor expression and activating cathelicidin antimicrobial peptide expression. In another study, the frog-derived HDP Temporin-Sha and its analogs effectively killed 90–100% of *H. pylori* in vitro while demonstrating low or no toxicity to human gastric cells and tissues, indicating its potential as an anti-*H. pylori* therapeutic peptide [63]. However, natural HDPs often face limitations such as rapid metabolism, poor gastrointestinal stability, and low bioavailability [64, 65]. To overcome the aforementioned limitations of natural HDPs, such as proteolytic instability, potential cytotoxicity, and poor pharmacokinetics, researchers are actively developing a multifaceted arsenal of innovative strategies. These approaches can be broadly categorized into three interconnected areas: First, the rational redesign of peptides, which involves modifying natural HDP sequences using D-amino acids or cyclization to enhance stability and reduce toxicity; second, the engineering of novel delivery systems, such as nanoparticles and engineered probiotics, to protect HDPs from degradation, facilitate targeted delivery to the infection site, and provide sustained release; third, the application of bioinformatics and high-throughput screening to mine vast databases for novel HDPs with optimal properties. The convergence of these advanced technologies is accelerating the translation of HDPs from the laboratory into viable therapeutic candidates.

For example, the recombinant PGLa-AM1 peptide demonstrated excellent anti-*H. pylori* activity both in vitro and in vivo [66]. Similarly, a cathelicidin antimicrobial peptide analog, CBF-K16, significantly reduced gastric *H. pylori* colonization in mice and alleviated inflammatory responses by suppressing IL-8 expression [67]. Wendler et al. hydrolyzed hβD1 to generate an 8-mer peptide and modified it using nonnatural D-amino acids. This peptide variant exhibited antimicrobial activity in vitro against microorganisms including *Escherichia coli*, *Pseudomonas aeruginosa*, and *Candida albicans* [68]. Furthermore, a pH-responsive antimicrobial peptide developed by Xiong et al. underwent conformational changes in the acidic gastric environment of mice and effectively killed infecting *H. pylori* [69]. Xie et al. developed an HDP-mimetic antimicrobial polymer, poly(2-oxazoline), which remains inactive under physiological conditions but is selectively activated at the infection site, enabling targeted antimicrobial therapy. This HDP-mimetic polymer demonstrated potent antibacterial efficacy against various Gram-negative bacteria in multiple animal infection models without inducing drug resistance [70]. Fonseca et al. utilized a nanoparticle-based microfluidic system to deliver HDPs, achieving faster and more effective eradication of *H. pylori* [71]. Zhang et al. employed engineered Cry3Aa crystals to deliver HDPs, protecting the peptides from degradation while promoting targeted release to the infection site [72]. In another study, oral administration of *Lactococcus lactis* encoding mouse cathelicidin significantly elevated mucosal cathelicidin levels in mice, concurrently reducing TNF-α and IL-6 expression, and mitigating *H. pylori* colonization and gastric mucosal inflammation [73]. Furthermore, Marcelo et al. computationally screened 1773 human metagenomes and identified 323 candidate sequences, from which they synthesized 78 potential antimicrobial peptides. Remarkably, 70.5% of these peptides demonstrated antimicrobial activity in mouse infection models and represent previously unreported structures [74]. This high-throughput screening strategy validates the feasibility of systematically mining novel antimicrobial peptides from extensive metagenomic datasets, while also providing robust methodological support for discovering therapeutic HDPs from diverse biological sources.

## 14.5 Probiotics

Probiotic therapy represents an emerging strategy for managing *H. pylori* infection. Probiotics such as *Lactobacillus*, *Bifidobacterium*, and *Bacillus* species exert anti-*H. pylori* effects through multiple mechanisms. On one hand, probiotics inhibit the colonization and survival of *H. pylori* via nonimmune pathways, including the secretion of antimicrobial substances, modulation of gastric pH, and competitive exclusion from adhesion sites [75–79]. For instance, *Lactobacillus* strains isolated from infant oral cavities and feces demonstrate antimicrobial effects by secreting substances such as small heat-stable peptides (SHSPs). Concurrently, *Lactobacillus* can competitively bind to adhesion sites on gastric mucosal epithelial cells through auto-aggregation, thereby reducing *H. pylori* colonization [80, 81]. Furthermore, various probiotics including *B. subtilis* inhibit *H. pylori* by secreting different bacteriocins, primarily through mechanisms involving inhibition of nucleic acid or cell wall protein synthesis, or by increasing bacterial membrane permeability [82, 83]. In clinical trials, oral administration of *Lactobacillus* not only reduced *H. pylori* load in volunteers but also alleviated discomfort symptoms such as diarrhea and mitigated inflammatory responses [84].

On the other hand, probiotics can also influence *H. pylori* infection through immunomodulatory pathways. For example, they regulate the activity and phenotype of immune cells such as natural killer (NK) cells and dendritic cells (DCs), and modulate the secretion of inflammatory cytokines, thereby shaping the host immune response to suppress *H. pylori* infection [85]. Specifically, probiotics promote the secretion of anti-inflammatory cytokines (e.g., IL-10) while inhibiting the production of pro-inflammatory cytokines (e.g., IL-6, IL-1β, and IFN-γ), leading to attenuated inflammation [86]. Studies have shown that specific probiotic strains (e.g., *Lactobacillus reuteri*) significantly reduce *H. pylori*-mediated levels of inflammatory factors, including TNF-α and IL-8, while promoting the secretion of the anti-inflammatory cytokine IL-10 [87]. In healthy volunteers, oral administration of *Bifidobacterium infantis* (*B. infantis*) upregulated Foxp3 expression in peripheral blood $CD4^+$ T cells and increased IL-10 levels. In vitro experiments further demonstrated that coculture of *B. infantis* with DCs promoted the generation of $Foxp3^+CD4^+$ T cells by blocking the activation of TLR2 and DC-SIGN and enhancing retinoic acid metabolism [88]. Probiotics effectively suppressed the activation of pro-inflammatory signaling pathways—such as NF-κB, IL-17, and TNF-α—in *H. pylori*-infected mice [89]. In another study, monotherapy or combination therapy with six probiotic strains promoted the infiltration of lymphocytes, plasma cells, and neutrophils in a mouse model; enhanced IFN-γ secretion; and consequently alleviated inflammation and accelerated *H. pylori* clearance [90]. These immunoregulatory functions play a significant role in mitigating chronic gastric inflammation induced by *H. pylori*. Currently, probiotic-based monotherapy or adjuvant therapy has demonstrated potential in clinical trials for anti-*H. pylori* regimens, evidenced by improved eradication rates and reduced antibiotic-associated side effects [91].

## 14.6 Summary

*H. pylori* infection represents a major global health burden and is a well-established causative agent for a spectrum of gastrointestinal pathologies, including chronic gastritis, peptic ulcer disease, and gastric cancer. For decades, conventional antibiotic-based therapies have formed the mainstay of clinical eradication regimens. However, the escalating prevalence of antibiotic resistance has substantially compromised treatment efficacy, highlighting the critical need for novel nonantibiotic alternatives.

Immunotherapy, which harnesses the host's immune system to combat diseases, has emerged as a transformative approach in medical therapeutics. Its application to infectious diseases offers a novel paradigm for managing *H. pylori* infection. This chapter systematically reviews recent advances in immunotherapeutic strategies against *H. pylori*, categorizing them into two principal modalities: active immunization and immunomodulation. In the domain of active immunization, diverse vaccine platforms—including recombinant subunit vaccines, multi-epitope vaccines, DNA vaccines, live-attenuated vector vaccines, and outer membrane vesicle (OMV) vaccines—have demonstrated the feasibility of inducing protective immune responses in preclinical models. These platforms elicit robust mucosal, humoral, and cellular immune responses, particularly Th1/Th17-polarized immunity, capable of suppressing or clearing *H. pylori* colonization. Simultaneously, immunomodulatory approaches using PRR agonists, HDPs, and probiotics have shown great potential in reshaping the hostile gastric microenvironment to favor bacterial clearance and resolution of inflammation, often with the added benefit of mitigating antibiotic-induced side effects. Notwithstanding these advancements, the clinical translation of immunotherapy for *H. pylori* continues to face considerable challenges. In vaccine development, key obstacles include inducing durable and broad-spectrum protection against genetically diverse strains, while optimizing antigen selection, adjuvant systems, and delivery platforms to maximize efficacy without compromising safety. Although immunomodulatory strategies show promise, their application remains predominantly adjunctive and often suffers from limited targeting specificity. Concerns regarding specificity and potential systemic toxicity necessitate further rigorous validation of their safety and efficacy profiles. A critical future challenge lies in precisely regulating the complex and dynamic immune network within the gastric mucosa. The ultimate objective is to achieve a delicate balance: eliciting a response sufficiently robust to eradicate established *H. pylori* infection, while adequately restraining it to prevent collateral tissue damage from excessive inflammation. Furthermore, the substantial heterogeneity in both host immune status and bacterial genetics underscores the imperative to develop individualized and precision immunotherapeutic approaches.

In conclusion, while challenges persist, the strategic shift from directly targeting the bacterium with antibiotics to empowering the host immune system offers a viable and exciting pathway forward. Future endeavors should be directed toward deepening our understanding of the dynamic interplay between *H. pylori* and host immunity, while fostering interdisciplinary collaborations to accelerate the translation of innovative immunotherapeutic strategies from bench to bedside.

## References


1. Zhang W, Chen Q, Liang X, Liu W, Xiao S, et al. Bismuth, lansoprazole, amoxicillin and metronidazole or clarithromycin as first-line *H. pylori* therapy. Gut. 2015;64(11):1715–20. https://doi.org/10.1136/gutjnl-2015-309900.
2. Marzhoseyni Z, Mousavi MJ, Ghotloo S. *H. pylori* antigens as immunomodulators of immune system. Helicobacter. 2024;29(1):e13058. https://doi.org/10.1111/hel.13058.
3. Gong J, Wang Q, Chen X, Lu J. *H. pylori* vaccine: mechanism of pathogenesis, immune evasion and analysis of vaccine types. Vaccines (Basel). 2025;13(5) https://doi.org/10.3390/vaccines13050526.
4. Sirit IS, Peek RM Jr. Decoding the ability of *H. pylori* to evade immune recognition and cause disease. Cell Mol Gastroenterol Hepatol. 2025;19(5):101470. https://doi.org/10.1016/j.jcmgh.2025.101470.
5. Zeng M, Mao XH, Li JX, Tong WD, Wang B, et al. Efficacy, safety, and immunogenicity of an oral recombinant *H. pylori* vaccine in children in China: a randomised, double-blind, placebo-controlled, phase 3 trial. Lancet. 2015;386(10002):1457–64. https://doi.org/10.1016/S0140-6736(15)60310-5.
6. Paydarnia N, Mansoori B, Esmaeili D, Kazemi T, Aghapour M, et al. *H. pylori* recombinant CagA regulates Th1/Th2 balance in a BALB/c murine model. Adv Pharm Bull. 2020;10(2):264–70. https://doi.org/10.34172/apb.2020.031.
7. Liu M, Zhong Y, Chen J, Liu Y, Tang C, et al. Oral immunization of mice with a multivalent therapeutic subunit vaccine protects against *H. pylori* infection. Vaccine. 2020;38(14):3031–41. https://doi.org/10.1016/j.vaccine.2020.02.036.
8. Malfertheiner P, Selgrad M, Wex T, Romi B, Borgogni E, et al. Efficacy, immunogenicity, and safety of a parenteral vaccine against *H. pylori* in healthy volunteers challenged with a Cag-positive strain: a randomised, placebo-controlled phase 1/2 study. Lancet Gastroenterol Hepatol. 2018;3(10):698–707. https://doi.org/10.1016/s2468-1253(18)30125-0.
9. Pilewski KA, Kramer KJ, Georgiev IS. Simultaneous immunization with multiple diverse Immunogens alters development of antigen-specific antibody-mediated immunity. Vaccines (Basel). 2021;9(9) https://doi.org/10.3390/vaccines9090964.
10. Jebali A, Esmaeilzadeh A, Esmaeilzadeh MK, Shabani S. Immunoinformatics design and synthesis of a multi-epitope vaccine against *H. pylori* based on lipid nanoparticles. Sci Rep. 2024;14(1):17910. https://doi.org/10.1038/s41598-024-68947-x.
11. Ru Z, Yu M, Zhu Y, Chen Z, Zhang F, et al. Immmunoinformatics-based design of a multi-epitope vaccine with CTLA-4 extracellular domain to combat *H. pylori*. FASEB J. 2022;36(4):e22252. https://doi.org/10.1096/fj.202101538RR.
12. Khan M, Khan S, Ali A, Akbar H, Sayaf AM, et al. Immunoinformatics approaches to explore *H. pylori* proteome (virulence factors) to design B and T cell multi-epitope subunit vaccine. Sci Rep. 2019;9(1):13321. https://doi.org/10.1038/s41598-019-49354-z.
13. Pan X, Ke H, Niu X, Li S, Lv J, et al. Protection against *H. pylori* infection in BALB/c mouse model by oral administration of multivalent epitope-based vaccine of cholera toxin B subunit-HUUC. Front Immunol. 2018;9:1003. https://doi.org/10.3389/fimmu.2018.01003.
14. Lv X, Song H, Yang J, Li T, Xi T, et al. A multi-epitope vaccine CTB-UE relieves *H. pylori*-induced gastric inflammatory reaction via up-regulating microRNA-155 to inhibit Th17 response in C57/BL6 mice model. Hum Vaccin Immunother. 2014;10(12):3561–9. https://doi.org/10.4161/hv.36096.
15. Guo L, Hong D, Wang S, Zhang F, Tang F, et al. Therapeutic protection against *H. pylori* infection in Mongolian gerbils by Oral immunization with a tetravalent epitope-based vaccine with polysaccharide adjuvant. Front Immunol. 2019;10:1185. https://doi.org/10.3389/fimmu.2019.01185.
16. Ansari H, Tahmasebi-Birgani M, Bijanzadeh M. DNA vaccine containing Flagellin a gene induces significant immune responses against *H. pylori* infection: an in vivo study. Iran J Basic Med Sci. 2021;24(6):796–804. https://doi.org/10.22038/ijbms.2021.54415.12227.

17. Chehelgerdi M, Doosti A. Effect of the cagW-based gene vaccine on the immunologic properties of BALB/c mouse: an efficient candidate for *H. pylori* DNA vaccine. J Nanobiotechnol. 2020;18(1):63. https://doi.org/10.1186/s12951-020-00618-1.
18. Azami Z, Farahmand M, Kavousi M. A new multi-epitope DNA vaccine against *H. pylori* infection in a BALB/c mouse model. Heliyon. 2024;10(21):e39433. https://doi.org/10.1016/j.heliyon.2024.e39433.
19. Kaveh-Samani A, Dalali S, Kaviani F, Piri-Gharaghie T, Doosti A. Oral administration of DNA alginate nanovaccine induced immune-protection against *H. pylori* in Balb/C mice. BMC Immunol. 2024;25(1):11. https://doi.org/10.1186/s12865-024-00602-6.
20. Katsande PM, Nguyen VD, Nguyen TLP, Nguyen TKC, Mills G, et al. Prophylactic immunization to *H. pylori* infection using spore vectored vaccines. Helicobacter. 2023;28(4):e12997. https://doi.org/10.1111/hel.12997.
21. Wang S, Ma J, Ji Q, Liu Q. Evaluation of an attenuated listeria monocytogenes as a vaccine vector to control *H. pylori* infection. Immunol Lett. 2021;238:68–74. https://doi.org/10.1016/j.imlet.2021.07.010.
22. Nie L, Huang Y, Cheng Z, Luo H, Zhan Y, et al. An intranasal influenza virus vector vaccine protects against *H. pylori* in mice. J Virol. 2024;98(3):e0192323. https://doi.org/10.1128/jvi.01923-23.
23. Ghasemi A, Wang S, Sahay B, Abbott JR, Curtiss R 3rd. Protective immunity enhanced salmonella vaccine vectors delivering *H. pylori* antigens reduce *H. pylori* stomach colonization in mice. Front Immunol. 2022;13:1034683. https://doi.org/10.3389/fimmu.2022.1034683.
24. Zhang F, Ni L, Zhang Z, Luo X, Wang X, et al. Recombinant L. lactis vaccine LL-plSAM-WAE targeting four virulence factors provides mucosal immunity against *H. pylori* infection. Microb Cell Factories. 2024;23(1):61. https://doi.org/10.1186/s12934-024-02321-4.
25. Cen Q, Gao T, Ren Y, Lu X, Lei H. Immune evaluation of a Saccharomyces cerevisiae-based oral vaccine against *H. pylori* in mice. Helicobacter. 2021;26(1):e12772. https://doi.org/10.1111/hel.12772.
26. Ru Z, Yu M, Zhu Y, Ding J, Li J, et al. Construction of a multi-epitope vaccine against *H. pylori* fused with the CTLA-4 extracellular domain and expressed in BCG. Gene. 2025;964:149634. https://doi.org/10.1016/j.gene.2025.149634.
27. Kashyap D, Panda M, Baral B, Varshney N, et al. Outer membrane vesicles: an emerging vaccine platform. Vaccines (Basel). 2022;10(10) https://doi.org/10.3390/vaccines10101578.
28. Bonnington KE, Kuehn MJ. Protein selection and export via outer membrane vesicles. Biochim Biophys Acta. 2014;1843(8):1612–9. https://doi.org/10.1016/j.bbamcr.2013.12.011.
29. Maisonneuve C, Bertholet S, Philpott DJ, De Gregorio E. Unleashing the potential of NOD- and Toll-like agonists as vaccine adjuvants. Proc Natl Acad Sci USA. 2014;111(34):12294–9. https://doi.org/10.1073/pnas.1400478111.
30. Zariri A, van Dijken H, Hamstra HJ, van der Flier M, Vidarsson G, et al. Expression of human CEACAM1 in transgenic mice limits the Opa-specific immune response against meningococcal outer membrane vesicles. Vaccine. 2013;31(47):5585–93. https://doi.org/10.1016/j.vaccine.2013.07.069.
31. Song Z, Li B, Zhang Y, Li R, Ruan H, et al. Outer membrane vesicles of *H. pylori* 7.13 as adjuvants promote protective efficacy against *H. pylori* infection. Front Microbiol. 2020;11:1340. https://doi.org/10.3389/fmicb.2020.01340.
32. Liu Q, Li B, Ma J, Lei X, Ma J, et al. Development of a recombinant outer membrane vesicles (OMVs)-based vaccine against *H pylori* infection in mice. J Extracell Vesicles. 2025;14(5):e70085. https://doi.org/10.1002/jev2.70085.
33. Liu Q, Shang Y, Shen L, Yu X, Cao Y, et al. Outer membrane vesicles from genetically engineered salmonella enterica serovar Typhimurium presenting *H. pylori* antigens UreB and CagA induce protection against *H. pylori* infection in mice. Virulence. 2024;15(1):2367783. https://doi.org/10.1080/21505594.2024.2367783.
34. Kronsteiner B, Bassaganya-Riera J, Philipson C, Viladomiu M, Carbo A, et al. Systems-wide analyses of mucosal immune responses to *H. pylori* at the interface between pathogenicity and symbiosis. Gut Microbes. 2016;7(1):3–21. https://doi.org/10.1080/19490976.2015.1116673.

35. Müller A, Oertli M, Arnold IC. *H. pylori* exploits and manipulates innate and adaptive immune cell signaling pathways to establish persistent infection. Cell Commun Signal. 2011;9(1):25. https://doi.org/10.1186/1478-811x-9-25.
36. Cheok YY, Tan GMY, Lee CYQ, Abdullah S, Looi CY, et al. Innate immunity crosstalk with *H. pylori*: pattern recognition receptors and cellular responses. Int J Mol Sci. 2022;23(14) https://doi.org/10.3390/ijms23147561.
37. Matsuura M. Structural modifications of bacterial lipopolysaccharide that facilitate gram-negative bacteria evasion of host innate immunity. Front Immunol. 2013;4:109. https://doi.org/10.3389/fimmu.2013.00109.
38. Nogueira AM, Marques T, Soares PC, David L, Reis CA, et al. Lewis antigen expression in gastric mucosa of children: relationship with *H. pylori* infection. J Pediatr Gastroenterol Nutr. 2004;38(1):85–91. https://doi.org/10.1097/00005176-200401000-00019.
39. den Dunnen J, Gringhuis SI, Geijtenbeek TB. Innate signaling by the C-type lectin DC-SIGN dictates immune responses. Cancer Immunol Immunother. 2009;58(7):1149–57. https://doi.org/10.1007/s00262-008-0615-1.
40. Gewirtz AT, Yu Y, Krishna US, Israel DA, Lyons SL, et al. *H. pylori* flagellin evades toll-like receptor 5-mediated innate immunity. J Infect Dis. 2004;189(10):1914–20. https://doi.org/10.1086/386289.
41. Andersen-Nissen E, Smith KD, Strobe KL, Barrett SL, Cookson BT, et al. Evasion of toll-like receptor 5 by flagellated bacteria. Proc Natl Acad Sci USA. 2005;102(26):9247–52. https://doi.org/10.1073/pnas.0502040102.
42. Wen SH, Hong ZW, Chen CC, Chang HW, Fu HW. *H. pylori* neutrophil-activating protein directly interacts with and activates toll-like receptor 2 to induce the secretion of Interleukin-8 from neutrophils and ATRA-induced differentiated HL-60 cells. Int J Mol Sci. 2021;22(21) https://doi.org/10.3390/ijms222111560.
43. Uno K, Kato K, Atsumi T, Suzuki T, Yoshitake J, et al. Toll-like receptor (TLR) 2 induced through TLR4 signaling initiated by *H. pylori* cooperatively amplifies iNOS induction in gastric epithelial cells. Am J Physiol Gastrointest Liver Physiol. 2007;293(5):G1004–12. https://doi.org/10.1152/ajpgi.00096.2007.
44. Sun X, Zhang M, El-Zataari M, Owyang SY, Eaton KA, et al. TLR2 mediates *H. pylori*-induced tolerogenic immune response in mice. PLoS One. 2013;8(9):e74595. https://doi.org/10.1371/journal.pone.0074595.
45. Dooyema SDR, Noto JM, Wroblewski LE, Piazuelo MB, Krishna U, et al. *H. pylori* actively suppresses innate immune nucleic acid receptors. Gut Microbes. 2022;14(1):2105102. https://doi.org/10.1080/19490976.2022.2105102.
46. Suarez G, Romero-Gallo J, Piazuelo MB, Wang G, Maier RJ, et al. Modification of *H. pylori* peptidoglycan enhances NOD1 activation and promotes cancer of the stomach. Cancer Res. 2015;75(8):1749–59. https://doi.org/10.1158/0008-5472.Can-14-2291.
47. Sun M, Liu Y, Ni X, Tan R, Wang Y, et al. Intranasal immunization with poly I:C and CpG ODN adjuvants enhances the protective efficacy against *H. pylori* infection in mice. Microbes Infect. 2025;27(3):105433. https://doi.org/10.1016/j.micinf.2024.105433.
48. Rynkiewicz D, Rathkopf M, Sim I, Waytes AT, Hopkins RJ, et al. Marked enhancement of the immune response to BioThrax® (anthrax vaccine adsorbed) by the TLR9 agonist CPG 7909 in healthy volunteers. Vaccine. 2011;29(37):6313–20. https://doi.org/10.1016/j.vaccine.2011.05.047.
49. Cooper CL, Davis HL, Angel JB, Morris ML, Elfer SM, et al. CPG 7909 adjuvant improves hepatitis B virus vaccine seroprotection in antiretroviral-treated HIV-infected adults. AIDS. 2005;19(14):1473–9. https://doi.org/10.1097/01.aids.0000183514.37513.d2.
50. Lu W, Cui C, Wang Y, Sun X, Wang S, et al. CpG ODN as an adjuvant arouses the vigor of B cells by relieving the negative regulation of surface TLR9 to enhance the antibody response to vaccine. Appl Microbiol Biotechnol. 2021;105(10):4213–24. https://doi.org/10.1007/s00253-021-11316-9.

51. Heineman TC, Cunningham A, Levin M. Understanding the immunology of Shingrix, a recombinant glycoprotein E adjuvanted herpes zoster vaccine. Curr Opin Immunol. 2019;59:42–8. https://doi.org/10.1016/j.coi.2019.02.009.
52. Zhang X, He Y, Zhang X, Fu B, Song Z, et al. Sustained exposure to *H. pylori* induces immune tolerance by desensitizing TLR6. Gastric Cancer. 2024;27(2):324–42. https://doi.org/10.1007/s10120-023-01461-7.
53. Malanovic N, Lohner K. Antimicrobial peptides targeting gram-positive bacteria. Pharmaceuticals (Basel). 2016;9(3) https://doi.org/10.3390/ph9030059.
54. Hancock REW, Alford MA, Haney EF. Antibiofilm activity of host defence peptides: complexity provides opportunities. Nat Rev Microbiol. 2021;19(12):786–97. https://doi.org/10.1038/s41579-021-00585-w.
55. Windham IH, Servetas SL, Whitmire JM, Pletzer D, Hancock REW, et al. *H. pylori* biofilm formation is differentially affected by common culture conditions, and proteins play a central role in the biofilm matrix. Appl Environ Microbiol. 2018;84(14) https://doi.org/10.1128/aem.00391-18.
56. Hancock RE, Haney EF, Gill EE. The immunology of host defence peptides: beyond antimicrobial activity. Nat Rev Immunol. 2016;16(5):321–34. https://doi.org/10.1038/nri.2016.29.
57. Svensson D, Nilsson BO. Human antimicrobial/host defense peptide LL-37 may prevent the spread of a local infection through multiple mechanisms: an update. Inflamm Res. 2025;74(1):36. https://doi.org/10.1007/s00011-025-02005-8.
58. Pero R, Coretti L, Nigro E, Lembo F, Laneri S, et al. β-Defensins in the fight against *H. pylori*. Molecules. 2017;22(3) https://doi.org/10.3390/molecules22030424.
59. Hase K, Murakami M, Iimura M, Cole SP, Horibe Y, et al. Expression of LL-37 by human gastric epithelial cells as a potential host defense mechanism against *H. pylori*. Gastroenterology. 2003;125(6):1613–25. https://doi.org/10.1053/j.gastro.2003.08.028.
60. Nuding S, Gersemann M, Hosaka Y, Konietzny S, Schaefer C, et al. Gastric antimicrobial peptides fail to eradicate *H. pylori* infection due to selective induction and resistance. PLoS One. 2013;8(9):e73867. https://doi.org/10.1371/journal.pone.0073867.
61. Patel SR, Smith K, Letley DP, Cook KW, Memon AA, et al. *H. pylori* downregulates expression of human β-defensin 1 in the gastric mucosa in a type IV secretion-dependent fashion. Cell Microbiol. 2013;15(12):2080–92. https://doi.org/10.1111/cmi.12174.
62. Bauer B, Pang E, Holland C, Kessler M, Bartfeld S, et al. The *H. pylori* virulence effector CagA abrogates human β-defensin 3 expression via inactivation of EGFR signaling. Cell Host Microbe. 2012;11(6):576–86. https://doi.org/10.1016/j.chom.2012.04.013.
63. Olleik H, Baydoun E, Perrier J, Hijazi A, Raymond J, et al. Temporin-SHa and its analogs as potential candidates for the treatment of *H. pylori*. Biomolecules. 2019;9(10) https://doi.org/10.3390/biom9100598.
64. Luong HX, Thanh TT, Tran TH. Antimicrobial peptides – advances in development of therapeutic applications. Life Sci. 2020;260:118407. https://doi.org/10.1016/j.lfs.2020.118407.
65. Li X, Zuo S, Wang B, Zhang K, Wang Y. Antimicrobial mechanisms and clinical application prospects of antimicrobial peptides. Molecules. 2022;27(9) https://doi.org/10.3390/molecules27092675.
66. Zhang X, Jiang A, Wang G, Yu H, Qi B, et al. Fusion expression of the PGLa-AM1 with native structure and evaluation of its anti-*H. pylori* activity. Appl Microbiol Biotechnol. 2017;101(14):5667–75. https://doi.org/10.1007/s00253-017-8302-9.
67. Jiang M, Ma L, Huang Y, Wu H, Dou J, et al. Antimicrobial activities of peptide Cbf-K(16) against drug-resistant *H. pylori* infection in vitro and in vivo. Microb Pathog. 2020;138:103847. https://doi.org/10.1016/j.micpath.2019.103847.
68. Wendler J, Schroeder BO, Ehmann D, Koeninger L, Mailänder-Sánchez D, et al. Proteolytic degradation of reduced human Beta Defensin 1 generates a novel antibiotic Octapeptide. Sci Rep. 2019;9(1):3640. https://doi.org/10.1038/s41598-019-40216-2.
69. Xiong M, Bao Y, Xu X, Wang H, Han Z, et al. Selective killing of *H. pylori* with pH-responsive helix-coil conformation transitionable antimicrobial polypeptides. Proc Natl Acad Sci USA. 2017;114(48):12675–80. https://doi.org/10.1073/pnas.1710408114.

70. Xie J, Zhou M, Cong Z, Xiao X, Liu L, et al. A host defense peptide–mimicking prodrug activated by drug-resistant gram-negative bacterial infections. Sci Transl Med. 2025;17(801):eadl4870.
71. Fonseca DR, Alves PM, Neto E, Custódio B, Guimarães S, et al. One-pot microfluidics to engineer chitosan nanoparticles conjugated with antimicrobial peptides using "Photoclick" chemistry: validation using the gastric bacterium *H. pylori*. ACS Appl Mater Interfaces. 2024;16(12):14533–47. https://doi.org/10.1021/acsami.3c18772.
72. Zhang W, Yang Z, Zheng J, Fu K, Wong JH, et al. A bioresponsive genetically encoded antimicrobial crystal for the oral treatment of *H. pylori* infection. Adv Sci (Weinh). 2023;10(30):e2301724. https://doi.org/10.1002/advs.202301724.
73. Zhang L, Yu J, Wong CC, Ling TK, Li ZJ, et al. Cathelicidin protects against *H. pylori* colonization and the associated gastritis in mice. Gene Ther. 2013;20(7):751–60. https://doi.org/10.1038/gt.2012.92.
74. Torres MDT, Brooks EF, Cesaro A, Sberro H, Gill MO, et al. Mining human microbiomes reveals an untapped source of peptide antibiotics. Cell. 2024;187(19):5453–67.e15. https://doi.org/10.1016/j.cell.2024.07.027.
75. Czerucka D, Rampal P. Diversity of saccharomyces boulardii CNCM I-745 mechanisms of action against intestinal infections. World J Gastroenterol. 2019;25(18):2188–203. https://doi.org/10.3748/wjg.v25.i18.2188.
76. Wu S, Xu Y, Chen Z, Chen Y, Wei F, et al. Lactiplantibacillus plantarum ZJ316 reduces *H. pylori* adhesion and inflammation by inhibiting the expression of Adhesin and urease genes. Mol Nutr Food Res. 2023;67(18):e2300241. https://doi.org/10.1002/mnfr.202300241.
77. Boyanova L, Gergova G, Markovska R, Yordanov D, Mitov I. Bacteriocin-like inhibitory activities of seven lactobacillus delbrueckii subsp. bulgaricus strains against antibiotic susceptible and resistant *H. pylori* strains. Lett Appl Microbiol. 2017;65(6):469–74. https://doi.org/10.1111/lam.12807.
78. Huang R, Wu F, Zhou Q, Wei W, Yue J, et al. Lactobacillus and intestinal diseases: mechanisms of action and clinical applications. Microbiol Res. 2022;260:127019. https://doi.org/10.1016/j.micres.2022.127019.
79. Ji J, Yang H. Using probiotics as supplementation for *H. pylori* antibiotic therapy. Int J Mol Sci 2020;21(3). DOI:https://doi.org/10.3390/ijms21031136.
80. Li J, Peng F, Huang H, Xu X, Guan Q, et al. Characterization, mechanism and in vivo validation of *H. pylori* antagonism by probiotics screened from infants' feces and oral cavity. Food Funct. 2024;15(3):1170–90. https://doi.org/10.1039/d3fo04592g.
81. Holz C, Busjahn A, Mehling H, Arya S, Boettner M, et al. Significant reduction in *H. pylori* load in humans with non-viable lactobacillus reuteri DSM17648: a pilot study. Probiotics Antimicrob Proteins. 2015;7(2):91–100. https://doi.org/10.1007/s12602-014-9181-3.
82. van Zyl WF, Deane SM, Dicks LMT. Molecular insights into probiotic mechanisms of action employed against intestinal pathogenic bacteria. Gut Microbes. 2020;12(1):1831339. https://doi.org/10.1080/19490976.2020.1831339.
83. Qureshi N, Li P, Gu Q. Probiotic therapy in *H. pylori* infection: a potential strategy against a serious pathogen? Appl Microbiol Biotechnol. 2019;103(4):1573–88. https://doi.org/10.1007/s00253-018-09580-3.
84. Yang H, Lin Y, Ma Y, Li J, Li J, et al. Screening probiotics for anti-*H. pylori* and investigating the effect of probiotics on patients with *H. pylori* infection. Foods. 2024;13(12) https://doi.org/10.3390/foods13121851.
85. Sanders ME, Merenstein DJ, Reid G, Gibson GR, Rastall RA. Probiotics and prebiotics in intestinal health and disease: from biology to the clinic. Nat Rev Gastroenterol Hepatol. 2019;16(10):605–16. https://doi.org/10.1038/s41575-019-0173-3.
86. Forooghi Nia F, Rahmati A, Ariamanesh M, Saeidi J, Ghasemi A, et al. The anti-*H. pylori* effects of Limosilactobacillus reuteri strain 2892 isolated from camel milk in C57BL/6 mice. World J Microbiol Biotechnol. 2023;39(5):119. https://doi.org/10.1007/s11274-023-03555-x.
87. Talebi G, Nabavi-Rad A, Sadeghloo Z, Doulberis M, Zali MR, et al. Inhibitory effects of lactobacillus reuteri strain I300 against *H. pylori* adhesion, invasion, and inflammatory response

in gastric epithelial cells in vitro. Folia Microbiol (Praha). 2025; https://doi.org/10.1007/s12223-025-01263-7.
88. Konieczna P, Groeger D, Ziegler M, Frei R, Ferstl R, et al. Bifidobacterium infantis 35624 administration induces Foxp3 T regulatory cells in human peripheral blood: potential role for myeloid and plasmacytoid dendritic cells. Gut. 2012;61(3):354–66. https://doi.org/10.1136/gutjnl-2011-300936.
89. He C, Peng C, Xu X, Li N, Ouyang Y, et al. Probiotics mitigate *H. pylori*-induced gastric inflammation and premalignant lesions in INS-GAS mice with the modulation of gastrointestinal microbiota. Helicobacter. 2022;27(4):e12898. https://doi.org/10.1111/hel.12898.
90. Shen S, Ren F, Qin H, Bukhari I, Yang J, et al. Lactobacillus acidophilus NCFM and Lactiplantibacillus plantarum Lp-115 inhibit *H. pylori* colonization and gastric inflammation in a murine model. Front Cell Infect Microbiol. 2023;13:1196084. https://doi.org/10.3389/fcimb.2023.1196084.
91. Baryshnikova NV, Ilina AS, Ermolenko EI, Uspenskiy YP, Suvorov AN. Probiotics and autoprobiotics for treatment of *H. pylori* infection. World J Clin Cases. 2023;11(20):4740–51. https://doi.org/10.12998/wjcc.v11.i20.4740.

# Clinical Research and Challenges of Immunotherapy

# 15

Verena Friedrich, Dietmar Pfeiffer, Paolo Palma,
Pierre van Damme, and Markus Gerhard

**Executive Summary**

*Helicobacter pylori* causes substantial gastroduodenal morbidity, yet rising antimicrobial resistance, treatment intolerance, and reinfection highlight the need for non-antibiotic strategies. This chapter evaluates immunotherapeutic approaches, active and passive, with an emphasis on disease-blocking rather than sterilizing immunity (Fig. 15.1).

The immunopathogenesis of *H. pylori* is characterized by a Th1/Th17-skewed response coexisting with regulatory pathways, epithelial checkpoint induction, and intimate adhesion with effector translocation. These features support host-directed interventions that attenuate tissue injury while maintaining bacterial containment. We consider (i) active immunization strategies that target virulence determinants (e.g., urease, VacA, NAP, adhesins) and are developed with mucosal safety constraints; (ii) passive immunotherapy using pathogen-specific antibodies to neutralize urease and toxins and to inhibit adhesion/effector delivery; (iii) microbiome-modulating adjuncts; and (iv) innate-mimetic bioactives.

V. Friedrich · D. Pfeiffer
School of Medicine and Health, Institute for Medical Microbiology, Immunology and Hygiene, Technical University of Munich (TUM), Munich, Germany

P. Palma · P. van Damme
Centre for the Evaluation of Vaccination (CEV), Vaccine and Infectious Disease Institute, Faculty of Medicine and Health Sciences, Antwerp, Belgium

M. Gerhard (✉)
TUM School of Medicine and Health, Department Preclinical Medicine, Institute of Medical Microbiology, Immunology and Hygiene, Technical University of Munich, Munich, Germany

German Center for Infection Research (DZIF), Partner Site Munich, Munich, Germany
e-mail: markus.gerhard@tum.de



J. Gao, Y. Du (eds.), *Non-Antibiotic Therapies for Helicobacter pylori*,
https://doi.org/10.1007/978-981-95-9598-3_15

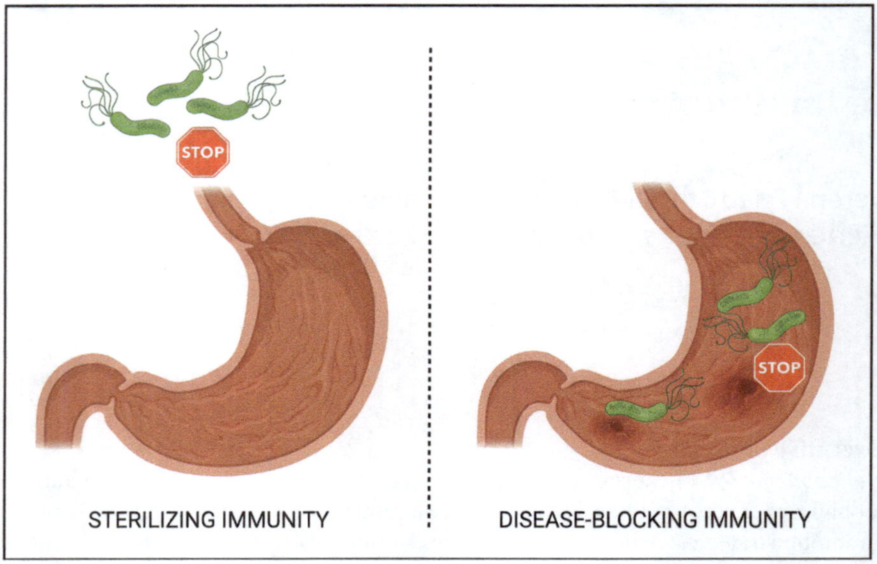


**Fig. 15.1** Schematic concept of sterilizing versus disease-blocking immunity. (Created using BioRender.com, used with permission)

In contrast, checkpoint pathway modulation (PD-1/PD-L1) is discussed only as a mechanistic concept and should be regarded as highly experimental, given persistent gastric safety concerns, rather than as a therapeutic option equivalent to vaccines or antibodies.

Passive antibodies require careful selection of route: oral delivery offers local action and excellent safety but needs protection from gastric degradation, whereas systemic delivery may access glandular niches more reliably yet requires gastric pharmacokinetics (PK)/pharmacodynamics (PD) confirmation and systemic safety monitoring. Furthermore, passive immunization may provide only temporary protection, and its effects may wane over time, requiring repeated administration based on clinical indications.

Translational recommendations prioritize biopsy-based pharmacodynamics and pathology-centered efficacy endpoints. Early-phase trials should, in addition to safety endpoints , employ blinded histologic scores, peptic-ulcer healing metrics, and validated symptom indices as primary outcomes, with urea breath test (UBT) and stool antigen as secondary endpoints. Safety monitoring must include scheduled endoscopy with targeted biopsies and predefined rules for halting in the event of worsening gastritis or ulceration. Adaptive designs are suitable for optimizing dose, route, and adjuvant/formulation.

Key challenges include antigenic/strain diversity, mucosal adjuvant tolerability, antibody stability and delivery, and balancing bacterial pressure with avoidance of excessive inflammation. Mitigation strategies comprise multivalent or conserved-epitope antigen design (including rational and computational selection), staged escalation of safer mucosal adjuvants, microencapsulation/enteric protection or systemic administration to ensure gastric exposure, and combination regimens (e.g., immunotherapy with proton-pump inhibitor (PPI) or bismuth; coadministration with probiotics).

The appendix provides tables summarizing antigens, passive formats, adjuvants, and trial designs; a schematic contrasting sterilizing versus disease-blocking aims; and a practical checklist to support protocol development. Collectively, these elements outline clinically and regulatorily consistent pathways for developing immunotherapies that target *H. pylori* pathogenesis while maintaining gastric safety.

## 15.1 Introduction: Clinical Need and Scope

*H. pylori* remains one of the world's most consequential chronic infections, driving peptic ulcer disease, gastric mucosa–associated lymphoid tissue (MALT) lymphoma, and gastric adenocarcinoma. Standard care still relies on multidrug antibiotic regimens combined with acid suppression. Yet clinical experience increasingly confronts the limits of this approach: antibiotic resistance (notably to clarithromycin, metronidazole, and fluoroquinolones) undermines efficacy across regions; tolerability and adherence suffer under complex dosing and gastrointestinal (GI) adverse effects. Furthermore, reinfection/recrudescence persists in certain populations despite initial clearance [1–3]. These constraints motivate exploration of non-antibiotic strategies, with a particular focus on immunotherapeutics that can complement or, in select contexts, substitute for antibiotics while supporting antimicrobial stewardship.

In this chapter, we intentionally focus on immunotherapeutic strategies based on a host–pathogen immunobiology perspective. Immunotherapy refers exclusively to immunologically mediated approaches, including active vaccination and passive or immunomodulatory strategies. Non-immunological anti-virulence or small-molecule approaches are addressed in other chapters and are not within the scope here. Active immunotherapy here refers to immune-mediated strategies that aim to induce or amplify antigen-specific protective responses; in the *H. pylori* field, this historically includes prophylactic or therapeutic vaccination concepts, though durable success in humans remains unproven [4–6]. Passive immunotherapy refers to the exogenous provision of functional antibodies (e.g., orally delivered egg yolk antibody IgY or bovine antibodies) designed to neutralize key bacterial factors directly within the gastric lumen, thereby offering a non-absorbed, reiterable intervention with a favorable safety profile.

A second definitional axis distinguishes prophylactic from therapeutic intent. Prophylactic approaches seek to prevent establishment of infection or disease in at-risk individuals, whereas therapeutic approaches intervene in established infection to reduce bacterial burden and, crucially, to blunt the pathways that connect colonization to disease. This leads to the core endpoint distinction of the chapter: sterilizing versus non-sterilizing outcomes. Several terms have been used to describe the non-sterilizing concept, including pathology-centered immunity, non-sterilizing protection, or pathogenesis/disease-blocking immunity; we will use the term disease-blocking immunity throughout this chapter. Sterilizing endpoints demand complete eradication. In contrast, disease-blocking endpoints accept that organisms may persist at some level but prioritize clinically meaningful benefits: reduction in histologic gastritis activity, ulcer healing, mitigation of symptoms, prevention of progression to atrophy or intestinal metaplasia, and lower relapse risk (Fig. 15.1). Given the immunopathogenesis of *H. pylori*, where chronic, dysregulated responses contribute as much to tissue injury as the pathogen itself, disease-blocking strategies can be not only adequate but preferable in terms of benefit–risk profile [4, 5]. Accordingly, the sections that follow focus on passive oral antibodies, blockade of defined adhesion/effector interfaces, mucosal tone modulation (including microbiota-directed adjuncts and innate mimetics), and targeted delivery technologies, while briefly treating vaccination for a mechanistic context. This approach aligns therapeutic goals with realistic biology, clinical feasibility, and antimicrobial stewardship in the era of rising resistance [1, 3].

## 15.2 Immunopathogenesis of *H. pylori* Relevant to Immunotherapeutic Strategies

### 15.2.1 Host–Pathogen Interactions and Chronicity

At the gastric surface, *H. pylori* meets a formidable but tightly regulated mucosal immune system. Pattern recognition by epithelial cells and innate leukocytes (e.g., via TLRs and NOD-like receptors) triggers chemokine release and recruitment of neutrophils and macrophages, establishing chronic active gastritis. Secretory IgA, and, to a lesser extent, luminally transuded IgG, bind bacteria and virulence factors, but neither prevents long-term carriage in most hosts, such as *Neisseria meningitidis* (nasopharyngeal carriage despite mucosal antibody presence [7]) or *Streptococcus pneumoniae* (persistent colonization in the nasopharynx even with sIgA responses [8]). The adaptive response is dominated by $CD4^+$ T cells with a characteristic Th1/Th17 skew that coexists with expanded regulatory T cells (Treg). This paradox of robust inflammation, without sterilizing clearance, helps explain why tissue injury (epithelial damage, atrophy, ulceration) can progress despite immune activation [9, 10]. In brief, Th1/Th17 effectors promote bacteriostatic pressure and pathology, whereas Tregs constrain excessive damage at the cost of permitting persistence. Evidence from primary immunodeficiency supports a direct link between discrete immune compartments and *H. pylori* outcome in humans. The clearest associations are observed in defects affecting the Th17 axis and mucosal B-cell/IgA immunity. STAT3 loss-of-function in autosomal dominant Hyper-IgE syndrome impairs Th17 differentiation and epithelial IL-17 signaling, and these patients display persistent *H. pylori* colonization with attenuated gastric responses, demonstrating that Th17 cells are necessary for mucosal containment rather than eradication [11, 12]. Likewise, common variable immunodeficiency, and particularly IgA deficiency within this spectrum, is associated with high *H. pylori* prevalence, nodular lymphoid hyperplasia and increased gastric carcinoma risk, indicating that mucosal antibodies limit epithelial remodeling and neoplastic progression rather than bacterial clearance [13]. In contrast, regulatory T-cell failure in FOXP3-deficient IPEX syndrome leads to severe destructive gastritis despite bacterial persistence, demonstrating that Treg restraint protects the mucosa but is not sufficient for sterilization [14]. From a therapeutic view, this suggests a requirement for precise rebalancing rather than global immune activation: the selective neutralization of pathogenic bacterial functions and local tuning of the Th1/Th17 ↔ Treg axis to reduce disease without requiring complete bacterial eradication [9, 10]. Collectively, these "experiments of nature" show that persistence reflects a regulated equilibrium: Th17 constrains expansion, Th1 drives pathology, B-cell/IgA pathways influence long-term mucosal remodeling, and Tregs prevent sterilizing-by-destruction inflammation.

### 15.2.2 Virulence Determinants That Mediate Colonization and Pathology

Several bacterial factors are pivotal for colonization, niche adaptation, and tissue injury—each offering nonantibiotic targets to interrupt pathogenesis:

- Urease buffers periplasmic pH by hydrolyzing urea into ammonia and $CO_2$, enabling motility and survival in acidic mucus. Neutralizing urease can hinder initial colonization and attenuate mucosal injury without necessarily sterilizing the stomach [15].
- Adhesins (e.g., blood group antigen–binding adhesin (BabA), sialic acid-binding adhesin (SabA), adhesion-like protein A/B (AlpA/B), *H. pylori* adhesin A (HpaA)) secure intimate attachment to fucosylated or sialylated glycans and to laminin/integrins, stabilizing the bacteria–epithelium interface and promoting downstream signaling. Blocking these interactions reduces adhesion, inflammation, and effector delivery [16, 17].
- Vacuolating cytotoxin A (VacA) disrupts endolysosomal trafficking, induces epithelial vacuolation, and modulates immune cells (e.g., T-cell suppression), thereby amplifying persistence and injury. Neutralization of VacA can blunt tissue damage even if organisms persist [15].
- Cytotoxin-associated gene A (CagA) (delivered by the cag type IV secretion system) perturbs host signaling (e.g., SHP2, β-catenin), drives pro-inflammatory cascades, and contributes to carcinogenic evolution; limiting CagA translocation or activity can decouple colonization from oncogenic/inflammatory consequences [15, 17].
- Neutrophil-activating protein (NAP) recruits and activates neutrophils, shaping a Th1-biased milieu. Modulating NAP's effects may reduce excessive neutrophil-mediated damage without necessarily lowering total bacterial counts [16].

Collectively, these determinants highlight receptor–ligand and toxin-neutralization strategies, via passive antibodies, decoy receptors, or small biologics, to block pathogenesis at the mucosal interface. Additional adhesin–receptor pairs not detailed here offer similar opportunities for disease-focused targeting.

### 15.2.3 Mechanistic Rationale for "Disease-Blocking" Versus "Sterilizing" Endpoints

Because *H. pylori* disease results from the interplay of persistent colonization and specific virulence-driven host responses, immunotherapies can aim to uncouple colonization from clinical harm. Neutralizing urease may impair mucosal fitness, blocking adhesins may limit intimate contact and downstream effector delivery, inhibiting VacA or CagA may dampen epithelial injury and oncogenic/inflammatory signaling, and modulating NAP may reduce neutrophil-driven tissue damage. Each of these can lower gastritis severity, ulcer risk, and progression to atrophy/metaplasia without strictly eliminating every bacterium. Consequently, primary endpoints for such approaches should prioritize pathology and patient benefit, validated histologic scores, ulcer healing, and symptom burden, while bacterial load (e.g., breath test, antigen assays) serves as a supportive or secondary outcome. This aligns with the mucosal immunology of the infection and acknowledges the risks of indiscriminate immune activation in a delicate tissue.

### 15.2.4 Immune Correlates of Protection in Animal Models and Limits in Humans

Experimental models have identified correlates such as mucosal IgA, Th1/Th17 effector responses, and reduced bacterial density following targeted immunization or passive antibody strategies in animals [5, 18–20]. These studies show that enhancing antigen-specific responses or blocking key virulence functions can curb pathology and reduce colonization. However, translation to humans is constrained by species differences in gastric physiology and glycan expression, variance in *H. pylori* strain virulence, and the human tendency toward long-lived regulatory circuits that blunt sterilizing immunity [9, 20]. Recent work has now identified gastric tissue-resident memory T cells (TRM) as a bona fide correlate of sterilizing protection, with CagA-dependent induction of Hobit$^{+}$ TRM, providing complete protection from reinfection in vivo [21], indicating that protection is governed by mucosal positioning of memory rather than amplification of systemic inflammation. In parallel, advanced human in vitro systems, such as gastric epithelial organoids [22, 23], air–liquid interface cultures [24], and epithelial–immune co-culture models [25, 26], enable mechanistic dissection of TRM licensing signals, epithelial cytokine cues, and virulence-factor imprinting in a physiologically relevant architecture. These systems preserve epithelial polarity, mucin organization, and host glycosylation patterns that determine *H. pylori* niche behavior and allow analysis of epithelial cytokine programs such as IL-17-driven antimicrobial defense and IFN-γ–induced stress signaling. Thus, while animal data validate mechanistic targets and disease-blocking concepts, human trials should emphasize safety and mucosal pharmacodynamics (e.g., epithelial cytokine signatures, regulated effector activity) and consider clinical benefit without full eradication as a legitimate and meaningful outcome [5, 19].

## 15.3 Conceptual Approach: Therapeutic Immunization to Block Disease

### 15.3.1 Definition and Scope: What "Blocking Disease" Means Operationally

Therapeutic immunization for *H. pylori* may require defining success in terms of clinically meaningful attenuation of disease, rather than compulsory eradication. Operationally, "disease blockade" denotes reductions in gastric inflammation, accelerated peptic ulcer healing, lower risk of progression to atrophy or intestinal metaplasia, and improvement in symptoms, even if bacteria persist at low or fluctuating levels. This shift aligns with the immunobiology of *H. pylori*, in which chronic, dysregulated host responses, rather than sheer bacterial biomass, drive tissue damage. As such, disease-blocking benefit can be both realistic and preferable when it minimizes mucosal injury without incurring the ecological and tolerability costs of repeated broad-spectrum antibiotics [4, 5, 27]. In practice, pathogenesis-blocking strategies prioritize local mechanisms (luminal neutralization, adhesion interference, effector inhibition, and calibrated immune modulation) that reduce the harm of the bacterium–host interaction, rather than demanding complete microbial elimination.

### 15.3.2 Mechanistic Routes to Block Pathogenesis by Immunization

Multiple immunologic routes can uncouple colonization from clinical harm, offering targets for therapeutic vaccination and, importantly, passive immunotherapy (oral antibodies), which can be administered safely and combined with standard care.

#### 15.3.2.1 Neutralization of Secreted Toxins (VacA)

VacA perturbs epithelial endolysosomal trafficking, induces vacuolation, and suppresses T-cell functions, potentiating persistence and epithelial injury. Antigen-specific responses that neutralize VacA, or passive antibodies that bind VacA in the gastric lumen, can mitigate cytotoxicity and immune dysregulation, thereby reducing gastritis severity without requiring complete clearance [15, 17].

#### 15.3.2.2 Inhibition of Adhesins to Reduce Epithelial Interaction

Adhesins such as BabA, SabA, AlpA/B, and HpaA mediate tight attachment to fucosylated/sialylated glycans and extracellular matrix, anchoring the bacterium for nutrient access and effector delivery. Immunization (active or passive) that blocks these interfaces is expected to shorten epithelial dwell time, limit microenvironmental signaling, and reduce downstream inflammation, again focusing on disease endpoints rather than sterilization [16, 17].

#### 15.3.2.3 Inhibition of Urease to Impair Acid Adaptation

Urease plays a central role in acid resistance and motility through the mucus layer. Neutralizing antibodies can decrease periplasmic buffering and stress the organism in situ, attenuating colonization fitness and epithelial irritation without necessarily eliminating all organisms [15].

#### 15.3.2.4 Opsonization to Reduce Local Inflammatory Signaling

At the mucosal surface, enhanced opsonization can facilitate bacterial aggregation, impede intimate epithelial contact, and modulate innate receptor engagement. While classical Fc-mediated clearance is constrained in the gastric lumen, opsonization can still modulate inflammatory cues and lessen tissue damage [5].

#### 15.3.2.5 Immune Modulation to Attenuate Tissue-Damaging Responses While Preserving Containment

Because Th1/Th17 effectors coexist with regulatory circuits, carefully calibrated immunization may rebalance the Th17↔Treg axis toward controlled containment rather than hyperinflammation. Here, formulations and routes matter: mucosa-targeted strategies that avoid excessive bystander activation are favored to prevent iatrogenic gastritis [9, 10].

Together, these mechanisms define a set of mechanistic approaches for pathogenesis blockade, in which therapeutic vaccination is one option, but passive, luminal antibodies are often a practical initial approach, given their safety, reversibility, and compatibility with reduced antibiotic exposure regimens.

### 15.3.3 Evidence Supporting Disease-Blocking Immunity

Across models (mouse, gerbil, ferret, dog), experimental immunization has repeatedly shown discordance between bacterial counts and disease severity: pathology improves even when culture or PCR detects residual organisms. Early work demonstrated that antigen-specific mucosal responses correlate with reduced gastritis and ulceration, with variable effects on colonization density [5, 18, 19]. In some settings, targeting selected virulence factors (e.g., urease subunits, adhesins, and VacA fragments) yielded histologic benefit and lower inflammatory scores independent of complete clearance, supporting the thesis that neutralizing key functions can decouple disease from persistent low-level carriage [15, 17]. Notably, protection in animals often involves composite immune profiles (mucosal IgA with contributory

Th1/Th17 signatures), but excessive Th1/Th17 activation can aggravate tissue injury, an important caution for translation [9, 10]. Collectively, the animal literature substantiates disease-blocking benefit as a coherent goal, while underscoring that dose, route, and antigen choice must be tuned to avoid over-inflammation [5, 18–20].

### 15.3.4 Implications for Clinical Endpoints and Regulatory Discussions

If the therapeutic intention is to block disease, primary outcomes in human studies should reflect patient-relevant gastric health, such as reduction of gastritis, ulcer recurrence, or progression toward atrophy and cancer, rather than binary eradication.

Given the risk of postimmunization gastritis with mucosal activation, immune biomarkers will serve as exploratory mechanistic readouts, while pathology-first endpoints will govern dose selection and success.

While CagA-imprinted Hobit gastric TRM correlates with sterilizing protection in vivo, its induction may not be necessary or sufficient for pathology mitigation and will not be used as a gating criterion for dose escalation or clinical success in therapeutic trials.

We propose a hierarchy:

Primary endpoints (pathology-focused):

- Histologic gastritis activity using validated scoring systems (neutrophils, mononuclear infiltrate, atrophy, intestinal metaplasia).
- Peptic ulcer outcomes (healing rates and time to healing on endoscopy, time to recurrence).
- Patient-reported symptom scales (pain, dyspepsia indices) with predefined minimal clinically important differences.
- Biomarkers of mucosal injury/repair, such as serum pepsinogens, and gastric pharmacodynamic readouts (e.g., epithelial cytokine signatures), when feasible in early-phase studies.

Key secondary endpoints (microbiologic/supportive):

- Bacterial load metrics (UBT) delta over baseline, stool antigen indices, qPCR copy number in biopsies to contextualize clinical benefit.
- Relapse/recrudescence rates over 6–12 months to determine efficacy in a disease-blocking approach.

Exploratory immune biomarker: quantify CagA-imprinted Hobit$^+$ gastric TRM to contextualize mechanism; primary pharmacodynamic success is defined by blinded histologic improvement and epithelial safety signatures under a disease-blocking, pathology-first framework.

Safety surveillance is central, with prespecified monitoring for postimmunization gastritis, upper GI bleeding/ulcer complications, and dyspeptic adverse events, especially when strategies might shift Th1/Th17 tone [9, 10]. From a regulatory perspective, developers should justify disease-blocking endpoints by linking them to recognized clinical benefits (ulcer healing, histologic improvement, symptom relief) and by demonstrating biological plausibility through targeted mechanisms (e.g., neutralization of VacA; interference with adhesins; urease inhibition) [15–17]. Importantly, trial designs can adopt a combination-first stance, adding passive antibodies or therapeutic immunization to standard care at reduced antibiotic exposure, to show reduction in antibiotic exposure value while de-risking safety and preserving feasibility [4, 5].

In sum, therapeutic immunization aimed at pathogenesis blockade reframes *H. pylori* care around clinically tangible, pathology-centered outcomes. By targeting the bacterial functions and host interactions that actually drive disease, such strategies can deliver benefit without insisting on complete eradication—an approach that is both biologically sound and developmentally feasible [4, 5, 18–20].

## 15.4 Immunotherapeutic Modalities (Classification and Mechanisms)

### 15.4.1 Active Immunotherapy

For blocking disease by active immunotherapy, the most compelling antigens are those that relate to disease mechanisms rather than those chosen purely for their ability to reduce colony counts in animal models. Canonical examples include urease (UreA/UreB), whose buffering activity induces mucosal fitness and epithelial irritation, VacA fragments that drive epithelial vacuolation and T-cell dysfunction, NAP, a potent neutrophil-activating and Th1-skewing factor, and adhesins such as BabA, SabA, AlpA/B, and HpaA that stabilize intimate epithelial contact and facilitate downstream signaling. Prior work has already shown that responses against these targets can blunt pathology and alter the host–pathogen interface; multivalent or multi-epitope constructs further hedge strain variability while preserving a mechanistic link to disease [16, 17, 28–30].

Because therapeutic intent privileges local benefit over sterilization, adjuvants and delivery routes must be selected for mucosal targeting and safety. Double-mutant heat-labile toxin (dmLT) and related constructs have strengthened gastric or proximal mucosal immunogenicity in preclinical vaccination while maintaining a wider safety margin than classical enterotoxins; Cholera toxin B subunit (CTB) has served as a carrier/adjuvant in several constructs; chitosan-based systems can improve residence time and uptake; α-GalCer leverages invariant natural killer T cells to help potentiate balanced Th1 responses. Still, the gastric environment is unforgiving: sublingual and intragastric routes can elicit robust local responses but have also revealed postimmunization gastritis signals in some settings, underscoring the need for careful dose escalation, endoscopic monitoring, and prespecified stopping rules in therapeutic trials [19, 31–33].

Preclinical studies across mice, gerbils, dogs, and nonhuman primates consistently demonstrate discordance between bacterial counts and disease severity:

vaccination can reduce histologic gastritis, glandular damage, or ulceration even when residual organisms remain detectable. Urease-based formulations and multivalent constructs have repeatedly improved pathology, and adhesin-targeted strategies have reduced epithelial interaction and downstream inflammation, while VacA-focused approaches block cytotoxicity and immune dysregulation. These effects support therapeutic, pathology-first objectives in humans rather than eradication as a singular endpoint [28, 34, 35].

The human experience, though still limited, reinforces both feasibility and caution. Early trials with recombinant urease showed immunogenicity in primates and highlighted the need for vigilant gastric safety surveillance when mucosal routes are pursued. Later summaries of pediatric oral subunit programs similarly emphasize the importance of clinical endpoints beyond eradication and the careful management of reactogenicity [10, 36].

In practice, therapeutic active immunization, if developed, should be selectively targeted. Appropriate candidates include patients with recurrent or refractory disease after guideline therapy, and highly selected individuals with high-risk precancerous lesions where pathogenesis-blocking could plausibly slow progression. In both scenarios, stringent endoscopic monitoring and conservative dose-finding are mandatory. Trials should adopt pathology-centered primary endpoints (validated histologic activity scores, ulcer healing, symptom scales, and gastric pharmacodynamic biomarkers) and treat bacterial load and eradication as secondary or exploratory measures [19, 31].

### 15.4.2 Passive Immunotherapy

Passive immunotherapy for *H. pylori* can be delivered orally as luminally acting antibodies or systemically as circulating antibodies that reach the gastric mucosa via the lamina propria and transudation. Orally administered preparations, such as polyclonal IgY or bovine antibodies, act topically to neutralize urease, adhesins, and VacA in the mucus layer, thereby interrupting epithelial engagement and dampening injurious signaling with a favorable local safety profile; they are repeatable, compatible with reduction in antibiotic exposure regimens, and supported by in vitro and clinical signals of symptom improvement and reduced urea breath test deltas [37–39]. However, their efficacy hinges on gastric stability and, critically, on access to the gastric glands and deep mucus niches where *H. pylori* resides, a point that remains uncertain and likely varies with formulation and dosing. In contrast, systemically administered monoclonal antibodies (e.g., IgG or engineered IgA/IgA-like formats) achieve tissue-level exposure through vascular delivery to the lamina propria, with subsequent movement into the mucus and glandular regions; this route may more reliably reach the sites of colonization, at the cost of higher manufacturing complexity, potential systemic adverse events, and the need to carefully manage Fc-mediated inflammation. From a development perspective, oral delivery emphasizes formulation science (encapsulation, mucoadhesion, timing with PPIs) and batch potency control for polyclonals [40], whereas systemic delivery emphasizes isotype/format engineering, pharmacokinetics, and gastric tissue pharmacodynamics to confirm on-target exposure [37, 40]. In both cases, the therapeutic aim remains pathogenesis blockade rather than compulsory eradication, with pathology-centered endpoints guiding dose and schedule.

#### 15.4.2.1 Multiple Molecular Formats and Sources Are Available

Polyclonal egg-yolk IgY is scalable and cost-effective, naturally lacks human Fc-mediated inflammatory effector functions, and can be raised against urease, adhesins, and VacA. Bovine colostral or milk-derived antibodies provide polyclonal breadth and a manufacturing pathway familiar with food-grade production. Increasingly, monoclonal antibodies and derivatives (including IgA/IgY formats, scFv and VHH/nanobody fragments) allow epitope-precise neutralization of defined virulence interfaces and can be engineered for gastric stability while keeping pro-inflammatory Fc engagement minimal [37, 40].

Mechanistically, passive antibodies enable prevention of pathogenesis: neutralizing urease reduces periplasmic buffering and bacterial survival, targeting adhesins limits intimate epithelial contact and the signaling that follows, and VacA neutralization attenuates epithelial damage and T-cell suppression.

A particularly compelling host–pathogen interaction is HopQ-carcinoembryonic antigen-related cell adhesion molecule (CEACAM): the *H. pylori* outer membrane protein HopQ binding to epithelial and immune CEACAMs stabilizes adhesion, promotes CagA translocation, and shapes noncanonical NF-κB signaling. Antibodies or small molecules that block HopQ–CEACAM offer a clear, druggable route to reduce attachment and effector delivery without broad antimicrobial pressure, and they can be layered with urease or adhesin-directed IgY to achieve complementary effects. In addition, antibody-mediated aggregation at the surface may modulate innate receptor engagement and lower pro-inflammatory signaling, indirectly favoring a less damaging mucosal milieu [17, 30, 37].

#### 15.4.2.2 Formulation Remains the Principal Challenge

Gastric acidity and proteases rapidly degrade proteins, so success depends on microencapsulation, enteric protection, and mucoadhesive carriers, often combined with acid suppression to extend luminal exposure. For polyclonals, lot-to-lot epitope coverage and potency must be controlled through standardized in vitro neutralization assays. Urease activity inhibition, adhesion blocking, and VacA vacuolation assays are commonly used [40–42].

Evidence supports clinical plausibility: In vitro, antigen-matched IgY and bovine antibodies robustly neutralize urease, block adhesin–receptor binding and reduce VacA-mediated cytotoxicity. In vivo, disease metrics improve in multiple models even when organisms are not fully cleared, mirroring the disease-blocking aims of this chapter [28, 35]. Early human studies and randomized evaluations report improvements in symptoms, reductions in urea breath test deltas, and lower bacterial density with oral anti-*H. pylori* antibodies, with generally favorable tolerability.

#### 15.4.2.3 Development and Regulatory Pathways Depend on Intended Claims

Where disease modification is asserted (e.g., reduced gastritis activity or accelerated ulcer healing), programs typically follow therapeutic-biologic routes and must anchor endpoints to recognized clinical benefits. Where products are framed as functional/medical foods, expectations for clinical evidence, chemistry, manufacturing, and controls (CMC), and labeling differ, but consistency of potency remains crucial. In both cases, companion assays that relate batch activity to clinical pharmacodynamics will be valuable for bridging [40].

Clinically, passive immunotherapy is attractive for refractory infection after failed antibiotic therapy, and for symptom suppression and disease mitigation when eradication is not feasible or antibiotics are contraindicated. These scenarios leverage the modality's strengths—local action, repeatability, and safety, while aligning outcomes with pathology-centered measures rather than solely focusing on eradication.

#### 15.4.2.4 Microbiome and Immunomodulatory Adjuncts

Selective probiotics and recombinant probiotic vectors can modulate mucosal immunity, improve tolerability of standard regimens, and yield modest efficacy gains; their most coherent role is as adjuncts that shift immune tone and epithelial resilience. Combinations with vaccines or IgY are mechanistically complementary: probiotics soften inflammatory milieu while antibodies blunt virulence at the surface [38, 43, 44].

#### 15.4.2.5 Checkpoint Axis Modulation (PD-1/PD-L1)

Data on PD-1/PD-L1 in the context of *H. pylori* arise largely from oncology settings; targeted checkpoint modulation as a primary therapeutic strategy for *H. pylori* remains conceptual and is not currently supported by clinical evidence. *H. pylori* induces PD-L1 on gastric epithelium and immune cells, a configuration that plausibly fosters T-cell exhaustion and persistence. While this makes checkpoint modulation mechanistically appealing, human gastric safety remains uncertain, and there is a plausible risk of exacerbating gastritis or ulceration. Any exploration should prioritize local delivery, dose minimization, and dense mucosal safety monitoring; at present, this path is investigational and not ready for broad clinical deployment [45–47].

#### 15.4.2.6 Innate-Mimetic and Bioactive Adjuncts (Antimicrobial Peptides, Lactoferrin)

Antimicrobial peptides and bovine lactoferrin are mentioned here as they exert measurable immunomodulatory effects but do not offer direct antimicrobial activity. These effects include iron sequestration and pattern-recognition signaling, and have shown improved eradication or reduced adverse events as add-ons in some trials. Signals are promising but heterogeneous, likely reflecting strain differences, background therapy, and dose/formulation. Within a disease-blocking approach, these agents are best positioned as supportive add-ons to antibody-based strategies or reduced-intensity antibiotic regimens, where they may modulate the inflammatory milieu and promote mucosal healing [48, 49].

## 15.5 Preclinical to Clinical Translation: Design, Endpoints, and Regulatory Considerations

The following sections outline translational and regulatory considerations for immunotherapy development and serve as a methodological framework for later clinical chapters. Designing therapeutic vaccine and passive immunotherapy trials for *H. pylori* requires embracing disease-blocking, pathology-oriented endpoints from the outset and selecting models and readouts that reliably bridge those endpoints in humans. Rodent systems remain indispensable for mechanism and screening, but larger animals and early human biomarker strategies are needed to mitigate translation risks.

### 15.5.1 Preclinical Models, Their Limits, and Bridging Strategies

Mice, and to some extend Mongolian gerbils, are still the models of choice for evaluating therapeutic vaccination, adjuvants, and delivery routes. Across these systems, urease-based constructs, multivalent epitope vaccines, and adjuvant comparisons (including mucosal adjuvants) have repeatedly shown discordance between bacterial counts and gastric pathology: histologic gastritis and glandular injury can decline even when residual organisms are detectable (a signal consistent with disease-blocking intent). Classic and contemporary exemplars include therapeutic intragastric vaccination that converted chronic carriage to protection on rechallenge in mice, epitope-based vaccines that delivered therapeutic benefit in gerbils, and iterative BALB/c programs articulating prophylactic and therapeutic effects with CTB-fused antigens [50–53]. The same platform logic extends to multivalent designs (e.g., UreB/HpaA/LTB) that improved pathology in C57BL/6 mice [54]. Nevertheless, rodents differ from humans in gastric physiology, mucosal glycan landscapes, and immunoregulation; they often overestimate sterilizing potential and underestimate reactogenicity. Practical bridges include: (i) pathology-first readouts in animals (validated gastritis scoring, ulcer presence) alongside colony forming unit (CFU)/qPCR; (ii) gastric PD panels (cytokines, epithelial signaling markers) that can be mirrored in early human biopsies; and (iii) dose-finding anchored to PD rather than titer alone [50–54].

### 15.5.2 Larger Animals and Nonhuman Primates (NHPs)

Dogs and ferrets provide valuable intermediate physiology for gastric kinetics, formulation behavior, and endoscopic histology; NHPs supply the most human-like mucosal architecture and immune compartmentalization. Across these species, immunization has elicited robust systemic and mucosal responses and sometimes improved gastric pathology or surrogate markers; however, sterilizing immunity is infrequent and variable, reinforcing the need to prioritize disease endpoints for translation [34, 35, 55]. Early NHP studies with recombinant urease documented immunogenicity and feasibility, and together with human safety summaries, highlighted the recurring signal of postimmunization gastritis after some mucosal regimens—an issue that must be prospectively managed in first-in-human studies of therapeutic candidates [34, 36]. Consequently, large-animal work is best positioned to: (i) validate endoscopic/histologic scoring workflows intended for trials; (ii) test mucosal delivery technologies and residence times; and (iii) prespecify safety monitoring (endoscopy, bleeding risk, ulcer surveillance) to be carried forward unchanged into early clinical phases.

### 15.5.3 Biomarkers and Clinical Endpoints for Disease-Blocking Trials

For therapeutic vaccination and passive immunotherapy alike, primary endpoints should be pathology-centered. Standardized histologic gastritis grading (activity and chronicity components; atrophy and intestinal metaplasia where applicable) provides an objective tissue anchor. Endoscopic outcomes, such as the presence, severity, or healing of ulcers and gastritis, together with healing rates and time-to healing, are direct markers of clinical benefit. Patient-reported symptom scales (e.g., validated dyspepsia indices) should be incorporated with predefined minimal clinically important differences to detect a significant benefit to patients. Circulating gastric injury/repair biomarkers can complement tissue readouts and reduce biopsy frequency in later phases. In this framework, bacterial load measures, $^{13}$C-UBT deltas, stool antigen indices, or qPCR copy numbers, are key secondary endpoints used to contextualize clinical improvement rather than to define success alone. Longitudinal relapse/recrudescence at 6–12 months remains important to demonstrate durability of disease control without insisting on complete eradication. These outcome sets align with contemporary immunobiology and the clinically relevant goal of pathogenesis blockade [10, 56].

### 15.5.4 Trial Design for Therapeutic Vaccination

Early-phase studies should prioritize dose and route that optimize gastric PD while minimizing reactogenicity; sublingual or intragastric strategies deserve careful graded dose escalation schedules with predefined stopping rules and dense endoscopic/histologic monitoring in sentinel cohorts. Randomization against optimized standard of care or an appropriate placebo should use pathology endpoints as primary, with eradication relegated to secondary/exploratory tiers. Where risk of postimmunization gastritis is material, adaptive designs can gate dose escalation on PD and safety thresholds observed in gastric biopsies [52, 57].

Trial design should explicitly consider both oral and systemic antibody delivery, because the route determines gastric exposure, safety profile, and operational complexity. For oral antibodies, early studies can pair changes in validated histologic activity with functional gastric PD, for example, ex vivo urease inhibition or adhesion-blocking assays on paired biopsies, while using $^{13}$C-UBT and stool antigen as supportive outcomes. Dosing and scheduling should be informed by in vitro neutralization potency, formulation-dependent gastric residence, and coadministration of PPIs to extend luminal exposure. Lot-to-lot potency for polyclonals requires standardized release assays [40]. For systemically administered antibodies, phase 1/2 trials should enrich patients with active gastritis or ulceration and incorporate gastric tissue pharmacokinetics (antibody levels in biopsies), spatial PD (immunostaining for target engagement within glands and deep mucus), and the same pathology-first primary endpoints used for oral products. Because systemic dosing may provide more reliable access to glandular niches, it warrants comparative evaluation against oral regimens in factorial or head-to-head designs where feasible; at the same time, systemic exposure necessitates careful safety surveillance for off-target effects and Fc-mediated inflammation, with isotype and Fc-engineering choices prespecified to minimize pro-inflammatory signaling. Across both routes, randomization against optimized standard of care should treat histologic gastritis activity, ulcer healing, and symptom indices as primary outcomes, reserving eradication for secondary/exploratory analysis and assessing durability at 6–12 months. Declaring the intended regulatory pathway upfront remains essential: orally delivered polyclonals framed as therapeutic biologics versus functional/medical foods face different evidentiary and Chemistry, Manufacturing, and Controls (CMC) requirements, while systemically administered monoclonals will follow conventional biologics pathways; in each case, bridging mechanistic potency assays to gastric PD strengthens the overall package [40].

### 15.5.5 Controlled Human Infection Models and Feasibility in *H. pylori* Immunotherapy

Controlled human infection models (CHIM) for *H. pylori* were first clinically established in a rigorous manner (with dose escalation) in the early 2000s, demonstrating reproducible colonization, gastric inflammation, and cytokine induction in healthy adult volunteers using an antibiotic-susceptible strain under predefined rescue therapy [58]. This confirmed that deliberate infection can be induced safely and reversibly and that epithelial inflammatory readouts can be measured over a defined timeline—a key prerequisite for immunotherapeutic evaluation. Renewed interest in CHIM has been driven by the identification of CagA-imprinted Hobit$^{+}$ gastric TRM as a causal and sterilizing correlate of protection [21], which for the first time provides a mechanistic endpoint that can be prospectively measured in human gastric biopsies. In parallel, next-generation human in vitro models (gastric organoids, air–liquid interface systems, and immune–epithelial co-cultures) now allow controlled interrogation of TRM licensing signals, epithelial cytokine responses, and safety parameters prior to in vivo exposure. Accordingly, modern CHIM designs would not rely on eradication as the primary endpoint, but on the induction of a protective residency program in the gastric mucosa, with changes in epithelial stress or metaplastic trajectories as secondary readouts. With antibiotic-sensitive challenge strains, predefined rescue therapy, and structured histological monitoring, CHIM might constitute a credible bridge between mechanistic immunology and early experimental medicine trials in humans.

In summary, credible translation of *H. pylori* immunotherapies depends less on driving CFUs to zero than on demonstrating tissue and patient benefit with acceptable safety. This entails selecting animal models and PD readouts that closely mirror human gastric biology, defining pathology-first clinical endpoints, and devising regulatory narratives centered on disease mitigation rather than eradication. Done this way, both therapeutic vaccination and passive antibody strategies can be advanced on evidentiary tracks that are biologically plausible and clinically meaningful.

## 15.6 Clinical and Translational Evidence from Large Animal and Nonhuman Primate Studies

### 15.6.1 Trial Phases and Safety Monitoring

The clinical translation of *H. pylori* immunotherapies must prioritize gastric safety (and overall safety). Early mucosal vaccine experiments identified a reproducible signal of postimmunization gastritis, in which heightened local immune activity coincided with protection in animals and also with transient epithelial injury—a phenomenon that maps to human risk and mandates prospectively defined surveillance in trials [19, 59]. Accordingly, phase 1–2 studies, whether evaluating therapeutic vaccines or passive antibodies, should incorporate sentinel endoscopy with targeted biopsies, blinded histologic activity scoring, and ulcer surveillance for vulnerable subgroups (e.g., prior ulcer, atrophy/metaplasia). Early protocols should include predefined stopping rules for worsening histology, bleeding, or significant symptom exacerbation, reflecting lessons from preclinical hyper-reactogenic regimens and human safety narratives [19, 57]. Biopsy-linked gastric pharmacodynamic (PD) readouts, such as local cytokine panels, pepsinogen indices, and ex vivo functional assays (urease inhibition, adhesion blockade), are essential both to titrate dose/route and to de-risk advancement before larger efficacy trials [60].

#### 15.6.1.1 Statistical Considerations

Because disease-blocking benefit, not eradication, is the intended goal, studies must be powered for histologic and symptom outcomes that exhibit different variance structures than binary culture/UBT endpoints. Pragmatically, early phase 2 designs can adopt co-primary endpoints, for example, change in validated histologic activity score plus a dyspepsia scale with a predefined minimal clinically important difference, while treating changes in $^{13}$C-UBT delta or stool antigen index as key secondary measures [60]. Adaptive and factorial designs are attractive in this space: they allow dose and adjuvant comparison for vaccines and route/formulation comparison for antibodies (oral vs. systemic), with interim looks gated on PD and safety thresholds to avoid exposing larger cohorts to overly reactogenic regimens [19, 57].

#### 15.6.1.2 Regulatory Path

Regulatory classification is driven by claims and mechanisms. Systemically administered monoclonals and therapeutic oral antibody products that claim disease modification fall under therapeutic biologic pathways and will need pathology-centered primary endpoints (histology, ulcer healing, validated symptom scales) plus standard quality, nonclinical, and PK/PD packages. In contrast, some IgY or bovine antibody preparations have been advanced as functional foods or medical foods with supportive symptom or microbial load claims; this path entails different evidentiary standards but still demands batch-to-batch potency and stability controls (e.g., in vitro urease and adhesion neutralization), given the gastric environment [40, 60]. Developers should declare the intended pathway from the outset and align endpoints and controls accordingly.

#### 15.6.1.3 Human Safety and Immunogenicity Studies

Bridging NHPs to humans established feasibility and immunogenicity for recombinant urease vaccines, while underscoring the need for cautious mucosal dosing and gastric monitoring [36]. Subsequent translational programs explored multivalent epitope vaccines incorporating NAP, urease, heat shock protein 60 (Hsp60), and HpaA in Mongolian gerbils, demonstrating therapeutic benefit that informed human dosing logic [61]. Work with refined mucosal adjuvants such as dmLT showed effective augmentation of local responses but also reinforced that human programs must be designed to detect and manage postimmunization gastritis [19]. Conceptual and historical context for these efforts is synthesized in Svennerholm & Lundgren's review, which emphasizes both the promise and the translational constraints of active immunization in *H. pylori* [4].

#### 15.6.1.4 Randomized and Controlled Trials of Passive Antibody Formulations

Passive antibody approaches have advanced the furthest in humans. Bovine antibody-containing milk reduced intragastric colonization in a randomized trial and improved clinical parameters, demonstrating the feasibility of oral passive immunotherapy as an adjunctive strategy [62]. A randomized double-blind study in type O individuals showed a clearance effect with bovine anti-*H. pylori* antibody milk, again with favorable tolerability [63]. Clinical IgY food products, including functional yogurt with *H. pylori*-specific IgY, suppressed infection markers and symptoms in small controlled studies [64]. More recently, multivalent dietary IgY increased eradication rates and improved symptoms in a controlled clinical study, and a 2024 real-world analysis reported the effectiveness and safety of a tetravalent IgY in refractory infection, highlighting applicability in contexts where antibiotic therapy is limited or contraindicated [38, 39]. Collectively, these data support pathogenesis-blocking benefits, symptom relief, improved histology surrogates, and reduced bacterial density, without insisting on sterilizing clearance.

#### 15.6.1.5 Outcomes Measured in Clinical Studies

Human trials report a consistent panel of microbiological and clinical readouts. Microbiologic measures include $^{13}C$-UBT delta, stool antigen indices, and occasionally culture/qPCR from biopsies; these are highly useful as secondary endpoints that contextualize disease improvement [60, 63]. Primary pathology-oriented outcomes include histologic gastritis scores (activity/chronicity, atrophy/metaplasia where relevant), ulcer healing rates, and patient-reported symptoms on validated dyspepsia scales. Passive antibody randomized controlled trials and controlled food-product studies commonly demonstrate symptom improvement and reductions in UBT/stool antigen, with eradication rates improved in some settings (especially with multivalent IgY) but not universally, a pattern consistent with a disease-blocking therapeutic goal [38]. Safety monitoring systematically collects adverse events (GI symptoms, bleeding) and includes per-protocol endoscopy/biopsy when feasible in early phases.

#### 15.6.1.6 Reports of Adverse Events and Postimmunization Gastritis

Mechanistic animal work linked local sIgA and heightened gastric immune activation to protection and also to post-immunization gastritis. This duality likely reflects the narrow therapeutic window for mucosal activation in the stomach [59]. Follow-on immunology in mice described expanded T-cell help and reduced regulation in protective vaccine settings, a pattern that, if recapitulated injudiciously in humans, could exacerbate symptoms or lesions [57]. Translational studies designed for human application with dmLT and other mucosal strategies have therefore adopted gradual dose escalation, route comparisons (sublingual vs. intragastric), and intensive histologic surveillance, making gastric safety a co-primary endpoint in first-in-human therapeutic studies [19]. These experiences collectively argue for designs that can separate on-target virulence neutralization (benefit) from excess bystander inflammation (risk) during clinical development.

#### 15.6.1.7 Synthesis and Practical Implications

Taken together, the human and translational evidence from large-animal and NHP studies already supports a disease-blocking, pathology-first path for *H. pylori* immunotherapies. Passive antibody strategies, with accumulating randomized and real-world data, show the capacity to improve symptoms and reduce surrogate load with favorable safety, particularly as adjuncts or maintenance in refractory cases [38, 39, 62, 63]. Active immunization remains promising but must progress under stringent gastric safety oversight, leveraging NHP bridging and modern mucosal adjuvants while accepting disease-blocking endpoints as primary markers of success [4, 19, 36]. For both modalities, regulators will expect clearly articulated claims, robust batch potency controls (especially for polyclonal IgY/bovine products), and statistical plans powered for histology/symptom outcomes with microbial measures as supportive evidence [40, 60].

## 15.7 Key Translational Challenges and Mitigation Strategies

Translating *H. pylori* immunotherapies into practice hinges on solving a set of persistent challenges that recur across programs. Four key challenges stand out and require specific measures:

### 15.7.1 Antigenic and Strain Diversity

Global *H. pylori* populations are highly heterogeneous in virulence repertoires (e.g., cytotoxin-associated gene pathogenicity island (cagPAI) integrity; variable adhesins such as BabA/SabA; divergent VacA and urease epitopes). This diversity hinders the development of generalizable vaccines and poses risks to geographic efficacy cliffs. The practical response is twofold: design multivalent constructs that combine complementary virulence targets and emphasize conserved epitope mapping to retain breadth across clades. Contemporary reverse vaccinology and AI-assisted epitope engineering can prioritize conserved, surface-exposed, and functionally constrained regions, while flagging escape-prone epitopes for de-prioritization. In therapeutic (disease-blocking) constructs, weighting antigens that directly drive disease, such as urease, VacA, NAP, and conserved adhesin interfaces, adds mechanistic insurance even if low-level colonization persists [15, 29, 65, 66].

### 15.7.2 Mucosal Adjuvant Safety

Potent mucosal adjuvants can not only raise effective gastric immunity but also risk postimmunization gastritis and ulcer exacerbation. A mitigation package should: (i) use safer adjuvants (e.g., optimized-dose dmLT; chitosan formulations) with stepwise, biopsy-anchored escalation; (ii) apply heterologous prime–boost and routes (sublingual versus intragastric) that localize responses while permitting tight reactogenicity control; and (iii) embed predefined stopping criteria tied to blinded histology and symptoms in first-in-human work. Programmatically, adjuvant choice should be justified with gastric PD data (cytokine panels, epithelial signaling) rather than serum titers alone [19, 31–33].

### 15.7.3 Antibody Stability and Delivery for Passive Immunotherapy

For oral IgY or bovine preparations, acid and protease degradation and variable access to glandular niches are the principal barriers. Three engineering levers repeatedly help: microencapsulation to shield protein cargo, enteric/mucoadhesive coatings to increase residence at the gastric surface, and coadministration with PPIs or buffering agents to decrease proteolysis during the exposure window. Release testing should mirror mechanism, urease inhibition, adhesion blockade, and VacA neutralization to ensure lot-to-lot potency. In parallel, systemic monoclonal antibodies deserve explicit consideration: by reaching the lamina propria via the bloodstream and transuding into mucus and glands, they may deliver more reliable exposure to the sites where *H. pylori* resides, at the cost of higher CMC complexity and potential Fc-mediated adverse events. Early comparative studies should therefore include tissue PK/PD (biopsy antibody levels, spatial target engagement) to rationally choose the route and format [40–42].

### 15.7.4 Balancing Clearance and Inflammation

Because gastric injury in *H. pylori* stems from both bacterial effectors and dysregulated host responses, indiscriminate immune activation can worsen disease. Mitigation begins with endpoint design: prioritize pathology-centered measures, validated histologic activity, ulcer healing, symptom scales, and pepsinogen I/II, as primary outcomes, with microbiologic readouts supportive, not determinative. Combine immunotherapies with injury-limiting agents (e.g., PPIs, bismuth, selected probiotics, or lactoferrin) during induction phases to narrow the inflammatory window, then taper as pathology improves. Finally, execute adaptive dose-finding gated on gastric PD and safety thresholds to separate on-target virulence neutralization from bystander damage before scaling [10, 57, 59, 60].

In summary, a translationally relevant strategy combines recent antigen design with cautious mucosal adjuvants, deploys oral or systemic antibodies using formulation or format engineering to ensure gastric exposure, and conducts biopsy-anchored, pathology-first trials that preempt and manage inflammation. This integrated approach aligns the biology of persistence with achievable clinical benefits, offering a path to meaningful disease control even when complete eradication is not the primary endpoint.

## 15.8 Trial Design Recommendations and Pragmatic Pathways for Clinical Development

### 15.8.1 Design Elements to Prioritize

Early-phase studies should center on gastric safety and biologic activity. Phase I trials ought to define explicit safety endpoints (endoscopic evidence of new or worsened gastritis or ulceration; bleeding; serious GI adverse events) alongside pharmacodynamic and immunogenicity readouts from gastric biopsies (cytokine panels, epithelial signaling markers, local antibody levels), rather than relying on serum titers alone [5, 60]. Inclusion criteria should enrich individuals with active noncomplicated gastritis or recent ulcer history when therapeutic intent is pathology mitigation; exclusion criteria should exclude patients with high bleeding risk, uncontrolled comorbidities, or recent high-dose nonsteroidal anti-inflammatory drug use. Stratification by *H. pylori* genotype (e.g., cagPAI status, VacA s/i/m types, dominant adhesin profiles) is advisable because virulence repertoires influence both pathogenesis and the theoretical responsiveness to targeted immunotherapies [67]. For prophylactic concepts, if pursued, populations at high exposure risk (household contacts, high-prevalence settings) are rational, but pediatric involvement requires a distinct ethical and safety framework [4, 36]. Sample size calculations should be powered for histology and symptom co-primary endpoints (e.g., change in validated gastritis activity score and a dyspepsia index with a prespecified minimal clinically important difference), with UBT and stool antigen treated as key secondary measures [60]. Adaptive designs can efficiently compare doses, routes, and adjuvant choices (for vaccines), or oral versus systemic delivery and formulations (for antibodies), with interim decisions gated on mucosal PD and safety [5, 61].

#### 15.8.1.1 Safety Monitoring and Dose Escalation

Given historical signals of postimmunization gastritis in mucosal vaccine settings, stepwise dose escalation with sentinel cohorts is recommended, incorporating scheduled endoscopy with targeted biopsies, blinded histologic scoring, and ulcer surveillance at predefined time points [19, 57]. Adjuvant selection should prioritize platforms with acceptable gastric safety margins (e.g., optimized-dose dmLT, chitosan-based systems), and routes should be chosen to favor local efficacy while controlling reactogenicity (sublingual vs. intragastric), with clear stopping rules for worsening histology or clinically significant symptom exacerbation [19]. Pediatric prophylactic strategies, if contemplated, should proceed only after adult proof of gastric and overall safety, with independent data monitoring, conservative dose-finding, and long-term surveillance for growth and nutritional outcomes [4].

#### 15.8.1.2 Prophylactic Versus Therapeutic Endpoints

Prophylactic programs, where ethically and practically justified, should use incident infection (serial UBT/stool antigen conversion) as a principal endpoint with preplanned analyses of symptomatic disease and histology in incident cases. Therapeutic programs should designate pathology-focused endpoints as primary (histologic activity and chronicity; ulcer healing rates/time to healing; validated symptom scales) and microbiological measures as secondary outcomes contextualizing disease improvement [5, 60].

#### 15.8.1.3 Combination Approaches

Given the multifactorial pathogenesis of *H. pylori*, combination regimens are scientifically justified. Vaccine in combination with probiotic approaches aim to enhance mucosal homeostasis and reduce adverse events while maintaining immunogenicity; several preclinical and early clinical reports support modest efficacy and improved tolerability [43, 44]. Vaccine plus IgY is rational where passive antibodies neutralize key virulence functions (urease, adhesins, VacA) while active immunization provides longer-lived immune pressure; careful sequencing (e.g., IgY induction then vaccine boost) may optimize safety. Vaccine or IgY combined with PPI or bismuth can reduce inflammatory injury and improve local exposure, aligning with a disease-blocking objective [38]. Combination trials should be factorial when feasible, with interaction terms prespecified and interim analyses focused on histologic and symptom co-primary endpoints.

#### 15.8.1.4 Pragmatic Pathways

For passive antibodies, an efficient path begins with single-arm Phase Ib studies to confirm gastric exposure and functional pharmacodynamics (ex vivo urease inhibition, adhesion blockade on paired biopsies), followed by randomized Phase II trials in refractory or antibiotic-intolerant patients, powered for histology and symptom improvement, with UBT/stool antigen as supportive endpoints. For therapeutic vaccines, proceed with cautious mucosal dosing, biopsy-anchored pharmacodynamics, and co-primary pathology endpoints; expansion should depend on reproducible safety and PD signals in two independent cohorts [19, 57]. Throughout, stratification by bacterial genotype and predefined rescue therapy rules are essential, and pediatric work should trail adult evidence with heightened ethical oversight [4, 36, 67].

## 15.9 Practical Recommendations for Authors, Clinicians, and Investigators

### 15.9.1 Antigen Selection Strategy

Prioritize antigens mechanistically linked to pathogenesis (urease, VacA, CagA, conserved adhesins), use multivalent designs and conserved-epitope mapping, stratify or prespecify analyses by *H. pylori* virulence genotype (e.g., cagPAI, VacA s/i/m types), and consider conservation and strain variability. Early protocols should justify antigen choices with pathology-focused preclinical data rather than CFU alone [5, 67].

### 15.9.2 Preclinical Packages

Combine rodent efficacy (histology readouts) with bridging in larger species or NHPs to assess mucosal delivery, residency, and safety. Incorporate gastric pharmacodynamic biomarkers (biopsy cytokines, epithelial signaling, local antibody levels) intended to be translated into FIH trials [36, 61].

### 15.9.3 Adjuvant Selection and Dosing

Favor mucosal adjuvants with acceptable gastric safety margins (e.g., optimized-dose dmLT, CTA1-DD, chitosan-based systems). Implement stepwise dose escalation with sentinel endoscopy/biopsy and blinded histology, using predefined stopping rules for postimmunization gastritis or ulcer exacerbation [19, 57].

### 15.9.4 Biomarkers and Endpoints

Use co-primary pathology outcomes (validated histologic activity/chronicity; ulcer healing/time to healing) with a patient-reported symptoms index; use UBT and stool antigen as key secondary endpoints. Specify minimal clinically important differences and power the trial accordingly; plan adaptive features for dose/route/adjuvant comparisons [5, 60].

### 15.9.5 Safety Monitoring

For therapeutic vaccination and antibody trials, schedule surveillance endoscopy with targeted biopsies and monitor systematically for bleeding, worsening gastritis, ulceration, and systemic reactogenicity. Pediatric prophylaxis, if contemplated, should follow adult proof of gastric safety, with independent monitoring and long-term follow-up [19, 36].

### 15.9.6 Passive Immunotherapy—Evidence and Positioning

Cite controlled studies and real-world data supporting oral antibody safety and symptom/microbiologic benefits when positioning passive products as adjuncts or for refractory settings. For systemic monoclonals, plan tissue PK/PD (antibody in biopsies, spatial target engagement) and Fc-engineering choices to minimize pro-inflammatory signaling [62, 63].

### 15.9.7 Initial Patient Populations

For therapeutic trials, enroll adults with refractory infection, intolerance/contraindication to antibiotics, and persistent symptomatic gastritis/ulceration despite standard care. For prophylactic concepts, consider high-exposure settings only after adult safety is established; pediatric studies require stringent ethical safeguards and prolonged surveillance [4, 36].

### 15.9.8 Combination Strategies

Where justified by preclinical or early clinical data, consider factorial designs testing immunotherapy ± probiotic, or adjunct IgY with PPI/bismuth to reduce inflammatory injury; predefine interaction analyses and retain pathology endpoints as primary [38, 43, 44].

These recommendations align study conduct with disease mechanisms, emphasize safety, and support regulatory strategy centered on clinically meaningful pathology outcomes rather than eradication alone (Tables 15.1, 15.2, 15.3, and 15.4).

**Data Annex: Suggested Immunologic Assays and Sample Handling**

- Gastric biopsies with standardized site mapping; histology scoring; immunostaining; ex vivo urease/adhesion assays.
- Mucosal immune readouts: Local cytokine panels; tissue antibody levels; sIgA in gastric juice/saliva.
- Systemic assays: Antigen-specific ELISpot and serum ELISA (contextual).
- Microbiologic endpoints: 13C-UBT delta, stool antigen; culture/qPCR from biopsies in early phases.
- Timing: Baseline, week 2–4 (PD), week 8–12 (primary), and 6–12-month relapse surveillance.

**Table 15.1** Candidate antigens for therapeutic/disease-blocking immunization

| Antigen/target | Pathogenesis role | Therapeutic mechanism (rationale) | Preclinical evidence (pathology-oriented) | Human data/status |
|---|---|---|---|---|
| Urease (UreA/UreB) | Acid adaptation; epithelial irritation | Neutralize urease activity to impair mucosal fitness and lessen epithelial injury | Murine protection with reduced gastritis using recombinant subunits and multivalent constructs | Human immunogenicity shown; therapeutic efficacy unproven; safety requires gastric monitoring |
| gGT | Glutamine metabolism, T-cell inhibition | Neutralize gGT activity to impair mucosal fitness and overcome immune evasion | gGT-deficient mutants show colonization defects. Immunization with gGT induces partial protection | Human safety and immunogenicity shown; therapeutic efficacy unproven |
| VacA (fragments) | Epithelial vacuolation; T-cell dysregulation | Toxin neutralization to decrease epithelial injury and immune suppression | Disease attenuation in models despite residual organisms | Human data limited to immunogenicity concepts/reviews |
| NAP | Neutrophil recruitment; Th1 skew | Reduce neutrophil-mediated damage without full clearance | Protection with reduced inflammatory injury in models | No therapeutic efficacy yet in humans |
| Adhesins (BabA, SabA, AlpA/B, HpaA) | Intimate epithelial attachment; effector delivery | Block adhesion to shorten epithelial dwell and limit downstream signaling | Decreased inflammation/colonization density in small-animal studies; multi-adhesin epitope designs | Early translational interest; no definitive human efficacy |
| Hsp60 | Chaperone; immunomodulation | Broaden responses; possible modulation of mucosal immunity | Included in multivalent epitope constructs with therapeutic signals in gerbils | Human efficacy unknown |
| Multivalent/multi-epitope vaccines | Covers complementary functions | Combine neutralization of virulence and anti-adhesion | Therapeutic benefit in Mongolian gerbils; CTB-fused epitope regimens active in mice | Inform human dosing logic; clinical proof pending |

Adapted from: [15, 16, 28, 30, 61]

**Table 15.2** Passive immunotherapy formats and clinical/preclinical evidence

| Format/source | Advantages | Limitations | Stability/formulation strategies | Representative evidence |
|---|---|---|---|---|
| Oral polyclonal IgY (egg yolk) | Local/topical action; scalable; minimal Fc-mediated inflammation | Acid/protease degradation; uncertain access to glands | Microencapsulation; enteric/mucoadhesive coatings; PPI coadministration | Symptom/UBT benefits in controlled studies; real-world tetravalent IgY effective/safe in refractory infection |
| Oral bovine colostral/milk antibodies | Broad polyclonal coverage; food-grade experience | Batch consistency; gastric stability | Similar to IgY; standardized potency QC (neutralization assays) | Randomized controlled trials show decreased colonization markers/ symptoms. |
| Systemic monoclonal antibodies (IgG/ engineered IgA) | Tissue exposure via lamina propria; likely access to glands/mucus | Systemic AEs; Fc-mediated inflammation; higher CMC burden | Isotype/Fc engineering; tissue PK/PD confirmation in gastric biopsies | Translational rationale strong; human efficacy studies needed |
| Recombinant fragments (scFv, VHH) | Epitope precision; potential stability | Shorter half-life; delivery constraints | Mucoadhesive carriers; fusion to stability tags | Preclinical neutralization/ adhesion-blocking data; human studies pending |
| Quality/CMC (cross-cutting) | – | – | In vitro potency panels: Urease inhibition, adhesion blockade, VacA neutralization | Manufacturing/ QC guidance for polyclonals |

Adapted from: [37–40, 62, 63]

**Table 15.3** Potency versus safety in adjuvants and mucosal delivery platforms

| Platform | Potency (models) | Human/clinical status and safety notes | Comments/route considerations |
|---|---|---|---|
| CTB (carrier/adjuvant) | Enhances mucosal responses in mice | Historical use; reactogenicity depends on dose/route | Often as fusion to epitopes; sublingual vs. intragastric |
| LT/dmLT | High potency; dmLT effective with improved safety margin | Human mucosal programs under careful monitoring; gastric safety essential | Candidate for therapeutic regimens with biopsy-anchored escalation |
| Chitosan formulations | Improved residence/uptake; vaccine efficacy in models | Human data limited; generally acceptable safety in small studies | Useful for oral/sublingual delivery; can carry epitopes |
| α-GalCer | Drives invariant natural killer T cell help; supports Th1 components | Translational interest; staged dosing required | Watch for inflammation; route optimization |
| Alum/systemic adjuvants | Systemic immunization protective in models | Not mucosa-selective; gastric targeting limited | Reference for immunogenicity; less suitable for therapeutic gastric focus |

Adapted from: [19, 32, 33, 53, 68]

**Table 15.4** Recommended clinical trial designs for disease-blocking endpoints

| Phase | Population | Design features | Primary endpoints | Key secondary endpoints | Safety/biomarkers |
|---|---|---|---|---|---|
| I (vaccine or antibody; oral or systemic) | Adults with active noncomplicated gastritis/ulcer; exclude high bleeding risk | Sentinel cohorts; stepwise dose/route escalation; blinded histology | Histologic activity/chronicity change; endoscopic ulcer healing/time-to-healing | UBT delta; stool antigen; biopsy qPCR | Endoscopy/biopsy; cytokine panels; pepsinogen I/II; ex vivo urease/adhesion assays |
| II (therapeutic) | Refractory or antibiotic-intolerant; genotype-stratified | Randomized vs. optimized SOC; factorial for adjuvant/route where feasible | Histology + dyspepsia index (co-primary with MCID) | UBT/stool antigen; relapse at 6–12 month | Safety as above; predefined stopping rules |
| II (prophylactic; selective settings) | High-exposure cohorts (adults first) | Incident infection (serial UBT/stool) as endpoint; pediatric after adult safety | Incident infection rate | Symptom burden; histology in incident cases | Pediatric ethics framework; long-term follow-up |

Adapted from: [5, 36, 60, 61, 67]

## References


1. Suzuki S, Esaki M, Kusano C, Ikehara H, Gotoda T. Development of *H. pylori* treatment: how do we manage antimicrobial resistance? World J Gastroenterol. 2019;25(16):1907–12. https://doi.org/10.3748/wjg.v25.i16.1907.
2. Lahaie RG, Gaudreau C. *H. pylori* antibiotic resistance: trends over time. Can J Gastroenterol. 2000;14(10):895–9. https://doi.org/10.1155/2000/218256.
3. Li S, Zhao W, Xia L, Kong L, Yang L. How long will it take to launch an effective *H. pylori* vaccine for humans? Infect Drug Resist. 2023;16:3787–805. https://doi.org/10.2147/IDR.S412361.
4. Svennerholm AM, Lundgren A. Progress in vaccine development against *H. pylori*. FEMS Immunol Med Microbiol. 2007;50(2):146–56. https://doi.org/10.1111/j.1574-695X.2007.00237.x.
5. Mueller A, O'Rourke J, Chu P, Kim CC, Sutton P, et al. Protective immunity against helicobacter is characterized by a unique transcriptional signature. Proc Natl Acad Sci USA. 2003;100(21):12289–94. https://doi.org/10.1073/pnas.1635231100.
6. Agarwal K, Agarwal S. *H. pylori* vaccine: from past to future. Mayo Clin Proc. 2008;83(2):169–75. https://doi.org/10.4065/83.2.169.
7. Horton RE, Stuart J, Christensen H, Borrow R, Guthrie T, et al. Influence of age and carriage status on salivary IgA to Neisseria meningitidis. Epidemiol Infect. 2005;133(5):883–9. https://doi.org/10.1017/S0950268805004097.
8. Beissegulova G, Ramazanova B, Mustafina K, Begadilova T, Koloskova Y, et al. Prevalence of nasopharyngeal streptococcus Pneumoniae carriage in infants: a systematic review and meta-analysis of cohort studies and randomized controlled trials. PLoS One. 2024;19(12):e0315461. https://doi.org/10.1371/journal.pone.0315461.
9. Velin D, Michetti P. Immunology of *H. pylori* infection. Digestion. 2006;73(2–3):116–23. https://doi.org/10.1159/000094043.
10. Velin D, Straubinger K, Gerhard M. Inflammation, immunity, and vaccines for *H. pylori* infection. Helicobacter. 2016;21(Suppl 1):26–9. https://doi.org/10.1111/hel.12336.
11. Holland SM, DeLeo FR, Elloumi HZ, Hsu AP, Uzel G, et al. STAT3 mutations in the hyper-IgE syndrome. N Engl J Med. 2007;357(16):1608–19. https://doi.org/10.1056/NEJMoa073687.
12. Milner JD, Brenchley JM, Laurence A, Freeman AF, Hill BJ, et al. Impaired T(H)17 cell differentiation in subjects with autosomal dominant hyper-IgE syndrome. Nature. 2008;452(7188):773–6. https://doi.org/10.1038/nature06764.
13. Agarwal S, Cunningham-Rundles C. Gastrointestinal manifestations and complications of primary immunodeficiency disorders. Immunol Allergy Clin N Am. 2019;39(1):81–94. https://doi.org/10.1016/j.iac.2018.08.006.
14. Barzaghi F, Amaya Hernandez LC, Neven B, Ricci S, Kucuk ZY, et al. Long-term follow-up of IPEX syndrome patients after different therapeutic strategies: an international multicenter retrospective study. J Allergy Clin Immunol. 2018;141(3):1036–49. https://doi.org/10.1016/j.jaci.2017.10.041.
15. Covacci A, Telford JL, Del Giudice G, Parsonnet J, Rappuoli R. *H. pylori* virulence and genetic geography. Science. 1999;284(5418):1328–33. https://doi.org/10.1126/science.284.5418.1328.
16. Satin B, Del Giudice G, Della Bianca V, Dusi S, Laudanna C, et al. The neutrophil-activating protein (HP-NAP) of *H. pylori* is a protective antigen and a major virulence factor. J Exp Med. 2000;191(9):1467–76. https://doi.org/10.1084/jem.191.9.1467.
17. Stein M, Ruggiero P, Rappuoli R, Bagnoli F. *H. pylori* CagA: from pathogenic mechanisms to its use as an anti-cancer vaccine. Front Immunol. 2013;4:328. https://doi.org/10.3389/fimmu.2013.00328.
18. Ermak TH, Giannasca PJ, Nichols R, Myers GA, Nedrud J, et al. Immunization of mice with urease vaccine affords protection against *H. pylori* infection in the absence of antibodies and is mediated by MHC class II-restricted responses. J Exp Med. 1998;188(12):2277–88. https://doi.org/10.1084/jem.188.12.2277.

19. Sjokvist Ottsjo L, Flach CF, Clements J, Holmgren J, Raghavan S. A double mutant heat-labile toxin from Escherichia coli, LT(R192G/L211A), is an effective mucosal adjuvant for vaccination against *H. pylori* infection. Infect Immun. 2013;81(5):1532–40. https://doi.org/10.1128/IAI.01407-12.
20. Nedrud JG. Animal models for gastric helicobacter immunology and vaccine studies. FEMS Immunol Med Microbiol. 1999;24(2):243–50. https://doi.org/10.1111/j.1574-695X.1999.tb01290.x.
21. Gong R, Huang B, Ralser A, Friedrich V, Mibus C, et al. CagA-dependent Hobit(+) gastric tissue-resident memory T cells confer full protection from *H. pylori* reinfection. Gut. 2025;74(11):1792–803. https://doi.org/10.1136/gutjnl-2025-334781.
22. Bartfeld S, Bayram T, van de Wetering M, Huch M, Begthel H, et al. In vitro expansion of human gastric epithelial stem cells and their responses to bacterial infection. Gastroenterology. 2015;148(1):126–36. https://doi.org/10.1053/j.gastro.2014.09.042.
23. Schlaermann P, Toelle B, Berger H, Schmidt SC, Glanemann M, et al. A novel human gastric primary cell culture system for modelling *H. pylori* infection in vitro. Gut. 2016;65(2):202–13. https://doi.org/10.1136/gutjnl-2014-307949.
24. Pang MJ, Burclaff JR, Jin R, Adkins-Threats M, Osaki LH, et al. Gastric organoids: progress and remaining challenges. Cell Mol Gastroenterol Hepatol. 2022;13(1):19–33. https://doi.org/10.1016/j.jcmgh.2021.09.005.
25. Sebrell TA, Hashimi M, Sidar B, Wilkinson RA, Kirpotina L, et al. A novel gastric spheroid co-culture model reveals chemokine-dependent recruitment of human dendritic cells to the gastric epithelium. Cell Mol Gastroenterol Hepatol. 2019;8(1):157–71. https://doi.org/10.1016/j.jcmgh.2019.02.010.
26. Schumacher MA, Aihara E, Feng R, Engevik A, Shroyer NF, et al. The use of murine-derived fundic organoids in studies of gastric physiology. J Physiol. 2015;593(8):1809–27. https://doi.org/10.1113/jphysiol.2014.283028.
27. Bugaytsova JA, Piddubnyi A, Tkachenko I, Rakhimova L, Edlund JO, et al. Vaccination with *H. pylori* attachment proteins protects against gastric cancer. *bioRxiv*. 2023. https://doi.org/10.1101/2023.05.25.542131.
28. Ferrero RL, Thiberge JM, Huerre M, Labigne A. Recombinant antigens prepared from the urease subunits of helicobacter spp.: evidence of protection in a mouse model of gastric infection. Infect Immun. 1994;62(11):4981–9. https://doi.org/10.1128/iai.62.11.4981-4989.1994.
29. Ji Q, Ma J, Wang S, Liu Q. Embedding of exogenous B cell epitopes on the surface of UreB structure generates a broadly reactive antibody response against *H. pylori*. Immunology. 2024;171(2):212–23. https://doi.org/10.1111/imm.13703.
30. Del Giudice G, Covacci A, Telford JL, Montecucco C, Rappuoli R. The design of vaccines against *H. pylori* and their development. Annu Rev Immunol. 2001;19:523–63. https://doi.org/10.1146/annurev.immunol.19.1.523.
31. Sjokvist Ottsjo L, Jeverstam F, Yrlid L, Wenzel AU, Walduck AK, et al. Induction of mucosal immune responses against *H. pylori* infection after sublingual and intragastric route of immunization. Immunology. 2017;150(2):172–83. https://doi.org/10.1111/imm.12676.
32. Longet S, Abautret-Daly A, Davitt CJH, McEntee CP, Aversa V, et al. An oral alpha-galactosylceramide adjuvanted *H. pylori* vaccine induces protective IL-1R- and IL-17R-dependent Th1 responses. NPJ Vaccines. 2019;4:45. https://doi.org/10.1038/s41541-019-0139-z.
33. Gong Y, Tao L, Wang F, Liu W, Jing L, et al. Chitosan as an adjuvant for a *H. pylori* therapeutic vaccine. Mol Med Rep. 2015;12(3):4123–32. https://doi.org/10.3892/mmr.2015.3950.
34. Dubois A, Lee CK, Fiala N, Kleanthous H, Mehlman PT, et al. Immunization against natural *H. pylori* infection in nonhuman primates. Infect Immun. 1998;66(9):4340–6. https://doi.org/10.1128/IAI.66.9.4340-4346.1998.
35. Rossi G, Ruggiero P, Peppoloni S, Pancotto L, Fortuna D, et al. Therapeutic vaccination against *H. pylori* in the beagle dog experimental model: safety, immunogenicity, and efficacy. Infect Immun. 2004;72(6):3252–9. https://doi.org/10.1128/IAI.72.6.3252-3259.2004.
36. Stadtlander CT, Gangemi JD, Khanolkar SS, Kitsos CM, Farris HE Jr, et al. Immunogenicity and safety of recombinant *H. pylori* urease in a nonhuman primate. Dig Dis Sci. 1996;41(9):1853–62. https://doi.org/10.1007/BF02088757.

37. Zhang L, Xiao Y, Ji L, Lin M, Zou Y, et al. Potential therapeutic effects of egg yolk antibody (IgY) in *H. pylori* infections horizontal line a review. J Agric Food Chem. 2021;69(46):13691–9. https://doi.org/10.1021/acs.jafc.1c05398.
38. Guo R, Wu S, Guan L, Xie Y, Yang X, et al. Dietary multivalent anti-*H. pylori* immunoglobulin Y significantly increase the *H. pylori* eradication and improve the clinical symptoms in patients. Helicobacter. 2021;26(5):e12843. https://doi.org/10.1111/hel.12843.
39. Hao N, Liu B, Zhao M, Lu M, Chen F, et al. Real-world evidence of a novel tetravalent immunoglobulin Y effectiveness and safety in patients with the refractory *H. pylori* infection. BMC Infect Dis. 2024;24(1):647. https://doi.org/10.1186/s12879-024-09498-4.
40. Schade R, Calzado EG, Sarmiento R, Chacana PA, Porankiewicz-Asplund J, et al. Chicken egg yolk antibodies (IgY-technology): a review of progress in production and use in research and human and veterinary medicine. Altern Lab Anim. 2005;33(2):129–54. https://doi.org/10.1177/026119290503300208.
41. Chang H, Lee Y, Chen C, Tu Y. Microencapsulation protects immunoglobulin in yolk (IgY) specific against *H. pylori* urease. J Food Sci. 2006;67(1):15–20. https://doi.org/10.1111/j.1365-2621.2002.tb11351.x.
42. Jung S, Kim H, Lee S, Lee N. Stability of anti-*H. pylori* IgY under some condition. J Anim Sci Technol. 2005;47(5):851–6. https://doi.org/10.5187/JAST.2005.47.5.851.
43. Gu Q, Song D, Zhu M. Oral vaccination of mice against *H. pylori* with recombinant Lactococcus lactis expressing urease subunit B. FEMS Immunol Med Microbiol. 2009;56(3):197–203. https://doi.org/10.1111/j.1574-695X.2009.00566.x.
44. Liu W, Tan Z, Xue J, Luo W, Song H, et al. Therapeutic efficacy of oral immunization with a non-genetically modified Lactococcus lactis-based vaccine CUE-GEM induces local immunity against *H. pylori* infection. Appl Microbiol Biotechnol. 2016;100(14):6219–29. https://doi.org/10.1007/s00253-016-7333-y.
45. Oster P, Vaillant L, Riva E, McMillan B, Begka C, et al. *H. pylori* infection has a detrimental impact on the efficacy of cancer immunotherapies. Gut. 2022;71(3):457–66. https://doi.org/10.1136/gutjnl-2020-323392.
46. Che H, Xiong Q, Ma J, Chen S, Wu H, et al. Association of *H. pylori* infection with survival outcomes in advanced gastric cancer patients treated with immune checkpoint inhibitors. BMC Cancer. 2022;22(1):904. https://doi.org/10.1186/s12885-022-10004-9.
47. Gong X, Shen L, Xie J, Liu D, Xie Y, et al. *H. pylori* infection reduces the efficacy of cancer immunotherapy: a systematic review and meta-analysis. Helicobacter. 2023;28(6):e13011. https://doi.org/10.1111/hel.13011.
48. Ciccaglione AF, Di Giulio M, Di Lodovico S, Di Campli E, Cellini L, et al. Bovine lactoferrin enhances the efficacy of levofloxacin-based triple therapy as first-line treatment of *H. pylori* infection: an in vitro and in vivo study. J Antimicrob Chemother. 2019;74(4):1069–77. https://doi.org/10.1093/jac/dky510.
49. Li H, Niu J, Wang X, Niu M, Liao C. The contribution of antimicrobial peptides to immune cell function: a review of recent advances. Pharmaceutics. 2023;15(9) https://doi.org/10.3390/pharmaceutics15092278.
50. Ghiara P, Rossi M, Marchetti M, Di Tommaso A, Vindigni C, et al. Therapeutic intragastric vaccination against *H. pylori* in mice eradicates an otherwise chronic infection and confers protection against reinfection. Infect Immun. 1997;65(12):4996–5002. https://doi.org/10.1128/iai.65.12.4996-5002.1997.
51. Guo L, Hong D, Wang S, Zhang F, Tang F, et al. Therapeutic protection against *H. pylori* infection in Mongolian gerbils by oral immunization with a tetravalent epitope-based vaccine with polysaccharide adjuvant. Front Immunol. 2019;10:1185. https://doi.org/10.3389/fimmu.2019.01185.
52. Guo L, Liu K, Xu G, Li X, Tu J, et al. Prophylactic and therapeutic efficacy of the epitope vaccine CTB-UA against *H. pylori* infection in a BALB/c mice model. Appl Microbiol Biotechnol. 2012;95(6):1437–44. https://doi.org/10.1007/s00253-012-4122-0.
53. Guo L, Yin R, Liu K, Lv X, Li Y, et al. Immunological features and efficacy of a multi-epitope vaccine CTB-UE against *H. pylori* in BALB/c mice model. Appl Microbiol Biotechnol. 2014;98(8):3495–507. https://doi.org/10.1007/s00253-013-5408-6.

54. Xie W, Zhao W, Zou Z, Kong L, Yang L. Oral multivalent epitope vaccine, based on UreB, HpaA, CAT, and LTB, for prevention and treatment of *H. pylori* infection in C57BL/6 mice. Helicobacter. 2021;26(3):e12807. https://doi.org/10.1111/hel.12807.
55. Ikewaki J, Nishizono A, Goto T, Fujioka T, Mifune K. Therapeutic oral vaccination induces mucosal immune response sufficient to eliminate long-term *H. pylori* infection. Microbiol Immunol. 2000;44(1):29–39. https://doi.org/10.1111/j.1348-0421.2000.tb01243.x.
56. Sutton P, Wilson J, Kosaka T, Wolowczuk I, Lee A. Therapeutic immunization against *H. pylori* infection in the absence of antibodies. Immunol Cell Biol. 2000;78(1):28–30. https://doi.org/10.1046/j.1440-1711.2000.00881.x.
57. Becher D, Deutscher ME, Simpfendorfer KR, Wijburg OL, Pederson JS, et al. Local recall responses in the stomach involving reduced regulation and expanded help mediate vaccine-induced protection against *H. pylori* in mice. Eur J Immunol. 2010;40(10):2778–90. https://doi.org/10.1002/eji.200940219.
58. Graham DY, Opekun AR, Osato MS, El-Zimaity HM, Lee CK, et al. Challenge model for *H. pylori* infection in human volunteers. Gut. 2004;53(9):1235–43. https://doi.org/10.1136/gut.2003.037499.
59. Goto T, Nishizono A, Fujioka T, Ikewaki J, Mifune K, et al. Local secretory immunoglobulin a and postimmunization gastritis correlate with protection against *H. pylori* infection after oral vaccination of mice. Infect Immun. 1999;67(5):2531–9. https://doi.org/10.1128/IAI.67.5.2531-2539.1999.
60. Patel SK, Pratap CB, Jain AK, Gulati AK, Nath G. Diagnosis of *H. pylori*: what should be the gold standard? World J Gastroenterol. 2014;20(36):12847–59. https://doi.org/10.3748/wjg.v20.i36.12847.
61. Guo L, Yang H, Tang F, Yin R, Liu H, et al. Oral immunization with a multivalent epitope-based vaccine, based on NAP, urease, HSP60, and HpaA, provides therapeutic effect on *H. pylori* infection in Mongolian gerbils. Front Cell Infect Microbiol. 2017;7:349. https://doi.org/10.3389/fcimb.2017.00349.
62. den Hoed CM, de Vries AC, Mensink PB, Dierikx CM, Suzuki H, et al. Bovine antibody-based oral immunotherapy for reduction of intragastric *H. pylori* colonization: a randomized clinical trial. Can J Gastroenterol. 2011;25(4):207–13. https://doi.org/10.1155/2011/672093.
63. Hu D, Zhang F, Zhou J, Xu B, Zhang H, et al. The clearance effect of bovine anti-*H. pylori* antibody-containing milk in O blood group *H. pylori*-infected patients: a randomized double-blind clinical trial. J Transl Med. 2015;13:205. https://doi.org/10.1186/s12967-015-0558-1.
64. Horie K, Horie N, Abdou AM, Yang JO, Yun SS, et al. Suppressive effect of functional drinking yogurt containing specific egg yolk immunoglobulin on *H. pylori* in humans. J Dairy Sci. 2004;87(12):4073–9. https://doi.org/10.3168/jds.S0022-0302(04)73549-3.
65. Tu HF, et al. Virus-like particle vaccine displaying an external, membrane adjacent MUC16 epitope elicits ovarian cancer-reactive antibodies. J Ovarian Res. 2024;17(1):19. https://doi.org/10.1186/s13048-023-01325-9.
66. Haghighi E, Abolmaali SS, Dehshahri A, Mousavi Shaegh SA, Azarpira N, et al. Navigating the intricate in-vivo journey of lipid nanoparticles tailored for the targeted delivery of RNA therapeutics: a quality-by-design approach. J Nanobiotechnol. 2024;22(1):710. https://doi.org/10.1186/s12951-024-02972-w.
67. Kabir S. The current status of *H. pylori* vaccines: a review. Helicobacter. 2007;12(2):89–102. https://doi.org/10.1111/j.1523-5378.2007.00478.x.
68. Gottwein JM, Blanchard TG, Targoni OS, Eisenberg JC, Zagorski BM, et al. Protective anti-helicobacter immunity is induced with aluminum hydroxide or complete Freund's adjuvant by systemic immunization. J Infect Dis. 2001;184(3):308–14. https://doi.org/10.1086/322032.

# The Future of Immunotherapy 16

Xiaobo He

## 16.1 Introduction

The efficacy of antibiotic therapy is increasingly compromised by the growing and formidable challenge of bacterial antimicrobial resistance, which has created an urgent and pressing need for the development of effective alternative strategies to combat these resilient pathogens. In this context, immunotherapy has emerged as a promising alternative approach, specifically targeting the elimination of *Helicobacter pylori* by modulating the host immune system to enhance its ability to fight off the infection.

In recent years, significant strides have been made in the preclinical and clinical development of various immunotherapies, which primarily include innovative vaccines alongside immunomodulatory adjuncts such as antimicrobial peptides (AMPs) and probiotics. However, despite these advancements, these strategies have not yet demonstrated satisfactory efficacy as standalone treatments for *H. pylori* infection, with most remaining in the preclinical stage or being utilized only as adjuvant therapies. This situation reflects the substantial scientific challenges that persist within the field of immunotherapy for *H. pylori*.

Specifically, the development of effective vaccines is hampered by the genetic diversity and high mutation rate of *H. pylori*, which complicates efforts to induce broad-spectrum and durable immune protection. Moreover, host genetic diversity leads to inter-individual heterogeneity in the intensity, type, and clinical outcome of vaccine-induced immune responses. Furthermore, the bacterium's unique adaptations for gastric mucosal colonization impose additional demands on vaccines, necessitating the elicitation of effective local immune responses to successfully

X. He (✉)
National Key Laboratory of Immunity and Inflammation & Institute of Immunology, College of Basic Medical Sciences, Naval Medical University, Shanghai, China
e-mail: hexiaobo@immunol.org



J. Gao, Y. Du (eds.), *Non-Antibiotic Therapies for Helicobacter pylori*,
https://doi.org/10.1007/978-981-95-9598-3_16

combat the infection. The complex interactions between the bacterium and the host immune system also present significant challenges for achieving precise immunomodulation through various strategies, making the task of developing effective therapies even more daunting.

To address these multifaceted challenges, this chapter undertakes a comprehensive review of current strategies and explores promising future directions in the field. It focuses on the optimization of vaccine antigens to enhance their effectiveness, the development of novel mucosal adjuvants and delivery systems that can improve immune responses, the identification of potential immunomodulatory targets that could be leveraged for therapeutic benefit, and the design of synergistic combination therapies that may offer a more robust approach to tackling *H. pylori* infections.

## 16.2 Next-Generation Vaccines

From past to present, the development of an *H. pylori* vaccine has remained an ongoing endeavor, yet most attempts have not advanced beyond the preclinical stage. A limited number of vaccine candidates have progressed into clinical trials and have shown promising preliminary results. For example, an oral *H. pylori* vaccine based on UreB and labile enterotoxin B subunit (LtB) components entered Phase III clinical trials and demonstrated a 71.8% reduction in infection incidence 1 year after vaccination. However, its protective efficacy declined to 55% after 2 years, indicating insufficient long-term durability [1]. Other related clinical studies have ultimately been unsuccessful due to failures in achieving complete bacterial clearance, providing satisfactory protection, or sustaining long-term efficacy [2–4]. Nevertheless, the experience accumulated throughout this process has yielded valuable insights to guide future vaccine development. These include the selection of highly conserved and diverse antigens to counter bacterial genetic variability and high mutation rates, the use of tailored adjuvants and delivery systems to elicit potent yet safe immune protection, and the combination of immunomodulators targeting *H. pylori* immune evasion mechanisms to enhance vaccine effectiveness. Furthermore, elucidating the mechanisms by which host genetic polymorphisms influence vaccine-induced immunity will further inform the optimization of vaccine design, personalized vaccination strategies, and precision interventions in the future.

### 16.2.1 Vaccine Antigens

*H. pylori* possesses a complex antigenic profile, including vacuolating cytotoxin A (VacA), cytotoxin-associated gene A (CagA), and blood group antigen-binding adhesin A (BabA) [5]. Most candidate vaccines have been developed based on one or several of these protein antigens. However, studies have revealed substantial genetic diversity among different *H. pylori*-associated antigenic proteins such as VacA and CagA. This diversity is widely observed within a single infected individual, among different individuals, and across populations in different geographical regions [6–8]. Such genetic variability is associated with high-frequency DNA mutation and recombination events [9]. Therefore, the genetic diversity and antigenic variability of *H. pylori* are critical considerations for the durability of vaccine-induced immune protection and its broad-spectrum effectiveness across diverse populations. In selecting vaccine antigens, enhancing antigen conservation and diversity, as well as identifying highly immunogenic antigens, is crucial to addressing the challenges posed by bacterial genetic diversity and high mutation rates, thereby enabling the induction of potent and broad-spectrum immune protection.

Multi-epitope vaccines are regarded as a promising strategy to enhance vaccine efficacy, as they achieve broader antigenic coverage by fusing dominant immunogenic epitopes derived from different antigenic proteins. Moreover, multiple studies have indicated that both cell-mediated and antibody-mediated immune responses play significant roles in vaccine-induced immune protection [10]. By incorporating both T-cell and B-cell epitopes, multi-epitope vaccine designs can elicit coordinated cellular and humoral immune responses, leading to more balanced and robust immune protection against *H. pylori*. In one study, a Lactobacillus-based vaccine composed of T-cell and B-cell epitopes from urease subunits effectively induced urease-specific antibody responses and reduced *H. pylori* colonization in a Mongolian gerbil model, demonstrating the potential of this multi-epitope approach [11].

Recent advances in bioinformatics have significantly accelerated the development of multi-epitope vaccines based on reverse vaccinology. Unlike traditional vaccine development, reverse vaccinology focuses on predicting potential protective antigens through analysis of pathogen genomic sequences, thereby expediting early-stage antigen identification. The general workflow involves: obtaining sequences of immunodominant proteins from databases such as NCBI; identifying conserved regions through sequence analysis; predicting and screening potential T-cell and B-cell epitopes based on comprehensive criteria, including immunogenicity and allergenicity; constructing multi-epitope immunogens using suitable linkers; and evaluating and refining the antigen design through physicochemical property analysis, molecular docking simulations, and immune response modeling. Finally, codon optimization is often applied to enhance the immunogenicity of the vaccine antigen [12–14]. Following this approach, Biniam et al. developed a novel and stable multi-epitope vaccine antigen using highly conserved virulence factors of *H. pylori*—protective surface antigen D15 and Cag pathogenicity island protein Cag11. By employing immunoinformatic tools, they demonstrated the antigen's ability to effectively induce immune responses via immune simulation [12]. In another recent representative study, researchers designed an in silico self-amplifying RNA-based multi-epitope peptide vaccine. This design incorporated five antigenic proteins (UreB, BabA, HpaA, CagA, and VacA) and predicted 12 epitopes covering B-cell, cytotoxic T-cell, and helper T-cell epitopes based on immunogenicity, allergenicity, toxicity, and cytokine induction potential. Further molecular docking and post-translational modification analyses supported the strong immunogenic profile of the construct. Moreover, by targeting conserved regions across *H. pylori* strains from diverse geographical origins, the vaccine achieved broad coverage against heterologous strains [15]. Similarly, other studies have selected B-cell, cytotoxic T-cell, and helper T-cell epitopes from six *H. pylori* antigens (CagA, GGT, NapA, PatA, urease, and VacA) to construct an mRNA-based multi-epitope vaccine [16]. These studies illustrate that reverse vaccinology, supported by diverse immunoinformatic tools, offers novel perspectives and optimization strategies for *H. pylori* vaccine antigen design. On one hand, the inclusion of diverse B- and T-cell epitopes enables precise induction of coordinated humoral and cellular immunity. On the other hand, epitopes derived from multiple antigens facilitate comprehensive and specific immune responses capable of addressing the challenge of genetic diversity. These potential advantages have stimulated growing research interest in this area [17, 18]. Despite offering a promising direction for antigen optimization, this strategy still faces several challenges. For instance, while in silico predictions can reveal potential immunogenicity, they cannot fully recapitulate the complexity of the human immune microenvironment. Moreover, variations in the screening criteria and analytical perspectives among different bioinformatic tools may lead to contradictory results, underscoring the need for careful selection and integrated interpretation of computational outcomes. These limitations also indicate important directions for future refinement of this strategy.

### 16.2.2 Induction of Effective Mucosal Immune Responses

The gastric mucosa serves as a critical immunological barrier within the human digestive system, playing a pivotal role in defending against pathogen invasion. *H. pylori* exhibits distinct mucosal tropism, primarily mediated by surface adhesins such as HpaA, BabA, and SabA, which specifically recognize and bind to glycostructures like Lewis antigens on gastric epithelial cells [19]. To survive the highly acidic gastric environment, *H. pylori* utilizes urease to hydrolyze urea and generate ammonia, thereby neutralizing local acidity and creating a favorable pH niche for colonization [20]. Recent studies further reveal that the acid-responsive two-component system ArsRS regulates SabA expression in a pH-dependent manner [21]. Specifically, under acidic conditions, phosphorylated ArsR suppresses sabA transcription, enabling bacterial adaptation to pH variations across different gastric regions and optimizing colonization sites. These findings illustrate the sophisticated environmental adaptation mechanisms employed by *H. pylori* to facilitate mucosal colonization.

Upon infection, the intricate mucosal immune system of the host promptly initiates a series of multi-layered defenses designed to exert potent antimicrobial effects against invading pathogens. In the initial phase of the innate mucosal immune response, once the pathogen successfully breaches the physical barrier that is primarily composed of epithelial cells and a protective layer of mucus, it is met with defense from innate immune cells, including macrophages and mast cells. These immune cells play a crucial role in mediating the clearance of the pathogen through processes such as phagocytosis or by activating immune responses via pattern recognition receptors (PRRs). Among the various PRRs, Toll-like receptors (TLRs)—notably TLR2, TLR4, TLR5, and TLR9—are particularly significant as they are instrumental in recognizing specific molecular patterns associated with *H. pylori* and subsequently triggering innate immune activation along with the release of pro-inflammatory cytokines. However, extensive research has revealed that *H. pylori* has developed a range of sophisticated strategies to subvert or evade PRR-mediated immune recognition [22]. The adaptive mucosal immune response is activated upon the presentation of antigens, a process that is primarily mediated by antigen-presenting cells (APCs), including macrophages and dendritic cells (DCs), as illustrated in Fig. 16.1. When naïve T cells become activated, they undergo a differentiation process that leads to the formation of various subsets, notably Th1, Th2, Th17, and Treg cells. Th1 and Th2 cells are particularly significant in the immune system, as they play crucial roles in the elimination of intracellular and extracellular pathogens, respectively, thus ensuring a robust defense against a wide array of infectious agents [23]. Th17 cells, on the other hand, contribute to the activation of the innate immune response by recruiting neutrophils to the site of infection and secreting a variety of chemokines and pro-inflammatory cytokines that amplify the immune response [24]. Additionally, Th17 cells produce interleukin (IL)-2 and IL-17, which serve to activate Th1 responses, highlighting the synergistic functions of these T cell subsets in orchestrating a comprehensive immune response

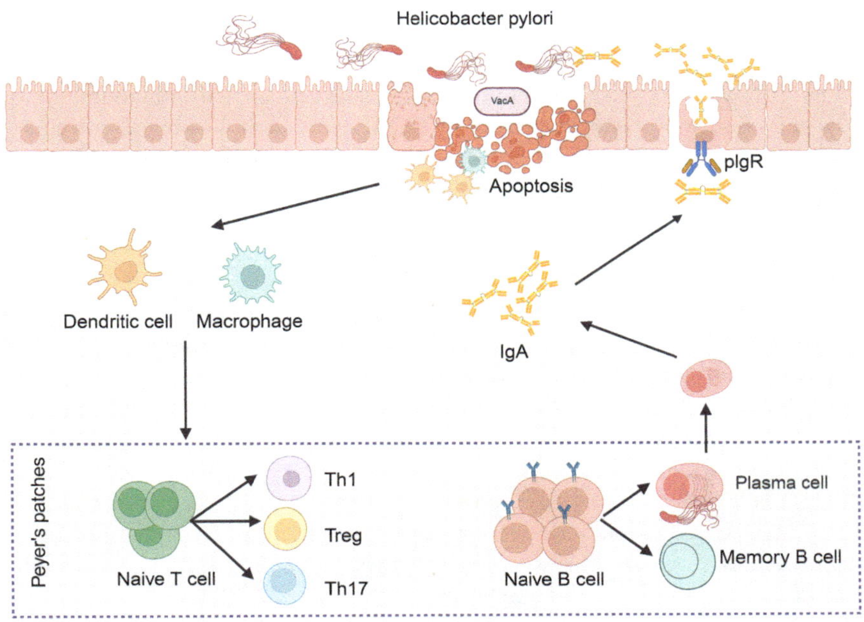


**Fig. 16.1** Adaptive mucosal immune response. (Image created with BioRender.com, with permission)

[25, 26]. Furthermore, Th17 cells have the capacity to upregulate polymeric immunoglobulin receptors through the action of IL-17, which, in turn, promotes the production of secretory immunoglobulin A (sIgA). As a pivotal component of mucosal immunity, sIgA plays a vital role in counteracting the colonization of *H. pylori*. It achieves this by directly blocking the binding of adhesins, such as SabA, to their respective receptors, thereby facilitating the aggregation of pathogens and their subsequent clearance from the mucosal surface—a mechanism referred to as immune exclusion. In addition to its protective functions, sIgA also contributes to the maintenance of mucosal homeostasis by inducing immune tolerance; for instance, sIgA immunocomplexes can be captured by DCs through DC-SIGN (Dendritic Cell-Specific Intercellular adhesion molecule-3-Grabbing Non-integrin). This interaction leads to a reduction in DC maturation and the secretion of inflammatory cytokines, ultimately promoting the differentiation of regulatory T cells (Treg), which further supports the delicate balance of the immune system in the mucosal environment [27]. Therefore, activating mucosal immunity represents a potential strategy for promoting the clearance of *H. pylori*. However, excessive mucosal immune responses may exacerbate inflammation and lead to more severe pathological damage. For example, Th1 cells secrete IFN-$\gamma$ and rely on the transcription factor T-bet to drive chronic inflammation. Th17 cells play a complex dual role, as they may aggravate gastric inflammation and recruit neutrophils through IL-17 production. In Treg-deficient mouse models, Th2-mediated immune responses are closely associated with the development of autoimmune gastritis. Thus, the regulation of mucosal immunity must consider not only its beneficial role in bacterial clearance but also its potential risks in driving inflammation and tissue pathology [24, 28, 29].

In vaccinology, the successful induction of coordinated cellular immunity, mediated by Th1, Th2, and Th17 cells, along with the establishment of humoral immunity—particularly through the promotion of sIgA production—may contribute to enhanced clearance of bacterial pathogens [30–32]. Achieving this goal involves addressing several challenges. Conventional vaccination routes, such as intramuscular injections, while capable of eliciting systemic immunity, frequently fall short in generating strong mucosal immune responses [33]. Mucosal-targeted delivery routes, such as oral administration, more effectively induce mucosal immunity but require additional consideration of complex factors such as the gastrointestinal environment. The development of mucosal vaccines that leverage novel delivery systems emerges as a promising strategy aimed at improving both mucosal targeting and the stability of antigens, thereby facilitating a more robust induction of mucosal immune responses. Additionally, the advancement of mucosal adjuvants is of paramount importance for fostering coordinated and potent mucosal immune responses that are crucially induced by these vaccines. However, it should be noted that the claims of improved efficacy in this chapter are primarily grounded in preclinical models and surrogate immunologic outcomes, such as sIgA and Th1/Th17 responses.

### 16.2.3 Adjuvants and Vaccine Delivery

Vaccine antigens alone often fail to induce sufficient protective immunity; therefore, adjuvants are widely employed in vaccine development to ensure a more effective immunological reaction [34]. Aluminum-based adjuvants, the first licensed for human use, remain the most extensively utilized. Common forms include aluminum phosphate and aluminum hydroxide. They function not only as delivery systems by creating an antigen depot that significantly enhances the uptake of antigens by APCs, but they also act as immunostimulants by activating the NLRP3 inflammasome. This activation leads to a Th2-biased immune response and the subsequent production of cytokines such as IL-1β, which are vital for orchestrating the immune response [35]. For instance, *H. pylori* vaccines that are formulated with aluminum hydroxide have been shown to effectively induce IL-5 secretion and robust T-cell responses [36]. In addition to aluminum-based adjuvants, emulsion-based adjuvants, such as MF59 and AS03, represent another class that is commonly used in vaccines for influenza and other epidemic diseases. These emulsion-based adjuvants enhance antigen delivery through a sustained-release mechanism, which improves antigen presentation to APCs and thereby boosts the immune response. While they are effective at inducing a Th2-biased immunity, their capacity to stimulate Th1 responses is relatively limited, which may affect the overall immune profile elicited by the vaccine [37]. Furthermore, conventional adjuvants are typically administered via non-mucosal routes, such as intramuscular injection, and they are generally ineffective at eliciting robust, site-specific mucosal immunity in the gastric mucosa, highlighting a limitation in their application and effectiveness.

The use of mucosal adjuvants offers a viable solution. These are substances administered via mucosal routes (e.g., oral and nasal) that significantly enhance immune responses to co-delivered antigens [38, 39]. Bacterial toxins and their derivatives, including cholera toxin (CT) and *Escherichia coli* heat-labile enterotoxin (LT), have been extensively investigated for this purpose. For instance, a recombinant multi-epitope vaccine has been developed by fusing the CT B subunit (CTB) with specific antigenic fragments derived from *H. pylori* urease subunits. This innovative approach has been shown to induce remarkably high levels of immunoglobulin A (IgA) and elicit a mixed immune response characterized by Th1, Th2, and Th17 cell activation, with the CTB component significantly enhancing the protective efficacy of the vaccine [40]. In another study, researchers utilized the *E. coli* heat-LT B subunit (LtB) as an adjuvant, demonstrating that this particular adjuvant not only increased fecal sIgA levels but also effectively reduced *H. pylori* colonization in the gastrointestinal tract. Notably, LtB was found to enhance the protective effect of an oral vaccine by promoting leukocyte migration and facilitating mucosal leakage, thereby contributing to a more robust immune response [41]. However, the potential toxicity associated with these toxin-derived adjuvants in human subjects poses significant limitations to their clinical application, which must be carefully considered [42]. The strategic combination of whole-cell bacteria or recombinant antigens with mucosal adjuvants is critical for achieving optimal vaccine efficacy. In a study conducted by Sukanya et al., the whole-cell vaccine adjuvanted with a multiple mutant CT (mmCT) led to a remarkable reduction in *H. pylori* colonization within mouse stomachs, achieving reductions ranging from 50- to 125-fold [43]. The mmCT adjuvant retains immunostimulatory potency that is comparable to that of wild-type CT, yet it exhibits an improved safety profile [44], offering a valuable strategy for optimizing other bacterial toxin-based adjuvants.

PRR agonists represent another promising class of mucosal adjuvants. Different PRRs recognize pathogen-associated molecular patterns (PAMPs) and cooperatively activate innate immunity, thereby shaping adaptive immune responses. Their agonists are increasingly explored as vaccine adjuvants to enhance immunogenicity [45]. For example, Rasmus et al. demonstrated that a combination of the TLR9 agonist CpG2006 [45] and a Mincle agonist effectively induced robust Th1/Th17 responses. Moreover, T cells from immunized mice exhibited strong expansion and mucosal recall responses upon antigen re-exposure 6 months later [46], underscoring the potential of PRR agonist-based adjuvants in promoting durable mucosal immunity and memory. Other studies have investigated particulate adjuvants, such as polysaccharides and virus-like particles (VLPs) [47, 48]. For instance, an *H. pylori* protein vaccine formulated with chitosan induced enhanced humoral immunity and a mixed Th1/Th2 response, improving bacterial clearance [49]. This study also found that chitosan exhibited adjuvant efficacy comparable to CT, supporting its potential as a mucosal adjuvant. Similarly, plant polysaccharides have been shown to elevate sIgA levels and promote Th1/Th17 responses [50]. Cytokines, which play central roles in bridging innate and adaptive immunity, are also considered effective adjuvants for enhancing mucosal vaccine potency. Examples include IL-1, IL-2, IL-7, IL-15, and interferons [51, 52]. However, their application requires careful consideration of potential dose-dependent toxicity (Fig. 16.2).

In addition to mucosal adjuvants, growing research attention has been directed toward novel delivery systems—particularly nanoparticles—for their advantageous

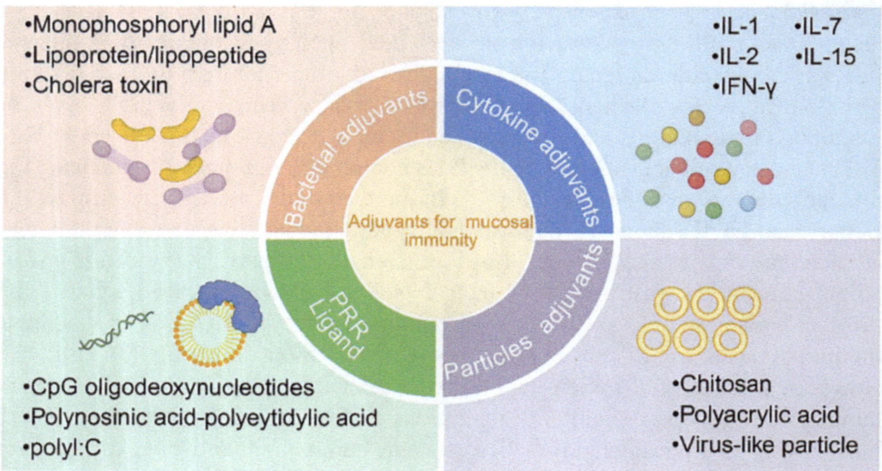


**Fig. 16.2** Adjuvants for mucosal immunity. (Image created with BioRender.com, with permission)

role in inducing mucosal immune responses. An ideal vaccine delivery strategy must overcome several key challenges: resisting the harsh acidic environment of the stomach, avoiding enzymatic degradation, penetrating the mucosal barrier, and enhancing targeting efficiency to mucosal tissues [53]. Nanoparticles have been extensively investigated as vaccine carriers, typically by covalently conjugating or encapsulating pathogen antigens. For example, Wang et al. developed vaccines based on PLGA (poly(lactic-co-glycolic acid)) microspheres and PLGA–chitosan microspheres encapsulating *H. pylori* proteins. These microspheres demonstrated excellent sustained antigen release in simulated gastric and intestinal fluids in vitro [54]. In another study, PLGA nanoparticles encapsulating a recombinant *H. pylori* antigen exhibited significant acid resistance in gastric conditions. In a mouse model, this vaccine elicited high levels of urease-specific antibodies and a Th1/Th17-polarized response, achieving complete protection against *H. pylori* challenge in 43% of the immunized animals [55]. To address the potential limitation of poor bioadhesion of plain PLGA particles, chitosan is often incorporated to improve mucosal adhesion and enhance antigen-specific mucosal immunity [56]. Chitosan itself exhibits inherent antibacterial activity and has been widely studied as a mucosal vaccine delivery vehicle [49, 57, 58]. In another approach, alginate-based nanocarriers delivering a DNA vaccine encoding UreH elicited strong mucosal immunity and systemic Th1 responses in mice. Lipid nanoparticles, which have demonstrated remarkable success in mRNA COVID-19 vaccine delivery, represent another potent nanocarrier platform [59]. Furthermore, nanoemulsion (NE)-based systems have been applied for intranasal vaccination against *H. pylori*. Such NEs not only prolong antigen release but also enhance antigen capture by APCs, thereby promoting potent Th1 responses [60]. Notably, several nanocarriers—including liposomes, chitosan, VLPs, and outer membrane vesicles—exhibit intrinsic self-adjuvanting properties, providing an additional advantage for enhancing mucosal immunity [49, 61–64].

Enhancing vaccine-induced mucosal immunity is promising for effective clearance of *H. pylori*. This section has highlighted a range of valuable strategies that involve the use of mucosal adjuvants and innovative delivery systems designed to optimize the immune response. These approaches strengthen mucosal immune induction through multiple mechanisms—such as amplifying local immune activation, improving mucosal antigen delivery, and prolonging antigen release—offering vital support and future directions for overcoming the limited efficacy of current *H. pylori* vaccine candidates. However, given the documented toxicity of toxin-derived and cytokine adjuvants, along with the dual role of Th17 cells in *H. pylori* pathogenesis [24], precise immunomodulation requires an integrated safety framework.

### 16.2.4 Host Genetic Polymorphisms

Host genetic polymorphism represents one of the pivotal factors contributing to inter-individual heterogeneity in vaccine responses. Multiple studies have demonstrated that genetic variation influences both the efficacy and durability of vaccine-induced protection. Among the most extensively investigated genetic determinants are the major histocompatibility complex (MHC) genes. Specifically, alleles and specific amino acid variants of HLA class II genes, such as HLA-DRB1 and HLA-DPB1, have been robustly associated with varying levels of antibody responses elicited by different vaccines [65]. These associations are attributed to differences in the efficiency of antigen presentation by APCs to $CD4^+$ T cells. Similarly, HLA class I genes have been linked to $CD8^+$ T cell-mediated immunity in response to vaccines such as influenza and measles [66, 67]. A deficiency in the complement component C4A gene has also been implicated in non-responsiveness to the hepatitis B vaccine, underscoring the role of the complement system in B cell activation, survival, or antibody class switching [68]. Beyond HLA, variations in genes encoding cytokines and their receptors involved in immune regulation further contribute to heterogeneous vaccine outcomes. For example, single-nucleotide polymorphisms (SNPs) in the IFNL3/IFNL4 locus have been correlated with seroconversion rates and antibody titers following influenza vaccination, with proposed mechanisms involving modulation of the Th1/Th2 balance [69]. Additionally, polymorphisms in genes of inflammatory cytokines such as IL-10 and IL-12B have been associated with differential vaccine responses [70, 71]. Genetic variants in PRRs, including TLR2, TLR4, and TLR8, as well as in downstream signaling molecules like MAL/TIRAP, may modulate vaccine-induced adaptive immunity by altering innate immune sensing [72, 73]. Similarly, specific combinations of killer-cell immunoglobulin-like receptors and their HLA ligands have been linked to variations in total IgA levels, highlighting the intricate crosstalk between innate and adaptive immunity in shaping vaccine responsiveness [74, 75]. Currently, genome-wide association studies serve as an important approach for identifying genetic variants associated with vaccine responses. Future research integrating multi-omics data, such as transcriptomics and proteomics, will further elucidate the molecular pathways through which SNPs exert their effects. In summary, host genetic polymorphisms—spanning HLA, cytokine and receptor genes, and innate immune regulators—constitute a fundamental source of inter-individual variability in vaccine immunogenicity and efficacy. A deeper understanding of these genetic determinants will be essential for advancing the development of more effective and personalized *H. pylori* vaccine strategies.

## 16.3 Targeted Modulation of Immune Evasion Mechanisms

*H. pylori* employs a diverse array of molecular strategies to dysregulate both innate and adaptive immune cells, thereby subverting host antimicrobial responses and facilitating persistent colonization. Therapeutic interventions that target these specific molecules or signaling pathways to reverse immune suppression represent a promising strategy for restoring effective anti-infective immunity. While agonists of PRRs constitute a relatively well-explored immunomodulatory approach, their application has been largely confined to vaccine adjuvants. Moreover, the non-specific immune activation induced by many PRR agonists raises concerns about potential excessive inflammation, necessitating a careful balance between effective immune stimulation and inflammatory pathology.

Beyond evading PRR recognition, *H. pylori* has been demonstrated to manipulate the function and phenotype of key immune cells—including macrophages, DCs, and T cells—through multiple mechanisms to foster a chronic infection state. A recent single-cell RNA sequencing study further illuminated the immunosuppressive landscape within the human gastric ecosystem during *H. pylori* infection, characterized by impaired antigen presentation and restricted proliferation of mucosa-infiltrating T cells. Consequently, targeting key nodes in these immunomodulatory pathways to counteract the negative regulation imposed by *H. pylori* and bolster host defense represents a viable strategy for precise immune intervention. This section outlines potential immunomodulatory targets and pathways in macrophages, DCs, and T cells. The application of specific inhibitors or immune cell-targeting modulators to reverse *H. pylori*-mediated suppression holds significant therapeutic promise and provides a rational foundation for developing precise and effective immunomodulatory strategies.

### 16.3.1 Macrophages

Macrophages play a pivotal role in the immune response to *H. pylori* infection by: [1] phagocytosing the pathogen, processing its antigens, and presenting antigenic peptides via MHC class II molecules to $CD4^+$ T cells to initiate adaptive immunity; and [2] recognizing PAMPs through PRRs to trigger the production of inflammatory mediators and drive innate immune responses. However, *H. pylori* subverts these macrophage functions to facilitate immune evasion and persistence. For instance, the virulence factor VacA interferes with phagosome–lysosome fusion, thereby impairing bacterial clearance and promoting intracellular survival [76]. Phosphorylated CagA upregulates heme oxygenase-1 expression, polarizing macrophages toward an anti-inflammatory M2 phenotype [77]. Phospholipase A disrupts macrophage membrane integrity, simultaneously attenuating autophagy and inducing RIP1-dependent apoptosis [78]. Additionally, outer membrane proteins such as HP1286 and HP1173 contribute to macrophage death by promoting tumor necrosis factor (TNF)-α and IL-8 secretion and activating caspase-dependent apoptotic pathways [79, 80].

### 16.3.2 Dendritic Cells

DCs play a critical role in the initiation of adaptive immunity, primarily by presenting antigenic peptides through MHC molecules and providing essential co-stimulatory signals for the activation and proliferation of T cells. However, in the context of *H. pylori* infection, these DCs are driven toward a tolerogenic phenotype, which is characterized by a significant downregulation of crucial co-stimulatory molecules, such as CD80, CD86, and CD40, as well as a reduction in the expression of MHC class II molecules. This shift toward a tolerogenic state in DCs leads to the promotion of immunosuppressive regulatory T cells (Tregs), which further suppress the activation of effector T cells, ultimately undermining the adaptive immune response that is necessary to combat the bacterium effectively. The mechanisms underlying this functional subversion are complex and multifactorial. The outer membrane protein OipA has been shown to play a significant role in this process by downregulating the expression of key molecules, including CD40, CD86, and MHC class II on DCs, thereby contributing to the overall immunosuppressive environment induced by *H. pylori* infection [81]. VacA impairs DC maturation by suppressing the expression of co-stimulatory molecules and the secretion of pro-inflammatory cytokines such as IL-1 and TNF-α following LPS stimulation, an effect mediated through the transcription factor E2F1 [82]. Phosphorylated CagA activates SHP-2, which, in turn, inhibits the kinase activity of TBK-1, leading to reduced IRF3 phosphorylation and nuclear translocation, ultimately suppressing IFN-γ production and impairing DC-mediated immune activation [83]. Furthermore, Lewis antigen-positive *H. pylori* strains engage the DC-SIGN receptor, inhibiting the ability of DCs to drive Th1 polarization [84]. The bacterial enzyme γ-glutamyl transpeptidase (GGT) promotes tolerogenic DC differentiation by modulating cAMP (Cyclic Adenosine Monophosphate) signaling and suppressing IL-6 secretion [85].

### 16.3.3 T Cell

*H. pylori* directly suppresses T-cell function through multiple virulence determinants. VacA induces cell cycle arrest at the G1/S phase, inhibiting T-cell proliferation and activation [86]. It also promotes the biased differentiation of immunosuppressive $CD4^{+}$ $CD25^{high}$ $Foxp3^{+}$ regulatory T cells (Tregs) [87, 88]. Similarly, GGT disrupts T-cell metabolism by depleting extracellular glutamine, leading to cell cycle arrest in the G1 phase and downregulation of c-Myc and IRF4, which collectively impair T-cell effector functions [89].

The CagA protein contributes to the suppression of T-cell activity through a variety of pathways. CagA upregulates squalene epoxidase, which increases palmitoyl-CoA levels. This enhances programmed death-ligand 1 (PD-L1) palmitoylation, reduces its ubiquitination, and ultimately stabilizes PD-L1 protein to suppress T-cell activity [90]; Additionally, CagA downregulates the expression of the co-stimulatory molecule B7-H2 within the gastric mucosa, a process that effectively suppresses the Th17 immune responses, which are crucial for combating certain infections and maintaining immune balance [91]. In addition, *H. pylori* enhances the release of soluble NKG2D (Natural Killer Group 2, member D) ligands from gastric epithelial cells in a VacA- and CagA-dependent manner, directly impairing the NKG2D-mediated cytotoxicity of effector T cells and NK cells [92]. Furthermore, this pathogenic bacterium induces the expression of PD-L1 on gastric epithelial cells through the actions of CagA and urease, a process that may lead to T-cell exhaustion, diminishing their ability to respond effectively to infections [93]. Furthermore, *H.*

*pylori*-derived ADP-heptose indirectly attenuates Th1 immunity by inhibiting DC activation [94].

## 16.4 Synergistic Combination Therapy Strategies

The eradication of *H. pylori* infection relies primarily on antibiotic regimens. However, the rising prevalence of antibiotic resistance has led to increasing treatment failure rates [95]. In this context, combination therapies integrating multiple treatment modalities have been proposed to overcome antibiotic resistance and enhance the efficacy of *H. pylori* clearance [96]. As immunotherapeutic approaches against *H. pylori* continue to advance, various combinations built around immunotherapy offer promising avenues for improving clinical outcomes.

The common combinatorial approach involves vaccines coupled with other immunomodulatory adjuncts. Immune stimulants frequently serve as vaccine adjuvants or delivery systems to exert synergistic anti-*H. pylori* effects. Agonists of TLRs are among the most widely studied adjuvants in this context [97]. TLR activation not only triggers innate immune responses but also shapes adaptive immunity, making them particularly suitable for enhancing vaccine-induced protection [98]. Given that *H. pylori* has evolved mechanisms to evade PRR-mediated recognition, TLR agonists can specifically counterbalance this immune escape. For example, adjuvants such as poly(I:C) and CpG ODN promote mucosal sIgA production and Th1/Th17 responses, acting synergistically with vaccine antigens [99]. Probiotics have also been engineered as live attenuated vaccine vectors. A recombinant *Lactococcus lactis* strain expressing multiple epitopes from urease, HpaA, HSP60, and NAP effectively targeted M cells and induced high levels of serum IgG and mucosal sIgA [53]. Similarly, a recombinant *Lactobacillus acidophilus* expressing adhesin *Hp0410* achieved comparable immunogenicity in mice [100]. In an advanced approach, an engineered probiotic vector co-expressing IL-2 and urease antigens further enhanced antigen-specific antibody responses and cytokine secretion [101].

Given that antibiotic therapy remains the most effective strategy for eradicating *H. pylori*, many studies have initially explored the synergistic effects of combining antibiotics with immunomodulatory adjuncts. Substantial evidence indicates that co-administration of probiotics with antibiotics can improve eradication rates, reduce side effects, and alleviate disease-related symptoms [102–104]. Mechanistically, probiotics help reinforce the gastrointestinal mucosal barrier, suppress the expression of pro-inflammatory cytokines such as IL-1β, and mitigate inflammatory responses [105]. They also restore *H. pylori*-induced metabolic disturbances in amino acid and fatty acid pathways, promoting immunometabolic homeostasis [106]. Clinical studies have shown that combining probiotics with triple antibiotic therapy significantly reduces serum levels of procalcitonin and IL-8, leading to higher eradication rates and concurrent amelioration of gastric inflammation [107]. Furthermore, this strategy not only improves eradication success but also allows for reduced antibiotic dosage, thereby curbing the emergence of resistance.

Beyond probiotics, AMPs also demonstrate synergistic potential when combined with antibiotics. For instance, the joint application of the AMP TP4 and conventional antibiotics enhances bacterial clearance, restores Th17/Treg balance, and improves gastric mucosal histology [108]. Thus, adjuvant treatments such as probiotics and AMPs not only exert direct antibacterial effects that complement antibiotic action but also modulate immune responses by lowering pro-inflammatory mediators such as IL-8 and TNF-α, thereby alleviating tissue inflammation.

In conclusion, the future of *H. pylori* eradication lies in sophisticated, multipronged combination strategies that simultaneously target the bacterium and recalibrate the host immune response. The synergistic integration of antibiotics with immunotherapeutic tools—ranging from vaccines to novel immunomodulators—holds the promise of not only overcoming antibiotic resistance but also achieving durable remission and potentially preventing reinfection. Notably, validated correlates of protection that reliably link immune surrogates to quantitative bacterial clearance remain to be established for *H. pylori*, underscoring the need for coordinated clinical studies. Success in this endeavor will therefore depend not only on a deeper understanding of host–pathogen–immune interactions, but also on the rational design of clinical trials capable of validating these innovative combinations and establishing clinically meaningful immune endpoints.

## 16.5 Summary

The development of immunotherapy for *H. pylori* has evolved through several phases: initial optimistic exploration, followed by a period of critical reflection, and more recently, a renewed effort driven by mechanistic insights and technological advances. Despite the significant progress that has been made in the field of immunotherapy, several persistent bottlenecks continue to hinder the effective translation of these promising immunotherapeutic strategies from the laboratory setting into clinical practice, where they can benefit patients. Most notably, vaccine candidates, which represent the most extensively studied category of immunotherapies, have demonstrated limited protective efficacy in real-world applications. A central challenge that remains is the induction of potent, durable, and broadly effective mucosal immunity that can be achieved across diverse populations with varying genetic backgrounds and immune responses. Meanwhile, immunomodulatory approaches are frequently employed as adjunct therapies; however, achieving precise and efficient immune modulation without compromising patient safety represents a crucial direction for future breakthroughs in this field.

To effectively address these multifaceted challenges, several strategic directions appear to hold considerable promise: [1] Optimization of Vaccine Antigens. Conventional vaccines that rely on one or just a few antigens are particularly vulnerable to the high genetic diversity and rapid mutation rate of *H. pylori*, which can undermine their effectiveness. Therefore, the development of multi-epitope vaccine designs, informed by reverse vaccinology to identify highly conserved and immunogenic epitopes, offers a viable path toward eliciting broad-spectrum and coordinated humoral and cellular immune responses that can withstand the challenges posed by this pathogen [2]. Induction and Maintenance of Mucosal Immunity. Given that the gastric mucosa is not a classical inductive site for immune responses, the development of innovative mucosal adjuvants and advanced delivery systems—such as nanoparticle-based carriers—becomes critical for ensuring precise antigen delivery, robust immune activation, and the establishment of long-term immunological memory that can effectively combat infections. However, current adjuvant discussions predominantly emphasize immunogenicity, while comprehensive safety endpoints—including gastric histopathology, systemic cytokine profiling, and clinical toxicity metrics—have not been uniformly applied. These parameters should be incorporated into future comparative studies [3]. Balancing and Regulating Immune Responses. *H. pylori* establishes persistent infection by driving immune cell dysfunction and disrupting the delicate balance of mucosal immune homeostasis. Therefore, targeted immunomodulatory strategies that aim to restore an antibacterial immune state, while simultaneously balancing effective immune activation against excessive inflammation, are essential for achieving therapeutic success in the treatment of infections caused by this resilient pathogen. Moreover, the balancing and regulation of immune responses must be operationalized through standardized safety endpoints to define therapeutic windows and ensure patient safety [4]. The integration of combination therapies represents a promising frontier in the treatment of *H. pylori* infections. By harnessing the unique strengths of various therapeutic modalities—such as vaccines, immunomodulatory adjuncts, and conventional antibiotics—into comprehensive multi-modal treatment regimens, we may achieve synergistic effects that significantly enhance treatment efficacy and potentially provide a solution to the growing challenge of antibiotic resistance.

Looking ahead, the overarching goal of *H. pylori* immunotherapy transcends the simple eradication of the bacteria itself. Instead, it seeks to implement precise immune interventions that can effectively reestablish a healthy gastric mucosal microenvironment. This restoration is crucial, as it aims to diminish the risk of developing associated gastrointestinal diseases, including the serious threat of gastric cancer. Immunotherapy for *H. pylori* holds promise in three key directions:

- Optimizing next-generation vaccine design: By refining adjuvants, antigens, and delivery systems, and leveraging insights from population genetic polymorphisms, it aims to achieve sustained, potent, and safe personalized immune protection.
- Blocking pathogenesis via immunization: A deeper understanding of bacterial immune evasion mechanisms will enable precise interventions targeting key virulence factors or signaling pathways. This approach seeks to promote effective bacterial clearance while avoiding tissue damage caused by excessive inflammatory responses.
- Developing synergistic combined treatment strategies: Constructing immunotherapy-based synergistic regimens requires in-depth investigation into the mechanisms of combined efficacy for vaccines, immunomodulators, antibiotics, and other therapies to overcome the limitations of monotherapies.

Despite the numerous challenges that persist in translating these innovative approaches from laboratory research to clinical practice, a more profound understanding of the intricate host–pathogen interactions occurring within the gastric niche—when combined with advancements in fields such as nanotechnology, immunoinformatics, and immune engineering—holds tremendous promise for the future. In due course, immunotherapy may emerge as a pivotal alternative or complementary strategy for *H. pylori* control, providing novel solutions to combat the escalating issue of antibiotic resistance that we face today.

## References


1. Zeng M, Mao XH, Li JX, Tong WD, Wang B, et al. Efficacy, safety, and immunogenicity of an oral recombinant *H. pylori* vaccine in children in China: a randomised, double-blind, placebo-controlled, phase 3 trial. Lancet. 2015;386(10002):1457–64. https://doi.org/10.1016/s0140-6736(15)60310-5.
2. Michetti P, Kreiss C, Kotloff KL, Porta N, Blanco JL, et al. Oral immunization with urease and Escherichia coli heat-labile enterotoxin is safe and immunogenic in *H. pylori*-infected adults. Gastroenterology. 1999;116(4):804–12. https://doi.org/10.1016/s0016-5085(99)70063-6.
3. Kotloff KL, Sztein MB, Wasserman SS, Losonsky GA, DiLorenzo SC, et al. Safety and immunogenicity of oral inactivated whole-cell *H. pylori* vaccine with adjuvant among volunteers with or without subclinical infection. Infect Immun. 2001;69(6):3581–90. https://doi.org/10.1128/iai.69.6.3581-3590.2001.
4. Malfertheiner P, Selgrad M, Wex T, Romi B, Borgogni E, et al. Efficacy, immunogenicity, and safety of a parenteral vaccine against *H. pylori* in healthy volunteers challenged with a Cag-positive strain: a randomised, placebo-controlled phase 1/2 study. Lancet Gastroenterol Hepatol. 2018;3(10):698–707. https://doi.org/10.1016/s2468-1253(18)30125-0.
5. Soluri MF, et al. Defining the *H. pylori* disease-specific antigenic repertoire. Front Microbiol. 2020;11:1551. https://doi.org/10.3389/fmicb.2020.01551.
6. Wilkinson DJ, Dickins B, Robinson K, Winter JA. Genomic diversity of *H. pylori* populations from different regions of the human stomach. Gut Microbes. 2022;14(1):2152306. https://doi.org/10.1080/19490976.2022.2152306.
7. Keikha M, Karbalaei M. EPIYA motifs of *H. pylori* cagA genotypes and gastrointestinal diseases in the Iranian population: a systematic review and meta-analysis. New Microbes New Infect. 2021;41:100865. https://doi.org/10.1016/j.nmni.2021.100865.
8. Hamza D, Elhelw R, Elhariri M, Ragab E. Genotyping and antimicrobial resistance patterns of *H. pylori* in human and dogs associated with A2142G and A2143G point mutations in clarithromycin resistance. Microb Pathog. 2018;123:330–8. https://doi.org/10.1016/j.micpath.2018.07.016.
9. Calado CRC. Antigenic and conserved peptides from diverse *H. pylori* antigens. Biotechnol Lett. 2022;44(3):535–45. https://doi.org/10.1007/s10529-022-03238-x.
10. Hasanzadeh Haghighi F, Menbari S, Mohammadzadeh R, Pishdadian A, Farsiani H. Developing a potent vaccine against *H. pylori*: critical considerations and challenges. Expert Rev Mol Med. 2024;27:e12. https://doi.org/10.1017/erm.2024.19.
11. Li X, Xing Y, Guo L, Lv X, Song H, et al. Oral immunization with recombinant Lactococcus lactis delivering a multi-epitope antigen CTB-UE attenuates *H. pylori* infection in mice. Pathog Dis. 2014;72(1):78–86. https://doi.org/10.1111/2049-632x.12173.
12. Moges Eskeziyaw B, Waihenya R, Maina N, Muuo Nzou S. Immunoinformatics-based designing of novel and potent multi-epitope PSA D15 and Cag11 immunogens for *H. pylori* immunodiagnostic assay development. Helicobacter. 2024;29(3):e13104. https://doi.org/10.1111/hel.13104.

13. Parvizpour S, Pourseif MM, Razmara J, Rafi MA, Omidi Y. Epitope-based vaccine design: a comprehensive overview of bioinformatics approaches. Drug Discov Today. 2020;25(6):1034–42. https://doi.org/10.1016/j.drudis.2020.03.006.
14. Basmenj ER, Pajhouh SR, Ebrahimi Fallah A, Naijian R, Rahimi E, et al. Computational epitope-based vaccine design with bioinformatics approach; a review. Heliyon. 2025;11(1):e41714. https://doi.org/10.1016/j.heliyon.2025.e41714.
15. Razzak A, Ahmed F, Mahmud MT. Development of a multi-epitope vaccine against *H. pylori* using a novel saRNA technology through an immunoinformatics approach. Sci Rep. 2025;15(1):33753. https://doi.org/10.1038/s41598-025-99512-9.
16. Barua A, Masum MHU, Mahdeen AA. A reverse vaccinology and Immunoinformatic approach for the designing of a novel mRNA vaccine against stomach cancer targeting the potent pathogenic proteins of *H. pylori*. Bioinform Biol Insights. 2025;19:11779322251331104. https://doi.org/10.1177/11779322251331104.
17. Chaleshtori ZA, Rastegari AA, Nayeri H, Doosti A. Use of immunoinformatics and the simulation approach to identify *H. pylori* epitopes to design a multi-epitope subunit vaccine for B- and T-cells. BMC Biotechnol. 2023;23(1):42. https://doi.org/10.1186/s12896-023-00814-5.
18. Khan M, Khan S, Ali A, Akbar H, Sayaf AM, et al. Immunoinformatics approaches to explore *H. pylori* proteome (virulence factors) to design B and T cell multi-epitope subunit vaccine. Sci Rep. 2019;9(1):13321. https://doi.org/10.1038/s41598-019-49354-z.
19. Camilo V, Sugiyama T, Touati E. Pathogenesis of *H. pylori* infection. Helicobacter. 2017;22(Suppl):1. https://doi.org/10.1111/hel.12405.
20. Sidebotham RL, Worku ML, Karim QN, Dhir NK, Baron JH. How *H. pylori* urease may affect external pH and influence growth and motility in the mucus environment: evidence from in-vitro studies. Eur J Gastroenterol Hepatol. 2003;15(4):395–401. https://doi.org/10.1097/00042737-200304000-00010.
21. Åberg A, Gideonsson P, Bhat A, Ghosh P, Arnqvist A. Molecular insights into the fine-tuning of pH-dependent ArsR-mediated regulation of the SabA adhesin in *H. pylori*. Nucleic Acids Res. 2024;52(10):5572–95. https://doi.org/10.1093/nar/gkae188.
22. Cheok YY, Tan GMY, Lee CYQ, Abdullah S, Looi CY, et al. Innate immunity crosstalk with *H. pylori*: pattern recognition receptors and cellular responses. Int J Mol Sci. 2022;23(14) https://doi.org/10.3390/ijms23147561.
23. Kao JY, Zhang M, Miller MJ, Mills JC, Wang B, et al. *H. pylori* immune escape is mediated by dendritic cell-induced Treg skewing and Th17 suppression in mice. Gastroenterology. 2010;138(3):1046–54. https://doi.org/10.1053/j.gastro.2009.11.043.
24. Dixon B, Hossain R, Patel RV, Algood HMS. Th17 cells in *H. pylori* infection: a dichotomy of help and harm. Infect Immun. 2019;87(11) https://doi.org/10.1128/iai.00363-19.
25. Shi Y, Liu XF, Zhuang Y, Zhang JY, Liu T, et al. *H. pylori*-induced Th17 responses modulate Th1 cell responses, benefit bacterial growth, and contribute to pathology in mice. J Immunol. 2010;184(9):5121–9. https://doi.org/10.4049/jimmunol.0901115.
26. Fu HW, Lai YC. The role of *H. pylori* neutrophil-activating protein in the pathogenesis of *H. pylori* and beyond: from a virulence factor to therapeutic targets and therapeutic agents. Int J Mol Sci. 2022;24(1) https://doi.org/10.3390/ijms24010091.
27. Li Y, Jin L, Chen T. The effects of secretory IgA in the mucosal immune system. Biomed Res Int. 2020;2020:2032057. https://doi.org/10.1155/2020/2032057.
28. Ikuse T, Blanchard TG, Czinn SJ. Inflammation, immunity, and vaccine development for the gastric pathogen *H. pylori*. Curr Top Microbiol Immunol. 2019;421:1–19. https://doi.org/10.1007/978-3-030-15138-6_1.
29. Nie S, Yuan Y. The role of gastric mucosal immunity in gastric diseases. J Immunol Res. 2020;2020:7927054. https://doi.org/10.1155/2020/7927054.
30. Pan X, Ke H, Niu X, Li S, Lv J, et al. Protection against *H. pylori* infection in BALB/c mouse model by Oral Administration of Multivalent Epitope-Based Vaccine of cholera toxin B subunit-HUUC. Front Immunol. 2018;9:1003. https://doi.org/10.3389/fimmu.2018.01003.

31. Kaveh-Samani A, Dalali S, Kaviani F, Piri-Gharaghie T, Doosti A. Oral administration of DNA alginate nanovaccine induced immune-protection against *H. pylori* in Balb/C mice. BMC Immunol. 2024;25(1):11. https://doi.org/10.1186/s12865-024-00602-6.
32. Lv X, Song H, Yang J, Li T, Xi T, et al. A multi-epitope vaccine CTB-UE relieves *H. pylori*-induced gastric inflammatory reaction via up-regulating microRNA-155 to inhibit Th17 response in C57/BL6 mice model. Hum Vaccin Immunother. 2014;10(12):3561–9. https://doi.org/10.4161/hv.36096.
33. Neutra MR, Kozlowski PA. Mucosal vaccines: the promise and the challenge. Nat Rev Immunol. 2006;6(2):148–58. https://doi.org/10.1038/nri1777.
34. Zhao T, Cai Y, Jiang Y, He X, Wei Y, et al. Vaccine adjuvants: mechanisms and platforms. Signal Transduct Target Ther. 2023;8(1):283. https://doi.org/10.1038/s41392-023-01557-7.
35. Zhang T, He P, Guo D, Chen K, Hu Z, et al. Research Progress of aluminum phosphate adjuvants and their action mechanisms. Pharmaceutics. 2023;15(6) https://doi.org/10.3390/pharmaceutics15061756.
36. Gottwein JM, Blanchard TG, Targoni OS, Eisenberg JC, Zagorski BM, et al. Protective anti-helicobacter immunity is induced with aluminum hydroxide or complete Freund's adjuvant by systemic immunization. J Infect Dis. 2001;184(3):308–14. https://doi.org/10.1086/322032.
37. Huang Z, Gong H, Sun Q, Yang J, Yan X, et al. Research progress on emulsion vaccine adjuvants. Heliyon. 2024;10(3):e24662. https://doi.org/10.1016/j.heliyon.2024.e24662.
38. Correa VA, Portilho AI, De Gaspari E. Vaccines, adjuvants and key factors for mucosal immune response. Immunology. 2022;167(2):124–38. https://doi.org/10.1111/imm.13526.
39. Song Y, Mehl F, Zeichner SL. Vaccine strategies to elicit mucosal immunity. Vaccines (Basel). 2024;12(2) https://doi.org/10.3390/vaccines12020191.
40. Yang J, Dai LX, Pan X, Wang H, Li B, et al. Protection against *H. pylori* infection in BALB/c mice by oral administration of multi-epitope vaccine of CTB-UreI-UreB. Pathog Dis. 2015;73(5) https://doi.org/10.1093/femspd/ftv026.
41. Peng X, Zhang R, Wang C, Yu F, Yu M, et al. E. coli enterotoxin LtB enhances vaccine-induced anti-*H. pylori* protection by promoting leukocyte migration into gastric mucus via inflammatory lesions. Cells. 2019;8(9) https://doi.org/10.3390/cells8090982.
42. Anderl F, Gerhard M. *H. pylori* vaccination: is there a path to protection? World J Gastroenterol. 2014;20(34):11939–49. https://doi.org/10.3748/wjg.v20.i34.11939.
43. Holmgren J, Nordqvist S, Blomquist M, Jeverstam F, Lebens M, et al. Preclinical immunogenicity and protective efficacy of an oral *H. pylori* inactivated whole cell vaccine and multiple mutant cholera toxin: a novel and non-toxic mucosal adjuvant. Vaccine. 2018;36(41):6223–30. https://doi.org/10.1016/j.vaccine.2018.07.073.
44. Lundgren A, Bourgeois L, Carlin N, Clements J, Gustafsson B, et al. Safety and immunogenicity of an improved oral inactivated multivalent enterotoxigenic Escherichia coli (ETEC) vaccine administered alone and together with dmLT adjuvant in a double-blind, randomized, placebo-controlled phase I study. Vaccine. 2014;32(52):7077–84. https://doi.org/10.1016/j.vaccine.2014.10.069.
45. Sharma M, Wagh P, Shinde T, Trimbake D, Tripathy AS. Exploring the role of pattern recognition receptors as immunostimulatory molecules. Immun Inflamm Dis. 2025;13(5):e70150. https://doi.org/10.1002/iid3.70150.
46. Woodworth JS, Contreras V, Christensen D, Naninck T, Kahlaoui N, et al. MINCLE and TLR9 agonists synergize to induce Th1/Th17 vaccine memory and mucosal recall in mice and non-human primates. Nat Commun. 2024;15(1):8959. https://doi.org/10.1038/s41467-024-52863-9.
47. Vacher G, Kaeser MD, Moser C, Gurny R, Borchard G. Recent advances in mucosal immunization using virus-like particles. Mol Pharm. 2013;10(5):1596–609. https://doi.org/10.1021/mp300597g.
48. Malik A, Gupta M, Gupta V, Gogoi H, Bhatnagar R. Novel application of trimethyl chitosan as an adjuvant in vaccine delivery. Int J Nanomedicine. 2018;13:7959–70. https://doi.org/10.2147/ijn.S165876.

49. Gong Y, Tao L, Wang F, Liu W, Jing L, et al. Chitosan as an adjuvant for a *H. pylori* therapeutic vaccine. Mol Med Rep. 2015;12(3):4123–32. https://doi.org/10.3892/mmr.2015.3950.
50. Liu C, Luo J, Xue RY, Guo L, Nie L, et al. The mucosal adjuvant effect of plant polysaccharides for induction of protective immunity against *H. pylori* infection. Vaccine. 2019;37(8):1053–61. https://doi.org/10.1016/j.vaccine.2018.12.066.
51. Cao X, Fu Y-X, Peng H. Promising cytokine adjuvants for enhancing tuberculosis vaccine immunity. Vaccine. 2024;12(5):477. https://doi.org/10.3390/vaccines12050477.
52. Wang X, Meng D. Innate endogenous adjuvants prime to desirable immune responses via mucosal routes. Protein Cell. 2014;6(3):170–84. https://doi.org/10.1007/s13238-014-0125-1.
53. Zhang F, Ni L, Zhang Z, Luo X, Wang X, et al. Recombinant L. lactis vaccine LL-plSAM-WAE targeting four virulence factors provides mucosal immunity against *H. pylori* infection. Microb Cell Factories. 2024;23(1):61. https://doi.org/10.1186/s12934-024-02321-4.
54. Chen YZ, Liu RJ, Yang L, Li MY, Suolangsizhu, et al. Preparation and release properties of *H. pylori* recombinant protein PLGA microspheres and PLGA-chitosan microspheres. Sichuan Da Xue Xue Bao Yi Xue Ban. 2021;52(5):794–8. https://doi.org/10.12182/20210960208.
55. Tan Z, Liu W, Liu H, Li C, Zhang Y, et al. Oral *H. pylori* vaccine-encapsulated acid-resistant HP55/PLGA nanoparticles promote immune protection. Eur J Pharm Biopharm. 2017;111:33–43. https://doi.org/10.1016/j.ejpb.2016.11.007.
56. Pawar D, Mangal S, Goswami R, Jaganathan KS. Development and characterization of surface modified PLGA nanoparticles for nasal vaccine delivery: effect of mucoadhesive coating on antigen uptake and immune adjuvant activity. Eur J Pharm Biopharm. 2013;85(3 Pt A):550–9. https://doi.org/10.1016/j.ejpb.2013.06.017.
57. Li D, Fu D, Kang H, Rong G, Jin Z, et al. Advances and potential applications of chitosan nanoparticles as a delivery carrier for the mucosal immunity of vaccine. Curr Drug Deliv. 2017;14(1):27–35. https://doi.org/10.2174/1567201813666160804121123.
58. Arif M, Ahmad R, Sharaf M, Samreen, Muhammad J, et al. Antibacterial and antibiofilm activity of mannose-modified chitosan/PMLA nanoparticles against multidrug-resistant *H. pylori*. Int J Biol Macromol. 2022;223(Pt A):418–32. https://doi.org/10.1016/j.ijbiomac.2022.10.265.
59. Baden LR, El Sahly HM, Essink B, Kotloff K, Frey S, et al. Efficacy and safety of the mRNA-1273 SARS-CoV-2 vaccine. N Engl J Med. 2021;384(5):403–16. https://doi.org/10.1056/NEJMoa2035389.
60. Yang Y, Chen L, Sun HW, Guo H, Song Z, et al. Epitope-loaded nanoemulsion delivery system with ability of extending antigen release elicits potent Th1 response for intranasal vaccine against *H. pylori*. J Nanobiotechnology. 2019;17(1):6. https://doi.org/10.1186/s12951-019-0441-y.
61. Zhang H, Liu Z, Li Y, Tao Z, Shen L, et al. Adjuvants for *H. pylori* vaccines: outer membrane vesicles provide an alternative strategy. Virulence. 2024;15(1):2425773. https://doi.org/10.1080/21505594.2024.2425773.
62. Chen X, Lin Z, Wang N, Zhou Y, Cheng L, et al. *H. pylori*-derived outer membrane vesicles: pathogenic roles, microbiota interactions, and biomedical applications. J Adv Res. 2025; https://doi.org/10.1016/j.jare.2025.09.055.
63. Kiaie SH, Majidi Zolbanin N, Ahmadi A, Bagherifar R, Valizadeh H, et al. Recent advances in mRNA-LNP therapeutics: immunological and pharmacological aspects. J Nanobiotechnol. 2022;20(1):276. https://doi.org/10.1186/s12951-022-01478-7.
64. Keikha R, Daliri K, Jebali A. The use of nanobiotechnology in immunology and vaccination. Vaccines (Basel). 2021;9(2) https://doi.org/10.3390/vaccines9020074.
65. Hammer C, Begemann M, McLaren PJ, Bartha I, Michel A, et al. Amino acid variation in HLA class II proteins is a major determinant of humoral response to common viruses. Am J Hum Genet. 2015;97(5):738–43. https://doi.org/10.1016/j.ajhg.2015.09.008.
66. Ovsyannikova IG, Pankratz VS, Vierkant RA, Jacobson RM, Poland GA. Human leukocyte antigen haplotypes in the genetic control of immune response to measles-mumps-rubella vaccine. J Infect Dis. 2006;193(5):655–63. https://doi.org/10.1086/500144.

67. Gonçalves E, Bonduelle O, Soria A, Loulergue P, Rousseau A, et al. Innate gene signature distinguishes humoral versus cytotoxic responses to influenza vaccination. J Clin Invest. 2019;129(5):1960–71. https://doi.org/10.1172/jci125372.
68. Milich DR. Influence of C4A deficiency on nonresponse to HBsAg vaccination: a new immune response gene. J Hepatol. 2002;37(3):396–9. https://doi.org/10.1016/s0168-8278(02)00237-4.
69. Egli A, Santer DM, O'Shea D, Barakat K, Syedbasha M, et al. IL-28B is a key regulator of B- and T-cell vaccine responses against influenza. PLoS Pathog. 2014;10(12):e1004556. https://doi.org/10.1371/journal.ppat.1004556.
70. Gröndahl-Yli-Hannuksela K, Vahlberg T, Ilonen J, Mertsola J, He Q. Polymorphism of IL-10 gene promoter region: association with T cell proliferative responses after acellular pertussis vaccination in adults. Immunogenetics. 2016;68(9):733–41. https://doi.org/10.1007/s00251-016-0923-0.
71. Poland GA, Ovsyannikova IG, Jacobson RM. Immunogenetics of seasonal influenza vaccine response. Vaccine. 2008;26(Suppl 4):D35–40. https://doi.org/10.1016/j.vaccine.2008.07.065.
72. Pellegrino P, Falvella FS, Cheli S, Perrotta C, Clementi E, et al. The role of toll-like receptor 4 polymorphisms in vaccine immune response. Pharmacogenomics J. 2016;16(1):96–101. https://doi.org/10.1038/tpj.2015.21.
73. Poland GA, Ovsyannikova IG, Kennedy RB. Personalized vaccinology: a review. Vaccine. 2018;36(36):5350–7. https://doi.org/10.1016/j.vaccine.2017.07.062.
74. Ovsyannikova IG, Schaid DJ, Larrabee BR, Haralambieva IH, Kennedy RB, et al. A large population-based association study between HLA and KIR genotypes and measles vaccine antibody responses. PLoS One. 2017;12(2):e0171261. https://doi.org/10.1371/journal.pone.0171261.
75. Scepanovic P, Alanio C, Hammer C, Hodel F, Bergstedt J, et al. Human genetic variants and age are the strongest predictors of humoral immune responses to common pathogens and vaccines. Genome Med. 2018;10(1):59. https://doi.org/10.1186/s13073-018-0568-8.
76. Abdullah M, Greenfield LK, Bronte-Tinkew D, Capurro MI, Rizzuti D, et al. VacA promotes CagA accumulation in gastric epithelial cells during *H. pylori* infection. Sci Rep. 2019;9(1):38. https://doi.org/10.1038/s41598-018-37095-4.
77. Gobert AP, Verriere T, Asim M, Barry DP, Piazuelo MB, et al. Heme oxygenase-1 dysregulates macrophage polarization and the immune response to *H. pylori*. J Immunol. 2014;193(6):3013–22. https://doi.org/10.4049/jimmunol.1401075.
78. Sit WY, Cheng ML, Chen TJ, Chen CJ, Chen BN, et al. *H. pylori* PldA modulates TNFR1-mediated p38 signaling pathways to regulate macrophage responses for its survival. Gut Microbes. 2024;16(1):2409924. https://doi.org/10.1080/19490976.2024.2409924.
79. Tavares R, Pathak SK. *H. pylori* secreted protein HP1286 triggers apoptosis in macrophages via TNF-independent and ERK MAPK-dependent pathways. Front Cell Infect Microbiol. 2017;7:58. https://doi.org/10.3389/fcimb.2017.00058.
80. Tavares R, Pathak SK. Induction of TNF, CXCL8 and IL-1β in macrophages by *H. pylori* secreted protein HP1173 occurs via MAP-kinases, NF-κB and AP-1 signaling pathways. Microb Pathog. 2018;125:295–305. https://doi.org/10.1016/j.micpath.2018.09.037.
81. Teymournejad O, Mobarez AM, Hassan ZM, Moazzeni SM, Ahmadabad HN. In vitro suppression of dendritic cells by *H. pylori* OipA. Helicobacter. 2014;19(2):136–43. https://doi.org/10.1111/hel.12107.
82. Kim JM, Kim JS, Yoo DY, Ko SH, Kim N, et al. Stimulation of dendritic cells with *H. pylori* vacuolating cytotoxin negatively regulates their maturation via the restoration of E2F1. Clin Exp Immunol. 2011;166(1):34–45. https://doi.org/10.1111/j.1365-2249.2011.04447.x.
83. Tanaka H, Yoshida M, Nishiumi S, Ohnishi N, Kobayashi K, et al. The CagA protein of *H. pylori* suppresses the functions of dendritic cell in mice. Arch Biochem Biophys. 2010;498(1):35–42. https://doi.org/10.1016/j.abb.2010.03.021.
84. Bergman MP, Engering A, Smits HH, van Vliet SJ, van Bodegraven AA, et al. *H. pylori* modulates the T helper cell 1/T helper cell 2 balance through phase-variable interaction between lipopolysaccharide and DC-SIGN. J Exp Med. 2004;200(8):979–90. https://doi.org/10.1084/jem.20041061.

85. Käbisch R, Semper RP, Wüstner S, Gerhard M, Mejías-Luque R. *H. pylori* γ-Glutamyltranspeptidase induces tolerogenic human dendritic cells by activation of glutamate receptors. J Immunol. 2016;196(10):4246–52. https://doi.org/10.4049/jimmunol.1501062.
86. Gebert B, Fischer W, Weiss E, Hoffmann R, Haas R. *H. pylori* vacuolating cytotoxin inhibits T lymphocyte activation. Science. 2003;301(5636):1099–102. https://doi.org/10.1126/science.1086871.
87. Wu YY, Hsieh CT, Tsay GJ, Kao JT, Chiu YM, et al. Recruitment of CCR6(+) Foxp3(+) regulatory gastric infiltrating lymphocytes in *H. pylori* gastritis. Helicobacter. 2019;24(1):e12550. https://doi.org/10.1111/hel.12550.
88. Altobelli A, Bauer M, Velez K, Cover TL, Müller A. *H. pylori* VacA targets myeloid cells in the gastric lamina Propria to promote peripherally induced regulatory T-cell differentiation and persistent infection. MBio. 2019;10(2) https://doi.org/10.1128/mBio.00261-19.
89. Wüstner S, Mejías-Luque R, Koch MF, Rath E, Vieth M, et al. *H. pylori* γ-glutamyltranspeptidase impairs T-lymphocyte function by compromising metabolic adaption through inhibition of cMyc and IRF4 expression. Cell Microbiol. 2015;17(1):51–61. https://doi.org/10.1111/cmi.12335.
90. Liu S, Zhang N, Ji X, Yang S, Zhao Z, et al. *H. pylori* CagA promotes gastric cancer immune escape by upregulating SQLE. Cell Death Dis. 2025;16(1):17. https://doi.org/10.1038/s41419-024-07318-w.
91. Lina TT, Pinchuk IV, House J, Yamaoka Y, Graham DY, et al. CagA-dependent downregulation of B7-H2 expression on gastric mucosa and inhibition of Th17 responses during *H. pylori* infection. J Immunol. 2013;191(7):3838–46. https://doi.org/10.4049/jimmunol.1300524.
92. Anthofer M, Windisch M, Haller R, Ehmann S, Wrighton S, et al. Immune evasion by proteolytic shedding of natural killer group 2, member D ligands in *H. pylori* infection. Front Immunol. 2024;15:1282680. https://doi.org/10.3389/fimmu.2024.1282680.
93. Aydın EM, Demir TD, Seymen N, Said SS, Oktem-Okullu S, et al. The crosstalk between *H. pylori* virulence factors and the PD1:PD-L1 immune checkpoint inhibitors in progression to gastric cancer. Immunol Lett. 2021;239:1–11. https://doi.org/10.1016/j.imlet.2021.06.009.
94. Neuper T, Frauenlob T, Dang HH, Krenn PW, Posselt G, et al. ADP-heptose attenuates *H. pylori*-induced dendritic cell activation. Gut Microbes. 2024;16(1):2402543. https://doi.org/10.1080/19490976.2024.2402543.
95. Tshibangu-Kabamba E, Yamaoka Y. *H. pylori* infection and antibiotic resistance – from biology to clinical implications. Nat Rev Gastroenterol Hepatol. 2021;18(9):613–29. https://doi.org/10.1038/s41575-021-00449-x.
96. Krzyżek P, Paluch E, Gościniak G. Synergistic therapies as a promising option for the treatment of antibiotic-resistant *H. pylori*. Antibiotics (Basel). 2020;9(10) https://doi.org/10.3390/antibiotics9100658.
97. Ghasemi A, Mohammad N, Mautner J, Karsabet MT, Ardjmand A, et al. Immunization with recombinant FliD confers protection against *H. pylori* infection in mice. Mol Immunol. 2018;94:176–82. https://doi.org/10.1016/j.molimm.2018.01.001.
98. Yang JX, Tseng JC, Yu GY, Luo Y, Huang CF, et al. Recent advances in the development of toll-like receptor agonist-based vaccine adjuvants for infectious diseases. Pharmaceutics. 2022;14(2) https://doi.org/10.3390/pharmaceutics14020423.
99. Sun M, Liu Y, Ni X, Tan R, Wang Y, et al. Intranasal immunization with poly I:C and CpG ODN adjuvants enhances the protective efficacy against *H. pylori* infection in mice. Microbes Infect. 2025;27(3):105433. https://doi.org/10.1016/j.micinf.2024.105433.
100. Hongying F, Xianbo W, Fang Y, Yang B, Beiguo L. Oral immunization with recombinant lactobacillus acidophilus expressing the adhesin Hp0410 of *H. pylori* induces mucosal and systemic immune responses. Clin Vaccine Immunol. 2014;21(2):126–32. https://doi.org/10.1128/cvi.00434-13.
101. Zhang HX, Qiu YY, Zhao YH, Liu XT, Liu M, et al. Immunogenicity of oral vaccination with Lactococcus lactis derived vaccine candidate antigen (UreB) of *H. pylori* fused with the human interleukin 2 as adjuvant. Mol Cell Probes. 2014;28(1):25–30. https://doi.org/10.1016/j.mcp.2013.08.003.

102. Viazis N, Argyriou K, Kotzampassi K, Christodoulou DK, Apostolopoulos P, et al. A four-probiotics regimen combined with a standard *H. pylori*-eradication treatment reduces side effects and increases eradication rates. Nutrients. 2022;14(3) https://doi.org/10.3390/nu14030632.
103. Zhou BG, Chen LX, Li B, Wan LY, Ai YW. Saccharomyces boulardii as an adjuvant therapy for *H. pylori* eradication: a systematic review and meta-analysis with trial sequential analysis. Helicobacter. 2019;24(5):e12651. https://doi.org/10.1111/hel.12651.
104. Wang Y, Wang X, Cao XY, Zhu HL, Miao L. Comparative effectiveness of different probiotics supplements for triple *H. pylori* eradication: a network meta-analysis. Front Cell Infect Microbiol. 2023;13:1120789. https://doi.org/10.3389/fcimb.2023.1120789.
105. Shen S, Ren F, Qin H, Bukhari I, Yang J, et al. Lactobacillus acidophilus NCFM and Lactiplantibacillus plantarum Lp-115 inhibit *H. pylori* colonization and gastric inflammation in a murine model. Front Cell Infect Microbiol. 2023;13:1196084. https://doi.org/10.3389/fcimb.2023.1196084.
106. Lin CC, Huang WC, Su CH, Lin WD, Wu WT, et al. Effects of multi-strain probiotics on immune responses and metabolic balance in *H. pylori*-infected mice. Nutrients. 2020;12(8) https://doi.org/10.3390/nu12082476.
107. Liao Q, Pan H, Zheng Z, Yu Y. Inclusion of probiotics to triple antibiotic therapy improves the outcome of anti-*H. pylori*-induced peptic ulcer. Curr Top Nutraceut Res. 2023; https://doi.org/10.37290/ctnr2641-452x.21.
108. Narayana JL, Huang HN, Wu CJ, Chen JY. Efficacy of the antimicrobial peptide TP4 against *H. pylori* infection: in vitro membrane perturbation via micellization and in vivo suppression of host immune responses in a mouse model. Oncotarget. 2015;6(15):12936–54. https://doi.org/10.18632/oncotarget.4101.

# Part IV

# Probiotic Therapies and Microbial Interventions

# Probiotic Mechanisms in *H. pylori* Infection

# 17

Weiliang Hou and Xueting Li

## 17.1 Introduction to *H. pylori* and the Therapeutic Challenge

*Helicobacter pylori* is a Gram-negative, microaerophilic, spiral-shaped bacterium that has coexisted with humans for thousands of years. Its remarkable ability to survive in the acidic gastric environment is largely attributed to the activity of urease, an enzyme that hydrolyzes urea into ammonia and carbon dioxide [1]. The released ammonia neutralizes gastric acid in the immediate vicinity, forming a microenvironment that enables the bacterium to persist on the gastric mucosa. However, this persistent colonization is not innocuous. Continuous stimulation of the host immune system results in chronic gastric inflammation, which may progress to peptic ulcer disease, gastric mucosa-associated lymphoid tissue lymphoma, or even gastric adenocarcinoma. Thus, eradication of *H. pylori* infection is a critical clinical goal for preventing long-term complications [2].

For decades, the mainstay of therapy has been triple or quadruple regimens combining antibiotics with a proton pump inhibitor (PPI). These approaches were once highly successful, achieving eradication rates above 85%. However, in recent years, their efficacy has declined sharply in many parts of the world, with success rates often falling below 70% [3]. The primary reason for this decline is the rapid emergence of antibiotic resistance, particularly against clarithromycin, metronidazole, and levofloxacin. This trend has made treatment more complex, often requiring longer courses or more drug combinations, which are associated with increased adverse events such as nausea, diarrhea, taste disturbances, and abdominal discomfort [4].

W. Hou (✉) · X. Li
Department of Gastroenterology, Shanghai Institute of Pancreatic Diseases, National Key Laboratory of Immunity and Inflammation, Changhai Clinical Research Unit, Changhai Hospital, Naval Medical University, Shanghai, China
e-mail: houweiliang@tongji.edu.cn


J. Gao, Y. Du (eds.), *Non-Antibiotic Therapies for Helicobacter pylori*,
https://doi.org/10.1007/978-981-95-9598-3_17

These side effects not only impair quality of life but also reduce patient adherence, further compromising therapeutic outcomes.

Another important but less frequently highlighted challenge is the disruption of the gastrointestinal microbiota caused both by *H. pylori* infection itself and by the antibiotics used in its treatment. Infection-induced dysbiosis alters microbial diversity and metabolic activity, influencing immune regulation and compromising the mucosal barrier [5]. Antibiotic therapy exacerbates this imbalance by suppressing commensal bacteria, thereby increasing the risk of gastrointestinal distress and opportunistic infections such as *Clostridium difficile* [6]. Together, these factors create an unfavorable environment that both facilitates pathogen persistence and diminishes the effectiveness of antimicrobial strategies.

Against this backdrop, probiotics have attracted considerable interest as an adjunctive therapeutic option. Defined by the FAO/WHO Food and Agriculture Organization / World Health Organization: FAO and WHO work closely together in the fields of food security and nutrition, particularly in the development of food safety standards and nutritional guidelines. Through joint efforts, the two organizations ensure that people worldwide have access to safe and nutritious food, so as to improve health outcomes and quality of life. as "live microorganisms which, when administered in adequate amounts, confer a health benefit on the host," probiotics are not typically considered as monotherapy for eradication but have demonstrated significant value in combination with standard regimens. Evidence from both experimental and clinical studies indicates that probiotics can exert beneficial effects through multiple mechanisms: they may compete with *H. pylori* for adhesion sites, secrete antimicrobial substances, enhance mucus production, stabilize epithelial barrier function, and modulate immune responses to reduce excessive inflammation [7, 8]. Furthermore, probiotic supplementation has been associated with improved tolerability of eradication therapy, reducing the incidence of antibiotic-associated side effects and improving patient compliance [9].

In this context, probiotics represent a promising nonantibiotic strategy for managing *H. pylori* infection (Fig. 17.1). The following sections of this chapter will systematically explore the direct inhibitory mechanisms of probiotics, their role in immune modulation, their ability to reinforce the gastric epithelial barrier, and their contribution to reducing antibiotic-associated adverse effects.

## 17.2 Direct Inhibitory Mechanisms of Probiotics

The ability of probiotics to exert direct antagonistic effects against *H. pylori* constitutes one of the most thoroughly investigated aspects of their biological activity. Unlike indirect mechanisms that depend on host immune responses or barrier enhancement, direct inhibition targets the pathogen itself, interfering with its ability to survive, colonize, and proliferate in the gastric environment. These effects include the secretion of bacteriocins and antimicrobial metabolites, competition for adhesion sites on epithelial cells, and suppression of urease activity, a crucial virulence factor of *H. pylori*. Together, these processes

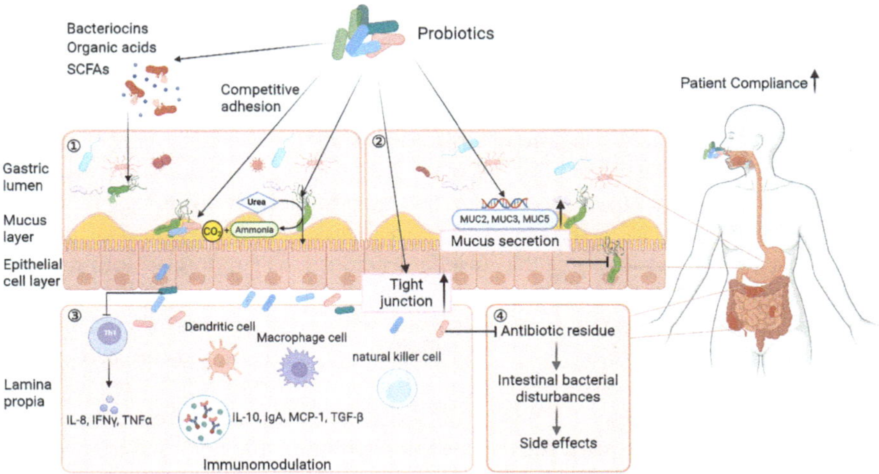


**Fig. 17.1** Inhibition mechanisms of probiotics on *H. pylori*. (1) Probiotics competitively inhibit *H. pylori* adhesion to the gastric epithelium and release antimicrobial metabolites—including bacteriocins, organic acids, and short-chain fatty acids (SCFAs)—that suppress bacterial growth and interfere with urease-mediated ammonia production. (2) Probiotics enhance epithelial barrier integrity by upregulating tight junction proteins and stimulating the expression of mucin genes (e.g., MUC2, MUC3, and MUC5), thereby reinforcing the mucus layer and limiting *H. pylori* penetration. (3) Within the lamina propria, probiotics modulate host immune responses by influencing dendritic cells (DCs), macrophages, and natural killer (NK) cells, shifting immune balance toward reduced pro-inflammatory cytokines [interleukin (IL)-8, interferon-gamma (IFN-γ), and tumor necrosis factor (TNF)-α] and increased regulatory mediators [IL-10, immunoglobulin A, MCP-1(Monocyte Chemoattractant Protein-1), and transforming growth factor-beta (TGF-β)]. (4) When used alongside antibiotics, probiotics help mitigate treatment-associated dysbiosis, lower intestinal side effects, and ultimately improve patient tolerance and compliance. Collectively, these actions highlight the synergistic potential of probiotics in controlling *H. pylori* infection through antimicrobial, barrier-protective, and immunomodulatory pathways. (Image created with BioRender.com, with permission)

significantly reduce bacterial load, attenuate pathogenicity, and create conditions less favorable for long-term colonization.

### 17.2.1 Production of Bacteriocins and Other Antimicrobial Compounds

Among the most important weapons in the probiotic arsenal are bacteriocins, ribosomally synthesized antimicrobial peptides that possess targeted activity against closely related bacterial species [10, 11]. Lactic acid bacteria, particularly strains of *Lactobacillus* and *Bifidobacterium*, have been shown to produce bacteriocins capable of inhibiting *H. pylori* growth [12]. Although their spectrum is generally narrower than that of antibiotics, bacteriocins act through unique mechanisms such as pore formation in bacterial membranes or interference with essential cellular processes, thereby reducing the likelihood of resistance development [13].

In addition to bacteriocins, probiotics also produce a variety of organic acids, including lactic, acetic, and propionic acids, as end-products of carbohydrate fermentation [14]. The acidic gastric environment is already a challenging niche for microbial survival; further acidification by probiotic-derived metabolites lowers the local pH to levels that *H. pylori* finds increasingly intolerable. Although *H. pylori* can resist acid stress to some extent through urease activity, its optimal growth occurs near neutral pH, meaning that sustained acidification imposes a considerable metabolic burden [15]. Importantly, undissociated forms of these organic acids penetrate the bacterial cell membrane, dissociate within the cytoplasm, and disrupt enzymatic activity and proton motive force, ultimately impairing energy production and growth [16].

Emerging evidence suggests that novel probiotics, such as *Weizmannia coagulans*, also secrete antimicrobial substances and alter the gastric microenvironment to the disadvantage of *H. pylori*. For example, animal models have demonstrated that *W. coagulans* supplementation not only suppresses bacterial colonization but also enhances host antioxidant defenses by increasing levels of superoxide dismutase and catalase, thereby indirectly limiting the damage induced by infection [17, 18]. Furthermore, isolates from nontraditional sources, such as lactic acid bacteria from fermented cocoa juice, have shown significant anti-*H. pylori* activity, underscoring the diverse potential reservoir of probiotic strains [19].

Collectively, these findings highlight the multifaceted antimicrobial potential of probiotics, which extends beyond traditional lactic acid bacteria and includes novel candidates capable of producing unique bacteriocins and bioactive metabolites.

### 17.2.2 Competitive Exclusion for Adhesion Sites

The colonization of the gastric epithelium is a prerequisite for *H. pylori* persistence, and adhesion to specific host receptors represents a critical step in this process. *H. pylori* recognizes and binds to molecules such as Lewis b antigens on the gastric epithelial surface [20]. Probiotics interfere with this essential step by competing for the same adhesion sites, thereby preventing *H. pylori* from establishing a stable niche.

This phenomenon, known as competitive exclusion, may operate through multiple mechanisms. Certain probiotic strains adhere to epithelial cells themselves, physically occupying binding sites that would otherwise be available to *H. pylori* [21]. Moreover, co-aggregation—a process in which probiotic cells bind directly to *H. pylori*—creates large clumps of bacteria that are more easily cleared from the stomach by peristalsis and gastric emptying [22].

Experimental studies confirm the clinical significance of this mechanism. For example, preincubation of gastric epithelial cells with *Lactobacillus* or *Bifidobacterium* significantly reduces subsequent adhesion of *H. pylori* [22, 23].

Strains such as *Lactobacillus rhamnosus* SD11 and *L. paracasei* CNCM I-1572 demonstrate robust co-aggregation capacity, while *L. rhamnosus MN45* not only aggregates with *H. pylori* but also suppresses urease activity and virulence factor expression, thereby reducing both bacterial survival and pathogenic potential [21, 24].

Through these mechanisms, probiotics effectively function as "biological decoys," intercepting *H. pylori* before it can attach to host cells and begin the colonization process.

### 17.2.3 Inhibition of Urease Activity

Urease is arguably the most critical virulence factor of *H. pylori*, enabling it to neutralize gastric acid and persist in an otherwise inhospitable environment. By hydrolyzing urea into ammonia and carbon dioxide, urease elevates the local pH and protects the bacterium from acid-induced damage [25]. Inhibition of urease, therefore, represents a direct and highly effective means of impairing *H. pylori* survival.

Experimental evidence supports this concept. *Lactobacillus acidophilus* NCFM and *Lactiplantibacillus plantarum* Lp-115, for example, exhibit anti-adhesion and anti-urease properties that significantly reduce colonization [26]. Similarly, *Lactiplantibacillus plantarum* ZJ316 has been shown to suppress the expression of adhesin and urease genes, thereby curtailing both bacterial adhesion and acid tolerance [27]. Even less common species, such as *Latilactobacillus sakei* LZ217, have demonstrated strong anti-*H. pylori* effects, including suppression of urease gene expression and interference with amino acid metabolism [28].

The inhibition of urease by probiotics not only compromises the bacterium's primary defense against gastric acidity but also sensitizes it to the host environment, reducing colonization density and enhancing the efficacy of conventional antimicrobial therapies.

## 17.3 Modulation of the Host Immune Response

While the direct antimicrobial activity of probiotics represents a critical defense against *H. pylori*, an equally important dimension of their benefit lies in their ability to regulate host immunity. The persistence of *H. pylori* is not solely due to its virulence factors but also to the complex interplay with the host immune system. The bacterium elicits a strong yet dysregulated immune response: pro-inflammatory mediators are chronically elevated, but sterilizing immunity is never achieved, resulting in persistent inflammation and mucosal injury. Probiotics can reshape this dynamic by attenuating excessive inflammation, enhancing regulatory pathways, and optimizing the balance between innate and adaptive responses. Through these immunomodulatory actions, probiotics contribute to bacterial clearance, protect gastric tissue from damage, and create conditions less conducive to long-term colonization.

### 17.3.1 Regulation of Cytokine Production

One of the hallmarks of *H. pylori* infection is a skewed T helper 1 (Th1)-dominated immune profile, characterized by heightened secretion of pro-inflammatory cytokines such as IL-1β, IL-6, IL-8, TNF-α, and IFN-γ. Among these, IL-8 plays a pivotal role by recruiting neutrophils, thereby perpetuating inflammation and tissue injury [29]. Chronic stimulation of these pathways contributes to gastritis, ulcer formation, and an increased risk of malignant transformation.

Probiotics have demonstrated the ability to modulate this cytokine milieu. Strains such as *Lactobacillus casei Shirota*, *L. acidophilus*, and *L. rhamnosus GG* have been shown in both in vitro and in vivo models to suppress IL-8 production by gastric epithelial cells in response to *H. pylori* [30–32]. Mechanistically, this effect is mediated through inhibition of the nuclear factor-kappa B (NF-κB) signaling cascade. Normally, NF-κB is activated upon *H. pylori* recognition, leading to the transcription of pro-inflammatory genes [33, 34].

In parallel, probiotics often enhance the production of anti-inflammatory cytokines such as IL-10 and TGF-β. These molecules are crucial for maintaining immune tolerance and preventing excessive tissue damage. The induction of IL-10 in particular has been linked to reduced gastric inflammation and improved mucosal healing [35]. Thus, probiotics not only mitigate harmful inflammation but also actively promote regulatory immune responses, restoring a healthier balance in the gastric mucosal environment.

### 17.3.2 Influence on Innate Immunity

The innate immune system constitutes the first line of defense against pathogens, relying on pattern recognition receptors such as Toll-like receptors (TLRs) to detect microbial components [36]. *H. pylori* interacts extensively with these receptors, often triggering inflammatory cascades that contribute to pathogenesis [37]. Probiotics modulate this interaction by engaging with the same receptors in a manner that promotes tolerance rather than hyperactivation.

For example, probiotic-derived ligands binding to TLR2 and TLR4 can shift downstream signaling toward anti-inflammatory outcomes, reducing the overproduction of cytokines and chemokines [38]. In addition, probiotics influence the activity of DCs, which serve as crucial antigen-presenting cells linking innate and adaptive immunity. Probiotic-stimulated DCs tend to adopt a "tolerogenic" phenotype, characterized by increased expression of IL-10 and reduced expression of IL-12 [39]. This phenotype is less likely to perpetuate chronic inflammation and more likely to promote the differentiation of regulatory T cells (Tregs), thereby containing excessive immune activation.

Probiotics also enhance innate defense by increasing the production of antimicrobial peptides and supporting the function of NK cells. These innate components can directly contribute to bacterial clearance while limiting collateral damage to host tissues. Thus, probiotics help recalibrate the innate immune system, ensuring that it remains effective against pathogens without becoming pathologically overactive.

### 17.3.3 Modulation of Adaptive Immunity

Adaptive immunity in *H. pylori* infection is often imbalanced, with an exaggerated Th1 response and inadequate regulatory mechanisms. This imbalance not only fails to eradicate the bacterium but also perpetuates chronic gastritis. Probiotics can modulate adaptive immunity in several ways, thereby shifting the balance toward more protective and less harmful responses.

One of the most significant effects is the promotion of Treg cell development. Tregs are instrumental in maintaining immune tolerance, largely through the secretion of IL-10 and TGF-β [40]. By expanding the Treg population, probiotics suppress excessive Th1-driven inflammation and prevent immune-mediated mucosal damage. This mechanism may also explain the reduced histological severity of gastritis observed in animal models supplemented with probiotics [41].

In addition, probiotics influence the humoral arm of adaptive immunity. Several studies have demonstrated enhanced production of secretory immunoglobulin A (sIgA) in the gastric mucosa following probiotic supplementation. sIgA serves as the frontline defense of the mucosal immune system, binding to pathogens and preventing their adhesion to epithelial surfaces. Increased sIgA levels can therefore improve immune surveillance and clearance of *H. pylori* [42].

Interestingly, some probiotics also attenuate the expression of *H. pylori* virulence factors such as vacuolating cytotoxin A and cytotoxin-associated gene A (CagA) by modulating host immune responses [41]. The suppression of these virulence factors contributes to reduced pathogenicity and lower risk of severe gastric disease.

In summary, probiotics exert profound effects on host immunity during *H. pylori* infection. By regulating cytokine production, modulating innate signaling pathways, promoting Treg differentiation, and enhancing mucosal antibody responses, they reshape the host immune landscape in a way that favors pathogen clearance while minimizing tissue injury. This immunological rebalancing not only contributes to the efficacy of eradication therapy but also reduces the risk of chronic gastritis and malignant progression.

## 17.4 Reinforcement of the Gastric Epithelial Barrier

The gastric epithelium forms the first physical and biochemical line of defense against *H. pylori*. Successful colonization and persistence of the pathogen depend in part on its ability to penetrate or compromise this barrier. Disruption of the epithelial layer exposes the underlying tissue to bacterial products and host-derived inflammatory mediators, amplifying mucosal damage and disease progression [43]. Probiotics contribute to host protection by reinforcing epithelial barrier integrity through two principal mechanisms: enhancement of mucus production and stabilization of intercellular tight junctions. Together, these effects strengthen the host's natural defenses and reduce susceptibility to chronic infection and tissue injury.

### 17.4.1 Enhancement of Mucus Production

The mucus layer, composed primarily of mucins secreted by gastric epithelial cells, serves as a viscous shield that prevents direct bacterial contact with the mucosal surface. *H. pylori*, however, has evolved strategies to degrade or penetrate this protective layer, thereby gaining access to epithelial cells [44]. Probiotics can counteract this process by upregulating mucin gene expression and increasing mucus secretion, thereby thickening and reinforcing the barrier.

Experimental studies have demonstrated that strains such as *Lactobacillus plantarum 299v* and *L. rhamnosus GG* enhance the expression of mucin genes, including MUC2, MUC3, and MUC5AC. Increased mucin production creates a more robust physical barrier, entrapping bacteria and facilitating their clearance from the stomach [45]. Additionally, probiotic-stimulated mucus secretion may contain bioactive molecules with antimicrobial properties, further hindering *H. pylori* survival [21].

In animal models, supplementation with *Bifidobacterium bifidum BF-1* significantly improved expression of MUC5AC and reduced gastric injury, highlighting the potential for probiotic-mediated mucosal protection [46]. These findings underscore the role of probiotics not only as antimicrobial agents but also as enhancers of host secretory defenses, which limit bacterial colonization and subsequent pathology.

### 17.4.2 Strengthening of Tight Junctions

Beyond the mucus layer, epithelial cells are connected by tight junctions, specialized protein complexes that regulate paracellular permeability and maintain mucosal barrier function. *H. pylori* infection disrupts these junctions, leading to increased epithelial permeability, bacterial translocation, and heightened exposure of the immune system to microbial antigens [47]. Such disruption contributes to chronic inflammation and tissue damage.

Probiotics counteract this deleterious effect by stabilizing and restoring tight junction integrity. They upregulate the expression of key junctional proteins such as zonula occludens-1 (ZO-1), occludin, and claudins, thereby reinforcing the paracellular barrier [48]. This restoration prevents the leakage of luminal contents into underlying tissues, maintaining epithelial homeostasis.

For example, *Lactobacillus paragasseri LPG-9*, isolated from the gastric mucosa, was shown to upregulate ZO-1 and occludin in infected tissues, alleviating gastritis in animal models [15]. Similarly, *Limosilactobacillus reuteri* 2892 isolated from camel milk had a protective effect on *H. pylori*-induced gastritis in C57BL/6 mice, significantly downregulated the expression of the virulence factor cagA gene, and increased the expression of tight junction molecules [49]. Probiotic-derived SCFAs also contribute to this effect by modulating signaling pathways that enhance tight junction assembly and repair [50].

### 17.4.3 Broader Implications for Gastric Health

Strengthening of the epithelial barrier by probiotics has implications that extend beyond *H. pylori* infection. Disruption of gastric or intestinal barrier function is a central feature of various gastrointestinal disorders, including inflammatory bowel disease, irritable bowel syndrome, and infections by other pathogens. Thus, the ability of probiotics to reinforce epithelial defenses highlights their potential as broad-spectrum agents for maintaining mucosal integrity across a range of conditions [51]. In the specific case of *H. pylori*, barrier enhancement complements direct antimicrobial activity and immune modulation, creating a multi-layered defense strategy that is more resilient to pathogen persistence.

In conclusion, probiotics play a pivotal role in maintaining the structural and functional integrity of the gastric epithelial barrier. By stimulating mucus secretion and strengthening tight junctions, they prevent bacterial adherence, limit epithelial permeability, and reduce mucosal inflammation. These actions are integral to the overall protective effects of probiotics, highlighting their value not only as adjuncts to antimicrobial therapy but also as agents that preserve host tissue health and resilience in the face of persistent infection.

## 17.5 Reduction of Antibiotic-Associated Side Effects

Although antibiotics remain indispensable for *H. pylori* eradication, their use is inevitably accompanied by adverse effects. Standard triple or quadruple therapies often include clarithromycin, amoxicillin, metronidazole, or levofloxacin, in combination with a PPI. While effective against the pathogen, these regimens exert collateral damage on the host's commensal microbiota, disrupt gastrointestinal homeostasis, and provoke a range of side effects such as diarrhea, bloating, abdominal pain, nausea, and altered taste perception. These adverse reactions represent a significant barrier to treatment adherence and, consequently, to eradication success. Probiotics have emerged as valuable adjuncts in this context, helping to mitigate antibiotic-associated side effects by restoring microbial balance, suppressing opportunistic pathogens, and enhancing treatment tolerability.

### 17.5.1 Antibiotic-Induced Dysbiosis

The gastrointestinal microbiota is a diverse ecosystem that plays an essential role in nutrient metabolism, immune regulation, and barrier function [52]. Antibiotic therapy, while targeting *H. pylori*, indiscriminately affects commensal bacteria, leading to reduced microbial diversity and shifts in community composition. This dysbiosis allows opportunistic organisms such as *Clostridium difficile* or *Candida* species to overgrow, causing secondary infections or gastrointestinal distress. Furthermore, disruption of normal microbial fermentation processes alters the production of metabolites like SCFAs, which are critical for mucosal health.

Probiotics help to counteract these disturbances by replenishing beneficial bacteria and competing with opportunistic pathogens. Their presence stabilizes the microbial ecosystem, reduces the risk of pathogen overgrowth, and restores normal metabolic activity, thereby alleviating many of the side effects associated with antibiotics [53, 54].

### 17.5.2 Alleviation of Gastrointestinal Symptoms

One of the most consistent clinical findings is the ability of probiotics to reduce the incidence and severity of antibiotic-associated diarrhea. This effect has been documented across multiple randomized controlled trials and meta-analyses. Probiotic strains such as *L. rhamnosus* GG, *Saccharomyces boulardii*, and multi-strain formulations have demonstrated significant efficacy in lowering rates of diarrhea, nausea, and bloating during eradication therapy. The protective effect is thought to arise from several mechanisms, including competitive inhibition of pathogenic bacteria, production of antimicrobial compounds, and reinforcement of mucosal barrier function [55].

Beyond diarrhea, probiotics also improve other gastrointestinal symptoms. Patients receiving probiotic supplementation frequently report reduced abdominal discomfort, less frequent nausea, and greater overall tolerance of therapy [56]. These improvements contribute not only to patient well-being but also to adherence, an essential determinant of eradication success.

### 17.5.3 Enhancement of Patient Compliance and Treatment Success

Adverse events are a leading cause of treatment discontinuation in *H. pylori* eradication regimens. Even mild gastrointestinal disturbances can lead patients to interrupt or abandon therapy, compromising eradication rates and contributing to the development of further antibiotic resistance. By reducing the frequency and severity of these adverse effects, probiotics play a crucial role in enhancing compliance.

Clinical studies support this relationship: patients receiving probiotics alongside standard eradication therapy demonstrate higher completion rates and greater overall treatment success. In some trials, probiotic supplementation has been associated with absolute increases of 5–10% in eradication rates compared with antibiotics alone [57, 58]. While these improvements may seem modest, they are clinically meaningful in the face of rising antibiotic resistance and declining global eradication success.

### 17.5.4 Strain-Specific and Formulation Considerations

It is important to note that not all probiotics confer equal protection against antibiotic-associated side effects. The efficacy of supplementation depends on factors such as strain specificity, dosage, and duration of administration. For instance,

*S. boulardii* is particularly effective in preventing diarrhea, while *L. rhamnosus* GG and multi-strain combinations provide broader benefits across a spectrum of gastrointestinal symptoms. Synbiotic formulations, combining probiotics with prebiotics, may further enhance efficacy by supporting the survival and activity of administered strains.

Thus, careful selection of probiotic strains and delivery strategies is essential for maximizing therapeutic benefit in clinical practice.

## 17.6 Conclusion

The mechanism of action of probiotics against *H. pylori* is not singular but a symphony of concerted effects. They function through direct antibacterial actions, including the production of inhibitory compounds and competitive exclusion; through immunomodulation by tempering the pro-inflammatory response and promoting regulatory pathways; and through enhancing the host's own defensive barriers by stimulating mucus production and tightening epithelial junctions. While they are not yet a standalone cure, their role as adjuvants to standard antibiotic therapy is well-supported by clinical evidence. They enhance eradication rates and, perhaps just as importantly, improve the patient's quality of life during treatment by minimizing side effects. Future research should focus on identifying the most potent strains and optimal combinations (synbiotics) and elucidating their precise molecular interactions with the pathogen and the host to develop next-generation probiotic therapeutics.

## References

1. Crowe SE. *H. pylori* infection. N Engl J Med. 2019;380(12):1158–65. https://doi.org/10.1056/NEJMcp1710945.
2. Camilo V, Sugiyama T, Touati E. Pathogenesis of *H. pylori* infection. Helicobacter. 2017;22(Suppl):1. https://doi.org/10.1111/hel.12405.
3. Tshibangu-Kabamba E, Yamaoka Y. *H. pylori* infection and antibiotic resistance – from biology to clinical implications. Nat Rev Gastroenterol Hepatol. 2021;18(9):613–29. https://doi.org/10.1038/s41575-021-00449-x.
4. Zhang M. High antibiotic resistance rate: a difficult issue for *H. pylori* eradication treatment. World J Gastroenterol. 2015;21(48):13432–7. https://doi.org/10.3748/wjg.v21.i48.13432.
5. Chen CC, Liou JM, Lee YC, Hong TC, El-Omar EM, Wu MS. The interplay between *H. pylori* and gastrointestinal microbiota. Gut Microbes. 2021;13(1):1–22. https://doi.org/10.1080/19490976.2021.1909459.
6. Engelsberger V, Gerhard M, Mejias-Luque R. Effects of *H. pylori* infection on intestinal microbiota, immunity and colorectal cancer risk. Front Cell Infect Microbiol. 2024;14:1339750. https://doi.org/10.3389/fcimb.2024.1339750.
7. Keikha M, Karbalaei M. Probiotics as the live microscopic fighters against *H. pylori* gastric infections. BMC Gastroenterol. 2021;21(1):388. https://doi.org/10.1186/s12876-021-01977-1.
8. Zhu XY, Liu F. Probiotics as an adjuvant treatment in *H. pylori* eradication therapy. J Dig Dis. 2017;18(4):195–202. https://doi.org/10.1111/1751-2980.12466.

9. Liang B, Yuan Y, Peng XJ, Liu XL, Hu XK, Xing DM. Current and future perspectives for *H. pylori* treatment and management: from antibiotics to probiotics. Front Cell Infect Microbiol. 2022;12:1042070. https://doi.org/10.3389/fcimb.2022.1042070.
10. Echegaray N, Yilmaz B, Sharma H, Kumar M, Pateiro M, Ozogul F, et al. A novel approach to Lactiplantibacillus plantarum: from probiotic properties to the omics insights. Microbiol Res. 2023;268:127289. https://doi.org/10.1016/j.micres.2022.127289.
11. Rwubuzizi R, Carneiro KO, Holzapfel WH, Vaz-Velho M, Todorov SD. Bacteriocin and antioxidant production, a beneficial properties of lactic acid bacteria isolated from fermented vegetables of Northwest Bulgaria. Probiotics Antimicrob Proteins. 2025;17(1):103–22. https://doi.org/10.1007/s12602-023-10140-z.
12. Kingkaew E, Woraprayote W, Booncharoen A, Niwasabutra K, Janyaphisan T, Vilaichone RK, et al. Functional genome analysis and anti-*H. pylori* activity of a novel bacteriocinogenic Lactococcus sp. NH2-7C from Thai fermented pork (Nham). Sci Rep. 2023;13(1):20362. https://doi.org/10.1038/s41598-023-47687-4.
13. Selle K, Klaenhammer TR. Genomic and phenotypic evidence for probiotic influences of lactobacillus gasseri on human health. FEMS Microbiol Rev. 2013;37(6):915–35. https://doi.org/10.1111/1574-6976.12021.
14. Teame T, Wang A, Xie M, Zhang Z, Yang Y, Ding Q, et al. Paraprobiotics and postbiotics of probiotic lactobacilli, their positive effects on the host and action mechanisms: a review. Front Nutr. 2020;7:570344. https://doi.org/10.3389/fnut.2020.570344.
15. Xu B, Kong J, Lin Y, Tang Z, Liu J, Chen Z, et al. Anti-*H. pylori* activity and gastroprotective effects of human stomach-derived lactobacillus paragasseri strain LPG-9. Food Funct. 2023;14(24):10882–95. https://doi.org/10.1039/d3fo03562j.
16. Yoon JH, Oh MS, Lee SY. Effectiveness of organic acids for inactivating pathogenic bacteria inoculated in laboratory media and foods: an updated minireview. Food Sci Biotechnol. 2024;33(12):2715–28. https://doi.org/10.1007/s10068-024-01618-9.
17. Wang Y, Gu Z, Zhang S, Li P. Complete genome sequencing revealed the potential application of a novel Weizmannia coagulans PL-W production with promising Bacteriocins in food preservative. Foods (Basel, Switzerland). 2023;12(1) https://doi.org/10.3390/foods12010216.
18. Zhai S, Gao Y, Jiang Y, Li Y, Fan Q, Tie S, et al. Weizmannia coagulans BC99 affects valeric acid production via regulating gut microbiota to ameliorate inflammation and oxidative stress responses in *H. pylori* mice. J Food Sci. 2024;89(12):9985–10002. https://doi.org/10.1111/1750-3841.17514.
19. Kouitcheu Mabeku LB, Ngue S, Bonsou Nguemo I, Leundji H. Potential of selected lactic acid bacteria from Theobroma cacao fermented fruit juice and cell-free supernatants from cultures as inhibitors of *H. pylori* and as good probiotic. BMC Res Notes. 2020;13(1):64. https://doi.org/10.1186/s13104-020-4923-7.
20. Hage N, Howard T, Phillips C, Brassington C, Overman R, Debreczeni J, et al. Structural basis of Lewis(b) antigen binding by the *H. pylori* adhesin BabA. Sci Adv. 2015;1(7):e1500315. https://doi.org/10.1126/sciadv.1500315.
21. Juntarachot N, Sunpaweravong S, Kaewdech A, Wongsuwanlert M, Ruangsri P, Pahumunto N, et al. Characterization of adhesion, anti-adhesion, co-aggregation, and hydrophobicity of *H. pylori* and probiotic strains. J Taibah Univ Med Sci. 2023;18(5):1048–54. https://doi.org/10.1016/j.jtumed.2023.02.017.
22. Yang H, Lin Y, Ma Y, Li J, Li J, Huo Z, et al. Screening probiotics for anti-*H. pylori* and investigating the effect of probiotics on patients with *H. pylori* infection. Foods (Basel, Switzerland). 2024;13(12) https://doi.org/10.3390/foods13121851.
23. Zuo F, Appaswamy A, Gebremariam HG, Jonsson AB. Role of Sortase a in lactobacillus gasseri Kx110A1 adhesion to gastric epithelial cells and competitive exclusion of *H. pylori*. Front Microbiol. 2019;10:2770. https://doi.org/10.3389/fmicb.2019.02770.
24. Hua Q, Hu J, Zhang X, He J, Wang P, Duan Z, et al. Postbiotics suppress *H. pylori* adhesion and survival through a Coaggregation mechanism. J Agric Food Chem. 2025;73(35):21976–86. https://doi.org/10.1021/acs.jafc.5c07004.

25. Guzel-Akdemir O, Akdemir A. Urease inhibitors for the treatment of *H. pylori*. Expert Opin Ther Pat. 2025;35(1):17–30. https://doi.org/10.1080/13543776.2024.2423004.
26. Shen S, Ren F, Qin H, Bukhari I, Yang J, Gao D, et al. Lactobacillus acidophilus NCFM and Lactiplantibacillus plantarum Lp-115 inhibit *H. pylori* colonization and gastric inflammation in a murine model. Front Cell Infect Microbiol. 2023;13:1196084. https://doi.org/10.3389/fcimb.2023.1196084.
27. Wu S, Xu Y, Chen Z, Chen Y, Wei F, Xia C, et al. Lactiplantibacillus plantarum ZJ316 reduces *H. pylori* adhesion and inflammation by inhibiting the expression of Adhesin and urease genes. Mol Nutr Food Res. 2023;67(18):e2300241. https://doi.org/10.1002/mnfr.202300241.
28. Xia C, Chen Z, Chen Y, Wei F, Wu S, Zhou Q, et al. Effects of Latilactobacillus sakei LZ217 on gastric mucosal colonization, metabolic interference, and urease expression in *H. pylori* infection. Helicobacter. 2024;29(4):e13130. https://doi.org/10.1111/hel.13130.
29. Dominguez-Martinez DA, Fontes-Lemus JI, Garcia-Regalado A, Juarez-Flores A, Fuentes-Panana EM. IL-8 secreted by gastric epithelial cells infected with *H. pylori* CagA positive strains is a chemoattractant for Epstein-Barr virus infected B lymphocytes. Viruses. 2023;15(3) https://doi.org/10.3390/v15030651.
30. Zhou Q, Xue B, Gu R, Li P, Gu Q. Lactobacillus plantarum ZJ316 attenuates *H. pylori*-induced gastritis in C57BL/6 mice. J Agric Food Chem. 2021;69(23):6510–23. https://doi.org/10.1021/acs.jafc.1c01070.
31. Jauvain M, Lepied G, Benejat L, Roudier N, Dussert C, Lehours P, et al. Effect of lactobacillus gasseri BIO6369 and Lacticaseibacillus rhamnosus BIO5326 on gastric carcinogenesis induced by *H. pylori* infection. Helicobacter. 2024;29(4):e13108. https://doi.org/10.1111/hel.13108.
32. Kang S, Guo Y, Rao J, Jin H, You HJ, Ji GE. In vitro and in vivo inhibition of *H. pylori* by lactobacillus plantarum pH3A, monolaurin, and grapefruit seed extract. Food Funct. 2021;12(21):11024–32. https://doi.org/10.1039/d1fo01480c.
33. Chen Z, Tang Z, Li W, Deng X, Yu L, Yang J, et al. Weizmannia coagulans BCF-01: a novel gastrogenic probiotic for *H. pylori* infection control. Gut Microbes. 2024;16(1):2313770. https://doi.org/10.1080/19490976.2024.2313770.
34. Yu HJ, Liu W, Chang Z, Shen H, He LJ, Wang SS, et al. Probiotic BIFICO cocktail ameliorates *H. pylori* induced gastritis. World J Gastroenterol. 2015;21(21):6561–71. https://doi.org/10.3748/wjg.v21.i21.6561.
35. Yu Z, Cao M, Peng J, Wu D, Li S, Wu C, et al. Lacticaseibacillus casei T1 attenuates *H. pylori*-induced inflammation and gut microbiota disorders in mice. BMC Microbiol. 2023;23(1):39. https://doi.org/10.1186/s12866-023-02782-4.
36. Ansaldo E, Farley TK, Belkaid Y. Control of immunity by the microbiota. Annu Rev Immunol. 2021;39:449–79. https://doi.org/10.1146/annurev-immunol-093019-112348.
37. Maldonado RF, Sa-Correia I, Valvano MA. Lipopolysaccharide modification in gram-negative bacteria during chronic infection. FEMS Microbiol Rev. 2016;40(4):480–93. https://doi.org/10.1093/femsre/fuw007.
38. Kaur H, Ali SA. Probiotics and gut microbiota: mechanistic insights into gut immune homeostasis through TLR pathway regulation. Food Funct. 2022;13(14):7423–47. https://doi.org/10.1039/d2fo00911k.
39. Wiese-Szadkowska M, Helmin-Basa A, Eljaszewicz A, Gackowska L, Januszewska M, Motyl I, et al. Selected commensal bacteria change profiles of *H. pylori*-induced T cells via dendritic cell modulation. Helicobacter. 2019;24(6):e12614. https://doi.org/10.1111/hel.12614.
40. Kronsteiner B, Bassaganya-Riera J, Philipson C, Viladomiu M, Carbo A, Abedi V, et al. Systems-wide analyses of mucosal immune responses to *H. pylori* at the interface between pathogenicity and symbiosis. Gut Microbes. 2016;7(1):3–21. https://doi.org/10.1080/19490976.2015.1116673.
41. Boltin D. Probiotics in *H. pylori*-induced peptic ulcer disease. Best Pract Res Clin Gastroenterol. 2016;30(1):99–109. https://doi.org/10.1016/j.bpg.2015.12.003.
42. Lin CC, Huang WC, Su CH, Lin WD, Wu WT, Yu B, et al. Effects of multi-strain probiotics on immune responses and metabolic balance in *H. pylori*-infected mice. Nutrients. 2020;12(8) https://doi.org/10.3390/nu12082476.

43. Keilberg D, Ottemann KM. How *H. pylori* senses, targets and interacts with the gastric epithelium. Environ Microbiol. 2016;18(3):791–806. https://doi.org/10.1111/1462-2920.13222.
44. Vllahu M, Voli A, Licursi V, Zagami C, D'Amore A, Traulsen J, et al. Inflammation promotes stomach epithelial defense by stimulating the secretion of antimicrobial peptides in the mucus. Gut Microbes. 2024;16(1):2390680. https://doi.org/10.1080/19490976.2024.2390680.
45. Qureshi N, Li P, Gu Q. Probiotic therapy in *H. pylori* infection: a potential strategy against a serious pathogen? Appl Microbiol Biotechnol. 2019;103(4):1573–88. https://doi.org/10.1007/s00253-018-09580-3.
46. Gomi A, Harima-Mizusawa N, Shibahara-Sone H, Kano M, Miyazaki K, Ishikawa F. Effect of Bifidobacterium bifidum BF-1 on gastric protection and mucin production in an acute gastric injury rat model. J Dairy Sci. 2013;96(2):832–7. https://doi.org/10.3168/jds.2012-5950.
47. Caron TJ, Scott KE, Fox JG, Hagen SJ. Tight junction disruption: *H. pylori* and dysregulation of the gastric mucosal barrier. World J Gastroenterol. 2015;21(40):11411–27. https://doi.org/10.3748/wjg.v21.i40.11411.
48. Wu S, Wei F, Chen Y, Chen Z, Luo Y, Fan J, et al. Lactiplantibacillus plantarum ZJ316 alleviates *H. pylori*-induced intestinal inflammation by sustaining intestinal homeostasis. Probiotics Antimicrob Proteins. 2025; https://doi.org/10.1007/s12602-025-10474-w.
49. Forooghi Nia F, Rahmati A, Ariamanesh M, Saeidi J, Ghasemi A, Mohtashami M. The anti-*H. pylori* effects of Limosilactobacillus reuteri strain 2892 isolated from camel milk in C57BL/6 mice. World J Microbiol Biotechnol. 2023;39(5):119. https://doi.org/10.1007/s11274-023-03555-x.
50. Chen J, Chen X, Ho CL. Recent development of probiotic Bifidobacteria for treating human diseases. Front Bioeng Biotechnol. 2021;9:770248. https://doi.org/10.3389/fbioe.2021.770248.
51. Latif A, Shehzad A, Niazi S, Zahid A, Ashraf W, Iqbal MW, et al. Probiotics: mechanism of action, health benefits and their application in food industries. Front Microbiol. 2023;14:1216674. https://doi.org/10.3389/fmicb.2023.1216674.
52. Zmora N, Suez J, Elinav E. You are what you eat: diet, health and the gut microbiota. Nat Rev Gastroenterol Hepatol. 2019;16(1):35–56. https://doi.org/10.1038/s41575-018-0061-2.
53. He C, Xie Y, Zhu Y, Zhuang K, Huo L, Yu Y, et al. Probiotics modulate gastrointestinal microbiota after *H. pylori* eradication: a multicenter randomized double-blind placebo-controlled trial. Front Immunol. 2022;13:1033063. https://doi.org/10.3389/fimmu.2022.1033063.
54. Han X, Liu H, Zhang K, Zeng Y, Liu P, Gan L, et al. Lactiplantibacillus plantarum MH-301 reduces *H. pylori* treatment-related adverse events via gut-vaginal axis: a randomized, double-blind, placebo-controlled trial. J Infect. 2025;91(2):106539. https://doi.org/10.1016/j.jinf.2025.106539.
55. Tokunaga E, Inoue J, Asaji N, Oyama K, Soga A, Shimamoto Y, et al. Impact of enterococcus faecium 129 BIO 3B-R on *H. pylori* eradication therapy side effects in adult patients: a randomized, double-blind, placebo-controlled study. Sci Rep. 2025;15(1):32085. https://doi.org/10.1038/s41598-025-15864-2.
56. Zhao Y, Niu X, Zhang Y, Zhao L, Zhang L, He J, et al. Impact of supplementing Limosilactobacillus fermentum MN-LF23 on the eradication of *H. pylori* with 14-day standard quadruple therapy: a randomized, double-blind, placebo-controlled trial. Nutr J. 2025;24(1):106. https://doi.org/10.1186/s12937-025-01124-6.
57. Viazis N, Argyriou K, Kotzampassi K, Christodoulou DK, Apostolopoulos P, Georgopoulos SD, et al. A four-probiotics regimen combined with a standard *H. pylori*-eradication treatment reduces side effects and increases eradication rates. Nutrients. 2022;14(3) https://doi.org/10.3390/nu14030632.
58. Niu Y, Li J, Qian H, Liang C, Shi X, Bu S. Evaluation of efficacy and safety of Lacticaseibacillus rhamnosus LRa05 in the eradication of *H. pylori*: a randomized, double-blind, placebo-controlled trial. Front Immunol. 2024;15:1450414. https://doi.org/10.3389/fimmu.2024.1450414.

# Probiotics in *H. pylori* Adjunctive Therapy

# 18

Jiulong Zhao and Yunyi Yang

## 18.1 Introduction

Current standard eradication therapies face substantial challenges due to increasing antibiotic resistance, significant drug side effects, and suboptimal patient adherence. Research has demonstrated that specific probiotic strains can markedly reduce the incidence of antibiotic-induced gastrointestinal adverse reactions, thereby effectively improving treatment tolerance and patient compliance. However, evidence regarding the efficacy of probiotics in enhancing *Helicobacter pylori (H. pylori)* eradication rates remains heterogeneous and subject to ongoing debate. The therapeutic effects of probiotics exhibit notable strain-specific variability, and optimal administration protocols—including timing, dosage, and treatment duration—have yet to be standardized, leading to inconsistent outcomes across clinical studies. Major international consensus guidelines acknowledge the role of specific probiotic strains in improving treatment tolerance but maintain a cautious stance toward their routine use for increasing eradication rates while emphasizing the importance of selecting clinically validated strains. In conclusion, probiotics represent a valuable adjunctive approach in *H. pylori* eradication therapy; however, their clinical application requires evidence-based strain selection and warrants further research to establish standardized treatment protocols.

J. Zhao (✉) · Y. Yang
Department of Gastroenterology, Shanghai Institute of Pancreatic Diseases, National Key Laboratory of Immunity and Inflammation, Changhai Clinical Research Unit, Changhai Hospital, Naval Medical University, Shanghai, China



J. Gao, Y. Du (eds.), *Non-Antibiotic Therapies for Helicobacter pylori*,
https://doi.org/10.1007/978-981-95-9598-3_18

## 18.2 Probiotic Overview and Frequently Studied Strains

Probiotics are defined as "live microorganisms that, when administered in adequate amounts, confer a health benefit on the host" [1]. Commonly used probiotic strains in humans include Lactobacillus species (e.g., *L. rhamnosus GG*, *L. casei Shirota*, *L. acidophilus* La-5, *L. plantarum*, and *L. gasseri*), Bifidobacterium species (e.g., *B. animalis* subsp. *Lactis Bb-12*, *B. breve*, and *B. longum*), *Streptococcus thermophilus*, *Enterococcus faecium*, *Bacillus clausii*, *Escherichia coli* Nissle 1917, and the yeast *Saccharomyces boulardii*. These microorganisms exert their beneficial effects through multiple mechanisms, including competitive inhibition of pathogen colonization, enhancement of intestinal epithelial barrier function, modulation of immune responses, and production of bioactive metabolites (such as short-chain fatty acids and bacteriocins), thereby improving the host metabolic status and maintaining the microbial balance [2]. Importantly, the effects of probiotics are strain-specific, indicating that different strains may exert distinct physiological influences. A growing body of evidence suggests that various probiotic strains are associated with the prevention or adjuvant management of a range of diseases, including acute and antibiotic-associated diarrhea, functional gastrointestinal disorders, and inflammatory bowel disease [3]. In recent years, increasing attention has been given to the adjunctive use of probiotics in the treatment of *H. pylori* infection. Among these strains, those belonging to the genera Lactobacillus, Bifidobacterium, and Saccharomyces are the most commonly used as adjuncts in anti-*H. pylori* therapy. These probiotics have been shown to improve the eradication rate of *H. pylori* while reducing gastrointestinal adverse effects associated with antibiotic treatment, thereby increasing patient compliance and overall clinical outcomes [4].

## 18.3 Efficacy of Probiotics in *H. pylori* Treatment

Standard eradication therapy is facing significant challenges, with the core dilemma being the increasingly severe problem of antibiotic resistance. In particular, the high global resistance rate to clarithromycin has directly led to the failure of classic triple therapy. Moreover, post-treatment management is notably inadequate: only a minority of patients undergo confirmatory eradication testing, and many failed cases are inappropriately retreated with the same regimen, further exacerbating resistance. Additionally, the complexity of multidrug combination regimens and their associated side effects—such as gastrointestinal reactions and *Clostridium difficile* infection—significantly impact patient adherence and treatment completion. Limitations in the diagnostic process, including interference from proton pump inhibitors with test results and the inability of serological tests to distinguish active infection, also pose additional obstacles to precise treatment and efficacy evaluation. Together, these factors constitute the major bottlenecks in current eradication efforts [5, 6]. Therefore, probiotics have gained significant attention as promising adjunctive therapeutic strategies. Current research indicates that probiotics exert a dual mechanism in *H. pylori* eradication therapy: they not only effectively mitigate antibiotic-induced gastrointestinal adverse reactions but also synergistically increase the clearance rate of the pathogen [7–9]. Building on these findings, researchers are actively exploring various application models for probiotics, including their use as standalone therapeutic regimens or in combination with conventional antibiotic protocols, with the aim of overcoming the current treatment challenges.

### 18.3.1 Probiotics as Single Therapies in *H. pylori* Treatment

The potential for eradicating *H. pylori* through probiotic monotherapy has been explored in several studies. Early work by Bhatia provided initial in vitro evidence that *Lactobacillus acidophilus* could inhibit the growth of *H. pylori* [10]. A small-scale clinical trial reported that oral administration of *L. acidophilus* alone led to successful eradication in 6 out of 14 patients, highlighting its potential antibacterial activity [11].

Research has further evaluated the efficacy of various probiotic strains—including single-strain formulations such as *Lactobacillus rhamnosus* [12], *Lactobacillus johnsonii La1* [13], and *Bifidobacterium bifidum BF-1* [14], as well as multistrain combinations—for example, *Lactobacillus reuteri* DSM 17938 plus *Lactobacillus reuteri* ATCC PTA 6457 [15] and a probiotic blend containing *Lactobacillus rhamnosus* GG, *Lactobacillus rhamnosus* LC705, *Propionibacterium freudenreichii* JS, and *Bifidobacterium animalis* Bb12 [16]. These studies consistently demonstrated a significant reduction in the gastric bacterial load, as reflected by decreased $^{13}$C-urea breath test (UBT) values, regardless of the probiotic strain or combination used.

However, the results from later randomized controlled trials (RCTs) have tempered earlier optimism. One double-blind study revealed that a composite probiotic containing *Lactobacillus acidophilus*, *Lactobacillus rhamnosus*, and *Lactobacillus sporogenes* had no significant inhibitory effect on *H. pylori* and could not replace antibiotic therapy [17]. Similarly, a clinical study involving 150 patients reported intention-to-treat eradication rates of only 18–32.5% with *Clostridium butyricum* and/or *Bacillus coagulans* monotherapy. Although these regimens exhibit favorable safety profiles, their efficacy remains substantially inferior to that of standard antibiotic therapy, precluding their use as substitutes [18].

A systematic review with pooled data analysis of 11 clinical trials concluded that probiotic monotherapy yields a low overall eradication rate of 14%, which is insufficient to recommend it as a stand-alone alternative to antibiotics. Nevertheless, the same analysis affirmed a direct—albeit modest—anti-*H. pylori* effect, as evidenced by a significantly higher eradication rate than that of the placebo (odds ratio (OR) = 7.91) and consistent reductions in UBT delta values, alongside a safety profile comparable to that of the placebo [19].

In summary, while probiotics exhibit a direct inhibitory effect against *H. pylori*, the magnitude of this activity is weak. Current evidence supports their investigation in specific clinical scenarios where conventional therapies are unsuitable, but their role remains primarily adjunctive. At present, probiotic monotherapy may be considered only for noncurative management of *H. pylori* colonization or for general gastric microbiota modulation, with clear clinical indications required to avoid misinterpretation of its therapeutic scope.

### 18.3.2 Probiotics as Adjuvant Therapy in Standard Eradication Protocols

The coadministration of probiotics with standard *H. pylori* eradication therapy represents the most extensively studied and evidence-supported adjuvant approach. The proposed benefits include enhancing eradication efficacy and improving treatment tolerability, although these effects remain subjects of debate.

A considerable body of evidence, including multiple RCTs and meta-analyses, supports the role of probiotics in improving eradication outcomes. One meta-analysis incorporating 30 RCTs (involving over 4500 patients) demonstrated that probiotic supplementation alongside standard triple therapy significantly improved the eradication rate by 12.2% in the per-protocol analysis and 14.1% in the intention-to-treat analysis. These benefits were consistent across adults and children, as well as in Asian and non-Asian populations. Additionally, probiotics significantly reduce the incidence of treatment-related adverse effects—such as nausea, vomiting, diarrhea, and epigastric pain—thereby increasing patient compliance and treatment tolerability [7].

Similarly, a broader meta-analysis of 140 studies (20,215 patients) systematically evaluated the effects of various probiotics in *H. pylori* eradication regimens. The results indicated that, compared with the placebo, most probiotics significantly increased the overall eradication rate (from 70.5% to 84.1%) and reduced adverse reactions (from 30.1% to 14.4%). However, in the context of standard triple therapy, no single probiotic type—such as *Lactobacillus* or *Saccharomyces boulardii*—or their combination clearly improved eradication rates. The combination regimens did not outperform single-strain supplementation. The efficacy of probiotics also appears to vary with treatment duration; for example, *Lactobacillus acidophilus* showed slightly better outcomes in 7-day regimens, whereas *S. boulardii* performed better in 10-day regimens [8].

In contrast, another meta-analysis of 21 RCTs reported that probiotic supplementation did not significantly improve eradication rates compared with placebo when it was added to standard triple therapy. Neither the intention-to-treat nor the per-protocol analyses revealed statistically significant benefits (OR = 1.21 and 1.28, respectively). Nonetheless, probiotics significantly reduce the frequency of adverse effects such as diarrhea and nausea [20]. A key source of heterogeneity among studies lies in the strain-level identity of probiotics. The effects of probiotics are highly strain-specific and cannot be generalized across taxonomic groups—or even among different isolates of the same species. The inconsistency in the conclusions of the current meta-analyses is likely attributable to the inclusion of various probiotics across the studies. To address this, an ideal solution would be to pool data from specific single strains and conduct meta-analyses.

For example, a 2015 systematic review and meta-analysis reported that supplementation with *S. boulardii* significantly increased the eradication rate from 71% to 80% and reduced the overall risk of side effects by approximately 56%, with notable reductions in diarrhea (49%) and nausea (40%) [21]. However, not all clinical trials have confirmed these positive findings. In a 2010 large-scale RCT by Song et al. ($n = 991$), patients were randomized into three groups: standard triple therapy, triple therapy plus *S. boulardii*, and triple therapy plus *S. boulardii* and a mucoprotective agent. While the intention-to-treat analysis revealed a significant improvement in eradication with *S. boulardii* (from 71.6% to 80.0%), the per-protocol analysis revealed no statistically significant differences among the groups (80.0% vs. 85.4% vs. 84.9%). Importantly, a previous study confirmed that *S. boulardii* significantly reduces overall side effects, particularly diarrhea and epigastric pain [22]. Similarly, a 2013 RCT by Zojaji et al. in Iran ($n = 160$) compared standard triple therapy with or without *S. boulardii*. Although the probiotic group presented a numerically greater eradication rate (87.5% vs. 81.2%), the difference was not statistically significant. The study highlighted, however, that *S. boulardii* significantly alleviated common gastrointestinal side effects—such as nausea, diarrhea, abdominal discomfort, and bloating—especially during the first 2 weeks of treatment [23]. Lactobacillus is another type of probiotic commonly used in clinical practice. A 2013 meta-analysis involving nine high-quality RCTs (involving a total of 1163 patients) revealed that probiotic supplementation containing Lactobacillus significantly increased the eradication rate of *H. pylori* (RR = 1.14 and NNT = 10) as an adjunctive therapy. This effect was more pronounced, particularly when Lactobacillus was used alone (a 17% increase in eradication rate, NNT = 6), and Lactobacillus was effective in both adults and children. However, probiotic supplementation did not significantly reduce overall treatment-related side effects (RR = 0.88) [24]. A recent systematic review and meta-analysis, through a comprehensive assessment of 26 high-quality RCTs (involving a total of 2314 patients), confirmed that supplementation with Lactobacillus as an adjunct therapy can significantly increase the standard eradication efficacy against *H. pylori* and improve tolerance during treatment. The analysis results revealed that, in the overall population, *Lactobacillus* supplementation significantly increased the eradication rate (RR = 1.063), and this benefit was more pronounced in children (RR = 1.223) than in adults (RR = 1.050). Furthermore, through in-depth subgroup analysis, the adjunctive effect of *Lactobacillus* varied depending on the treatment regimen and bacterial strain. Its enhancing effect was clear for triple therapy (RR = 1.124) but was not significant for quadruple therapy. At the strain level, *Lactobacillus reuteri* (RR = 1.049) demonstrated positive potential, whereas *Lactobacillus GG* did not have a significant effect. With respect to treatment duration, supplementation regimens of 14 days and 28–30 days were confirmed to be the most effective. In addition to improving efficacy, Lactobacillus can significantly reduce gastrointestinal side effects caused by antibiotic therapy, notably providing significant preventive effects against diarrhea (RR = 0.324), nausea/vomiting (RR = 0.566), and abdominal pain (RR = 0.438) [25].

Despite the standardization of probiotic strains, our meta-analysis continues to reveal inconsistent outcomes. We hypothesize that this persistent heterogeneity may be attributed to variations in administration protocols and demographic–geographic factors. The dosing regimens for probiotics in clinical trials are remarkably heterogeneous [26]. Timing strategies vary widely, ranging from the simultaneous administration of probiotics with antibiotics—termed "full-course concomitant" therapy—to sequential approaches where probiotics are administered as a pretreatment or as a posteradication stabilizer. Furthermore, the treatment duration differs significantly across studies, typically ranging from 2 to 8 weeks. The daily dosage also covers a broad spectrum, commonly ranging from $10^7$ to $10^{11}$ colony-forming units (CFUs). These methodological inconsistencies directly undermine comparability across trials and challenge the validity of pooling their results. Geodemographic variations introduce another layer of complexity. The efficacy of probiotic supplementation is modulated by a combination of region-specific factors, including local *H. pylori* antibiotic resistance patterns, host genetic polymorphisms, prevailing dietary habits, and the composition of the baseline gut microbiota.

In contrast to the ongoing debate over eradication efficacy, the capacity of specific probiotic strains to mitigate therapy-related adverse events is supported by robust and reproducible evidence. Numerous RCTs and subsequent meta-analyses have consistently concluded that adjunctive probiotic supplementation significantly reduces the overall incidence of antibiotic-associated adverse effects. The most pronounced protective effects are observed against gastrointestinal symptoms, particularly diarrhea and bloating.

This safety benefit holds direct clinical significance by enhancing patient adherence. A substantial proportion of eradication failures are not due to bacterial resistance but are a consequence of premature treatment discontinuation caused by intolerable side effects. By significantly improving the tolerability of eradication regimens, probiotics play an indirect yet crucial role. They facilitate the completion of the full course of antibiotic therapy, thereby making an essential contribution to achieving the ultimate therapeutic goal of *H. pylori* eradication.

### 18.3.3 The Application of Probiotics in Children with *H. pylori* Infection

Several studies have investigated probiotics as supplements for *H. pylori* eradication therapy in children. An RCT published in 2014 examined the effect of supplementing standard triple therapy with *Lactobacillus acidophilus* and *Bifidobacterium bifidum* in children with *H. pylori* infection. Compared with triple therapy alone, probiotic supplementation significantly increased the *H. pylori* eradication rate (83.7% vs. 64.4%). Furthermore, real-time quantitative polymerase chain reaction analysis revealed that probiotic supplementation effectively maintained the balance of the intestinal flora, preventing the decrease in Bifidobacterium and *Lactobacillus acidophilus* numbers and the increase in *Escherichia coli* numbers typically caused by standard antibiotic therapy. These findings indicate that probiotics, as adjunctive therapies, play a positive role in improving the eradication rate and protecting the intestinal microecology [27]. Moreover, a randomized, double-blind clinical trial conducted in Iran investigated the effect of adjuvant probiotic therapy on *H. pylori* infection in children. A total of 66 *H. pylori*-positive children aged 3–14 years were randomly assigned to receive either a probiotic containing seven strains or a placebo while receiving standard triple therapy. The results revealed that the probiotic group had a significantly greater *H. pylori* eradication rate (90.09% vs. 69.69%) and significantly lower incidences of nausea/vomiting (6.06% vs. 27.27%) and diarrhea (6.06% vs. 24.24%) during treatment. These studies indicate that probiotic supplementation can increase the eradication rate of *H. pylori* in children and reduce antibiotic-related gastrointestinal side effects [28]. A 2017 network meta-analysis systematically evaluated the efficacy and safety of probiotic-supplemented triple therapy for *H. pylori* infection in children. The analysis included 29 clinical trials (3122 pediatric patients) involving 17 probiotic regimens. Compared with placebo, probiotic supplementation significantly increased the *H. pylori* eradication rate and reduced the overall risk of side effects. Furthermore, *Lactobacillus casei* was the most effective for improving the eradication rate, whereas the multistrain combination of *Lactobacillus acidophilus* and *Lactobacillus rhamnosus* was the most effective for reducing overall side effects. Additionally, the study identified the most effective probiotic strains or combinations for specific side effects, such as diarrhea and nausea/vomiting [29]. However, some researchers have reported contrary findings. A study conducted in Turkey revealed that supplementing standard triple therapy with a specific probiotic combination (containing *Lactobacillus casei* 2401, *Lactobacillus acidophilus* 2027, and *Bifidobacterium lactis* 2211) for treating *H. pylori* infection in children demonstrated no additional benefits compared with standard triple therapy alone. The results revealed no statistically significant differences between the two groups in either the *H. pylori* eradication rate (66.6% in the probiotic group vs. 68.9% in the standard therapy group) or the incidence of drug side effects. The authors speculated that the high local antibiotic resistance rate might be an important reason for the generally low eradication rate (approximately 68%) and the inability of probiotics to further increase treatment efficacy [30]. These inconsistent findings may be constrained by the limited number of subgroup analyses conducted specifically in pediatric populations.

## 18.4 Guidelines for the Clinical Use of Probiotics

Several international authoritative guidelines have evaluated the adjunctive role of probiotics in the treatment of *H. pylori* infection. The Italian guidelines state that certain probiotics, such as Lactobacillus, Bifidobacterium, and *Saccharomyces boulardii*, when used as an adjunct to *H. pylori* eradication therapy, can reduce antibiotic-associated adverse effects—particularly diarrhea—but their role in increasing eradication rates remains unclear; thus, they are recommended to improve treatment tolerability, with a level of evidence of 3a and a grade of recommendation of B [31]. According to the Toronto Consensus, the routine use of probiotics is not recommended for eradication therapy for *H. pylori* in adults. The consensus states that the current quality of evidence is very low; probiotics have not been proven to effectively increase the eradication rate, nor are they recommended for routine addition to reduce treatment side effects. Although some individual studies suggest potential benefits, the expert panel strongly advises against their routine use as an adjunctive therapy owing to the lack of standardization in probiotic formulations and the increased cost and complexity they add to the treatment regimen [32]. The Maastricht V/Florence Consensus states that specific probiotic strains—such as Lactobacillus, Bifidobacterium, and Saccharomyces—can significantly reduce gastrointestinal side effects during *H. pylori* eradication therapy, but marked interstrain differences exist; therefore, only strains with proven clinical efficacy should be chosen. Any improvement in eradication rates remains uncertain and, if present, is probably an indirect consequence of better treatment tolerability [33].

## 18.5 Clinical Challenges in Implementing Probiotics as Adjunctive Therapies for *H. pylori* Infection

Despite a growing evidence base, the routine integration of probiotics into *H. pylori* eradication practices continues to face critical hurdles.

### 18.5.1 Strain Selection and Uneven Support

Numerous studies and meta-analyses have indicated that probiotics such as Lactobacillus, Saccharomyces, and Bifidobacterium exhibit certain beneficial effects in eradication therapy [7, 34]. However, on the basis of the current clinical evidence, it remains unclear which probiotic is most effective in eradicating *H. pylori*. In clinical practice, one of the main challenges faced by physicians is selecting the appropriate type of probiotic. To date, only a limited number of strains or combinations—primarily specific strains of Lactobacillus and Bifidobacterium—have been supported by substantial evidence from large, reproducible, high-quality RCTs [25, 35]. Furthermore, some studies suggest that multistrain probiotic compounds may be more effective than single-strain preparations because of synergistic effects, although this conclusion still requires further validation through additional research [36].

### 18.5.2 Absence of a Standardized Administration Protocol

Even when an evidence-based strain is chosen, no authoritative "prescription" exists. The optimal timing (pretreatment, concomitant, or post-treatment), daily dose (CFUs per day), and total duration (weeks to months) remain undefined, resulting in heterogeneous clinical practices that compromise reproducibility and generalizability. For example, a meta-analysis revealed that the adjunct dosing regimens of *Saccharomyces boulardii* vary, with doses ranging from 100 to 1000 mg per day and treatment durations ranging from 1 to 4 weeks. Among these, the regimen of 500 mg daily for 2 weeks has been more frequently studied; however, no consensus has been reached regarding the optimal protocol [21]. Furthermore, some studies have proposed a sequential strategy distinct from concurrent administration, which involves the use of probiotics either before initiation or after the completion of eradication therapy. For example, one study demonstrated that a 2-week probiotic pretreatment followed by standard eradication therapy achieved an eradication rate exceeding 90% [37]. A meta-analysis further indicated that supplementing with probiotics before antibiotic treatment (pretreatment) or continuing supplementation after the course of therapy significantly improved *H. pylori* eradication rates, whereas concurrent administration yielded limited efficacy. Additionally, extending the supplementation period beyond 2 weeks improved therapeutic outcomes [38]. These findings suggest that "avoiding concurrent administration and prolonging the probiotic course" may be a key strategy for optimizing the adjunctive efficacy of probiotics.

In conclusion, variations in probiotic strains, dosages, and treatment durations have led to inconsistent and even contradictory outcomes in *H. pylori* eradication, making the findings difficult to interpret. Nevertheless, adjuvant probiotic therapy remains a valuable addition to current eradication regimens, as it can increase eradication rates and reduce treatment-related side effects through direct or indirect anti-*H. pylori* mechanisms. However, the optimal application strategy has yet to be clearly defined and warrants further investigation.

## 18.6 Conclusion

On the basis of current evidence, probiotics have considerable potential as adjunctive therapies in *H. pylori* eradication regimens. Clinical benefits are primarily manifested in two aspects: specific strains moderately increase eradication rates while significantly reducing antibiotic-associated gastrointestinal adverse effects, thereby improving treatment adherence. Nevertheless, several challenges impede standardized clinical implementation. Substantial interstrain variability in efficacy and the absence of established administration protocols—including optimal strain selection, timing, dosage, and treatment duration—contribute to inconsistent clinical outcomes and unreliable therapeutic effects. Current international guidelines consistently recognize the value of probiotics in improving treatment tolerance while remaining cautious about their role in increasing eradication rates due to insufficient evidence. In clinical practice, evidence-based probiotic strains should be selectively considered, particularly for patients with poor tolerance or high risk of treatment failure. Future research must move beyond asking *whether* probiotics work and instead focus on defining *which* specific strains, at doses and timings, are effective for *which* patient populations. Large, RCTs based on standardized, evidence-based protocols are essential to translate this promising adjunctive potential into reliable clinical practice.

## References


1. Baryshnikova NV, Ilina AS, Ermolenko EI, Uspenskiy YP, Suvorov AN. Probiotics and autoprobiotics for treatment of *H. pylori* infection. World J Clin Cases. 2023;11(20):4740–51. https://doi.org/10.12998/wjcc.v11.i20.4740.
2. Markowiak P, Śliżewska K. Effects of probiotics, prebiotics, and Synbiotics on human health. Nutrients. 2017;9(9) https://doi.org/10.3390/nu9091021.
3. Sanders ME, Merenstein DJ, Reid G, Gibson GR, Rastall RA. Probiotics and prebiotics in intestinal health and disease: from biology to the clinic. Nat Rev Gastroenterol Hepatol. 2019;16(10):605–16. https://doi.org/10.1038/s41575-019-0173-3.
4. Compare D, Sgamato C, Nardone OM, Rocco A, Coccoli P, Laurenza C, et al. Probiotics in gastrointestinal diseases: all that glitters is not gold. Dig Dis. 2022;40(1):123–32. https://doi.org/10.1159/000516023.
5. Yang JC, Lu CW, Lin CJ. Treatment of *H. pylori* infection: current status and future concepts. World J Gastroenterol. 2014;20(18):5283–93. https://doi.org/10.3748/wjg.v20.i18.5283.
6. Crowe SE. *H. pylori* infection. N Engl J Med. 2019;380(12):1158–65. https://doi.org/10.1056/NEJMcp1710945.
7. Lau CS, Ward A, Chamberlain RS. Probiotics improve the efficacy of standard triple therapy in the eradication of *H. pylori*: a meta-analysis. Infect Drug Resist. 2016;9:275–89. https://doi.org/10.2147/idr.S117886.
8. Wang F, Feng J, Chen P, Liu X, Ma M, Zhou R, et al. Probiotics in helicobacter pylori eradication therapy: systematic review and network meta-analysis. Clin Res Hepatol Gastroenterol. 2017;41(4):466–75. https://doi.org/10.1016/j.clinre.2017.04.004.
9. Goderska K, Agudo Pena S, Alarcon T. *H. pylori* treatment: antibiotics or probiotics. Appl Microbiol Biotechnol. 2018;102(1):1–7. https://doi.org/10.1007/s00253-017-8535-7.
10. Bhatia SJ, Kochar N, Abraham P, Nair NG, Mehta AP. Lactobacillus acidophilus inhibits growth of campylobacter pylori in vitro. J Clin Microbiol. 1989;27(10):2328–30. https://doi.org/10.1128/jcm.27.10.2328-2330.1989.
11. Canducci F, Cremonini F, Armuzzi A, Di Caro S, Gabrielli M, Santarelli L, et al. Probiotics and *H. pylori* eradication. Dig Liver Dis. 2002;34:S81–S3. https://doi.org/10.1016/S1590-8658(02)80172-4.
12. Chen MJ, Chen CC, Huang YC, Tseng CC, Hsu JT, Lin YF, et al. The efficacy of lactobacillus acidophilus and rhamnosus in the reduction of bacterial load of *H. pylori* and modification of gut microbiota-a double-blind, placebo-controlled, randomized trial. Helicobacter. 2021;26(6):e12857. https://doi.org/10.1111/hel.12857.
13. Gotteland M, Cruchet S. Suppressive effect of frequent ingestion of lactobacillus johnsonii La1 on *H. pylori* colonization in asymptomatic volunteers. J Antimicrob Chemother. 2003;51(5):1317–9. https://doi.org/10.1093/jac/dkg227.
14. Miki K, Urita Y, Ishikawa F, Iino T, Shibahara-Sone H, Akahoshi R, et al. Effect of Bifidobacterium bifidum fermented milk on *H. pylori* and serum pepsinogen levels in humans. J Dairy Sci. 2007;90(6):2630–40. https://doi.org/10.3168/jds.2006-803.
15. Francavilla R, Polimeno L, Demichina A, Maurogiovanni G, Principi B, Scaccianoce G, et al. Lactobacillus reuteri strain combination in *H. pylori* infection: a randomized, double-blind, placebo-controlled study. J Clin Gastroenterol. 2014;48(5):407–13. https://doi.org/10.1097/mcg.0000000000000007.
16. Myllyluoma E, Kajander K, Mikkola H, Kyrönpalo S, Rasmussen M, Kankuri E, et al. Probiotic intervention decreases serum gastrin-17 in *H. pylori* infection. Dig Liver Dis. 2007;39(6):516–23. https://doi.org/10.1016/j.dld.2007.02.015.
17. Lee CY, Shih HC, Yu MC, Lee MY, Chang YL, Lai YY, et al. Evaluation of the potential inhibitory activity of a combination of L. acidophilus, L. rhamnosus and L. sporogenes on *H. pylori*: a randomized double-blind placebo-controlled clinical trial. Chin J Integr Med. 2017;23(3):176–82. https://doi.org/10.1007/s11655-016-2531-0.

18. Zhang J, Guo J, Li D, Chen M, Liu J, Feng C, et al. The efficacy and safety of clostridium butyricum and Bacillus coagulans in *H. pylori* eradication treatment: an open-label, single-arm pilot study. Medicine (Baltimore). 2020;99(45):e22976. https://doi.org/10.1097/md.0000000000022976.
19. Losurdo G, Cubisino R, Barone M, Principi M, Leandro G, Ierardi E, et al. Probiotic monotherapy and *H. pylori* eradication: a systematic review with pooled-data analysis. World J Gastroenterol. 2018;24(1):139–49. https://doi.org/10.3748/wjg.v24.i1.139.
20. Lu C, Sang J, He H, Wan X, Lin Y, Li L, et al. Probiotic supplementation does not improve eradication rate of *H. pylori* infection compared to placebo based on standard therapy: a meta-analysis. Sci Rep. 2016;6(1):23522. https://doi.org/10.1038/srep23522.
21. Szajewska H, Horvath A, Kołodziej M. Systematic review with meta-analysis: saccharomyces boulardii supplementation and eradication of *H. pylori* infection. Aliment Pharmacol Ther. 2015;41(12):1237–45. https://doi.org/10.1111/apt.13214.
22. Song MJ, Park DI, Park JH, Kim HJ, Cho YK, Sohn CI, et al. The effect of probiotics and mucoprotective agents on PPI-based triple therapy for eradication of *H. pylori*. Helicobacter. 2010;15(3):206–13. https://doi.org/10.1111/j.1523-5378.2010.00751.x.
23. Zojaji H, Ghobakhlou M, Rajabalinia H, Ataei E, Jahani Sherafat S, Moghimi-Dehkordi B, et al. The efficacy and safety of adding the probiotic saccharomyces boulardiito standard triple therapy for eradication of H. pylori: a randomized controlled trial. Gastroenterol Hepatol Bed Bench. 2013;6(Suppl 1):S99–s104.
24. Zheng X, Lyu L, Mei Z. Lactobacillus-containing probiotic supplementation increases *H. pylori* eradication rate: evidence from a meta-analysis. Rev Esp Enferm Dig. 2013;105(8):445–53. https://doi.org/10.4321/s1130-01082013000800002.
25. Mishra V, Dash D, Panda AK, Pathak SK. Efficacy of lactobacillus spp. supplementation in *H. pylori* eradication: a systematic meta-analysis of randomized controlled trials with trial sequential analysis. Helicobacter. 2024;29(6):e70006. https://doi.org/10.1111/hel.70006.
26. Homan M, Orel R. Are probiotics useful in *H. pylori* eradication? World J Gastroenterol. 2015;21(37):10644–53. https://doi.org/10.3748/wjg.v21.i37.10644.
27. Wang Y-h, Huang Y. Effect of lactobacillus acidophilus and Bifidobacterium bifidum supplementation to standard triple therapy on *H. pylori* eradication and dynamic changes in intestinal flora. World J Microbiol Biotechnol. 2014;30(3):847–53. https://doi.org/10.1007/s11274-013-1490-2.
28. Ahmad K, Fatemeh F, Mehri N, Maryam S. Probiotics for the treatment of pediatric *H. pylori* infection: a randomized double blind clinical trial. Iran J Pediatr. 2013;23(1):79–84.
29. Feng JR, Wang F, Qiu X, McFarland LV, Chen PF, Zhou R, et al. Efficacy and safety of probiotic-supplemented triple therapy for eradication of *H. pylori* in children: a systematic review and network meta-analysis. Eur J Clin Pharmacol. 2017;73(10):1199–208. https://doi.org/10.1007/s00228-017-2291-6.
30. Akcam M, Koca T, Salman H, Karahan N. The effects of probiotics on treatment of *H. pylori* eradication in children. Saudi Med J. 2015;36(3):286–90. https://doi.org/10.15537/smj.2015.3.10124.
31. Zagari RM, Romano M, Ojetti V, Stockbrugger R, Gullini S, Annibale B, et al. Guidelines for the management of *H. pylori* infection in Italy: the III working group consensus report 2015. Dig Liver Dis. 2015;47(11):903–12. https://doi.org/10.1016/j.dld.2015.06.010.
32. Fallone CA, Chiba N, van Zanten SV, Fischbach L, Gisbert JP, Hunt RH, et al. The Toronto consensus for the treatment of *H. pylori* infection in adults. Gastroenterology. 2016;151(1):51–69. https://doi.org/10.1053/j.gastro.2016.04.006.
33. Malfertheiner P, Megraud F, O'Morain CA, Gisbert JP, Kuipers EJ, Axon AT, et al. Management of *H. pylori* infection-the Maastricht V/Florence consensus report. Gut. 2017;66(1):6–30. https://doi.org/10.1136/gutjnl-2016-312288.
34. Wang Y, Wang X, Cao XY, Zhu HL, Miao L. Comparative effectiveness of different probiotics supplements for triple *H. pylori* eradication: a network meta-analysis. Front Cell Infect Microbiol. 2023;13:1120789. https://doi.org/10.3389/fcimb.2023.1120789.

35. Jiang X, Xu C, Liu B, Chen P, Xu Q, Zhang L. Efficacy and safety of bifidobacterium quadruple viable tablets in the treatment of *H. pylori*-infected peptic ulcer or gastritis patients: a systematic review and meta-analysis. BMC Infect Dis. 2023;23(1):313. https://doi.org/10.1186/s12879-023-08211-1.
36. Chapman CM, Gibson GR, Rowland I. Health benefits of probiotics: are mixtures more effective than single strains? Eur J Nutr. 2011;50(1):1–17. https://doi.org/10.1007/s00394-010-0166-z.
37. Dajani AI, Abu Hammour AM, Yang DH, Chung PC, Nounou MA, Yuan KY, et al. Do probiotics improve eradication response to *H. pylori* on standard triple or sequential therapy? Saudi J Gastroenterol. 2013;19(3):113–20. https://doi.org/10.4103/1319-3767.111953.
38. Lv Z, Wang B, Zhou X, Wang F, Xie Y, Zheng H, et al. Efficacy and safety of probiotics as adjuvant agents for *H. pylori* infection: a meta-analysis. Exp Ther Med. 2015;9(3):707–16. https://doi.org/10.3892/etm.2015.2174.

# The Future of Probiotic Therapy

# 19

Weiliang Hou and Xueting Li

## 19.1 Introduction

The management of *Helicobacter pylori* infection remains a formidable global health challenge. Despite remarkable advances in antimicrobial therapy, the persistent prevalence of this pathogen underscores the limitations of current treatment strategies. Standard eradication regimens—typically involving combinations of antibiotics and proton pump inhibitors—were once highly effective. However, the rise of antibiotic resistance, treatment-related adverse effects, and difficulties with patient adherence have significantly reduced their success rates in many regions of the world [1]. These obstacles highlight the urgent need for alternative or adjunctive therapeutic approaches that are both effective and sustainable.

Probiotics have emerged as a promising candidate in this regard. Defined as live microorganisms that, when administered in adequate amounts, confer a health benefit on the host, probiotics have traditionally been associated with gastrointestinal well-being [2]. In the context of *H. pylori* infection, probiotics offer a multifaceted therapeutic potential [3]. They can exert direct antimicrobial effects against *H. pylori* through the secretion of organic acids, hydrogen peroxide, and bacteriocins. They may also compete with the pathogen for adhesion sites on the gastric mucosa, thereby limiting colonization [4]. Beyond these direct effects, probiotics are increasingly recognized for their ability to modulate host immune responses, reduce gastric inflammation, and restore microbial balance disrupted by antibiotic therapy [5]. Collectively, these mechanisms provide a compelling rationale for investigating

W. Hou (✉) · X. Li
Department of Gastroenterology, Shanghai Institute of Pancreatic Diseases, National Key Laboratory of Immunity and Inflammation, Changhai Clinical Research Unit, Changhai Hospital, Naval Medical University, Shanghai, China
e-mail: houweiliang@tongji.edu.cn



J. Gao, Y. Du (eds.), *Non-Antibiotic Therapies for Helicobacter pylori*,
https://doi.org/10.1007/978-981-95-9598-3_19

probiotics as both adjuncts to conventional eradication regimens and as potential stand-alone strategies in selected populations.

The clinical evidence accumulated thus far supports the notion that probiotics can improve treatment outcomes. Several randomized controlled trials and meta-analyses have demonstrated that adding probiotics to standard triple or quadruple therapy can modestly increase eradication rates while significantly reducing common side effects, such as diarrhea, nausea, and abdominal discomfort [6]. These benefits improve patient adherence, which is often a critical determinant of treatment success. However, despite these encouraging findings, probiotics have not yet been universally adopted as a standard component of *H. pylori* therapy. The variability in strain selection, dosage, duration, and study design across trials has led to inconsistent results, preventing the establishment of firm clinical guidelines.

Looking to the future, the role of probiotics in *H. pylori* management is poised to expand in several directions. Advances in microbiome science, genomic technologies, and bioengineering are creating opportunities to identify novel probiotic strains with superior anti-*H. pylori* properties or to design genetically modified microorganisms capable of delivering targeted therapeutic molecules. Innovative delivery systems, such as microencapsulation and nanotechnology-based carriers, promise to enhance the survival and efficacy of probiotics in the hostile gastric environment. Meanwhile, the integration of precision medicine approaches—guided by microbiome profiling, host genetics, and computational analytics—offers the prospect of tailoring probiotic regimens to individual patients, thereby maximizing therapeutic benefit.

At the same time, it is important to recognize the challenges that remain. The safety of long-term probiotic use in diverse populations must be rigorously assessed, particularly among immunocompromised individuals. Regulatory frameworks are inconsistent across countries, creating obstacles to the clinical translation of research findings. Ensuring the quality, authenticity, and functional stability of probiotic products is another pressing issue, as variability among commercial formulations undermines both clinical efficacy and patient confidence. Finally, the cost-effectiveness of advanced probiotic formulations must be addressed if they are to play a meaningful role in global *H. pylori* control, especially in low- and middle-income countries where disease burden is highest.

Against this backdrop, the future development of probiotic therapy for *H. pylori* represents both a scientific and a clinical frontier. This chapter will explore the emerging directions that define this evolving field, including innovations in strain selection, delivery systems, personalized approaches, synergistic combinations with other therapies, and the regulatory frameworks necessary for safe and effective implementation (Fig. 19.1). Together, these advancements point toward a future in which probiotics may transition from optional adjuncts to integral components of *H. pylori* management, contributing to improved patient outcomes and alleviating the global burden of antibiotic resistance.

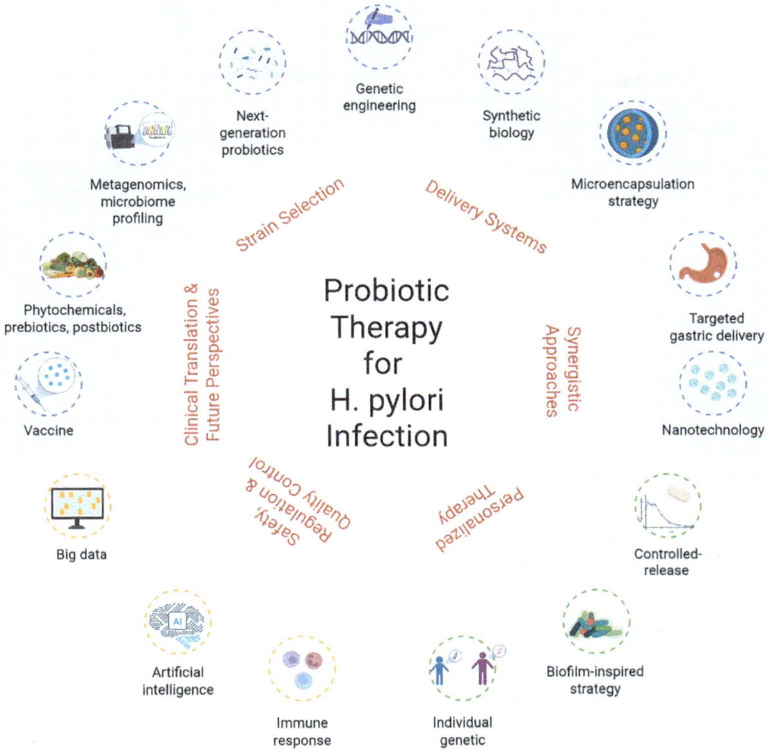


**Fig. 19.1** Future development direction of probiotic therapy. (Image created with BioRender.com, with permission)

## 19.2 Advances in Probiotic Strain Selection

The efficacy of probiotic therapy in *H. pylori* infection depends critically on the strains employed. Early investigations often treated probiotics as a broad, interchangeable category, but it has since become clear that probiotic effects are highly strain-specific. Some microorganisms show significant anti-*H. pylori* activity, while others exhibit minimal or no benefit. Consequently, progress in this field requires a shift from generic use to rational, evidence-based strain selection, drawing on advances in microbiology, genomics, and synthetic biology. This section discusses emerging directions in strain selection, organized into three interrelated themes: the identification of novel strains, the development of multi-strain and next-generation probiotics (NGPs), and the application of genetic engineering and synthetic biology.

### 19.2.1 Identification of Novel Probiotic Strains with Anti-*H. pylori* Activity

The discovery of probiotics with activity against *H. pylori* has historically relied on empirical screening. Strains such as *Lactobacillus rhamnosus GG*, *Lactobacillus reuteri*, and *Bifidobacterium animalis* have repeatedly demonstrated inhibitory effects on *H. pylori* growth, as well as the ability to reduce gastrointestinal side effects associated with antibiotic therapy [7, 8]. Other organisms, including the yeast *Saccharomyces boulardii*, have shown promise in improving eradication rates and patient adherence when used as adjuncts to standard therapy [9].

However, advances in metagenomics and microbiome profiling are revolutionizing the search for new strains. By sequencing gastric and intestinal microbial communities in infected and uninfected individuals, researchers are identifying previously unrecognized organisms that may naturally antagonize *H. pylori* [10–12].

Moreover, strains isolated from the gastric niche itself may have particular advantages. Unlike conventional probiotics sourced from the intestine or fermented foods, gastric-adapted microorganisms are more likely to withstand low pH and compete effectively with *H. pylori* for mucosal niches [13]. Early exploratory studies suggest that the human stomach harbors a wider range of cultivable microbes than previously thought, opening new possibilities for strain discovery [14]. Such as BCF-01, a novel gastric-derived probiotic, can effectively inhibit *H. pylori* infection [15].

Another promising approach involves functional screening based on specific mechanisms rather than broad survival assays. For example, identifying strains capable of binding urease—the key enzyme that enables *H. pylori* to neutralize gastric acid—could provide highly targeted inhibition [16]. Similarly, probiotics that interfere with virulence factors such as cytotoxin-associated gene A (CagA) or vacuolating cytotoxin A (VacA) may directly attenuate the pathogenicity of *H. pylori* [17].

The future of probiotic strain discovery thus lies at the intersection of systems biology and mechanistic microbiology. By combining high-throughput sequencing with functional assays, researchers are moving toward a rational pipeline for selecting candidate strains with the highest therapeutic potential.

### 19.2.2 Multi-strain Probiotics and NGPs

Single-strain probiotics have provided valuable insights, but their clinical benefits remain modest in most trials [18]. This has fueled interest in multi-strain formulations designed to harness the complementary properties of different organisms. Multi-strain consortia can act on multiple pathogenic pathways simultaneously: one strain may reduce adhesion, another may enhance mucosal immunity, while a third modulates inflammation [19]. Such combinations may also increase ecological resilience within the gastric niche, reducing the risk of *H. pylori* re-colonization.

Meta-analyses suggest that multi-strain probiotics often outperform single-strain probiotics in improving eradication rates and reducing antibiotic-associated side effects. For example, combinations of Lactobacillus, Bifidobacterium, and Saccharomyces species have shown greater consistency in clinical outcomes than monotherapies [20, 21]. Nonetheless, the challenge lies in selecting the optimal blend of strains, as antagonistic interactions can occur if species compete for the same resources. Advanced modeling of microbial interactions may help optimize such consortia for maximal efficacy.

In parallel, the concept of NGPs is gaining attention. Unlike traditional strains, which are typically selected for safety and a long history of human use, NGPs include commensals with unique functional traits that may not yet be widely commercialized [22]. Examples include *Akkermansia muciniphila*, noted for its mucin-degrading capacity and immune-modulatory properties [23], and *Faecalibacterium prausnitzii*, recognized for anti-inflammatory activity [24]. Although not directly studied in *H. pylori* eradication, these organisms could theoretically modulate the gastric environment to make colonization less favorable.

Next-generation consortia may also incorporate postbiotics (non-viable microbial components with health benefits) and parabiotics (inactivated microbial cells), offering additional safety advantages for vulnerable populations. Together, multi-strain and next-generation approaches represent a major leap toward a new generation of probiotic therapeutics.

### 19.2.3 Genetic Engineering and Synthetic Biology Approaches

Perhaps the most transformative advance in probiotic strain selection lies in synthetic biology. Traditional probiotics act indirectly, modulating the environment or immune responses in ways that are sometimes unpredictable. In contrast, engineered strains can be designed to deliver specific therapeutic functions with precision.

One approach involves engineering Lactobacillus or *Escherichia coli Nissle* 1917 strains to express antimicrobial peptides (AMPs) or bacteriocins specifically targeting *H. pylori*. For instance, pretreatment with bioengineered *Lactococcus lactis* that actively secretes cathelicidin significantly increased mucosal cathelicidin levels and reduced *H. pylori* infection and the associated inflammation [25].

Another promising avenue is the expression of neutralizing agents against virulence factors. Strains engineered to secrete antibody fragments or decoy receptors that bind CagA or VacA could blunt *H. pylori* pathogenicity without necessarily eradicating the bacterium, thus reducing disease risk while minimizing selective pressure for resistance.

Synthetic probiotics could also serve as live biotherapeutic delivery platforms. For example, engineered microbes could produce anti-inflammatory cytokines, mucosal protectants, or biofilm-disrupting enzymes directly at the site of infection. Such multifunctional strains would go far beyond the capabilities of conventional probiotics [26].

The challenges in this field are considerable. Regulatory acceptance of genetically modified organisms remains limited, particularly for live microbial therapies intended for human consumption. Safety concerns—including the risk of horizontal gene transfer or unintended ecological effects—must be rigorously addressed. Nonetheless, the rapid progress in Clustered Regularly Interspaced Short Palindromic Repeats (CRISPR)-based genome editing and synthetic microbial circuits makes it plausible that engineered probiotics will move from experimental systems to clinical trials within the next decade [27, 28].

## 19.3 Innovative Delivery Systems

The therapeutic potential of probiotics in *H. pylori* infection is often constrained by the harsh gastric environment. Acidic pH, digestive enzymes, and bile salts significantly reduce probiotic viability, while the pathogen itself occupies mucosal niches that are difficult for exogenous microbes to colonize. Consequently, innovative delivery systems have become central to enhancing probiotic survival, colonization, and efficacy. Advances in encapsulation, nanotechnology, and biofilm-inspired systems are reshaping the field.

### 19.3.1 Microencapsulation and Targeted Gastric Delivery

Microencapsulation is one of the most established strategies to protect probiotics from gastric acidity and enzymatic degradation. Encapsulation materials such as alginate, chitosan, and cellulose derivatives create a protective barrier around microbial cells. This increases survival rates during gastric transit and allows for gradual release in the stomach or small intestine [29, 30].

Recent developments focus on targeted gastric delivery, ensuring that viable probiotics reach the stomach—the primary site of *H. pylori* colonization. By modifying capsule coatings to resist degradation until reaching acidic pH, researchers have improved delivery precision [31]. Some encapsulated formulations also include prebiotic carriers, which both shield the microbes and enhance their growth upon release [32].

### 19.3.2 Nanotechnology-Based Probiotic Formulations

Nanotechnology offers new opportunities for enhancing probiotic delivery. Nanoencapsulation enables precise control over particle size, release kinetics, and surface properties [33]. Nanocarriers can be engineered to respond to specific physiological triggers, such as gastric pH or mucosal enzymes, thereby releasing probiotics only when and where they are most effective [34].

Functionalization with mucosal adhesion ligands further improves colonization by enabling probiotics to bind tightly to gastric epithelial surfaces, increasing their persistence in competition with *H. pylori* [35, 36]. Additionally, nanoparticle systems can be designed to co-deliver synergistic agents, such as AMPs or phytochemicals, alongside probiotics. These multifunctional platforms may significantly enhance therapeutic outcomes, reducing reliance on antibiotics.

### 19.3.3 Controlled-Release and Biofilm-Inspired Strategies

Another emerging approach is the development of controlled-release systems, which allow probiotics to be delivered gradually over extended periods. Such systems maintain a stable microbial population at the infection site, avoiding rapid clearance and providing sustained therapeutic activity [37].

Biofilm-inspired systems mimic the natural protective matrix of microbial communities. By embedding probiotics within biofilm-like structures or polysaccharide matrices, researchers can enhance both their survival and adherence to the gastric mucosa [38, 39]. This design not only improves colonization but may also promote ecological resilience, reducing the likelihood of *H. pylori* re-colonization after therapy.

Although still experimental, these strategies show promise for transforming probiotics into long-acting, site-specific therapeutics.

## 19.4 Personalized and Precision Probiotic Therapy

While conventional probiotic therapy has largely adopted a "one-size-fits-all" approach, emerging evidence shows that responses to probiotics vary significantly between individuals. Factors such as host genetics, microbiome composition, immune status, and lifestyle all influence therapeutic outcomes. To maximize efficacy in *H. pylori* infection, future probiotic therapy must transition toward personalized and precision medicine.

### 19.4.1 Role of Host Microbiome Profiling in Therapy Selection

Each individual harbors a unique gut and gastric microbiome, which dynamically interacts with *H. pylori*. Some microbial communities are more resistant to colonization, while others provide niches that favor persistence. High-throughput sequencing now makes it possible to profile a patient's microbiome before treatment [40].

Such profiling can guide strain selection: for example, patients with reduced Lactobacillus populations may benefit most from supplementation with specific Lactobacillus strains. Similarly, individuals with disrupted microbial balance after multiple antibiotic courses might require multi-strain or next-generation formulations.

Microbiome-guided therapy could also help predict treatment success. Patients with microbial signatures linked to antibiotic resistance or poor probiotic colonization might be directed toward alternative regimens. This precision approach would avoid unnecessary therapies while improving overall eradication rates.

### 19.4.2 Individual Genetic and Immune Response Differences

Host genetics play a crucial role in shaping responses to both *H. pylori* and probiotics. Polymorphisms in genes regulating gastric acid secretion, mucosal immunity, or inflammatory pathways influence infection outcomes [41, 42]. For instance, variations in interleukin-1β and tumor necrosis factor-α genes are associated with higher gastric inflammation and cancer risk [43]. Probiotics that modulate immune signaling may provide greater benefit to genetically susceptible individuals.

Immune status also affects therapeutic success. Immunocompromised patients may require non-viable preparations such as postbiotics to avoid infection risks while still gaining anti-inflammatory benefits. Conversely, immunocompetent patients may tolerate live strains that stimulate protective immune responses, including enhanced IgA secretion or regulatory T-cell activation.

Thus, tailoring probiotic therapy to genetic and immune profiles represents a promising step toward precision medicine, ensuring both safety and maximal benefit.

### 19.4.3 AI- and Big Data-Driven Personalization

The integration of artificial intelligence (AI) and big data analytics is transforming the landscape of personalized probiotic therapy. By analyzing vast datasets that include microbiome profiles, host genetics, clinical history, and treatment outcomes, machine learning algorithms can identify complex patterns not evident through traditional analysis [44].

Such models could be used to predict the most effective probiotic regimen for an individual patient, including strain selection, dosage, and delivery method. For instance, AI could suggest that a patient with specific microbial and genetic features would benefit most from a Lactobacillus–Bifidobacterium combination delivered via encapsulation, while another patient might be better suited to a yeast-based formulation.

Beyond guiding clinical practice, AI systems could also facilitate real-time monitoring of therapy. Wearable biosensors or digital health platforms could track biomarkers of inflammation or microbiome changes, feeding data back into adaptive algorithms that refine treatment over time. This would represent a major step toward dynamic, patient-centered probiotic care.

## 19.5 Synergistic Approaches

The complexity of *H. pylori* infection suggests that no single intervention is likely to be universally effective. Probiotics, while beneficial, typically provide modest improvements when used alone. Future development is therefore moving toward synergistic strategies, combining probiotics with other therapies to enhance efficacy, reduce side effects, and mitigate antibiotic resistance.

### 19.5.1 Combination of Probiotics with Antibiotics for Resistance Mitigation

Antibiotic resistance remains the major barrier to successful *H. pylori* eradication. Clarithromycin- and levofloxacin-resistant strains are increasingly common, with eradication rates falling below acceptable thresholds in many regions. Probiotics can play an important supportive role in this context [18].

When co-administered with antibiotics, probiotics reduce gastrointestinal side effects such as diarrhea and bloating, which often compromise adherence. Improved adherence directly correlates with higher eradication success [45, 46]. Moreover, some strains—particularly *Lactobacillus reuteri* and *Saccharomyces boulardii*—appear to reduce the emergence of resistant *H. pylori* subpopulations by limiting bacterial load and suppressing virulence gene expression.

Future strategies may involve designing probiotic-antibiotic regimens tailored to regional resistance patterns, maximizing both efficacy and tolerability. This synergy offers a pragmatic bridge between current antibiotic-dependent protocols and future microbiome-based therapies.

### 19.5.2 Probiotics with Phytochemicals, Prebiotics, and Postbiotics

Beyond antibiotics, probiotics may act synergistically with plant-derived compounds (phytochemicals), prebiotics, and postbiotics. Many phytochemicals, such as polyphenols from green tea, curcumin, and sulforaphane, demonstrate anti-*H. pylori* activity through inhibition of urease or suppression of bacterial adhesion [47–50]. When combined with probiotics, these natural compounds may enhance colonization resistance and reduce gastric inflammation.

Prebiotics—non-digestible food ingredients that selectively promote the growth of beneficial microbes—can further support probiotic colonization and persistence. For example, inulin or fructooligosaccharides may create a favorable niche for probiotic strains, amplifying their effects against *H. pylori* [51, 52].

Postbiotics, consisting of microbial metabolites, cell wall components, or inactivated cells, also offer potential. They retain immunomodulatory and antimicrobial activity without requiring live organisms, making them safer for vulnerable populations. Combining probiotics with postbiotics could create layered protection, leveraging both live and inactivated microbial functions [53].

### 19.5.3 Vaccine-Probiotic Combined Strategies

Despite decades of research, no licensed vaccine against *H. pylori* is currently available. Challenges include antigenic variability, immune evasion strategies, and the difficulty of inducing robust mucosal immunity. Nevertheless, vaccine research continues, and probiotics may serve as valuable adjuvants or delivery vehicles.

Certain probiotic strains can enhance antigen presentation, stimulate mucosal IgA production, and reduce vaccine-associated inflammation [54]. Engineered probiotics could even deliver *H. pylori* antigens directly to the gastric mucosa, functioning simultaneously as both vaccine vectors and colonization inhibitors.

The integration of vaccines and probiotics offers a long-term vision: a preventive approach in which probiotics not only complement therapeutic regimens but also support population-level control of infection through enhanced immunization efficacy.

## 19.6 Safety, Regulation, and Quality Control

For probiotics to transition from supportive adjuncts to recognized therapeutic agents against *H. pylori*, their safety, regulatory status, and quality control must be addressed systematically. Unlike antibiotics, which undergo rigorous clinical testing and standardized manufacturing, probiotic products remain highly variable in formulation, labeling, and oversight across regions. Ensuring their safe, consistent, and reliable application is therefore essential for future development.

### 19.6.1 Ensuring Probiotic Safety in Long-Term Use

Most probiotics currently used in *H. pylori* studies—such as Lactobacillus, Bifidobacterium, and Saccharomyces—have long histories of safe use in food and supplements. However, safety cannot be assumed for all strains, particularly for NGPs or genetically engineered organisms.

Potential risks include the following: infection in vulnerable populations, such as immunocompromised or critically ill patients; horizontal gene transfer, where antibiotic resistance genes from probiotic strains could spread to pathogenic bacteria; and metabolic side effects, if engineered probiotics produce unintended metabolites.

To mitigate risks, safety evaluation must include genomic analysis (to screen for virulence or resistance genes), toxicological studies, and long-term monitoring of colonization. For engineered probiotics, built-in "kill switches" or containment strategies may be necessary to prevent uncontrolled persistence in the host or environment.

### 19.6.2 Regulatory Challenges and Global Standardization

Regulatory frameworks for probiotics vary widely. In some regions, probiotics are classified as dietary supplements, subject to minimal oversight; in others, they fall under the category of live biotherapeutic products, requiring pharmaceutical-level evaluation. This lack of global harmonization complicates research, commercialization, and clinical use.

Key challenges include the following:

Classification ambiguity: Should probiotics for *H. pylori* be regulated as foods, supplements, or drugs?

Clinical trial requirements: Few probiotics have undergone large-scale, multi-center randomized trials comparable to those required for antibiotics.

International inconsistency: A strain recognized as safe in Europe may face stricter barriers in the United States or Asia.

Global health agencies, such as the World Health Organization and the International Scientific Association for Probiotics and Prebiotics, are calling for standardized guidelines on probiotic evaluation, focusing on safety, efficacy, and labeling. For probiotics to become integral in *H. pylori* therapy, convergence toward such standards will be essential.

### 19.6.3 Quality Control in Probiotic Production

Quality control is another critical barrier. Studies have shown that many commercial probiotic products contain fewer viable organisms than claimed, or even entirely different strains. Inconsistent production undermines both clinical efficacy and patient trust.

Key quality control priorities include the following: strain authentication: using genomic sequencing to confirm that the product contains the intended organism; viability testing: ensuring that therapeutic doses survive throughout shelf life and after gastric transit; stability assurance: maintaining efficacy during storage, transport, and variable environmental conditions; and good manufacturing practice: applying pharmaceutical-grade production standards to probiotic products intended for clinical use.

Advanced technologies, such as microencapsulation at production scale and real-time viability monitoring, may further improve product reliability. In addition, transparent labeling of strain identity, viable count, and expiration date should become mandatory to ensure informed use in clinical settings.

## 19.7 Clinical Translation and Future Perspectives

Probiotic therapy for *H. pylori* has advanced considerably in preclinical and early clinical research, yet its translation into routine practice remains limited. Bridging the gap between laboratory findings and real-world implementation requires addressing clinical trial evidence, public health relevance, and long-term integration into healthcare systems.

### 19.7.1 Current Status of Clinical Trials in *H. pylori* Infection

Clinical evidence increasingly supports the role of probiotics as adjunctive agents in *H. pylori* therapy. Randomized controlled trials and meta-analyses have shown that probiotics, when combined with standard triple or quadruple antibiotic regimens, can modestly increase eradication rates and significantly reduce side effects such as diarrhea, bloating, and abdominal discomfort [21, 55–57]. These improvements enhance treatment compliance, which is crucial for successful eradication.

However, differences in probiotic strains, dosage, and study design have led to inconsistent results, and no single formulation has proven universally effective. Conventional strains such as Lactobacillus and Bifidobacterium remain the most studied, yet their clinical outcomes vary among populations. To advance clinical translation, future trials should adopt standardized methodologies, larger sample sizes, and include microbiome-based analyses to clarify strain-specific mechanisms and host responses.

### 19.7.2 Potential for Reducing Antibiotic Resistance Burden

Antibiotic resistance has become a critical barrier to *H. pylori* eradication worldwide. Probiotics may help mitigate this burden by complementing antibiotic action and reducing the need for high-dose or prolonged antibiotic exposure. As mentioned above, certain probiotic strains produce bacteriocins, organic acids, and hydrogen peroxide that directly inhibit *H. pylori* growth, while others improve antibiotic tolerance by stabilizing gut microbiota and preventing dysbiosis. These effects collectively decrease selective pressure on bacteria and limit the spread of resistance genes within the gastrointestinal environment.

Additionally, probiotics enhance treatment adherence by alleviating drug-related side effects, indirectly preventing partial antibiotic courses that favor resistant strain selection. Although current evidence is promising, large-scale clinical trials specifically designed to assess the impact of probiotics on antibiotic resistance patterns are still needed to confirm their broader public health benefits.

### 19.7.3 Future Outlook: Toward Routine Clinical Application

Looking ahead, the successful integration of probiotics into routine clinical practice for *H. pylori* management will rely on several converging developments. Foremost, the establishment of robust evidence-based guidelines is essential. Large-scale, multi-center clinical trials and comprehensive meta-analyses are required to define the most effective probiotic strains, appropriate dosages, and optimal treatment durations, thereby transforming the current empirical use into standardized clinical protocols. At the same time, the advancement of personalized medicine will enable more targeted probiotic therapies through the integration of microbiome profiling and precision-based approaches. Such strategies will allow clinicians to tailor probiotic interventions to individual microbial and immune profiles, optimizing treatment efficacy across diverse patient populations.

Equally critical is the development of harmonized global regulatory frameworks that classify probiotics intended for *H. pylori* therapy as therapeutic products rather than dietary supplements, ensuring consistent standards for safety, quality, and efficacy. From an innovation standpoint, progress must also focus on creating advanced yet affordable probiotic formulations. Scalable delivery systems and engineered strains should be designed to maintain viability and targeted activity while remaining cost-effective, particularly in regions where *H. pylori* prevalence and antibiotic resistance are high, but healthcare resources are limited. In the broader context of infection control, probiotics may eventually integrate with preventive health strategies such as vaccines or dietary interventions, shifting the emphasis from treatment to long-term prevention and population-level protection.

Collectively, these developments suggest that the future of probiotics in *H. pylori* therapy will unfold gradually, initially as adjuncts that enhance antibiotic regimens, then as precision-guided therapeutic tools, and ultimately as independent or preventive modalities contributing to sustainable management of *H. pylori* infection worldwide.

## 19.8 Conclusion

The future of probiotic therapy in *H. pylori* infection lies in its transformation from an empirical supplement-based approach to a precision-guided, evidence-based medical strategy. Recent studies show that probiotics not only alleviate antibiotic side effects but also exert direct antimicrobial, immunomodulatory, and ecological actions that support *H. pylori* control. Advances in microencapsulation, nanotechnology, and biofilm-inspired systems now enhance probiotic survival and targeted delivery, while synthetic biology enables the design of engineered strains with specific antimicrobial or immune-modulating functions. Parallel progress in personalized medicine, integrating microbiome profiling, genetics, and AI, will further refine probiotic selection and dosing, ensuring optimal efficacy across diverse patient populations.

At the same time, the therapeutic potential of probiotics will depend on synergistic integration and strict regulation. Combining probiotics with antibiotics, prebiotics, phytochemicals, or vaccines can enhance eradication outcomes, reduce resistance, and restore microbial homeostasis. Yet, the absence of harmonized global standards for safety, labeling, and clinical evaluation remains a major barrier. Future success will require robust quality control, genomic safety screening, and clear regulatory pathways. Ultimately, probiotics are poised to evolve from adjunctive aids into personalized, preventive biotherapeutics, offering a sustainable, microbiome-centered paradigm for *H. pylori* management.

## References

1. Tshibangu-Kabamba E, Yamaoka Y. *H. pylori* infection and antibiotic resistance – from biology to clinical implications. Nat Rev Gastroenterol Hepatol. 2021;18(9):613–29. https://doi.org/10.1038/s41575-021-00449-x.
2. Suez J, Zmora N, Segal E, Elinav E. The pros, cons, and many unknowns of probiotics. Nat Med. 2019;25(5):716–29. https://doi.org/10.1038/s41591-019-0439-x.
3. Liang B, Yuan Y, Peng XJ, Liu XL, Hu XK, Xing DM. Current and future perspectives for *H. pylori* treatment and management: from antibiotics to probiotics. Front Cell Infect Microbiol. 2022;12:1042070. https://doi.org/10.3389/fcimb.2022.1042070.
4. Zuo F, Appaswamy A, Gebremariam HG, Jonsson AB. Role of Sortase a in lactobacillus gasseri Kx110A1 adhesion to gastric epithelial cells and competitive exclusion of *H. pylori*. Front Microbiol. 2019;10:2770. https://doi.org/10.3389/fmicb.2019.02770.
5. Lin CC, Huang WC, Su CH, Lin WD, Wu WT, Yu B, et al. Effects of multi-strain probiotics on immune responses and metabolic balance in *H. pylori*-infected mice. Nutrients. 2020;12(8) https://doi.org/10.3390/nu12082476.
6. Viazis N, Argyriou K, Kotzampassi K, Christodoulou DK, Apostolopoulos P, Georgopoulos SD, et al. A four-probiotics regimen combined with a standard *H. pylori*-eradication treatment reduces side effects and increases eradication rates. Nutrients. 2022;14(3) https://doi.org/10.3390/nu14030632.
7. Westerik N, Reid G, Sybesma W, Kort R. The probiotic lactobacillus rhamnosus for alleviation of *H. pylori*-associated gastric pathology in East Africa. Front Microbiol. 2018;9:1873. https://doi.org/10.3389/fmicb.2018.01873.
8. Shen S, Ren F, Qin H, Bukhari I, Yang J, Gao D, et al. Lactobacillus acidophilus NCFM and Lactiplantibacillus plantarum Lp-115 inhibit *H. pylori* colonization and gastric inflammation in a murine model. Front Cell Infect Microbiol. 2023;13:1196084. https://doi.org/10.3389/fcimb.2023.1196084.
9. Dirjayanto VJ, Audrey J, Simadibrata DM. Vonoprazan-amoxicillin dual regimen with saccharomyces boulardii as a rescue therapy for *H. pylori*: current perspectives and implications. World J Gastroenterol. 2024;30(10):1280–6. https://doi.org/10.3748/wjg.v30.i10.1280.
10. Mao LQ, Zhou YL, Wang SS, Chen L, Hu Y, Yu LM, et al. Impact of *H. pylori* eradication on the gastric microbiome. Gut Pathog. 2021;13(1):60. https://doi.org/10.1186/s13099-021-00460-2.
11. Guo Y, Cao XS, Guo GY, Zhou MG, Yu B. Effect of *H. pylori* eradication on human gastric microbiota: a systematic review and meta-analysis. Front Cell Infect Microbiol. 2022;12:899248. https://doi.org/10.3389/fcimb.2022.899248.
12. Yu T, Lu T, Deng W, Yao D, He C, Luo P, et al. Microbiome and function alterations in the gastric mucosa of asymptomatic patients with *H. pylori* infection. Helicobacter. 2023;28(3):e12965. https://doi.org/10.1111/hel.12965.

13. Liu Z, Zhang D, Chen S. Unveiling the gastric microbiota: implications for gastric carcinogenesis, immune responses, and clinical prospects. J Exp Clin Cancer Res. 2024;43(1):118. https://doi.org/10.1186/s13046-024-03034-7.
14. Bik EM, Eckburg PB, Gill SR, Nelson KE, Purdom EA, Francois F, et al. Molecular analysis of the bacterial microbiota in the human stomach. Proc Natl Acad Sci USA. 2006;103(3):732–7. https://doi.org/10.1073/pnas.0506655103.
15. Chen Z, Tang Z, Li W, Deng X, Yu L, Yang J, et al. Weizmannia coagulans BCF-01: a novel gastrogenic probiotic for *H. pylori* infection control. Gut Microbes. 2024;16(1):2313770. https://doi.org/10.1080/19490976.2024.2313770.
16. Wu S, Xu Y, Chen Z, Chen Y, Wei F, Xia C, et al. Lactiplantibacillus plantarum ZJ316 reduces *H. pylori* adhesion and inflammation by inhibiting the expression of Adhesin and urease genes. Mol Nutr Food Res. 2023;67(18):e2300241. https://doi.org/10.1002/mnfr.202300241.
17. Ansari S, Yamaoka Y. *H. pylori* virulence factor Cytotoxin-associated Gene a (CagA)-mediated gastric pathogenicity. Int J Mol Sci. 2020;21(19) https://doi.org/10.3390/ijms21197430.
18. Ji J, Yang H. Using probiotics as supplementation for *H. pylori* antibiotic therapy. Int J Mol Sci. 2020;21(3) https://doi.org/10.3390/ijms21031136.
19. Wei Z, Peng Z, Li J, Tang H, Qiao B, Madjirebaye P, et al. Screening, characterization, and mode of action of three lactic acid bacteria against *H. pylori*. Probiotics Antimicrob Proteins. 2025; https://doi.org/10.1007/s12602-025-10620-4.
20. Hodzhev V, Dzhambazov K, Sapundziev N, Encheva M, Todorov S, Youroukova V, et al. High-dose probiotic mix of lactobacillus spp., Bifidobacterium spp., Bacillus coagulans, and saccharomyces boulardii to prevent antibiotic-associated diarrhea in adults: a multicenter, randomized, double-blind, placebo-controlled trial (SPAADA). Open Forum Infect Dis. 2024;11(11):ofae615. https://doi.org/10.1093/ofid/ofae615.
21. Wang Y, Wang X, Cao XY, Zhu HL, Miao L. Comparative effectiveness of different probiotics supplements for triple *H. pylori* eradication: a network meta-analysis. Front Cell Infect Microbiol. 2023;13:1120789. https://doi.org/10.3389/fcimb.2023.1120789.
22. Khan MT, Dwibedi C, Sundh D, Pradhan M, Kraft JD, Caesar R, et al. Synergy and oxygen adaptation for development of next-generation probiotics. Nature. 2023;620(7973):381–5. https://doi.org/10.1038/s41586-023-06378-w.
23. Cani PD, Depommier C, Derrien M, Everard A, de Vos WM. Akkermansia muciniphila: paradigm for next-generation beneficial microorganisms. Nat Rev Gastroenterol Hepatol. 2022;19(10):625–37. https://doi.org/10.1038/s41575-022-00631-9.
24. Shi Z, Li M, Zhang C, Li H, Zhang Y, Zhang L, et al. Butyrate-producing Faecalibacterium prausnitzii suppresses natural killer/T-cell lymphoma by dampening the JAK-STAT pathway. Gut. 2025;74(4):557–70. https://doi.org/10.1136/gutjnl-2024-333530.
25. Zhang L, Wu WK, Gallo RL, Fang EF, Hu W, Ling TK, et al. Critical role of antimicrobial peptide cathelicidin for controlling *H. pylori* survival and infection. J Immunol. 2016;196(4):1799–809. https://doi.org/10.4049/jimmunol.1500021.
26. Yang Y, Yu P, Huang Y, Zhang W, Nie Y, Gao C. Metabolic engineering of lactobacilli spp. for disease treatment. Microb Cell Factories. 2025;24(1):53. https://doi.org/10.1186/s12934-025-02682-4.
27. Hidalgo-Cantabrana C, O'Flaherty S, Barrangou R. CRISPR-based engineering of next-generation lactic acid bacteria. Curr Opin Microbiol. 2017;37:79–87. https://doi.org/10.1016/j.mib.2017.05.015.
28. Goh YJ, Barrangou R. Harnessing CRISPR-Cas systems for precision engineering of designer probiotic lactobacilli. Curr Opin Biotechnol. 2019;56:163–71. https://doi.org/10.1016/j.copbio.2018.11.009.
29. Guo Q, Tang J, Li S, Qiang L, Chang S, Du G, et al. Lactobacillus plantarum 21805 encapsulated by whey protein isolate and dextran conjugate for enhanced viability. Int J Biol Macromol. 2022;216:124–31. https://doi.org/10.1016/j.ijbiomac.2022.06.207.
30. Zhang C, Chen H, Zou Y, Ling X, Lai Y, Ou L, et al. Computationally-guided multifunctional nanoplatform for enhanced oral delivery and targeted therapy of *H. pylori* using herbal bioactives. Biomaterials. 2026;324:123517. https://doi.org/10.1016/j.biomaterials.2025.123517.

31. Li C, Wang ZX, Xiao H, Wu FG. Intestinal delivery of probiotics: materials, strategies, and applications. Adv Mater. 2024;36(32):e2310174. https://doi.org/10.1002/adma.202310174.
32. Khan S, Sharaf M, Ahmed I, Khan TU, Shabana S, Arif M, et al. Potential utility of nano-based treatment approaches to address the risk of *H. pylori*. Expert Rev Anti-Infect Ther. 2022;20(3):407–24. https://doi.org/10.1080/14787210.2022.1990041.
33. Jain P, Jain S, Prasad KN, Jain SK, Vyas SP. Polyelectrolyte coated multilayered liposomes (nanocapsules) for the treatment of *H. pylori* infection. Mol Pharm. 2009;6(2):593–603. https://doi.org/10.1021/mp8002539.
34. Kamankesh M, Yadegar A, Llopis-Lorente A, Liu C, Haririan I, Aghdaei HA, et al. Future nanotechnology-based strategies for improved Management of *H. pylori* infection. Small (Weinheim an der Bergstrasse, Germany). 2024;20(3):e2302532. https://doi.org/10.1002/smll.202302532.
35. Sharma B, Chaurasia A, Rajput DS, Gupta N. Mucoadhesive microspheres: a promising delivery system for *H. pylori* (*H. pylori*) treatment. Zhongguo Ying Yong Sheng Li Xue Za Zhi. 2024;40:e20240006. https://doi.org/10.62958/j.cjap.2024.006.
36. Grosso R, Benito E, Carbajo-Gordillo AI, Diaz MJ, Garcia-Martin MG, de-Paz MV. Advanced interpenetrating polymer networks for innovative gastroretentive formulations targeting *H. pylori* gastric colonization. Eur J Pharm Sci. 2024;200:106840. https://doi.org/10.1016/j.ejps.2024.106840.
37. Zhu X, Su T, Wang S, Zhou H, Shi W. New advances in nano-drug delivery systems: *H. pylori* and gastric cancer. Front Oncol. 2022;12:834934. https://doi.org/10.3389/fonc.2022.834934.
38. Xu C, Ban Q, Wang W, Hou J, Jiang Z. Novel nano-encapsulated probiotic agents: encapsulate materials, delivery, and encapsulation systems. J Control Release. 2022;349:184–205. https://doi.org/10.1016/j.jconrel.2022.06.061.
39. Wala SJ, Ragan MV, Sajankila N, Volpe SG, Purayil N, Dumbauld Z, et al. Probiotics and novel probiotic delivery systems. Semin Pediatr Surg. 2023;32(3):151307. https://doi.org/10.1016/j.sempedsurg.2023.151307.
40. Xiao Y, Zhai Q, Zhang H, Chen W, Hill C. Gut colonization mechanisms of lactobacillus and Bifidobacterium: an argument for personalized designs. Annu Rev Food Sci Technol. 2021;12:213–33. https://doi.org/10.1146/annurev-food-061120-014739.
41. Shah SC, et al. Host genetic determinants associated with *H. pylori* eradication treatment failure: a systematic review and meta-analysis. Gastroenterology. 2021;161(5):1443–59. https://doi.org/10.1053/j.gastro.2021.07.043.
42. Sterbenc A, Jarc E, Poljak M, Homan M. *H. pylori* virulence genes. World J Gastroenterol. 2019;25(33):4870–84. https://doi.org/10.3748/wjg.v25.i33.4870.
43. Sugimoto M, Furuta T, Yamaoka Y. Influence of inflammatory cytokine polymorphisms on eradication rates of *H. pylori*. J Gastroenterol Hepatol. 2009;24(11):1725–32. https://doi.org/10.1111/j.1440-1746.2009.06047.x.
44. Coveney P, Highfield R, Stahlberg E, Vazquez M. Digital twins and big AI: the future of truly individualised healthcare. NPJ Digit Med. 2025;8(1):494. https://doi.org/10.1038/s41746-025-01874-x.
45. Goderska K, Agudo Pena S, Alarcon T. *H. pylori* treatment: antibiotics or probiotics. Appl Microbiol Biotechnol. 2018;102(1):1–7. https://doi.org/10.1007/s00253-017-8535-7.
46. Cekin AH, Sahinturk Y, Akbay Harmandar F, Uyar S, Yolcular BO, Cekin Y. Use of probiotics as an adjuvant to sequential *H. pylori* eradication therapy: impact on eradication rates, treatment resistance, treatment-related side effects, and patient compliance. Turk J Gastroenterol. 2017;28(1):3–11. https://doi.org/10.5152/tjg.2016.0278.
47. Deng G, Wu Y, Song Z, Li S, Du M, Deng J, et al. Tea polyphenol liposomes overcome gastric mucus to treat *H. pylori* infection and enhance the intestinal microenvironment. ACS Appl Mater Interfaces. 2022;14(11):13001–12. https://doi.org/10.1021/acsami.1c23342.
48. Hidalgo A, Bravo D, Soto C, Maturana G, Cordero-Machuca J, Zuniga-Lopez MC, et al. The anti-oxidant curcumin solubilized as oil-in-water nanoemulsions or chitosan nanocapsules effectively reduces *H. pylori* growth, bacterial biofilm formation, gastric cell adhesion and internalization. Antioxidants. 2023;12(10) https://doi.org/10.3390/antiox12101866.

49. Sarkar A, De R, Mukhopadhyay AK. Curcumin as a potential therapeutic candidate for *H. pylori* associated diseases. World J Gastroenterol. 2016;22(9):2736–48. https://doi.org/10.3748/wjg.v22.i9.2736.
50. He S, Sun L, Chen J, Li Y, Pan Y, Su A, et al. Therapeutic effects of Sulforaphane on *H. pylori*-infected mice: insights from high-coverage metabolomics and Lipidomics analyses of serum and liver. Int J Mol Sci. 2025;26(16) https://doi.org/10.3390/ijms26167791.
51. Fayed B, Jagal J, Cagliani R, Kedia RA, Elsherbeny A, Bayraktutan H, et al. Co-administration of amoxicillin-loaded chitosan nanoparticles and inulin: a novel strategy for mitigating antibiotic resistance and preserving microbiota balance in *H. pylori* treatment. Int J Biol Macromol. 2023;253(Pt 2):126706. https://doi.org/10.1016/j.ijbiomac.2023.126706.
52. Singh RS, Singh RP, Kennedy JF. Recent insights in enzymatic synthesis of fructooligosaccharides from inulin. Int J Biol Macromol. 2016;85:565–72. https://doi.org/10.1016/j.ijbiomac.2016.01.026.
53. Hua Q, Hu J, Zhang X, He J, Wang P, Duan Z, et al. Postbiotics suppress *H. pylori* adhesion and survival through a Coaggregation mechanism. J Agric Food Chem. 2025;73(35):21976–86. https://doi.org/10.1021/acs.jafc.5c07004.
54. Haghighi FH, Farsiani H. Is Lactococcus lactis a suitable candidate for use as a vaccine delivery system against *H. pylori*? Curr Microbiol. 2024;82(1):30. https://doi.org/10.1007/s00284-024-03994-1.
55. Zhang M, Zhang C, Zhao J, Zhang H, Zhai Q, Chen W. Meta-analysis of the efficacy of probiotic-supplemented therapy on the eradication of *H. pylori* and incidence of therapy-associated side effects. Microb Pathog. 2020;147:104403. https://doi.org/10.1016/j.micpath.2020.104403.
56. Ivashkin V, Maev I, Poluektova E, Sinitsa A, Avalueva E, Mnatsakanyan M, et al. Efficacy and safety of Postbiotic contained inactivated lactobacillus reuteri (Limosilactobacillus reuteri) DSM 17648 as adjuvant therapy in the eradication of *H. pylori* in adults with functional dyspepsia: a randomized double-blind placebo-controlled trial. Clin Transl Gastroenterol. 2024;15(9):e1. https://doi.org/10.14309/ctg.0000000000000750.
57. Zhang Y, Lu B, Dong Y, Zhang Y, Du Q, Chen Y, et al. Saccharomyces boulardii combined with triple therapy alter the microbiota in the eradication of *H. pylori* infection. Sci Rep. 2024;14(1):13152. https://doi.org/10.1038/s41598-024-63894-z.

# Part V

# Traditional Medicine and Natural Product Based Therapies

# The Theoretical Basis of Traditional Chinese Medicine for Treating *Helicobacter pylori*

20

Ziyou Zhong and Meicun Yao

## 20.1 Introduction

In humanity's long pursuit of health, different civilizations have developed their own distinctive systems of medical theory. Traditional Chinese Medicine (TCM), a treasure of Chinese civilization, has been refined through millennia of practice and accumulation. It has not only made immense contributions to the flourishing and well-being of the Chinese people, but its profound philosophy and rich practical experience have also deeply influenced neighboring traditions—evident, for example, in Japan's Kampo and Traditional Korean Medicine. At both the theoretical and practical levels, TCM has amassed a vast corpus of knowledge and experience, forming a profound repository of wisdom that provides an inexhaustible source of inspiration and valuable insights for modern medical research and clinical practice.

In digestive diseases, chronic gastric conditions are particularly prevalent; yet it was not until the discovery of *Helicobacter pylori* in 1982 that modern medicine attributed the primary etiology of many such disorders to infection by this microorganism. Within the traditional theoretical framework of TCM, there is no explicit concept of specific microorganisms like *H. pylori*; nevertheless, this has not impeded TCM's recognition and treatment of related illnesses. The symptom patterns systematically recorded in TCM classics—such as *weiwan tong* (epigastric pain), *pi man* (epigastric fullness), *cao za* (gastric upset), and *tun suan* (acid regurgitation)—correlate closely with the clinical manifestations of chronic gastritis and peptic ulcer disease associated with *H. pylori* in modern medicine. Synthesizing millennia

Z. Zhong
Gastroenterology Department, Shenzhen Traditional Chinese Medicine Hospital, Shenzhen, China

M. Yao (✉)
School of Pharmaceutical Sciences (Shenzhen), Sun Yat-sen University, Shenzhen, China
e-mail: Lssymc@mail.sysu.edu.cn


J. Gao, Y. Du (eds.), *Non-Antibiotic Therapies for Helicobacter pylori*,
https://doi.org/10.1007/978-981-95-9598-3_20

of clinical practice and the collective wisdom of physicians, TCM theory situates pathogens such as *H. pylori* within the broader category of external *xie qi* (pathogenic factors) [1], while advancing a deeper insight that does not confine itself to the pathogen alone. As articulated in the *Huangdi Neijing* (Yellow Emperor's Inner Canon), "where pathogenic factors accumulate, the body's righteous qi must be deficient," and through the diagnostic–therapeutic paradigm of *bian zheng lun zhi* (syndrome differentiation and treatment), the pathogenesis is succinctly summarized as *ben xu biao shi* (root deficiency with branch excess), thereby establishing the fundamental principle of *fu zheng qu xie* (supporting the righteous qi and dispelling pathogenic factors). Interpreted in contemporary scientific terms, *qu xie* corresponds to inhibiting or eradicating the pathogen, while *fu zheng* corresponds to regulating and enhancing host immune function; together, these approaches aim to restore the body's intrinsic balance and resistance.

This chapter systematically elucidates the TCM theoretical framework for the prevention and treatment of *H. pylori*. We first examine TCM's macroscopic understanding of this pathogen, analyzing why it is classified under *shi re du xie* (damp-heat-toxin pathogen). We then show how TCM transcends a pathogen-centric perspective by emphasizing *pi wei xu ruo* (spleen–stomach weakness) as the internal condition that permits colonization and disease, thereby establishing the core pathogenesis of *zheng xu xie fan* (righteous qi deficiency and pathogenic invasion). Building on this foundation, the chapter details the fundamental therapeutic principles—*fu zheng qu xie*; *biao ben jian zhi* (treating both the branch and the root)—and outlines the framework of *fen xing lun zhi* (treatment based on syndrome typing), which embodies individualized therapy. We further synthesize core consensus concepts consolidated through long-term contemporary practice, including the co-management philosophy of the "human–microbe–environment" triad focused on modulating the internal milieu, and the *zhi wei bing* (preventive treatment of disease) principle that permeates the entire prevention-and-therapy continuum. Taken together, these discussions demonstrate that in addressing *H. pylori*, TCM provides a self-consistent, comprehensive cognitive framework and strategy independent of simple bactericidal logic, thereby laying a solid theoretical foundation for the clinical practice presented in subsequent chapters.

## 20.2 TCM'S Theoretical Understanding of *H. pylori*

From the microscopic perspective of modern medicine, *H. pylori* is a pathogenic microorganism with specific biological characteristics. By contrast, within TCM's macroscopic framework of syndrome differentiation, the pathogen is not named explicitly but is characterized by a distinct *xiang* (phenomenal manifestation); accordingly, TCM's understanding of *H. pylori* is inferred from the constellation of clinical signs and symptoms it elicits. At present, there is a broad consensus within TCM regarding the etiology and pathogenesis of *H. pylori* infection: the primary *bing wei* (disease location) resides in the stomach and is closely linked to the functional states of the spleen and liver. This theoretical understanding can be organized along four dimensions: the nature of the pathogen, the fundamental cause of disease, the course of pathological evolution, and the major influencing factors.

I. Core Pathogenic Perspective: Damp-Heat-Toxin Pathogen Theory

Classifying the pathogenic nature of *H. pylori* as *shi re du xie* is the prevailing, widely accepted view; some scholars also refer to it as *zhuo du* (turbid toxin). This is not a mere terminological mapping, but a synthesis grounded in the strong concordance between extensive clinical phenomenology and classical theoretical constructs [2].

First, from a theoretical perspective, the pathogenic characteristics of *H. pylori* are highly consistent with the nature of *shi re* (damp-heat). TCM theory holds that pathogenic damp-heat often enters the body through the mouth and nose and is closely related to diet and environment. Modern medicine, in turn, recognizes that the primary transmission routes for *H. pylori* are oral-oral and fecal-oral. Its route of invasion and pathogenic characteristics align perfectly with the features of a damp-heat pathogen. *H. pylori* colonizes the stomach, causing inflammation, ulceration, and other conditions, just as the distinguished *wen bing* (warm disease) physician Zhang Xugu of the Qing dynasty stated: "Although the pathogen of damp-heat is initially received from the exterior, it ultimately settles in the spleen–stomach."

Its "damp" nature is characterized by stickiness and a lingering, protracted course. Pathogenic dampness is heavy, turbid, viscous, and greasy, and readily obstructs the flow of *qi ji* (qi dynamics). This corresponds precisely to the characteristics of *H. pylori* infection: a protracted disease course, recurrent flare-ups, and difficulty in achieving a swift recovery. *H. pylori* adheres to the gastric mucosa, disrupting the normal ascending and descending functions of the stomach qi. This leads to symptoms such as epigastric fullness and distention, loss of appetite, and a sticky or greasy sensation in the mouth. The persistent, lingering nature of the disease course is a direct reflection of the "sticky and viscous" characteristic of dampness [3].

Its *re* (heat) nature is manifested in upward-flaring and corrosive burning. Heat is a yang pathogen that readily consumes body fluids (*jin ye*) and corrodes flesh and blood. This is highly consistent with the pathological changes induced by *H. pylori* infection—namely, inflammatory responses of the gastric mucosa, congestion/hyperemia, edema, erosion, and even ulceration. Common clinical signs such as a burning sensation in the epigastrium, bitter taste in the mouth, acid regurgitation, a red tongue body, and a yellow, greasy tongue coating are all manifestations of the "heat" aspect of damp-heat accumulation in the stomach.

Its *du* (toxic) nature is manifested by severity and injury to the collaterals (*luo*). Here, "*du*" denotes a pathogen with strong virulence and the capacity to inflict profound harm. As stated in *Shi Re Tiao Bian*, "prolonged stagnation of damp-heat often transforms into toxin." The vacuolating cytotoxin gene A and cytotoxin-associated gene A produced by *H. pylori* can directly damage gastric mucosal epithelial cells and disrupt the gastric wall barrier; this accords with the properties of a *du xie* (toxic pathogen) that injures the collaterals and corrupts bodily substance. Over the long term, the persistent presence of this "*du*" pathogen is a key factor leading to serious outcomes such as gastric mucosal atrophy, intestinal metaplasia, and even carcinogenesis.

Furthermore, from a clinical standpoint, evidence from modern epidemiological surveys corroborates this theory. A single-center study of 2818 *H. pylori*–positive patients found that *pi wei shi re zheng* (spleen–stomach damp-heat syndrome) was the most common pattern, accounting for 32.69% of cases [4]. Likewise, a literature analysis spanning nearly 30 years and encompassing 5731 cases of *H. pylori*-associated ulcers showed that *pi wei shi re zheng* was the most frequently observed syndrome type [5].

II. The Fundamental Pathogenic View: Righteous-Qi Deficiency With Pathogenic Invasion

TCM's core pathogenic perspective does not consider *xie qi* in isolation but emphasizes the dynamic interplay between *zheng qi* (righteous qi) and *xie qi*. The onset and progression of *H. pylori* infection result precisely from the combined effects of *zheng qi kui xu* (deficiency of righteous qi) and *xie qi qin xi* (pathogenic invasion).

*Zheng xu xie fan* constitutes the fundamental *bing ji* (pathogenic mechanism). As stated in the *Huangdi Neijing*, "When righteous qi is present within, pathogenic factors cannot invade," and "Where pathogenic factors accumulate, qi must be deficient," incisively revealing the internal laws governing disease onset. Among the various contributors to *zheng xu* (righteous-qi deficiency), *pi wei xu ruo* is regarded as the root and internal cause predisposing to *H. pylori* infection. The *Pi Wei Lun* (Treatise on the spleen–stomach) similarly notes: "When the spleen–stomach are injured within, a hundred diseases arise." As the *hou tian zhi ben* (foundation of the post-heaven constitution) and the source of the generation and transformation of qi and blood, the spleen–stomach maintain digestion and absorption. When dietary irregularities, overwork, or emotional disturbance impair their function—that is, when *pi wei xu ruo* occurs—transportation and transformation of fluids and dampness become dysregulated, generating internal damp turbidity and providing the "soil" for *H. pylori* colonization.

This pathogenic view explains why, despite the widespread presence of *H. pylori* in the environment, not all exposed individuals become ill, and why clinical presentations range from asymptomatic carriage to severe ulceration: outcomes hinge on the strength of the individual's *pi wei zheng qi*. It also cautions that focusing solely on *qu xie* (dispelling the pathogen—i.e., bactericidal actions) without concurrently *fu zheng* (supporting the righteous qi—i.e., improving spleen–stomach function) is akin to weeding without improving the soil; recurrence is then highly likely.

III. The Evolution of the Core Pathogenesis: The Theory of Blood Stasis Obstructing the Collaterals

When the disease course becomes prolonged—on the foundation of damp-heat and qi deficiency—the key link in its pathological evolution lies in *yu xue zu luo* (blood stasis obstructing the collaterals). TCM theory holds that *jiu bing ru luo* (a prolonged illness will enter the collaterals) and *jiu bing bi yu* (a prolonged illness will inevitably lead to blood stasis). The long-term, chronic nature of *H. pylori* infection causes the Stomach's *qi ji* to remain persistently dysregulated; with enduring stagnation, the movement of blood is impeded, and *yu xue* (blood stasis) ultimately forms, becoming deeply embedded within the stomach collaterals.

This theory is crucial for explaining the transition of *H. pylori* infection from early functional disturbance to the organic lesions seen in the mid-to-late stages. The precancerous gastric lesions recognized by modern medicine in the later phase of *H. pylori* infection—gastric mucosal gland atrophy, intestinal metaplasia, and dysplasia—are highly consistent in pathological essence with the TCM understanding of *yu xue zu luo*. Typical *xiang* (manifestations) under gastroscopy—such as a dim and dusky mucosal color, clearly visible vessels, and an uneven surface with irregular relief—are classic signs of stasis and obstruction within the stomach collaterals that have lost proper nourishment. Therefore, *yu xue* is both a pathological product resulting from long-standing *H. pylori* infection and a key pathogenic factor driving worsening disease and progression toward malignant transformation. In treating patients in the mid-to-late stages, whether to incorporate *huo xue hua yu* (invigorating blood and transforming stasis) becomes a critical consideration for intercepting further transmission of the disease and for cancer prevention.

IV. An Important Influencing Factor: Liver–Stomach Disharmony

In addition to the inherent weakened state of the spleen–stomach, emotional factors are the most common influences that induce and exacerbate *H. pylori*-related gastric disease. According to TCM's Five Elements (*Wu Xing*) theory, the liver belongs to wood and the spleen–stomach to Earth; under physiological conditions, they coordinate, but in pathology, the counteracting relationship of "*mu ke tu*" (wood overacting on Earth) becomes manifest.

When a person remains in adverse emotional states—mental tension, anxiety, or depression—liver-qi stagnation (*gan qi yu jie*) readily develops; the constrained liver qi then "transversely invades" the stomach (*heng ni fan wei*), producing *gan wei bu he* (liver–stomach disharmony). This disrupts the Stomach's functions of receiving and descending, giving rise to sharp or distending epigastric pain, chest and hypochondriac distension, frequent belching, and acid regurgitation. Clinically, among patients with *H. pylori* infection, it is a very common phenomenon that symptoms are markedly aggravated or recur in response to emotional fluctuations. One clinical study showed that liver–stomach disharmony syndrome (*gan wei bu he* zheng) accounted for 21.62% of *H. pylori*-infected cases, second only to spleen–stomach damp-heat as the most prevalent pattern [4]. These findings indicate that regulating emotions and *shu gan li qi* (soothing the liver and regulating qi) is not only a crucial link for symptom relief but also an essential therapeutic dimension for controlling disease progression and preventing recurrence. This perspective also fully reflects TCM's holistic view of *xin shen tong zhi*—treating mind and body together.

## 20.3 Therapeutic Principles Guided by TCM Theory

Building on the preceding, in-depth understanding of the etiology and pathogenesis of *H. pylori* infection, TCM has developed a distinctive set of therapeutic principles centered on holistic regulation, aiming to restore the body's inherent balance. The essence of this approach is not a "bactericidal action" directed at a single pathogen; rather, it treats the human being as an integrated whole. By harmonizing the qi and blood of the *zang-fu* organs (viscera–bowel system), it seeks to modify the internal environment upon which *H. pylori* depends for survival, thereby achieving the ultimate goal that "when righteous qi is present within, pathogenic factors cannot invade."

I. Fundamental Principle: *Fu zheng qu xie*, Treating Both Branch and Root

*Fu zheng qu xie*—supporting the righteous qi and dispelling pathogenic factors—is TCM's root method for treating disease, and it is especially crucial for *H. pylori* infection, a pattern characterized by *ben xu biao shi* (root deficiency with branch excess). The *Suwen · Sanbu Jiuhou Lun* emphasizes: "One must first assess whether the body is stout or thin to regulate the deficiency or excess of qi. If there is excess, reduce it; if there is deficiency, supplement it." This classic statement establishes the two core therapeutic laws of *fu zheng* (supporting the righteous) and *qu xie* (dispelling the pathogenic).

Although in fundamental pathogenesis, *zheng xu* (righteous-qi deficiency) is the root (*ben*) and *xie shi* (pathogenic excess) is the branch (*biao*), when confronted with a clearly defined, continuously damaging pathogen such as *H. pylori*, TCM strategy emphasizes *ji ze zhi qi biao* (in acute conditions, treat the branch). In practice, the first priority is to control or eliminate the pathogenic factor to halt ongoing gastric-mucosal damage, thereby creating the conditions and space for subsequent functional repair (*fu zheng*). As with a city under siege, reconstruction cannot proceed while enemy troops still occupy the streets; only on the foundation of effectively "removing the enemy" does *fu zheng* attain its fullest significance.

*Qu xie* (Dispelling Pathogenic Factors). The primary targets are the *shi re du xie* generated in *H. pylori* infection and pathological products such as *qi zhi xue yu* (qi stagnation and blood stasis). Specific methods include clearing heat and transforming dampness, resolving toxins and expelling turbidity, and regulating qi while invigorating the blood, all aimed at eliminating the pathogen and improving the local gastric environment.

*Fu zheng* (Supporting the Righteous Qi). The principal focus is the internal root—*pi wei xu ruo*. Specific methods include *jian pi yi qi* (strengthening the spleen and supplementing qi) and *yang yin he wei* (nourishing yin and harmonizing the stomach). The emphasis is on restoring spleen–stomach function to enhance the body's intrinsic disease resistance and reparative capacity.

In the course of treatment, the two approaches must operate complementarily—neither should be pursued to the exclusion of the other. If one focuses only on *fu zheng* (supporting the righteous) without *qu xie* (dispelling the pathogenic), there is a risk of *bi men liu kou* ("trapping the pathogen by closing the door"), paradoxically causing the illness to become protracted and lingering. Conversely, if one only dispels the pathogenic without supporting the righteous, the *zheng qi* becomes further depleted, allowing the pathogenic factor to persist or recur. Therefore, only by dynamically adjusting the proportion between supporting the righteous and dispelling the pathogenic—based on the relative strengths of the patient's *zheng* versus *xie*—can the ideal therapeutic goal of *biao ben jian zhi* be achieved.

II. The Core Strategy: The Theoretical Framework of Treatment Based on Syndrome Typing

*Fen xing lun zhi*, also called *bian zheng lun zhi*, is the essence of individualized, precision therapy in TCM. It requires that patients with *H. pylori* not be viewed as a single category; rather, based on each patient's specific clinical manifestations—*she xiang* (tongue presentation) and *mai xiang* (pulse condition) included—patients are classified into distinct "syndrome types" (*zheng hou lei xing*), and corresponding therapeutic principles are determined. This embodies the high flexibility of *tong bing yi zhi* (treating the same disease with different methods).

Below are several of the most common syndrome types in *H. pylori* infection, with their therapeutic principles and representative formulas:

(I) Spleen–stomach damp-heat syndrome (*Pi Wei Shi Re Zheng*).
  Core pathogenesis: damp-heat accumulating and binding in the middle *jiao*.
  Therapeutic principle: clear heat and transform dampness, regulate qi, and harmonize the Middle (*li qi he zhong*).
  Representative formula: *Lian Po Yin*.

(II) Spleen–stomach weakness syndrome (*Pi Wei Xu Ruo* Zheng).
  This pattern represents the "root" of the disease.
  Core pathogenesis: insufficiency of middle qi (*zhong qi bu zu*) with impaired transportation and transformation (*yun hua shi jian*).
  Therapeutic principle: strengthen the spleen and supplement qi (*jian pi yi qi*); harmonize the stomach and quiet the middle (*he wei an zhong*).
  Representative formula: *Xiang Sha Liu Jun Zi Tang*.

(III) Liver–stomach disharmony syndrome (*Gan Wei Bu He* Zheng).
  A common pattern driven by emotional factors.
  Core pathogenesis: failure of the liver's free coursing (*gan shi shu xie*) with transverse invasion of the stomach (*heng ni fan wei*).
  Therapeutic principle: soothe the liver, regulate qi, and harmonize the stomach (*shu gan li qi he wei*).
  Representative formula: *Chai Hu Shu Gan San*.

(IV) Cold–heat complex syndrome (*Han Re Cuo Za Zheng*).
  A mixed deficiency–excess pattern.
  Core pathogenesis: disordered ascending–descending of spleen–stomach with cold and heat mutually binding in the middle.
  Therapeutic principle: open with acrid and descend with bitter (*xin kai ku jiang*); harmonize the Stomach and dissipate focal distention (*he wei xiao pi*).
  Representative formula: *Ban Xia Xie Xin Tang*.

Note: Detailed composition, pharmacological mechanisms, and clinical modifications of the above formulas will be elaborated in the next chapter.

III. Core Therapeutic Philosophies

Behind the therapeutic principles described above lie two core philosophical concepts that run through the entire approach; together, they constitute TCM's unique worldview in responding to *H. pylori* infection.

(I) Holistic Co-Management of the "Human–Microbe–Environment"

The starting point of TCM treatment is not to directly target *H. pylori* itself, but to focus on the human who is infected and the internal environment. As an external *xie qi*, whether *H. pylori* can cause disease depends on the strength of the body's *zheng qi*—in particular, the functional state of the spleen–stomach. Therefore, the key to treatment lies in altering the gastric microenvironment of "damp-heat-toxin accumulation" (*shi re du xie*) on which *H. pylori* relies for survival. By employing methods such as strengthening the spleen (*jian pi*), transforming dampness (*hua shi*), and regulating qi (*li qi*), the internal environment is changed, making it impossible for the pathogen to colonize and inflict damage. This holistic regulatory model—co-managing the human–microbe–environment triad—aims to restore the body's homeostatic stability and is the fundamental strategy for achieving long-term health and security. This philosophy is highly aligned with the "host–pathogen–microbiota" three-element model emphasized in modern medicine, reflecting the forward-looking nature of TCM theory in holistic regulation and its synergy with contemporary biology.

(II) The Forward-Looking Perspective of the *Zhi Wei Bing* Philosophy

The philosophy of *zhi wei bing* is an essence of TCM theory and provides forward-looking guidance for the prevention and treatment of *H. pylori* infection. It encompasses three levels:

1. *Wei Bing Xian Fang* (Preventing Disease Before Its Onset).

This refers to the stage before infection occurs or during asymptomatic carriage. By regulating the spleen–stomach and improving poor lifestyle habits, it seeks to eliminate, at the source, susceptible constitutions (e.g., those prone to damp-heat), thereby removing the "soil" on which *H. pylori* colonizes.

2. *Ji Bing Fang Bian* (Preventing Progression Once Disease Is Present).

This denotes active intervention after infection to prevent progression to serious outcomes such as atrophic gastritis, intestinal metaplasia, or even carcinogenesis. For example, when a *shi re du xie* is seen stagnating in the stomach, one should anticipate its potential to damage the stomach collaterals; therefore, while clearing heat and transforming dampness, promptly apply methods of supplementing qi and nourishing yin (*yi qi yang yin*) and invigorating the blood and transforming stasis (*huo xue hua yu*) to intercept the transmission and progression of disease.

3. *Chai Hou Fang Fu* (Preventing Recurrence After Recovery).

This refers to the period after *H. pylori* eradication and resolution of clinical symptoms, which is not the end of treatment. At this stage, the patient's underlying spleen–stomach weakness (*pi wei xu ruo*) is not yet fully reversed, and the gastric environment still carries a risk of re-harboring pathogens. The focus is on consolidating therapeutic effects and preventing relapse through long-term spleen-strengthening and qi-supplementing therapy, conditioning and maintenance, and sustained healthy lifestyle and dietary habits. The goal is to thoroughly transform the "soil" on which *H. pylori* depends, thereby achieving long-term health and stability and effectively reducing long-term recurrence.

(III) The Syndrome-Differentiation View of Individualized and Precise Treatment

Treating *H. pylori* with TCM is not a one-size-fits-all approach. It strictly follows *bian zheng lun zhi* to achieve highly individualized, precise therapy—continuing the diagnostic–therapeutic spirit of the *Shang Han Lun* (Treatise on Cold Damage): "observe the pulse and signs, discern what has been violated, and treat according to the syndrome."

As noted above, *fen xing lun zhi* provides the primary framework for individualization. A higher tier of individualization, however, requires that even within the same syndrome type, therapeutic principles and formulas be finely adjusted and modified (*jia jian*) according to a patient's age, constitution, regional context, and seasonal conditions.

For example, among two patients with the same *Pi Wei Shi Re Zheng* (spleen–stomach damp-heat syndrome), if one also exhibits *qi xu* (qi deficiency), qi-supplementing (*yi qi*) herbs should be added on top of clearing-heat and transforming-dampness. If the presentation occurs in a humid summer, the potency of damp-transforming herbs should be increased. This philosophy of "suiting treatment to the individual, the time, and the place" (*yin ren, yin shi, yin di zhi yi*) achieves a deeper level of precise targeting of the person, rather than merely delivering a generic framework aimed at the "disease syndrome."

## 20.4 The Theoretical Advantages of Integrating Chinese and Western Medicine

Although TCM possesses a self-consistent, comprehensive theoretical system for addressing *H. pylori* infection, we must objectively recognize that modern, antibiotic-centered therapies have unparalleled advantages in *qu xie* (dispelling pathogenic factors)—that is, directly, rapidly, and forcefully eradicating the pathogen—hence their status as the dominant global strategy for *H. pylori* eradication. Against this clinical reality, scenarios in which TCM serves as the primary modality for *H. pylori* are presently concentrated mainly in TCM hospitals or TCM departments within comprehensive hospitals in China; in broader clinical practice, TCM more often functions as an important adjunct and complementary therapy, working synergistically with modern medicine to form a distinctive model of "Integrative Chinese and Western Medicine [6]". This integration is not a simple superposition of drugs, but a strategic complement and theoretical fusion grounded in a nuanced understanding of each system's strengths and limitations, with the aim of delivering more effective and safer care. Its theoretical advantages are reflected chiefly in the following three aspects:

I. Complementarity of Therapeutic Targets

Modern medical therapy—centered on antibiotics—employs a singular and highly precise target: it acts directly on *H. pylori* itself, with the aim of eradicating the pathogen.

By contrast, TCM theory proposes a broader, multidimensional target—the infected human as an integrated whole. This includes modulating the gastric microenvironment on which *H. pylori* depends (changing the "soil") and, more critically, directly enhancing host immune function. Through *fu zheng* (supporting the righteous qi), TCM seeks to activate and regulate the body's immune system, forging an active, durable defense network capable of continuously resisting pathogenic factors.

Accordingly, at the theoretical level, integrative Chinese–Western medicine constructs a triple-target model of "pathogen–host environment–host immunity." Western medicine executes the precision strike against the "external enemy," while Chinese medicine transforms the internal terrain and fortifies the "internal guard." In terms of therapeutic targets, the two approaches perfectly complement each other—superior to either system operating alone.

II. Synergy of Therapeutic Effects

Grounded in the complementarity of therapeutic targets, the two therapeutic systems can, in theory, generate synergistic effects in clinical outcomes [7]. This synergy is chiefly manifested in two aspects: "enhancing efficacy" (*zeng xiao*) and "reducing toxicity" (*jian du*).

**Enhancing Efficacy (Dual Synergism)**

Direct synergism in *qu xie* (dispelling the pathogen): Many Chinese herbs themselves contain active constituents that inhibit or kill *H. pylori*—for example, berberine in *Huang Lian* (*Rhizoma Coptidis*). These components can act in direct synergy with antibiotics, jointly accomplishing the task of *qu xie* [8].

Indirect synergism via *fu zheng* (supporting the righteous qi): A key modern biological interpretation of *fu zheng* is regulating and enhancing host immune function. By activating and modulating the body's immune surveillance and clearance, it works together with the antibiotics' "external aid" to form a "pincer attack from within and without." This collaboration jointly clears the pathogen and lays a solid immunologic foundation for lowering the long-term recurrence rate.

**Reducing Toxicity (Attenuating Side Effects)**

While eradicating *H. pylori*, antibiotics are often accompanied by gastrointestinal adverse reactions and gut-flora dysbiosis, which can undermine patient compliance.

The *fu zheng* approach in TCM—especially *jian pi he wei* (strengthening the spleen and harmonizing the stomach)—can precisely and targetedly regulate gastrointestinal function, protect the gastric mucosa, and maintain intestinal microecological balance, thereby, in theory, effectively mitigating the toxic and side effects of Western pharmaceuticals.

III. The Holistic Nature of Therapeutic Goals

The core indicator for evaluating the efficacy of modern medical treatment is the eradication rate of *H. pylori*—a microscopic, disease-centered metric. In contrast, the therapeutic goal of integrating Chinese and Western medicine is broader and more holistic. It not only pursues clearance of the pathogenic cause, but it is also dedicated to restoring the health state of the person.

This higher-level goal includes not only resolution of clinical symptoms and improvement in quality of life but also reconstruction of immune defenses and stabilization of the internal environment's homeostasis. The aim is to help patients achieve a long-term, steady state of health that resists future pathogen invasion. This elevates treatment from a purely biomedical model to a more advanced biopsychosocial medical model, realizing a paradigm shift from pathogen-centered to patient-centered therapy.

## 20.5 Conclusion

This chapter has systematically elucidated the TCM theoretical framework for the prevention and treatment of *H. pylori*. We first clarified that TCM's cognitive logic does not originate from the microscopic "pathogen," but it is rooted in the macroscopic "holistic person."

At the level of theoretical understanding, we established a four-dimensional cognitive model: (1) the core pathogenic nature of *H. pylori* is defined as *shi re du xie*; (2) *zheng xu xie fan*—with *pi wei xu ruo* at its core—is identified as the fundamental cause of disease onset; (3) *yu xue zu luo* is revealed as the key pathological link underlying a protracted course and deepening progression; and (4) *gan wei bu he* is emphasized as an important emotion-related factor that triggers and aggravates the condition.

Building on this framework, we further expounded the corresponding therapeutic principles and core philosophies. We established the fundamental principle of *fu zheng qu xie*; *biao ben jian zhi* and, in light of *H. pylori*'s characteristics, stressed the clinical priority of *ji ze zhi qi biao*. Under this overarching principle, we adopted *fen xing lun zhi* as the core strategy and distilled three core therapeutic philosophies: the holistic co-management of the "human–microbe–environment" triad; the *zhi wei bing* philosophy—covering *wei bing xian fang* (preventing disease before onset), *ji bing fang bian* (preventing progression once present), and *chai hou fang fu* (preventing recurrence after recovery); and an individualized, precise approach tailored to the person.

Finally, grounded in clinical reality, we discussed the inherent theoretical advantages of integrating Chinese and Western medicine. By analyzing the complementarity of therapeutic targets, the synergy of therapeutic effects (enhancing efficacy and reducing toxicity), and the holistic nature of therapeutic goals, we show that Chinese–Western integration represents a strategic fusion aimed at achieving superior clinical outcomes.

## 20.6 Outlook

The TCM theoretical system systematically expounded in this chapter is not a classical philosophy to be admired from afar, but an action blueprint for guiding future clinical practice and scientific research. It provides a firm theoretical basis for answering "Why use TCM?" and "How is TCM effective?" Looking ahead, the pathway for development is both clear and expansive:

First, there must be the continued deepening of basic research. With the aid of modern technology, we should further clarify the modern biological connotations of TCM syndromes such as *shi re du xie* and *pi wei xu ruo*, and precisely analyze the multi-component, multi-target synergistic mechanisms of effective formulas and herbs [9]. This will provide an ever more solid scientific basis for clinical practice and drive the continuous optimization of therapeutic solutions.

Second, there should be continuous optimization of clinical solutions. Through large-sample, high-quality evidence-based studies, we can develop integrated Chinese–Western diagnostic and treatment plans and clinical pathways with high levels of evidence and strong generalizability, and actively promote their inclusion in authoritative guidelines, thereby enhancing their clinical uptake and impact.

Finally, the line of prevention must be advanced forward. We should fully leverage TCM's unique advantage of *zhi wei bing*: vigorously carry out constitutional conditioning and early interventions for populations at high risk of *H. pylori* infection, and provide long-term maintenance and care for patients after eradication. These measures can effectively control infection rates and interrupt progression toward atrophy, intestinal metaplasia, and even gastric cancer.

Next, in the following chapter (Chap. 21), guided by this theoretical system, we will move from "theory" to "practice." We will systematically describe the modern research advances of representative single herbs and classic compound formulas that embody the above effects, as well as their specific application strategies in clinical practice. Through the close integration of theory and practice, we will jointly witness how this theoretical system—rooted in millennia of accumulated wisdom—can, with the support of modern science and technology, contribute its unique and irreplaceable value to addressing the global challenge of *H. pylori* infection.

## References

1. Sun R, Du Y, Cui J, et al. Prevention and treatment strategies for *H. pylori*-associated gastritis in ancient and modern Chinese medicine literature. Lishizhen Med Mater Med Res. 2021;32(11):2735–7.
2. Lv G, Lao S. An analysis of the relationship between *H. pylori* infection and spleen–stomach damp-heat. Liaoning J Tradit Chin Med. 2004;31(1):16–7.
3. Chen Y, Ye H, Liu Y, et al. Treating *H. pylori* infection from the perspective of spleen–stomach damp-heat. J Basic Chin Med. 2019;25(2):195–7.
4. Fu C, Huang C, Lv W, et al. Study on the distribution of traditional Chinese medicine syndrome types in 2818 patients with *H. pylori* associated gastritis. J Li-Shizhen Tradit Chin Med. 2025;36(15):2896–9.
5. Liu J, Peng T, Yi M, et al. Analysis of Chinese medicine syndrome types and syndrome element distribution in *H. pylori*-associated ulcers based on modern literature. New Chin Med. 2025;57(13):164–72.
6. Hu F, Zhang S, Zhang X, et al. Expert consensus on integrative medicine for *H. pylori* related "diseases and syndromes" in China (2024 edition). Chin J Gastroenterol Hepatol. 2024;33(12):1581–8.
7. Zhang X, Wei W, Lan Y, et al. Chinese medicine and Western medicine collaborative expert consensus on diagnosis and treatment of *H. pylori* caused gastritis in adults (2020, Beijing). J Tradit Chin Med. 2020;61(22):2016–24.
8. Wan H, Li C, Li W, et al. Anti-*H. pylori* mechanism of Chinese medicine: a review. Chin J Exp Tradit Med Formulae. 2023;29(3):203–12.
9. Zhang P-P, Li L, Qu H-Y, et al. Traditional Chinese medicine in the treatment of *H. pylori*-related gastritis: the mechanisms of signalling pathway regulations. World J Gastroenterol. 2025;31(3):96582.

# Traditional Chinese Medicine Pharmacology and Clinical Practice in *Helicobacter pylori*

21

Chuqiu Zhang and Meicun Yao

## 21.1 Introduction

The previous chapter systematically elucidated the TCM framework for preventing and treating *Helicobacter pylori (H. pylori)* infection. Building on that foundation, this chapter integrates theory with practice and presents the concrete pharmacological studies and clinical application strategies derived from that framework.

Across millennia of confronting disorders such as *weiwan tong* (epigastric pain) and *pi man* (epigastric fullness), TCM has not only amassed a vast body of clinical experience but has also developed a distinctive, practice-oriented system grounded in the material basis of Chinese materia medica. At the heart of this system lies not merely the use of single herbs, but—more fundamentally—the Chinese herbal formula (*zhong yao fu fang*).

A formula is the highest expression of TCM therapeutic thought. It is not a simple admixture of multiple herbs; rather, it follows rigorous principles of compatibility (*pei wu*). Each herb carries its own theoretical attributes—such as *si qi wu wei* (four natures and five flavors) and *gui jing* (channel tropism)—and when they are combined, the strict rule of *jun chen zuo shi* (sovereign–minister–assistant–envoy) is observed: the *jun* (sovereign) herb targets the principal disorder; the *chen* (minister) herbs reinforce its action; the *zuo* (assistant) herbs address concurrent patterns or moderate the intensity of the *jun* and *chen*; and the *shi* (envoy) herbs guide the effects to specific channels or harmonize the whole prescription.

It is precisely this sophisticated compatibility wisdom that enables formulas to exert multi-target, multi-pathway synergy, thereby achieving the overarching

C. Zhang · M. Yao (✉)
School of Pharmaceutical Sciences (Shenzhen), Sun Yat-sen University, Shenzhen, China
e-mail: Lssymc@mail.sysu.edu.cn



J. Gao, Y. Du (eds.), *Non-Antibiotic Therapies for Helicobacter pylori*,
https://doi.org/10.1007/978-981-95-9598-3_21

regulatory aim of enhancing efficacy while reducing toxicity—an advantage unattainable with single herbs or isolated extracts alone.

Based on the pathomechanistic theories outlined in the previous chapter, TCM has, in practice, established a set of core therapeutic principles: employing *qing re jie du, hua shi xie zhuo* (clearing heat and resolving toxins; transforming dampness and expelling turbidity) to eliminate pathogenic factors; *jian pi yi qi, fu zheng gu ben* (strengthening the Spleen and supplementing qi; supporting the righteous and consolidating the root) to modify the "soil" that fosters disease; and *li qi he wei, huo xue hua yu* (regulating qi and harmonizing the Stomach; invigorating blood and transforming stasis) to improve the local milieu. Ultimately, these principles are implemented through concrete prescriptions and medications (*fang yao*).

Accordingly, this chapter moves from the simple to the complex and from basic research to clinical application, and will be presented on three levels:

First, we examine modern pharmacological evidence for single Chinese herbs with anti-*H. pylori* activity and discuss their limitations. Second, we analyze in depth how Chinese herbal formulas and Chinese patent medicines (*zhong cheng yao*) capitalize on holistic regulation through sophisticated compatibility. Finally, we focus on the current mainstream model in clinical practice—integrative Chinese and Western medicine (*zhong xi yi jie he*)—and explore its concrete synergistic strategies and clinical value.

## 21.2 Core Therapeutic Principles and Modern Interpretations

The previous chapter systematically expounded a four-dimensional cognitive model in TCM for *H. pylori* infection. It illuminates etiology and pathogenesis across four axes: the pathogenic nature of *shi re du xie* (damp-heat-toxin pathogen), the fundamental cause of onset as *zheng xu xie fan* (righteous-qi deficiency with pathogenic invasion), the pathological evolution of *yu xue zu luo* (blood stasis obstructing the collaterals), and the key influencing factor of *gan wei bu he* (Liver–Stomach disharmony).

Theory, however, finds its ultimate value in guiding practice. These pathomechanistic insights must be translated into concrete therapeutic principles (*zhi ze*). In view of the complex pattern of *H. pylori* infection—marked by *ben xu biao shi* (root deficiency with branch excess) and *xu shi jia za* (intermingling of deficiency and excess)—the core TCM principles are articulated along four main dimensions:

I. ***Qing re hua shi, jie du qu zhuo*—Clearing Heat and Transforming Dampness; Resolving Toxins and Expelling Turbidity**

**Theoretical Basis** This principle directly addresses the core pathogenic nature of *H. pylori* as a *shi re du xie* (damp-heat-toxin pathogen).

- Interpretation *Qing re* (clearing heat) aims to suppress the heat-toxin inflammatory reaction of the gastric mucosa induced by *H. pylori*.

- *Hua shi* (transforming dampness) seeks to remove the damp-turbid gastric environment—the "soil" that allows *H. pylori* to colonize.
- *Jie du qu zhuo* (resolving toxins and expelling turbidity) targets the direct mucosal injury caused by the various toxins produced by *H. pylori*.

Taken together, these three facets constitute TCM's most direct approach to *qu xie* (dispelling the pathogenic factor) in the context of *H. pylori*.

From a modern pharmacological perspective, herbs with these actions typically show broad-spectrum antibacterial and anti-inflammatory effects and can inhibit the activity of *H. pylori* toxins [9].

II. ***Jian pi yi qi, fu zheng gu ben*—Strengthening the Spleen and Supplementing Qi; Supporting the Righteous and Consolidating the Root**

**Theoretical Basis** This principle addresses the fundamental paradigm of *zheng xu xie fan* (righteous-qi deficiency with pathogenic invasion), with particular focus on *pi wei xu ruo* (spleen–stomach weakness).

**Interpretation** This is the defining feature that distinguishes TCM management of *H. pylori* from purely antibacterial therapy and is the concentrated expression of the *fu zheng* (supporting the righteous) philosophy. The aim is not only to passively enhance gastric-mucosal barrier and repair, but also—more importantly—to strengthen spleen–stomach function so as to regulate and activate the host immune system, enabling active recognition, phagocytosis, and clearance of the pathogen. This constitutes a higher-level, indirect strategy of *qu xie* (dispelling the pathogenic factor): eliminating the pathogen by arming the body's own "troops." Only when *zheng qi cun nei* (righteous qi is present within) can the conditions on which *H. pylori* depends be fundamentally removed, thereby effectively reducing long-term recurrence [1].

III. ***Li qi he wei, shu gan jie yu*—Regulating Qi and Harmonizing the Stomach; Soothing the Liver and Relieving Constraint**

**Theoretical Basis** This principle primarily targets the key influencing factor of *gan wei bu he* (liver–stomach disharmony).

**Interpretation** The manifestations of *H. pylori* infection are often exacerbated by emotional factors. The underlying mechanism is *gan qi yu jie* (liver-qi stagnation) that *heng ni fan wei* (transversely invades the stomach). Thus, *shu gan li qi* (soothing the liver and regulating qi) and restoring the smooth operation of the *qi ji* (qi dynamics) are critical to re-establishing the Stomach's normal ascending–descending functions. This principle embodies the TCM notion of *xin shen tong zhi* (treating mind and body together). From a modern medical perspective, it corresponds to modulating the brain–gut axis, improving gastrointestinal motility, and relieving stress-related digestive symptoms [37].

IV. ***Huo xue hua yu, tong luo xiu fu*****—Invigorating Blood and Transforming Stasis; Unblocking the Collaterals and Promoting Repair**

**Theoretical Basis** This principle corresponds to the late-stage pathogenic evolution of *yu xue zu luo* (blood stasis obstructing the collaterals) in *H. pylori* infection.

**Interpretation** In patients with a protracted course who already exhibit precancerous changes—such as gastric mucosal atrophy or intestinal metaplasia—simple heat-clearing or spleen-strengthening is insufficient to reverse the trajectory. At this juncture, *huo xue hua yu* (invigorating blood and transforming stasis) should be added to improve gastric-mucosal microcirculation, promote glandular regeneration, and inhibit fibrotic progression, thereby repairing deep injury and intercepting the pathway toward gastric carcinogenesis. This is a concrete embodiment of the *ji bing fang bian* (preventing progression once disease is present) strategy within the TCM philosophy of *zhi wei bing* (preventive treatment of disease) [36].

In summary, these four principles are not applied in isolation. In clinical practice, they are organically combined and dynamically adjusted to the patient's specific syndrome (*zheng hou*) within the framework of *bian zheng lun zhi* (syndrome differentiation and treatment). Together, they constitute an integrated, multi-dimensional strategy network, providing clear guidance for the concrete pharmacologic applications discussed in the next section.

## 21.3 Modern Research on Single Chinese Herbs

While Chinese herbal formulas dominate clinical practice, pharmacological studies on single herbs (*dan wei zhong yao*) underpin our understanding of how TCM acts against *H. pylori* and provide a vital source for new drug discovery. Since the bacterium's discovery, researchers have used modern scientific methods to screen and validate traditional herbs historically used for conditions such as epigastric pain. By incomplete tally, hundreds of Chinese herbs have been reported to show varying degrees of in vitro anti-*H. pylori* activity. Selected representative studies are summarized below to illustrate the scope of this work (Table 21.1).

Next, we select several representative herbs from different functional categories for detailed discussion.

**Table 21.1** Selected representative single herbs with in vitro anti-*H. pylori* activity

| Chinese name (Pinyin) | Latin name (*italic*) | Representative active constituent | MIC range (mg/mL) | Primary action/ classification | References |
|---|---|---|---|---|---|
| 黄连 (Huang Lian) | *Coptis chinensis* | 小檗碱 (Berberine) | 3.9–15.6 | Clear heat, resolve toxins | [12, 24, 33] |
| 黄芩 (Huang Qin) | *Scutellaria baicalensis* | 黄芩素 (Baicalin) | 0.125 | Clear heat, resolve toxins | [14, 43] |
| 地榆 (Di Yu) | *Sanguisorba officinalis* | 鞣花酸 (Ellagic acid) | 0.16–0.32 | Clear heat, cool blood | [2, 31] |
| 蒲公英 (Pu Gong Ying) | *Taraxacum mongolicum* | 绿原酸 (Chlorogenic acid) | 0.5–4 | Clear heat, resolve toxins | [4] |
| 丁香 (Ding Xiang) | *Syzygium aromaticum* | 丁香酚 (eugenol) | 0.16–0.32 | Aromatically transform dampness | [5, 28] |
| 厚朴 (Hou Pu) | *Magnolia officinalis* | 厚朴酚 (Magnolol) | ≤0.1 | Aromatically transform dampness | [15, 32] |
| 木香 (Mu Xiang) | *Aucklandia lappa* | 木香烃内酯 (Costunolide) | 40 | Aromatically regulate *qi* | [39] |
| 诃子 (He Zi) | *Terminalia chebula* | 诃子酸 (Chebulinic acid) | 0.02–0.16 | Ethnic medicine/ astringent | [26, 27] |
| 毛诃子 (Mao He Zi) | *Terminalia bellirica* | 鞣质 (Tannins) | 0.16 | Ethnic medicine/ astringent | [53] |
| 五指毛桃 (Wu Zhi Mao Tao) | *Ficus hirta* | 补骨脂素 (psoralen) | 0.64 | Support righteousness, strengthen spleen | [30] |
| 甘草 (Gan Cao) | *Glycyrrhiza uralensis* | 甘草酸 (Glycyrrhizic acid) | ≥0.1 | Support righteousness, strengthen spleen | [17, 23] |
| 白术 (Bai Zhu) | *Atractylodes macrocephala* | 苍术酮 (Atractylone) | >0.1 | Support righteousness, strengthen spleen | [10, 42] |

*Note*. MIC (minimum inhibitory concentration) values are compiled from multiple reports. Differences in strains, extraction/processing, and experimental protocols may lead to variability; the ranges above are for reference

I. **Herbs That Clear Heat and Resolve Toxins (*qing re jie du yao*)**

This category receives the most attention in anti-*H. pylori* research because many herbs exhibit direct antibacterial or bactericidal activity—an intuitive embodiment of the TCM principle of *qu xie* (dispelling the pathogenic factor).

***Huang Lian (Coptidis Rhizoma)***
As a flagship heat-clearing, toxin-resolving herb, Huang Lian has been most extensively investigated for anti-*H. pylori* effects [52]. Its primary active constituent, berberine, exerts multiple actions against *H. pylori*: inhibiting bacterial growth, suppressing urease activity, and interfering with adhesion [33]. Beyond berberine, other bioactive alkaloids in Huang Lian—such as palmatine and coptisine—also show in vitro anti-*H. pylori* activity, though via differing mechanisms [12, 35]. This illustrates that even a "single" herb often works through the synergy of multiple constituents rather than a lone ingredient.

Clinically, however, Huang Lian is intensely bitter and cold (*da ku da han*), prone to injuring the spleen–stomach, and thus, it is unsuitable as a long-term or high-dose single agent. It is commonly used as the *jun yao* (sovereign herb) and paired with spleen-fortifying or acrid-warm substances to temper its cold nature and protect stomach qi—embodying the principle of "purging fire without injuring the Stomach." For example, in the classic formula *Ban Xia Xie Xin Tang*, Huang Lian and Huang Qin (both bitter-cold) are combined with Gan Jiang and Ban Xia (both acrid-warm) to "open with acridity and descend with bitterness" (*xin kai ku jiang*) [19, 20].

***Pu Gong Ying (Taraxaci Herba; Dandelion)***
This herb's notable advantage is its "medicine–food homology" (*yao shi tong yuan*), which confers high safety and broad public acceptance. Extracts not only inhibit *H. pylori* but also effectively disrupt its biofilm [4, 8]. Current studies suggest that the anti-*H. pylori* effect relates to its richness in chlorogenic acid, caffeic acid, polysaccharides, and other constituents. No single compound has shown exceptionally strong, standalone activity, again pointing to multi-component synergy. Because its nature is bitter–cold (*ku han*), it should be used cautiously in patients with spleen–stomach deficiency-cold (*pi wei xu han*).

***Di Yu (Radix Sanguisorbae)***
Traditionally used to clear heat and cool blood (*qing re liang xue*) and to astringe and stop bleeding (*shou se zhi xue*), Di Yu has recently shown substantial potential against *H. pylori*. Both its aqueous extract and the active constituent ellagic acid display strong in vitro inhibitory effects against multiple clinical isolates of *H. pylori* [31]. Moreover, in a completed project supported by the National Natural Science Foundation of China, preliminary clinical observations were encouraging: as monotherapy, a Di Yu preparation relieved symptoms in >95% of patients, with an *H. pylori* eradication rate approaching 40%. Unlike intensely bitter–cold agents such as Huang Lian, Di Yu is moderately bitter–cold (*ku han zhong du*) with an astringent property, causing less irritation to the gastric mucosa. It therefore shows better prospects for applications requiring longer courses to improve the gastric environment.

II. **Herbs That Aromatically Transform Dampness (*fang xiang hua shi yao*)**

This group, characterized by a distinctive fragrance, "awakens the Spleen and harmonizes the Stomach, transforms dampness, and promotes the movement of stagnation" (*xing pi he wei, hua shi xing zhi*). They are essential for patterns in which damp-heat obstructs the middle *jiao* (*zhong jiao shi zu*) and the qi mechanism is impeded (*qi ji bu chang*). Because pathogenic dampness (*shi xie*) is viscous and stagnant, it readily blocks the spleen–stomach qi mechanism. The acrid, fragrant, dispersing, and mobilizing nature (*xin xiang zou cuan*) of aromatic herbs invigorates spleen-yang and frees the qi mechanism—like sunlight dispelling clammy gloom—thereby achieving "transformation of dampness" (*hua shi*).

***Ding Xiang (Syzygium aromaticum)—Clove***

An outstanding representative of aromatic herbs, clove warms the middle and directs counterflow downward (*wen zhong jiang ni*), and transforms dampness to awaken the Spleen (*hua shi xing pi*) [25]. Traditionally, eugenol was regarded as its principal active constituent against *H. pylori*. Recent studies, however, have found very strong anti-*H. pylori* activity (MIC 160–320 μg/mL) even in aqueous extracts containing <0.15% eugenol [28]. Mechanistically, actions span multiple links: urease inhibition, disruption of bacterial ATP metabolism, and modulation of key bacterial protein expression. These findings strongly support that clove's anti-*H. pylori* effect arises from multi-component, multi-target synergy, rather than eugenol alone.

III. **Herbs That Support Righteousness and Strengthen the Spleen (*fu zheng jian pi yao*)**

This category embodies the *fu zheng* (supporting the righteous) philosophy. By enhancing host functions, these herbs achieve indirect anti-pathogen effects—a strategy that addresses both branch and root.

***Huang Qi (Astragali Radix) and Wu Zhi Mao Tao (Ficus Hirta Radix)***

Huang Qi, renowned as the "chief herb for supplementing qi" (*bu qi zhi chang*), is known as "Bei Qi" (Northern Astragalus); Wu Zhi Mao Tao, a representative herb from the Lingnan region, is called "Nan Qi" (Southern Astragalus). Both supplement qi and strengthen the spleen (*jian pi yi qi*), enhance host immunity, and help repair the gastric mucosa [29, 48]. Notably, components in Wu Zhi Mao Tao, such as psoralen, show a direct inhibitory effect on *H. pylori* (MIC up to 6.25 μg/mL) [47]. This finding enriches the traditional view of "supporting the righteous" by adding a concomitant "dispelling the pathogenic factor" (*qu xie*) dimension.

***Gan Cao (Glycyrrhizae Radix)—Licorice***

One of the most frequently used herbs in TCM formulas—hence the honorific "*guo lao*" (state elder)—Gan Cao's core function is to *tiao he zhu yao* (harmonize the actions of the other herbs). In *H. pylori* therapy, its role is likewise indispensable [17]. Although its main constituents, glycyrrhizic acid and glycyrrhetinic acid, show limited direct anti-*H. pylori* activity, they provide strong anti-inflammatory effects, promote mucus secretion, and protect the gastric mucosa—counteracting damage from *H. pylori* infection and from drugs such as antibiotics [23]. Consequently, in anti-*H. pylori* prescriptions, licorice often serves as the "*hou qin guan*" (logistics officer) and "*he shi lao*" (peacemaker): consolidating defenses, repairing injury, and moderating the harsh properties of other ingredients.

IV. **Representative Herbs With Ethnic Minority Characteristics**

These medicines are grounded in the distinctive theoretical frameworks of various ethnic medical traditions. Differences in lifestyle, environment, and culture shape divergent views on pharmacological actions and clinical applications. This experience- and context-driven diversity has greatly enriched the broader field of pharmacology and offers fresh perspectives and valuable resources for anti-*H. pylori* research.

***He Zi (Chebulae Fructus)***

A treasured medicine in Tibetan practice, He Zi—and its related herb Mao He Zi (*Bellirica Fructus*)—shows unique value against *H. pylori*. Studies indicate that He Zi extracts not only directly inhibit *H. pylori* (MIC 20–160 μg/mL) by disrupting the bacterial cell membrane and suppressing urease activity but also effectively downregulate inflammatory mediators (e.g., interleukin (IL)-8) induced in gastric epithelial cells by *H. pylori* infection [27]. In parallel, its abundant tannins astringe the gastric mucosa and promote ulcer healing [26]. Together, these features illustrate an ethnic-medicine advantage that combines "direct antibacterial action" with "improvement of the local gastric environment."

V. **Limitations of Single-Herb Application**

Research on single Chinese herbs has yielded valuable mechanistic insights, but direct clinical use of single herbs shows clear limitations. A single target is unlikely to address the complex, intertwined pathomechanisms comprehensively; therapeutic strength and stability can be inadequate, and there is also a risk of resistance. Consequently, TCM's wisdom is better expressed through the synergistic compatibility of multiple herbs—namely, Chinese herbal formulas (*zhong yao fu fang*).

## 21.4 Application of Chinese Herbal Formulas

The limitations of single-agent herbs highlight the substantial advantages of **Chinese herbal formulas** (*zhong yao fu fang*). Confronted with the complex pathogenesis of *H. pylori* infection—characterized by *ben xu biao shi* (root deficiency with branch excess) and *xu shi jia za* (intermingling of deficiency and excess)—formulas follow the rigorous compatibility rules of *jun chen zuo shi* (sovereign, minister, assistant, and envoy). By organically combining herbs with different medicinal properties, they achieve holistic regulation of the body through multi-component, multi-target synergy.

This section will present the application of formulas in the anti-*H. pylori* field from several dimensions, including Chinese patent medicines (*zhong cheng yao*, standardized finished TCM preparations), representative formulas guided by *bian zheng* (syndrome differentiation), as well as classic formulas (*jing fang*) and empirically derived formulas (*yan fang*).

First, Table 21.2 provides an overview of several representative formulas that have been most extensively studied against *H. pylori*.

I. **Standardized Use of Modern Chinese Patent Medicines (*zhong cheng yao*)**

Chinese patent medicines are a major outcome of the modernization of TCM. Having passed approval by the national drug authority, they possess clear quality standards and legal status, making them the most readily—and lawfully—accessible TCM preparations for clinicians (including Western-trained physicians in general hospitals) and thus the most widely used in practice. Their chief advantages are standardization, convenience, and broad availability. When classified as OTC (over-the-counter) products, the public can purchase them directly. Their shortcomings, however, are also evident: fixed formulas sacrifice the individualized flexibility of *sui zheng jia jian* (modifying according to *bian zheng*), and fluctuations along the industrial production chain—from raw-material sourcing to

**Table 21.2** Selected representative formulas for anti-*H. pylori* and their research progress

| Name | Category | Primary therapeutic principle | Research stage |
|---|---|---|---|
| (胃复春) | Proprietary Chinese medicine | Strengthen spleen and supplement *qi*; invigorate blood and resolve toxins | Large-scale clinical research + expert consensus |
| Sanjiu Weitai (三九胃泰) | Proprietary Chinese medicine | Clear heat and dry dampness; move *qi* and invigorate blood | Clinical research + OTC (over-the-counter) |
| Moluodan (摩罗丹) | Proprietary Chinese medicine | Harmonize stomach and direct *qi* downward; strengthen spleen and reduce distention | Clinical research |
| Ban Xia Xie Xin Tang (半夏泻心汤) | Classic formula (*jing fang*) | Open with acrid and descend with bitter; use cold and heat together | Preclinical + clinical research |
| Huang Lian Jie Du Tang (黄连解毒汤) | Classic formula (*jing fang*) | Clear heat and purge fire; resolve toxins and dry dampness | Preclinical + clinical research |
| Lian Pu Yin (连朴饮) | Classic formula (*shi fang*) | Clear heat and transform dampness; regulate *qi* and harmonize the middle | Preclinical + clinical research |
| Xiang Sha Liu Jun Zi Tang (香砂六君子汤) | Classic formula (*shi fang*) | Strengthen spleen and supplement *qi*; harmonize stomach and quiet the middle | Preclinical + clinical research |
| Chai Hu Shu Gan San (柴胡疏肝散) | Classic formula (*shi fang*) | Soothe liver, regulate *qi*, and harmonize stomach | Preclinical + clinical research |
| Ping Wei San (平胃散) | Classic formula (*shi fang*) | Dry dampness and fortify spleen; move *qi* and harmonize stomach | Preclinical + clinical research |
| Dan Shen Yin (丹参饮) | Classic formula (*shi fang*) | Invigorate blood and dispel stasis; move *qi* and stop pain | Preclinical + clinical research |

manufacturing processes—may lead to inter-batch variability in quality, which can in turn affect clinical efficacy.

**Weifuchun (胃复春)**
This product is a landmark example of in-depth research on late-stage pathological changes related to *H. pylori* infection within the field of Chinese patent medicines.

**Origin and Composition** Co-developed by the Zhejiang Academy of Traditional Chinese Medicine and Hangzhou Hu Qing Yu Tang Pharmaceutical Co., Ltd. The formula consists of Hong Shen (Radix Ginseng Rubra), Xiang Cha Cai (*Isodon* spp.), and Zhi Qiao (Aurantii Fructus).

**Functions and Indications** *Jian pi yi qi* (strengthen the spleen and supplement qi) and *huo xue jie du* (invigorate blood and resolve toxins). It is mainly indicated for precancerous gastric lesions, as adjuvant therapy after gastric cancer surgery, and for chronic superficial gastritis of the *spleen–stomach deficiency-weakness* pattern.

**Modern Research** According to the Expert Consensus on the Clinical Application of Weifuchun for Precancerous Lesions of Atrophic Gastritis, evidence-based studies show that Weifuchun significantly increases the pathological remission rates of atrophy and intestinal metaplasia (Level B evidence, strong recommendation) and effectively relieves clinical symptoms such as epigastric discomfort and distention (Level B evidence, strong recommendation). This consensus provides high-quality evidence supporting the precise application of this Chinese patent medicine for the late course of *H. pylori* infection characterized by *zheng qi deficiency with lingering pathogens* (*zheng xu xue yu*) [19].

**Sanjiu Weitai Granules (三九胃泰颗粒)**
A widely used OTC Chinese patent medicine that exemplifies clearing and resolving *"damp-heat"* (*shi re*).

**Origin and Composition** Derived from a military empirical formula and composed of San Ya Ku (*Radix et Ramus Evodiae*), Jiu Li Xiang (*Folium et Cacumen Murrayae*), Bai Shao (*Radix Paeoniae Alba*), Di Huang (*Radix Rehmanniae*), Mu Xiang (*Radix Aucklandiae*), and related herbs.

**Functions and Indications** *Qing re zao shi* (clear heat and dry dampness), *xing qi huo xue* (move qi and invigorate blood), and *rou gan zhi tong* (soften the liver and relieve pain).

**Modern Research** Clinical studies show that, for symptoms such as epigastric burning pain, bitter taste in the mouth, and epigastric fullness caused by the common *spleen–stomach damp-heat* pattern during *H. pylori* infection, Sanjiu Weitai provides convenient and effective symptomatic relief and can be used as an adjunctive medication to improve symptoms [46, 51].

**Moluodan (摩罗丹)**
A representative Chinese patent medicine used in the treatment of chronic atrophic gastritis.

**Origin and Composition** Derived from a folk empirical prescription and composed of 18 Chinese herbs, including *Bai He* (Bulbus Lilii), *Fu Ling* (Poria), *Xuan Shen* (Radix Scrophulariae), *Mai Dong* (Radix Ophiopogonis), *Shi Hu* (Herba Dendrobii), *San Qi* (Radix et Rhizoma Notoginseng), *Di Yu* (Radix Sanguisorbae), and others.

**Functions and Indications** *He wei jiang ni* (harmonize the stomach and direct qi downward), *jian pi xiao zhang* (strengthen the spleen and reduce distention), and *tong luo ding tong* (unblock collaterals and relieve pain).

**Modern Research** The formula contains multiple components and integrates several therapeutic approaches—*zi yang wei yin* (nourishing stomach yin), *jian pi he wei* (strengthening the spleen and harmonizing the stomach), and *huo xue hua yu* (invigorating blood and transforming stasis)—reflecting a comprehensive regulation strategy for the complex pathogenesis of chronic gastric disease. Clinical studies show that Moluodan effectively alleviates symptoms such as dull epigastric pain and poor appetite in patients with chronic atrophic gastritis, and it also has a certain effect on reversing gastric mucosal atrophy and intestinal metaplasia during or after *H. pylori*-related disease courses [3, 16].

II. **Representative Formulas Based on Syndrome Differentiation**

In the therapeutic principles section of the previous chapter, we established corresponding principles and representative formulas for different syndrome types (*zheng xing*). These formulas are concrete embodiments of the TCM diagnostic thinking of syndrome differentiation (*bian zheng lun zhi*), are recommended by relevant national consensus documents, and form the basis for individualized treatment by TCM physicians. Unlike the fixed compositions of Chinese patent medicines, these representative formulas are commonly used clinically as core base prescriptions; physicians then modify them (*sui zheng jia jian*) according to each patient's specific condition and constitutional features to achieve optimal individualized outcomes.

***Representative Formula for Spleen–Stomach Damp-Heat (pi wei shi re) Syndrome: Lian Pu Yin (连朴饮)***

Originating from Wang Mengying's *Huo Luan Lun* (Treatise on Cholera), this formula is composed of Huang Lian (*Rhizoma Coptidis*), Hou Pu (*Cortex Magnoliae Officinalis*), Shi Chang Pu (*Rhizoma Acori Tatarinowii*), Zhi Ban Xia (*Rhizoma Pinelliae Praeparatum*), and related herbs. Its key compatibility feature is the combination of bitter-cold Huang Lian to *clear heat and dry dampness* (*qing re zao shi*) with acrid-bitter-warm Hou Pu to *move qi and transform dampness* (*xing qi hua shi*), thereby jointly addressing the two major pathogenic factors of "dampness" and "heat." Modern studies show that Lian Pu Yin not only inhibits *H. pylori* but also significantly reduces gastric tissue inflammatory mediators (e.g., tumor necrosis factor-α and IL-6), demonstrating favorable efficacy for damp-heat-related gastric mucosal inflammation [34].

***Representative Formula for Spleen–Stomach Weakness (pi wei xu ruo) Syndrome: Xiang Sha Liu Jun Zi Tang (香砂六君子汤)***

Originating from the Qing dynasty *Gu Jin Ming Yi Fang Lun* (Formulas of Famous Physicians Past and Present), this prescription is built on *Liu Jun Zi Tang* (Six Gentlemen Decoction) with the addition of *Mu Xiang* (Radix *Aucklandiae*) and *Sha Ren* (Fructus *Amomi*). It centers on "*fu zheng*" (supporting the righteous) with *Ren Shen* (Radix et Rhizoma *Ginseng*) and *Bai Zhu* (Rhizoma *Atractylodis Macrocephalae*), and complements with aromatic qi-regulating herbs (*fang xiang li qi*) such as *Mu Xiang* and *Sha Ren* to "*qu xie*" (dispel the pathogen). Clinical studies show that, when combined with standard quadruple therapy, the *H. pylori* eradication rate can be raised from ~75% to >90%, while significantly reducing adverse events such as abdominal distension and poor appetite [18].

***Representative Formula for Liver–Stomach Disharmony (gan wei bu he) Syndrome: Chai Hu Shu Gan San (柴胡疏肝散)***

Recorded in the Ming dynasty *Yi Xue Tong Zhi* (Comprehensive Medicine), this formula contains *Chai Hu* (Radix *Bupleuri*), *Bai Shao* (Radix *Paeoniae Alba*), *Zhi Ke* (Fructus *Aurantii*), and *Xiang Fu* (Rhizoma *Cyperi*), among others. Its core action is to soothe the liver and relieve constraint (*shu gan jie yu*), restoring the liver's free-coursing function. Clinical research indicates that in *H. pylori*-infected patients with pronounced emotional fluctuations, adjunctive use of *Chai Hu Shu Gan San* significantly improves anxiety and depression scores, increases the *H. pylori* eradication rate, and reduces symptom recurrence [7, 41].

***Representative Formula for Cold–Heat Complex (han re cuo za) Syndrome: Ban Xia Xie Xin Tang (半夏泻心汤)***

A classic formula from Zhang Zhongjing's *Shang Han Lun* (Treatise on Cold Damage), it comprises *Ban Xia* (Rhizoma *Pinelliae*), *Huang Qin* (Radix *Scutellariae*), *Huang Lian* (Rhizoma *Coptidis*), *Gan Jiang* (Rhizoma *Zingiberis Siccatum*), *Ren Shen* (Radix et Rhizoma *Ginseng*), and so on. By employing both cold and warm agents together, it simultaneously attacks the pathogen and supports deficiency (*gong bu jian shi*). Foundational studies suggest that it acts by modulating the gut microbiota and repairing the gastric mucosal barrier [20]. Clinical data indicate potential value for refractory *H. pylori*, with improvements in long-standing symptoms such as epigastric fullness (*pi man*) and belching (*ai qi*) [6].

III. Inheritance and Exploration of Classic Formulas (*jing fang*) and Empirical Formulas (*yan fang*)

Beyond Chinese patent medicines and consensus-recommended formulas, TCM clinical practice preserves a vast, ocean-like corpus of traditional prescriptions—an inexhaustible treasury of therapeutic wisdom. By incomplete estimates, hundreds of thousands of ancient and contemporary formulas exist. Broadly, they fall into two groups: *jing fang* (classic formulas) and *yan fang* (empirical formulas), together representing both the depth and the frontier of individualized TCM therapy.

***Classic Formula (jing fang): Huang Lian Jie Du Tang (黄连解毒汤)***

Originating from Ge Hong's Eastern Jin text *Zhou Hou Bei Ji Fang* (Handbook of Prescriptions for Emergencies), this formula comprises four intensely bitter-cold herbs—*Huang Lian* (Rhizoma *Coptidis*), *Huang Qin* (Radix *Scutellariae*), *Huang Bai* (Cortex *Phellodendri Chinensis*), and *Zhi Zi* (Fructus *Gardeniae*). It typifies the method of clearing heat, purging fire, and resolving toxins (*qing re xie huo jie du*). With a concise yet forceful composition that directly targets the "fire-heat-toxin pathogen" (*huo re du xie*), modern studies show the decoction's water extract has strong in vitro anti-*H. pylori* activity and significantly suppresses *H. pylori*-induced IL-6 secretion in gastric epithelial cells [44]. Clinically, for *H. pylori*-infected patients whose syndrome pattern corresponds to exuberant fire-heat, short courses of this formula can yield favorable outcomes [50].

***Renowned Physician's Empirical Formula (ming jia yan fang/jing yan fang)***

The late National Master of Chinese Medicine Zhu Liangchun frequently and skillfully used *dui yao* (herb pairs) when treating epigastric pain (*weiwan tong*). A core trio he favored combines Di Yu (*Radix Sanguisorbae*), Yan Hu Suo (*Rhizoma Corydalis*), and Hai Piao Xiao (*Endoconcha Sepiae seu Sepiellae*).

- **Hai Piao Xiao** serves as the first choice for neutralizing acid and relieving pain (*zhi suan zhi tong*), helping neutralize gastric acid while astringing and protecting the mucosa.
- **Yan Hu Suo**, lauded as the "holy medicine for analgesia" (*zhi tong sheng yao*), moves qi, invigorates blood, and stops pain (*xing qi huo xue zhi tong*); modern studies also show a strong direct anti-*H. pylori* effect.
- **Di Yu** provides both notable bactericidal activity and immune-enhancing effects.

Used together, the three herbs act synergistically—simultaneously offering direct antibacterial action, symptom improvement, and immune modulation.

IV. **Reflections on Inheritance and Exploration**

*Jing fang* (classic formulas) and *yan fang* (empirical formulas) are invaluable treasures of TCM. Yet their further development faces several challenges. Many formulas in China have not been widely developed into *zhong cheng yao* and remain largely within the domain of individualized modification (*jia jian*) by TCM physicians. In addition, most empirical formulas have a relatively narrow scope of use and lack large-scale clinical validation; their efficacy and safety still require further confirmation. Although some empirical prescriptions can be prepared as "*yuan nei zhi ji*" (in-hospital preparations), their circulation and use are strictly limited and thus cannot benefit a broader patient population. Determining how to validate and develop these high-potential formulas through modern scientific methods is a key task for the future development of TCM.

In summary, Chinese herbal formulas, including Chinese patent medicines, are the core armamentarium of TCM for treating *H. pylori* infection. Their most prominent characteristics lie in their holistic nature and synergistic action: through the scientific compatibility of multiple herbs, they can simultaneously act on multiple targets and pathological links, balancing "*qu xie*" (dispelling pathogenic factors) and "*fu zheng*" (supporting the righteous). While inhibiting the pathogen, they also regulate the body's internal environment and improve clinical symptoms.

However, their complexity and non-standardization (particularly referring to decoctions) also present challenges, such as difficulty in fully elucidating mechanisms of action and significant hurdles in quality control. It is precisely these challenges, coupled with modern medicine's pursuit of precision and standardization, that have jointly given rise to the currently most widely applied model in clinical practice—integrative Chinese and Western medicine therapy (*zhong xi yi jie he zhi liao*).

## 21.5 Clinical Strategies for Integrating Chinese and Western Medicine

Given current clinical realities, standardized antibiotic-based regimens still have an unmatched advantage for directly and rapidly eradicating *H. pylori*. Consequently, in most real-world settings, integrative Chinese–Western therapy has become the mainstream strategy to improve efficacy and patient experience.

In practice, two dominant models are seen:

1. **General hospitals (gastroenterology):** Nationally approved Chinese patent medicines are combined with standardized regimens.
2. **TCM hospitals (spleen–stomach specialties):** Decoctions tailored by syndrome differentiation are favored for more individualized care.

Although reports of the second model often show good outcomes, non-standardized formulations and variable preparation processes limit reproducibility and weaken the evidence base. This chapter synthesizes both models and outlines practical clinical strategies for integrative care.

First, Table 21.3 summarizes representative Chinese patent medicines and formulas with comparatively strong clinical support in integrated therapy.

**Table 21.3** Selected representative formulas/proprietary medicines commonly used in integrated Chinese–Western treatment for *H. pylori*

| Name (English Pinyin) | Category | Synergistic role |
|---|---|---|
| 荆花胃康胶丸 (Jing Hua Wei Kang Jiao Wan) | Proprietary Chinese medicine | Enhance efficacy (*Zeng Xiao*, dispel pathogen), reduce toxicity (*Jian Du*, clear heat) |
| 胃苏颗粒 (Wei Su Ke Li) | Proprietary Chinese medicine | Regulate *qi* and harmonize stomach, and improve motility |
| 养胃舒颗粒 (Yang Wei Shu Ke Li) | Proprietary Chinese medicine | Support righteousness, nourish *Yin*, and repair mucosa |
| 猴头菌片 (Hou Tou Jun Pian) | Proprietary Chinese medicine | Protect gastric mucosa and repair damage |
| 参苓白术散 (Shen Ling Bai Zhu San) | Classic formula (*Shi Fang*) | Enhance efficacy (*Zeng Xiao*, support righteousness), reduce toxicity (*Jian Du*, stop diarrhea) |
| 温脾汤 (Wen Pi Tang) | Classic formula (*Shi Fang*) | Support righteousness and warm *Yang*, dispel cold, and unblock stagnation |
| 黄芪建中汤 (Huang Qi Jian Zhong Tang) | Classic formula (*Jing Fang*) | Support righteousness and consolidate root, repair mucosa |
| 丹参饮 (Dan Shen Yin) | Classic formula (*Shi Fang*) | Improve mucosal blood flow, repair damage |
| 左金丸 (Zuo Jin Wan) | Classic formula (*Shi Fang*) | Soothe liver and purge fire, descend rebellious *qi*, and stop vomiting |
| 良附丸 (Liang Fu Wan) | Classic formula (*Shi Fang*) | Warm middle and dispel cold, move *qi*, and stop pain |

I. **Enhance Efficacy and Reduce Toxicity (*zeng xiao jian du*)**

This is the core and most widely studied model of integrated therapy, aiming to achieve a true "1 + 1 > 2" synergy in both eradication rates and patient experience.

### Synergistic Efficacy Enhancement

Abundant clinical evidence shows that adding Chinese medicine to standard quadruple therapy significantly increases *H. pylori* eradication.

- Chinese patent medicine: A meta-analysis of 10 randomized controlled trials ($n$ = 1181) found a significantly higher *H. pylori* eradication rate in the group receiving quadruple therapy combined with Jinghua Weikang Capsule compared to the quadruple therapy alone group (risk ratio = 1.16, 95% confidence interval 1.10–1.23, $P < 0.00001$) [21].
- Decoction guided by *bian zheng lun zhi*: When Xiang Sha Liu Jun Zi Tang is combined with Western therapy, systematic reviews indicate an additional 10–15% absolute increase in eradication ([11]; [45]).

### Synergistic Toxicity Reduction

Standard quadruple regimens are often accompanied by nausea, diarrhea, and other adverse events; Chinese medicine contributes a clear "detoxifying" advantage.

- Shen Ling Bai Zhu San (with Renshen, Baizhu, and Fuling as core herbs) reduced antibiotic-associated diarrhea from 28.6% to 5% in one clinical study [13].
- Wei Su Ke Li (e.g., Zisu Geng and Xiang Fu) markedly relieves abdominal distension and belching during treatment, with reported overall symptom–response rates exceeding 90% [22].

II. **"Peri-Treatment Period" Management (*wei zhi liao qi guan li*): Full-Course Intervention to Consolidate Efficacy**

The value of TCM lies not only in its synchronous use with antibiotics; more importantly, it embodies the throughout-the-course philosophy of *zhi wei bing* ("preventive treatment of disease").

**Pre-treatment Conditioning ("Battle Preparations")**

For patients with obvious constitutional imbalance, a short course of TCM conditioning (about 1–2 weeks) may be initiated before antibiotic therapy.

- Example (damp-heat internal accumulation, *shi re nei yun*): Give Lian Pu Yin (Huang Lian, Hou Pu, etc.) first to clear and transform damp-heat (*qing hua shi re*).
- Rationale/Evidence: Studies indicate that such formulas can significantly reduce gastric-mucosal inflammatory levels, thereby—*theoretically*—creating the optimal physiological conditions for subsequent antibiotics to exert their maximal effect [49].

**Post-treatment Repair ("Post-Battle Reconstruction")**

After eradication of *H. pylori*, gastric-mucosal injury and spleen–stomach often have not fully recovered. At this stage, applying TCM for "after-care" (*shan hou*) is crucial.

- Example: Huang Qi Jian Zhong Tang (Huang Qi, Gui Zhi, Yi Tang, etc.).
- Clinical findings: Continuing this formula after *H. pylori* eradication can significantly improve pathological scores for gastric-mucosal atrophy and intestinal metaplasia; the overall effective rate has been reported to reach >85%, far exceeding that of control groups [40].

III. **Remedial Treatment After Eradication Failure**

With rising antibiotic resistance, the number of patients who fail eradication therapy continues to grow. A notable clinical phenomenon is that many patients, after aggressive antibiotic regimens fail, exhibit a shift in syndrome pattern (*zheng hou*): the initial presentation dominated by *"shi re"* (damp-heat) often transforms into a *"han re cuo za"* (cold–heat complex) or even *"pi wei xu han"* (spleen–stomach deficiency-cold). In response to this transformation, treatment based on syndrome differentiation (*bian zheng lun zhi*) offers great flexibility.

***Cold–Heat Complex (han re cuo za)***

For patients whose pattern has shifted to a cold–heat complex, reports indicate that remedial therapy can be attempted with Ban Xia Xie Xin Tang, composed of *Ban*

*Xia* (Rhizoma Pinelliae), *Huang Qin* (Radix Scutellariae), *Huang Lian* (Rhizoma Coptidis), *Gan Jiang* (Rhizoma Zingiberis Siccatum), and so on. A study in patients with refractory *H. pylori* infection—those who had experienced multiple prior failures—showed that combining this formula with sensitive antibiotics achieved a remedial eradication rate > 89% [6].

***Spleen–Stomach Deficiency-Cold (pi wei xu han)***
For patients whose pattern has shifted to spleen–stomach deficiency-cold, one may select formulas such as Liang Fu Wan, composed of *Gao Liang Jiang* (Rhizoma Alpiniae Officinarum) and *Xiang Fu* (Rhizoma Cyperi). By warming the middle and dispelling cold (*wen zhong san han*) and moving qi and relieving pain (*xing qi zhi tong*), this approach aims to alter the Stomach's "cold-damp" environment (*han shi*), thereby creating favorable conditions for subsequent eradication therapies against *H. pylori* [38].

IV. **Advantages and Disadvantages of the Integrated Chinese–Western Medicine Strategy**

**Advantages** The integrated strategy combines the precision and potent pathogen-clearing (*qu xie*) effects of Western medicine with the holistic, harmonizing, and righteousness-supporting (*fu zheng*) approach of TCM. This not only increases the eradication rate of *H. pylori* but also improves clinical symptoms and quality of life, making it the most practical and optimal solution for managing the complex condition of *H. pylori* infection at present.

**Disadvantages** At this time, integrated approaches still lack unified, widely accepted standardized guidelines. The evidence level supporting most Chinese patent medicines (*zhong cheng yao*) in combination therapy needs to be strengthened. In addition, decoctions based on syndrome differentiation (*bian zheng lun zhi* tang ji) are inherently individualized, which makes large-scale, standardized, and reproducible validation difficult. Potential interactions between Chinese and Western medicines (e.g., effects on drug metabolism) also require further basic research to clarify their mechanisms.

## 21.6 Conclusion and Outlook

### 21.6.1 Conclusion

This chapter systematically outlined TCM practice against *H. pylori*, closely integrating the theoretical framework from the previous chapter with modern pharmacological research and clinical application. Following a path "from bench to bedside" and "from the simple to the complex," we sketched a panorama of multi-level, multi-strategy TCM responses to *H. pylori* infection.

First, we examined modern research on single Chinese herbs, highlighting the breadth of anti-*H. pylori* work and summarizing the mechanisms of representative herbs such as *Huang Lian* (Rhizoma Coptidis) and *Di Yu* (Radix Sanguisorbae). Pharmacological studies confirm that many single herbs exhibit clear bioactivity; however, most evidence remains preclinical (in vitro or animal models). Their real-world efficacy and safety in humans still require validation through more high-quality clinical trials.

Next, we explored the core of TCM practice—formulas (*fu fang*) and Chinese patent medicines. We discussed them at three levels: modern patent preparations, representative formulas based on syndrome differentiation (*bian zheng lun zhi*), and classic (*jing fang*) versus empirical formulas (*yan fang*), affirming their respective value in standardization, individualization, and heritage-driven exploration. Nonetheless, much of the clinical evidence comes from single-center, small-sample studies. There is an urgent need for rigorously designed, multi-center, large-sample randomized controlled trials to provide higher-grade evidence.

Finally, we focused on the model most widely used in current clinical practice—integrated Chinese–Western medicine (*zhong xi yi jie he*). We analyzed concrete strategies such as "enhancing efficacy and reducing toxicity" (*zeng xiao jian du*) and "peri-treatment period management" (*wei zhi liao qi*). Although clinical benefits are evident, the underlying synergistic mechanisms remain insufficiently defined. Future work should strengthen basic research to elucidate pharmacokinetics, pharmacodynamics, and potential interactions when Chinese and Western medicines are used together, thereby providing a more robust scientific basis for clinical efficacy and safety.

### 21.6.2 Outlook

Building on the current practice and evidence summarized in this chapter, the future of TCM in the anti-*H. pylori* arena is both clear and promising, with work prioritized in the following areas:

1. **From active constituents to novel drugs.** Research on single herbs has already identified multiple compounds with definite anti-*H. pylori* activity (e.g., berberine and ellagic acid). The next step is to optimize these leads and advance pharmacology/toxicology programs so as to deliver new anti-*H. pylori* agents with independent intellectual property, stronger efficacy, and fewer adverse effects.
2. **From empirical know-how to clarified mechanisms.** The multi-component, multi-target nature of Chinese formulas is both an advantage and a research challenge. Multi-omics—systems biology, network pharmacology, and gut microbiome science—should be leveraged to resolve, at molecular levels, the science behind *jun–chen–zuo–shi* compatibility and the synergistic mechanisms of "*zeng xiao jian du*" in integrated TCM–Western therapy.
3. **From basic research to high-level clinical evidence.** A key bottleneck is the relative lack of high-quality clinical data. Rigorously designed, large-sample,

multi-center RCTs that follow international standards are needed—not only to verify the effectiveness of formulas and integrated care models but also to provide the bridge that moves single-herb findings from preclinical benches to clinical application.

4. **From traditional decoctions to modern preparations.** A major task is transforming validated classic and empirical formulas into modern, standardized TCM products—using contemporary manufacturing—to ensure stable quality, convenient use, and compliance with international norms. This will expand clinical reach and help bridge the gap between individualization and standardization, with Chinese patent medicines serving as a crucial vehicle.

## References


1. Cai G. Practical Chinese medicine gastroenterology. Shanghai: Shanghai University of Traditional Chinese Medicine Press; 1996.
2. Chen P, Chen M, Peng C, et al. *In vitro* anti-bacterial activity and its preliminary mechanism of action of the non-medicinal parts of *Sanguisorba officinalis* L. against *H. pylori* infection. J Ethnopharmacol. 2024;318(Pt B):116981.
3. Ding Z. Advances in research on pharmacological effects and clinical application of MoLuodan. China J Pharm Econ. 2018;9:127–9.
4. Dong H, Qiao J, Hou S, et al. Potentialities of dandelion (*Taraxacum Mongolicum* Hand.-Mazz.) flower extracts on gastric protection against *H. pylori* and characterization of its bioactive constituents. Chem Biodivers. 2024;21(6):e202400140.
5. El-Shouny WA, Ali SS, Hegazy HM, et al. *Syzygium aromaticum* L.: traditional herbal medicine against cagA and vacA toxin genes-producing drug resistant *H. pylori*. J Tradit Complement Med. 2020;10(4):366–77.
6. Fang C, Huang Y, Liu L, et al. Clinical study of staged comprehensive treatment of Banxia Xiexin decoction in treating refractory *H. pylori* infection. Acta Chin Med Pharmacol. 2021;49(11):78–82.
7. Guo Z, Long T, Yao J, et al. Potential antidepressant effects of traditional Chinese botanical drug formula Chaihu-Shugan-San and its active ingredients. Front Pharmacol. 2024;15:1337876.
8. Han SH, Hwang JK, Park SN, et al. Potential effect of solvent fractions of *Taraxacum mongolicum* Hand.-Mazz. on protection of gastric mucosa. Korean. J Food Sci Technol. 2005;37(1):84–9.
9. Hu F, Zhang S, Zhang X, et al. Expert consensus on integrative medicine for *H. pylori* related "Diseases and Syndromes" in China (2024 Edition). Chin J Gastroenterol Hepatol. 2024;33(12):1581–8.
10. Jeong SY, Kang DM, Kim HJ, et al. Anti-*H. pylori* and anti-inflammatory Sesquiterpenoids from the Rhizoma of Atractylodes macrocephala. Molecules. 2025;30(15):3142.
11. Jiang H. Effects of Xiangsha Liujunzi decoction combined with conventional quadruple therapy on patients with *H. pylori*-positive duodenal ulcer. Med J Chin People's Health. 2025;37(8):119–22.
12. Jung J, Choi JS, Jeong CS. Inhibitory activities of palmatine from coptis chinensis against *Helicobactor pylori* and gastric damage. Toxicol Res. 2014;30(1):45–8.
13. Lao H. Efficacy observation of Shenling Baizhu powder in treating antibiotic-associated Diarrhea. Shandong J Tradit Chin Med. 2009;28(01):23–4.
14. Lee B, Park IH, Yim D, et al. Comprehensive evaluation of the anti- *H. pylori* activity of Scutellariae radix. Nat Prod Sci. 2017;23(1):46.
15. Lee HA, Hong S, Yoo JH, et al. Anti-*H. pylori* activity and inhibition of gastritis by Magnolia officinalis extract. Thai. J Vet Med. 2018;48(2):203–10.

16. Li C. A clinical study on Moluodan (condensed pill) in treating chronic atrophic gastritis. Paper presented at the 20th National Conference on gastroenterology of Chinese medicine by the spleen and stomach disease branch of China Association of Chinese Medicine, Handan Pharmaceutical Co., Ltd., Handan, 1 September 2008.
17. Li Y, Xu C, Zhang Q, et al. *In vitro* anti-*H. pylori* action of 30 Chinese herbal medicines used to treat ulcer diseases. J Ethnopharmacol. 2005;98(3):329–33.
18. Li S, Liu H, Tang Y. Clinical observation on 48 cases of *H. pylori*-induced peptic ulcer with spleen-stomach deficiency treated with Xiangsha Liujunzi decoction combined with quadruple therapy. J Tradit Chin Med. 2016;57(21):5.
19. Li J, Liu Z, Han X. Expert consensus on the clinical application of Weifuchun for managing gastric precancerous lesions in atrophic gastritis. J Tradit Chin Med. 2023a;64(2):212–6.
20. Li XH, Xu JY, Wang X, et al. BanXiaXieXin decoction treating gastritis mice with drug-resistant *H. pylori* and its mechanism. World J Gastroenterol. 2023b;29(18):2818–35.
21. Liu Q, Hu J, Li B, et al. Jinghuaweikang capsules combined with quadruple therapy in the treatment of *H. pylori* associated gastritis: a systematic evaluation and meta-analysis. Chin J Pharmacovigil. 2023;20(9):1039–45.
22. Liu N, Gao Y, Cui J. Recent advances in the clinical application and research of Weisu granules. J Clin Ration Drug Use. 2024a;17(31):177–80.
23. Liu Y, Liu L, Wang J, et al. Improvement effects of glycyrrhizin on *H. pylori*-associated gastritis in rats and its mechanism. China Pharm. 2024b;35(2):166–71.
24. Ma F, Chen Y, Li J, et al. Screening test for anti-*H. pylori* activity of traditional Chinese herbal medicines. World J of Gastroenterol. 2010;16(44):5629.
25. Otunola GA. Culinary spices in food and medicine: an overview of Syzygium aromaticum (L.) Merr and L M Perry Myrtaceae. Front Pharmacol. 2021;12:793200.
26. Ou L, Hao Y, Liu H, et al. Chebulinic acid isolated from aqueous extracts of *Terminalia chebula* Retz inhibits *H. pylori* infection by potential binding to cag a protein and regulating adhesion. Front Microbiol. 2024a;15:1416794.
27. Ou L, Liu HR, Shi XY, et al. *Terminalia chebula* Retz. aqueous extract inhibits the *H. pylori*-induced inflammatory response by regulating the inflammasome signaling and ER-stress pathway. J Ethnopharmacol. 2024b;320:117428.
28. Peng C, Sang S, Shen X, et al. *In vitro* anti-*H. pylori* activity of *Syzygium aromaticum* and the preliminary mechanism of action. J Ethnopharmacol. 2022;288:114995.
29. Quan T, Zhou F, Chen H, et al. *Ficus hirta* Vahl. Ameliorates nonalcoholic fatty liver disease through regulating lipid metabolism and gut microbiota. Oxid Med Cell Longev. 2022;2022:3474723.
30. Sang S, Zhang C, Lu B. Mechanistic insights and synergistic effects of qi-Fu-Qing-you formulation in combating *H. pylori* infection and maintaining gut microbial homeostasis. J Funct Foods. 2025;128:106824.
31. Shen X, Zhang W, Peng C, et al. *In vitro* anti-bacterial activity and network pharmacology analysis of *Sanguisorba officinalis* L. against *H. pylori* infection. Chin Med. 2021;16(1):33.
32. Shi DH, Liu YW, Liu WW, et al. Inhibition of urease by extracts derived from 15 Chinese medicinal herbs. Pharm Biol. 2011;49(7):752–5.
33. Song D, Hao J, Fan D. Biological properties and clinical applications of berberine. Front Med. 2020;14(5):564–82.
34. Tan Y. Efficacy observation of modified Lianpu Yin in treating *H. pylori*-positive chronic gastritis. J Sichuan Tradit Chin Med. 2016;34(5):142–4.
35. Tang Q, Ma Z, Tang X, et al. Coptisine inhibits *H. pylori* and reduces the expression of CagA to alleviate host inflammation *in vitro* and *in vivo*. J Ethnopharmacol. 2023;314:116618.
36. Wang J, Jin M, Wang H, et al. Compilation of clinical experiences from distinguished veteran practitioners of traditional Chinese medicine. Beijing: China Press Tradit Chin Med; 2006.
37. Wei B, Zhang W, Chen Z. Integrated traditional Chinese and Western medicine gastroenterology. Beijing: People's Med. Publ House; 2003.

38. Wei L, Tian Y, Xia J. Clinical comparative study on Liangfu Wan herbal granules versus traditional decoction in the treatment of epigastric pain. Lishizhen Med Mater Med Res. 2009;20(3):612–4.
39. Wei H, Yan L, Feng W, et al. Research progress on active ingredients and pharmacologic properties of Saussurea lappa. J Asian Nat Prod Res. 2014;10(11):1045–53.
40. Xia D. Effect analysis of Huangqi Jianzhong decoction combined with conventional Western medicine in treating chronic gastritis of spleen-stomach deficiency-cold type. J Chin Foreign Med Pharm Res. 2025;4(07):63–5.
41. Xu Y. Clinical observation on Chaihu Shugan powder combined with standard triple therapy for *H. pylori*-associated chronic gastritis. Hubei: Hubei University of Chinese Medicine; 2012.
42. Yu M, Wang X, Ling F, et al. Atractylodes lancea volatile oils attenuated *H. pylori* NCTC11637 growth and biofilm. Microb Pathog. 2019;135:103641.
43. Yu Z, Sheng WD, Yin X, et al. Coptis, Pinellia, and Scutellaria as a promising new drug combination for treatment of *H. pylori* infection. World J Clin Cases. 2022;10(34):12500–14.
44. Zhang Y, Lei J. Efficacy of modified Huanglian Jiedu decoction in treating *H. pylori*-positive gastric ulcer and its effect on serum inflammatory factor levels. Mod J Integr Tradit Chin West Med. 2017;26(29):3270–1.
45. Zhang Z, Yan H. Efficacy observation on treating 55 cases of gastritis related with *H. pylori* with modified Xiangsha Liujunzi decoction. Clin Res Tradit Chin Med. 2013;5(5):99–100.
46. Zhang Z, Zhu H. Clinical observation on 37 cases of superficial gastritis with Qi stagnation and blood stasis pattern treated with Sanjiu Weitai granules. Mod Dig Interv. 2012;17(2):116–7.
47. Zhang BL, Fan CQ, Dong L, et al. Structural modification of a specific antimicrobial lead against *H. pylori* discovered from traditional Chinese medicine and a structure–activity relationship study. Eur J Med Chem. 2010;45(11):5258–64.
48. Zhang PP, Tang JN, Xiang BY, et al. Unlocking the potential of radix Astragali and its active ingredients in gastric ulcer therapy. J Asian Nat Prod Res. 2025a:1–15.
49. Zhang X, Wang L, Huang L, et al. Potential mechanisms by which Jiawei Lianpu Yin inhibits *H. pylori* colonization and alleviates gastric mucosal inflammation and damage: integrated transcriptomics, network pharmacology, and experimental validation. J Ethnopharmacol. 2025b;348:119793.
50. Zhao X, Li C, Zhang Y, et al. Efficacy observation of Huanglian Jiedu decoction and Sijunzi decoction combined with quadruple therapy for *H. pylori* infection. Pract Clin J Integr Tradit West Med. 2025;25(3):51–3.
51. Zhu Z, Yang J, Shi Z. Clinical efficacy observation of Sanjiu Weitai granules in treating non-atrophic gastritis. Paper presented at the 26th National Academic Conference on digestive system diseases in integrated traditional Chinese and Western medicine, digestive system diseases Committee of Chinese Association of integrative medicine, Guangzhou, 1–4 August 2014
52. Zhu X, Zeng M, Song H, et al. To investigate the molecular mechanism underlying Rhei radix et Rhizoma and Coptidis Rhizoma herb pair against *H. pylori* infection based on core genes and key micrornas. Chin J Integr Tradit West Med Dig. 2021;29(1):36–44.
53. Zou Y, Lu B, Feng Z, et al. Anti *H. pylori* activity and gastrointestinal protective effects of *Terminalia bellirica*: mechanistic insights from *in vitro* and *in vivo* studies. J Ethnopharmacol. 2025;345:119569.

# Natural Product Therapy Against *H. pylori*

22

Yanyan Shi and Meicun Yao

As outlined in the preceding chapters, traditional Chinese medicine (TCM), including single herbal medicine and compound formulas, has accumulated a wealth of effective drug resources and treatment strategies for *Helicobacter pylori* infection in long-term clinical practice. However, the complex composition, variations in quality, and difficulty in fully elucidating the compound mechanism of TCM pose challenges to its modernization and promotion. In this context, the isolation and identification of natural products (NPs) with clear pharmacological activities from TCM has emerged as a pivotal pathway bridging traditional experience with modern drug development.

## 22.1 The Role of NPs and Ethnomedicine

NPs are chemical entities derived from living organisms, including plants, animals, and microorganisms, renowned for their diverse pharmacological properties and critical role in the discovery and development of therapeutic agents. Advancements in research techniques and analytical technologies have further spurred the exploration of NPs, with a significant proportion of currently utilized pharmaceutical agents now being either directly derived from or inspired by NPs.

These remarkable pharmacological properties are rooted in the unique structural characteristics of NPs, which distinguish them from synthetic compounds. For example, NPs typically possess a greater proportion of $sp^3$-hybridized atoms and chiral centers compared with synthetic small-molecule drugs [1]. Moreover, they

Y. Shi (✉)
Peking University Third Hospital, Peking, China
e-mail: shiyanyan@bjmu.edu.cn

M. Yao
School of Pharmaceutical Sciences (Shenzhen), Sun Yat-sen University, Shenzhen, China



J. Gao, Y. Du (eds.), *Non-Antibiotic Therapies for Helicobacter pylori*,
https://doi.org/10.1007/978-981-95-9598-3_22

often exhibit a lower nitrogen content and a higher oxygen content. These structural features confer NPs with scaffold architectures that have been evolutionarily optimized to achieve potent biological activity and high target selectivity. As such, the drug-likeness and synthetic feasibility of NPs markedly differ from those of traditional synthetic compounds, calling for distinct approaches in their development and optimization.

Their relevance is particularly acute in combating *H. pylori*. Current antibiotic therapies are increasingly hindered by rising resistance, notable side effects, patient non-adherence, contraindications, and disruption of the gut microbiome [2, 3]. NPs, as previously outlined, offer a unique combination of exceptional biocompatibility and remarkable structural diversity, allowing them to engage in a wide range of interactions with biological targets, thereby exhibiting potent bioactivity. In the field of antibacterial therapy, NPs combat pathogens by disrupting cell wall integrity, altering membrane permeability, and interfering with critical metabolic processes [4]. These characteristics are especially pertinent when addressing the significant challenge of *H. pylori* eradication, which is crucial for promoting gastric mucosal healing and preventing the progression of peptic diseases. Therefore, NPs offer a promising therapeutic alternative, providing a potential solution to overcome these challenges and enhance the efficacy of eradication treatments.

In this section, traditional ethnomedicine is highlighted as a valuable source for identifying novel anti-*H. pylori* agents. Ethnomedicine refers to the study of natural remedies based on the traditional knowledge and practices of ethnic groups [5], often recorded in historical texts or passed down orally through generations. Rooted in empirical observations and cultural beliefs, ethnomedicine differs from scientific medicine in its approach. While some elements of these practices still await scientific validation, ethnomedicine has long been essential for primary health care in many communities [6, 7], providing a wealth of testable hypotheses and naturally optimized compounds that can contribute to modern drug discovery [8].

## 22.2 The Mechanisms of NPs in *H. pylori* Eradication

NPs are bioactive compounds synthesized by living organisms, typically serving as primary and secondary metabolites in various biochemical pathways. These compounds can be classified into diverse groups, such as alkaloids, flavonoids, saponins, terpenes, and phenols, among others. NPs often serve as lead compounds for the development of novel molecular structures. Through subsequent structural modifications, these compounds can give rise to new drug candidates that retain the core properties of the parent molecules, while offering enhanced efficacy, improved affinity, and fewer side effects.

Primary metabolites serve as fundamental building blocks within organisms and participate directly in vital physiological processes; however, few exhibit antimicrobial properties. Research indicates that certain polysaccharides can inhibit the adhesion of *H. pylori*, an effect attributed to the disruption of adhesin binding to host gastric epithelial cells [9–11]. In contrast, secondary metabolites are bioactive compounds that often possess a broader spectrum of activities and unique mechanisms of action. Consequently, they are regarded as having significant application potential, thus attracting considerable research interest for their pharmacological value. In the combat against *H. pylori*, several key classes of NPs play prominent roles.

Terpenoids are a major class of plant-derived secondary metabolites composed of five-carbon isoprene units ($C_5H_8$). Their structural diversity arises primarily from variations in the number of isoprene units they contain. Based on this structural characteristic, terpenoids are classified into various categories, including: monoterpenes ($C_{10}$), sesquiterpenes ($C_{15}$), diterpenes ($C_{20}$), diheterospermas ($C_{25}$), triterpenes ($C_{30}$), tetraterpenes ($C_{40}$), and polycyclic terpenes ($C_n$, $n > 40$) [12, 13]. As widely distributed compounds in plants, terpenoids have been identified for their broad range of biological activities, such as anticancer, anti-inflammatory, antibacterial, antiviral, antioxidant, and immunomodulatory effects.

Equally important are polyphenols, a major class of plant secondary metabolites, which are characterized by aromatic rings with one or more hydroxyl groups. Widely present in fruits, vegetables, and grains, they are categorized into phenolic acids, flavonoids, and other nonflavonoids based on the structure of their aglycones. Their multiple hydroxyl groups enable potent ROS scavenging and antioxidant effects [14].

A third major class with significant anti-*H. pylori* potential is alkaloids, a structurally diverse class of nitrogen-containing NPs. They are widely distributed across more than 300 plant families, with over 10,000 compounds identified to date. They exhibit a broad spectrum of pharmacological activities, including antimalarial, anticancer, antibacterial, anti-inflammatory, and hypoglycemic effects. Among medicinal alkaloids, compounds such as berberine, coptisine, and various quinolone-type alkaloids have shown promising antimicrobial potential against *H. pylori* [15]. While current research has largely focused on berberine and coptisine, the anti-*H. pylori* mechanisms of many other alkaloids remain insufficiently studied.

Despite the ongoing challenges in NP development and clinical translation, NPs remain a significant source of oral therapeutics, often surpassing conventional drug development frameworks such as Lipinski's Rule of Five. This indicates their potential to address complex biological targets, including those involved in *H. pylori* eradication [16, 17]. NPs sourced from traditional medicine are extensively regarded as active components for combating *H. pylori* infection, and innovative research is vigorously advancing anti-*H. pylori* drug discovery. This advanced understanding of the pathogen's biological characteristics and disease-causing mechanisms, gained through decades of investigation, has positioned ethnomedicine as a cornerstone for identifying efficacious NPs and alternative therapies [18], particularly for antibiotic-resistant cases. Some patented medicines have been included in clinical guidelines as adjuvant treatments. Consequently, in countries where traditional medicine is prevalent, systems such as TCM and Kampo are now recommended treatment options [19, 20]. The following section summarizes recent advances in the study of secondary metabolites for eradicating *H. pylori*.

### 22.2.1 Direct Antibacterial Activity

The TCM principles of "*clearing heat and detoxifying*" and "*expelling pathogenic factors*" align closely with modern pharmacological findings, particularly in relation to direct antibacterial mechanisms. NPs that operate through these principles represent a key strategy for combating *H. pylori*, as they primarily target the bacterium itself—either inflicting lethal structural damage to the cell envelope or disrupting essential metabolic pathways, ultimately resulting in bacterial death or growth inhibition. Bitter, cold-natured herbs such as *Coptis chinensis* and *Phellodendron amurense* exemplify this mechanism by compromising bacterial membrane integrity. Their major bioactive constituents, particularly berberine from *Coptis chinensis*, are among the most extensively studied compounds with well-established antibacterial properties. Representative NPs exhibiting direct antibacterial activity are summarized in Table 22.1.

Monoterpenoids, a prominent subclass of terpenoids composed of two isoprene units, are major constituents of plant essential oils (e.g., pine, lemon, and thyme) and exhibit direct antibacterial activity in vitro, often accompanied by gastroprotective effects in vivo [21]. Among them, eugenol stands out as one of the most potent natural compounds against *H. pylori* due to *its* antimembrane activity, with a minimum inhibitory concentration (MIC) ranging from 23.0 to 51.0 μg/mL [22]. Similarly, carvone (MIC50/90 31.3/62.5 mg/L) found in essential oils exhibits inhibitory effects without inducing acquired resistance [23, 24]. It is hypothesized that certain volatile compounds can penetrate bacterial cells and induce extensive cellular damage, thereby contributing to their antimicrobial properties. Thymol and carvacrol, classified as terpenoid phenols, display stronger efficacy with MICs of 64–128 and 16–64 μg/mL, respectively [25]. Their primary antimicrobial mechanism involves membrane disruption: their hydrophobicity and phenolic hydroxyl groups enable insertion into the bacterial membrane, increasing permeability, causing ion leakage, and impairing (Adenosine Triphosphate) ATP synthesis [26]. Sesquiterpenoids also demonstrate direct antibacterial effects, as observed in compounds such as atractylenolide I and III, which suppress bacterial growth at levels comparable to metronidazole [27]. Among triterpenoids, triterpene glycosides (saponins) show notable anti-*H. pylori* activity. Glycyrrhizin, a major saponin from *Glycyrrhiza glabra*, acts through its metabolite glycyrrhetinic acid to rapidly inhibit *H. pylori* growth in vitro [28]. However, prolonged intake is associated with an increased risk of gastric mucosal damage and therefore requires careful evaluation [29].

Phenolic acids, a major class of nonflavonoid polyphenols, are categorized into benzoic acid and cinnamic acid derivatives. Several dietary nonflavonoid polyphenols, including resveratrol in grapes and red wine and ellagitannins in berries and nuts, demonstrate health benefits. Syringic and cinnamic acids specifically inhibit both *H. pylori* growth and $H^+/K^+$-ATPase activity. Monomeric hydrolyzable tannins such as tellimagrandin I and II display strong antibacterial activity through membrane disruption [30]. Ellagitannins also show direct anti-*H. pylori* effects with MIC values below 25 μM, as exemplified by castalagin and vescalagin from *Castanea sativa Mill.* leaves [31].

**Table 22.1** Representative NPs with direct anti-*H. pylori* activity

| Compounds name | Natural herb source | MIC (μg/mL) | MBC (μg/mL) | *H. pylori strains* | Mechanism | Research phase |
|---|---|---|---|---|---|---|
| Thymol 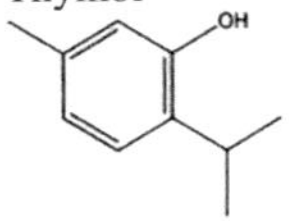 | *Thymus kotschyanus* | 64–128 | / | Eight strains of *H. pylori* (one reference and seven clinical isolates) | Ability of pH hemostasis modulation ↓;<br>Damage *H. pylori*;<br>Membrane integrity;<br>ATP synthesis ↓;<br>Ability of biofilm production ↓. | In vitro;<br>In vivo: animal models; limited human pilot studies |
| Carvacrol 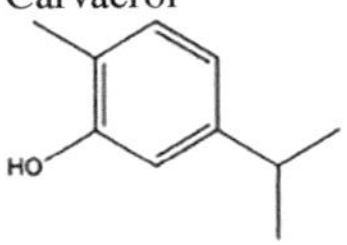 | *Origanum vulgare* | 16–64 | 32–64 | Nine strains of *H. pylori* (one reference strain and eight clinical isolates) | Ability of pH hemostasis modulation ↓;<br>Damage *H. pylori*;<br>Membrane integrity;<br>ATP synthesis↓;<br>Ability of biofilm production↓. | In vitro;<br>In vivo: animal models. |
| Eugenol 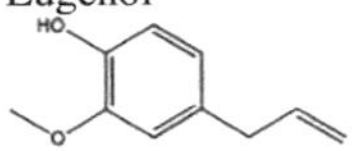 | *Eugenia caryophillis* | 23.0–51.0 | / | *H. pylori* isolates and standard strain NCTC 11637 | Antibiofilm ↓;<br>Gene expression of virulent factors ↓; | In vitro;<br>In vivo: animal models. |
| Glycyrrhetinic acid 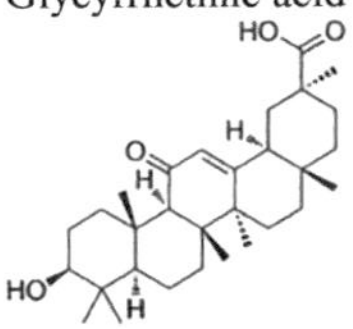 | *Glycyrrhiza glabra* | 50 | / | 29 different *H. pylori* strains | Inhibition of HMGB1;<br>Prevention of bacterial attachment to gastric epithelial cells;<br>Anti-inflammatory effects | In vitro;<br>In vivo: animal models |

(continued)

**Table 22.1** (continued)

| Compounds name | Natural herb source | MIC (μg/mL) | MBC (μg/mL) | *H. pylori strains* | Mechanism | Research phase |
|---|---|---|---|---|---|---|
| Tellimagrandin 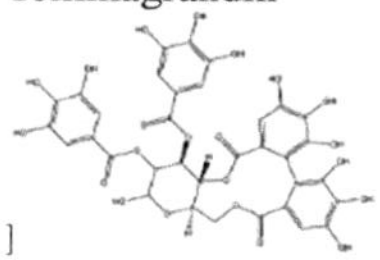 ] | *Cornus canadensis* | 12.5 | / | NCTC 11638, ATCC 43504, A-13, and A-19 | Damage *H. pylori*;<br>Membrane integrity | In vitro;<br>In vivo: animal models |
| Coptisine 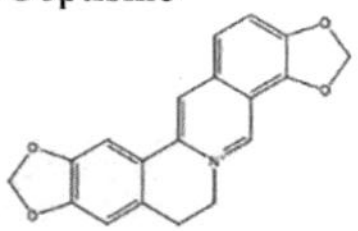 | *Coptis chinensis* | 16–32 | 32–64 | ATCC 43504, 700392, and PMSS1 | Urease ↓;<br>Cag expression ↓;<br>DNA fragmentation ↑;<br>Phosphatidylserine exposure ↑;<br>Membrane stability ↓ | In vitro;<br>In vivo: animal models |
| Berberine 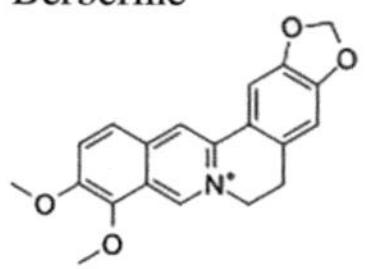 | *Coptis chinensis*/*Phellodendron amurense* | 16–40 | 40–80 | NCTC 26695 | Antibiofilm ↓;<br>Immunomodulation↑;<br>Gastric mucosal protection. | In vitro;<br>In vivo: Animal models;<br>Clinical trial |

In addition, alkaloids exert direct antibacterial effects through multitarget mechanisms. Compounds such as coptisine and squalamine compromise bacterial membrane integrity, potentially facilitating the entry of conventional antibiotics and enhancing the efficacy of combination therapies. Berberine possesses both inhibitory and bactericidal properties against *H. pylori*. Its mechanism is distinct from that of standard antibacterial drugs, characterized by a synergistic activity when used alongside antibiotics that are primarily effective against the bacterium in the logarithmic growth phase, which has been supported by clinical evidence.

### 22.2.2 Inhibition of Virulence and Pathogenesis

Beyond direct antibacterial activity, alternative therapeutic strategies align more closely with the TCM concept of "*transforming dampness and turbidity*," which focuses on eliminating the pathogenic environment in the stomach and thereby preventing *H. pylori* colonization. Herbs that exemplify this principle—such as *Magnolia officinalis* and *Atractylodes macrocephala*—have demonstrated the ability to disrupt key virulence factors of the bacterium. Bioactive compounds from these herbs, including magnolol, are known to inhibit urease activity and interfere with bacterial adhesion. By targeting essential virulence mechanisms such as urease function and adhesion pathways, these NPs hinder the bacterium's ability to establish itself within the host and initiate tissue damage. Thus, suppression of virulence and adhesion represents an important therapeutic approach for preventing *H. pylori* colonization and its associated pathological effects (Table 22.2).

Monoterpenoids, including thymol and carvacrol, inhibit bacterial carbonic anhydrase, consequently disrupting pH homeostasis and biofilm formation [25]. Eugenol also reduces virulence by inhibiting the expression of related factors [32]. In the same way, components of cedarwood oil (e.g., α-cedrene and cedrol) suppress urease activity [33]. The sesquiterpenoid patchouli alcohol (PA) enhances cellular defense by activating Transcription Factor EB (TFEB) to promote lysosomal function. It additionally displays strong synergy with clarithromycin, attributed to its inhibition of bacterial efflux pumps. Separate evidence indicates that its antibacterial property may also involve DNA polymerase inhibition, as demonstrated in *Streptococcus mutans* [34], providing new insight into the potent activity of sesquiterpenoids.

Flavonoids, comprising more than 9000 identified compounds characterized by a core structure of two benzene rings linked by a heterocyclic ring, exhibit direct antibacterial effects partly through urease inhibition [35]. For example, quercetin, a widely consumed flavonoid with strong antioxidant capacity, inhibits *H. pylori* through noncompetitive urease inhibition, with structure–activity analyses, indicating that its hydroxyl groups are essential for this effect. Similarly, epigallocatechin, which is a major catechin in green tea, shares this urease-inhibiting mechanism. Notably, both compounds retain their inhibitory activity under anaerobic conditions, highlighting their potential relevance in the gastric environment. Myricetin, a well-known antivirulence compound widely distributed in plants such as bayberry

**Table 22.2** Representative NPs that inhibit *H. pylori* virulence and pathogenesis

| Compounds name | Natural source | MIC (μg/mL) | MBC (μg/mL) | *H. pylori strains* | Mechanism | Research phase |
|---|---|---|---|---|---|---|
| β-Pinene 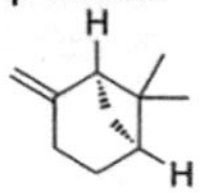 | *Picea asperata* | 500 | / | ATCC 43504 | Urease ↓;<br>Affect cell permeability | In vitro;<br>In vivo: animal models |
| Quercetin 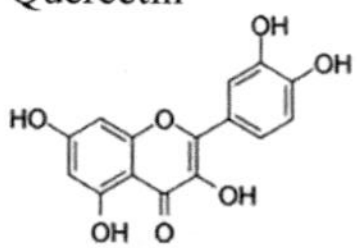 | Tea; apple; broccoli; tomato; onion; grape; berries | 25.5–51 | 42.5–68 | Nine strains of *H. pylori* | Urease ↓;<br>Cell membrane stability ↓ | In vitro;<br>In vivo: animal models |
| 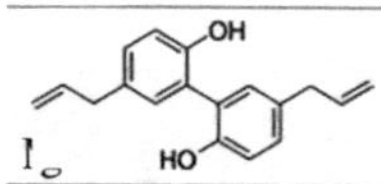 | *Magnolia officinalis* | 10–40 | / | Six strains of *H. pylori* | Vac A ↓. | In vitro |
| Myricetin 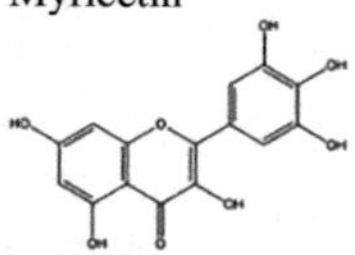 | *Myrica* | 160 | 320 | ATCC 700824, 51932 | Immune system detection ↑;<br>Antibiotic sensitivity ↑;<br>Biofilm↓ | In vitro |
| Sanguinarine 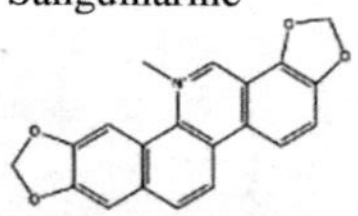 | *Zanthoxylum nitidum* | 6.25–50 | / | A total of 14 clinical isolates | Urease↓;<br>Cell lysis | In vitro;<br>In vivo: animal models |
| Epiberberine 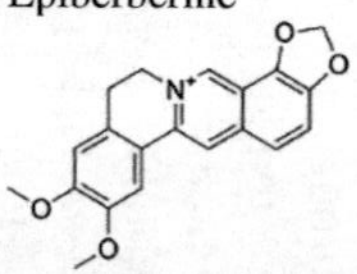 | *Rhizoma Coptidis* | 64–100 | / | ATCC 43504, 26,695, and PMSS1 | Inhibit urease activity;<br>UreB expression ↓ | In vitro;<br>In vivo: animal models |

bark, vine tea leaves and stems, onions, and various berries, has been shown to inhibit the morphological transition of *H. pylori* [36]. Luteolin and its derivatives (e.g., luteolin 5-O-β-glucoside) also show strong urease inhibitory and antiulcer effects, although their precise mechanisms remain elusive. According to reports, magnolol extracted from *Magnolia officinalis* exerts antibacterial effects by inhibiting the formation and secretion of the *H. pylori* virulence factor vacuolating cytotoxin A (VacA). Collectively, flavonoids combat *H. pylori* through multiple mechanisms, including inhibition of key bacterial enzymes (e.g., urease and HpPDF), disruption of membrane integrity, and interference with bacterial adhesion [37].

Urease inhibition is also a common anti-virulence mechanism among alkaloids. Sanguinarine, isolated from *Zanthoxylum nitidum*, suppresses urease activity by chelating $Ni^{2+}$ ions and interacting with thiol groups, as evidenced by reduced activity in the presence of sulfhydryl reagents or competitive $Ni^{2+}$ binders. Likewise, coptisine targets both the $Ni^{2+}$ active site and thiol residues of urease while also reducing cytotoxin-associated gene A (CagA) expression and inducing DNA breakage in *H. pylori*. Chalcones and their derivatives interfere with bacterial–epithelial adhesion [38], whereas anthocyanins reduce bacterial attachment and increase antibiotic sensitivity without harming gastric cells. Epiberberine, a constituent of *Rhizoma Coptidis*, attenuates host cell apoptosis and inflammation by inhibiting urease activity and downregulating UreB expression.

### 22.2.3 Immunomodulation and Anti-Inflammatory Effects

The therapeutic intent of "*strengthening and consolidating body resistance*" and "*strengthening the spleen and replenishing qi*" in TCM is systematically embodied in the immunomodulatory strategy against *H. pylori* infection. This principle is frequently implemented through the application of herbs renowned for their tonifying effects, with *Glycyrrhiza uralensis* acting as a prominent representative, its bioactive extracts being particularly noted for their immunomodulatory effects. Rather than directly targeting the bacterium or its virulence factors, this approach modulates the host's inflammatory response, thereby mitigating infection-induced tissue damage. By attenuating excessive inflammation and enhancing endogenous defense mechanisms, it provides a modern pharmacological interpretation of "*supporting the right qi*" in TCM. Representative NPs with demonstrated immunomodulatory and anti-inflammatory activities are summarized in Table 22.3.

Monoterpenoid compounds regulate oxidative stress and inflammatory responses in gastric mucus, demonstrated by their ability to decrease pro-inflammatory cytokines (e.g., Myeloperoxidase (MPO), nuclear factor (NF)-κB, tumor necrosis factor (TNF)-α, interleukin (IL)-6, and IL-1β) while increasing the anti-inflammatory cytokine IL-10 [39]. The sesquiterpenoid PA alleviates *H. pylori* urease-induced damage in gastric epithelial cells by reducing apoptosis, mitochondrial dysfunction, and inflammation [40]. Recent studies indicate that the triterpenoid glycyrrhizin restores lysosomal function by inhibiting HMGB1 (High-Mobility Group Box 1), modulating inflammation and ROS (Reactive Oxygen Species) production to counteract *H. pylori* infection [41]. This lysosomal acidification capacity highlights its

**Table 22.3** Representative NPs with immunomodulatory and anti-inflammatory activities

| Compounds name | Natural source | MIC (μg/mL) | MBC (μg/mL) | *H. pylori strains* | Mechanism | Research phase |
|---|---|---|---|---|---|---|
| Resveratrol 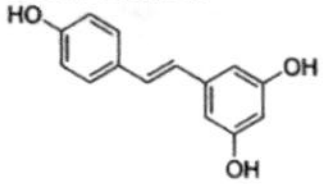 | *Veratrum grandiflorum* | 64 | / | ATCC 26695 | Inhibit type F1 ATPase, ribosomal function, and outer membrane protein (OMP) expression;<br>Inhibit multiple inflammatory factors and pathways. | In vitro;<br>In vivo: Animal models |
| β-carotene 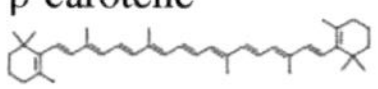 | Orange-colored fruits and vegetables | / | / | / | ROS levels ↓;<br>Inflammatory signaling inhibition | In vitro;<br>In vivo: *animal models* |
| Astaxanthin 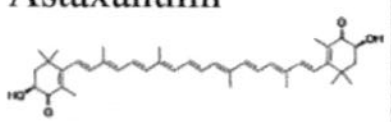 | Microalgae, fungi, complex plant, and crustaceans | / | / | / | Inhibit multiple inflammatory factors and pathways | In vitro;<br>In vivo: animal models |
| Kaempferol 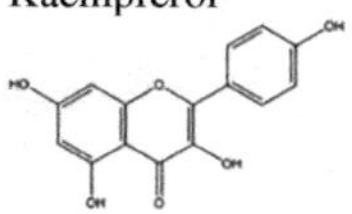 | *Kaempferol galanga* | >600 | / | ATCC 43504 and 26,695 | CagA and VacA transport to host cells ↓;<br>Stability of membrane ↓;<br>Interfere with key metabolic pathways | In vitro;<br>In vivo: animal models |
| Chalcone 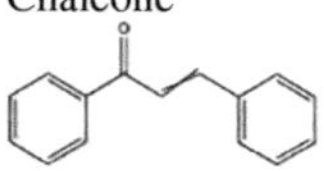 | Species of the *Leguminosae, Asteraceae, and Moraceae* families | 12.5–50 | / | 17 *H. pylori* strains including clinical ones | Interactions between *H. pylori* and gastric epithelium ↓;<br>Motility ↓;<br>Urease ↓ | In vitro;<br>In vivo: animal models |
| Curcumin 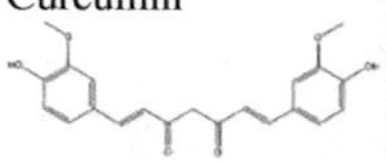 | *Turmeric* | 30 | 40 | ATCC 700392 | Urease ↓;<br>Cell membrane stability ↓;<br>Inhibition of virulence factors;<br>Immunomodulation ↑. | In vitro;<br>In vivo: animal models;<br>Clinical trial (effect on functional dyspepsia and peptic ulcer) |

potential as an autophagy inducer. Despite these benefits, licorice consumption may lead to glucocorticoid-like effects, electrolyte imbalances, edema, and hypertension, necessitating strict safety evaluation when developing anti-*H. pylori* products containing licorice extracts. Tetraterpenoids, characterized by a $C_{40}$ structure, are known for their robust antioxidant properties. Carotenoids such as β-carotene, astaxanthin, and lycopene are noted for their potent antioxidant properties and therapeutic potential in *H. pylori*-related gastric disorders [42]. For instance, β-carotene suppresses ROS production by inhibiting (Nicotinamide Adenine Dinucleotide Phosphate) NADPH oxidase and enhancing antioxidant enzyme activity, while downregulating (Inducible Nitric Oxide Synthase) iNOS and COX-2 (Cyclooxygenase-2) expression. These actions collectively block ROS-mediated inflammatory pathways, including (Mitogen-Activated Protein Kinase) MAPK and NF-κB signaling [43–45], yet β-carotene's role in anti-*H. pylori* remains unconfirmed.

Astaxanthin, a xanthophyll derived from marine crustaceans, exhibits approximately tenfold stronger antioxidant capacity than β-carotene. In (A subline of the BALB/c inbred mouse strain, widely used in biomedical research) BALB/cA mice, astaxanthin-rich diets reduced *H. pylori* colonization and gastric inflammation [46–48]. It attenuates oxidative stress and suppresses ROS, shifting Th1/Th2 immune responses toward enhanced bacterial clearance [49, 50]. Additionally, astaxanthin inhibits JAK1/STAT3 (Janus Kinase 1 / Signal Transducer and Activator of Transcription 3) activation and integrin α5-mediated cell adhesion and migration in gastric epithelial cells [51].

Kaempferol is a flavonoid and polyphenolic compound found in a wide range of edible plants, such as broccoli, strawberries, onions, and papaya, as well as in various medicinal herbs. It exhibits anti-inflammatory activity by inhibiting the translocation of the CagA and VacA proteins, thereby downregulating pro-inflammatory cytokine expression [52]. This effect is attributed to its ability to downregulate the transcription of components of the type IV secretion system. Certain nonflavonoid polyphenols suppress IL-8 release via inhibition of the NF-κB pathway. Resveratrol possesses broad-spectrum antibacterial activity against *H. pylori* [53]. It modulates multiple targets, including sirtuins, NF-κB, COX-2, and proinflammatory cytokines (TNF-α, IL-1β, and IL-6), and reduces *H. pylori*-induced biomarkers such as IL-8, iNOS, and Nrf2 [54]. However, its clinical application is constrained by low oral bioavailability, prompting the development of novel delivery systems. Apigenin reduces *H. pylori*-induced inflammation in MKN45 cells by suppressing NF-κB activation [55]. In silico data revealed the potential of anti-*H. pylori*, anti-inflammatory, and MMP-9 inhibition for the chalcones.

Curcumin significantly decreases *H. pylori* colonization in mice by lowering MPO and urease activity while enhancing immune regulation [56]. Clinical trials demonstrate that curcumin as adjuvant therapy improves dyspepsia symptoms, reduces oxidative stress, and increases eradication rates [57]. It synergistically targets Ureα/β subunits to inhibit bacterial survival. Given its poor solubility and bioavailability, chitosan-based delivery systems hold promise for enhancing its therapeutic efficacy.

Beyond direct antibacterial effects, alkaloids exert substantial anti-inflammatory effects. They reduce proinflammatory cytokines (e.g., IL-6, IL-17, and CXCL1 (C-X-C Motif Chemokine Ligand 1)), modulate macrophage activity, and inhibit NF-κB signaling. Clinical evidence supports the translational potential of alkaloids. Meta-analyses confirm that berberine as an adjunct to standard triple therapy significantly improves eradication rates, promotes ulcer healing, and reduces adverse events [58]. Moreover, newly developed berberine derivatives have been developed that target OMP transport, increase bacterial membrane permeability, or interact directly with *H. pylori* DNA [59, 60], highlighting their promise as multi-target anti-*H. pylori* candidates with favorable safety profiles.

### 22.2.4 Gastric Mucosal Protection and Repair

The TCM principle of "*strengthening body resistance*" is supported by modern pharmacological findings showing that certain NPs can restore and reinforce the gastric mucosal barrier. Herbs traditionally used for this purpose, such as *Astragalus membranaceus* and *Atractylodes macrocephala*, are particularly noteworthy. For example, atractylone—a sesquiterpene oxide derivative found in the essential oils of *Atractylodes macrocephala*—has been reported to promote effective repair of gastric mucosal injury. These NPs enhance both the structural and functional recovery of the gastric mucosa, thereby strengthening endogenous defense mechanisms and mitigating damage caused by *H. pylori* and other pathogenic factors. This mechanism provides a modern biological basis for the TCM concept of "*strengthening body resistance*" and further highlights the therapeutic importance of natural medicines in maintaining gastric homeostasis (Table 22.4).

Key components of *Citrus lemon* (Rutaceae) essential oil, such as limonene and β-pinene, are recognized for their anti-*H. pylori* properties. In vitro studies have shown that these compounds possess antibacterial effects against *H. pylori*. Notably, limonene promotes mucus secretion by maintaining optimal levels of prostaglandin E2 ($PGE_2$), a mechanism that helps inhibit *H. pylori* colonization and enhances the gastric mucosa's resistance to ulcer-inducing factors [61]. Additionally, β-pinene promotes mucus production and alleviates oxidative stress and inflammation, thereby offering protective effects on the gastric mucosa [39]. A comparable mechanism is observed in α-pinene, which protects the gastric mucosa by modulating oxidative stress, $PGE_2$ levels, and histamine production [62]. β-Myrcene exhibits anti-*H. pylori* activity (MIC: 500 μg/mL) [63] and provides complementary gastroprotection by strengthening the gastric antioxidant defense system, as evidenced by increased glutathione peroxidase, glutathione reductase, and total glutathione levels [64]. Atractylone, a sesquiterpene isolated from the rhizomes of *Atractylodes macrocephala*, has been shown to attenuate *H. pylori*-induced gastric mucosal injury in zebrafish. It also reduces oxidative stress-related gastric damage and alleviates ulcer formation, in part by modulating and reshaping the gut microbiota [65]..

Phenolic acids, including gallic acid, syringic acid, cinnamic acid, catechin, and chlorogenic acid, exhibit gastroprotective effects against *H. pylori*. Beyond direct

**Table 22.4** Representative NPs with gastric mucosal protective and repair effects

| Compounds name | Natural source | MIC (μg/mL) | MBC (μg/mL) | *H. pylori strains* | Mechanism | Research phase |
|---|---|---|---|---|---|---|
| Limonene 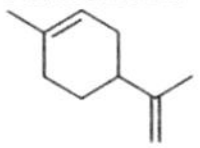 | *Citrus lemon* essential oils | 75 | / | ATCC 43504 | Damage bacterial membrane; Gastric mucosal defense; Regulate the balance of inflammatory factors | In vitro; In vivo: animal models |
| α-pinene 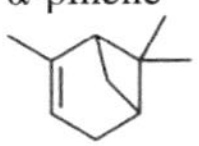 | *Citrus* fruit; Edible essential oils | / | 100 | P1 | Modulation of oxidative stress and $PGE_2$ and histamine levels | In vitro; In vivo: animal models; Clinical trial |
| Atractylone I 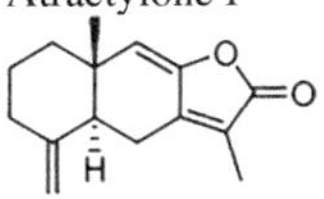 | *Atractylodes macrocephala* | 3.13–12.5 | 54.9–94.3 | Strain 51, 26,695, and 43,504 | Gastric mucosal protection; Anti-inflammatory | In vitro; In vivo |
| β-Myrcene 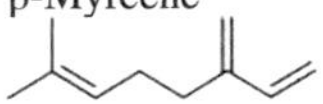 | *Pinus sylvestris var. Mongolica*; Microorganism | 500 | / | ATCC 43504 | Gastroprotection; Urease ↓ | In vitro |
| Palmatine 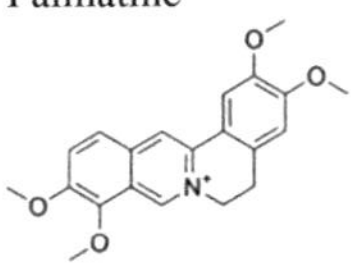 | *Phellodendron amurense* | 50 | / | NCTC 26695 and SS1 | Gastric mucosal protection; Urease expression and activity↓ | In vitro; In vivo: Animal models |
| | | 32–64 | 64–128 | ATCC 43504, 700,392, and PMSS1 | | |
| Jatrorrhizine 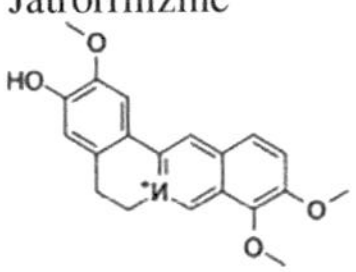 | *Tinospora sagittata* (Oliv.) Gagnep; *Coptis chimensis* Franch | 100 | / | NCTC 11637 | Inhibit multiple inflammatory factors and pathways; Elevate protective gastric mucosal factors | In vitro |

antibacterial effects, alkaloids exert gastric protective actions on the gastric mucosa. Jatrorrhizine, sharing the same protoberberine molecular skeleton with berberine, suppresses *H. pylori*-induced inflammation by inhibiting the NLRP3 inflammasome and NF-κB pathway while elevating protective gastric mucosal factors.

Palmatine, a plant-derived isoquinoline alkaloid first isolated from *Coptidis Rhizoma*, has been proposed as a potential non-antibiotic therapeutic agent capable of attenuating (Matrix Metalloproteinase 10) MMP-10-dependent inflammatory responses in the gastric mucosa by blocking (A Disintegrin and Metalloproteinase 17 / Epidermal Growth Factor Receptor) ADAM17/EGFR signaling, thereby alleviating *H. pylori*-induced histological damage [66].

Carveol, which contains a p-menthane skeleton, is primarily found in the essential oils of *Cymbopogon giganteus*, *Illicium pachyphyllum*, and *Carum carvi*. It is also present in the essential oils of plants from the Lamiaceae and Labiatae families. An in vitro study reported a MIC to be 0.6 mg/mL, although the specific mechanism underlying its potent antibacterial activity remains to be elucidated [67].

## 22.3 Clinical Evidence and Research Status of Anti-*H. pylori* NPs

The translation of NPs from traditional remedies or laboratory discoveries into clinically validated therapies is a multi-stage process. This progression spans from initial antimicrobial screening and mechanistic studies to large-scale clinical trials, with each phase generating increasingly rigorous evidence. A systematic evaluation of their bioactivity, mechanisms, and developmental status across structural classes is therefore essential to identify lead compounds with genuine clinical potential and recognize major developmental challenges.

The systematic profiling of representative NPs highlights both the potential and challenges in developing anti-*H. pylori* therapies. Terpenoids such as thymol, carvacrol, and eugenol exhibit multi-target antibacterial activity, while certain monoterpenes such as limonene and α-pinene demonstrate complementary gastroprotective and anti-inflammatory effects. Nevertheless, most terpenoids remain at early research stages with limited clinical evidence. Polyphenols act through both direct antibacterial activity and modulation of host inflammatory responses. Curcumin, one of the few NPs to enter clinical trials, has been studied mainly for symptomatic relief in functional dyspepsia and peptic ulcers rather than as an eradication agent. However, poor bioavailability and chemical instability impede their clinical advancement. Among the structural classes, alkaloids have progressed furthest toward clinical application. Berberine, the most extensively studied, has entered clinical trials as an adjunct to conventional antibiotics, improving eradication rates and reducing adverse effects.

In summary, the development of NPs against *H. pylori* remains largely preclinical, with limited clinical trial data and narrow outcome measures that fall short of fully validating their therapeutic potential. Advancement beyond the preclinical stage requires overcoming fundamental limitations in physicochemical properties.

A major translational bottleneck lies in bridging the gap between in vitro activity and in vivo efficacy, often due to poor stability, solubility, and oral bioavailability in the gastric environment. Research on traditional compound formulations faces additional challenges, such as the complexity of "syndrome differentiation" theory and the multi-component, multi-target nature of these remedies, which complicate standardization, mechanistic studies, and regulatory approval. Consequently, the therapeutic value of many NPs may lie not in monotherapy but in synergistic combinations with antibiotics, underscoring the need to establish optimal dosing and avoid antagonism.

Given that combining NPs with antibiotics shows synergistic potential, further rigorous investigation is required to clarify their molecular interactions, long-term safety, and optimal clinical protocols in order to improve eradication rates and mitigate antibiotic resistance. Moving forward, efforts should prioritize lead compounds with favorable pharmacokinetic profiles and advance them through well-designed preclinical and clinical studies. Modern technological approaches will be essential for accelerating their transition from laboratory discovery to clinical application.

## 22.4 Translational Challenges and Future Prospects of NPs

The structural complexity and evolutionary optimization of NPs make them an indispensable resource for drug discovery. Although intact NP fragments are rarely found in clinical drugs, they serve as the structural foundation for most core molecular scaffolds in pharmaceuticals. Notably, these scaffolds exhibit a markedly lower risk of CYP450-mediated interactions and hepatotoxicity compared to their synthetic counterparts ($P < 0.001$) [68, 69] and are increasingly represented across clinical trial phases, underscoring their high translational potential. Crucially, their ability to interact with multiple cellular targets simultaneously significantly lowers the risk of resistance development, which is a critical advantage in combating drug-resistant *H. pylori*.

The exploration of NPs against *H. pylori* is powerfully guided by ethnomedicine. Herbal formulations have historically been used to treat conditions such as gastritis, dyspepsia, and peptic ulcers, all of which are now recognized as being associated with *H. pylori* infection [70–72]. This underscores the importance of NPs in combating *H. pylori* and the need to assess their anti-*H. pylori* properties within traditional treatments. Such evaluations could identify potentially effective medicinal plants that might otherwise be overlooked [70]. Consequently, an ethnobotanical approach focused on plants used for gastric disorders increases the likelihood of discovering active compounds. Research increasingly shows that the effectiveness of these remedies often stems from synergistic interactions within their complex mixtures, a strategy that aligns with the need to tackle antibiotic resistance.

To translate this potential into therapies, innovative methodologies are being employed. Innovative methodologies and a shift toward precision and sustainability are revitalizing the field of NP research for *H. pylori* treatment. In recent years, high-throughput screening has rapidly identified active lead compounds from extensive libraries, while molecular biology techniques are clarifying their complex mechanisms. Bioinformatics, including spanning genomics, proteomics, and metabolomics, along with artificial intelligence, is increasingly applied in biosynthesis and drug innovation. Another promising direction involves the valorization of agricultural and industrial waste. For instance, terpenoids have been isolated from unused parts of *Phoenix dactylifera*. Repurposing such waste streams lowers production costs and environmental impacts and promotes sustainable and cost-effective sourcing.

However, the clinical translation of potent NPs is hindered by inherent limitations such as poor solubility, low bioavailability, and nonspecific targeting. To address these challenges, novel delivery systems are being developed. For example, terpene-based therapeutic eutectic solvents disrupt bacterial membrane integrity via microfluidic technology, offering a sustainable strategy for targeted anti-*H. pylori* therapy [73]. Similarly, fucoidan-modified, quercetin-loaded glycyrrhizic acid nanovesicles (Fu-GaLip@KP) enhance intracellular bacterial clearance while repairing gastric mucosa and alleviating inflammation [41]. In vivo studies show that such nanoparticles achieve efficacy comparable to triple therapy and help restore gastrointestinal homeostasis. Overall, nanoparticle delivery systems, chemical modifications, and bio-enhancers endow new formulations with improved mucosal penetration, biofilm disruption, and sustained release, potentially enabling dose reduction and fewer side effects.

Despite these advances, significant obstacles remain. Ethnomedicine, grounded in syndrome differentiation, lacks a universally accepted disease-pattern classification, impeding the standardization of NPs and mechanistic studies. Bridging this gap will require integrating clinical experience and expert consensus to define syndrome patterns in *H. pylori*-related diseases. Moreover, the transition from extracts to viable drug candidates suffers from high attrition due to unpredictable pharmacokinetics or toxicity. Although NP–antibiotic combinations improve eradication rates and mitigate resistance, their synergistic mechanisms, pharmacodynamic interactions, and long-term safety require further investigation. Variability in the source, composition, and extraction methods of NPs also poses challenges for quality consistency and large-scale production. There is an urgent need to establish standardized, rigorous evaluation protocols to ensure reproducible efficacy and quality across studies.

Future efforts should adopt an integrated strategy to translate theoretical promise into clinical reality. In-depth mechanistic studies of compounds and their combinations will provide a foundation for rational drug design, biomarker identification, and validation of combination therapies. More multicenter, double-blind, randomized controlled trials are essential to confirm the safety and efficacy of natural formulations against *H. pylori*. Research should also systematically optimize the combination of NPs and antibiotics, determine the synergistic ratio, administration sequence, and fixed-dose formula to eradicate the disease to the greatest extent while minimizing drug resistance and adverse reactions. Finally, leveraging artificial intelligence and bioinformatics will accelerate development, deepen mechanistic understanding, and advance the application of NPs in treating *H. pylori*-related diseases.

## References


1. Henkel T, Brunne RM, Müller H, Reichel F. Statistical investigation into the structural complementarity of natural products and synthetic compounds. Angew Chem Int Ed Engl. 1999;38(5):643–7.
2. Wang L, Li Z, Tay CY, Marshall BJ, Gu B. Multicentre, cross-sectional surveillance of *H. pylori* prevalence and antibiotic resistance to clarithromycin and levofloxacin in urban China using the string test coupled with quantitative PCR. Lancet Microbe. 2024;5(6):e512–e3.
3. Rocha GR, Lemos FFB, Silva LGO, Luz MS, Correa Santos GL, Rocha Pinheiro SL, et al. Overcoming antibiotic-resistant *H. pylori* infection: current challenges and emerging approaches. World J Gastroenterol. 2025;31(10):102289.
4. Zhao Y, Wei J, Li C, Ahmed AF, Liu Z, Ma C. A comprehensive review on mechanism of natural products against Staphylococcus aureus. J Future Foods. 2022;2(1):25–33.
5. Lai J, Tang Y, Yang F, Chen J, Huang FH, Yang J, et al. Targeting autophagy in ethnomedicine against human diseases. J Ethnopharmacol. 2022;282:114516.
6. Buenz EJ, Verpoorte R, Bauer BA. The ethnopharmacologic contribution to bioprospecting natural products. Annu Rev Pharmacol Toxicol. 2018;58:509–30.
7. Deng R, Chen X, Zhao S, Zhang Q, Shi Y. The effects and mechanisms of natural products on *H. pylori* eradication. Front Cell Infect Microbiol. 2024;14:1360852.
8. Süntar I. Importance of ethnopharmacological studies in drug discovery: role of medicinal plants. Phytochem Rev. 2020;19(5):1199–209.
9. Gottesmann M, Goycoolea FM, Steinbacher T, Menogni T, Hensel A. Smart drug delivery against *H. pylori*: pectin-coated, mucoadhesive liposomes with antiadhesive activity and antibiotic cargo. Appl Microbiol Biotechnol. 2020;104(13):5943–57.
10. Menchicchi B, Hensel A, Goycoolea FM. Polysaccharides as bacterial antiadhesive agents and "smart" constituents for improved drug delivery systems against *H. pylori* infection. Curr Pharm Des. 2015;21(33):4888–906.
11. Loke MF, Lui SY, Ng BL, Gong M, Ho B. Antiadhesive property of microalgal polysaccharide extract on the binding of *H. pylori* to gastric mucin. FEMS Immunol Med Microbiol. 2007;50(2):231–8.
12. Hoshino Y, Villanueva L. Four billion years of microbial terpenome evolution. FEMS Microbiol Rev. 2023;47(2)
13. Khan MSA, Khundmiri SUK, Khundmiri SR, Al-Sanea MM, Mok PL. Fruit-derived polysaccharides and terpenoids: recent update on the Gastroprotective effects and mechanisms. Front Pharmacol. 2018;9:569.
14. Bonacorsi C, Raddi MS, da Fonseca LM, Sannomiya M, Vilegas W. Effect of Byrsonima crassa and phenolic constituents on *H. pylori*-induced neutrophils oxidative burst. Int J Mol Sci. 2012;13(1):133–41.
15. Wen J, Wu S, Ma X, Zhao Y. Zuojin pill attenuates *H. pylori*-induced chronic atrophic gastritis in rats and improves gastric epithelial cells function in GES-1 cells. J Ethnopharmacol. 2022;285:114855.
16. Chopra B, Dhingra AK. Natural products: a lead for drug discovery and development. Phytother Res. 2021;35(9):4660–702.
17. Doak BC, Over B, Giordanetto F, Kihlberg J. Oral druggable space beyond the rule of 5: insights from drugs and clinical candidates. Chem Biol. 2014;21(9):1115–42.
18. Vale FF, Oleastro M. Overview of the phytomedicine approaches against *H. pylori*. World J Gastroenterol. 2014;20(19):5594–609.
19. Zhou L, Lu H, Song Z, Lyu B, Chen Y, Wang J, et al. Chinese national clinical practice guideline on *H. pylori* eradication treatment. Chin Med J (Engl). 2022;135(24):2899–910.
20. Miwa H, Nagahara A, Asakawa A, Arai M, Oshima T, Kasugai K, et al. Evidence-based clinical practice guidelines for functional dyspepsia 2021. J Gastroenterol. 2022;57(2):47–61.
21. Rozza AL, Moraes Tde M, Kushima H, Tanimoto A, Marques MO, Bauab TM, et al. Gastroprotective mechanisms of citrus lemon (Rutaceae) essential oil and its majority com-

pounds limonene and β-pinene: involvement of heat-shock protein-70, vasoactive intestinal peptide, glutathione, sulfhydryl compounds, nitric oxide and prostaglandin $E_2$. Chem Biol Interact. 2011;189(1–2):82–9.
22. Elbestawy MKM, El-Sherbiny GM, Moghannem SA. Antibacterial, Antibiofilm and anti-inflammatory activities of eugenol clove essential oil against resistant *H. pylori*. Molecules. 2023;28(6)
23. Piasecki B, Korona-Głowniak I, Kiełtyka-Dadasiewicz A, Ludwiczuk A. Composition and anti-*H. pylori* properties of essential oils obtained from selected mentha cultivars. Molecules. 2023;28(15)
24. Ohno T, Kita M, Yamaoka Y, Imamura S, Yamamoto T, Mitsufuji S, et al. Antimicrobial activity of essential oils against *H. pylori*. Helicobacter. 2003;8(3):207–15.
25. Grande R, Carradori S, Puca V, Vitale I, Angeli A, Nocentini A, et al. Selective inhibition of *H. pylori* carbonic anhydrases by Carvacrol and Thymol could impair biofilm production and the release of outer membrane vesicles. Int J Mol Sci. 2021;22(21)
26. Ultee A, Bennik MH, Moezelaar R. The phenolic hydroxyl group of carvacrol is essential for action against the food-borne pathogen Bacillus cereus. Appl Environ Microbiol. 2002;68(4):1561–8.
27. Jeong SY, Kang DM, Kim HJ, Yeon SW, Lee HH, Kim MH, et al. Anti-*H. pylori* and anti-inflammatory Sesquiterpenoids from the Rhizoma of Atractylodes macrocephala. Molecules. 2025;30(15)
28. Wittschier N, Faller G, Hensel A. Aqueous extracts and polysaccharides from liquorice roots (Glycyrrhiza glabra L.) inhibit adhesion of *H. pylori* to human gastric mucosa. J Ethnopharmacol. 2009;125(2):218–23.
29. Périco LL, Heredia-Vieira SC, Beserra FP, de Cássia Dos Santos R, Weiss MB, Resende FA, et al. Does the gastroprotective action of a medicinal plant ensure healing effects? An integrative study of the biological effects of Serjania marginata Casar. (Sapindaceae) in rats. J Ethnopharmacol. 2015;172:312–24.
30. Funatogawa K, Hayashi S, Shimomura H, Yoshida T, Hatano T, Ito H, et al. Antibacterial activity of hydrolyzable tannins derived from medicinal plants against *H. pylori*. Microbiol Immunol. 2004;48(4):251–61.
31. Piazza S, Martinelli G, Fumagalli M, Pozzoli C, Maranta N, Giavarini F, et al. Ellagitannins from Castanea sativa Mill. leaf extracts impair *H. pylori* viability and infection-induced inflammation in human gastric epithelial cells. Nutrients. 2023;15(6)
32. Hu Q, Zhou M, Wei S. Progress on the antimicrobial activity research of clove oil and eugenol in the food antisepsis field. J Food Sci. 2018;83(6):1476–83.
33. Korona-Glowniak I, Glowniak-Lipa A, Ludwiczuk A, Baj T, Malm A. The in vitro activity of essential oils against *H. pylori* growth and urease activity. Molecules. 2020;25(3)
34. Takao Y, Kuriyama I, Yamada T, Mizoguchi H, Yoshida H, Mizushina Y. Antifungal properties of Japanese cedar essential oil from waste wood chips made from used sake barrels. Mol Med Rep. 2012;5(5):1163–8.
35. González A, Casado J, Lanas Á. Fighting the antibiotic crisis: flavonoids as promising antibacterial drugs against *H. pylori* infection. Front Cell Infect Microbiol. 2021;11:709749.
36. Krzyżek P, Migdał P, Paluch E, Karwańska M, Wieliczko A, Gościniak G. Myricetin as an Antivirulence compound interfering with a morphological transformation into coccoid forms and potentiating activity of antibiotics against *H. pylori*. Int J Mol Sci. 2021;22(5)
37. Widelski J, Okińczyc P, Suśniak K, Malm A, Bozhadze A, Jokhadze M, et al. Correlation between chemical profile of Georgian Propolis extracts and their activity against *H. pylori*. Molecules. 2023;28(3)
38. Michalkova R, Kello M, Cizmarikova M, Bardelcikova A, Mirossay L, Mojzis J. Chalcones and gastrointestinal cancers: experimental evidence. Int J Mol Sci. 2023;24(6)
39. de Souza MC, Vieira AJ, Beserra FP, Pellizzon CH, Nóbrega RH, Rozza AL. Gastroprotective effect of limonene in rats: influence on oxidative stress, inflammation and gene expression. Phytomedicine. 2019;53:37–42.

40. Xie J, Lin Z, Xian Y, Kong S, Lai Z, Ip S, et al. (−)-Patchouli alcohol protects against *H. pylori* urease-induced apoptosis, oxidative stress and inflammatory response in human gastric epithelial cells. Int Immunopharmacol. 2016;35:43–52.
41. Zhang S, Ma S, Hao S, Pan J, Li Y, Yuan G, et al. Fucoidan-modified antibiotic-free nanovesicles: a multidimensional approach to eradicate intracellular and extracellular *H. pylori* and restore gastrointestinal homeostasis. Int J Biol Macromol. 2025;307(Pt 2):141786.
42. Kim JH, Lee J, Choi IJ, Kim YI, Kwon O, Kim H, et al. Dietary carotenoids intake and the risk of gastric cancer: a case-control study in Korea. Nutrients. 2018;10(8)
43. Jang SH, Lim JW, Kim H. Beta-carotene inhibits *H. pylori*-induced expression of inducible nitric oxide synthase and cyclooxygenase-2 in human gastric epithelial AGS cells. J Physiol Pharmacol. 2009;60(Suppl 7):131–7.
44. Park Y, Lee H, Lim JW, Kim H. Inhibitory effect of β-Carotene on *H. pylori*-induced TRAF expression and hyper-proliferation in gastric epithelial cells. Antioxidants (Basel). 2019;8(12)
45. Bae S, Lim JW, Kim H. β-Carotene inhibits expression of matrix Metalloproteinase-10 and invasion in *H. pylori*-infected gastric epithelial cells. Molecules. 2021;26(6)
46. Bennedsen M, Wang X, Willén R, Wadström T, Andersen LP. Treatment of *H. pylori* infected mice with antioxidant astaxanthin reduces gastric inflammation, bacterial load and modulates cytokine release by splenocytes. Immunol Lett. 1999;70(3):185–9.
47. Naguib YM. Antioxidant activities of astaxanthin and related carotenoids. J Agric Food Chem. 2000;48(4):1150–4.
48. Wang X, Willén R, Wadström T. Astaxanthin-rich algal meal and vitamin C inhibit *H. pylori* infection in BALB/cA mice. Antimicrob Agents Chemother. 2000;44(9):2452–7.
49. Kim SH, Lim JW, Kim H. Astaxanthin inhibits mitochondrial dysfunction and Interleukin-8 expression in *H. pylori*-infected gastric epithelial cells. Nutrients. 2018;10(9)
50. Davinelli S, Melvang HM, Andersen LP, Scapagnini G, Nielsen ME. Astaxanthin from shrimp cephalothorax stimulates the immune response by enhancing IFN-γ, IL-10, and IL-2 secretion in Splenocytes of *H. pylori*-infected mice. Mar Drugs. 2019;17(7)
51. Woo J, Lim JW, Kim H. Astaxanthin inhibits integrin α5 expression by suppressing activation of JAK1/STAT3 in *H. pylori*-stimulated gastric epithelial cells. Mol Med Rep. 2023;27(6)
52. Yeon MJ, Lee MH, Kim DH, Yang JY, Woo HJ, Kwon HJ, et al. Anti-inflammatory effects of Kaempferol on *H. pylori*-induced inflammation. Biosci Biotechnol Biochem. 2019;83(1):166–73.
53. Trautmann D, Suazo F, Torres K, Simón L. Antitumor effects of resveratrol opposing mechanisms of *H. pylori* in gastric cancer. Nutrients. 2024;16(13)
54. Spósito L, Morais-Silva G, Fonseca D, Neves MM, Vieira Silva M, Bauab TM, et al. Nano-in-microparticles approach: targeted gastric ulcer therapy using trans-resveratrol nanoparticles encapsulated in hyaluronic acid and alginate microparticles. Int J Biol Macromol. 2025;305(Pt 2):141010.
55. Wang YC, Huang KM. In vitro anti-inflammatory effect of apigenin in the *H. pylori*-infected gastric adenocarcinoma cells. Food Chem Toxicol. 2013;53:376–83.
56. De R, Kundu P, Swarnakar S, Ramamurthy T, Chowdhury A, Nair GB, et al. Antimicrobial activity of curcumin against *H. pylori* isolates from India and during infections in mice. Antimicrob Agents Chemother. 2009;53(4):1592–7.
57. Judaki A, Rahmani A, Feizi J, Asadollahi K, Hafezi Ahmadi MR. Curcumin in combination with triple therapy regimes ameliorates oxidative stress and histopathologic changes in chronic gastritis-associated *H. pylori* infection. Arq Gastroenterol. 2017;54(3):177–82.
58. Li Z, Wang Y, Xu Q, Ma J, Li X, Yan J, et al. Berberine and health outcomes: an umbrella review. Phytother Res. 2023;37(5):2051–66.
59. Guo X, Zhao X, Lu X, Zhao L, Zeng Q, Chen F, et al. A deep learning-driven discovery of berberine derivatives as novel antibacterial against multidrug-resistant *H. pylori*. Signal Transduct Target Ther. 2024;9(1):183.
60. Zhang X, Wang G, Kuang W, Xu L, He Y, Zhou L, et al. Discovery and evolution of berberine analogues as anti-*H. pylori* agents with multi-target mechanisms. Bioorg Chem. 2024;151:107628.

61. Moraes TM, Kushima H, Moleiro FC, Santos RC, Rocha LR, Marques MO, et al. Effects of limonene and essential oil from Citrus aurantium on gastric mucosa: role of prostaglandins and gastric mucus secretion. Chem Biol Interact. 2009;180(3):499–505.
62. Al-Sayed E, Gad HA, El-Kersh DM. Characterization of four piper essential oils (GC/MS and ATR-IR) coupled to Chemometrics and their anti-*H. pylori* activity. ACS. Omega. 2021;6(39):25652–63.
63. Wei A, Shibamoto T. Antioxidant/lipoxygenase inhibitory activities and chemical compositions of selected essential oils. J Agric Food Chem. 2010;58(12):7218–25.
64. Bonamin F, Moraes TM, Dos Santos RC, Kushima H, Faria FM, Silva MA, et al. The effect of a minor constituent of essential oil from Citrus aurantium: the role of β-myrcene in preventing peptic ulcer disease. Chem Biol Interact. 2014;212:11–9.
65. Li L, Du Y, Wang Y, He N, Wang B, Zhang T. Atractylone alleviates ethanol-induced gastric ulcer in rat with altered gut microbiota and metabolites. J Inflamm Res. 2022;15:4709–23.
66. Chen X, Wang R, Bao C, Zhang J, Zhang J, Li R, et al. Palmatine ameliorates *H. pylori*-induced chronic atrophic gastritis by inhibiting MMP-10 through ADAM17/EGFR. Eur J Pharmacol. 2020;882:173267.
67. Aslam H, Khan AU, Qazi NG, Ansari SF, Minhas AM, Alvi AM, et al. Plausible role of carveol in gastrointestinal disorders: mechanistic insights via H(+)/K(+)-ATPase and voltage-gated calcium channel inhibition. Biochem Biophys Res Commun. 2025;758:151615.
68. Bharate SB, Lindsley CW. Natural products driven medicinal chemistry. J Med Chem. 2024;67(23):20723–30.
69. Cragg GM, Newman DJ, Snader KM. Natural products in drug discovery and development. J Nat Prod. 1997;60(1):52–60.
70. Dinat S, Orchard A, Van Vuuren S. A scoping review of African natural products against gastric ulcers and *H. pylori*. J Ethnopharmacol. 2023;301:115698.
71. Takagi K, Harada M. Pharmacological studies on herb paeony root. II. Anti-inflammatory effect, inhibitory effect on gastric juice secretion, preventive effect on stress ulcer, antidiuretic effect of paeoniflorin and combined effects with licorice component Fm 100. Yakugaku Zasshi. 1969;89(7):887–92.
72. Escobedo-Hinojosa WI, Del Carpio JD, Palacios-Espinosa JF, Romero I. Contribution to the ethnopharmacological and anti-*H. pylori* knowledge of Cyrtocarpa procera Kunth (Anacardiaceae). J Ethnopharmacol. 2012;143(1):363–71.
73. Brzeziński M, Chmiela M, Picchio M, Calderón M, Gonciarz W. Terpene-based eutectic solvent microdroplets: a strategy to combat antibiotic-resistant *H. pylori*. Langmuir. 2025;41(23):14915–24.

# Part VI

# Innovative Non-antibiotic Therapies and Clinical Translation

# Phage Therapy for *Helicobacter pylori* 23

Jinlong Ru, Jinling Xue, Raquel Mejías Luque,
Markus Gerhard, and Li Deng

## 23.1 Introduction and Rationale for Phage Therapy

### 23.1.1 The Clinical Challenge of *Helicobacter pylori* in the Era of Antibiotic Resistance

*Helicobacter pylori*, a Gram-negative, spiral-shaped bacterium, is uniquely adapted to chronically colonize the human gastric epithelium. It is one of the most successful human pathogens, infecting an estimated 50% of the global population, with prevalence rates exceeding 80% in some developing nations [1]. While many infections remain asymptomatic, chronic gastritis is a universal consequence among infected individuals. A significant subset, estimated at 10–20% of those infected,

J. Ru · J. Xue
Technical University of Munich, School of Life Sciences, Freising, Germany

TUM Central Institute of Infection Prevention (ZIP), Freising, Germany

Institute of Virology, Helmholtz Centre Munich—German Research Centre for Environmental Health, Neuherberg, Germany

R. M. Luque · M. Gerhard
TUM School of Medicine and Health, Department Preclinical Medicine, Institute of Medical Microbiology, Immunology and Hygiene, Technical University of Munich, Munich, Germany

German Center for Infection Research (DZIF), Partner Site Munich, Munich, Germany

L. Deng (✉)
Technical University of Munich, School of Life Sciences, Freising, Germany

TUM Central Institute of Infection Prevention (ZIP), Freising, Germany

Institute of Virology, Helmholtz Centre Munich—German Research Centre for Environmental Health, Neuherberg, Germany

German Center for Infection Research (DZIF), Munich, Germany
e-mail: li.deng@tum.de


J. Gao, Y. Du (eds.), *Non-Antibiotic Therapies for Helicobacter pylori*,
https://doi.org/10.1007/978-981-95-9598-3_23

will develop more severe pathologies, including peptic ulcer disease [2]. More alarmingly, *H. pylori* is the primary etiological agent for the majority of non-cardia gastric cancers and is classified as a Group 1 carcinogen by the International Agency for Research on Cancer [3]. It is also the leading risk factor for gastric mucosa-associated lymphoid tissue (MALT) lymphoma, a malignancy that, in some cases, can enter complete remission following bacterial eradication alone [2].

The pathogenesis of *H. pylori* is complex, driven by a sophisticated arsenal of virulence factors that enable its survival and persistence in the hostile gastric lumen. Its potent urease enzyme hydrolyzes urea to produce ammonia, creating a neutralized microenvironment around the bacterium to buffer against gastric acid [1]. Motility, provided by flagella, allows it to penetrate the viscous gastric mucus layer and adhere to epithelial cells. Key virulence factors such as the cytotoxin-associated gene A and the vacuolating cytotoxin A are directly involved in host cell damage, disruption of cellular processes, and the induction of chronic inflammation. Moreover, *H. pylori* forms biofilms on gastric surfaces; these structured microbial communities enhance persistence, reduce antimicrobial penetration, and contribute to treatment recalcitrance [4].

Standard eradication treatment for *H. pylori* consists of complex multidrug regimens, typically administered for 10–14 days [3]. These regimens combine a potent acid-suppressing agent, such as a proton pump inhibitor (PPI) or a potassium-competitive acid blocker (P-CAB), with two or three antibiotics chosen from a limited arsenal including clarithromycin, amoxicillin, metronidazole, and tetracycline. However, the foremost challenge in managing *H. pylori* is the escalating crisis of antibiotic resistance. The efficacy of standard therapies has been severely compromised, with eradication rates that should ideally exceed 90% having fallen to as low as 57–75% in many regions [3]. Resistance to cornerstone antibiotics is now widespread. In the United States, for example, resistance rates are approximately 31% for clarithromycin and 42% for metronidazole [3]. This therapeutic "arms race" against a highly adaptable pathogen underscores the urgent and critical need for fundamentally new treatment approaches.

The clinical and public-health implications are substantial. Empiric multidrug regimens impose high pill burdens and complex dosing schedules that challenge adherence, increase adverse events (e.g., antibiotic-associated diarrhea and taste disturbance), and disrupt the gut microbiome, which are costs that accumulate with repeated attempts at eradication. Drug–drug interactions (notably with macrolides and metronidazole), contraindications in specific populations (e.g., pregnancy, pediatrics, and comorbidities), and variable access to susceptibility testing complicate individualized care. Although guideline bodies increasingly advocate susceptibility-guided therapy where feasible, diagnostic access and turnaround are uneven, and resistance continues to outpace incremental modifications of antibiotic cocktails [5].

Collectively, these trends argue for a strategic pivot: complementing and, where possible, substituting antibiotic approaches with alternative and adjunctive modalities that (i) bypass established resistance mechanisms, (ii) function within the unique gastric microenvironment (acid, mucus, biofilm), (iii) minimize collateral microbiome damage, and (iv) integrate with stewardship principles. Candidate nonantibiotic strategies span immuno- and vaccine-based approaches, targeted microbiome modulation, mucolytics/biofilm disruptors, anti-adhesive agents, and bacteriophage therapy. Phages offer intrinsic specificity, the potential for self-amplification at the site of infection, modular engineering to overcome mucus/biofilm barriers, and combinability with sub-therapeutic antibiotic backbones to restore efficacy. In the context of rising antimicrobial resistance (AMR) and stagnating eradication rates, systematically developing such alternatives is no longer optional but essential to achieve durable, safe, and globally scalable control of *H. pylori* infection [1–4].

### 23.1.2 Bacteriophages as a Targeted "Living" Antibiotic

Bacteriophage (phage) therapy, an approach that utilizes viruses to specifically target and eliminate pathogenic bacteria, has re-emerged as a compelling alternative to conventional antibiotics. Phages are the natural predators of bacteria and are the most abundant biological entities on Earth [6]. Their therapeutic potential resides primarily in their lytic replication cycle, a ruthlessly efficient process of bacterial destruction. The cycle begins when a lytic phage specifically recognizes and attaches to receptor molecules on the surface of a target bacterium. The phage then injects its genetic material, hijacking the host cell's metabolic and replicative machinery to synthesize hundreds of new phage components. These components self-assemble into progeny virions, which are ultimately liberated through the action of phage-encoded enzymes, such as holins and endolysins, that degrade the bacterial cell wall from within, causing the cell to lyse [7].

This mechanism confers several key theoretical advantages over traditional antibiotics. First, their high target specificity minimizes disruption to the host's commensal microbiota, thereby reducing the risk of collateral damage such as antibiotic-associated diarrhea or secondary opportunistic infections [7]. Furthermore, their ability to self-replicate in situ allows a small initial dose to multiply exponentially where susceptible host bacteria are present, a stark contrast to conventional antibiotics, which are consumed during their action [6]. Finally, because their mechanism of killing is fundamentally different from that of any antibiotic, phages are often effective against bacteria that have developed resistance to multiple classes of conventional drugs, making them a vital tool in the fight against antimicrobial resistance [8].

### 23.1.3 Chapter Objectives

The growing crisis of antibiotic-resistant *H. pylori* provides a powerful rationale for exploring phage therapy. However, the application of this strategy to a pathogen residing in the unique and challenging niche of the human stomach is still in its infancy [2]. The objective of this chapter is therefore to provide a comprehensive and critical assessment of the current state of *H. pylori* phage therapy. We will review the landscape of isolated and characterized phages, summarize the existing preclinical evidence from in vitro studies, and, most importantly, dissect the significant biological, technical, and environmental hurdles that currently impede its translation into a viable clinical therapy. By examining both the promise and the peril, this chapter aims to offer a realistic perspective on the future directions required to harness the potential of bacteriophages against this major global pathogen.

## 23.2 Current State-of-the-Art of *H. pylori* Phage Research

The exploration of bacteriophages as a therapeutic modality for *H. pylori* is a field in its nascent stages, characterized by a small but growing body of research. Progress has been methodical, reflecting the significant technical challenges associated with isolating and propagating phages for a fastidious, microaerophilic bacterium like *H. pylori* [2]. The current landscape is defined by the successful isolation of a handful of candidate phages, promising but preliminary in vitro efficacy data, and an overarching need for in vivo validation. A notable feature of the field is the apparent predominance of temperate (lysogenic) phages found in association with *H. pylori*, with strictly lytic phages proving more elusive [2].

### 23.2.1 Isolation and Characterization of Candidate Phages

The search for therapeutic phages against *H. pylori* has spanned diverse environments, from clinically relevant sources such as patient gastric biopsies to environmental reservoirs such as wastewater and human feces. The process involves meticulous laboratory work using standard characterization techniques, including plaque assays to quantify lytic activity, transmission electron microscopy to determine morphology, and host range analysis to assess the breadth of their activity. The limited number of characterized phages can be broadly categorized into those with demonstrated lytic cycles and the more commonly found temperate phages.

The pursuit of strictly lytic phages, the preferred candidates for therapy, has yielded a few key isolates. For instance, the phage Hpφ was isolated directly from human gastric biopsies, a finding of significant clinical relevance as it confirms the presence of active phages at the site of infection. Detailed characterization revealed its lytic parameters, including a latent period of approximately 45 min and a burst size of around 80 plaque-forming units (PFUs) per infected cell, providing quantitative measures of its bactericidal efficiency [9]. In contrast to clinical sources, environmental reservoirs have also proven fruitful. The lytic phages ΦHPE1 and ΦHPE2 were isolated from wastewater and represent two distinct morphological families. ΦHPE1 is a member of the *Podoviridae* family, with a characteristic icosahedral head and a short tail, while ΦHPE2 belongs to the *Siphoviridae* family, possessing a longer, non-contractile tail. This morphological diversity suggests that *H. pylori* is susceptible to different types of phages, which could be advantageous for developing phage cocktails [10].

A significant, and perhaps defining, feature of *H. pylori* phage biology is the prevalence of temperate phages, which are often found as dormant prophages integrated into the genomes of clinical isolates [2]. While their lysogenic potential raises safety concerns for therapy, their study has provided invaluable insights. Among the most thoroughly characterized is HPy1R, a temperate podovirus induced from a clinical *H. pylori* strain using UV radiation. Its 31.2 kbp linear dsDNA genome was fully sequenced, and crucially, in silico analysis confirmed the absence of any known virulence or antibiotic resistance genes, a critical safety checkpoint for any therapeutic candidate [11]. Another well-studied temperate phage, KHP30, highlights the morphological novelty within this group. It is a spherical, lipid-containing phage, representing a new phage type. Interestingly, despite possessing an integrase gene, KHP30 is maintained as an episome (an extrachromosomal plasmid) within its host, suggesting an alternative lysogenic strategy to chromosomal integration [12]. Other temperate phages, such as 1961P [13] and PhiHp33 (isolated from a MALT lymphoma patient strain) [2], have further illustrated the role of prophages in driving the genetic diversity and evolution of *H. pylori*, for instance, through processes like generalized transduction.

The key characteristics of these representative phages are summarized in Table 23.1.

Of particular therapeutic relevance is the remarkable acid tolerance observed in phages such as KHP30 [12] and HPy1R [11]. Their ability to remain viable at pH levels as low as 2.5–3.0 is a significant finding, strongly suggesting a degree of co-evolution with their acidophilic host. This inherent stability is a major advantage for oral therapy and could potentially reduce the complexity of protective formulations required to survive gastric transit, a challenge faced by phages isolated from pH-neutral environments.

### 23.2.2 Summary of In Vitro Efficacy

The foundational evidence supporting the potential of phage therapy against *H. pylori* comes exclusively from in vitro studies. These experiments have focused on three critical parameters: lytic activity, stability under simulated gastric conditions, and host range.

The lytic phage Hpφ demonstrated potent antimicrobial activity in cell culture models. When combined with lactoferrin and hydroxyapatite nanoparticles, its efficacy was enhanced fourfold. This formulation led to a significant reduction in *H. pylori* colonization of AGS human gastric adenocarcinoma cells and a mitigation of the associated inflammatory response, all without exhibiting cytotoxicity to the human cells [9].

The temperate phages have also shown surprising in vitro promise. HPy1R was able to suppress *H. pylori* populations for up to 24 h, even at low multiplicities of

**Table 23.1** Characteristics of key investigated *H. pylori* phages

| Phage name | Type | Family | Source of isolation | Key morphological and genomic features | Stability and lytic parameters | References |
|---|---|---|---|---|---|---|
| Hpφ | Lytic | Not specified | Human gastric biopsies | Not specified in detail | Latent period: ~45 min; burst size: ~80 PFU/cell. Stable with nanoparticle formulation | [9] |
| ΦHPE1 | Lytic | Podoviridae | Wastewater | Icosahedral head (~62 nm) and short tail | Good thermal and acid stability. Titer: $10^9$ PFU/mL | [10] |
| ΦHPE2 | Lytic | Siphoviridae | Wastewater | Isometric head (~92.5 nm) and long tail | Good thermal and acid stability. Titer: $10^{10}$ PFU/mL | [10] |
| HPy1R | Temperate | Podoviridae | UV induction from a clinical strain | Linear dsDNA genome (31.2 kbp); integrase gene present. No known virulence or resistance genes | Stable at pH 3–11 and 37 °C. Showed a 2.2–2.3 log reduction in simulated gastric fluid | [11] |
| KHP30 | Temperate | New spherical type | Spontaneous release from a clinical strain | Spherical, lipid-containing (~69 nm); integrase gene present but maintained as an episome | Stable at pH 2.5–10. Latent period: ~140 min; burst size: ~13 PFU/cell | [12] |
| 1961P | Temperate | Podoviridae | Lysate of a clinical strain | dsDNA genome (26.8 kbp); capable of generalized transduction | Not specified in detail | [13] |
| PhiHp33 | Temperate | Siphoviridae | UV induction from a MALT lymphoma strain | Genome size of 24.6 kbp; inducible prophage | Not specified in detail | [2] |

infection. Its broad activity, demonstrated by the formation of inhibition haloes against nearly 79% of 76 clinical isolates, is particularly noteworthy. The discrepancy between its broad ability to inhibit growth and its limited ability to form plaques (indicative of full replication) in most strains suggests a possible mechanism of "lysis from without," where bacterial killing occurs due to a high concentration of phage adsorption without a completed replicative cycle [11]. Should this prove to be a dominant mechanism, it would have significant implications for therapeutic dosing, as the benefit of in situ phage amplification would be limited, likely requiring higher initial concentrations to achieve a clinical effect.

Stability is arguably the most crucial in vitro finding for this therapeutic area. The ability of HPy1R and KHP30 to withstand highly acidic conditions is a vital prerequisite for any orally administered phage [11, 12]. Further validating this, HPy1R was tested in a simulated gastric digestion model. While it showed encouraging stability, it still experienced a 2.2–2.3 log reduction in titer after 1–2 h [11]. This result highlights that even for inherently acid-stable phages, some form of protective formulation may be necessary to maximize the viable dose delivered to the gastric mucosa.

Collectively, these in vitro results provide solid proof of concept. They confirm that phages capable of killing *H. pylori* can be isolated, that some possess the requisite stability for the gastric environment, and that they can effectively reduce bacterial populations in laboratory settings. However, these promising laboratory findings must be interpreted with caution, as they represent the full extent of the current evidence base.

### 23.2.3 Novel Methodologies: The Advent of Viral Tagging

A significant methodological innovation that promises to overcome the slow pace of traditional phage discovery is viral tagging (VT), a technique demonstrated in the foundational work by Deng et al. [14]. This culture-independent technique involves fluorescently labeling a mixed population of environmental viruses and co-incubating them with a specific "bait" host. Host cells that become fluorescently "tagged" by an adsorbing virus can then be isolated using fluorescence-activated cell sorting. Subsequent sequencing of the viral DNA from these sorted cells directly links unknown viruses to their specific host. In its initial application, a single VT experiment discovered 42 new, uncultured phage types for one marine cyanobacterium, dramatically increasing the known viral diversity for that host. This powerful approach bypasses the need for conventional, and often difficult, phage culturing and plaque assays. While originally developed in marine systems, the principle has been successfully adapted to the far more complex human gut microbiome [15]. If adapted for the difficult-to-culture *H. pylori* and the challenging gastric environment, VT could revolutionize the discovery of therapeutically relevant phages, rapidly expanding our limited arsenal of candidates.

## 23.3 Challenges and Limitations for *H. pylori* Phage Therapy Development

The promising in vitro findings detailed in the preceding section represent the current frontier of *H. pylori* phage research. However, a stark and defining gap exists between these laboratory successes and any demonstrable therapeutic effect in vivo. There is a conspicuous lack of published data from established animal models validating the efficacy of phage therapy against an active *H. pylori* infection. This absence is not due to a shortage of such models but rather signifies the immense difficulty in translating laboratory potential into a living system. This gap underscores that the journey from petri dish to patient is impeded by a series of formidable, interconnected obstacles. To translate laboratory potential into a clinical reality, each challenge must be addressed in sequence, starting with the hostile gastric environment itself.

### 23.3.1 The Gastric Environment: The Foremost Physiological Barrier

The human stomach is an organ optimized for digestion and for the exclusion of invading microbes. *H. pylori* is one of the very few bacterial species that have adapted to this extreme niche; by contrast, exogenously administered bacteriophages must traverse a gauntlet of overlapping defenses before they can even encounter their target. This multilayered barrier begins in the lumen, where acidity and proteolysis act in concert, and culminates in the viscoelastic mucus layer that shields the epithelium and harbors bacterial communities within glandular niches. From a therapeutic perspective, these features collectively define the baseline attrition that any oral phage preparation must overcome to exert an effect in vivo.

#### 23.3.1.1 Gastric Acidity and Proteolytic Enzymes

The most immediate challenge is the extreme acidity of the gastric lumen, where the pH can drop to between 1.5 and 3.5. Bacteriophages, as biological entities composed primarily of proteins, are inherently susceptible to denaturation at such low pH levels. The phage capsid, a precisely assembled protein structure, can be irreversibly damaged, rendering the phage non-infectious. Compounding this is the presence of potent proteolytic enzymes, chiefly pepsin, which functions optimally in this acidic milieu. Pepsin can readily digest the proteinaceous capsid of phages, leading to their degradation [9]. While the discovery of phages such as KHP30 and HPy1R with remarkable acid stability in vitro is encouraging, it does not guarantee survival. The observation that HPy1R still suffered a significant loss of titer in a simulated gastric fluid model containing enzymes underscores that the combined assault of acid and proteases is more detrimental than acidity alone [11]. These observations argue strongly for formulation approaches that either transiently buffer the microenvironment or physically shield phages during gastric transit, particularly in the absence of concomitant acid suppression.

#### 23.3.1.2 The Gastric Mucus Layer as a Physical Obstacle

Beyond the lumen, *H. pylori* establishes its primary residence within or beneath the thick, viscous layer of mucus that coats the gastric epithelium [4]. This mucus layer, composed of mucin glycoproteins, serves as a protective sanctuary for the bacteria. For a bacteriophage, however, this same layer constitutes a significant physical and chemical barrier. The viscous nature of mucus can physically trap phages, severely impeding their diffusion and preventing them from reaching their bacterial targets residing deep within the gastric pits [16]. Microscopic and biophysical studies in mucin-rich systems consistently show that particle size, surface chemistry, and local gel architecture govern transport; for phages, this translates into a requirement for either intrinsic "mucus-inert" properties or extrinsic assistance (e.g., transient reduction of viscosity) to achieve therapeutically relevant access to the epithelial niche. Therefore, a successful therapeutic phage must not only survive transit through gastric acid but must also possess the ability to efficiently penetrate this dense mucus barrier to access the bacterial population at the mucosal surface.

#### 23.3.1.3 Spatial Heterogeneity and Niche Protection

Finally, the gastric environment is not uniform: pH, mucus viscosity, and bacterial density vary between the antrum and the corpus and along the glandular axes. *H. pylori* exploits these gradients to occupy protected microenvironments that are intermittently replenished and physically segregated. For phage therapy, such spatial heterogeneity imposes two linked constraints, demanding that the dose be sufficient to saturate multiple compartments and that delivery ensure that infectious particles reach regions where adsorption and replication are feasible. Absent these conditions, even acid- and protease-resistant phages may fail to translate into measurable bacterial clearance in vivo, providing a parsimonious explanation for the persisting gap between encouraging in vitro findings and the lack of validated efficacy in animal models.

### 23.3.2 Inherent Biological and Technical Hurdles

The development of phage therapy for *H. pylori* is constrained as much by biology as by technique. Unlike enteric pathogens that can be targeted in relatively accessible environments, *H. pylori* occupies a specialized, spatially structured niche at the gastric epithelium and exhibits pronounced strain diversity with dynamic surface phenotypes. These features reduce the likelihood that any single phage will display broad, durable activity across clinical isolates and complicate reproducible cultivation and testing under laboratory conditions. At the same time, the current catalog of *H. pylori* phages remains limited, and the methods used to discover, propagate, and characterize them are not yet standardized across laboratories. As a result, datasets are often difficult to compare, and promising activity in one setting may not translate elsewhere. Two issues, in particular, dominate early discovery and preclinical selection: the difficulty of isolating strictly lytic phages with a broad host range, and the persistent risk posed by temperate biology and associated gene flow.

#### 23.3.2.1 Difficulty in Isolating Broad-Spectrum Lytic Phages

A major bottleneck in the field is the scarcity of strictly lytic phages that are active against *H. pylori* [2]. The high specificity of phages, while a key advantage, becomes a practical limitation when confronting a genetically diverse pathogen like *H. pylori*. Most isolated phages exhibit a narrow host range, meaning they can only infect a small subset of clinical strains. This necessitates the development of "phage cocktails." Such formulations contain multiple distinct phages that recognize different bacterial receptors, thereby providing broad enough coverage for empirical therapy. However, developing, characterizing, and manufacturing such cocktails adds significant complexity and regulatory burden. Beyond the sheer scarcity of strictly lytic candidates, breadth is systematically overestimated when discovery proceeds on narrow, convenience panels. *H. pylori* displays extensive surface heterogeneity and phase-variable expression of outer-membrane structures that function as phage receptors; adsorption phenotypes observed against a small number of laboratory or local clinical isolates therefore provide, at best, an optimistic upper bound on true coverage. As a result, putatively "broad-spectrum" isolates will contract in range once evaluated across geographically diverse strain collections and under assay conditions that more faithfully recapitulate gastric chemistry and mucin exposure. In practical terms, host-range claims require validation on large, representative panels and under gastric-mimetic conditions before they can inform cocktail design or clinical feasibility.

#### 23.3.2.2 The Risks of Temperate Phages

The majority of phages identified in association with *H. pylori* to date are temperate. While more readily isolated, the use of temperate phages in therapy is fraught with significant risks and inefficiencies [17]. Their life cycle includes the potential to integrate their genome into the host bacterium's chromosome as a prophage, which poses several dangers. Prophages can carry genes that alter the phenotype of the host bacterium, potentially enhancing its virulence or conferring antibiotic resistance via transduction (lysogenic conversion). Furthermore, a bacterium that contains a prophage often becomes immune to subsequent infection by related phages (superinfection immunity), potentially rendering the therapy ineffective. Finally, the primary goal of therapy is rapid bacterial eradication, which is not guaranteed by temperate phages that can establish a dormant lysogenic state instead of proceeding to immediate lysis [6].

### 23.3.3 Potential for *H. pylori* Resistance to Phages

Just as *H. pylori* has masterfully evolved resistance to a wide array of antibiotics, it must be assumed that it will also develop resistance to bacteriophages [18]. The bacterium's remarkable genetic plasticity, high mutation rate, and capacity for recombination suggest that phage resistance could emerge rapidly under selective pressure. Mechanisms of bacterial resistance to phages are well-documented and include modification or loss of the specific surface receptors that phages use for attachment; production of an extracellular matrix, as in biofilms, that physically prevents phage access; and the use of innate or adaptive defense systems, such as restriction-modification and Clustered Regularly Interspaced Short Palindromic Repeats (CRISPR)-Cas systems, which recognize and degrade foreign phage DNA [19]. Therefore, any long-term therapeutic strategy must proactively account for the emergence of phage resistance.

Mitigation must be built into product design and dosing from the outset. Cocktails composed of strictly lytic phages that use orthogonal receptors raise the genetic barrier to escape and limit the fitness advantage of single-step receptor changes. Sequential or alternating schedules that pair brief mucolysis with phage dosing, followed, when indicated, by low-intensity antibiotic backbones, can suppress emergent resistant subpopulations without recreating the collateral damage of full-dose antibiotic therapy. Synergy with selected antibiotics, bismuth-based regimens, or probiotic strategies that destabilize *H. pylori* biofilms and alter niche occupancy should be explored empirically in gastric-mimetic systems, with rapid adaptation of cocktail composition as resistance signatures appear.

Operationally, resistance management requires surveillance. Longitudinal sampling of gastric isolates during and after therapy, coupled to host-range retesting and whole-genome sequencing, can detect early escape and guide timely updates to the cocktail. In this framework, combinations are not adjuncts but core components: by reducing bacterial effective population size, disrupting protective matrices, and diversifying selective pressures, they convert the high evolvability of *H. pylori* from a fatal liability into a manageable risk for durable phage-based control.

### 23.3.4 Preclinical Evaluation: Demonstrating In Vivo Efficacy and Safety

A decisive obstacle to clinical progress in *H. pylori* phage therapy is the persistent gap between promising in vitro observations and demonstrable efficacy in vivo. In culture, phages encounter bacteria under conditions that maximize adsorption and lytic replication: neutral pH, abundant nutrients, and limited physical impediments. In the stomach, the same virions must survive acid and pepsin, traverse viscoelastic mucus, and locate sparsely distributed targets in spatially heterogeneous niches while bacterial physiology is shaped by urea metabolism, microaerophily, and biofilm architecture. Bridging these environmental discontinuities requires preclinical systems that are not merely convenient but biologically faithful, and a study logic that couples exposure with response in a way that survives out-of-sample testing.

The first requirement is model fidelity. Rodent systems have proven useful for studying colonization biology and host responses, yet many commonly used protocols prioritize ease of infection over recapitulation of human gastric ecology. Short infection windows before intervention, reliance on single laboratory strains, and endpoints focused on presence/absence rather than quantitative burden all inflate apparent effect sizes. A credible efficacy model must enforce stable, gland-associated colonization prior to treatment, sample anatomically defined regions of the stomach to capture antral–corpus heterogeneity, and quantify bacterial load using both culture and molecular assays calibrated against one another. Histopathology should be scored by blinded assessment with predefined criteria because improvements in inflammation and epithelial integrity are mechanistically informative even when eradication is not achieved. Where feasible, larger-animal models can contribute confirmatory data on dosing logistics, gastric residence times, and the practicalities of formulation and administration that rodents cannot resolve.

Pharmacokinetic (PK) and pharmacodynamic (PD) characterization in a relevant biological context is the second pillar. The dose delivered to the mouth is not the dose that reaches the epithelial niche. The fate of phages in the stomach spans rapid luminal clearance, partial sequestration within the mucus, and uncertain localization at the cell surface where *H. pylori* resides. Preclinical evaluation should therefore measure infectious titers over time in distinct gastric compartments, including the luminal contents, mucus, and epithelial scrapings, after single and repeated dosing, ideally under the same acid-suppression regimen anticipated clinically. These exposure data must be linked to bacterial burden in order to determine whether observed killing reflects productive replication in situ or non-replicative adsorption-mediated effects at high local concentrations. This distinction is relevant: regimens predicated on self-amplification will fail if replication is physiologically constrained, whereas regimens built around direct-kill kinetics will underestimate efficiency if phages do, in fact, amplify once the epithelial niche is reached.

Safety assessment must proceed in parallel and with the same insistence on context. Oral phage therapy destined for the stomach poses a particular combination of local and systemic questions. Locally, repeated exposure of an inflamed mucosa to high-titer viral protein assemblies and formulation excipients necessitates careful monitoring for exacerbation of gastritis, disruption of barrier function, or unexpected alterations in mucus architecture. Systemically, even when absorption is negligible, immune recognition can generate neutralizing antibodies that curtail efficacy on repeat dosing; profiling humoral responses and their functional impact is therefore part of the efficacy story as much as of safety. Product quality is integral to safety: complete genomic annotation to exclude lysogeny modules or mobilizable cargo, rigorous removal of host-cell impurities and endotoxin, and stability testing under clinically realistic storage and handling conditions are prerequisites rather than embellishments. Finally, given the central role of the microbiome in gastric and intestinal health, off-target ecological effects should be benchmarked against standard antibiotic regimens to substantiate the often-invoked selectivity advantage of phages.

Study design should formalize these considerations into analyses that are both adequately powered and resistant to bias. Infection must be established at a stable plateau before randomization; allocation and outcome assessment should be blinded; and a primary endpoint, typically the log-transformed reduction of *H. pylori* burden in prespecified gastric regions, should be defined a priori with an effect size grounded in power calculations. Secondary endpoints ought to include histopathology, inflammatory mediators, and durability of response after a washout period, because early post-treatment rebounds are common in gastric infection and inform mechanisms. Dosing schedules should compare monotherapy to rational combinations that address known barriers, such as brief mucolytic pretreatment to reduce viscoelastic impedance or low-intensity antibiotic backbones designed to suppress escape without recreating the selection pressures of full-dose therapy. Crucially, all interventions should be paired with contemporaneous measurements of phage titers in gastric compartments to deconvolute failures of exposure from failures of intrinsic activity.

Interpreting negative or equivocal results is as important. A study that demonstrates high luminal titers but negligible signal at the epithelial surface argues for improved formulation or delivery rather than abandonment of the modality. Conversely, adequate exposure with minimal bacterial impact points to host-range mismatches, biofilm-mediated tolerance, or life-cycle constraints that diminish replication under physiologic stress; these cases motivate cocktail redesign, incorporation of depolymerases, or a pivot to dosing strategies that do not assume in situ amplification. In either case, transparent reporting of methods and raw data, including details on the exact media, atmospheric conditions, pH control, mucin sources, sampling maps, and assay calibration, enables replication and meta-analysis across groups, which the field urgently needs to move beyond single-laboratory claims.

Taken together, a credible preclinical package for *H. pylori* phage therapy will demonstrate three things. First, that candidate phages or cocktails reach and persist within the gastric mucus and at the epithelial interface at concentrations predicted to be effective under physiologic constraints. Second, that these exposures translate into reproducible, quantitative reductions in bacterial burden and improvements in mucosal pathology in models that reflect human disease rather than laboratory convenience. Third, that the intervention is locally and systemically well tolerated, genomically and chemically clean, and operationally compatible with the dosing contexts foreseeable in clinical use. Only when these conditions are met does the in vitro promise of *H. pylori* phage therapy become a clinically plausible proposition.

### 23.3.5 Formulation, Delivery, and Manufacturing Obstacles

The hostile gastric environment dictates that simply administering a liquid phage suspension orally is unlikely to be effective. This necessitates the development of sophisticated formulation and delivery strategies.

#### 23.3.5.1 Challenges in Protective Formulation and Delivery

To shield phages from acid and pepsin, encapsulation within protective matrices is essential. Potential strategies include acid-resistant micro- or nanoparticles, enteric-coated capsules, or embedding phages within mucoadhesive polymers to prolong gastric residence time [20]. Beyond protection, the formulation must facilitate delivery to the site of infection. This may involve using mucus-penetrating technologies, such as particles coated with hydrophilic polymers, or the co-administration of mucolytic agents to locally decrease mucus viscosity [21].

### 23.3.5.2 Manufacturing and Standardization Hurdles

Translating *H. pylori* phage therapy into the clinic hinges on chemistry, manufacturing, and controls as much as on biology. Upstream, production must deliver high-titer, genetically stable, strictly lytic phages under Good Manufacturing Practice (GMP). Because *H. pylori* grows slowly under microaerophilic conditions, seed trains and bioreactors require tight control of atmosphere, pH, and nutrients, with periodic genomic confirmation to prevent drift. Downstream, purification must remove bacterial debris, host proteins, residual DNA, and endotoxin to pharmaceutically acceptable limits without damaging infectivity. Scalable trains typically combine tangential-flow filtration for clarification/concentration with gentle nuclease treatment and chromatography or membrane adsorbers, all tuned to minimize shear and adsorption losses.

Quality control needs to be stability-indicating and standardized. Classical plaque assays remain the primary measure of potency but are method-sensitive; orthogonal quantitative polymerase chain reaction (qPCR) of genome copies helps detect degradation and, combined with PFU counts, yields a meaningful PFU:genome ratio. Release testing should include sterility/bioburden, residual DNA/protein, and endotoxin, with container–closure assessments to avoid titer loss from surface adsorption or interfacial stress. Reference materials and cross-site proficiency testing are essential to make data comparable over time.

Formulation is the second critical pillar. Products must retain activity through storage and survive gastric transit. Lyophilized formats with protective excipients can extend shelf-life and ease distribution; liquid presentations generally require cold-chain control and freeze–thaw protection. For oral delivery, enteric or pH-triggered microencapsulation can shield phages from acid and pepsin, particularly with PPI/P-CAB co-therapy, yet excessive encapsulation or mucoadhesion may hinder diffusion through gastric mucus to the epithelial niche. Conversely, "mucus-penetrating" designs favor transport but may reduce residence time; these trade-offs should be optimized using assays that include simulated gastric fluid with pepsin and native mucins.

Cocktail products add further complexity. Constituents can differ in stability and interact during manufacture and storage; identity and potency of each phage must be verified pre- and post-blend, with comparability plans for inevitable process changes. The field would benefit from harmonized standards specific to *H. pylori*: representative reference strain panels, gastric-mimetic potency conditions, and minimal release specifications. Without such standardization, manufacturing remains bespoke and cross-study comparisons confounded by method variance rather than biology.

## 23.3.6 Regulatory and Clinical Translation Hurdles

Clinical translation of *H. pylori* phage therapy is gated as much by regulation and quality systems as by biology. Current frameworks were built for small molecules and conventional biologics; for dynamically composed, living agents, the pathways are often fragmented or ill-defined. Regulators will expect the same fundamentals: clear mechanism of action, robust Chemistry, Manufacturing, and Controls (CMC), and reproducible clinical performance. However, they must also adjudicate questions unique to phages: iterative cocktail updates, lot-to-lot genomic stability, and the acceptability of adaptive composition in response to resistance. Early, sustained scientific advice with authorities is therefore essential, not merely to secure an Investigational New Drug/Medicinal Product authorization, but to agree on principles for comparability when manufacturing improves or when cocktail members are added, replaced, or retired.

Product characterization is crucial for securing regulatory approval. Sponsors must present strictly lytic, genome-resolved phages with curated annotations excluding lysogeny modules, toxins, and antimicrobial-resistance genes; demonstrate genetic stability across the master/working bank system; and validate GMP processes that deliver consistent potency while controlling host-derived impurities and endotoxin. Release specifications should pair infectivity (PFU) with genome copy measurements, provide sterility/bioburden and residual DNA/protein limits, and include stability-indicating assays under clinically realistic storage and handling. Because formulation directly determines gastric survival and mucus traversal, performance testing in simulated gastric fluids with pepsin and native mucins becomes part of the CMC story, not an optional add-on.

Clinical trial design remains insufficiently standardized for this indication. Phage-specific protocols should prespecify gastric-context endpoints and logistics: enrollment after confirmed colonization; background acid suppression harmonized across arms; primary outcomes based on urea breath or stool antigen tests corroborated by quantitative culture/qPCR from anatomically mapped biopsies; and durability assessed after a washout period to detect recrudescence. Blinding can be challenged by distinct dosing schedules (e.g., mucolytic pretreatments or multiple daily doses) and must be engineered accordingly. Because host range and resistance are central, trials should incorporate baseline and on-therapy susceptibility testing and prospectively defined rules for adaptive cocktail adjustments, with statistical controls that preserve interpretability. Safety monitoring must address local gastric tolerability, systemic immunogenicity (including neutralizing antibodies), and microbiome effects benchmarked against antibiotic standards.

Ultimately, regulators will judge *H. pylori* phage therapy on the same translational triad as other biologics: well-characterized product, consistent manufacturing, and convincingly designed trials that demonstrate clinically meaningful, durable eradication with an acceptable safety profile. Establishing consensus reference methods, such as strain panels, gastric-mimetic potency assays, and minimal release specifications, would not only de-risk individual programs but also make the evidence base coherent across sponsors, accelerating a credible path from bench to bedside.

## 23.4 Future Directions and Recommendations

The nascent state of *H. pylori* phage therapy, defined by both its immense theoretical promise and its substantial practical hurdles, necessitates a strategic, innovative, and multidisciplinary path forward. Overcoming the challenges detailed in the previous section requires moving beyond traditional discovery paradigms and embracing cutting-edge technologies in bioengineering, formulation science, and computational biology. The following directions represent a potential roadmap for translating the concept of *H. pylori* phage therapy into a clinical reality.

### 23.4.1 Enhancing Phage Discovery and AI-Driven Engineering

To address the critical scarcity of suitable lytic phages and their narrow host range (as discussed in Sect. 23.3.2.1), future efforts must be directed toward expanding the arsenal of therapeutic candidates through both systematic discovery and rational design.

#### 23.4.1.1 High-Throughput Discovery and Phage Biobanking

The field must transition from fortuitous isolation toward systematic, high-throughput screening of diverse environmental and clinical sources, such as sewage, gastric biopsies, and fecal samples from colonized individuals. The establishment of comprehensive, well-annotated phage biobanks is a crucial infrastructural goal. These repositories should contain phages that are fully sequenced, phenotypically characterized for host range and stability, and certified free of undesirable genes, serving as a ready resource for formulating patient-specific cocktails [22].

#### 23.4.1.2 Precision Phage Engineering

Genetic and synthetic biology offer the power to not just discover, but to *design* phages with superior therapeutic profiles. A key strategy, given the prevalence of temperate phages (Sect. 23.3.2.2), is their rational conversion into obligately lytic variants. Using precision gene-editing tools like CRISPR-Cas, genes essential for lysogeny, such as integrases and repressors, can be cleanly deleted, permanently locking the phage into a therapeutic lytic cycle [6]. Beyond this, phage genomes can be engineered to enhance efficacy. The modification of receptor-binding proteins (RBPs) could broaden the host range to target conserved receptors across diverse *H. pylori* strains. Capsid proteins could be engineered for enhanced stability against acid and pepsin. Phages can even be armed with additional therapeutic payloads, such as genes encoding exopolysaccharide depolymerases to degrade *H. pylori* biofilms, thereby addressing multiple barriers simultaneously.

#### 23.4.1.3 The Role of Computational Biology and Artificial Intelligence

The integration of artificial intelligence (AI) and machine learning can significantly accelerate these discovery and engineering processes. AI algorithms can be trained on vast metagenomic datasets to identify novel phage sequences in silico, predicting their lytic potential by identifying key genetic markers. Furthermore, AI-driven structural biology tools can predict the three-dimensional structures of phage RBPs. By computationally modeling the interactions of these proteins with known *H. pylori* outer membrane proteins, it becomes possible to predict phage-host pairings and host range with increasing accuracy, dramatically accelerating the identification of promising candidates [22].

### 23.4.2 Developing Intelligent Oral Delivery Systems

To overcome the formidable gastric barriers detailed in Sect. 23.3.1, the development of intelligent oral delivery systems is non-negotiable. Research must focus on sophisticated vehicles that can protect the phage payload during its perilous transit. This includes nano-encapsulation in biocompatible polymers, pH-responsive hydrogels that swell to buffer acid, and mucoadhesive systems that prolong gastric residence time. Concurrently, strategies to penetrate the mucus layer are essential. This could be achieved through co-formulation with mucolytic agents, such as N-acetylcysteine, that locally reduce mucus viscosity [23], or by designing "mucus-inert" nanoparticles (e.g., with a dense polyethylene glycol coating) that can diffuse through the mucus mesh to the site of infection [21]. The ultimate goal is a stable, solid oral dosage form suitable for widespread clinical use.

### 23.4.3 Investigating Synergistic Combination Therapies

Given the adaptability of *H. pylori* and the potential for resistance (Sect. 23.3.3), monotherapy of any kind may prove insufficient. The future of treatment likely lies in multipronged, synergistic attacks.

#### 23.4.3.1 Phage-Antibiotic Synergy

The combination of phages with antibiotics to which *H. pylori* retains susceptibility, such as amoxicillin and tetracycline, is a highly promising strategy. This approach creates a multitarget assault that makes the evolution of simultaneous resistance far less likely. Phage-antibiotic synergy may also manifest in other ways; for instance, phages could potentially "resensitize" resistant bacteria to antibiotics or allow for the use of lower, less toxic antibiotic doses [24].

#### 23.4.3.2 Phages with Probiotics and Engineered Microbiome Therapeutics

Combining phages with probiotics offers a compelling ecological benefit. Phages can specifically reduce the pathogen load, creating a niche that beneficial probiotics can more easily colonize to restore a healthy gastric microbiota and exert their own anti-*H. pylori* effects [25]. A more advanced concept involves the development of engineered probiotics, such as *Lactococcus lactis*, designed to act as in situ "living factories" that produce and secrete *H. pylori*-specific phages or phage-derived lytic proteins directly within the stomach, ensuring continuous, targeted delivery [26].

### 23.4.4 Establishing a Clear Roadmap for Clinical Translation

Bridging the vast gap between the laboratory and the clinic requires a deliberate and standardized approach to preclinical and clinical development.

#### 23.4.4.1 Standardization of Preclinical Models

The field urgently needs to conduct robust in vivo studies using standardized animal models. This requires a consensus on infection protocols, outcome measures (bacterial load and histological inflammation scores), and the application of PK/PD modeling to understand phage behavior in the gastric environment, including their distribution, amplification, and clearance [27].

#### 23.4.4.2 Proactive Regulatory Engagement and Personalized Medicine

A clear pathway for clinical trials must be established through early and continuous engagement with regulatory agencies such as the U.S. Food and Drug Administration (FDA) and the European Medicines Agency (EMA) [28]. This dialogue is essential for defining the requirements for manufacturing (GMP standards), quality control, and the design of scientifically sound clinical trials. The ultimate vision for *H. pylori* phage therapy is one of personalized medicine. In this paradigm, rapid genomic sequencing of a patient's infecting *H. pylori* strain could be fed into an AI system. The algorithm would then instantly select the most effective, bespoke phage cocktail from a vast, well-characterized biobank to treat that specific infection, marking a definitive shift from one-size-fits-all regimens to truly individualized antimicrobial therapy.

## References


1. Elbehiry A, Abalkhail A, Anajirih N, Alkhamisi F, Aldamegh M, Alramzi A, et al. *H. pylori*: routes of infection, antimicrobial resistance, and alternative therapies as a means to develop infection control. Diseases. 2024;12:311. https://doi.org/10.3390/diseases12120311.
2. Muñoz AB, Stepanian J, Trespalacios AA, Vale FF. Bacteriophages of *H. pylori*. Front Microbiol. 2020;11:549084. https://doi.org/10.3389/fmicb.2020.549084.
3. Aldhaleei WA, Wallace MB, Harris DM, Bi Y. *H. pylori*: a concise review of the latest treatments against an old foe. CCJM. 2024;91:481–7. https://doi.org/10.3949/ccjm.91a.24031.
4. Ali A, AlHussaini KI. *H. pylori*: a contemporary perspective on pathogenesis, diagnosis and treatment strategies. Microorganisms. 2024;12:222. https://doi.org/10.3390/microorganisms12010222.
5. Schulz C, Liou J-M, Alboraie M, Bornschein J, Campos Nunez C, Coelho LG, et al. *H. pylori* antibiotic resistance: a global challenge in search of solutions. Gut. 2025;74:1561–70. https://doi.org/10.1136/gutjnl-2025-335523.

6. Strathdee SA, Hatfull GF, Mutalik VK, Schooley RT. Phage therapy: from biological mechanisms to future directions. Cell. 2023;186:17–31. https://doi.org/10.1016/j.cell.2022.11.017.
7. Gordillo Altamirano FL, Barr JJ. Phage therapy in the postantibiotic era. Clin Microbiol Rev. 2019;32:e00066–18. https://doi.org/10.1128/CMR.00066-18.
8. Palma M, Qi B. Advancing phage therapy: a comprehensive review of the safety, efficacy, and future prospects for the targeted treatment of bacterial infections. Infect Dis Rep. 2024;16:1127–81. https://doi.org/10.3390/idr16060092.
9. Cuomo P, Papaianni M, Fulgione A, Guerra F, Capparelli R, Medaglia C. An innovative approach to control *H. Pylori*-induced persistent inflammation and colonization. Microorganisms. 2020;8:1214. https://doi.org/10.3390/microorganisms8081214.
10. Abdel-Haliem ME, Askora A. Isolation and characterization of bacteriophages of *H. pylori* isolated from Egypt. Future Virol. 2013;8:821–6. https://doi.org/10.2217/fvl.13.58.
11. Ferreira R, Sousa C, Gonçalves RFS, Pinheiro AC, Oleastro M, Wagemans J, et al. Characterization and genomic analysis of a new phage infecting *H. pylori*. Int J Mol Sci. 2022;23:7885. https://doi.org/10.3390/ijms23147885.
12. Uchiyama J, Takeuchi H, Kato S, Gamoh K, Takemura-Uchiyama I, Ujihara T, et al. Characterization of *H. pylori* Bacteriophage KHP30. Appl Environ Microbiol. 2013;79:3176–84. https://doi.org/10.1128/AEM.03530-12.
13. Luo C-H, Chiou P-Y, Yang C-Y, Lin N-T. Genome, integration, and transduction of a novel temperate phage of *H. pylori*. J Virol. 2012;86:8781–92. https://doi.org/10.1128/JVI.00446-12.
14. Deng L, Ignacio-Espinoza JC, Gregory AC, Poulos BT, Weitz JS, Hugenholtz P, et al. Viral tagging reveals discrete populations in Synechococcus viral genome sequence space. Nature. 2014;513:242–5. https://doi.org/10.1038/nature13459.
15. Džunková M, Low SJ, Daly JN, Deng L, Rinke C, Hugenholtz P. Defining the human gut Host–Phage network through single-cell viral tagging. Nat Microbiol. 2019;4(12):2192–2203. https://doi.org/10.1038/s41564-019-0526-2.
16. Pacios O, Blasco L, Ortiz Cartagena C, Bleriot I, Fernández-García L, López M, et al. Molecular studies of phages-Klebsiella pneumoniae in mucoid environment: innovative use of mucolytic agents prior to the administration of lytic phages. Front Microbiol Internet. 2023;14:1286046. https://doi.org/10.3389/fmicb.2023.1286046.
17. Monteiro R, Pires DP, Costa AR, Azeredo J. Phage therapy: going temperate? Trends Microbiol. 2019;27:368–78. https://doi.org/10.1016/j.tim.2018.10.008.
18. Hasanuzzaman M, Bang CS, Gong EJ. Antibiotic resistance of *H. pylori*: mechanisms and clinical implications. J Korean Med Sci. 2024;39(4):e44. https://doi.org/10.3346/jkms.2024.39.e44.
19. Malik S, Ahsan O, Muhammad K, Munawar N, Waheed Y. Phagetherapy updates: new frontiers against antibiotic resistance. EuJMI. 2025;15:1–12. https://doi.org/10.1556/1886.2024.00126.
20. Khan S, Ullah S, Bilal M, Hasnain M, Urooj F, Atif M, et al. A systematic and comprehensive review on recent trends in *H. pylori* eradication: current opinion and future perspective. BJSTR. 2022;44(3):35593–35612. https://doi.org/10.26717/BJSTR.2022.44.007069.
21. Malik DJ, Sokolov IJ, Vinner GK, Mancuso F, Cinquerrui S, Vladisavljevic GT, et al. Formulation, stabilisation and encapsulation of bacteriophage for phage therapy. Adv Colloid Interf Sci. 2017;249:100–33. https://doi.org/10.1016/j.cis.2017.05.014.
22. Paul S, Mukhopadhyay AK. Progress of bacteriophage as therapeutic agent against *H. pylori*, especially in India. Discov Appl Sci. 2025;7:399. https://doi.org/10.1007/s42452-025-06971-w.
23. Huynh HQ, Couper RTL, Tran CD, Moore L, Kelso R, Butler RN. N-Acetylcysteine, a novel treatment for *H. pylori* infection. Dig Dis Sci. 2004;49:1853–61. https://doi.org/10.1007/s10620-004-9583-2.
24. Anastassopoulou C, Ferous S, Petsimeri A, Gioula G, Tsakris A. Phage-based therapy in combination with antibiotics: a promising alternative against multidrug-resistant gram-negative pathogens. Pathogens. 2024;13:896. https://doi.org/10.3390/pathogens13100896.
25. Addissouky TA, Ali MMA, El Tantawy El Sayed I, Wang Y. Recent advances in diagnosing and treating *H. pylori* through botanical extracts and advanced technologies. Arch Pharmacol Ther. 2023;5(1):53–66 https://doi.org/10.33696/Pharmacol.4.045.

26. Choudhury A, Ortiz PS, Young M, Mahmud MT, Stoffel RT, Greathouse KL, et al. Control of *H. pylori* with engineered probiotics secreting selective guided antimicrobial peptides. Microbiol Spectrum. 2023;11:e02014–23. https://doi.org/10.1128/spectrum.02014-23.
27. Patil S, Yu S, Jobby R, Ravichandran V, Sarkar S. A critical review on In Vivo and Ex Vivo models for the investigation of *H. pylori* infection. Front Cell Infect Microbiol Internet. 2025;15:1516237. https://doi.org/10.3389/fcimb.2025.1516237.
28. Yang Q, Le S, Zhu T, Wu N. Regulations of phage therapy across the world. Front Microbiol Internet. 2023;14:1250848. https://doi.org/10.3389/fmicb.2023.1250848.

# Antibacterial Peptides Against *Helicobacter pylori*

24

Zelin Zheng, Yejiao Shi, and Honggang Hu

## 24.1 Conventional Treatments for the *H. pylori* Infection

*Helicobacter pylori* (*H. pylori*) is a Gram-negative bacterium that colonizes on the human gastric mucosa. Its infection is closely associated with the development of chronic gastritis, peptic ulcers, and even gastric cancer, which are globally concerning public health issues. In developing countries, the infection rate of *H. pylori* reaches almost 50%. As a well-known Class I carcinogen, its long-term infection is a significant contributor to gastric cancer. The effective inhibition and eradication of *H. pylori* is crucial for preventing the related malignancies and improving public health.

Currently, the first-line clinical treatments of *H. pylori* infection are primarily based on the triple therapy or quadruple therapy, which involve the combination of proton pump inhibitors with two or three antibiotics [1]. The proton pump inhibitors decrease the gastric acid secretion and thus increase the gastric pH, creating a more stable environment for the antibiotics. While the antibiotics commonly include clarithromycin, amoxicillin, metronidazole, and levofloxacin, these standardized therapies have been recommended over the past three decades, achieving eradication rates as high as 80–90% in 1990s. Nonetheless, their eradication rates are annually declining, and their limitations are apparently increasing.

First of all, the triple or quadruple therapy usually requires patients to take numerous tablets daily for 10–14 days. Such complex dosing could easily lead to the misuse or overuse of antibiotics, reducing the treatment effectiveness and

Z. Zheng · H. Hu (✉)
Institute of Translational Medicine, Shanghai University, Shanghai, China

School of Medicine, Shanghai University, Shanghai, China
e-mail: hhu66@shu.edu.cn

Y. Shi
Institute of Translational Medicine, Shanghai University, Shanghai, China



J. Gao, Y. Du (eds.), *Non-Antibiotic Therapies for Helicobacter pylori*,
https://doi.org/10.1007/978-981-95-9598-3_24

raising bacterial resistance. The emergence of resistance directly caused the decrease in eradication rate and the increase in treatment failure. In the past 10 years, these salvage therapies have still failed in approximately 10–30% patients. Besides, while the antibiotics in the triple or quadruple therapy kill *H. pylori*, they also attack the beneficial commensal bacteria indiscriminately, disrupting the balance among the intestinal microbiota. As such, the triple or quadruple therapy is usually associated with side effects such as diarrhea, bloating, and antibiotic-associated colitis. These side effects not only severely impact the medication compliance of patients but also severely impact their immune and metabolic health in the long term. Therefore, the conventional treatments for the *H. pylori* infection are stuck in a vicious cycle of increasing resistance, increasing dosage, and increasing resistance unless novel antibacterial therapies that are less prone to resistance are developed.

## 24.2 Antibacterial Peptides for the *H. pylori* Infection

The accelerating rise in bacterial resistance has spurred the discovery of novel antimicrobial agents and therapies. Since the first discovery of cecropin in moths by Hans Boman et al. in 1981, a class of peptides with antibacterial bioactivity has been increasingly found in a growing number of animals and plants [2]. These antibacterial peptides are short-chain molecules composed of 5–50 amino acids. Their molecular weights lie between those of the small molecular antibiotics and the large molecular defensins, endowing them with several distinct properties [3].

Compared to the conventional antibiotics, antibacterial peptides are less likely to induce bacterial resistance [4]. The conventional antibiotics usually attack a single target of bacteria, such as the specific enzyme or ribosome, to inhibit the bacterial element synthesis and thus terminate the bacterial life process. Therefore, the bacteria tend to generate genetic mutations to induce resistance. By contrast, the antibacterial peptides exert their bactericidal effects by nonspecific membrane disruption, predominated by multiple targets. These antibacterial peptides, typically with positive charges, can interact with the negatively charged bacterial membranes via electrostatic interactions, subsequently disrupting the structure of bacterial membranes and allowing the leakage of cellular contents. Therefore, it is difficult for the bacteria to alter their compositions and properties of their membranes, preventing the development of bacterial resistance. Moreover, antibacterial peptides can also affect the metabolism of the bacteria by entering the bacterial membranes and interfering with their nucleic acid and protein synthesis, or modulate the immunity of the host. These multiple bactericidal mechanisms of the antibacterial peptides hinder the evolution of bacterial resistance as well.

Besides, antibacterial peptides offer a versatile platform for molecular engineering. By the unnatural amino acid substitution, cyclization, glycosylation, or lipidation, the stability of the antibacterial peptides in the gastric environment can be significantly enhanced, preventing their degradation and guaranteeing their effectiveness for the treatment of *H. pylori* infections. By the rational design or modification, the specificity of the antibacterial peptides toward the *H. pylori* can also be remarkably optimized, minimizing their damage to the host cells and other probiotics and maintaining a healthy intestinal microbiota during the treatment of *H. pylori* infections.

The extremely low bacterial resistance induction, together with the remarkably high molecular designability of the antibacterial peptides, makes them the most promising alternatives as the next generation therapeutics for the effective and safe treatment of *H. pylori* infections.

## 24.3 Anti-infection Mechanisms of the Antibacterial Peptides

### 24.3.1 Direct Killing

The antibacterial peptides are enriched with a high proportion of lysine and arginine residues, conferring strong positive charges. These positive charges enable their initial attachment to the negatively charged outer membranes of the *H. pylori,* which are constituted by the lipopolysaccharides (LPSs), phosphatidylserine, and phosphatidylglycerol. Progressively, the hydrophobic regions of their amphipathic α-helical structures insert into the inner membranes of the *H. pylori,* inducing aggregation or forming transmembrane channels on their membranes. Consequently, the intracellular contents of *H. pylori* leaked, leading to its death ultimately.

Depending on the different membrane-disruptive processes, four models have been proposed, including the barrel-stave model, toroidal-pore model, carpet model, and aggregate channel model. In the barrel-stave model, multiple α-helical antibacterial peptides, such as human dermcidin (SSLLEKGLDGAKKAVGGLGKLGK DAVEDLESVGKGAVHDVKDVLDSVL) [5], assemble on the bacterial membranes and insert perpendicularly into the lipid bilayers. The hydrophobic domains of the antibacterial peptides interact with the hydrophobic core of the membrane lipids, while the hydrophilic regions of the antibacterial peptides align to form a central hydrophilic pore. The formed pore creates a continuous conduit across the membrane, resulting in the leakage of cytoplasmic components and the death of bacteria.

By contrast, in the toroidal-pore model, the antibacterial peptides, such as arenicin-1 (RWCVYAYVRVRGVLVRYRRCW) [6], align parallel to the lipid bilayers. The hydrophilic segments of the antibacterial peptides associate with the lipid head groups, while the hydrophobic regions embed into the acyl chain domains. As the peptide-to-lipid ratio rises, the lipid bilayers undergo curvature strain, leading to the formation of inward-bent pores. The surfactant-like action of the antibacterial peptide compromises the ability of the bacterial membranes to maintain osmotic homeostasis, ultimately causing bacterial death.

Disparately, the carpet model does not involve any defined transmembrane pores. Instead, the antibacterial peptides, such as aurein 1.2 (GLFDIIKKIAESF) [7], align parallel to and cover the membrane surface in a carpet-like manner. Upon reaching a critical concentration, the antibacterial peptides reorganize into micellar structures, which dissolve the bacterial membrane and lead to their disintegration.

In comparison, the aggregate channel model begins with the adsorption of antibacterial peptides, such as maculatin [8], onto the bacterial membrane, followed by the lipid binding and the formation of peptide–lipid aggregates. Increasing concentration of the antibacterial peptides elevates the local density of these complexes, inducing severe membrane distortion. The resulting mechanical tension promotes the formation of transient fissures, through which the bacterial cellular contents leak. These temporary openings may also facilitate the translocation of the antibacterial peptides into the bacterial cytoplasm, potentially enabling their interactions with the intracellular targets.

Apart from these membrane-disruptive mechanisms, the antibacterial peptides can also translocate into *H. pylori* and interact with its intracellular components such as nucleic acids and proteins, disrupting its essential metabolic processes and leading to its death. For instance, Buforin II, with 21-amino acids, is isolated from the gastric tissue of the Asian toad. It can initially permeabilize the bacterial membranes without causing any immediate lysis and subsequently enter the bacterial cytoplasm, where it exhibits high affinity for the DNA and RNA. These bindings significantly inhibit nucleic acid metabolism, leading to rapid bacterial death. Besides, pyrrhocoricin with 20 amino acids is derived from the European firebug *Pyrrhocoris apterus*. It can non-lytically enter the bacterial cytoplasm, specifically bind to the bacterial homolog of the eukaryotic heatshockprotein70 (Hsp70) chaperone DnaK. As a consequence, the ATPase activity of bacteria is inhibited, and the protein folding function of bacteria is abrogated, causing bacterial death. In addition to the nucleic acids and proteins, the antibacterial peptides can also interfere with the key enzymes inside bacteria. For instance, LL-37 has been reported to inhibit the activity of the lipid A palmitoyltransferase (PagP) of the gram-negative bacteria, thus disturbing the acylation of lipid and the integrity of bacterial membrane for exerting its bactericidal potency.

### 24.3.2 Indirect Protection

In addition to the direct killing of *H. pylori*, the indirect protection is another feasible way to inhibit or eradicate *H. pylori* infection, which is further developed from the extensive bacterial colonization. Since the colonization process is closely associated with the release of pathogenic adhesins and virulence factors, the antibacterial peptides can exert anti-adhesion or anti-virulence capacities as well for the treatment of *H. pylori* infection.

The attachment of *H. pylori* onto the mucosae of the host is the key step for its infection. There are various adhesins expressed by the bacteria interacting specifically with their receptors expressed by the host. By blocking these adhesins or receptors, the colonization of *H. pylori* can be restricted. For instance, based on the computer-aided design, three peptides derived from the defatted wheat germ protein were identified to bind to the blood group antigen-binding adhesin and sialic acid-binding adhesin, significantly inhibiting the *H. pylori* adherence [9]. Besides, a serine *O*-heptosylated peptide mimicking the bacterial adhesins was synthesized as well. It acted as an effective blocker to occupy the receptor on the host, remarkably inhibiting the *H. pylori* colonization [10]. Moreover, the antibacterial peptides can also enter the bacterial membrane and downregulate the bacterial genes that are related to the expression of adhesins. For instance, the HF-18 derived from the hagfish intestinal peptide HFIAP-1 was demonstrated to downregulate the transcription of adhesion genes alpA/alpB, thus reducing the expression of the adhesin cytotoxin-associated gene A, ultimately attenuating the bacterial colonization [11].

With its successful colonization onto the host, the *H. pylori* can secrete virulence factors to facilitate its infection and damage the host tissues. By neutralizing these virulence factors, both the *H. pylori* infection and its related damage effects can be confined. For instance, after attaching to the host mucosae, the *H. pylori* can secrete cytotoxin vacuolating toxin A (VacA). The VacA binds to the host receptors and inserts into the host membrane to form anion-selective channels, resulting in the intracellular ion imbalance and water accumulation. And the subsequent vacuolation will upregulate inflammation and cause apoptosis. Therefore, based on the phage display, the antibacterial peptide with high affinity to the VacA was identified. It was demonstrated could effectively neutralize the VacA and thus reduce the pathogenicity of the *H. pylori* [12].

### 24.3.3 Immune Regulation

In addition to their activities against bacteria for the inhibition and eradication of the *H. pylori* infection, the antibacterial peptides can also modulate the immune responses of the hosts to combat the *H. pylori*. At physiological concentrations, these antibacterial peptides elicit a broad spectrum of biological reactions related to inflammation, innate immunity, and adaptive immunity.

At the early stage of *H. pylori* infection, the antibacterial peptides act as danger signals. They can bind to the pattern recognition receptors on the surface of immune cells, activating these cells to secrete more pro-inflammatory cytokines. Thereby, these cells are initiated into the immune defensive stage with enhanced phagocytic and bactericidal capabilities. For instance, the muramyl dipeptide and glucosaminylmuramyl dipeptide can specifically bind to the Y-box binding protein 1, upregulating the nuclear factor (NF)-κB2 gene expression and initiating the innate immune responses. The well-known LL-37 has also been reported to bind to the formyl peptide receptor 2, thus activating the multiple downstream signaling pathways, particularly the mitogen-activated protein kinase/extracellular signal-regulated kinase cascade (MAPK/ERK). These activations significantly enhance the expression of various chemokines and cytokines. By engaging their cognate receptors, these signaling molecules promote the directed migration of neutrophils, monocytes, and T cells to sites of infection or inflammation, thereby playing a pivotal role in orchestrating immune defense.

Although these inflammatory responses are necessary for inhibiting and eradicating *H. pylori* infections, the excessive inflammation can also cause severe tissue damage. The LPS derived from Gram-negative bacteria is a key toxin that can trigger fatal systemic inflammatory responses such as sepsis. By neutralizing LPS after the early stage of *H. pylori* infection, the antimicrobial peptides can prevent the inflammatory storm as well, thus protecting the host from the harm of autoimmune reactions. For instance, the thrombin-derived host defense peptide GKY25 can directly bind to LPS, dampening the pathogen-associated molecular pattern (PAMP)-mediated immune activation and thus suppressing the pro-inflammatory effects. Studies also found the rationally designed IDR01018 could significantly reduce the expression of pro-inflammatory factors, such as interleukin (IL)-8, tumor necrosis factor-α, and IL-1β, by inhibiting the NF-κB/MAPK pathway. Meanwhile, the expression of toll-like receptor 2 (TLR2)/toll-like receptor 4 (TLR4) could be downregulated as well to reduce the binding of the LPS from *H. pylori* to the host receptors, thereby blocking the inflammatory cascade reactions (Fig. 24.1).

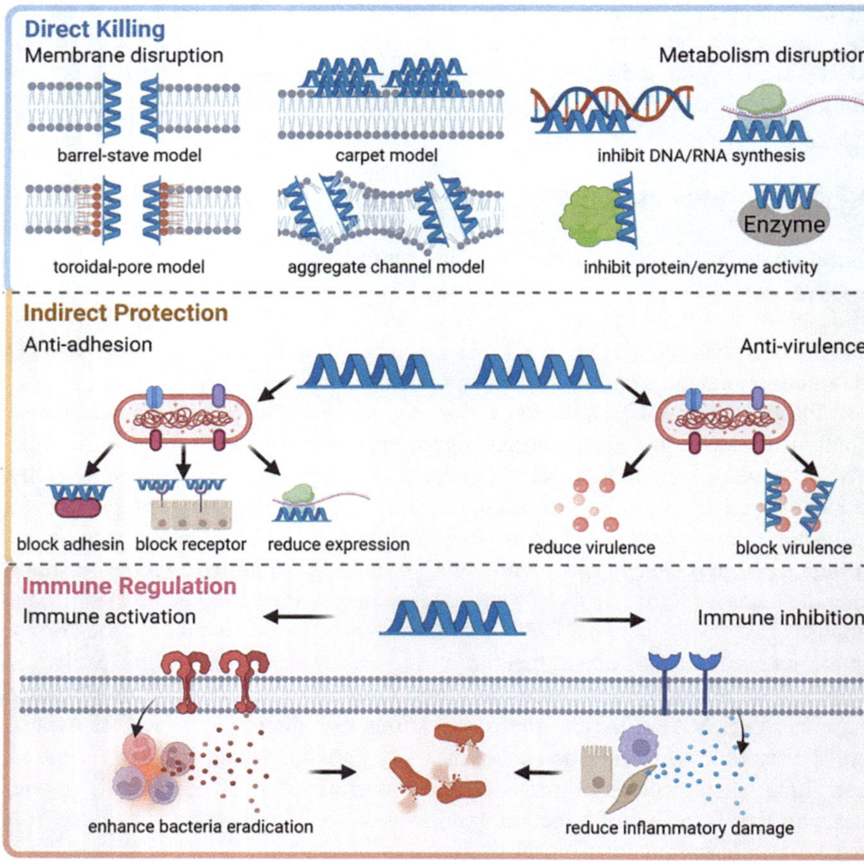


**Fig. 24.1** The mechanisms of the antibacterial peptides for the treatment of *H. pylori* infection. The antibacterial peptides can inhibit and eradicate the *H. pylori* infection mainly by direct killing, indirect protection, or immune regulation. (Created with BioRender.com)

## 24.4 Discovery Approaches of the Antibacterial Peptides

### 24.4.1 Natural Mining

Almost all organisms in nature can synthesize bioactive peptides, especially the antimicrobial peptides that serve as the core components of innate immunity to respond to the threats from the pathogenic microorganisms. As such, extracting and screening the specific tissues of mammals, plants, and microbes is the most direct way to mine the antibacterial peptides. After being extracted, the total proteins obtained are separated and purified using a series of chromatographic techniques. Then the relatively small peptide fractions are collected and screened for their antimicrobial activities. Ultimately, the mass spectrometry or Edman degradation is exploited to determine their exact molecular sequence and weight. Although the process is time-consuming and costly, it is still the most recognized and standardized approach to discovering novel antibacterial peptides.

Besides, optimizing the known peptides with identified functions is another approach for their applications in the *H. pylori* infection treatment. With a good understanding of the different anti-infection mechanisms, any peptides with the known capacities of membrane disruptive, LPS binding, immunomodulating, and so on could be rationally tailored for the inhibition and eradication of the *H. pylori* infection.

### 24.4.2 Rational Design

Based on the different anti-infection mechanisms, diverse anti-infection targets have been identified. Therefore, the antibacterial peptides can also be rationally designed to interact with a specific target. The computer-aided design has been proven to be capable of effectively exploring the conformational space and discovering new hidden pockets of these targets. It is crucial for guiding the subsequent structure design of the antibacterial peptides and experimental verification of their antibacterial activities. In computer-aided design, the structure of the antibacterial peptides that interact with the aimed targets can be predicted, with their interaction processes being simulated and their binding affinities being calculated. By sequentially performing the homology modeling, molecular docking, and virtual screening, the antibacterial peptide with the best affinity for the target can be screened out in the molecular library. Besides, the molecular dynamics simulation can also be indispensably utilized to map the ligand binding sites and analyze the binding modes.

Phage display is another commonly used technique that links the genotype and phenotype to discover antibacterial peptides. It involves inserting a random peptide library into the coat protein gene of a bacteriophage, thus allowing each phage to display a different peptide on its surface. The phage library displaying different peptides is then incubated with the aimed target, such as the virulence factor of *H. pylori*, and eluted. After multiple cycles, the antibacterial peptide with the highest affinity is enriched. Nonetheless, the enriched antibacterial peptides are typically

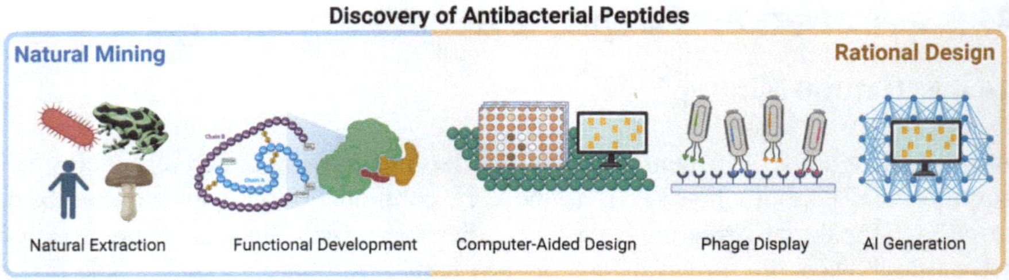


**Fig. 24.2** The discovery approaches of the antibacterial peptides for the treatment of *H. pylori* infection. The antibacterial peptides can be discovered by either natural mining or rational design. (Created with BioRender.com)

unmodified. To optimize their therapeutic parameters, including the solubility, stability, pharmacokinetics, and so on, the medicinal chemistry approaches are rationally exploited in combination for generating the desired agents treating *H. pylori* infection effectively and safely.

In addition, the high-throughput screening platform can also be built on the solid-phase peptide synthesis, which enables the automated, random, and rapid construction of the peptide library. The antibacterial activity, cytotoxicity, solubility, and stability of these peptides can then be evaluated in parallel, generating a promising sequence of peptides and a vast amount of data in a short period of time. The sequence of peptides can be modified or used directly for the treatment *of H. pylori* infection. Meanwhile, the vast amount of data can be further analyzed by machine learning. Key features of peptides that influence their anti-infection efficacy can be identified and summarized, which can be continuously utilized to guide the next generation of antibacterial peptides with optimized efficacy for treating *H. pylori* infection (Fig. 24.2).

## 24.5 Optimization Strategies of the Antibacterial Peptides

### 24.5.1 Chemical Structure Modification

Although several antibacterial peptides have entered the market for the clinical treatment of *H. pylori* infection, the inevitable weaknesses of these naturally occurring molecules still severely limit their further development. These weaknesses include the low efficacy caused by their random secondary structure, the poor bioavailability caused by their proteolytic instability, as well as the potential cytotoxicity caused by their unpredicted selectivity. To overcome these weaknesses, various chemical structure modification strategies have been exploited, including unnatural amino acid substitution, cyclization, and chemical chain modification. Since the antibacterial mechanism and physicochemical properties of the antibacterial peptides depend on their primary structure, specific chemical modifications allow for the tailored enhancement of both their stability and antibacterial efficacy.

#### 24.5.1.1 Unnatural Amino Acid Substitution

As naturally occurring molecules, the antibacterial peptides are predominantly composed of L-amino acids. The in vivo proteases with the evolved active sites exhibit high stereospecificity to recognize and cleave the peptide bonds formed by the L-amino acids. As such, the antibacterial peptides usually suffer from their poor in vivo stability for the effective treatment of bacterial infections. The unnatural amino acids with unique functional groups and spatial conformations are not encoded by the genetic machinery. Thereby, they possess distinct metabolic properties and can serve as valuable tools for stabilizing the antibacterial peptides in vivo. For instance, by replacing the unessential residues with their D-enantiomers, the Lip7 ($C_{12}$-RRW) was demonstrated to have enhanced resistance to the proteolytic hydrolysis without compromising its antibacterial potency [13].

In addition to D-amino acid substitution, the incorporation of halogenated amino acids into the sequence of the antibacterial peptides represents another promising strategy. The steric bulk imposed by the halogen atoms in the side chain can hinder the access of proteases to the peptide bond, thereby reducing the susceptibility of antibacterial peptides to hydrolysis. Besides, the high electronegativity of halogen atoms can also alter the charge distribution as well as the local environment of amino acids, further disrupting the recognition and cleavage efficiency of proteases. For instance, by modifying its hydrophobic leucine with the trifluoromethyl, both buforin II (TRSSRAGLQFPVGRVHRLLRK) and magainin II (GIGKFLHAAKKF AKAFVAEIMNS) with hexafluoroleucine exhibited enhanced resistance to tryptic degradation and improved antibacterial activity against *Bacillus subtilis* and *Escherichia coli* [14].

#### 24.5.1.2 Cyclization

Inspired by the quite a lot Food and Drug Administration (FDA) approved antibacterial peptides with cyclic and rigid structures, such as vancomycin, daptomycin, and polymyxin, the cyclization is also considered an effective strategy to optimize the antibacterial peptides. The cyclization can stabilize the conformation of the antibacterial peptides, thereby not only reducing their susceptibility to hydrolysis but also reducing the energy barrier for their membrane adaptation and protein interaction. Consequently, the rapid accumulation of the antibacterial peptides onto the bacterial surface is facilitated, with the enhanced antibacterial efficacy being achieved. Depending on the site of cyclization, three modes of modifications, including sidechain to sidechain, sidechain to backbone, and backbone to backbone, have been developed.

Typically, the sidechain to sidechain cyclization involves the formation of disulfide bonds between cysteine residues. It allows precise control over the cyclization site and ring size of the antibacterial peptides. For instance, the cycloCP-11 with the sequence ICLKKWPWWPWRRCK-$NH_2$ cyclized via disulfide bonding between the additionally inserted cysteines at the second and penultimate positions. Compared to its linear parent peptide, it demonstrated a longer half-life. After incubation with trypsin for 1.5 h, it retained the antibacterial potency that was comparable to the original one [15]. However, these disulfide bonds are usually not stable in the acidic environment. Thereby, the all-hydrocarbon stapled antibacterial peptides have emerged as a promising candidate with enhanced protease stability and antibacterial potency [16, 17]. These peptides were synthesized by firstly incorporating two unnatural amino acids bearing α-methyl and α-alkenyl groups during the solid-phase peptide synthesis (SPPS) and then inducing the all-hydrocarbon cross-link via the olefin metathesis reaction [18]. For instance, by stapling the truncated LL-37 peptide KR-12, an optimized molecule with reinforced α-helical conformation was generated. The reinforced α-helical conformation not only protected it from proteolytic hydrolysis but also improved its antibacterial efficacy via effective membrane perturbation and anti-inflammatory efficacy [19].

In comparison, sidechain to backbone cyclization usually involves the formation of amide bonds, such as lactam, lactone, or thiolactone linkages, between the backbone atoms and the side chain residues such as lysine, aspartic acid, glutamic acid, cysteine, or threonine. For instance, the antibacterial peptide microcin J25, in which a cyclic structure forms between its backbone amide and its side chain Glu8, displays a unique rotaxane-like topology known as a lasso structure. The rigid and distinctive structure confers its high stability under high temperatures and extreme pH conditions.

Similar to the sidechain to backbone cyclization, the backbone to backbone cyclization involves amide bond formation as well, but between the *C*-terminal carboxyl and *N*-terminal amino groups. The increased steric hindrance from the macrocyclic structure directly impedes protease access to the peptide backbone. For instance, inspired by the natural cyclic antibacterial peptides, the $R_4W_4$ (RRRRWWWW) was synthesized based on the head-to-tail cyclization. After 24 hours of incubation in plasma, approximately 60% of the synthesized cyclic peptide remained, while only about 21% of its linear counterpart was left, suggesting that its significantly improved plasma stability [20]. Besides, the cyclic peptide also demonstrated higher antibacterial potency than its linear analog, particularly against the gram-positive bacteria [20].

#### 24.5.1.3 Chemical Group Modification

Beyond the unnatural amino acid substitution and cyclization, chemical chain modification offers greater flexibility and diversity for improving the therapeutic parameters of antibacterial peptides. It allows for the selection of different modifying groups and sites to meet the specific requirements for the treatment of *H. pylori* infection. Among all these modifications, the three most commonly utilized ones are PEGylation, lipidation, and glycosylation.

PEGylation is one of the most commonly utilized strategies. By conjugating to the *C*-terminus, *N*-terminus, or side group, the hydrophilic PEG chain forms a protective layer around the antibacterial peptides. It increases the molecular weight and solubility of the antibacterial peptides, reducing their renal filtration, decreasing their reticuloendothelial system clearance, and thus extending their in vivo half-life successively. Besides, it also creates a steric barrier that hinders the protease's access to the antibacterial peptides, reducing their enzymatic catabolism and degradation risk. For instance, with PEGylating its *C*-terminus, the half-life of the antibacterial peptide Onc-112 extended from 0.5–12 h to 7–42 h in the mouse serum. In spite of infrequency, the PEGylation was reported to reduce the nonspecific interactions of antibacterial peptides with cell membranes as well, diminishing their cytotoxicity and adverse effects as a consequence.

Lipidation is another effective modification strategy involving the attachment of the fatty acid chains to the peptide chains via either the amide or ester bonds. The generated lipopeptides exhibited several advantages, including improved bioavailability, enhanced protease stability, and increased membrane permeability. For instance, with the *N*-terminal lipidation by the branched octanoic tail, the antibacterial peptide Buforin II exerted enhanced penetrating ability across the bacterial membranes. Besides, the remarkable penetrating capacity also enabled its antibiofilm activity. Both the inhibition and eradiation of *Porphyromonas gingivalis* biofilm formation, as effective as the commercialized polymyxin B, were observed for it [21]. Moreover, comprising the hydrophilic peptide moiety and the hydrophobic acyl chain, the lipopeptides are also endowed with the self-assembling capacity into nanostructures. Compared to the single antibacterial peptide molecule, the self-assembled antibacterial peptide nanofibers displayed significantly improved capacity to inhibit and eradicate the biofilm of both pathogenic bacterial *Pseudomonas aeruginosa* and fungus *Candida albicans* [22].

Glycosylation is the key post-translational modification in living organisms. It involves attaching the carbohydrate molecules to the peptide chains via asparagine side chains (*N*-glycosylation), serine or threonine side chains (*O*-glycosylation), or cysteine side chains (*S*-glycosylation) [23]. By glycosylating the antibacterial peptides, not only their hydrophilicity and bioavailability can be improved, but also their targeting and binding to the targeted bacteria can be enhanced. Although it may not always enhance the direct antibacterial potency of peptides due to the potential reductions in their hydrophobicity and positive charge, it can increase their resistance to the enzymatic degradation, prolonging their half-life. For instance, SLP-51 with Arg *N*-lactose demonstrated superior in vitro antimicrobial activities against drug-resistant Gram-positive or -negative clinical isolates and improved proteolytic stability [24].

### 24.5.2 Delivery System Engineering

As promising alternative therapeutics, the antibacterial peptides possess more rapid bacterial killing capacities and lower propensity to induce resistance. As for the clinical treatment of *H. pylori* infection, these peptides are still severely limited by their degradation susceptibility to gastric acid and proteases, as well as their insufficient targeting of the gastric mucosa. In addition to the chemical structure modifications, the nanostructure-based delivery systems have also been widely exploited for combating these limitations. These nanostructures can not only reduce the systemic exposure and the nonspecific toxicity of antibacterial peptides but also enable their active targeting and accumulation at the infection site of the gastric epithelial cells.

Among all the nanocarriers developed for the antibacterial peptide delivery, the pH-responsive polymer nanoparticle is the most representative one. It can protect the antibacterial peptides from degradation in the gastric fluids and release the antibacterial peptides on demand at the infection site in relatively lower pH [25]. For instance, a chitosan (CS)-based nanocarrier coated with concanavalin A was developed for the targeted delivery of the antimicrobial peptide CM11 to *H. pylori* infection sites. The acid resistance and pH-responsive release of CS help prevent drug degradation in the stomach, and the system demonstrated enhanced anti-*H. pylori* activity in vitro and effective anti-infection efficacy in vivo [26]. It maximizes the therapeutic efficacy of the antibacterial peptides and minimizes their potential side effects on the hosts' normal flora and intestinal mucosa.

Besides, the antibacterial peptides can also be rationally designed at the molecular or even atomic level to self-assemble into delivery nanostructures and spontaneously shield the antibacterial via the programmed non-covalent interactions, such as hydrogen bonding, electrostatic interactions, and hydrophobic interactions [27]. Stimuli-responsive domain can be designed to sense the specific signals within the gastric infection microenvironment, such as lower pH, bacterial secreted urease, or specific proteases, triggering their structural disassembly and thus enabling on-demand release of the antibacterial peptides [3]. These dynamic assembly/disassembly systems based on the covalent peptide chemistry have excellent performance in improving the pharmacokinetic properties. Their relatively stable covalent interactions can mitigate the burst release effect of the antibacterial peptides that are often associated with their non-covalent nanocarrier delivery systems. Moreover, multiple functional domains, such as targeting, penetrating, and responsive sequences, can be modularly designed into one self-assembling peptide molecule to achieve the integrated therapeutic effects. For instance, the Cry3Aa protein from *Bacillus thuringiensis* naturally forms submicron-sized crystals, which have been demonstrated to preserve the function of encapsulated cargo proteins, conferring enhanced stability and protection against proteolytic degradation. The antimicrobial peptide LL37 and the *H. pylori*-targeting peptide P17 were incorporated into these Cry3Aa crystals via genetic fusion. This design enables targeted delivery to the infection site, followed by pH-responsive release of the antimicrobial peptide, resulting in efficient killing of *H. pylori* [28].

Most notably, with the advanced biotechnologies, the probiotics have also been engineered as living nanocarriers for the antibacterial peptide delivery. For instance, the probiotic *Lactococcus lactis* was engineered to secrete a guided antimicrobial peptide that was a fusion protein comprising an *H. pylori*-binding peptide and a broad-spectrum antimicrobial peptide. Oral administration of the engineered probiotic significantly cleared the infection and rapidly restored microbial homeostasis in the stomach of mouse models [29]. Combining the probiotic therapy with the antibacterial peptides, the interdisciplinary approach represents a paradigm shift in paving the way for innovative solutions for the effective and safe treatment of *H. pylori* infection (Fig. 24.3).

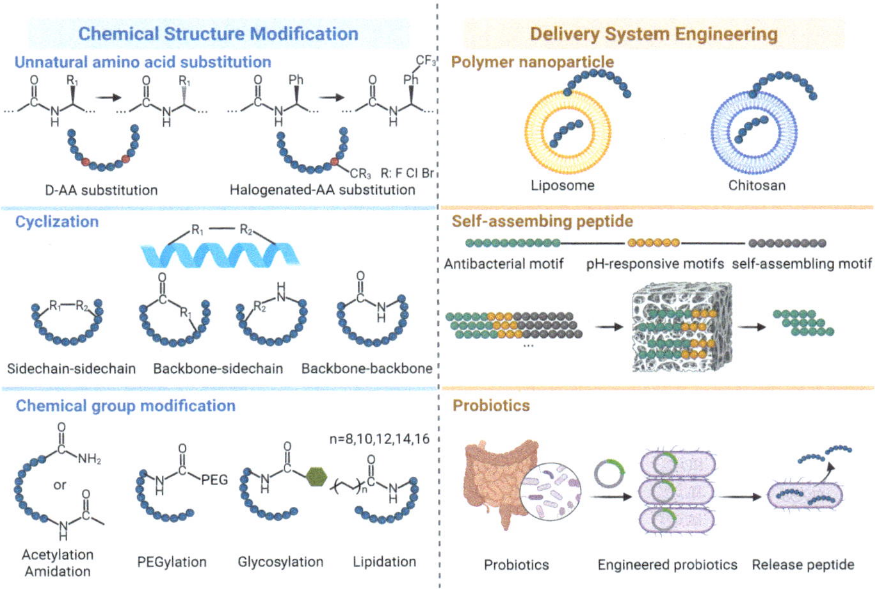


**Fig. 24.3** The optimization strategies of the antibacterial peptides for the effective and safe treatment of *H. pylori* infection. The antibacterial peptides can be optimized by either chemical structure modification or delivery system engineering. (Created with BioRender.com)

## 24.6 Challenges and Perspectives

Although holding significant potential for inhibiting and eradicating the *H. pylori* infections in preclinical studies, the translation of antibacterial peptides from the laboratory to the bedside remains challenging. Three major issues regarding their in vivo instability, high production and purification cost, as well as potential immunogenicity, are still waiting to be solved.

First of all, the antibacterial peptides with the featured biodegradable property have a short plasma half-life and a rapid systemic clearance upon intravenous administration. Meanwhile, they can also be degraded and inactivated in the acidic and enzymatic gastrointestinal environment upon oral administration. Their in vivo instability reduces their bioavailability, necessitating higher dosages. Diverse modification strategies and advanced delivery systems have been investigated to prolong their structural and functional integrities until they reach and act on the *H. pylori*. Nonetheless, a universal strategy that is suitable for most antibacterial peptides still needs to be proposed for the effective treatment of *H. pylori* infections.

These stability enhancement optimizations of the antibacterial peptides further exacerbate the high cost in their production and purification. By far, the large-scale peptide synthesis still relies heavily on the SPPS. To obtain the antibacterial peptides with the correct molecular structure and spatial conformation, rigorous SPPS and HPLC processes are required to ensure their strict quality control and batch-to-batch bioactivity consistency, which are extremely expensive. The biosynthesis via microbial fermentation may offer a cost-effective alternative to acquire the antibacterial peptides with longer sequences. However, the antibacterial peptides with optimized structures are not yet feasible to be biosynthesized. An advanced biosynthesis platform still needs to be established for the economical treatment of *H. pylori* infections.

Last but not least, the immunogenicity risk of the antibacterial peptides should also be taken into consideration during their translation for the treatment of *H. pylori* infection. Although being recognized as less immunogenic than the large-sized proteins, the antibacterial peptides can still trigger immune responses in patients. Such responses may cause allergic reactions or systemic inflammation. If they reassemble with the in vivo proteins, more severe autoimmunity may also be induced. The degree of immunogenicity depends on multiple factors of the antibacterial peptides, including their sequences, modifications, impurities in the formulations, and routes of administration. Molecular engineering strategies that well balance their antibacterial potency and immunogenicity are still in demand for the safe treatment of *H. pylori* infection.

To address these challenges, future breakthroughs will rely on interdisciplinary collaborations from microbiologists, chemists, materials scientists, and clinical researchers. Moreover, the appearance of artificial intelligence (AI) has also provided a new vitality to facilitate better *H. pylori* infection treatment by the antibacterial peptides. The AI features a series of computational models and algorithms so that the researchers can design and optimize the antibacterial peptides through computing rather than laboratory screening [30]. The well-developed AlphaFold2 by David Baker, Demis Hassabis, and John M. Jumper, who won the 2024 Nobel Prize in Chemistry, is capable of predicting protein and peptide structure with unprecedented accuracy via multiple sequence alignment, feature extraction, and three-dimensional structure prediction. Leveraging AI, the optimal sequences and structures of the antibacterial peptides could be achieved with high stability and activity, as well as low immunogenicity and toxicity. With the personalized formulations, the optimized antibacterial peptides could also be precisely delivered and released, exerting their potency in a tailored manner. These interdisciplinary collaborations are paving the way for the antibacterial peptides-based innovative therapies to effectively and safely cure *H. pylori* infection (Fig. 24.4).

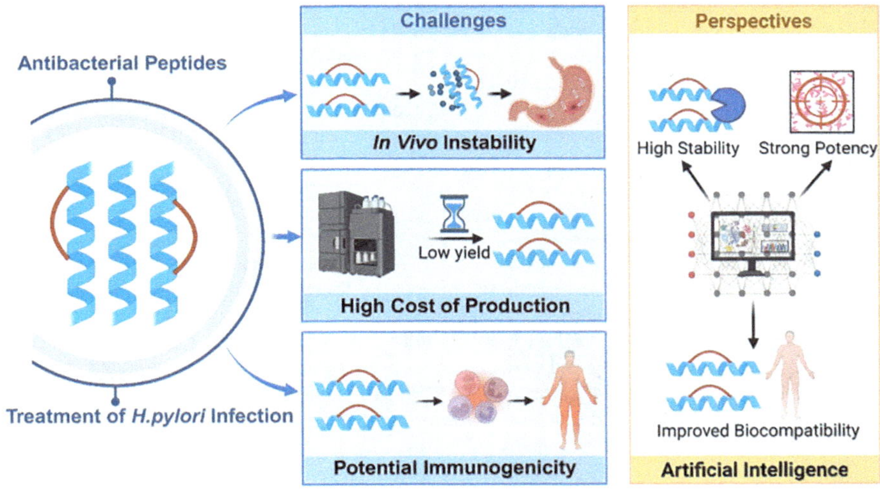


**Fig. 24.4** The challenges and perspectives of the antibacterial peptides for the effective and safe treatment of *H. pylori* infection. The antibacterial peptides can be empowered by the AI with high stability, strong potency, and improved biocompatibility. (Created with BioRender.com)

## References


1. Lee YC, Dore MP, Graham DY. Diagnosis and treatment of *H. pylori* infection. Annu Rev Med. 2022;73:183–95. https://doi.org/10.1146/annurev-med-042220-020814.
2. Boman HG, Wade D, Boman IA, Wåhlin B, Merrifield RB. Antibacterial and antimalarial properties of peptides that are cecropin-melittin hybrids. FEBS Lett. 1989;259(1):103–6. https://doi.org/10.1016/0014-5793(89)81505-4.
3. Shi Y, Wareham DW, Phee LM, Azevedo HS. Self-assembled peptide nanostructures for antibacterial applications. In: Guler MO, editor. Peptide-based biomaterials. London: The Royal Society of Chemistry; 2020. p. 395–428.
4. Henninot A, Collins JC, Nuss JM. The current state of peptide drug discovery: back to the future? J Med Chem. 2018;61(4):1382–414. https://doi.org/10.1021/acs.jmedchem.7b00318.
5. Song C, Weichbrodt C, Salnikov ES, Dynowski M, Forsberg BO, et al. Crystal structure and functional mechanism of a human antimicrobial membrane channel. Proc Natl Acad Sci USA. 2013;110(12):4586–91. https://doi.org/10.1073/pnas.1214739110.
6. Shenkarev ZO, Balandin SV, Trunov KI, Paramonov AS, Sukhanov SV, et al. Molecular mechanism of action of β-hairpin antimicrobial peptide arenicin: oligomeric structure in dodecylphosphocholine micelles and pore formation in planar lipid bilayers. Biochemistry. 2011;50(28):6255–65. https://doi.org/10.1021/bi200746t.
7. Fernandez DI, Le Brun AP, Whitwell TC, Sani MA, James M, et al. The antimicrobial peptide aurein 1.2 disrupts model membranes via the carpet mechanism. Phys Chem Chem Phys. 2012;14(45):15739–51. https://doi.org/10.1039/c2cp43099a.
8. Parton DL, Akhmatskaya EV, Sansom MS. Multiscale simulations of the antimicrobial peptide maculatin 1.1: water permeation through disordered aggregates. J Phys Chem B. 2012;116(29):8485–93. https://doi.org/10.1021/jp212358y.

9. Sun X, Li S, Okagu OD, Wang H, Zhang S, et al. Identification of peptides from defatted wheat germ proteins with dual functionality: emulsifying activity and anti-adhesive activity against *H. pylori*. Innovative Food Sci Emerg Technol. 2023;86:103367. https://doi.org/10.1016/j.ifset.2023.103367.
10. Li X, Liao C, Xu Y, Lu QH, Chen S, et al. Configuration-specific antibody for bacterial heptosylation: an Antiadhesion therapeutic strategy. J Am Chem Soc. 2023;145(1):322–33. https://doi.org/10.1021/jacs.2c09990.
11. Zhou C, Jiang M, Ye X, Liu X, Zhao W, et al. Antibacterial activities of peptide HF-18 against *H. pylori* and its virulence protein CagA. Int J Pept Res Ther. 2022;28(2):63. https://doi.org/10.1007/s10989-022-10372-7.
12. Ortiz PS, Young M, Mahmud T, Sohag MMH, Kearney CM. Generation of VacA-targeting guide peptides to increase specific antimicrobial peptide toxicity against *H. pylori*. J Biotechnol. 2025;403:17–29. https://doi.org/10.1016/j.jbiotec.2025.03.018.
13. Zou J, Wang J, Gao L, Xue W, Zhu J, et al. Ultra-short lipopeptides containing d-amino acid exhibiting excellent stability and antibacterial activity against gram-positive bacteria. Eur J Med Chem. 2025;287:117341. https://doi.org/10.1016/j.ejmech.2025.117341.
14. Meng H, Kumar K. Antimicrobial activity and protease stability of peptides containing fluorinated amino acids. J Am Chem Soc. 2007;129(50):15615–22. https://doi.org/10.1021/ja075373f.
15. Rozek A, Powers JP, Friedrich CL, Hancock RE. Structure-based design of an indolicidin peptide analogue with increased protease stability. Biochemistry. 2003;42(48):14130–8. https://doi.org/10.1021/bi035643g.
16. Shi Y, Luo G, Zhen B, Liu Z, Chen S, et al. Systematic all-hydrocarbon stapling analysis for Cecropin a generates a potent and stable antimicrobial peptide. J Med Chem. 2025;68(6):6372–85. https://doi.org/10.1021/acs.jmedchem.4c02852.
17. Zheng M, Chen H, Li X, Chen S, Shi Y, et al. Discovery of a novel antifungal agent: all-hydrocarbon stapling modification of peptide Aurein1.2. J Pept Sci. 2024;30(1):e3533. https://doi.org/10.1002/psc.3533.
18. Li X, Chen S, Zhang WD, Hu HG. Stapled helical peptides bearing different anchoring residues. Chem Rev. 2020;120(18):10079–144. https://doi.org/10.1021/acs.chemrev.0c00532.
19. Zhang Y, Zheng M, Wang Z, Liu Z, Chen S, et al. Discovery of novel antibacterial agent for the infected wound treatment: all-hydrocarbon stapling optimization of LL-37. Theranostics. 2024;14(3):1181–94. https://doi.org/10.7150/thno.87916.
20. Lohan S, Mandal D, Choi W, Konshina AG, Tiwari RK, et al. Small amphiphilic peptides: activity against a broad range of drug-resistant bacteria and structural insight into membranolytic properties. J Med Chem. 2022;65(1):665–87. https://doi.org/10.1021/acs.jmedchem.1c01782.
21. Shi Y, Chen H, Chen S, Kong X, Liu Z, et al. N-terminal Lipidation of Buforin II generates a potent and stable peptide antibiotic. J Med Chem. 2025;68(16):17624–37. https://doi.org/10.1021/acs.jmedchem.5c01258.
22. Lombardi L, Shi Y, Falanga A, Galdiero E, de Alteriis E, et al. Enhancing the potency of antimicrobial peptides through molecular engineering and self-assembly. Biomacromolecules. 2019;20(3):1362–74. https://doi.org/10.1021/acs.biomac.8b01740.
23. Xue J, Dou C, Shi Y, Fu Y, Du X, et al. Glyco-based building blocks for the chemical synthesis of glycoproteins. Int J Biol Macromol. 2025;313:144141. https://doi.org/10.1016/j.ijbiomac.2025.144141.
24. Li X, Ding Y, Xue J, Fu Y, Yan F, et al. Peptide double-stapling and arginine N-glycosylation triggered the development of therapeutic antimicrobial peptides capable of killing drug-resistant bacteria in mice. J Med Chem. 2025;68(4):4511–26. https://doi.org/10.1021/acs.jmedchem.4c02564.
25. Wang C, Hong T, Cui P, Wang J, Xia J. Antimicrobial peptides towards clinical application: delivery and formulation. Adv Drug Deliv Rev. 2021;175:113818. https://doi.org/10.1016/j.addr.2021.05.028.
26. Moosazadeh Moghaddam M, Bolouri S, Golmohammadi R, Fasihi-Ramandi M, Heiat M, et al. Targeted delivery of a short antimicrobial peptide (CM11) against *H. pylori* gastric infection

using concanavalin A-coated chitosan nanoparticles. J Mater Sci Mater Med. 2023;34(9):44. https://doi.org/10.1007/s10856-023-06748-w.
27. Shi Y, Lin R, Cui H, Azevedo HS. Multifunctional self-assembling peptide-based nanostructures for targeted intracellular delivery: design, physicochemical characterization, and biological assessment. Methods Mol Biol. 2018;1758:11–26. https://doi.org/10.1007/978-1-4939-7741-3_2.
28. Zhang W, Yang Z, Zheng J, Fu K, Wong JH, et al. A bioresponsive genetically encoded antimicrobial crystal for the oral treatment of *H. pylori* infection. Adv Sci. 2023;10(30):e2301724. https://doi.org/10.1002/advs.202301724.
29. Choudhury A, Ortiz Patrick S, Young M, Mahmud MT, Stoffel Ryan T, et al. Control of *H. pylori* with engineered probiotics secreting selective guided antimicrobial peptides. Microbiol Spectr. 2023;11(5):e02014–23. https://doi.org/10.1128/spectrum.02014-23.
30. Shi Y, Hu H. AI accelerated discovery of self-assembling peptides. Biomater Transl. 2023;4(4):291–3. https://doi.org/10.12336/biomatertransl.2023.04.008.

# Gene Editing Technology Against *Helicobacter pylori*

# 25

Mengfan Li, Jie Gao, and Tinglin Zhang

## 25.1 Introduction

*Helicobacter pylori* infects more than half of the world's population and is a leading cause of gastric cancer [1, 2]. Classified as a Group I carcinogen by the World Health Organization, *H. pylori*-related diseases result in over one million deaths annually. Current standard treatments combine antibiotics such as clarithromycin, amoxicillin, and metronidazole with proton pump inhibitors. However, widespread antibiotic misuse has led to rising resistance rates, contributing to increased treatment failure [3]. In some regions, clarithromycin resistance exceeds 30%, while metronidazole resistance rates surpass 50%. Moreover, broad-spectrum antibiotics disrupt the gut microbiota, potentially causing intestinal dysbiosis and metabolic disorders. Thus, there is an urgent need for precise, efficient, and microbiome-friendly anti-*H. pylori* strategies.

M. Li
Changhai Clinical Research Unit, Shanghai Changhai Hospital, Naval Medical University, Shanghai, China

College of Life Science, Mudanjiang Medical University, Mudanjiang, China

J. Gao
Changhai Clinical Research Unit, Shanghai Changhai Hospital, Naval Medical University, Shanghai, China

Shanghai Key Laboratory of Nautical Medicine and Translation of Drugs and Medical Devices, Shanghai, China

College of Life Science, Mudanjiang Medical University, Mudanjiang, China

T. Zhang (✉)
Changhai Clinical Research Unit, Shanghai Changhai Hospital, Naval Medical University, Shanghai, China

Shanghai Key Laboratory of Nautical Medicine and Translation of Drugs and Medical Devices, Shanghai, China



J. Gao, Y. Du (eds.), *Non-Antibiotic Therapies for Helicobacter pylori*,
https://doi.org/10.1007/978-981-95-9598-3_25

Gene editing technologies, especially the Clustered Regularly Interspaced Short Palindromic Repeats (CRISPR)-Cas system, have revolutionized life sciences. Initially identified as a bacterial adaptive immune mechanism, CRISPR-Cas has been repurposed into a highly efficient gene editing tool [4]. In *H. pylori* research, this technology is used not only to investigate bacterial pathogenesis but also to develop novel diagnostics, targeted therapies, and prophylactic vaccines, opening new avenues for precision management of *H. pylori* infection.

## 25.2 CRISPR-Cas Systems for Precise Diagnosis of *H. pylori*

Rapid and accurate diagnosis is essential for the effective treatment of *H. pylori*. Traditional methods such as the urea breath test, serology, and histology have limitations. Although quantitative polymerase chain reaction (qPCR) is highly sensitive, it requires specialized equipment and trained personnel, restricting its broad application. The combination of CRISPR-Cas systems with isothermal amplification techniques, such as recombinase polymerase amplification (RPA), offers a promising platform for next-generation molecular diagnostics.

Liu et al. developed a digital microfluidics (DMF)-based RPA-CRISPR-Cas12a system for rapid detection and genotyping of *H. pylori* [5]. His approach integrates the rapid amplification capability of RPA with the trans-cleavage activity of Cas12a [6] on an automated DMF chip, integrating them automatically onto a DMF chip. The system simultaneously detects a conserved *H. pylori* gene (*ureB*) and key virulence genes (*cagA* and vacuolating cytotoxin A (*vacA*)) within 30 minutes, with a detection limit as low as 10 copies per reaction. When tested on 80 clinical saliva samples, it showed 100% sensitivity and specificity compared to qPCR. This work highlights the high sensitivity, specificity, and potential for portability and automation of CRISPR-based diagnostics, supporting future point-of-care testing and personalized treatment guidance for *H. pylori* infection.

The CRISPR-Cas13 system (e.g., SHERLOCK) also shows promise for *H. pylori* detection. Upon binding target RNA, Cas13 exhibits nonspecific RNase activity, enabling signal detection via fluorescence or lateral flow strips. Although originally designed for RNA viruses, it can be adapted for DNA targets through reverse transcription, offering high sensitivity and multiplex detection capabilities [7]. Recent research efforts increasingly focus on integrating CRISPR diagnostics with microfluidics, smartphone-based imaging, and other technologies to further enhance its field applicability.

Precise diagnosis is critical because virulence gene genotypes such as *cagA* and *vacA* are closely linked to the pathogenicity of *H. pylori* strains and clinical outcomes [7]. Rapid determination of strain virulence helps clinicians assess patient risk and make informed decisions regarding eradication therapy, avoiding unnecessary treatment for low-virulence infections.

**Table 25.1** Major gene editing technologies and their applications in microbiological research

| Technology type | Mechanism of action | Application in microbiological research | References |
|---|---|---|---|
| ZFNs | Utilize zinc finger protein domains to recognize specific DNA sequences, fused with FokI nuclease to induce DNA double-strand breaks. | Used for generating gene knockout strains to study virulence gene function; less common in bacteria, predominantly used in eukaryotic systems. | [8] |
| TALENs | Employ TAL effectors to recognize DNA sequences, fused with FokI nuclease for site-specific cleavage. | Used for bacterial genome editing, studying resistance mechanisms; applicable in *H. pylori* for constructing specific virulence factor knockout models. | [9] |
| CRISPR-Cas9 | Utilizes single-guide RNA (sgRNA) to direct Cas9 nuclease to specific genomic loci, causing DNA double-strand breaks. | Widely applied in *H. pylori* for gene function studies, virulence factor identification, resistance gene screening, and vaccine development. | [4] |
| CRISPR-Cas12a | Guided by crRNA, cleaves DNA and exhibits trans-cleavage activity, suitable for diagnostics and multiplex gene editing. | Used for rapid molecular diagnosis of *H. pylori* (e.g., DMF-RPA-CRISPR system) and multiplex gene knockout. | [6] |
| CRISPR-Cas13 | Targets RNA rather than DNA, possesses trans-cleavage activity, suitable for RNA detection and regulation. | Used for pathogen RNA detection and gene expression regulation; applicable in *H. pylori* for studying non-coding RNA function. | [10] |

### 25.2.1 Overview of Major Gene Editing Technologies and Their Applications in Microbiological Research

Gene editing technologies have evolved from zinc finger nucleases (ZFNs) and transcription activator-like effector nucleases (TALENs) to CRISPR-Cas systems. These tools induce DNA double-strand breaks, activating cellular repair mechanisms to achieve gene knockout, insertion, or replacement. In microbiological research, particularly for pathogens like *H. pylori*, gene editing is indispensable for dissecting pathogenic mechanisms, identifying drug targets, and developing novel therapies. Table 25.1 summarizes the main gene editing technologies, their mechanisms, and applications in microbiological research.

## 25.3 Gene Editing Technology Facilitating Research into *H. pylori* Pathogenesis and Drug Target Discovery

Understanding the pathogenic mechanisms of *H. pylori* is fundamental to developing new drugs. CRISPR-Cas9 technology enables efficient knockout or modification of specific genes in *H. pylori*, facilitating the functional analysis of virulence factors at the molecular level.

For instance, cytotoxin-associated gene A (CagA) is one of the most critical virulence factors of *H. pylori*, injected into host gastric epithelial cells via a type IV secretion system, where it disrupts multiple intracellular signaling pathways, leading to aberrant cell proliferation [11]. Utilizing CRISPR-Cas9 to construct CagA knockout *H. pylori* strains and comparing them with wild-type strains allows researchers to precisely delineate the specific role of CagA in inducing gastritis, promoting epithelial cell proliferation, and carcinogenesis [12]. Similarly, gene editing of other key virulence factors such as VacA and urease (UreB) helps clarify their individual and synergistic effects in *H. pylori* colonization, immune evasion, and tissue damage [13].

In recent years, CRISPR interference (CRISPRi) technology has also been widely adopted in *H. pylori* research. Unlike complete knockout, CRISPRi utilizes a catalytically dead Cas9 fused to transcriptional repressor domains to achieve reversible suppression of target genes. This method is particularly suitable for studying the functions of essential genes, avoiding bacterial lethality resulting from complete gene deletion. For example, researchers employed CRISPRi to gradiently inhibit multiple genes related to acid adaptation in *H. pylori* (e.g., ureI and amiE), systematically evaluating their contribution to gastric colonization, thereby providing a theoretical basis for developing novel inhibitors targeting bacterial acid adaptation pathways [14].

These fundamental studies not only deepen our understanding of *H. pylori* pathogenicity but also, more importantly, provide clues for screening new drug targets. Drugs targeting these key virulence factors rather than essential bacterial genes may exert less selective pressure, potentially slowing the development of resistance. For instance, developing small molecule inhibitors that specifically block CagA injection or VacA pore-forming activity represents an attractive "anti-virulence" rather than bactericidal therapeutic strategy [9].

Furthermore, CRISPR-based genome-wide screening technology demonstrates considerable potential in *H. pylori* research. By constructing an sgRNA library covering the entire *H. pylori* genome coupled with high-throughput sequencing, researchers can systematically evaluate the role of each gene in *H. pylori*'s adaptation to the gastric environment, resistance to antibiotic stress, and host interaction in a single experiment. For example, one study using this technology identified several previously unknown genes associated with *H. pylori* biofilm formation, offering new targets for intervening in *H. pylori* community behavior and resistance development [15].

## 25.4 Direct Application of Gene Editing Technology in Developing Novel Antimicrobial Therapies

Beyond serving as a research tool, gene editing technology itself can be directly translated into an antimicrobial weapon. The core concept involves utilizing the CRISPR-Cas system to specifically target and cleave essential genes or antibiotic resistance genes in the pathogen, thereby achieving bactericidal effects or restoring antibiotic sensitivity.

### 25.4.1 Targeted Delivery Systems for Antimicrobials

Direct use of CRISPR plasmids or ribonucleoprotein complexes targeting *H. pylori* still faces challenges related to low in vivo delivery efficiency. An innovative strategy involves combining gene editing technology with advanced delivery systems. Zhang et al. reported a bio-responsive genetically encoded antimicrobial crystal (Cry3Aa-metal ion-induced auto-processing (MIIA)-LL37-P17), offering a new approach for oral antimicrobial peptide (AMP) therapy against *H. pylori* infection [16]. Although this particular study did not directly use CRISPR, the protein engineering and genetic encoding strategies employed align with the principles of genetic manipulation [17]. The researchers fused the gene for the AMP LL-37 with an *H. pylori*-targeting peptide (P17), an MIIA linker peptide, and the Cry3Aa crystal protein gene, expressing and forming protein crystals in Bacillus thuringiensis. These crystals protect LL-37 in the stomach acid and specifically release LL-37 in the localized elevated pH environment resulting from *H. pylori* urease activity, thereby precisely killing *H. pylori* while significantly reducing disruption to the gut microbiota. The concept of such "intelligent" delivery platforms can readily be integrated with CRISPR systems. For example, incorporating genetic elements encoding CRISPR-Cas systems targeting essential *H. pylori* genes into such crystals could enable oral delivery of targeted gene-editing therapeutics.

Another notable delivery strategy utilizes natural or synthetic nanoparticles as carriers for CRISPR-Cas systems. For instance, the successful application of lipid nanoparticles (LNPs) in mRNA vaccines provides a reliable platform for nucleic acid drug delivery. Researchers have attempted to encapsulate CRISPR-Cas9 mRNA targeting essential *H. pylori* genes into gastric mucosa-adhesive LNPs for oral delivery. Preliminary animal studies indicate local expression in the stomach and effective reduction of *H. pylori* load. Such non-viral delivery systems exhibit good biocompatibility and potential for scalable production, representing a significant direction for future oral gene-editing drugs [18].

### 25.4.2 Engineered Probiotics and Phage Therapy

Utilizing genetically edited probiotics or bacteriophages as vectors to deliver the CRISPR-Cas system to the site of *H. pylori* infection constitutes another promising strategy (1). Engineered Probiotics: Modifying probiotics capable of colonizing the gastric mucosa (e.g., certain *Lactobacillus* strains) to express and secrete CRISPR-Cas systems targeting essential *H. pylori* genes. These probiotics can act as "living factories" continuously producing antimicrobial effects at the infection site [19] (2). Engineered Bacteriophages: Bacteriophages are natural viruses of bacteria with high species specificity. Editing the phage genome via gene editing technology to carry CRISPR-Cas components can create "phage antibiotics" capable of precisely lysing *H. pylori* [20]. This therapy theoretically has minimal impact on the human microbiota.

Although phage therapy for *H. pylori* remains exploratory, progress in this field is rapid. Challenges include the isolation of *H. pylori*-specific phages and their limited host range, areas where gene editing technology provides tools for optimizing phage host range and efficacy.

## 25.5 Application of Gene Editing Technology in *H. pylori* Vaccine Design

Vaccines represent the most cost-effective means of preventing *H. pylori* infection. However, decades of *H. pylori* vaccine development have faced repeated setbacks, partly due to the complex immune evasion mechanisms and high antigenic diversity of *H. pylori*. Gene editing technology introduces new opportunities for screening and optimizing vaccine antigens and constructing delivery platforms.

### 25.5.1 Outer Membrane Vesicles as a Multi-antigen Delivery Platform

Outer membrane vesicles (OMVs) are nanometer-scale vesicles naturally secreted by Gram-negative bacteria, carrying various pathogen-associated molecular patterns and possessing inherent immunoadjuvant properties, making them potent vaccine delivery platforms [21]. In 2024, Liu et al. utilized genetically engineered OMVs from *Salmonella enterica* serovar Typhimurium to successfully present three key *H. pylori* antigens—UreB, CagA, and VacA—and evaluated their immunoprotective efficacy in a mouse model [22]. Further advancing this approach, Liu et al. (2025) shifted to directly utilizing *H. pylori*-derived OMVs and employed gene editing technology to modify their lipopolysaccharide (LPS) structure to counteract *H. pylori*'s immune evasion mechanisms [23]. This study used CRISPR-Cas9 to knock out key genes in *H. pylori* LPS synthesis (lpxE, lpxF, and futB), successfully converting the LPS from a low-immunogenicity tetra-acylated lipid A structure to a hexa-acylated form recognizable by TLR4, and removed the molecular mimicry of human Lewis antigens from the O-antigen, thereby significantly enhancing the immune-stimulatory capacity of the OMVs.

The researchers further utilized the hemoglobin protease autotransporter system to display major *H. pylori* antigenic proteins (UreB, VacA, and CagA) on the OMV surface. Experimental results demonstrated that OMVs with modified LPS (especially the ΔlpxE ΔlpxF ΔfutB triple knockout strain) effectively activated human and murine TLR4 signaling pathways, promoted dendritic cell maturation, and, upon oral immunization, induced robust serum IgG and gastric mucosal sIgA antibody responses, alongside a mixed Th1/Th2/Th17 cellular immune response dominated by Th17 cells.

In challenge experiments, this OMV vaccine significantly reduced gastric *H. pylori* colonization and urease activity in mice, with histopathology revealing intact gastric mucosal structure. Notably, the combination of UreB and CagA again demonstrated optimal protective efficacy, consistent with earlier findings using the Salmonella OMV platform. This study not only showcases the powerful capability of gene editing technology in directly optimizing pathogen-derived OMV vaccine platforms but also provides new insights for developing safe, effective, and potentially broad-spectrum *H. pylori* vaccines.

### 25.5.2 Live Attenuated Vaccines and Antigen Optimization

Gene editing technology can also be applied to construct live attenuated *H. pylori* vaccines [24]. Precise knockout of *H. pylori* virulence genes (e.g., CagA and VacA) or metabolic essential genes using CRISPR-Cas9 can yield safer candidate vaccine strains. These attenuated strains can transiently persist in the host, mimicking natural infection to elicit comprehensive and durable immune protection without causing disease [25].

Furthermore, gene editing technology can be used to optimize the antigen genes themselves. For instance, rational design can introduce point mutations into antigen genes like UreB to enhance their immunogenicity or stability; or epitopes from different antigens can be concatenated to construct multi-epitope fusion antigens, aiming to elicit stronger cross-protective immune responses.

## 25.6 Challenges and Future Perspectives

Although gene editing technology demonstrates immense potential in the realm of *H. pylori* therapy, its translation from laboratory to clinical application faces several challenges: (1) Delivery Efficiency and Specificity: Efficiently and specifically delivering CRISPR-Cas systems or therapeutics based on their principles to the *H. pylori* colonization niche within the gastric mucosa remains the primary technical bottleneck [26]. The extreme acidity of the stomach, the gastric mucosal barrier, and the ecological niche of *H. pylori* primarily beneath the mucus layer pose significant delivery hurdles. Novel delivery systems such as nanocarriers and engineered bacteria/phages require further optimization (2). Safety Considerations: For CRISPR-based antimicrobial therapies, off-target effects constitute a risk that must be rigorously assessed. Off-target cleavage could lead to unintended editing of the human host genome or beneficial gut microbiota, with unforeseeable consequences. Development of high-fidelity Cas variants and comprehensive safety evaluation systems is imperative (3). Emergence of Resistance: Similar to antibiotics, *H. pylori* may develop resistance to gene editing therapies, for example, through mutations in the CRISPR target sequence or the Cas protein recognition interface. Employing multiplex targeting strategies (simultaneously targeting multiple essential genes) or combination therapies may delay the emergence of resistance.

Looking ahead, future trends in the application of gene editing technology for *H. pylori* treatment may include: (1) Multifunctional Integrated Platforms: Combining diagnosis and treatment to develop "theragnostic" systems, for instance, intelligent systems that activate CRISPR antimicrobial activity only upon sensing the presence of specific *H. pylori* virulence factors [27] (2). Personalized Medicine: Rapid genotyping of clinically isolated *H. pylori* strains to "tailor" sequence-specific CRISPR drugs or select the most matched vaccine antigens (3). Modulating Host-Pathogen Interaction: Beyond directly targeting *H. pylori*, gene editing technology could be used to modulate host cell gene expression, for example, enhancing the defensive capacity of gastric epithelial cells or regulating immune responses, thereby creating a microenvironment less conducive to *H. pylori* colonization.

## 25.7 Conclusion

Gene editing technology, particularly the CRISPR-Cas system, is profoundly transforming the landscape of *H. pylori* research, diagnosis, and therapy. It provides unprecedented precision diagnostic tools, such as the DMF-based RPA-CRISPR detection platform, laying the foundation for rapid genotyping of *H. pylori* infection. Furthermore, it opens entirely novel therapeutic avenues, ranging from engineered crystals for the intelligent delivery of AMPs, to multi-valent antigen-displaying OMV vaccine platforms, and to highly promising targeted antimicrobial strategies (e.g., engineered probiotics/bacteriophages). These innovative strategies aim to overcome the two core limitations of conventional antibiotic therapies: resistance and microbiota disruption.

Despite persistent challenges related to delivery efficiency, safety, and resistance management, the precision, flexibility, and powerful functionality exhibited by gene editing technology position it as one of the most promising new weapons against the significant public health threat posed by *H. pylori*. With the continued maturation of relevant technologies and the refinement of safety assessments, it is reasonable to anticipate that gene editing-based precision medicine approaches will bring about revolutionary breakthroughs in the prevention and treatment of *H. pylori* infection in the foreseeable future.

## References

1. Hooi JKY, Lai WY, Ng WK, Suen MMY, Underwood FE, et al. Global prevalence of *H. pylori* infection: systematic review and meta-analysis. Gastroenterology. 2017;153(2):420–9. https://doi.org/10.1053/j.gastro.2017.04.022.
2. Malfertheiner P, Megraud F, Rokkas T, Gisbert JP, Liou JM, et al. Management of *H. pylori* infection: the Maastricht VI/Florence consensus report. Gut. 2022; https://doi.org/10.1136/gutjnl-2022-327745.
3. Savoldi A, Carrara E, Graham DY, Conti M, Tacconelli E. Prevalence of antibiotic resistance in *H. pylori*: a systematic review and meta-analysis in World Health Organization regions. Gastroenterology. 2018;155(5):1372–1382.e1317. https://doi.org/10.1053/j.gastro.2018.07.007.

4. Doudna JA, Charpentier E, Genome editing. The new frontier of genome engineering with CRISPR-Cas9. Science. 2014;346(6213):1258096. https://doi.org/10.1126/science.1258096.
5. Liu J, Fu R, Zhang S, Hou J, Ma H, et al. Rapid and multi-target genotyping of *H. pylori* with digital microfluidics. Biosens Bioelectron. 2024;256:116282. https://doi.org/10.1016/j.bios.2024.116282.
6. Chen JS, Ma E, Harrington LB, Da Costa M, Tian X, et al. CRISPR-Cas12a target binding unleashes indiscriminate single-stranded DNase activity. Science. 2018;360(6387):436–9. https://doi.org/10.1126/science.aar6245.
7. Gootenberg JS, Abudayyeh OO, Kellner MJ, Joung J, Collins JJ, et al. Multiplexed and portable nucleic acid detection platform with Cas13, Cas12a, and Csm6. Science. 2018;360(6387):439–44. https://doi.org/10.1126/science.aaq0179.
8. Urnov FD, Rebar EJ, Holmes MC, Zhang HS, Gregory PD. Genome editing with engineered zinc finger nucleases. Nat Rev Genet. 2010;11(9):636–46. https://doi.org/10.1038/nrg2842.
9. Joung JK, Sander JD. TALENs: a widely applicable technology for targeted genome editing. Nat Rev Mol Cell Biol. 2013;14(1):49–55. https://doi.org/10.1038/nrm3486.
10. Abudayyeh OO, Gootenberg JS, Konermann S, Joung J, Slaymaker IM, et al. C2c2 is a single-component programmable RNA-guided RNA-targeting CRISPR effector. Science. 2016;353(6299):aaf5573. https://doi.org/10.1126/science.aaf5573.
11. Hatakeyama M. *H. pylori* CagA and gastric cancer: a paradigm for hit-and-run carcinogenesis. Cell Host Microbe. 2014;15(3):306–16. https://doi.org/10.1016/j.chom.2014.02.008.
12. Tegtmeyer N, Wessler S, Backert S. Role of the cag-pathogenicity Island encoded type IV secretion system in *H. pylori* pathogenesis. FEBS J. 2011;278(8):1190–202. https://doi.org/10.1111/j.1742-4658.2011.08035.x.
13. Kao CY, Sheu BS, Wu JJ. *H. pylori* infection: an overview of bacterial virulence factors and pathogenesis. Biom J. 2016;39(1):14–23. https://doi.org/10.1016/j.bj.2015.06.002.
14. Wang T, Guan C, Guo J, Liu B, Wu Y, et al. Pooled CRISPR interference screening enables genome-scale functional genomics study in bacteria with superior performance. Nat Commun. 2018;9(1):2475. https://doi.org/10.1038/s41467-018-04899-x.
15. Liu X, Gallay C, Kjos M, Domenech A, Slager J, et al. High-throughput CRISPRi phenotyping identifies new essential genes in Streptococcus pneumoniae. Mol Syst Biol. 2017;13(5):931. https://doi.org/10.15252/msb.20167449.
16. Zhang W, Yang Z, Zheng J, Fu K, Wong JH, et al. A bioresponsive genetically encoded antimicrobial crystal for the oral treatment of *H. pylori* infection. Adv Sci (Weinh). 2023;10(30):e2301724. https://doi.org/10.1002/advs.202301724.
17. Bikard D, Euler CW, Jiang W, Nussenzweig PM, Goldberg GW, et al. Exploiting CRISPR-Cas nucleases to produce sequence-specific antimicrobials. Nat Biotechnol. 2014;32(11):1146–50. https://doi.org/10.1038/nbt.3043.
18. Qiu M, Glass Z, Chen J, Haas M, Jin X, et al. Lipid nanoparticle-mediated codelivery of Cas9 mRNA and single-guide RNA achieves liver-specific in vivo genome editing of Angptl3. Proc Natl Acad Sci USA. 2021;118(10):e2020401118. https://doi.org/10.1073/pnas.2020401118.
19. Citorik RJ, Mimee M, Lu TK. Sequence-specific antimicrobials using efficiently delivered RNA-guided nucleases. Nat Biotechnol. 2014;32(11):1141–5. https://doi.org/10.1038/nbt.3011.
20. Gomaa AA, Klumpe HE, Luo ML, Selle K, Barrangou R, et al. Programmable removal of bacterial strains by use of genome-targeting CRISPR-Cas systems. MBio. 2014;5(1):e00928–13. https://doi.org/10.1128/mBio.00928-13.
21. van der Pol L, Stork M, van der Ley P. Outer membrane vesicles as platform vaccine technology. Biotechnol J. 2015;10(11):1689–706. https://doi.org/10.1002/biot.201400395.
22. Liu Q, Shang Y, Shen L, Yu X, Cao Y, et al. Outer membrane vesicles from genetically engineered Salmonella enterica serovar Typhimurium presenting *H. pylori* antigens UreB and CagA induce protection against *H. pylori* infection in mice. Virulence. 2024;15(1):2367783. https://doi.org/10.1080/21505594.2024.2367783.
23. Liu Q, Li B, Ma J, Lei X, Ma J, et al. Development of a recombinant outer membrane vesicles (OMVs)-based vaccine against *H. pylori* infection in mice. J Extracell Vesicles. 2025;14(5):e70085. https://doi.org/10.1002/jev2.70085.

24. Sutton P, Boag JM. Status of vaccine research and development for *H. pylori*. Vaccine. 2019;37(50):7295–9. https://doi.org/10.1016/j.vaccine.2018.01.001.
25. Aebischer T, Bumann D, Epple HJ, Metzger W, Schneider T, et al. Correlation of T cell response and bacterial clearance in human volunteers challenged with *H. pylori* revealed by randomised controlled vaccination with Ty21a-based salmonella vaccines. Gut. 2008;57(8):1065–72. https://doi.org/10.1136/gut.2007.145839.
26. Gohil N, Bhattacharjee G, Lam NL, Perli SD, Singh V. CRISPR-Cas systems: challenges and future prospects. Prog Mol Biol Transl Sci. 2021;180:141–51. https://doi.org/10.1016/bs.pmbts.2021.01.008.
27. Adli M. The CRISPR tool kit for genome editing and beyond. Nat Commun. 2018;9(1):1911. https://doi.org/10.1038/s41467-018-04252-2.

# Potassium-Competitive Acid Blockers Against *Helicobacter pylori* 26

Zhiqiang Song

## 26.1 Role of Gastric Acid Suppression in *Helicobacter pylori* Eradication

In addition to antibiotic resistance and patient adherence, the intragastric acidic environment is a critical factor influencing the success of *Helicobacter pylori* eradication. Multiple studies have demonstrated that the efficacy of antibiotics or bismuth alone is suboptimal, whereas combination therapy with acid-suppressive agents significantly improves eradication rates. Furthermore, the *H. pylori* eradication rate is closely associated with both the degree and duration of gastric acid suppression [1–5].

### 26.1.1 *H. pylori* Eradication Regimens in Domestic and International Guidelines

Both domestic and international guidelines emphasize the necessity of adequate gastric acid suppression during *H. pylori* eradication therapy. This suppression enhances bacterial susceptibility to antibiotics and stabilizes antibiotic concentrations within the gastric environment. Consequently, acid-suppressive agents are considered essential components of all *H. pylori* eradication regimens. Primary recommended regimens include bismuth-containing quadruple therapy (one acid suppressant, two antibiotics, and bismuth), non-bismuth quadruple therapy (one acid suppressant and three antibiotics), triple therapy (one acid suppressant and two antibiotics), and dual therapy (one acid suppressant and one antibiotic). Although specific regimen recommendations differ among countries and regions (see Table 26.1), all guidelines incorporate a gastric acid suppressant in their treatment protocols.

Z. Song (✉)
Department of Gastroenterology, Peking University Third Hospital, Beijing, China



J. Gao, Y. Du (eds.), *Non-Antibiotic Therapies for Helicobacter pylori*,
https://doi.org/10.1007/978-981-95-9598-3_26

**Table 26.1** Comparison of recommendations for various *H. pylori* eradication regimens in guidelines from different countries/regions

| Eradication regimen | USA [6] | Europe [7] | China [8] | Japan [9] |
|---|---|---|---|---|
| Bismuth-containing quadruple therapy | Recommended | Recommended | Recommended | Not mentioned |
| Non-bismuth quadruple therapy | Not mentioned | Recommended | Not recommended | Not mentioned |
| Clarithromycin-containing triple therapy | Recommended | Recommended | Not recommended | Recommended |
| Levofloxacin-containing triple therapy | Recommended | Recommended | Not recommended | Not mentioned |
| Amoxicillin dual therapy | Recommended | Recommended | Recommended | Not mentioned |
| Rifabutin-containing triple therapy | Not mentioned | Recommended | Not recommended | Not mentioned |

Note: USA: *2024 ACG Clinical Guideline: Treatment of H. pylori Infection*; Europe: *Management of H. pylori infection: the Maastricht VI/Florence Consensus Report*; China: *2022 Chinese Guidelines for the Treatment of H. pylori Infection*; Japan: *2016 Japanese Guidelines for the Management of H. pylori Infection (Revised Edition)*

## 26.2 Mechanisms of Acid-Suppressive Agents in *H. pylori* Eradication

### 26.2.1 Enhanced Antibiotic Susceptibility of *H. pylori*

Elevated intragastric pH inhibits urease activity, reduces the metabolic activity of *H. pylori*, and impairs its colonization of the gastric mucosa [10]. At a gastric pH of 4–6, *H. pylori* enters a dormant state in which it remains viable but non-replicating and exhibits reduced susceptibility to eradication. Upon antibiotic withdrawal, the bacteria can resume growth and proliferation. When the gastric pH reaches 6–8, *H. pylori* transitions to a replicative state, during which the proliferating bacteria become more susceptible to antibiotics and are more readily eradicated. Therefore, acid-suppressive agents must increase the intragastric pH to ≥6 to induce an active growth phase in *H. pylori*, thereby enhancing its antibiotic susceptibility and facilitating eradication [11].

### 26.2.2 Enhanced Antibiotic Stability and Bioavailability

Most antibiotics demonstrate potent antimicrobial activity in vitro but are acid-labile and susceptible to degradation in the acidic gastric environment. Administering acid-suppressive agents to raise intragastric pH enhances the stability of acid-sensitive antibiotics in the stomach, thereby increasing their efficacy against *H. pylori*. Studies indicate that when the pH increases from 1 to 7, the half-lives

($T_{1/2}$) of metronidazole, amoxicillin, and clarithromycin in gastric fluid are extended by approximately two-fold, ten-fold, and 200-fold, respectively [12]. Therefore, elevating intragastric pH prolongs the in vivo $T_{1/2}$ of antibiotics and enhances their bioavailability.

### 26.2.3 Increased Antibiotic Concentration in the Mucosal Layer

*H. pylori* typically colonizes the gastric epithelial layer and the overlying mucus layer. Antibiotics must penetrate this mucus layer and achieve adequate concentrations to exert effective bactericidal activity. The physicochemical properties of the gastric mucus layer vary with pH levels, which influence antibiotic penetration. Under low pH conditions, antibiotic penetration is reduced. By raising the intragastric pH, acid-suppressive agents enhance antibiotic penetration through the mucus layer. This increase in penetration elevates antibiotic concentrations at the sites of *H. pylori* colonization, thereby enhancing bactericidal efficacy [13].

### 26.2.4 Increased Gastric Anti-*H. pylori* Immunoglobulin A Concentration

Anti-*H. pylori* immunoglobulin A (IgA) is a protective antibody generated following *H. pylori* infection and is susceptible to degradation by pepsin, an enzyme active at low pH value. Elevated intragastric pH reduces pepsin activity, thereby decreasing anti-*H. pylori* IgA degradation. The resulting increase in anti-*H. pylori* IgA levels enhance local immune defenses and may contribute to bacterial eradication. Studies have reported that *H. pylori* eradication can occur in some cases even with proton pump inhibitor (PPI) monotherapy, an effect potentially mediated by the rise in gastric anti-*H. pylori* IgA and subsequent immune clearance [14].

## 26.3 Factors Influencing the Efficacy of Acid-Suppressive Therapy in *H. pylori* Eradication

### 26.3.1 Acid Suppression Intensity

Acid-suppressive agents must maintain an intragastric pH above 6 to enhance the bactericidal effect of antibiotics against *H. pylori*. Inadequate acid suppression reduces antibiotic activity and leads to decreased eradication rates. An observational study reported that the median 24-h intragastric pH was significantly higher in patients with successful *H. pylori* eradication than in those without (6.4 vs. 5.2, $P = 0.0131$) [3]. A randomized controlled trial (RCT) further demonstrated a 100% eradication rate in patients with a median intragastric pH $\geq 6$, which was significantly higher than the 90.5% rate observed in patients with a median pH $<6$ ($P = 0.023$) [4].

### 26.3.2 Duration of Acid Suppression

In addition to the intensity of acid suppression, its duration is a critical determinant of successful *H. pylori* eradication [5]. Multiple studies have demonstrated that patients with successful eradication exhibit significantly higher 24-h and nocturnal intragastric pH levels than those with failed eradication [3, 4, 15]. Ke et al. reported that the holding time ratio (HTR) for pH >6 was significantly longer in the success group compared to the failure group ($P < 0.05$). They also observed that most *H. pylori* infections were eradicated when the 24-h median intragastric pH exceeded 5.7, confirming that adequate duration of acid suppression is essential for successful treatment [4]. Based on this evidence, the current recommended standard for acid suppression in *H. pylori* eradication therapy is maintenance of pH >6 with HTR >75% [16].

### 26.3.3 Stability of Acid Suppression

Nocturnal acid breakthrough (NAB) also influences the efficacy of *H. pylori* eradication. A randomized controlled study demonstrated a significantly shorter mean duration of NAB in the eradication group compared to the persistent infection group (99.3 vs. 293.8 min, $P < 0.05$). The eradication rate was notably lower in patients with NAB than in those without (61.3% vs. 83.3%, $P = 0.055$). These findings suggest that maintaining stable acid suppression and preventing NAB are equally important for successful *H. pylori* eradication.

## 26.4 Evolution of Acid-Suppressive Agents

Acid-suppressive agents have undergone extensive development through multiple iterations and currently comprise three main classes: $H_2$ receptor antagonists ($H_2$RAs), PPIs, and potassium-competitive acid blockers (P-CABs) [17]. The history of acid-suppressive therapy dates back to the early twentieth century, when early agents such as aluminum- and magnesium-based weak bases functioned by neutralizing gastric acid. In 1964, pharmacologist James W. Black pioneered research on $H_2$RAs, leading to the development of the first-generation agent cimetidine. However, due to its complex pharmacological profile and frequent adverse effects, cimetidine was rapidly superseded by the second-generation $H_2$RA ranitidine. Although ranitidine demonstrated fewer side effects than its predecessor, its acid-suppressive efficacy was limited, and it showed a tendency for rapid tachyphylaxis. With the advent of more potent acid suppressants, $H_2$RAs were progressively replaced by PPIs, marking the beginning of a new era in acid suppression therapy [18]. In recent years, increasing antimicrobial resistance in *H. pylori* has exposed the limitations of PPI-based regimens, leading to a gradual decline in successful eradication rates [19]. Inadequate acid suppression, which fails to induce a drug-susceptible replicative state in *H. pylori*, represents one of the main factors contributing to this decline [20]. The novel class of P-CABs has attracted considerable clinical interest due to their potent acid inhibition and capacity to address major limitations of PPIs. Currently, the *2022 Chinese Guidelines for the Treatment of H. pylori Infection* [8] recognize P-CABs as a new class of highly effective acid suppressants for *H. pylori* eradication therapy.

### 26.4.1 Types and Characteristics of Acid-Suppressive Agents

#### 26.4.1.1 $H_2$ Receptor Antagonists

$H_2$RAs represent the earliest class of acid-suppressive drugs. They reduce gastric acid secretion by reversibly binding to histamine $H_2$ receptors on parietal cells. Before PPIs became available, antibiotic regimens combined with $H_2$RAs served as a primary approach for *H. pylori* eradication. Early regimens that combined a ranitidine–bismuth complex with clarithromycin and amoxicillin achieved eradication rates of approximately 80% [21]. Owing to their relatively weak acid-suppressive efficacy, $H_2$RAs are no longer recommended for *H. pylori* eradication therapy.

#### 26.4.1.2 Proton Pump Inhibitors

In the late 1970s, Ganser, Forte, and Sachs first identified the $H^+/K^+$-ATPase (proton pump) structure in gastric parietal cells, which enabled chemists to subsequently develop PPI drugs. PPIs inhibit gastric acid secretion by irreversibly blocking the proton pump in parietal cells, thus preventing the final step of acid secretion, the exchange of $K^+$ for $H^+$ into the gastric lumen. Following 5 days of once-daily standard-dose PPI therapy, maximal gastric acid output decreases by approximately 66%. This treatment results in intragastric pH exceeding 3 for approximately 70.8% of a 24-h period and exceeding 5 for about 37.5% of the time [18]. Despite demonstrating superior acid suppression compared to $H_2$RAs, PPIs present several clinical limitations: (1) The onset of action and time to reach a steady state are relatively slow. As weakly basic prodrugs, PPIs form irreversible covalent bonds with $H^+/K^+$-ATPase. However, they only bind to actively secreting pumps, leaving resting parietal cells unaffected. This mechanism necessitates multiple doses over 3–5 days to achieve complete acid suppression. (2) The prodrug requires activation in the acidic environment of secretory canaliculi. Consequently, PPIs must be administered 30–60 min before meals to achieve optimal acid suppression [22]. (3) Most PPIs have a relatively short half-life (approximately 1–2 h), while the gastric $H^+/K^+$-ATPase has a half-life of about 50 h. With approximately one-third of proton pumps being replaced daily by newly synthesized enzymes, NAB may occur even with twice-daily dosing since these new pumps remain unexposed to PPI inhibition [23]. (4) Most PPIs are primarily metabolized by the cytochrome P450 (CYP) 2C19 enzyme, making their efficacy susceptible to genetic polymorphisms in CYP2C19 [23, 24].

**Table 26.2** Comparison of key characteristics between P-CABs and PPIs [18, 29]

| Characteristic | PPIs | P-CABs |
|---|---|---|
| Mechanism of action | Sulfenamide forms an irreversible covalent bond with cysteine residues in the sulfhydryl group of gastric $H^+$/$K^+$-ATPase | Competes with $K^+$ for the $K^+$ binding site on $H^+/K^+$-ATPase in ionized form via reversible binding |
| Molecular target | Inhibits only the activated state of $H^+$/$K^+$-ATPase | Inhibits both resting and activated states of $H^+/K^+$-ATPase |
| Physical properties | Low $pK_a$ (3.8–4.5), unstable in acidic environment | High $pK_a$ (6.0–9.0), stable in acidic environment |
| Half-life ($T_1/_2$) | Short (0.5–2.1 h) | Long (5–9 h) |
| Tissue distribution | Moderate accumulation in parietal cells (≈1000× plasma concentration) | High accumulation in parietal cells (≈100,000× plasma concentration) |
| Food effect | Acid suppression significantly affected by food intake | Acid suppression minimally affected by food intake |
| Genetic influence | Acid suppression significantly influenced by CYP2C19 polymorphisms | Acid suppression minimally influenced by CYP2C19 polymorphisms |
| Dosing for full effect | Requires multiple repeated doses to achieve full effect | Full effect achieved with first dose |

#### 26.4.1.3 Potassium-Competitive Acid Blockers

A new class of acid-suppressive agents, the P-CABs, emerged in the 1980s. P-CABs inhibit gastric acid secretion through competitive and reversible blockade of the potassium ion channel in $H^+/K^+$-ATPase. Although both P-CABs and PPIs target the proton pump, several key differences exist between them (see Table 26.2): (1) P-CABs are not prodrugs and do not require activation in an acidic environment. This property permits effective administration without regard to meals, thereby improving dosing convenience [25]. (2) P-CABs inhibit both resting and activated forms of $H^+/K^+$-ATPase, enabling a full inhibitory effect from the first dose [25]. (3) As lipophilic weak bases, P-CABs possess a higher acid dissociation constant ($pK_a$) than PPIs. This characteristic facilitates their accumulation within the acidic environment of the parietal cell secretory canaliculi, leading to more potent acid suppression [26]. (4) P-CABs exhibit a longer half-life ($T_1/_2$). For instance, vonoprazan has a half-life of up to 9 h, making P-CABs superior to conventional PPIs for suppressing nocturnal gastric acid secretion [27]. (5) P-CABs are primarily metabolized by CYP3A4. Their pharmacokinetics are not affected by CYP2C19 genotype, which minimizes interindividual variability and ensures consistent acid suppression across patients with different CYP2C19 genotypes [28].

### 26.4.2 Physicochemical Properties and Developmental History of P-CABs

The $H^+/K^+$-ATPase exists in two conformational states, E1 and E2 [30]. In the E1 form, the ion-binding sites face the cytoplasmic side of the membrane and demonstrate high affinity for $H^+$ but low affinity for $K^+$. In contrast, the E2 form presents its ion-binding sites toward the luminal side, showing high affinity for $K^+$ and low affinity for $H^+$. Upon binding $H^+$, the proton pump assumes the E1($H^+$) conformation. Subsequent hydrolysis of ATP provides energy for conversion to the phosphorylated E2($H^+$) conformation. This E2($H^+$) state then binds $K^+$ from the secretory canaliculi, enabling $K^+/H^+$ exchange. After releasing $H^+$ into the gastric lumen, the pump returns to the cytoplasm, releases $K^+$, and reverts to the E1 form, thus completing the acid secretion cycle. P-CABs inhibit gastric acid secretion by competitively binding to the $K^+$ site on the E2P conformation, which blocks $K^+/H^+$ exchange [31].

The initial P-CAB compounds featured chemical structures classified as imidazopyridine, imidazonaphthyridine, pyrimidine, and quinoline derivatives [30]. Revaprazan, a pyrimidine-based P-CAB, received approval in South Korea in 2007. However, its clinical adoption remained limited due to acid-suppressive efficacy comparable to PPIs, and it has not been introduced in the Chinese market [32]. In 2009, Japanese scientists developed vonoprazan, a novel P-CAB from the imidazopyridine class. Animal studies demonstrated that vonoprazan provided threefold greater acid inhibition than revaprazan and exhibited a significantly faster onset of action compared to *omeprazole* [33]. Vonoprazan was first approved in Japan in December 2014 and subsequently completed phase III clinical trials in China for reflux esophagitis (RE) treatment and *H. pylori* eradication. Following positive trial outcomes, it gained approval in China in 2020, becoming the first P-CAB launched in the Chinese market. Subsequently, tegoprazan and ceprazone have also been approved and marketed in China.

### 26.4.2.1 Vonoprazan

#### Molecular Structure

The Chinese chemical name of vonoprazan is 1–5-(2-Chlorophenyl)-1-(pyridin-3-ylsulfonyl)-1H-pyrrol-3-yl-N-methylmethanamine monofumarate. It belongs to the imidazopyridine class and features a pyrrolopyridine bicyclic structure as its core scaffold (see Fig. 26.1). The compound remains stable for at least 8 h in acidic gastric environments (pH 1–2). Studies show that under acidic conditions, the residual amounts of vonoprazan at 2, 4, 6, and 8 h are 99.2%, 98.3%, 97.2%, and 96%, respectively, demonstrating its sustained and stable inhibition of gastric acid secretion [31]. The methylamino and sulfonyl groups in vonoprazan form hydrogen bonds with Glu795 and Tyr799 residues of the $H^+/K^+$-ATPase protein target [34],

**Fig. 26.1** Molecular structure of vonoprazan

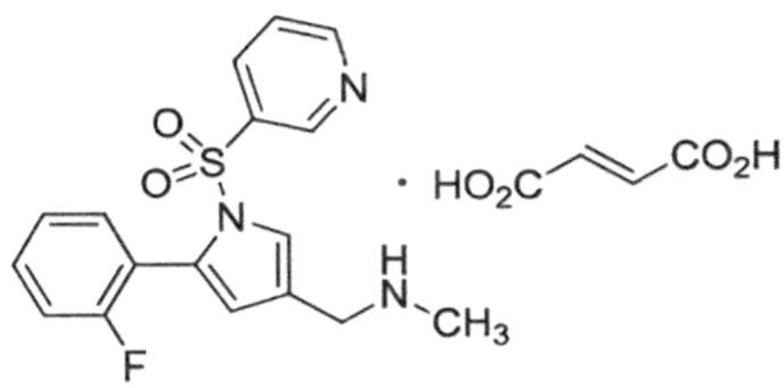

conferring high binding affinity and slow dissociation from the enzyme. In a 20 mmol/L potassium chloride solution at pH 7, the half-life of vonoprazan dissociation from $H^+/K^+$-ATPase is 7.5 h [34].

### Pharmacokinetics

Studies in healthy non-Chinese populations indicate that after a single 20 mg oral dose of vonoprazan, the median time to peak concentration ($T_{max}$) is ≤2.0 h, with a mean half-life ($T_{1/2}$) ranging from 7 to 9 h [35, 36]. In healthy Chinese males, $T_{max}$ values are 2.0 h under fasting conditions and 2.5 h when administered postprandially. Under repeated dosing conditions, $T_{max}$ values are 1.5 h in non-Chinese populations and 2.0 h in Chinese subjects, with no significant difference observed between groups [36]. In Chinese subjects, the peak plasma concentration ($C_{max}$) following a single fasting dose is 18.2 ng/mL, which increases to 25.2 ng/mL under fed conditions. The elimination half-life is 7.0 h after single-dose administration and extends to 7.6 h with repeated dosing [36].

Vonoprazan is predominantly metabolized by the hepatic cytochrome P450 enzyme *CYP3A4*, with additional contributions from *CYP2B6*, *CYP2C19*, *CYP2D6*, and sulfotransferase *SULT2A1* (in vitro) [36]. This metabolic profile enables vonoprazan to largely avoid the interindividual variability in acid suppression associated with *CYP2C19* rapid metabolizer genotypes. Consistent with this, studies have demonstrated that the pharmacokinetics and pharmacodynamics of vonoprazan remain unaffected by *CYP2C19* polymorphisms in both healthy male volunteers and patient populations, including those with RE [37].

### Pharmacodynamics

Vonoprazan is a basic compound with a $pK_a$ (9.3) higher than that of PPIs. It ionizes immediately upon exposure to an acidic environment and rapidly accumulates within the acidic secretory canaliculi. The pharmacodynamic activity of vonoprazan under neutral conditions (pH 7.5) is nearly identical to that under weakly acidic conditions (pH 6.5) [38]. Furthermore, vonoprazan demonstrates high acid stability. These characteristics promote its accumulation in gastric parietal cells, achieving intracellular concentrations up to 100 million times higher than plasma levels. This allows the drug to remain protonated and effectively bind to $H^+/K^+$-ATPase, thereby inhibiting acid secretion [35]. Clinical studies indicate that a single 20-mg dose of vonoprazan elevates intragastric pH to approximately 7 within 4 h after administration. After 7 days of once-daily 40-mg dosing, the percentage of time with intragastric pH ≥ 4 was 85.3% on Day 1 and 100.0% on Day 7 [35]. A multicenter, RCT enrolled 418 treatment-naïve adults with *H. pylori* infection who received vonoprazan 20 mg twice daily (after meals) plus amoxicillin 1000 mg three times daily (after meals). The median intragastric pH was 7.4 on Day 1 and rose to 7.8 by Day 7. Intragastric pH exceeded 6 within 4–5 h after the first dose and was consistently maintained between 7 and 9 throughout the study [39].

#### 26.4.2.2 Tegoprazan

**Molecular Structure**

The Chinese chemical name of tegoprazan is (S)-4-(5,7-difluorochroman-4-yl)oxy-N,N,2-trimethyl-1H-benzodimidazole-6-carboxamide. As a member of the benzimidazole class, it shares a core structural motif with PPIs. Animal studies demonstrate that tegoprazan produces reversible inhibition of porcine $H^+/K^+$-ATPase, with a half-maximal inhibitory concentration (IC ~ 50~) of 0.53 μM (Fig. 26.2).

**Pharmacokinetics**

In healthy Korean populations, the time to reach peak plasma concentration ($T_{max}$) ranges from 0.5 to 1.5 h. After administration of single doses at 50, 100, 200, and 400 mg, the corresponding peak plasma concentrations ($C_{max}$) were 669.9, 1526.9, 2639.2, and 5053.2 ng/mL, respectively, and the area under the plasma concentration–time curve ($AUC_{0-\infty}$) ranged from 2788.0 to 24,866.0 ng·h·mL$^{-1}$. Following once-daily administration of 100 mg or 200 mg tegoprazan for 7 days, the $C_{max}$ on Day 7 was slightly reduced compared to Day 1 [40]. In a single-dose escalation study involving healthy Chinese subjects who received 50, 100, and 200 mg of tegoprazan, the observed $T_{max}$ was 0.5 h. The corresponding Cmax values for each dose were 813.80, 1494.60, and 2829.00 ng/mL, respectively, and the elimination half-life ($T_{1/2}$) ranged from 3.87 to 4.57 h [41].

**Pharmacodynamics**

Tegoprazan exhibits a $pK_a$ of 5.1, which is moderately higher than that of PPIs (3.8–4.5) but lower than that of vonoprazan (9.3). Following once-daily administration of 50 mg tegoprazan, the percentage of time with intragastric pH > 4 over 24 h was 54.5% on Day 1 and 68.2% on Day 7 [42]. At a dose of 100 mg, the corresponding values were 62.3% on Day 1 and 70.4% on Day 7, while administration of 200 mg resulted in values of 76.8% on Day 1 and 94.6% on Day 7 [40].

**Fig. 26.2** Molecular structure of tegoprazan

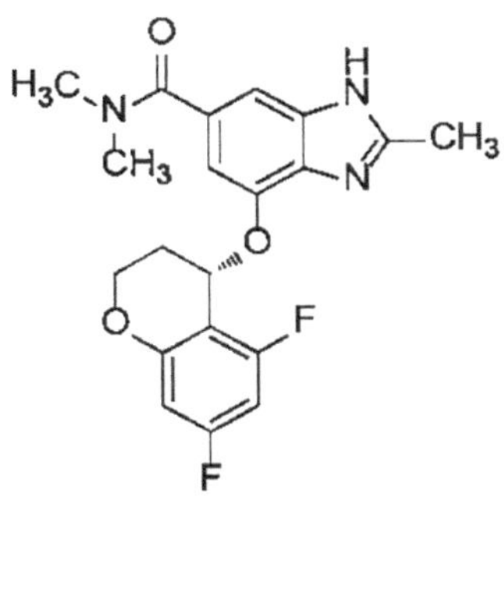

**Fig. 26.3** Molecular structure of ceprazone

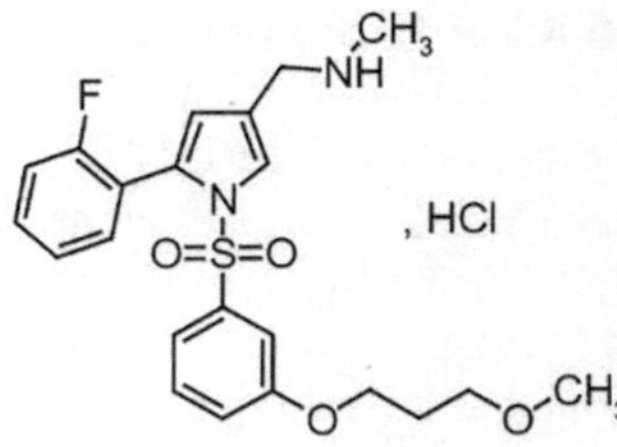

### 26.4.2.3 Ceprazone

**Molecular Structure**

The Chinese chemical name of ceprazone is 1–5-(2-Fluorophenyl)-1-1H-pyrrol-3-yl-N,N-dimethylmethanamine hydrochloride [43]. Under conditions of pH 6, the half-maximal inhibitory concentration (IC ~ 50~) of ceprazone is 0.029 μM [44] (Fig. 26.3).

**Pharmacokinetics**

Following single-dose administration of 10 mg and 20 mg ceprazone hydrochloride tablets to healthy subjects, the time to peak plasma concentration ($T_{max}$) was 1.25 h and 1.50 h, respectively, with peak plasma concentrations ($C_{max}$) of 12.4 ng/mL and 31.7 ng/mL. The $C_{max}$ of ceprazone is influenced by high-fat meals, which increase the systemic exposure of the parent drug by 35% and 27%, respectively [43]. The prescribing information recommends administration before breakfast. In vitro experiments using human liver CYP450 enzymes indicate that ceprazone is predominantly metabolized by *CYP3A4*, with minimal or negligible contributions from other isoenzymes, including *CYP1A2*, *CYP2B6*, *CYP2C8*, *CYP2C9*, *CYP2C19*, and *CYP2D6* [43].

**Pharmacodynamics**

After 7 days of repeated administration to healthy subjects, the 20-mg and 40-mg dose groups maintained intragastric pH > 3 for 99.6% and 100.0% of the 24-h period, respectively. The corresponding percentages were 98.3% and 100.0% for pH > 4 and 97.4% and 100.0% for pH > 5 [43] (Table 26.3).

## 26.5 Clinical Efficacy of P-CABs in *H. pylori* Eradication Therapy

All patients infected with *H. pylori* should receive eradication therapy; however, increasing antimicrobial resistance poses a major challenge to successful treatment [50, 51]. The earliest effective regimen for *H. pylori* infection consisted of bismuth, metronidazole, and tetracycline. With the emergence of metronidazole resistance, PPIs were incorporated, leading to the development of acid-suppressive bismuth-containing quadruple therapy. During the 1990s, PPI-based triple therapy (a PPI

**Table 26.3** Differences in pharmacological characteristics among P-CABs

| | Vonoprazan | Tegoprazan | Ceprazone |
|---|---|---|---|
| Chemical structure | Pyrrole-pyridine derivative [36] | Benzimidazole derivative | Pyridine derivative |
| Mechanism of action | Competitive $K^+$ binding via ionic bond (reversible) [36] | Competitive $K^+$ binding via ionic bond (reversible) [45] | Competitive $K^+$ binding via ionic bond (reversible) [43] |
| Tissue/plasma concentration | High accumulation in canaliculi (10 [8]-fold ionized form) [31] | 10 [4]-fold ionized form [31] | No data |
| Acid activation required | No (administration independent of meals) [31, 46] | No (administration independent of meals) [47, 48] | Swallow whole; take before breakfast [43] |
| $T_{1/2}$ | Long (7.7 h) [31] | Short (4.1 h) [48] | Intermediate (6.0 h) [43] |
| PKa | 9.331 | 5.147 | No published data |
| IC50 | 0.019 μM [38] | 0.53 μM [45] | 0.029 μM [44] |
| Pumps bound | Resting and active [49] | No data | No data |
| Binding sites | Multiple sites on proton pump [49] | No data | No data |
| Electrostatic barrier | Present (blocks back-diffusion) [49] | No data | No data |
| Membrane permeability | Good at pH 7.4 [31] | No data | No data |
| Solubility | Rapid dissolution [31] | No data | No data |
| Acid stability | Stable in acidic environment [31] | No data | No data |

plus two antibiotics) achieved eradication rates of approximately 90% and was established as the primary internationally recommended regimen. However, due to rising resistance, the efficacy of this triple therapy has declined to unacceptable levels (eradication rates below 80%) [52]. This prompted a shift in research focus toward bismuth quadruple therapy, which combines a PPI, bismuth, and two antibiotics. By 2012, PPI-based bismuth quadruple therapy had become the globally recommended first-line regimen for *H. pylori* eradication. Nonetheless, growing resistance to key antibiotics, particularly metronidazole (~45%), clarithromycin (~21%), and levofloxacin (~27%), has reduced the effectiveness of PPI-based quadruple therapy [53], underscoring the urgent need for new agents to address resistance issues. P-CABs, as a novel class of acid suppressants, offer several advantages over PPIs, including faster onset of action and more potent acid suppression. A meta-analysis demonstrated that vonoprazan-based quadruple therapy yields significantly higher eradication rates than PPI-based quadruple therapy (92.94% vs. 87.82%; risk ratio (RR) = 1.06, 95% confidence interval (CI): 1.03–1.09) [54]. This superior efficacy was observed both in the overall population (85.6% vs. 77.8%, $P < 0.001$) and among patients with clarithromycin-resistant strains (73.7% vs. 41.5%, $P = 0.020$) [55]. The *2022 Chinese Guidelines for the Treatment of H. pylori*

*Infection* [8] endorse both PPI- and P-CAB-based quadruple regimens as appropriate options for *H. pylori* eradication.

Bismuth quadruple therapy presents several limitations, including complex dosing regimens, suboptimal patient adherence, significant effects on gut microbiota, and limited access to effective antibiotics in some regions [56]. For these reasons, the Maastricht VI/Florence consensus incorporated high-dose PPI plus amoxicillin dual therapy into clinical practice to minimize unnecessary antibiotic exposure and curb the development of antimicrobial resistance [7]. However, PPIs are unable to provide potent and sustained acid suppression over a full 24-h period, which restricts the maximal antibacterial activity of co-administered antibiotics. To overcome this drawback, P-CAB-based dual therapy has been introduced. A clinical study conducted in China investigated the efficacy of vonoprazan plus amoxicillin dual therapy for *H. pylori* eradication. The results demonstrated that the vonoprazan–amoxicillin dual regimen, used as first-line treatment, was more effective than PPI-based quadruple therapy (eradication rate: 96.30% vs. 77.78%, P = 0.005) and was associated with a lower incidence of adverse events (5.66% vs. 22.00%, P = 0.033) [57]. The *2024 ACG Clinical Guideline: Treatment of* H. pylori *Infection* suggests P-CAB plus amoxicillin dual therapy as a first-line option for treatment-naïve patients (conditional recommendation; moderate quality of evidence). For those without prior *H. pylori* treatment and with unknown clarithromycin susceptibility, P-CAB–clarithromycin triple therapy is recommended over PPI–clarithromycin triple therapy (conditional recommendation; moderate quality of evidence) [6].

### 26.5.1 Clinical Efficacy of P-CAB-Based Quadruple Regimens

Bismuth-containing quadruple therapy represents the most widely used first-line regimen in China. Clinical investigations of P-CAB-based bismuth quadruple therapy have predominantly been conducted in Chinese populations. Research has primarily focused on vonoprazan-based quadruple therapy, followed by tegoprazan-based regimens, whereas only a single study has explored ceprazone-based quadruple therapy.

Multiple RCTs in Chinese populations have reported *H. pylori* eradication rates of 93.3–97.7% for vonoprazan-based quadruple therapy in per-protocol (PP) analyses [58–62]. A prospective, open-label, non-inferiority RCT enrolled 332 treatment-naïve patients with *H. pylori* infection and penicillin allergy [62]. Participants were randomly assigned in a 1:1 ratio to receive either a 14-day P-CAB quadruple regimen (vonoprazan 20 mg twice daily; colloidal bismuth 200 mg twice daily; doxycycline 100 mg twice daily; and metronidazole 400 mg three times daily) or a 14-day PPI quadruple regimen (rabeprazole 10 mg twice daily; the other three drugs identical to the P-CAB group). The study compared eradication rates, adverse events, and adherence between the two groups. The results demonstrated a significantly higher eradication rate in the P-CAB quadruple therapy group (90.4% vs. 71.1%, P = 0.013). No significant differences were observed in the overall incidence of adverse events (39.8% vs. 40.4%; P = 0.911) or treatment adherence (88.0% vs. 91.0%; P = 0.372) between the groups.

Another multicenter randomized clinical trial evaluated the treatment duration for vonoprazan-based quadruple therapy by enrolling 1300 patients with initial *H. pylori* infection. Participants were randomly assigned to 10-day or 14-day regimens consisting of vonoprazan 20 mg twice daily, bismuth 220 mg twice daily, amoxicillin 1000 mg twice daily, plus either clarithromycin 500 mg twice daily or tetracycline 500 mg four times daily. Eradication was assessed via [13]C-urea breath test at least 6 weeks post-treatment. PP analysis showed eradication rates of 93.2% for the 10-day regimen and 93.7% for the 14-day regimen (P < 0.001), while intention-to-treat (ITT) analysis demonstrated rates of 88.6% and 89.4%, respectively (P = 0.007). The 10-day group experienced fewer adverse events than the 14-day group (22.59% vs. 28.50%, P = 0.016), though no significant differences were found in treatment adherence or discontinuation due to severe adverse events. Three real-world studies further confirmed that vonoprazan-based quadruple regimens (containing vonoprazan, amoxicillin, clarithromycin/furazolidone/minocycline, and bismuth) achieved PP eradication rates above 93% [63–65].

A meta-analysis of East Asian populations included 10 clinical studies (seven RCTs and three cohort studies) comparing vonoprazan-based versus PPI-based quadruple therapy for *H. pylori* eradication. The results demonstrated significantly higher eradication rates with vonoprazan-based quadruple therapy (92.94% vs. 87.82%; RR = 1.06, 95% CI: 1.03–1.09) [54]. In a separate systematic review, Javier P. Gisbert et al. evaluated 77 studies (including 24 RCTs), primarily from Asian countries, with one study each from the United States/Europe and Australia. This analysis reported an overall eradication rate of 94% for vonoprazan-based quadruple regimens [66]. A network meta-analysis of 101 RCTs (21,745 patients) conducted between January 2000 and September 2021 used the surface under the cumulative ranking curve (SUCRA) to evaluate treatment efficacy. The vonoprazan-based bismuth-containing quadruple regimen achieved the highest SUCRA value, ranking first in effectiveness among all *H. pylori* eradication regimens [67].. Another network meta-analysis of 43 studies (11,043 patients) also identified the vonoprazan-based bismuth quadruple regimen as the most effective treatment, with a SUCRA value of 90.88%, superior to PPI-based triple and quadruple therapies as well as vonoprazan-based dual and triple regimens [68].

Only a limited number of RCTs have investigated tegoprazan-based quadruple regimens for *H. pylori* eradication. In a multicenter, randomized, double-blind, parallel-group comparative study, 555 Chinese patients with *H. pylori* infection were randomly assigned to receive either tegoprazan (50 mg) or esomeprazole (20 mg) for 14 days. In the full analysis set, eradication rates were 93.5% for the tegoprazan group versus 86.4% for the esomeprazole group (difference: 7%; 95% CI: 2.06–11.99%), with superiority testing confirming significantly higher efficacy with tegoprazan (P = 0.006). Secondary outcomes indicated consistently higher eradication rates with tegoprazan across all subgroups compared to esomeprazole, though these differences did not reach statistical significance (P > 0.05) [69].

Only one meta-analysis from Korea has directly compared the efficacy and safety of tegoprazan-based versus PPI-based regimens for *H. pylori* eradication. This analysis incorporated seven studies comprising 3200 patients. The ITT analysis demonstrated comparable eradication rates between tegoprazan-based and PPI-based quadruple therapy (77.3% vs. 76.4%, P = 0.68). Similarly, PP analysis showed nearly identical rates (84.3% vs. 84.2%, P = 0.69). These results indicate no statistically significant difference in eradication efficacy between tegoprazan-based and PPI-based quadruple regimens under either ITT or PP analysis [70].

Currently, only one randomized, double-blind, multicenter phase III clinical trial has evaluated ceprazone-based versus esomeprazole-based bismuth quadruple therapy for *H. pylori* eradication. This study enrolled 576 patients with confirmed *H. pylori* infection, randomly assigning them to receive either a bismuth quadruple regimen containing ceprazone 20 mg or an esomeprazole 20 mg-based regimen. In the ITT population, eradication was achieved in 87.8% of ceprazone-treated patients, demonstrating non-inferiority to the 82.52% rate observed with esomeprazole (difference: 5.29%; 95% CI: −0.55% to 11.18%; P = 0.075). PP analysis showed significantly higher eradication rates with ceprazone (93.5% vs. 88.2%; treatment difference: 5.25%, 95% CI: 0.29–10.45%; P = 0.038) [71].。 .

## 26.5.2 Clinical Efficacy of P-CAB-Based Triple Therapies

Most clinical evidence for P-CAB-based triple therapy originates from Japanese studies. Kusano et al. [72] investigated first-line *H. pylori* eradication using a vonoprazan-based triple regimen (vonoprazan 20 mg, amoxicillin 750 mg, and clarithromycin 200 mg, all twice daily for 7 days) in 118 Japanese junior high school students. Eradication rates were 81.4% by ITT analysis and 85.7% by PP analysis, reflecting good tolerability in this adolescent cohort. Multiple additional Japanese clinical studies evaluating 7-day vonoprazan triple therapy have consistently reported high eradication rates, surpassing those achieved with 7-day PPI-based triple therapy (containing amoxicillin and clarithromycin), along with maintained efficacy against clarithromycin-resistant strains [73–75]. In a randomized, parallel-group, single-blind trial, Maruyama et al. [76] directly compared 7-day vonoprazan-based and PPI-based triple regimens as first-line *H. pylori* treatment. Both ITT (95.8% vs. 69.6%, P = 0.00003) and PP analyses (95.7% vs. 71.4%, P = 0.0002) demonstrated significantly superior eradication rates with vonoprazan.

A Chinese three-arm RCT evaluated vonoprazan-based dual and triple therapies against PPI-based bismuth quadruple therapy for *H. pylori* eradication. The results demonstrated non-inferiority of the vonoprazan triple regimen to bismuth quadruple therapy in both ITT (95.9% vs. 92.0%, P < 0.001) and PP analyses (96.7% vs. 97.4%, P = 0.009) [77]. A systematic review by Jung et al. of 10 studies reported significantly higher eradication rates with vonoprazan-based triple therapy versus PPI-based triple therapy in ITT analysis (88.1% vs. 72.8%, RR = 1.19, P < 0.00001), with comparable adverse event incidence (approximately 8.0% each, P = 0.06) [78]. Another meta-analysis of 897 patients confirmed the superiority of vonoprazan-based triple therapy over PPI-based regimens (ITT eradication: 91.4% vs. 74.8%, OR = 3.68, P < 0.05) [79]. Accordingly, the Japanese Society of Gastroenterology's Practice Guidelines for Peptic Ulcer Disease recommend the vonoprazan–amoxicillin–clarithromycin triple regimen as first-line treatment for *H. pylori* eradication [77].

Two RCTs and four real-world studies have evaluated tegoprazan-based triple therapy for *H. pylori* eradication. A Korean randomized, double-blind, controlled phase III study enrolled 300 patients to assess a tegoprazan-containing triple regimen 80. ITT analysis showed eradication rates of 62.9% with tegoprazan versus 60.6% with lansoprazole (P = 0.009), while PP analysis demonstrated rates of 69.3% versus 67.3%, respectively, establishing non-inferiority of the tegoprazan-based regimen (P = 0.013) [80]. Another RCT involving 97 patients compared the efficacy and safety of different tegoprazan dosing regimens against vonoprazan-based triple therapy. The 50-mg tegoprazan triple regimen achieved a *H. pylori* eradication rate of 60.6%, significantly lower than the 100-mg tegoprazan regimen (78.8%) and the 20-mg vonoprazan regimen (84.9%). All treatments demonstrated good tolerability without significant safety differences [81]. These results indicate comparable efficacy between 100-mg tegoprazan and 20-mg vonoprazan, while 50-mg tegoprazan appears to provide suboptimal acid suppression. A retrospective study of 948 patients receiving tegoprazan triple therapy (50 mg twice daily plus amoxicillin 1000 mg and clarithromycin 500 mg twice daily) compared 7-day (n = 435) and 14-day (n = 513) treatment durations. Eradication rates were 70.5% with 7-day therapy versus 85.1% with 14-day therapy, with no significant difference in adverse event rates [82]. Currently, no meta-analysis has specifically addressed tegoprazan-based triple therapy, and studies evaluating ceprazone-based triple regimens remain unavailable.

### 26.5.3 Clinical Efficacy of P-CAB-Based Dual Therapies

Most clinical evidence for P-CAB-based dual therapy derives from vonoprazan studies. Zhang et al. [83] conducted an RCT among 250 *H. pylori*-infected patients, evaluating first-line eradication with vonoprazan–amoxicillin dual therapy. The dual-therapy group received vonoprazan 20 mg twice daily plus amoxicillin 1.0 g three times daily for 14 days, while the control group received bismuth quadruple therapy (lansoprazole 30 mg, clarithromycin 500 mg, amoxicillin 1000 mg, and bismuth potassium citrate 240 mg, all twice daily for 14 days). ITT analysis showed eradication rates of 92.0% versus 88.0%, and PP analysis demonstrated 95.8% versus 91.7% for dual versus quadruple therapy, respectively. The dual-therapy group had significantly fewer adverse events (11.2% vs. 20.8%, P = 0.038), establishing non-inferior efficacy with improved safety. A meta-analysis by Abosheaishaa H et al. of nine randomized trials [84] found comparable overall eradication between vonoprazan-based dual therapy and PPI-based bismuth quadruple regimens (OR: 1.44, 95% CI: 0.96–2.14; I [2] = 56%).

In sensitivity analysis excluding the study by Yang et al., dual therapy showed significantly superior *H. pylori* eradication compared to quadruple therapy (OR: 1.64, 95% CI: 1.12–2.41; I [2] = 43%, P = 0.01), with heterogeneity remaining stable. The dual-therapy group demonstrated significantly lower overall adverse events (OR: 0.32, 95% CI: 0.26–0.40; P = 0.0001), particularly for nausea and vomiting (OR: 0.42, 95% CI: 0.28–0.65; P = 0.0001). Treatment adherence was comparable between groups (OR: 1.16, 95% CI: 0.76–1.76; P = 0.50). These results consistently demonstrate that dual therapy achieves equivalent or superior eradication rates while substantially reducing adverse events compared to PPI-based bismuth quadruple therapy. In a cost-effectiveness analysis of 100 patients, Jian Wenhua et al. [85] found comparable eradication rates (94% vs. 88%, P > 0.05) but significantly fewer adverse events with vonoprazan dual therapy versus PPI-based quadruple therapy (12% vs. 54%, P < 0.05). Total costs were substantially lower for dual therapy (143.02 CNY vs. 384.41 CNY).

Numerous clinical studies have examined antibiotic dosing and treatment duration in dual therapy. Early trials in China and Japan evaluating 7-day vonoprazan dual therapy (20 mg twice daily plus amoxicillin 1000 mg, 750 mg, or 500 mg three times daily) failed to achieve the clinically acceptable 90% eradication threshold for *H. pylori*, potentially due to insufficient treatment duration or suboptimal amoxicillin dosing [86–89]. Subsequent RCTs investigated 10-day and 14-day vonoprazan dual therapy regimens. Pooled data indicate that with a daily amoxicillin dose of 2000 mg, the 10-day regimen produces suboptimal results, while the 14-day regimen achieves satisfactory eradication (> 76% ITT, > 90% PP) [90]. When daily amoxicillin exceeds 3000 mg, both 10-day and 14-day vonoprazan dual regimens demonstrate high efficacy (> 80% ITT and > 90% PP). A 2022 meta-analysis confirmed real-world eradication rates exceeding 90% for vonoprazan–amoxicillin dual therapy after excluding Western populations [86]. A recent multicenter RCT in China [39] compared 10-day versus 14-day vonoprazan–amoxicillin dual therapy (20 mg twice daily and 1000 mg three times daily, both after meals) for efficacy, safety, and adherence. The 14-day regimen (n = 209) demonstrated satisfactory efficacy with PP eradication >95% and ITT eradication of 88%, whereas the 10-day regimen (n = 209) showed suboptimal performance. A contemporary meta-analysis of 29 clinical studies on P-CAB dual therapies reported vonoprazan-based eradication rates of 82.4%, 90.7%, and 93.9% for 7-day, 10-day, and 14-day regimens, respectively [91].

A recent multicenter study in *The Lancet Microbe* [92] investigated 14-day vonoprazan dual therapy with low-dose (2000 mg/day) versus high-dose (3000 mg/day) amoxicillin for *H. pylori* eradication. The low-dose regimen demonstrated non-inferior efficacy to high-dose therapy while exerting minimal effects on gut microbiota composition and resistance gene abundance. In contrast, high-dose therapy significantly increased $\beta$-lactamase resistance genes, with levels remaining elevated during short-term follow-up. These findings suggest that extended observation is required to fully evaluate the ecological impact of high-dose regimens. The vonoprazan plus low-dose amoxicillin dual regimen appears to minimize effects on gut microbial ecology and resistance gene development while maintaining favorable safety.

Two single-center Chinese studies [93, 94] evaluated vonoprazan-based dual therapy in penicillin-allergic patients. Both the regimen combining vonoprazan 20 mg twice daily with tetracycline 500 mg three times daily (92.0% ITT and 95.1% PP) and the regimen combining vonoprazan 20 mg twice daily with minocycline 100 mg twice daily (90.7% ITT and 94.2% PP) demonstrated high eradication efficacy. Additionally, a real-world study in China reported 92.5% eradication using 14-day vonoprazan–amoxicillin dual therapy as rescue treatment for *H. pylori* infection [95].

Limited data exist regarding the safety and efficacy of tegoprazan and ceprazone in dual therapy regimens, with only three studies available for tegoprazan. Two RCTs [96, 97] reported comparable *H. pylori* eradication rates between tegoprazan-based dual therapy and PPI-based or tegoprazan-based quadruple therapy, alongside significantly reduced adverse events. A multicenter non-inferiority RCT assigned 368 *H. pylori*-infected patients to either tegoprazan-amoxicillin dual therapy (50 mg twice daily +750 mg four times daily) or esomeprazole-amoxicillin dual therapy (20 mg four times daily +750 mg four times daily) for 14 days. ITT analysis showed eradication rates of 84.2% versus 85.8% ($P = 0.77$), while PP analysis demonstrated 88.5% versus 88.2% ($P > 0.99$), confirming comparable efficacy between regimens [98].

## 26.6 Safety of P-CABs

Early P-CAB acid suppressants such as Scheme 28080 were primarily discontinued due to severe hepatotoxicity associated with long-term or repeated administration [99]. The newer-generation P-CAB vonoprazan was structurally optimized during development to minimize cytotoxic potential [34]. To date, no hepatotoxicity cases have been reported for vonoprazan, and multiple clinical studies confirm that 4–8 weeks of continuous treatment does not cause significant alterations in alanine aminotransferase or aspartate aminotransferase levels [34, 37, 100]. Across clinical trials, P-CABs demonstrate adverse event profiles similar to PPIs, with gastrointestinal symptoms (e.g., diarrhea and constipation) as the most frequently reported reactions. These are typically mild to moderate and resolve upon treatment discontinuation [67, 76, 79, 88]. Direct comparative safety studies between different P-CABs are currently unavailable. The following summarizes safety data for individual agents.

### 26.6.1 Vonoprazan [36]

Adverse reactions associated with vonoprazan are generally mild, primarily comprising nausea, vomiting, diarrhea, and rash. Commonly reported adverse reactions ($\geq 1/100$ to $<1/10$) in clinical studies include diarrhea and constipation. In trials combining vonoprazan with amoxicillin, clarithromycin, and bismuth citrate, frequently observed adverse reactions (> 1%) encompassed taste disturbance, palpitations, hyperlipidemia, proteinuria, elevated serum bilirubin, and elevated serum phosphorus. All reported events were mild in intensity, and none were observed during vonoprazan monotherapy.

### 26.6.2 Tegoprazan [101]

Two domestic clinical trials have documented adverse reactions to tegoprazan, showing elevated serum gastrin as very common (≥ 1/10), while hyperlipidemia, hyperuricemia, abnormal hepatic function, and headache are common (≥ 1/100 to <1/10). A domestic Phase II trial assessing the safety of tegoprazan combined with amoxicillin, clarithromycin, and bismuth in adults for *H. pylori* eradication included 278 patients receiving tegoprazan 50 mg, amoxicillin 1000 mg, clarithromycin 500 mg, and bismuth potassium citrate 240 mg twice daily. Safety outcomes from the 14-day treatment revealed the most frequent adverse reactions (> 5%) as taste disturbance (26.3%), elevated serum gastrin (23.7%), diarrhea (5.8%), and nausea (5.4%). Other common reactions (> 1%) included melena, vomiting, insomnia, rash, and QRS (Complex in electrocardiogram) axis deviation. These adverse events were not observed during tegoprazan monotherapy and were predominantly mild.

### 26.6.3 Ceprazone [43]

The safety profile of ceprazone hydrochloride tablets was assessed in two Phase III trials and one Phase II trial involving adults. Commonly reported adverse reactions (≥ 1/100 to <1/10) included hyperlipidemia, hyperuricemia, gastric polyps, diarrhea, hepatic function abnormalities, arrhythmia, sinus bradycardia, elevated transaminases, proteinuria, and ST-segment abnormalities on electrocardiogram.

A Phase III trial assessed the safety of ceprazone hydrochloride combined with amoxicillin, clarithromycin, and bismuth as quadruple therapy for *H. pylori* eradication in adults. Among 287 patients receiving ceprazone 20 mg twice daily with amoxicillin 1000 mg, clarithromycin 500 mg, and bismuth 240 mg twice daily for 14 days, the most frequent adverse reactions (incidence ≥5%) included taste disturbance (38.3%), melena (8.7%), hyperuricemia (6.3%), and nausea (5.9%). Other common reactions (≥ 1%) comprised dry mouth (2.4%), elevated gamma-glutamyl transferase (2.4%), abdominal bloating (1.7%), fatigue (1.7%), urinary tract infection (1.7%), dysgeusia (1.7%), vomiting (1.4%), and stool discoloration (1.4%). These events were not reported during ceprazone monotherapy and were predominantly mild.

## 26.7 Conclusion

Gastric acid suppression represents a crucial determinant for successful *H. pylori* eradication, and acid-suppressive agents constitute essential components of *H. pylori* treatment protocols recommended in both domestic and international guidelines. Novel P-CABs, exemplified by vonoprazan, significantly enhance eradication efficacy through their rapid-onset, stable, potent, and sustained acid inhibition. Growing recognition of adequate acid suppression requirements, supported by accumulating evidence, has driven increased adoption of P-CAB-based regimens in clinical practice. Currently, vonoprazan-based eradication regimens possess the most substantial evidence base and well-established clinical experience across quadruple, triple, and dual therapy strategies.

Historically, P-CABs have established a more robust foundation for acid suppression compared to PPIs, leading to a paradigm shift in *H. pylori* eradication from traditional "add-on" to contemporary "streamlined" therapeutic approaches. Owing to their superior eradication rates, simplified administration, and favorable safety profiles, P-CAB-based dual therapies have regained recognition as first-line regimens in major guidelines. The vonoprazan plus single-antibiotic dual regimen demonstrates particular promise, offering high efficacy, reduced adverse effects, enhanced compliance, minimal disruption to gut microbiota, and lower treatment costs, attributes that may expand its therapeutic benefits to broader patient populations. Future advances in personalized medicine will require large-scale clinical investigations to optimize tailored treatment strategies, including precise regimen selection, treatment duration, and dosing schedules for specific patient subgroups.

## References


1. Laine L, Johnson E, Suchower L, Ronca P, Hwang C, et al. US double-blind, controlled trials of omeprazole and amoxycillin for treatment of *H. pylori*. Aliment Pharmacol Ther. 1998;12(4):377–82. https://doi.org/10.1046/j.1365-2036.1998.00303.x.
2. Hu F. Evaluation of therapeutic regimens for *H. pylori* infection. Mod Dig Intervention. 2003;8(3)
3. Sugimoto M, Furuta T, Shirai N, Kodaira C, Nishino M, et al. Evidence that the degree and duration of acid suppression are related to *H. pylori* eradication by triple therapy. Helicobacter. 2007;12(4):317–23. https://doi.org/10.1111/j.1523-5378.2007.00508.x.
4. Ke H, Li J, Lu B, Yang C, Wang J, et al. The appropriate cutoff gastric pH value for *H. pylori* eradication with bismuth-based quadruple therapy. Helicobacter. 2021;26(1):e12768. https://doi.org/10.1111/hel.12768.
5. Sjöstedt S, Sagar M, Lindberg G, Wikström B, Nord CE, et al. Prolonged and profound acid inhibition is crucial in *H. pylori* treatment with a proton pump inhibitor combined with amoxicillin. Scand J Gastroenterol. 1998;33(1):39–43.
6. Chey WD, Howden CW, Moss SF, Morgan DR, Greer KB, et al. ACG clinical guideline: treatment of *H. pylori* infection. Am J Gastroenterol. 2024;119(9):1730–53. https://doi.org/10.14309/ajg.0000000000002968.
7. Malfertheiner P, Megraud F, Rokkas T, Gisbert JP, Liou JM, et al. Management of *H. pylori* infection: the Maastricht VI/Florence consensus report. Gut. 2022; https://doi.org/10.1136/gutjnl-2022-327745.

8. H. pylori Study Group of Chinese Society of Gastroenterology. 2022 Chinese guidelines for the treatment of *H. pylori* infection. Chin J Dig (Zhonghua Xiaohua Zazhi). 2022;42(11) https://doi.org/10.3760/cma.j.cn311367-20220929-00479.
9. Kato M, Ota H, Okuda M, Kikuchi S, Satoh K, et al. Guidelines for the management of *H. pylori* infection in Japan: 2016 Revised Edition. Helicobacter. 2019;24(4):e12597. https://doi.org/10.1111/hel.12597.
10. Zhang GX. Role of proton pump inhibitors in *H. pylori* eradication. 2016
11. Ierardi E, Losurdo G, Fortezza RF, Principi M, Barone M, et al. Optimizing proton pump inhibitors in *H. pylori* treatment: old and new tricks to improve effectiveness. World J Gastroenterol. 2019;25(34):5097–104. https://doi.org/10.3748/wjg.v25.i34.5097.
12. Erah PO, Goddard AF, Barrett DA, Shaw PN, Spiller RC. The stability of amoxycillin, clarithromycin and metronidazole in gastric juice: relevance to the treatment of *H. pylori* infection. J Antimicrob Chemother. 1997;39(1):5–12. https://doi.org/10.1093/jac/39.1.5.
13. Goddard AF, Spiller RC. The effect of omeprazole on gastric juice viscosity, pH and bacterial counts. Aliment Pharmacol Ther. 1996;10(1):105–9. https://doi.org/10.1111/j.1365-2036.1996.tb00183.x.
14. Meining A, Wick M, Miehlke S, Bayerdörffer E, Stolte M, et al. The presence of immunoglobulins in the gastric juice of patients infected with *H. pylori* is related to a reduced secretion of acid. Helicobacter. 2002;7(1):67–70. https://doi.org/10.1046/j.1523-5378.2002.00058.x.
15. Labenz J, Stolte M, Blum AL, Jorias I, Leverkus F, et al. Intragastric acidity as a predictor of the success of *H. pylori* eradication: a study in peptic ulcer patients with omeprazole and amoxicillin. Gut. 1995;37(1):39–43. https://doi.org/10.1136/gut.37.1.39.
16. Yuan YZ. Acid suppression criteria and drug selection in acid-related diseases. 2023.
17. Srebro J, Brniak W, Mendyk A. Formulation of dosage forms with proton pump inhibitors: state of the art, challenges and future perspectives. Pharmaceutics. 2022;14(10) https://doi.org/10.3390/pharmaceutics14102043.
18. Herszényi L, Bakucz T, Barabás L, Tulassay Z. Pharmacological approach to gastric acid suppression: past, present, and future. Dig Dis. 2020;38(2):104–11. https://doi.org/10.1159/000505204.
19. Lee JY, Park KS. Optimal first-line treatment for *H. pylori* infection: recent strategies. Gastroenterol Res Pract. 2016;2016:9086581. https://doi.org/10.1155/2016/9086581.
20. Sugano K. Effect of *H. pylori* eradication on the incidence of gastric cancer: a systematic review and meta-analysis. Gastric Cancer. 2019;22(3):435–45. https://doi.org/10.1007/s10120-018-0876-0.
21. Vondracek TG. Ranitidine bismuth citrate in the treatment of *H. pylori* infection and duodenal ulcer. Ann Pharmacother. 1998;32(6):672–9. https://doi.org/10.1345/aph.17247.
22. Chey WD, Mody RR, Izat E. Patient and physician satisfaction with proton pump inhibitors (PPIs): are there opportunities for improvement? Dig Dis Sci. 2010;55(12):3415–22. https://doi.org/10.1007/s10620-010-1209-2.
23. Strand DS, Kim D, Peura DA. 25 years of proton pump inhibitors: a comprehensive review. Gut Liver. 2017;11(1):27–37. https://doi.org/10.5009/gnl15502.
24. Ramani A, Merchant A, Cash BD. Review of the clinical development of fexuprazan for gastroesophageal reflux-related disease. Eur J Clin Pharmacol. 2023;79(8):1023–9. https://doi.org/10.1007/s00228-023-03521-4.
25. Scarpignato C, Hunt RH. The potential role of potassium-competitive acid blockers in the treatment of gastroesophageal reflux disease. Curr Opin Gastroenterol. 2019;35(4):344–55. https://doi.org/10.1097/mog.0000000000000543.
26. Wang P, Liu YT, Zhao Y, Li X. Research advances in potassium-competitive acid blockers for acid-related diseases. Chin J Prescription Drugs (Zhongguo Chufangyao). 2023;21(5)
27. Wang W, Yaseen HS, Li X, Ahmad MZ, Chen Y, et al. Tegoprazan: a novel, highly selective, and potent potassium-competitive acid blocker (P-CAB). Eur J Clin Pharmacol. 2025;81(8):1103–17. https://doi.org/10.1007/s00228-025-03850-6.

28. Kagami T, Sahara S, Ichikawa H, Uotani T, Yamade M, et al. Potent acid inhibition by vonoprazan in comparison with esomeprazole, with reference to CYP2C19 genotype. Aliment Pharmacol Ther. 2016;43(10):1048–59. https://doi.org/10.1111/apt.13588.
29. He JJ, Cheng NN. Vonoprazan: a potassium-competitive acid blocker. Chin J Clin Pharm (Zhongguo Linchuang Yaoxue Zazhi). 2019;28(3)
30. Geibel JP. Role of potassium in acid secretion. World J Gastroenterol. 2005;11(34):5259–65. https://doi.org/10.3748/wjg.v11.i34.5259.
31. Otake K, Sakurai Y, Nishida H, Fukui H, Tagawa Y, et al. Characteristics of the novel potassium-competitive acid blocker Vonoprazan fumarate (TAK-438). Adv Ther. 2016;33(7):1140–57. https://doi.org/10.1007/s12325-016-0345-2.
32. Kim HK, Park SH, Cheung DY, Cho YS, Kim JI, et al. Clinical trial: inhibitory effect of revaprazan on gastric acid secretion in healthy male subjects. J Gastroenterol Hepatol. 2010;25(10):1618–25. https://doi.org/10.1111/j.1440-1746.2010.06408.x.
33. Mori H, Tonai-Kachi H, Ochi Y, Taniguchi Y, Ohshiro H, et al. N-(2-hydroxyethyl)-N,2-dimethyl-8-{(4R)-5-methyl-3,4-dihydro-2H-chromen-4-ylamino}imidazo1,2-apyridine-6-carboxamide (PF-03716556), a novel, potent, and selective acid pump antagonist for the treatment of gastroesophageal reflux disease. J Pharmacol Exp Ther. 2009;328(2):671–9. https://doi.org/10.1124/jpet.108.146415.
34. Shin JM, Inatomi N, Munson K, Strugatsky D, Tokhtaeva E, et al. Characterization of a novel potassium-competitive acid blocker of the gastric H,K-ATPase, 1-5-(2-fluorophenyl)-1-(pyridin-3-ylsulfonyl)-1H-pyrrol-3-yl-N-methylmethanamine monofumarate (TAK-438). J Pharmacol Exp Ther. 2011;339(2):412–20. https://doi.org/10.1124/jpet.111.185314.
35. Jenkins H, Sakurai Y, Nishimura A, Okamoto H, Hibberd M, et al. Randomised clinical trial: safety, tolerability, pharmacokinetics and pharmacodynamics of repeated doses of TAK-438 (vonoprazan), a novel potassium-competitive acid blocker, in healthy male subjects. Aliment Pharmacol Ther. 2015;41(7):636–48. https://doi.org/10.1111/apt.13121.
36. <_Voque® (Vonoprazan Fumarate Tablets) Package Insert.pdf>.
37. Ashida K, Sakurai Y, Nishimura A, Kudou K, Hiramatsu N, et al. Randomised clinical trial: a dose-ranging study of vonoprazan, a novel potassium-competitive acid blocker, vs. lansoprazole for the treatment of erosive oesophagitis. Aliment Pharmacol Ther. 2015;42(6):685–95. https://doi.org/10.1111/apt.13331.
38. Hori Y, Imanishi A, Matsukawa J, Tsukimi Y, Nishida H, et al. 1-5-(2-Fluorophenyl)-1-(pyridin-3-ylsulfonyl)-1H-pyrrol-3-yl-N-methylmethanamine monofumarate (TAK-438), a novel and potent potassium-competitive acid blocker for the treatment of acid-related diseases. J Pharmacol Exp Ther. 2010;335(1):231–8. https://doi.org/10.1124/jpet.110.170274.
39. Song Z, Zuo X, Zhang Z, Wang X, Lin Y, et al. 1242: 10-day versus 14-day vonoprazan and amoxicillin dual therapy for the first-line eradication of *H. pylori* Infection: a Nationwide Multicenter, Randomized, Controlled Trial. Gastroenterology. 2025;169(1):S-307–8. https://doi.org/10.1016/s0016-5085(25)01655-5.
40. Han S, Choi HY, Kim YH, Nam JY, Kim B, et al. Randomised clinical trial: safety, tolerability, pharmacokinetics, and pharmacodynamics of single and multiple oral doses of tegoprazan (CJ-12420), a novel potassium-competitive acid blocker, in healthy male subjects. Aliment Pharmacol Ther. 2019;50(7):751–9. https://doi.org/10.1111/apt.15438.
41. He J, Cao G, Yu J, Wang J, Cheng N, et al. Safety, tolerability and pharmacokinetics of single ascending and multiple oral doses of Tegoprazan in healthy Chinese subjects. Clin Drug Investig. 2021;41(1):89–97. https://doi.org/10.1007/s40261-020-00986-4.
42. Warrington S, Baisley K, Boyce M, Tejura B, Morocutti A, et al. Effects of rabeprazole, 20 mg, or esomeprazole, 20 mg, on 24-h intragastric pH and serum gastrin in healthy subjects. Aliment Pharmacol Ther. 2002;16(7):1301–7. https://doi.org/10.1046/j.1365-2036.2002.01292.x.
43. _Ceprazone Access - Electronic Prescribing Information.pdf
44. Li CY, Su M, Yan YY, Zhou L, Ao LY, et al. KFP-H008 blocks gastric acid secretion through inhibiting H(+)-K(+)-ATPase. Eur J Pharmacol. 2017;810:112–9. https://doi.org/10.1016/j.ejphar.2017.06.020.

45. Takahashi N, Take Y. Tegoprazan, a novel potassium-competitive acid blocker to control gastric acid secretion and motility. J Pharmacol Exp Ther. 2018;364(2):275–86. https://doi.org/10.1124/jpet.117.244202.
46. Martinucci I, Blandizzi C, Bodini G, Marabotto E, Savarino V, et al. Vonoprazan fumarate for the management of acid-related diseases. Expert Opin Pharmacother. 2017;18(11):1145–52. https://doi.org/10.1080/14656566.2017.1346087.
47. Kim DK, Lee KH, Kim SJ, Kim SJ, Lee SJ, et al. Effects of Tegoprazan, a novel potassium-competitive acid blocker, on rat models of gastric acid-related disease. J Pharmacol Exp Ther. 2019;369(3):318–27. https://doi.org/10.1124/jpet.118.254904.
48. Yoon DY, Sunwoo J, Shin N, Kim AR, Kim B, et al. Effect of meal timing on pharmacokinetics and pharmacodynamics of tegoprazan in healthy male volunteers. Clin Transl Sci. 2021;14(3):934–41. https://doi.org/10.1111/cts.12958.
49. Scott DR, Munson KB, Marcus EA, Lambrecht NW, Sachs G. The binding selectivity of vonoprazan (TAK-438) to the gastric H+, K+ -ATPase. Aliment Pharmacol Ther. 2015;42(11–12):1315–26. https://doi.org/10.1111/apt.13414.
50. Sugano K, Tack J, Kuipers EJ, Graham DY, El-Omar EM, et al. Kyoto global consensus report on *H. pylori* gastritis. Gut. 2015;64(9):1353–67. https://doi.org/10.1136/gutjnl-2015-309252.
51. Megraud F, Bruyndonckx R, Coenen S, Wittkop L, Huang TD, et al. *H. pylori* resistance to antibiotics in Europe in 2018 and its relationship to antibiotic consumption in the community. Gut. 2021;70(10):1815–22. https://doi.org/10.1136/gutjnl-2021-324032.
52. Moss SF, Chey WD, Daniele P, Pelletier C, Jacob R, et al. Brief communication: global temporal trends in the efficacy of clarithromycin-based regimens for the treatment of *H. pylori* infection. Ther Adv Gastroenterol. 2023;16:17562848231167284. https://doi.org/10.1177/17562848231167284.
53. Kuo YT, Liou JM, El-Omar EM, Wu JY, Leow AHR, et al. Primary antibiotic resistance in *H. pylori* in the Asia-Pacific region: a systematic review and meta-analysis. Lancet Gastroenterol Hepatol. 2017;2(10):707–15. https://doi.org/10.1016/s2468-1253(17)30219-4.
54. Cheng Y, Yang L, Xu S, Zhang C. Vonoprazan is not inferior to proton pump inhibitors in bismuth-containing quadruple therapy for *H. pylori* eradication: a meta-analysis of 10 studies from East Asia. J Clin Gastroenterol. 2024;58(10):950–6. https://doi.org/10.1097/mcg.0000000000002001.
55. Jin T, Wu W, Zhang L, Xuan H, Zhang H, et al. The efficacy and safety of Vonoprazan and Tegoprazan in *H. pylori* eradication: a comprehensive systematic review and meta-analysis of randomized controlled trials. Ther Adv Gastroenterol. 2025;18:17562848251314801. https://doi.org/10.1177/17562848251314801.
56. Malfertheiner P, Megraud F, O'Morain CA, Gisbert JP, Kuipers EJ, et al. Management of *H. pylori* infection-the Maastricht V/Florence consensus report. Gut. 2017;66(1):6–30. https://doi.org/10.1136/gutjnl-2016-312288.
57. _Efficacy analysis of vonoprazan combined with high-dose amoxicillin dual therapy for first-line eradication of *H. pylori*.pdf. https://doi.org/10.3969/j.issn.1672-9463.2023.03.010
58. _ Efficacy Analysis of Vonoprazan, a Novel Potassium-Competitive Acid Blocker, in *H. pylori* Eradication Therapy.pdf. https://doi.org/10.3969/j.issn.1671-038X.2022.11.07
59. Ding YM, Duan M, Han ZX, Song XH, Zhang FL, et al. Bismuth-containing quadruple therapy for *H. pylori* eradication: a randomized clinical trial of 10 and 14 days. Dig Dis Sci. 2024;69(7):2540–7. https://doi.org/10.1007/s10620-024-08460-3.
60. Lu L, Wang Y, Ye J, Han Y, Lou G, et al. Quadruple therapy with vonoprazan 20 mg daily as a first-line treatment for *H. pylori* infection: a single-center, open-label, noninferiority, randomized controlled trial. Helicobacter. 2023;28(1):e12940. https://doi.org/10.1111/hel.12940.
61. Hu Y, Huang XH, Zhou B, Liu ML, Liu YF, et al. Vonoprazan and amoxicillin dual therapy for 14 days as the first-line treatment of *H. pylori* infection: a non-inferiority, randomized clinical trial. Helicobacter. 2024;29(1):e13045. https://doi.org/10.1111/hel.13045.
62. Yan T, Wang J, Zhu R, Ma D, Gao J, et al. Vonoprazan improves the efficacy of bismuth quadruple therapy containing doxycycline and metronidazole as first-line *H. pylori* treat-

ment in penicillin-allergic patients: a randomized controlled trial. J Antimicrob Chemother. 2025;80(4):927–34. https://doi.org/10.1093/jac/dkae467.
63. Zhou J, Jia L, Liu Z, Zhao W, Liu L, et al. Efficacy and safety of vonoprazan-based bismuth quadruple therapy for first-line *H. pylori* eradication: a large-scale, real-world study. Medicine (Baltimore). 2024;103(42):e40170. https://doi.org/10.1097/md.0000000000040170.
64. _ Efficacy of Vonoprazan-Containing Quadruple Therapy for Eradicating *H. pylori* Infection. pdf. https://doi.org/10.14109/j.cnki.xyylc.2024.03.10
65. He Q, Ou Y, Zhu H, Chen Z, Yang D, et al. Efficacy and safety of bismuth quadruple regimens containing minocycline and vonoprazan for eradication of *H. pylori*: real-world evidence. JGH Open. 2024;8(5):e13070. https://doi.org/10.1002/jgh3.13070.
66. Benito BM, Nyssen OP, Gisbert JP. Efficacy and safety of Vonoprazan in dual/triple/quadruple regimens both in first-line and rescue therapy for *H. pylori* eradication: a systematic review with meta-analysis. Helicobacter. 2024;29(6):e13148. https://doi.org/10.1111/hel.13148.
67. Li J, Shi H, Zhou F, Xie L, Lin R. The efficacy and safety of regimens for *H. pylori* eradication treatment in China: a systemic review and network meta-analysis. J Clin Gastroenterol. 2024;58(1):12–23. https://doi.org/10.1097/mcg.0000000000001902.
68. Zhang CG, et al. The efficacy and safety of Vonoprazan-based regimens for *H. pylori* eradication treatment_ a systemic review and network meta-analysis. 2024 ACG. .pdf.
69. Song Z, Wang W, Li P, Zuo X, Zhu Y, et al. Tegoprazan-based versus esomeprazole-based triple therapy plus bismuth for first-line *H. pylori* eradication: a nationwide, multicenter, double-blind, double-dummy, randomized controlled trial. Chin Med J (Engl). 2025; https://doi.org/10.1097/cm9.0000000000003780.
70. Cho JH, Jin SY, Park S. Comparison of tegoprazan and proton pump inhibitors for first-line *H. pylori* eradication: a systematic review with meta-analysis. Expert Rev Anti-Infect Ther. 2025;23(2–4):227–33. https://doi.org/10.1080/14787210.2025.2459722.
71. Tan N, Wu H, Lan C, Liu C, Liao A, et al. The efficacy of keverprazan-based quadruple therapy for *H. pylori* eradication: a phase III, randomised, double-blind, multicentre trial. Int J Antimicrob Agents. 2024;64(5):107320. https://doi.org/10.1016/j.ijantimicag.2024.107320.
72. Kusano C, Gotoda T, Suzuki S, Ikehara H, Moriyama M. Safety of first-line triple therapy with a potassium-competitive acid blocker for *H. pylori* eradication in children. J Gastroenterol. 2018;53(6):718–24. https://doi.org/10.1007/s00535-017-1406-2.
73. Okubo H, Akiyama J, Kobayakawa M, Kawazoe M, Mishima S, et al. Vonoprazan-based triple therapy is effective for *H. pylori* eradication irrespective of clarithromycin susceptibility. J Gastroenterol. 2020;55(11):1054–61. https://doi.org/10.1007/s00535-020-01723-6.
74. Tanabe H, Ando K, Sato K, Ito T, Goto M, et al. Efficacy of vonoprazan-based triple therapy for *H. pylori* eradication: a multicenter study and a review of the literature. Dig Dis Sci. 2017;62(11):3069–76. https://doi.org/10.1007/s10620-017-4664-1.
75. Sakurai K, Suda H, Ido Y, Takeichi T, Okuda A, et al. Comparative study: vonoprazan and proton pump inhibitors in *H. pylori* eradication therapy. World J Gastroenterol. 2017;23(4):668–75. https://doi.org/10.3748/wjg.v23.i4.668.
76. Maruyama M, Tanaka N, Kubota D, Miyajima M, Kimura T, et al. Vonoprazan-based regimen is more useful than PPI-based one as a first-line *H. pylori* eradication: a randomized controlled trial. Can J Gastroenterol Hepatol. 2017;2017:4385161. https://doi.org/10.1155/2017/4385161.
77. Kamada T, Satoh K, Itoh T, Ito M, Iwamoto J, et al. Evidence-based clinical practice guidelines for peptic ulcer disease 2020. J Gastroenterol. 2021;56(4):303–22. https://doi.org/10.1007/s00535-021-01769-0.
78. Jung YS, Kim EH, Park CH. Systematic review with meta-analysis: the efficacy of vonoprazan-based triple therapy on *H. pylori* eradication. Aliment Pharmacol Ther. 2017;46(2):106–14. https://doi.org/10.1111/apt.14130.
79. Lyu QJ, Pu QH, Zhong XF, Zhang J. Efficacy and safety of Vonoprazan-based versus proton pump inhibitor-based triple therapy for *H. pylori* eradication: a meta-analysis of randomized clinical trials. Biomed Res Int. 2019;2019:9781212. https://doi.org/10.1155/2019/9781212.

80. Choi YJ, Lee YC, Kim JM, Kim JI, Moon JS, et al. Triple therapy-based on tegoprazan, a new potassium-competitive acid blocker, for first-line treatment of *H. pylori* infection: a randomized, double-blind, phase III, clinical trial. Gut Liver. 2022;16(4):535–46. https://doi.org/10.5009/gnl220055.
81. Park JY, Choi IJ, Kim GH, Hong SJ, Shin SK, et al. Comparative efficacy of potassium-competitive acid blocker-based triple therapy with tegoprazan versus vonoprazan for *H. pylori* eradication: a randomized, double-blind, active-controlled pilot study. Gut Liver. 2025; https://doi.org/10.5009/gnl250067.
82. Jung YS, Kim S, Kim HY, Noh SJ, Park JH, et al. 7-day versus 14-day tegoprazan-based triple therapy to treat *H. pylori* infection: real-world evidence. J Gastroenterol Hepatol. 2022;37(10):1911–8. https://doi.org/10.1111/jgh.15939.
83. Zhang JY, Chen JH, Huang YL, Li J, Xu D, et al. Fourteen-day vonoprazan-amoxicillin dual therapy versus 14-day bismuth-based quadruple therapy for *H. pylori* treatment: a randomized clinical trial. Ther Adv Gastroenterol. 2025;18:17562848251354868. https://doi.org/10.1177/17562848251354868.
84. Abosheaishaa H, Elsayed H, Abdelhalim O, Haddad IME, Harfoush MK, et al. Vonoprazan and amoxicillin dual therapy versus bismuth-based therapy for *H. pylori* eradication: a systematic review and meta-analysis of randomized controlled trials. Proc (Bayl Univ Med Cent). 2025;38(5):716–21. https://doi.org/10.1080/08998280.2025.2491964.
85. Clinical Comparative Study of Modified Dual Therapy versus Standard Quadruple Therapy for *H. pylori* Infection.pdf
86. Wang Z, Wang F. Efficacy and safety comparison of *H. pylori* eradication between vonoprazan dual therapy versus triple therapy: a systematic review and meta-analysis. Ther Adv Gastroenterol. 2022;15:17562848221125308. https://doi.org/10.1177/17562848221125308.
87. Lin Y, Xu H, Yun J, Yu X, Shi Y, et al. The efficacy of vonoprazan combined with different dose amoxicillin on eradication of *H. pylori*: an open, multicenter, randomized clinical study. Ann Transl Med. 2022;10(18):987. https://doi.org/10.21037/atm-22-4133.
88. Ouyang Y, Wang M, Xu YL, Zhu Y, Lu NH, et al. Amoxicillin-vonoprazan dual therapy for *H. pylori* eradication: a systematic review and meta-analysis. J Gastroenterol Hepatol. 2022;37(9):1666–72. https://doi.org/10.1111/jgh.15917.
89. Suzuki S, Gotoda T, Kusano C, Ikehara H, Ichijima R, et al. Seven-day vonoprazan and low-dose amoxicillin dual therapy as first-line *H. pylori* treatment: a multicentre randomised trial in Japan. Gut. 2020;69(6):1019–26. https://doi.org/10.1136/gutjnl-2019-319954.
90. Li X, Jiang C, Su Y, Gao R, Yang P, et al. Efficacy and safety of vonoprazan-amoxicillin dual therapy versus bismuth-containing quadruple therapy for patients with *H. pylori* infection: a meta-analysis. Front Microbiol. 2025;16:1561749. https://doi.org/10.3389/fmicb.2025.1561749.
91. Hu Y, Du RC, Lai YK, Hu YX, Zhu YC, et al. 40 years after the discovery of *H. pylori*: performing optimized "subtraction" for clinical eradication. iMetaOmics. 2025;2(2) https://doi.org/10.1002/imo2.70017.
92. Hu Y, Zhang ZY, Wang F, Zhuang K, Xu X, et al. Effects of amoxicillin dosage on cure rate, gut microbiota, and antibiotic resistome in vonoprazan and amoxicillin dual therapy for *H. pylori*: a multicentre, open-label, non-inferiority randomised controlled trial. Lancet Microbe. 2025;6(3):100975. https://doi.org/10.1016/j.lanmic.2024.100975.
93. Gao W, Liu J, Wang X, Li J, Zhang X, et al. Simplified *H. pylori* therapy for patients with penicillin allergy: a randomised controlled trial of vonoprazan-tetracycline dual therapy. Gut. 2024;73(9):1414–20. https://doi.org/10.1136/gutjnl-2024-332640.
94. Wang X, Teng G, Dong X, Dai Y, Wang W. The efficacy and safety of a simple 14-day vonoprazan-minocycline dual therapy for *H. pylori* eradication: a retrospective pilot study. Ther Adv Gastroenterol. 2024;17:17562848241299734. https://doi.org/10.1177/17562848241299734.
95. Gao W, Teng G, Wang C, Xu Y, Li Y, et al. Eradication rate and safety of a "simplified rescue therapy": 14-day vonoprazan and amoxicillin dual regimen as rescue therapy on treatment of *H. pylori* infection previously failed in eradication: a real-world, retrospective clinical study in China. Helicobacter. 2022;27(5):e12918. https://doi.org/10.1111/hel.12918.

96. Liu HN, Wang R, Cao Y, Xian F, Bi XJ, et al. Comparison of the efficacy between the dual therapy of tegoprazan and the quadruple therapy of tegoprazan: a randomized controlled multicenter study. Clin Transl Gastroenterol. 2024;15(10):e1. https://doi.org/10.14309/ctg.0000000000000699.
97. Lin X, Huang H, Liu Y, Zeng Y, Lu S, et al. Tegoprazan-amoxicillin dual therapy for *H. pylori* eradication: a prospective, randomized, multicenter study in Fujian, China. Helicobacter. 2024;29(6):e13151. https://doi.org/10.1111/hel.13151.
98. Kong Q, Mirza IA, Zhang X, Song X, Li X, et al. Fourteen-day tegoprazan-amoxicillin dual therapy as the first-line treatment of *H. pylori* infection (SHARE2301): a Multicenter, Noninferiority, Randomized Clinical Trial. Helicobacter. 2024;29(3):e13098. https://doi.org/10.1111/hel.13098.
99. Parsons ME, Keeling DJ. Novel approaches to the pharmacological blockade of gastric acid secretion. Expert Opin Investig Drugs. 2005;14(4):411–21. https://doi.org/10.1517/13543784.14.4.411.
100. Ashida K, Iwakiri K, Hiramatsu N, Sakurai Y, Hori T, et al. Maintenance for healed erosive esophagitis: phase III comparison of vonoprazan with lansoprazole. World J Gastroenterol. 2018;24(14):1550–61. https://doi.org/10.3748/wjg.v24.i14.1550.
101. Tegoprazan Tablets - Updated Hp-Stamped Prescribing Information (October 21, 2024).pdf

# Nonantibiotic Clinical Trials for *Helicobacter pylori*

27

Yingli Gong, Jie Gao, and Tinglin Zhang

## 27.1 The Significance of Nonantibiotic Clinical Trials

*Helicobacter pylori* infection has a global prevalence of up to 50% and is a major risk factor for gastric cancer [1]. The International Agency for Research on Cancer (IARC) classifies *H. pylori* as a Group 1 carcinogen, posing a significant public health challenge [2]. On July 7, 2025, the IARC/World Health Organization (WHO) published a research paper entitled Global lifetime estimates of expected and preventable gastric cancers across 185 countries in *Nature Medicine* [3]. This study assessed lifetime estimates of expected and preventable gastric cancer cases across 185 countries and regions worldwide. The results indicate that among individuals born globally between 2008 and 2017, approximately 15.8 million are projected to

Y. Gong
Changhai Clinical Research Unit, Shanghai Changhai Hospital, Naval Medical University, Shanghai, China

College of Life Science, Mudanjiang Medical University, Mudanjiang, China

J. Gao
Changhai Clinical Research Unit, Shanghai Changhai Hospital, Naval Medical University, Shanghai, China

Shanghai Key Laboratory of Nautical Medicine and Translation of Drugs and Medical Devices, Shanghai, China

College of Life Science, Mudanjiang Medical University, Mudanjiang, China

T. Zhang (✉)
Changhai Clinical Research Unit, Shanghai Changhai Hospital, Naval Medical University, Shanghai, China

Shanghai Key Laboratory of Nautical Medicine and Translation of Drugs and Medical Devices, Shanghai, China


J. Gao, Y. Du (eds.), *Non-Antibiotic Therapies for Helicobacter pylori*, https://doi.org/10.1007/978-981-95-9598-3_27

develop gastric cancer at some point in their lives. Among these cases, 76% (11.9 million individuals) are associated with *H. pylori* infection.

The latest fifth edition of the national consensus recommends the following first-line eradication regimen for *H. pylori* infection: "any bismuth compound + any proton pump inhibitor (PPI) + amoxicillin + metronidazole," which is administered orally for a 14-day course. *The Maastricht VI/Florence Consensus Report* emphasized [4] that bismuth-containing quadruple therapy (comprising PPIs, bismuth compounds, tetracycline, and metronidazole) and clarithromycin triple therapy (comprising PPIs, clarithromycin, and amoxicillin) constitute the first-line treatments for *H. pylori* infection. Dual therapy regimens (such as PPIs combined with amoxicillin) are also commonly employed owing to their simplified medication regimen and reduced incidence of adverse effects. However, current primary antibiotic therapies face various challenges, as outlined below. The issue of antibiotic resistance in *H. pylori* infections is particularly acute. The latest antimicrobial resistance surveillance data from the Asia–Pacific region indicate that resistance rates for clarithromycin and levofloxacin approach 50%. The increase in resistance not only significantly increases the failure rate of standard treatment regimens but also exacerbates complex clinical phenomena such as heteroresistance [5–7], where both susceptible and resistant strains coexist within the same patient, making infection control and eradication increasingly challenging. Consequently, there is an urgent need to develop novel nonantibiotic strategies to address the challenge of resistance.

Current standard clinical treatment regimens rely primarily on broad-spectrum antibiotics; however, these antibiotics lack specific targeting capabilities against *H. pylori* within the stomach. Their systemic distribution prevents their selective action against the target pathogen, instead indiscriminately eliminating beneficial commensal flora in the gut and causing dysbiosis. Research indicates [8] that dysbiosis of the gut microbiota significantly impacts human health. Long-term antibiotic use to eradicate *H. pylori* may induce gut dysbiosis in patients, with some individuals exhibiting persistent and pronounced gut microbiota disruption for ≥12 months following treatment completion [9–12]. Therefore, current *H. pylori* treatment urgently requires not only effective pathogen eradication but also preservation of the gut microbial balance to the greatest extent possible.

Therefore, nonantibiotic therapies have gradually emerged as a research focus in *H. pylori* infection treatment because of their advantages in circumventing resistance, high safety profile, and broad applicability. Through mechanisms such as immunomodulation, microbiome intervention, natural antibacterial agents, traditional Chinese medicine (TCM) synergism, and targeted bactericidal action, these therapies not only serve as complementary adjuncts to antibiotic regimens but also demonstrate dual preventive and therapeutic potential in certain protocols. This chapter systematically examines the clinical significance, mechanisms of action, efficacy, and safety of nonantibiotic therapies while outlining future development directions to inform precision prevention and control strategies for *H. pylori* infection.

## 27.2 Principal Types of Nonantibiotic Clinical Trials

Nonantibiotic therapies are categorized into four types on the basis of their mechanism: immunomodulatory agents, microecological modulators, natural-origin products, and TCM. Each aims to control *H. pylori* infection safely and effectively.

### 27.2.1 Immunomodulatory Agents

Immunomodulatory therapies enhance immune responses or provide passive immunity to suppress or eradicate *H. pylori*. They include vaccines and yolk antibodies (immunoglobulin Y (IgY)), offering both preventive and therapeutic benefits.

#### 27.2.1.1 HPV Vaccines

Traditional mucosal vaccines utilize *H. pylori* whole-bacterial lysate, urease (Ure), or flagellar proteins (FlaA/B) as antigens, which are delivered orally or via nasogastric tubes and rely on the gastrointestinal mucosal immune response for efficacy. The first anti-*H. pylori* vaccine was successfully developed in China. Research published in The Lancet in 2015 by Professor Zou Quanming and his team at Army Medical University demonstrated [13] that the oral recombinant urease B vaccine demonstrated a 1-year protective efficacy of 71.8% and a 3-year efficacy of 65.8% in 4464 children aged 6–15 years. However, this vaccine type has limitations, including weaker immune responses in adults and susceptibility to degradation by gastric acid and pepsin, resulting in low antigen utilization. The clinical application of *H. pylori* vaccines currently remains predominantly focused on children, as they exhibit milder gastric mucosal inflammation and stronger immune responses. Vaccination provides long-term protection against infection and reduces the risk of gastric cancer in adulthood.

#### 27.2.1.2 Yolk Antibodies (IgY)

IgY, which is extracted from the egg yolk of immunized laying hens, is a specific immunoglobulin that targets *H. pylori*. This passive immunotherapy bypasses the need to activate the host's innate immune system, which relies directly on exogenous antibodies to exert its effects. It is particularly suitable for populations with compromised immune function, including children, elderly individuals, and immunocompromised individuals. Clinical research data indicate [14] that when administered orally as a polyvalent anti-*H. pylori* IgY (0.1 g per dose, three times daily for 2 weeks), the *H. pylori* eradication rate reached 30.6%, with a 56.0% reduction in urea breath test ( [13]C-UBT) values and an 87.3% improvement in clinical symptoms such as abdominal distension and acid reflux. When probiotics (*Lactobacillus acidophilus* LA-5, 10 [10] CFU daily) are combined [15], the eradication rate can be increased to 45.2% ($P < 0.05$), with an adverse reaction incidence of only 2.0–3.0% (primarily mild abdominal distension), which is significantly lower than that of antibiotic therapy (20–30%). Its primary indications include antibiotic intolerance, pediatric *H. pylori* infection, and symptom management for mild *H. pylori* infection ( [13]C-UBT value <4.0‰) [6].

### 27.2.2 Microbiome Modulation Therapies

Microecological modulation therapy demonstrates limited efficacy when employed as a monotherapy; however, when combined with antibiotic therapy, it significantly enhances eradication rates while reducing adverse reactions. It currently represents the most widely utilized nonantibiotic adjunctive treatment in clinical practice [16].

Probiotics are primarily employed to increase the eradication rate of conventional quadruple therapy regimens, particularly in regions with high antibiotic resistance rates. A meta-analysis incorporating 26 randomized controlled trials (n = 3878) evaluated the efficacy of probiotics combined with first-line quadruple therapy (PPI + two antibiotics + bismuth compound). The results demonstrated that [17] the improvement in the eradication rate was as follows: the probiotic combination group achieved an *H. pylori* eradication rate of 90.0% (1741/1935), which was significantly greater than that of the quadruple therapy alone group (81.8%; 1612/1970), representing an 8.2% increase (95% CI: 6.5–9.9%). Subgroup analysis revealed that in regions with clarithromycin resistance rates >20%, the improvement in the eradication rate was more pronounced in the combination group (from 75.0% to 86.5%), suggesting that probiotics may partially counteract the impact of clarithromycin resistance and reduce adverse reactions. The incidence of antibiotic-associated diarrhea in the combination group was 4.0% (77/1935), which was significantly lower than that in the quadruple therapy alone group (22.5%, 443/1970), representing a reduction of 18.5%.

In another clinical study involving children [18], 100 children aged 3–12 years with *H. pylori* infection were enrolled and randomized into two groups: amoxicillin + clarithromycin + PPI (n = 50, 2 weeks) and combination with *Bifidobacterium longum* BB-12 (n = 50, probiotic 10 [10]CFU daily, administered 2 h apart from antibiotics, for 4 weeks). The results revealed the following: the eradication rate of the combined group was 82.0% (41/50), which was significantly higher than that of the antibiotic-only group (68.0%, 34/50); the incidence of adverse reactions was 5.0% (2/50, mild abdominal distension) in the combination group versus 24.0% (12/50, diarrhea, rash) in the antibiotic-only group. Gut microbiota: In the combination group, bifidobacteria and lactobacilli abundances returned to normal levels within 2 weeks post-treatment cessation, whereas the antibiotic-only group required 4 weeks for recovery.

### 27.2.3 Natural-Origin Products

Lactoferrin and allicin derivatives are the two most effective categories among natural-source products; when combined with antibiotics, they significantly increase eradication rates. Lactoferrin combined with levofloxacin triple therapy regimen: A meta-analysis incorporating eight randomized controlled trials (n = 980) evaluated the efficacy of lactoferrin (300 mg/day) combined with levofloxacin triple therapy (PPI + levofloxacin + amoxicillin) [19]. The eradication rate was as follows: the eradication rate in the combination group reached 85.0% (418/490), which was significantly higher than that in the triple therapy alone group (70.0%, 343/490), representing a 15% improvement. In the drug-resistant subgroup, in regions with levofloxacin resistance rates exceeding 25%, the eradication rate in the combination group remained at 78.0%, which was significantly higher than that in the monotherapy group (60.0%). Safety: The incidence of adverse reactions in the combination group was <1%, representing a significant reduction compared with that in the monotherapy group (15–20%).

Allicin Derivatives Combined With Amoxicillin: In a clinical study [20], 60 *H. pylori*-infected subjects (metronidazole resistance rate > 40%) were enrolled and divided into two groups: the allicin-cyclodextrin inclusion complex (200 mg/day) + amoxicillin (1 g per dose, twice daily) group (n = 30, 2 weeks) and the amoxicillin monotherapy group (n = 30, 2 weeks). The results demonstrated the following: the eradication rate was 65.0% in the combination group (19/30), which was significantly higher than that in the amoxicillin-alone group (40.0%, 12/30); the degree of improvement in inflammation was 60–70% in the combination group, markedly exceeding that in the amoxicillin-alone group (30–40%); and the incidence of adverse reactions was 3.0% (1/30, mild nausea) in the combination group and 10.0% (3/30, diarrhea) in the monotherapy group.

Green Tea Catechins: Green tea catechins are polyphenolic compounds found in green tea (Camellia sinensis), with epigallocatechin gallate (EGCG) exhibiting the strongest antibacterial activity, accounting for 50–60% of the total catechins. In a clinical study (n = 45, adults) [21], daily oral administration of 800 mg EGCG (divided into two doses) reduced *H. pylori* colonization rates by 40–50% after 4 weeks, with [13]C-UBT values decreasing from 8.5‰ to 4.2‰. However, *H. pylori* colonization rates returned to baseline levels within 2 months after the intervention was discontinued, suggesting that EGCG is more suitable for controlling *H. pylori* colonization than for eradication therapy.

Sulforaphane (SFN) From Broccoli: SFN is the primary active compound in broccoli sprouts (*Brassica oleracea* var. botrytis) and is formed from glucosinolates via the action of myrosinase. It has antioxidant, anti-inflammatory, and antibacterial properties. Clinical studies [22](n = 30, adults) demonstrated that daily consumption of broccoli sprout juice containing 200 μmol SFN resulted in a 38.0% reduction in [13]C-UBT values and a 45% decrease in the gastric mucosal ammonia concentration after 8 weeks. However, these indicators returned to baseline levels 2 months after cessation of the intervention. This may be attributed to the elimination of SFN from the body. This resulted in a 38.0% reduction in [13]C-UBT values and a 45% decrease in the gastric mucosal ammonia concentration after 8 weeks. However, these indicators returned to baseline levels within 2 months after the intervention was discontinued. This may be attributed to the rapid metabolism of SFN in the body (with a half-life of approximately 2 h), which prevents the sustained inhibition of *H. pylori*. The advantage of SFN lies in its exceptional safety profile—no adverse reactions occur at daily intakes below 400 μmol—and its direct availability through broccoli sprout consumption, making it suitable for the routine prevention and management of *H. pylori* infection.

### 27.2.4 TCM Therapies

Jinghua Weikang Capsules: Formulated from *Chenopodium ambrosioides* and *Adina pilulifera*, this preparation regulates qi, dispels cold, clears heat, and resolves stasis. *H. pylori* infection presents with mixed cold–heat patterns characterized by epigastric distension and pain, a preference for warmth and pressure, acid regurgitation, and a bitter taste, with a white greasy coating on the tongue accompanied by yellow discoloration. The mechanism of action for Jinghua Weikang capsules involves disruption of biological membranes, anti-inflammatory effects, and mucosal repair. Clinical research [23] (n = 320, multicenter randomized controlled trial) demonstrated that Jinghua Guaikang capsules (160 mg per dose, three times daily) combined with a quadruple regimen containing furazolidone (PPI + furazolidone + amoxicillin + bismuth potassium citrate) for 2 weeks increased the eradication rate from 75% to 88%, with a 92.0% symptom relief rate, which was significantly higher than that of the quadruple regimen alone (72.0%). Adverse reactions were primarily oral irritation (incidence 8.0–10.0%), which was alleviated by administration after meals.

Liu Junzi Decoction: Comprising wood fragrant, amomum, codonopsis, atractylodes, poria, and liquorice, it effectively invigorates qi and strengthens the spleen, harmonizes the stomach, and alleviates pain. *H. pylori* infection can present with spleen–stomach deficiency syndrome, where patients exhibit epigastric dull pain, poor appetite, fatigue and lassitude, loose stools, and a pale tongue with white coating. The mechanism of action of Liu Junzi decoction involves primarily repairing the mucosal barrier and modulating immunity. Clinical studies (n = 50, adults) have demonstrated [24] that Liu Junzi Decoction (one dose daily, decocted and taken in two divided doses), combined with PPI therapy for *H. pylori*-associated chronic gastritis over 4 weeks, achieved an 88.0% symptom improvement rate and a 75% healing rate for gastric mucosal erosions. However, the eradication rate was approximately 45.0%, suggesting its greater suitability for symptom control and mucosal repair in *H. pylori* infection patients than direct *H. pylori* eradication.

## 27.3 Safety and Tolerability of Nonantibiotic Therapies

One of the core advantages of nonantibiotic therapies is their high safety profile. Compared with antibiotics (with an adverse reaction incidence rate of 20–30%), their adverse reactions are predominantly mild to moderate (incidence generally <10%) and typically resolve spontaneously, posing no serious life-threatening risks. However, safety characteristics vary across different therapeutic categories. A comprehensive assessment must integrate the mechanism of action, route of administration, and patient demographics while clearly defining contraindications and risk control measures to mitigate potential safety hazards. To maximize the safety of nonantibiotic therapies, clinical practice must establish a comprehensive management system encompassing pretreatment assessment, intratreatment monitoring, and adverse reaction management. The specific protocols are as follows.

### 27.3.1 Premedication Assessment

Liver/kidney function, immune status, allergies, and current medications should be evaluated. Stratify dosages for children, pregnant women, and the elderly.

### 27.3.2 Monitoring During Medication

Adverse reactions were recorded every 2 weeks. Special populations should be monitored weekly. Reassess [13]C-UBT or gastroscopy every 4–8 weeks. For long-term use, liver/kidney function was checked every 4 weeks.

### 27.3.3 Management of Adverse Reactions

1. Mild-to-moderate reactions (e.g., slight abdominal bloating and rash): No need to discontinue medication. The administration method (e.g., take after meals and reduce a single dose) should be adjusted, the patient should be observed for 2–3 days, and the patient should resume medication once symptoms subside.
2. Moderate reactions (e.g., persistent constipation and oral irritation): Medication should be suspended for 1–2 days. Once symptoms resolve, the original dose should be restarted at half the original dose. If reactions persist, change therapy (e.g., switching from IgY to lactoferrin).
3. Severe reactions (e.g., allergic rash and bacteremia): Discontinue immediately, administer symptomatic treatment (e.g., antihistamines and antibiotics), and after recovery, assess whether to resume nonantibiotic therapy (or switch to a treatment without cross-reactivity risk).

## 27.4 Research Progress and Future Directions in Nonantibiotic Clinical Trials

Nonantibiotic clinical trials for *H. pylori* infection represent a crucial strategy to address the global antibiotic resistance crisis, meet therapeutic needs in special populations, and achieve integrated prevention–treatment–consolidation control. However, nonantibiotic therapies still face challenges: single-therapy eradication rates remain below 50%, preventing the complete replacement of antibiotics; novel therapies (such as mRNA vaccines and bacteriophages) lack large-scale clinical data; and standardization systems for natural products and TCM remain incomplete. Future progress must advance nonantibiotic therapies from "supplementary adjuncts" to "core alternatives" through four pathways: "enhancing efficacy via novel formulation development," "overcoming resistance through optimized combination regimens," "ensuring quality via standardization," and "filling gaps through clinical translation."

In the future, as research has progressed, nonantibiotic clinical trial therapies will achieve three major breakthroughs: ① precision: selecting personalized therapies on the basis of the *H. pylori* genotype (e.g., CagA+/CagA-) and host immune status (e.g., sIgA levels), such as prioritizing mRNA vaccines targeting CagA for CagA+ strains; ② integration: establishing a comprehensive prevention and control system encompassing "vaccine prophylaxis + combination therapy treatment + probiotic consolidation," aligning closely with the WHO's target of reducing the incidence of gastric cancer by 10% by 2030; and ③internationalization: Chinese original therapies such as TCM and *H. pylori* vaccines becoming global public health tools for *H. pylori* prevention and control through international multicenter research.

## References


1. Yang L, Kartsonaki C, Yao P, de Martel C, Plummer M, et al. The relative and attributable risks of cardia and noncardia gastric cancer associated with *H. pylori* infection in China: a case-cohort study. Lancet Public Health. 2021;6(12):e888–96. https://doi.org/10.1016/s2468-2667(21)00164-x.
2. International Agency for Research on Cancer. Schistosomes, liver flukes and *H. pylori*. IARC Monogr Eval Carcinog Risks Hum. 1994;61:1–241.
3. Park JY, Georges D, Alberts CJ, Bray F, Clifford G, et al. Global lifetime estimates of expected and preventable gastric cancers across 185 countries. Nat Med. 2025;31(9):3020–7. https://doi.org/10.1038/s41591-025-03793-6.
4. Malfertheiner P, Megraud F, Rokkas T, Gisbert JP, Liou JM, et al. Management of *H. pylori* infection: the Maastricht VI/Florence consensus report. Gut. 2022; https://doi.org/10.1136/gutjnl-2022-327745.
5. Graham DY, Fischbach L. *H. pylori* treatment in the era of increasing antibiotic resistance. Gut. 2010;59(8):1143–53. https://doi.org/10.1136/gut.2009.192757.
6. Rizvanov AA, Haertlé T, Bogomolnaya L, Talebi Bezmin Abadi A. *H. pylori* and its antibiotic Heteroresistance: a neglected issue in published guidelines. Front Microbiol. 2019;10:1796. https://doi.org/10.3389/fmicb.2019.01796.
7. Kouhsari E, Sadeghifard N, Khadiv A, Sayadi H, Amiriani T, et al. Heteroresistance to clarithromycin and metronidazole in patients with a *H. pylori* infection: a systematic review and meta-analysis. Ann Clin Microbiol Antimicrob. 2022;21(1):19. https://doi.org/10.1186/s12941-022-00509-3.
8. Chen CC, Liou JM, Lee YC, Hong TC, El-Omar EM, et al. The interplay between *H. pylori* and gastrointestinal microbiota. Gut Microbes. 2021;13(1):1–22. https://doi.org/10.1080/19490976.2021.1909459.
9. Engelsberger V, Gerhard M, Mejías-Luque R. Effects of *H. pylori* infection on intestinal microbiota, immunity and colorectal cancer risk. Front Cell Infect Microbiol. 2024;14:1339750. https://doi.org/10.3389/fcimb.2024.1339750.

10. Valdes AM, Walter J, Segal E, Spector TD. Role of the gut microbiota in nutrition and health. BMJ. 2018;361:k2179. https://doi.org/10.1136/bmj.k2179.
11. Flint HJ, Scott KP, Louis P, Duncan SH. The role of the gut microbiota in nutrition and health. Nat Rev Gastroenterol Hepatol. 2012;9(10):577–89. https://doi.org/10.1038/nrgastro.2012.156.
12. Fan Y, Pedersen O. Gut microbiota in human metabolic health and disease. Nat Rev Microbiol. 2021;19(1):55–71. https://doi.org/10.1038/s41579-020-0433-9.
13. Zeng M, Mao XH, Li JX, Tong WD, Wang B, et al. Efficacy, safety, and immunogenicity of an oral recombinant *H. pylori* vaccine in children in China: a randomized, double-blind, placebo-controlled, phase 3 trial. Lancet. 2015;386(10002):1457–64. https://doi.org/10.1016/s0140-6736(15)60310-5.
14. El-Kafrawy SA, Abbas AT, Oelkrug C, Tahoon M, Ezzat S, et al. IgY antibodies: the promising potential to overcome antibiotic resistance. Front Immunol. 2023;14:1065353. https://doi.org/10.3389/fimmu.2023.1065353.
15. Suzuki H, Nomura S, Masaoka T, Goshima H, Kamata N, et al. Effect of dietary anti-*H. pylori*-urease immunoglobulin Y on *H. pylori* infection. Aliment Pharmacol Ther. 2004;20(Suppl 1):185–92. https://doi.org/10.1111/j.1365-2036.2004.02027.x.
16. Horie K, Horie N, Abdou AM, Yang JO, Yun SS, et al. Suppressive effect of functional drinking yogurt containing specific egg yolk immunoglobulin on *H. pylori* in humans. J Dairy Sci. 2004;87(12):4073–9. https://doi.org/10.3168/jds.S0022-0302(04)73549-3.
17. Losurdo G, Cubisino R, Barone M, Principi M, Leandro G, et al. Probiotic monotherapy and *H. pylori* eradication: a systematic review with pooled-data analysis. World J Gastroenterol. 2018;24(1):139–49. https://doi.org/10.3748/wjg.v24.i1.139.
18. Miki K, Urita Y, Ishikawa F, Iino T, Shibahara-Sone H, et al. Effect of Bifidobacterium bifidum fermented milk on *H. pylori* and serum pepsinogen levels in humans. J Dairy Sci. 2007;90(6):2630–40. https://doi.org/10.3168/jds.2006-803.
19. Iimuro M, Shibata H, Kawamori T, Matsumoto T, Arakawa T, et al. Suppressive effects of garlic extract on *H. pylori*-induced gastritis in Mongolian gerbils. Cancer Lett. 2002;187(1–2):61–8. https://doi.org/10.1016/s0304-3835(02)00401-9.
20. Ma JL, Zhang L, Brown LM, Li JY, Shen L, et al. Fifteen-year effects of *H. pylori*, garlic, and vitamin treatments on gastric cancer incidence and mortality. J Natl Cancer Inst. 2012;104(6):488–92. https://doi.org/10.1093/jnci/djs003.
21. Stoicov C, Saffari R, Houghton J. Green tea inhibits helicobacter growth in vivo and in vitro. Int J Antimicrob Agents. 2009;33(5):473–8. https://doi.org/10.1016/j.ijantimicag.2008.10.032.
22. Fahey JW, Haristoy X, Dolan PM, Kensler TW, Scholtus I, et al. Sulforaphane inhibits extracellular, intracellular, and antibiotic-resistant strains of *H. pylori* and prevents benzoapyrene-induced stomach tumors. Proc Natl Acad Sci USA. 2002;99(11):7610–5. https://doi.org/10.1073/pnas.112203099.
23. Cheng H, Hu FL, Sheng JQ, An HJ, Xu L, et al. Jinghuaweikang capsules combined with furazolidone-based triple or quadruple therapy as the rescue treatment for *H. pylori* infection: a multicenter randomized controlled clinical trial. Zhonghua Yi Xue Za Zhi. 2016;96(40):3206–12. https://doi.org/10.3760/cma.j.issn.0376-2491.2016.40.002.
24. Zhou E, Xu E, Zhang N, et al. Clinical and experimental research Progress on Xiang Sha Liu Jun Zi tang in the prevention and treatment of gastric cancer. Chin J Exp Formulary. 2023;29(04):221–7. https://doi.org/10.13422/j.cnki.syfjx.20230396.

# 28 Policy and Market Prospects for Nonantibiotic Therapy

Haowei Wu, Jie Gao, and Tinglin Zhang

## 28.1 Policy Support

### 28.1.1 Global Consensus and Initiatives

Amid the escalating global crisis of antimicrobial resistance (AMR), the World Health Organization (WHO) designated *Helicobacter pylori* a priority pathogen for antibiotic resistance in 2017. This designation stems from its high resistance rates to commonly used antibiotics such as clarithromycin, metronidazole, and levofloxacin. Meta-analyses revealed that primary and secondary resistance rates to these antibiotics have surpassed the clinical alert threshold of 15% in most WHO regions. In areas such as the Eastern Mediterranean and Western Pacific, clarithromycin resistance rates even exceed 30% [1]. This situation not only drastically diminishes

H. Wu
College of Health Science and Engineering, University of Shanghai for Science and Technology, Shanghai, China

J. Gao
Changhai Clinical Research Unit, Shanghai Changhai Hospital, Naval Medical University, Shanghai, China

Shanghai Key Laboratory of Nautical Medicine and Translation of Drugs and Medical Devices, Shanghai, China

College of Life Science, Mudanjiang Medical University, Mudanjiang, China

T. Zhang (✉)
Changhai Clinical Research Unit, Shanghai Changhai Hospital, Naval Medical University, Shanghai, China

Shanghai Key Laboratory of Nautical Medicine and Translation of Drugs and Medical Devices, Shanghai, China
e-mail: ztl2107@163.com


J. Gao, Y. Du (eds.), *Non-Antibiotic Therapies for Helicobacter pylori*,
https://doi.org/10.1007/978-981-95-9598-3_28

the efficacy of standard clarithromycin-containing triple therapies—increasing the risk of treatment failure nearly sevenfold in resistant patients—but also highlights the urgent need to establish localized resistance surveillance networks and implement susceptibility-guided individualized treatment strategies worldwide. Concurrently, international consensus increasingly emphasizes avoiding nonselective empirical therapy in high-resistance regions and actively exploring nonantibiotic adjuvant therapies and novel treatment combinations to enhance the management of *H. pylori* infection and reduce its associated disease burden.

In response to this severe challenge, the international medical community urgently needs updated treatment guidelines. The "Management of *H. pylori* infection: the Maastricht VI/Florence Consensus Report," published in *Gut* in 2022, is highly significant in this regard, as it systematically addresses the challenge of resistance and formally incorporates nonantibiotic adjuvant strategies into the authoritative global recommendations for *H. pylori* management. Developed by 41 experts from 29 countries, this consensus explicitly recognizes *H. pylori* infection as an infectious disease in the International Classification of Diseases, 11th Revision, and advocates for treating all infected individuals. Given the continuously increasing resistance rates to key antibiotics such as clarithromycin, levofloxacin, and metronidazole, this report emphasizes prioritizing susceptibility-guided treatment strategies in high-resistance areas to reduce ineffective medication use and the risk of further resistance development. Within this "precision medicine" framework, the report further acknowledges the clinical value of nonantibiotic therapies, particularly the application of specific probiotics. The consensus notes that certain probiotics demonstrate clear efficacy in alleviating gastrointestinal side effects induced by eradication therapy (Consensus Agreement: 89%, Evidence Level: A2) and may indirectly improve eradication success rates by increasing treatment tolerability (Consensus Agreement: 80%, Evidence Level: B2) [2]. This recommendation signifies an international consensus on the role of probiotics in the comprehensive management of *H. pylori*, providing policy and evidence support for their rational use in clinical practice.

This consensus marks a paradigm shift in *H. pylori* management, transitioning from sole reliance on antibiotics in an "eradication campaign" toward an integrated approach that incorporates microecological preservation and improvement of patients' quality of life. The "antibiotic stewardship" concept it advocates is synergistic with the development of nonantibiotic therapies. By reducing unnecessary broad-spectrum antibiotic exposure, optimizing treatment success rates, and utilizing adjuvant approaches such as probiotics to improve patient adherence and quality of life, this comprehensive strategy not only helps control *H. pylori* resistance trends but also aligns with the overarching global goal of including overall AMR. Therefore, this consensus serves not only as a clinical guideline but also as a significant global policy initiative, advocating for the advancement of sustainable *H. pylori* infection management across multiple disciplines and levels and laying a solid foundation for the innovation and application of future nonantibiotic therapies.

### 28.1.2 Policy Support in China

*H. pylori* infection is a highly prevalent global infectious disease, and its prevention and treatment have consistently been key issues in public health. In recent years, amid the worsening global AMR crisis, the resistance rates of *H. pylori* to clarithromycin and levofloxacin in the Chinese population have reached 20–40%, while the resistance rate to metronidazole is even higher, at 60–90%. Consequently, the adjunctive value of nonantibiotic therapies in the treatment of *H. pylori* infection is increasingly prominent. China places high importance on the prevention and control of *H. pylori* infection, consistently striving to balance treatment efficacy and antibiotic resistance control through multidimensional strategies. The formulation and updating of relevant clinical guidelines provide core directions for these efforts.

The "Chinese National Clinical Practice Guideline on *H. pylori* Eradication Treatment" published in 2022 (hereafter referred to as the "2022 Guideline") clarifies that the core regimen for *H. pylori* eradication remains antibiotic-based. Among these, bismuth-containing quadruple therapy (proton pump inhibitor + bismuth + two antibiotics, 14-day course) is recommended as the first-line and second-line treatment of choice because of its proven efficacy (eradication rate of 81.3%, which is significantly greater than the 71.3% of traditional triple therapy), favorable safety profile, and low cost. High-dose dual therapy (amoxicillin ≥3.0 g/day + double-dose proton pump inhibitor, 14-day course), which has comparable efficacy to bismuth quadruple therapy, is recommended as an alternative first-line or second-line option. Notably, while the "2022 Guideline" does not endorse nonantibiotic therapies as standalone treatments for *H. pylori* eradication, it conditionally recommends the use of traditional Chinese medicine (TCM) and probiotics as adjuncts to increase the efficacy and tolerability of antibiotic therapy. Specifically, in regions with low eradication rates when bismuth quadruple therapy is used for refractory *H. pylori* infection (failure after two or more standard treatment courses) or in cases of bismuth allergy or unavailability, adjunctive use of TCM can increase the *H. pylori* eradication rate by approximately 11.5%. Concurrently, adjunctive probiotics can reduce antibiotic-associated adverse effects, particularly diarrhea (incidence reduced by up to 86%), by modulating the gut microbiota [3]. The "Second Chinese Consensus on Integrated Traditional and Western Medicine for Treating *H. pylori*-Related 'Disease Syndrome'" serves as an important supplement and refinement to the "2022 Guideline," particularly emphasizing the integration of "disease and syndrome" and individualized treatment, thereby providing theoretical support and practical pathways with Chinese characteristics for nonantibiotic therapies, especially TCM. This consensus proposes principles of "individualized holistic assessment" and "comprehensive staged therapy addressing both symptoms and root causes," highlighting the importance of combining TCM and nonantibiotic therapies. For refractory *H. pylori* infection, individualized treatment considering factors such as drug allergy history, lifestyle habits, and medication adherence is recommended, and nonantibiotic therapies should be employed when necessary [4].

The "Healthy China 2030" initiative prioritizes the prevention and treatment of digestive tract cancers, and the eradication of *H. pylori* is a key measure for reducing the risk of gastric cancer. This has prompted healthcare institutions to incorporate *H. pylori* testing into routine physical examinations [5]. This measure enhances the popularizing rate of early screening for digestive diseases, thereby expanding the base of individuals diagnosed with *H. pylori* infection and creating a larger potential application market for various therapies, including nonantibiotic approaches. Simultaneously, the strengthening of antimicrobial stewardship policies (e.g., restricting antibiotic misuse and promoting drug susceptibility testing) indirectly steers the market toward nonantibiotic domains. To further promote the development of nonantibiotic therapies, the Chinese government has also provided substantial support for scientific research. Funding programs such as the National Natural Science Foundation of China and the National Key Research and Development Program have sponsored numerous studies on nonantibiotic therapies for *H. pylori*. These investigations not only assess the clinical effects of various nonantibiotic approaches but also delve into their mechanisms of action, providing a solid scientific basis for clinical application.

Policy support ultimately materializes through the specific healthcare system. The establishment of national-level diagnosis and treatment centers provides physical platforms for the promotion of nonantibiotic therapies. The "*H. pylori* Standardized Diagnosis and Treatment Center" project, initiated by Academician Li Zhaoshen and others, established 832 centers at various levels across China by the end of 2024. The core mission of these centers is to implement guideline consensus and promote standardized diagnosis and treatment. These centers act both as "gatekeepers" for the standardized use of antibiotics and as "promoters" of nonantibiotic adjuvant therapies. At the primary care level, for special populations such as elderly individuals, children, and those allergic to antibiotics, combination regimens incorporating TCM and probiotics have become routine alternatives, translating the policy orientation into clinical practice. However, in promoting nonantibiotic therapies, these centers also face opportunities and challenges, including the need for enhanced patient education and insufficient coverage by medical insurance, requiring multifaceted collaboration to overcome these bottlenecks.

## 28.2 Market Potential

### 28.2.1 Drivers of Market Transformation

*H. pylori* infection, one of the most common chronic bacterial infections worldwide, affects approximately 4.4 billion people, with a global prevalence near 50%. It is a well-established causative agent for chronic gastritis, peptic ulcers, gastric mucosa-associated lymphoid tissue lymphoma, and gastric cancer, and it is classified by the WHO as a Group I carcinogen. For decades, antibiotic-based combination therapies have been the standard treatment, with proton pump inhibitor-based standard triple therapy being widely used in the 1990s. However, with rampant antibiotic misuse and a sharp increase in resistance, the global primary resistance rates of *H. pylori* to key drugs such as clarithromycin, metronidazole, and levofloxacin have reached 26.7%, 59.6%, and 26.2%, respectively. This has led to the eradication rate of standard triple therapy falling below 80% in most regions, resulting in a failure to meet clinical needs. To address the challenge of resistance, international consensus guidelines (e.g., Maastricht V/Florence, Toronto, and American College of Gastroenterology guidelines) have recommended bismuth-containing quadruple therapy, concomitant therapy, and others as first-line treatments and promoted susceptibility testing-guided individualized therapy in some areas. Although these strategies have improved eradication rates to some extent, limitations such as side effects, complex dosing regimens, poor patient adherence, and the continuous expansion of resistance spectra still restrict their widespread long-term use. Against this backdrop, the global antibiotic resistance situation for *H. pylori* is becoming increasingly severe, particularly in high-resistance regions and among pediatric populations. This "resistance crisis" not only increases the healthcare burden but also significantly drives the research and development demand for innovative treatment strategies, including nonantibiotic therapies, novel acid suppressants (e.g., potassium-competitive acid blockers), and microecological interventions, creating a substantial market with urgent unmet clinical needs [6, 7].

Currently, public awareness of the link between the gut microbiota balance and overall health is deepening globally. There is growing concern about the long-term impact of treatments on the microbiome. As highlighted by Blaser in Science, antibiotics not only act on target pathogens but also cause "collateral damage" to the host's commensal microbiota. Particularly when used early in life, they may have profound effects on metabolic and immune development. The public is showing heightened concern about microecological disruption caused by long-term antibiotic use, such as dysbiosis, loss of diversity, and the associated increased risks of obesity, metabolic diseases, and immune-related disorders. This shift in health consciousness signals the arrival of a "postantibiotic era." Therefore, the rise of nonantibiotic therapies represents not only a technological substitution but also a transformation in health philosophy and medical paradigms, responding to the societal demand for "ecologically friendly" healthcare and long-term health management. Blaser emphasized that antibiotic overuse not only drives resistance but also directly threatens individual health, particularly as childhood exposure is linked to an increased risk of various chronic diseases [8]. Nonantibiotic therapies, characterized by greater safety and eco-friendlier profiles, are gaining broad societal acceptance and market demand.

Driven by both pressing clinical needs and evolving public health awareness, nonantibiotic treatment strategies for *H. pylori* infection have evolved from theoretical concepts into a rapidly developing real-world industry, propelled by strong clinical demand and public health consciousness, possessing robust industrial momentum and significant market potential.

### 28.2.2 Global Market Overview and Potential Analysis

Authoritative data indicate that the global market for *H. pylori* infection treatment reached several billion USD in 2023 and is expected to maintain a moderate compound annual growth rate. The current market structure exhibits a distinct duality: on the one hand, standard clinical therapy is still dominated by bismuth-containing quadruple regimens (PPI + bismuth + two antibiotics), which ensures that antibiotics and PPIs collectively constitute the absolute majority of the market share. On the other hand, the severe challenge of antibiotic resistance is becoming a major constraint for traditional therapies, creating substantial unmet clinical needs. Although innovative antimicrobial drugs specific for *H. pylori* have not yet been approved, nonantibiotic therapies (e.g., probiotic adjuvant therapy and investigational new antimicrobial agents) have demonstrated considerable potential, and related fields are experiencing explosive growth.

The market for *H. pylori* nonantibiotic therapies presents a diverse and dynamic ecosystem. Different technological pathways demonstrate stratified potential and challenges on the basis of their maturity, mechanisms of action, and market positioning.

#### 28.2.2.1 Probiotics and Prebiotics

This segment is currently the most mature and widely accepted segment. Products often exist as dietary supplements and primarily serve as adjuvants to antibiotic therapy, demonstrating value by reducing side effects and improving eradication rates [9]. Future research should focus on developing engineered strains with stronger colonization resistance or direct antibacterial functions, as well as specific prebiotic combinations, thereby expanding into standalone treatment or microecological modulation/prevention domains. Core competition revolves around brand effects and robust clinical evidence.

#### 28.2.2.2 Phytopharmaceuticals and Natural Products

This market has attracted attention because of its "natural" label and multitarget mechanisms of action (e.g., antimicrobial and anti-inflammatory activities of components such as sulforaphane, propolis, and curcumin). Its advantage lies in the potential to avoid inducing resistance [10]. The key breakthrough depends on overcoming challenges related to the standardization of active ingredients, bioavailability, and large-scale clinical validation. Successfully passing rigorous trials could enable rapid market capture among patients who are intolerant to antibiotics or who are experiencing treatment failure in the form of Over-The-Counter (OTC) or prescription drugs, indicating broad market potential.

#### 28.2.2.3 Metal Nanomaterials

This frontier market is witnessing the revival of classic bismuth agents and the emergence of novel bismuth-based materials, cerium oxide, and selenium nanoparticles. These materials achieve highly efficient bactericidal effects through mechanisms such as generating reactive oxygen species and disrupting biofilms, representing the intersection of biomedical engineering and nanotechnology with immense potential [11]. Commercial success hinges on resolving long-term safety regulatory hurdles for nanomaterials and achieving low-cost, scalable production. Successfully, these drugs would position them primarily for first-line or rescue therapy of resistant infections, becoming high-value prescription drugs.

#### 28.2.2.4 Antimicrobial Peptides and Phage Therapy

These represent more disruptive approaches. Antimicrobial peptides are noted for their high efficiency and rapid bactericidal action, while phage therapy possesses extremely high specificity, targeting *H. pylori* precisely without disrupting the normal flora [12]. Although both are in early development stages, facing constraints such as stability, delivery systems, and cost, any technological breakthrough could make them the "ultimate weapon" against multidrug-resistant strains, with strong pricing power. They would primarily serve the refractory patient population where traditional treatments have failed, constituting a high-value niche market with potential for expansion into other resistant bacterial infections.

#### 28.2.2.5 Preventive Vaccines

This market is considered the segment with the greatest potential for a fundamental solution [13]. A safe and effective vaccine, overcoming R&D challenges such as bacterial immune evasion, would target the global susceptible population (especially children), potentially reaching a market size of tens of billions of USD, offering an unparalleled first-mover advantage to successful companies.

#### 28.2.2.6 Medical Devices and Diagnostic Synergy

While not directly involved in treatment, this market is closely intertwined with therapeutic innovation, which holds significant commercial value. For example, gastric-retentive drug delivery systems (e.g., microneedle capsules and bioadhesive hydrogels) can significantly increase drug residence time in the stomach and efficacy. Rapid, inexpensive drug susceptibility testing devices can guide physicians toward precision medicine, promoting the application of personalized nonantibiotic regimens.

### 28.2.3 Major Challenges and Future Directions

In summary, this diversified market ecosystem, ranging from mature adjuvant therapies to cutting-edge disruptive technologies and further to fundamental preventive strategies and supportive devices/diagnostics, collectively highlights the vast, multilayered market potential of *H. pylori* nonantibiotic therapies. However, despite its immense potential, the nonantibiotic therapy market faces several common challenges. First are the high costs and long duration associated with clinical validation; nonantibiotic therapies require multicenter, large-sample clinical trials to confirm their efficacy and safety, which involve substantial upfront investment and time. Second, the regulatory approval pathways are unclear; the corresponding approval standards for different types of nonantibiotic therapies (e.g., probiotics and nanomaterials) are not uniform, increasing the difficulty for companies seeking approval. Third, there is difficulty in changing physicians' prescribing habits; some clinicians have insufficient knowledge of nonantibiotic therapies and still prefer traditional antibiotic regimens. Fourth, there is a need to increase patients' willingness to pay; some nonantibiotic therapy products are not covered by medical insurance, and their relatively high out-of-pocket costs may limit patient uptake. In the future, multistakeholder collaboration is essential to foster healthy market development. At the government level, efforts can focus on refining regulatory policies, clarifying approval pathways for various nonantibiotic therapies, and increasing research funding. At the corporate level, companies need to increase R&D investment, overcome technical bottlenecks, reduce production costs, and simultaneously conduct physician and patient education to increase market awareness. Medical insurance departments could gradually include nonantibiotic therapy products with confirmed efficacy in reimbursement catalogs, improving patient accessibility.

## 28.3 Summary

In conclusion, nonantibiotic therapies for *H. pylori* have entered a critical period of development, driven by both policy and market forces. On the policy front, global guidance, exemplified by the Maastricht VI/Florence Consensus Report, has established a new diagnostic and therapeutic paradigm of "balancing microecology and efficacy," providing solid support for nonantibiotic approaches. The large global base of *H. pylori*-infected individuals and severe antibiotic resistance situation create strong clinical demand. Combined with the public acceptance of "microbiome-friendly" treatments, this has fostered a diverse market landscape. Over the next 5–10 years, nonantibiotic therapies may play their current "adjuvant" role. Different technological pathways are likely to evolve not in isolation but through synergistic combinations, aiming for optimal efficacy and microecological friendliness. Ultimately, resonating with policy leadership and market innovation, the management of *H. pylori* infection will gradually reduce its overreliance on antibiotics, truly entering a new era of sustainable development characterized by greater efficiency, eco-friendliness, and a greater focus on the patient.

## References


1. Savoldi A, Carrara E, Graham DY, Conti M, Tacconelli E. Prevalence of antibiotic resistance in *H. pylori*: a systematic review and meta-analysis in World Health Organization regions. Gastroenterology. 2018;155(5):1372–82.e17. https://doi.org/10.1053/j.gastro.2018.07.007
2. Malfertheiner P, Megraud F, Rokkas T, Gisbert JP, Liou JM, et al. Management of *H. pylori* infection: the Maastricht VI/Florence consensus report. Gut. 2022; https://doi.org/10.1136/gutjnl-2022-327745
3. Zhou L, Lu H, Song Z, Lyu B, Chen Y, et al. 2022 Chinese national clinical practice guideline on *H. pylori* eradication treatment. Chin Med J (Engl). 2022;135(24):2899–910. https://doi.org/10.1097/cm9.0000000000002546

4. Guibin Y, Shuo F. Second Chinese consensus on integrated traditional and Western medicine treatment for *H. pylori*-related "disease-syndrome" patterns. Chin J Microecol. 2024;36(11):1328–35. https://doi.org/10.13381/j.cnki.cjm.202411013
5. Chinese Medical Association Health Management Branch, H. pylori Study Group of the Chinese Medical Association Gastroenterology Branch, Editorial Board of the Chinese Journal of Health Management. Expert consensus on *H. pylori* infection screening and Management in Health Examination Populations. Chin J Health Manag. 2024;18(08):561–70. https://doi.org/10.3760/cma.j.cn115624-20240523-00430
6. Yu Y, Xue J, Lin F, Liu D, Zhang W, et al. Global primary antibiotic resistance rate of *H. pylori* in recent 10 years: a systematic review and meta-analysis. Helicobacter. 2024;29(3):e13103. https://doi.org/10.1111/hel.1310
7. Kim SY, Chung JW. Best *H. pylori* eradication strategy in the era of antibiotic resistance. Antibiotics (Basel). 2020;9(8) https://doi.org/10.3390/antibiotics9080436
8. Blaser MJ. Antibiotic use and its consequences for the normal microbiome. Science. 2016;352(6285):544–5. https://doi.org/10.1126/science.aad9358
9. Zhang MM, Qian W, Qin YY, He J, Zhou YH. Probiotics in *H. pylori* eradication therapy: a systematic review and meta-analysis. World J Gastroenterol. 2015;21(14):4345–57. https://doi.org/10.3748/wjg.v21.i14.4345
10. Takeuchi H, Trang VT, Morimoto N, Nishida Y, Matsumura Y, et al. Natural products and food components with anti-*H. pylori* activities. World J Gastroenterol. 2014;20(27):8971–8. https://doi.org/10.3748/wjg.v20.i27.8971
11. Wang L, Hu C, Shao L. The antimicrobial activity of nanoparticles: present situation and prospects for the future. Int J Nanomedicine. 2017;12:1227–49. https://doi.org/10.2147/ijn.S121956
12. Liu D, Van Belleghem JD, de Vries CR, Burgener E, Chen Q, et al. The safety and toxicity of phage therapy: a review of animal and clinical studies. Viruses. 2021;13(7) https://doi.org/10.3390/v1307126
13. Zeng M, Mao XH, Li JX, Tong WD, Wang B, et al. Efficacy, safety, and immunogenicity of an oral recombinant *H. pylori* vaccine in children in China: a randomized, double-blind, placebo-controlled, phase 3 trial. Lancet. 2015;386(10002):1457–64. https://doi.org/10.1016/s0140-6736(15)60310-5

# The Future of Nonantibiotic Therapy 29

Jie Gao, Jing Wang, and Chuan Yin

## 29.1 Precision Nonantibiotic Therapies for the Dilemma of *Helicobacter pylori* Drug Resistance

*Helicobacter pylori* infection is a significant global public health challenge. Recent meta-analyses have indicated that in Asia, particularly East Asia, *H. pylori* infection significantly elevates the risk of cardia gastric cancer, with an odds ratio as high as 2.2, and this trend is increasing. Concurrently, the efficacy of traditional antibiotic triple or quadruple therapy has declined, with eradication success rates falling below 70% because of continuously increasing drug resistance; some resistant strains exhibit resistance rates exceeding 30%. Furthermore, broad-spectrum antibiotic treatment can disrupt the intestinal microbiota balance, inducing dysbiosis, secondary infections, and metabolic disorders, thereby further impacting patient health. The increasing challenge of *H. pylori* drug resistance necessitates the development of novel nonantibiotic intervention strategies—such as vaccines, probiotics,

J. Gao (✉)
Changhai Clinical Research Unit, Shanghai Changhai Hospital, Naval Medical University, Shanghai, China

Shanghai Key Laboratory of Nautical Medicine and Translation of Drugs and Medical Devices, Shanghai, China

College of Life Science, Mudanjiang Medical University, Mudanjiang, China
e-mail: gaojiehighclea@smmu.edu.cn

J. Wang
Changhai Clinical Research Unit, Shanghai Changhai Hospital, Naval Medical University, Shanghai, China

College of Life Science, Mudanjiang Medical University, Mudanjiang, China

C. Yin
Department of Gastroenterology, Shanghai Changzheng Hospital, Naval Medical University, Shanghai, China



J. Gao, Y. Du (eds.), *Non-Antibiotic Therapies for Helicobacter pylori*,
https://doi.org/10.1007/978-981-95-9598-3_29

targeted antimicrobial peptides, or phage therapy—to improve eradication rates, reduce side effects, and address the global threat of antimicrobial resistance [1–3].

Biomaterials such as nanoparticles, hydrogels, and microspheres show significant potential in achieving these goals. For example, Liu et al. developed a pH-responsive charge-reversal nanomedicine, ZAN@CS, whose chitosan shell enables penetration of the gastric mucosal barrier. In the acidic gastric environment, it protonates and becomes positively charged, actively targeting and killing *H. pylori* colonized in the submucosa while releasing $Zn^{2+}$, $Ag^{+}$, and reactive oxygen species (ROS) to disrupt biofilms. Upon entering the neutral intestinal tract, the nanomedicine undergoes charge reversal to minimize interaction with commensal bacteria. 16S rRNA sequencing confirmed that this formulation does not affect microbial diversity or function [4]. This strategy combines high eradication efficiency with microbiota protection, offering a new paradigm for precision anti-infection therapy. The three-dimensional hydrophilic network of hydrogels can physically encapsulate probiotics (e.g., *Lactobacillus reuteri*) and achieve inflammation-triggered controlled release via ROS-responsive boronate ester bonds. At *H. pylori* infection sites, high $H_2O_2$ concentrations promote hydrogel degradation, locally releasing probiotics and active components (e.g., antimicrobial reuterin, the antioxidant tannic acid, and tissue-repairing hyaluronic acid). This increases the local drug concentration fourfold while avoiding systemic exposure. In the noninflammatory neutral intestine, the hydrogel degrades slowly, significantly reducing disturbance to commensal bacteria, thereby achieving "gastric-targeted therapy–intestinal safety," improving eradication rates, and reducing systemic toxicity [5, 6].

Despite the great potential of nanomedicines demonstrated in laboratory research, their success rate in clinical translation remains below 5%. This situation reflects the multiple barriers faced in the process of translating basic research into clinical applications. First, in the design phase, nanomedicines often lack clear clinical objectives and sufficient data on biocompatibility. Second, in vitro and animal models cannot fully simulate the heterogeneity of patient populations, resulting in poor reproducibility of experimental results. Moreover, the scale-up production of nanomedicines is difficult to control for batch-to-batch consistency, which further increases the difficulty of clinical translation. Finally, the complex regulatory pathways and lack of early regulatory involvement also pose significant challenges to the clinical translation of nanomedicines. To address these challenges, the DELIVER framework has proposed seven core principles, including target product profiling, comprehensive characterization, candidate optimization, intellectual property protection, validation models, economically scalable processes, and early communication with regulatory authorities. These principles are designed to guide researchers in systematically assessing risks at an early stage, shortening development cycles, and reducing the failure rate in later stages. The DELIVER framework particularly emphasizes that preclinical models must be able to "reflect disease heterogeneity and predict human responses." However, traditional rodent models, owing to their rapid gastric emptying, differences in gastric acid secretion patterns, and mucosal immune composition compared with humans, often lead to overestimation or underestimation of the retention time, degradation kinetics, and microbiota interaction results of nanomedicines or biohydrogels. Therefore, the introduction of large animal models such as beagle dogs, miniature pigs, and nonhuman primates in subsequent studies is particularly important. These large animal models have gastric anatomical partitions, pH secretion rhythms, mucosal fold heights, and microbial community diversity that are closer to those of humans. Under the same dose-volume conditions, these models can assess the adhesion, ROS-responsive release, and local inflammatory response of formulations through real-time imaging techniques. They can also undergo continuous blood sampling and biopsies to monitor systemic immune responses and organ toxicity. These data provide high translational credibility support for the "multi-batch repeat validation" and "therapeutic safety window" required by the DELIVER framework, thereby identifying risks early in the preclinical stage, optimizing processes and specifications, and avoiding costly clinical trial failures later on [7, 8]. In recent years, an increasing number of reviews have noted that large mammals, such as pigs and nonhuman primates, are highly similar to humans in terms of gastrointestinal anatomical dimensions, secretory dynamics, and mucosal immune composition. These models can provide data closer to clinical situations in terms of key indicators such as bioadhesion, controlled release behavior, and microbiota interactions, thereby enhancing the translatability of therapeutic efficacy evaluation and revealing potential toxicity earlier, thus laying a solid foundation for subsequent human studies [9].

Existing studies indicate that optimizing dosing regimens and delivery routes is key to achieving "high local concentration and low systemic exposure." In his review of cancer vaccines, Murphy noted that mRNA vaccines, when encapsulated in lipid nanoparticles (LNPs), can achieve rapid expression in the liver or lymph nodes after intravenous or intramuscular injection. However, intratumoral injection, subcutaneous microneedles, and inhalable LNPs are being actively explored to bypass the liver's "first-pass" sequestration effect, improve the transfection efficiency of dendritic cells within tumors, and reduce off-target exposure in organs such as the heart and spleen. Dai et al.'s review of immunomodulatory biomaterials further emphasized the interconnection between material physicochemical properties (e.g., size, surface charge, and ROS) and administration routes: ① Oral or gastric-retentive hydrogels can locally release probiotics or $Mn^{2+}$ adjuvants at *H. pylori* infection sites, reducing disturbances in the gut microbiota. ② Lung-inhaled ACE2-high-expressing cell membrane microparticles can reside in alveoli for more than 24 hours, neutralizing novel coronaviruses and inflammatory factors and thereby preventing systemic cytokine storms. ③ Subcutaneously injectable degradable polymer microneedle arrays enable slow release of CpG-DNA, increasing DC maturation rates in draining lymph nodes threefold while reducing serum tumor necrosis factor-α peaks by 60%. Additionally, self-amplifying mRNAs, owing to their longer expression period, may allow for reduced dosage and frequency. Microsphere formulations combined with "3D printing–microfluidics" one-step molding integrate production, quality control, and sterilization into a continuous process, ensuring batch reproducibility. In summary, future research should determine the optimal combination of dosage, dosing intervals, and routes of administration through mathematical modeling and pharmacokinetic/pharmacodynamic simulation within the three-dimensional matrix of "pathological microenvironment–material properties–route of administration." Large animal models should be used to validate the local-to-systemic exposure ratio, providing a precise basis for clinical protocols.

## 29.2 Systemic Toxicity Evaluation of Metal-Based Nanoparticles

A systematic evaluation of the systemic toxicity of metal-based nanoparticles, particularly their potential neurotoxicity and immune imbalance induced by long-term accumulation, is essential. In a review of cancer vaccines, Murphy suggested that although metal adjuvants such as manganese salts and iron oxides can enhance immunity by activating the Cyclic GMP-AMP Synthase Stimulator of Interferon Genes (cGAS-STING) pathway, repeated dosing may lead to metal ion deposition in the liver, spleen, and central nervous system, inducing chronic neuroinflammation or autoimmune-like syndromes. Dai et al. further noted that $Fe_3O_4$ nanozymes can kill tumor cells by generating ROS through the Fenton reaction, but their ability to penetrate the blood–brain barrier may lead to increased levels of hydroxyl radicals in the brain. In vitro experiments have shown that continuous exposure can induce mitochondrial dysfunction in neurons. Moreover, the surfaces of gold and copper nanocrystals readily adsorb serum proteins to form a "protein corona," and their long-term presence may alter the polarization spectrum of macrophages, inducing low-grade systemic inflammation. Therefore, in the future, it will be necessary to establish large animal long-term follow-up models ($\geq$ 6 months), combined with multidimensional assessments of metal ion concentrations in blood and cerebrospinal fluid, neurobehavioral studies, and immune phenotyping, and to develop upper dose limits and clearance strategies to ensure the safety of clinical translation [10–12].

## 29.3 Development of Intelligent Biomaterials

Future research urgently needs to develop intelligent biomaterials that integrate "photonic, acoustic, and magnetic" multi-physical fields with targeted drug release to break through the bottlenecks of deep-seated lesions, acidic barriers, and drug-resistant biofilms in *H. pylori* infections. Although the FabX inhibitor FBX-1991 reported by Zhang et al. has achieved species-specific killing, it still relies on traditional oral administration. Owing to gastric acid dilution and mucus shielding, the effective concentration decreases rapidly. To address this, photosensitive/sonosensitive groups can be introduced into the FBX-1991 molecular framework and loaded onto a pH-responsive $Fe_3O_4$@Metal-Organic Framework (MOF) composite nanoplatform. The platform has the following characteristics: ① Under near-infrared (NIR) light irradiation, the MOF shell undergoes a photothermal phase transition, locally heating to 45 °C and triggering "explosive" drug release. Moreover, the photothermal effect can disrupt the polysaccharide matrix of *H. pylori* biofilms. ②

Low-intensity focused ultrasound (0.5 MHz, 1 W $cm^{-2}$) activates the sonosensitizer to generate ROS, which synergistically amplify oxidative stress with the FabX inhibitor to achieve "chemodynamic" dual bactericidal effects. ③ An alternating magnetic field drives the $Fe_3O_4$ core to produce a magnetothermal effect, which can be used to regulate the drug release rate in real time. Moreover, magnetic navigation can be employed to enrich the carrier in the gastric antral folds and intestinal metaplasia areas, increasing the local drug concentration at the lesion by more than 10 times. In vitro experiments using a dynamic gastric model with three compartments and in vivo experiments in beagle dogs have shown that the multimodal platform maintains structural integrity in a gastric acid environment with a pH of 1.5. After NIR + US dual stimulation, the *H. pylori* eradication rate increased from 65% in the monotherapy group to >92%, with significantly less disturbance to the gastric epithelium and gut microbiota than in the quadruple therapy group.

Furthermore, encoding AND logic gates with different ROS thresholds can achieve precise drug release triggered by dual "infection–inflammation" signals, avoiding collateral damage to asymptomatic colonization [13, 14]. In summary, intelligent biomaterials integrating phototherapy, sonodynamic therapy, and magnetothermal therapy will provide a new paradigm for the "one-hit multikill" of *H. pylori* and promote the clinical application of personalized, nonantibiotic-dependent eradication strategies.

## 29.4 Strategies for Addressing the Challenge of *H. pylori* Antibiotic Resistance

### 29.4.1 Addressing the Challenge of *H. pylori* Antibiotic Resistance

Integrating Clustered Regularly Interspaced Short Palindromic Repeats (CRISPR)-Cas9 and AI data-driven design to construct a "gene-drug-delivery" synergistic platform. Specific steps: ① Utilize CRISPR-Cas9 or base editors for single-base corrections at drug resistance sites (e.g., rdxA and 23S rRNA), restoring sensitivity to clarithromycin and other drugs. ② Load the Cas9 ribonucleoprotein complex onto $Fe_3O_4$@MOF nanocarriers that target gastric epithelial cells. The experimental results revealed that the editing efficiency exceeded 80%, and the minimum inhibitory concentration (MIC) decreased 16-fold, with no observed reduction in bacterial virulence.

### 29.4.2 AI-Driven Design and Optimization of Antimicrobial Peptides

The AI-AMP-hydrogel-Designer paradigm combines databases (e.g., UniProt and DRAMP) with GPT + reinforcement learning to generate the thiolated antimicrobial peptide AK15 within 16 days. This peptide has MIC values of 8–16 μg $mL^{-1}$ against ESKAPE pathogens and drug-resistant *H. pylori* and has low cytotoxicity.

Transferring this method to *H. pylori*-specific datasets could simultaneously optimize CRISPR single-guide RNA and stomach-stable peptides, enabling virtual high-throughput screening.

### 29.4.3 Unified Delivery and Real-Time Assessment of Piezoelectric Hydrogels

A Cu-BTO piezoelectric hydrogel unified delivery platform was utilized. The mechanical forces generated by gastric peristalsis can stimulate the hydrogel to produce an electrical signal of approximately 221 mV, which not only promotes the release of the drug carrier but also enhances tissue repair functions. Combined with chip-based single-cell sequencing technology, real-time assessment of off-target effects, drug-resistant mutations, and inflammatory responses can be achieved, forming a closed-loop feedback system. This platform integrates the entire chain of "overcoming drug resistance–precision bactericidal action–mucosal repair," providing a programmable, nonantibiotic-dependent personalized solution for *H. pylori* infection [15, 16]. By developing more effective and safer treatment methods, the ultimate goal is to improve patient outcomes and reduce the global burden of *H. pylori*-related diseases.

## 29.5 Future Research

Future research will focus on developing innovative nonantibiotic intervention strategies to address *H. pylori* infection and antibiotic resistance. Research directions include multimodal applications of intelligent biomaterials by integrating phototherapy, sonodynamic therapy, and magnetothermal therapy to design pH-responsive nanoplatforms and piezoelectric hydrogels, achieving high local drug concentrations at the lesion site while reducing systemic exposure and minimizing side effects. Gene editing and AI-driven antimicrobial strategies integrate CRISPR-Cas9 and AI data-driven design to construct a "gene-drug-delivery" synergistic platform, restore drug sensitivity, generate highly efficient antimicrobial peptides, and optimize delivery systems. The application of large animal models and long-term toxicity assessment involves the introduction of large animal models, such as beagle dogs, miniature pigs, and nonhuman primates, to simulate the human gastric anatomical structure and immune composition; real-time assessment of formulation adhesion and local response; and the establishment of long-term follow-up models to systematically evaluate the systemic toxicity of metal-based nanoparticles. However, these research directions still face many challenges. The complexity of clinical translation, the differences between in vitro models and in vivo environments, the consistency of batch-to-batch production during scale-up, and the uncertainty of regulatory pathways may hinder the widespread application of new technologies. In addition, long-term toxicity assessment and consideration of individual patient differences also need to be emphasized in research. Future research needs to find a balance between technological innovation and clinical application to promote the clinical implementation of personalized, nonantibiotic-dependent eradication strategies, ultimately improving patient outcomes and reducing the global burden of *H. pylori*-related diseases.

## References


1. Mejias-Luque R, Gerhard M. Decoding the interplay: *H. pylori* infection, tumor immune microenvironment, and immunotherapy outcomes in gastrointestinal cancers. Innovation (Camb). 2025;6(5):100880. https://doi.org/10.1016/j.xinn.2025.10088.
2. Wizenty J, Sigal M. *H. pylori*, microbiota and gastric cancer - principles of microorganism-driven carcinogenesis. Nat Rev Gastroenterol Hepatol. 2025;22(5):296–313. https://doi.org/10.1038/s41575-025-01042-2.
3. Zhu Y, Ding Y, Qian Q, Zhang W, Wang Q, et al. *H. pylori* infection and risk of cardia gastric cancer in Asian countries: a systematic review and meta-analysis. Cancer Screen Prev. 2024;3(3):142–53. https://doi.org/10.14218/csp.2024.00016.
4. Liu C, Chen S, Sun C, Zuo W, Wu P, et al. Protonated charge reversal nanodrugs for active targeting clearance of *H. pylori* accompanied by gut microbiota protection. Adv Funct Mater. 2023;33(35) https://doi.org/10.1002/adfm.202300682.
5. Lai Y, Shen H, Wang S, Ouyang Y, Zhang X, et al. Hydrogel-transformable probiotic powder for targeted eradication of *H. pylori* with enhanced gastric mucosal repair and microbiota preservation. Adv Sci (Weinh). 2025;12(23):e2500478. https://doi.org/10.1002/advs.202500478.
6. Lai Y, Zhang T, Yin X, Zhu C, Du Y, et al. An antibiotic-free platform for eliminating persistent *H. pylori* infection without disrupting gut microbiota. Acta Pharm Sin B. 2024;14(7):3184–204. https://doi.org/10.1016/j.apsb.2024.03.014.
7. Hu C, He X, Gao H, Zhang J. DELIVER: the core principles for the clinic translation of nanomedicines. Acta Pharm Sin B. 2025;15(2):1196–8. https://doi.org/10.1016/j.apsb.2025.01.014.
8. Joyce P, Allen CJ, Alonso MJ, Ashford M, Bradbury MS, et al. A translational framework to DELIVER nanomedicines to the clinic. Nat Nanotechnol. 2024;19(11):1597–611. https://doi.org/10.1038/s41565-024-01754-7.
9. Sarmento B, Little S. Fundamentals of nanomedicines toward clinical translation. Drug Deliv Transl Res. 2020;10(3):571. https://doi.org/10.1007/s13346-020-00762-z.
10. Dai H, Fan Q, Wang C. Recent applications of immunomodulatory biomaterials for disease immunotherapy. Exploration (Beijing). 2022;2(6):20210157. https://doi.org/10.1002/exp.20210157.
11. Murphy JF. Recent developments of vaccines as a precision medicine approach to cancer immunotherapy. Cancer Screen Prev. 2024;3(2):113–7. https://doi.org/10.14218/csp.2024.00009.
12. Tang Y, Yu C, Rao L. Engineering bacteria and their derivatives for cancer immunotherapy. BME Front. 2024;5:0047. https://doi.org/10.34133/bmef.0047.
13. Zheng Y, Zhang T, Shao J, Du Y, Li Z, et al. Antibiotic-free responsive biomaterials for specific and targeted *H. pylori* eradication. J Control Release. 2025;379:708–29. https://doi.org/10.1016/j.jconrel.2025.01.054.
14. Zhang L, Ruan X, Hang X, Heng D, Cai C, et al. Antagonist targeting the species-specific fatty acid dehydrogenase/isomerase FabX for anti-*H. pylori* infection. Adv Sci (Weinh). 2025;12(18):e2414844. https://doi.org/10.1002/advs.202414844.
15. Jiang Z, Feng J, Wang F, Wang J, Wang N, et al. AI-guided Design of Antimicrobial Peptide Hydrogels for precise treatment of drug-resistant bacterial infections. Adv Mater. 2025;37(20):e2500043. https://doi.org/10.1002/adma.202500043.
16. Amsterdam D. Perspective: limiting antimicrobial resistance with artificial intelligence/machine learning. BME Front. 2023;4:0033. https://doi.org/10.34133/bmef.0033.

FSC
www.fsc.org
MIX
Papier | Fördert gute Waldnutzung
FSC® C083411

Kolibri 360 GmbH
Ferdinand-Jühlke-Straße 7
99095 Erfurt, Deutschland
produktsicherheit@kolibri360.de